KINETICS

Coefficient of Restitution (Impact)

$$e = -\frac{(v_{Bf})_n - (v_{Af})_n}{(v_{Bi})_n - (v_{Ai})_n} = -\frac{(v_{B/A})_{fn}}{(v_{B/A})_{in}}$$

Principle of Angular Impulse amd Momentum

$$\mathbf{H}_{Oi} + \int_{t_i}^{t_f} \Sigma \mathbf{M}_O \, dt = \mathbf{H}_{Of}$$

$$\mathbf{H}_O = \mathbf{r}_{P/O} \times (m\mathbf{v})$$

(For a Particle– Relative to a Fixed Point O)

$$\mathbf{H}_{Gi} + \int_{t_i}^{t_f} \Sigma \mathbf{M}_G \, dt = \mathbf{H}_{Gf}$$

$$\mathbf{H}_G = -\omega I_{Gxz}\,\mathbf{i} - \omega I_{Gyz}\,\mathbf{j} + \omega I_{Gz}\,\mathbf{k}$$

(For the General Plane Motion of a Rigid Body Relative to an Axis Through the Mass Center G)

Planetary Motion

$$F = \frac{Gm_1 m_2}{r^2}$$

$$r^2 \dot{\theta} = h \quad \text{(a constant)}$$

$$r = \frac{h^2}{Gm_1} \frac{1}{1 + e \cos \theta}$$

$$v = \frac{Gm_1}{h} \sqrt{e^2 + 2e \cos \theta + 1}$$

$$T = \left[\frac{4\pi^2 a^3}{Gm_1}\right]^{1/2}$$

$$a = \frac{1}{2}(r_a + r_p) = \frac{h^2}{2Gm_1(1 - e)} + \frac{h^2}{2Gm_1(1 + e)} = \frac{h^2}{Gm_1(1 - e^2)}$$

$$G = 6.673(10^{-11}) \text{ m}^3/(\text{kg} \cdot \text{s}^2) = 3.439(10^{-8}) \text{ ft}^4/(\text{lb} \cdot \text{s}^4)$$

The Earth

Mass	5.976(10^24) kg	4.095(10^23) lb · s²/ft
Mean radius	6370 km	3960 mi

Vibrations

Equations of Motion – Canonical Form

$$m\ddot{x} + c\dot{x} + kx = P_0 \sin \Omega t$$

General Solution

$$x(t) = x_c(t) + x_p(t)$$

$$x_p(t) = D \sin(\Omega t - \psi_s)$$

$$D = \frac{P_0/k}{\sqrt{[1 - (\Omega/\omega_n)^2]^2 + (2\zeta\Omega/\omega_n)^2}}$$

$$\tan \psi_s = \frac{2\zeta\Omega/\omega_n}{1 - (\Omega/\omega_n)^2}$$

$$x_c(t) = D_1 e^{\lambda_1 t} + D_2 e^{\lambda_2 t} \qquad \zeta > 1$$

$$x_c(t) = (B + Ct) e^{-\omega_n t} \qquad \zeta = 1$$

$$x_c(t) = e^{-\zeta\omega_n t}(B \cos \omega_d t + C \sin \omega_d t) \qquad \zeta < 1$$

$$\omega_n = \sqrt{k/m} = 2\pi f_n$$

$$f_n = \frac{1}{\tau_n} = \frac{\omega_n}{2\pi}$$

$$\zeta = \frac{c}{2m\omega_n}$$

$$\lambda_{1,2} = -\zeta\omega_n \pm \omega_n \sqrt{\zeta^2 - 1} \qquad \zeta \geq 1$$

$$\omega_d = \omega_n \sqrt{1 - \zeta^2} \qquad \zeta < 1$$

ENGINEERING MECHANICS
DYNAMICS

SECOND EDITION

WILLIAM F. RILEY

Professor Emeritus
Iowa State University

LEROY D. STURGES

Iowa State University

JOHN WILEY & SONS, INC.

New York · Chichester · Brisbane · Toronto · Singapore

ACQUISITIONS EDITOR   Charity Robey
DEVELOPMENTAL EDITOR   Madalyn Stone
MARKETING MANAGER   Debra Riegert
PRODUCTION SERVICE   York Production Services
TEXT DESIGNER   Ann Marie Renzi and Karin Kincheloe
COVER DESIGNER   Karin Kincheloe
MANUFACTURING MANAGER   Mark Cirillo
PHOTO EDITOR   Hilary Newman
ILLUSTRATION EDITOR   Sigmund Malinowski
ILLUSTRATION DEVELOPMENT   Boris Starosta
ELECTRONIC ILLUSTRATIONS   Precision Graphics
COVER PHOTO   Tim Brown/ProFILES West, Inc.

This book was set in Palatino by York Graphic Services, and printed and bound by Von Hoffmann Press. The cover was printed by Phoenix Color.

Recognizing the importance of preserving what has been written, it is a policy of John Wiley & Sons, Inc. to have books of enduring value published in the United States printed on acid-free paper, and we exert our best efforts to that end.

The paper in this book was manufactured by a mill whose forest management programs include sustained yield harvesting of its timberlands. Sustained yield harvesting principles ensure that the number of trees cut each year does not exceed the amount of new growth.

Library of Congress Cataloging in Publication Data:

Riley, William F. (William Franklin). 1925–
    Engineering mechanics : dynamics / William F. Riley, Leroy D. Sturges. — 2nd ed.
        p.   cm.
    Includes index.
    ISBN 0–471–05339–2 (cloth : alk. paper)
    1. Dynamics.   I. Sturges, Leroy D.   II. Title.
TA352.R55 1996
620.1'04—dc20   95–46438
                                                                              CIP

Printed in the United States of America

10 9 8 7 6 5 4 3 2

PREFACE

Our purpose in writing this dynamics book, together with the companion statics book, was to present a fresh look at the subject and to provide a more logical order of presentation of the subject material. We believe our order of presentation will give students a greater understanding of the material and will better prepare students for future courses and later professional life.

INTRODUCTION

This text has been designed for use in undergraduate engineering programs. Students are given a clear, practical, comprehensible, and thorough coverage of the theory normally presented in introductory mechanics courses. Application of the principles of dynamics to the solution of practical engineering problems is demonstrated. This text can also be used as a reference book by practicing aerospace, automotive, civil, mechanical, mining, and petroleum engineers.

Extensive use is made in this text of prerequisite course materials in mathematics and physics. Students entering a dynamics course that uses this book should have a working knowledge of introductory differential and integral calculus, should have taken an introductory course in vector algebra, and should have taken or be enrolled in an introductory course in differential equations.

Vector methods do not always simplify solutions of two-dimensional problems in dynamics. For three-dimensional problems, however, vector algebra provides a systematic procedure that often eliminates errors that might occur with a less systematic approach. In this book, vector algebra is used wherever it provides an efficient solution to a problem. If vector algebra offers no advantages, a scalar approach is used. Likewise, students are encouraged to develop the ability to select the mathematical tools most appropriate for the particular problem they are attempting to solve.

FEATURES

Engineering Emphasis

Throughout this book, a strong emphasis has been placed on the engineering significance of the subject area in addition to the mathematical methods of analysis. Many illustrative example problems have been integrated into the main body of the text at points where the presentation of a method can be best reinforced by the immediate illustration of the method. Students are usually more enthusiastic about a subject if they can see and appreciate its value as they proceed into the subject.

We believe that students can progress in a mechanics course only by understanding the physical and mathematical principles jointly, not by mere memorization of formulas and substitution of data to obtain answers to simple problems. Furthermore, we think that it is better to teach a few fundamental principles for solving problems than to teach a large number of special cases and trick procedures. Therefore the text aims to develop in the student the ability to analyze a given problem in a simple and logical manner and to apply a few fundamental, well-understood principles to its solution.

A conscientious effort has been made to present the material in a simple and direct manner, with the student's point of view constantly in mind.

Coverage of Kinematics Before Kinetics

As we did in the first edition, we have purposefully organized this book to present the complete coverage of kinematics of particles and rigid bodies *before* the discussion of kinetics of particles and rigid bodies.

The best way to learn is to begin with simple situations and progress gradually to more complicated ones. That is the fundamental reason for the way we organize the book. Each principle is applied first to a particle, then to a system of particles, then to a rigid body subjected to a coplanar system of forces, and finally to the general case of a rigid body subjected to a three-dimensional system of forces.

We believe this approach represents both the logical and desirable organization of material for an introductory course in dynamics. Mastery of kinematics is essential for a successful study of kinetics. Kinetics, using Newton's Laws, is developed completely before the student is introduced to work-energy methods and impulse-momentum methods. This way, the student is introduced to each of the methods in a cohesive fashion.

In addition, this organization is also applied to the solving of kinetics problems. The three common methods of solving kinetics problems—(1) force, mass, and acceleration; (2) work and energy; (3) impulse and momentum—are developed, in turn, and applied first to particles, then to a system of particles, then to rigid bodies under plane motion, and finally to general three-dimensional situations.

The progression of material within major sections is from the slightly familiar (from a previous physics course) concepts of particle mechanics to the unfamiliar concepts of two-dimensional rigid-body mechanics to the more complex concepts of three-dimensional rigid-body motion. The more challenging topics are thus spread more evenly over the semester.

The more standard method of presentation found in other books on the market suffers from three difficulties. First, students find it hard to master simultaneously three methods of solution for a problem. Second, all the easy and familiar concepts (dealing with particle kinematics and kinetics) are covered in the first few weeks of the course and all the more difficult and less familiar concepts (dealing with rigid-body kinematics and kinetics) are dealt with in the last few weeks of the course. Third, if only the particle dynamics portion of the course is mastered, the student does not have the ability to perform in future courses requiring competency in rigid-body dynamics.

Although we think our organization makes more sense, we recognize that not all teachers will agree with us. The chapters dealing with particle dynamics

Chapter 13: Kinematics of Particles
Chapter 15: Kinetics of Particles: Newton's Laws
Chapter 17: Kinetics of Particles: Work and Energy Methods
Chapter 19: Kinetics of Particles: Impulse and Momentum

do not depend on the intervening chapters dealing with rigid body dynamics and they can be covered first if so desired. Chapter 21 dealing with mechanical vibrations could be covered anytime after Chapter 15 is completed.

Free-body Diagrams

Most engineers consider the free-body diagram to be the single most important tool for the solution of mechanics problems. The free-body diagram is just as important in dynamics as it is in statics. It is our approach that, whenever an equation of equilibrium or an equation of motion is written, it must be accompanied by a complete, proper free-body diagram.

Problem-solving Procedures

Success in engineering mechanics courses depends, to a surprisingly large degree, on a well-disciplined method of problem solving and on the solution of a large number of problems. The student is urged to develop the ability to reduce problems to a series of simpler component problems that can be easily analyzed and combined to give the solution of the initial problem. Along with an effective methodology for problem decomposition and solution, the ability to present results in a clear, logical, and neat manner is emphasized throughout the text. A first course in mechanics is an excellent place to begin development of this disciplined approach which is so necessary in most engineering work.

Homework Problems

This book contains a large selection of problems that illustrate the wide application of the principles of dynamics to the various fields of engineering. The problems in each set represent a considerable range of difficulty. And, in this edition, we have grouped the homework problems according to this range of difficulty. We believe that a student gains mastery of a subject through application of basic theory to the solution of problems that appear somewhat difficult. Mastery, in general, is not achieved by solving a large number of simple but similar problems. The problems in this text require an understanding of the principles of dynamics without demanding excessive time for computational work. We have replaced approximately 400 homework problems with new ones.

Significant Figures

Results should always be reported as accurately as possible. However, results should not be reported to 10 significant figures merely because the calculator displays that many digits. One of the tasks in all engineering work is to determine the accuracy of the given data and the expected accuracy of the final answer. Results should reflect the accuracy of the given data.

In a textbook, however, it is not possible for students to examine or question the accuracy of the given data. It is also impractical, in an introductory course, to give error bounds on every number. Therefore, since an accuracy greater than about 0.2 percent is seldom possible for practical engineering problems, all given data in Example Problems and Homework Problems, regardless of the number of figures shown, will be assumed sufficiently accurate to justify rounding off the final answer to approximately this degree of accuracy (three to four significant figures).

SI vs. USCS Units

Most large engineering companies deal in an international marketplace. In addition, the use of the International System of Units (SI) is gaining acceptance in the United States. As a result, most engineers must be proficient in both the SI system and the U.S. Customary System (USCS) of units. In response to this need, both U.S. Customary units and SI units are used in approximately equal proportions in the text for both illustrative examples and homework problems. As an aid to the instructor in problem selection, all odd-numbered problems are given in USCS units and even-numbered problems in SI units.

Chapter Summaries and Key Equations

As an aid to students we have written a summary that appears at the end of each chapter. These sections provide a synopsis of the major concepts that are explained in the chapter and can be used by students as a review or study aid. In this edition, we have highlighted key equations in color to help students distinguish them.

Answers Provided

Answers to about half of the problems are included in the back of the book. We believe that the first assignment on a given topic should include some problems for which the answers are given. Since the simpler problems are usually reserved for this first assignment, answers are provided for the first few problems of each article and thereafter are given for approximately half of the remaining problems. The problems whose answers are provided are indicated by an asterisk after the problem number.

DESIGN

Use of Color

In this edition we have continued to use a variety of colors. We believe that color will help students learn mechanics more effectively for two reasons: First, today's visually oriented students are more motivated by texts that depict the real world more accurately. Second, the careful color coding makes it easier for students to understand the figures and text.

Following are samples of figures found in the book. As you can see, force vectors are depicted as red arrows; velocity vectors are depicted as green arrows. Position vectors appear in blue; dimensions as a thin black line; and unit vectors in bold black. This pedagogical use of color is consistent throughout the book.

We have also used color to help students identify the most important study elements. For instance, example problems are always outlined in yellow and important equations appear in a blue box.

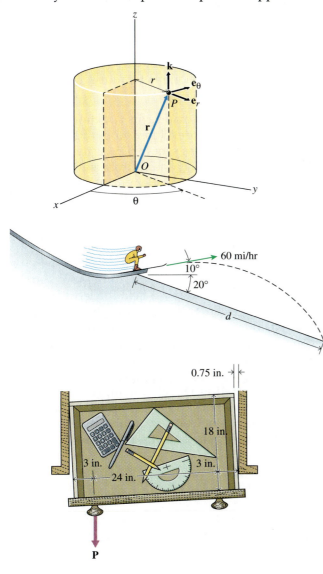

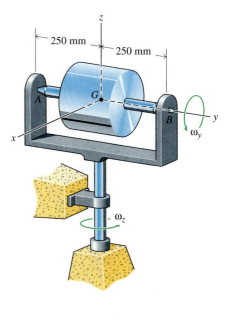

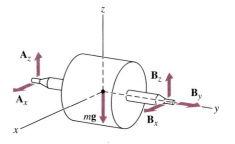

Illustrations

One of the most difficult things for students to do is to visualize engineering problems. Over the years, students have struggled with the lack of realism in mechanics books. We think that mechanics illustrations should be as colorful and three-dimensional as life is. To hold students' attention, we developed the text illustrations with this point in mind.

We started with a basic sketch. Then a specialist in technical illustration added detail. Then the art studio created the figures using *Adobe Illustrator*©. All of these steps enabled us to provide you with the most realistic and accurate illustrations on the market.

Accuracy

After many years of teaching, we appreciate the importance of an accurate text. We have made an extraordinary effort to provide you with an error-free book. Every problem in the text has been worked out at least twice independently; many of the problems have been worked out a third time independently.

DEVELOPMENT PROCESS

This book is the most extensively developed text ever published for the engineering market. The development process involved several steps.

1. **Input from Students.** We have made a special effort to involve our end-users, the students, in the development of this text. An extensive market research survey was sent to 525 current or prior users of the text. We also asked for suggestions from students on student comment cards, and we held two student focus groups, one at the University of Maryland and one at the State University of New York at Stony Brook.

2. **Input from Faculty**. Professors from the United States and Canada carefully reviewed each draft of this manuscript. In addition to these reviews, we asked 7 professors to keep diary reviews of the first edition as they were actually using it in class. To supplement the traditional reviewing process, we held a teleconference focus group with 8 professors across the country. The Wiley sales staff reported hundreds of comments and suggestions from professors across the country. All of these suggestions were carefully considered by the author and editorial team and incorporated whenever feasible.

3. **Manuscript and Illustration Development.** A development editor worked with the authors to hone both the manuscript and the art sketches to their highest potential. A special art developer worked with the authors and the art studio to enhance the illustrations.

SUPPLEMENTS FOR THE BOOK

Solutions Manual

After years of teaching, we realize the importance of an accurate solution manual that matches the quality of the text. For that reason, we have prepared the manual ourselves. The manual includes a complete solution for every problem in the book. Each solution appears with the original problem statement and, where appropriate, the problem figure. We do this for the convenience of the instructor, who no longer will have to refer to both book and solution manual in preparing for class. The manual also contains transparency masters for use in preparing overhead transparencies. In this edition we are making the solutions manual and overheads available in digital form (CD-ROM IBM compatibility) as well as in print form.

Software

For students who want extra help with this course, Wiley offers a package developed by NextGen Texts called *Dynamics Software for Students: Based on the BEST (Basic Engineering Software Tutorial) software developed by the University of Missouri at Rolla.* It is primarily for engineering students who want more practice with additional worked-out example problems. However, instructors will also find it useful for classroom demonstrations. The software requires IBM-PC and compatibles 386 and 486 running Windows 3.1.

ACKNOWLEDGMENTS

Many people participated directly and indirectly in the preparation of this book. In addition to the authors, many present and former colleagues and students contributed ideas concerning methods of presentation, example problems, and homework problems. Final judgments concerning organization of material and emphasis of topics, however, were made by the authors. We will be pleased to receive comments from readers and will attempt to acknowledge all such communications.

We'd like to thank the following people for their suggestions and encouragement throughout the reviewing process.

Jeff Arrington	Abilene Christian University
Leonard B. Baldwin	University of Wyoming
Mark V. Bower	University of Alabama-Huntsville
Phillip Bridges	Mississippi State University
Julie Chen	Boston University
Ravinder Chona	Texas A&M University
Jennifer Cordes	Stevens Institute of Technology
Bruce Dewey	University of Wyoming
Jack Forrester	University of Tennessee-Knoxville
Joseph Ianelli	University of Tennessee-Knoxville
John Kosmatka	University of California-San Diego
Charles Krousgrill	Purdue University
Norman Laws	University of Pittsburgh
Donald Lemke	University of Illinois-Chicago
Mark Lusk	Colorado School of Mines
John McKelliget	University of Massachusetts-Lowell
Robert E. Miller	University of Illinois-Urbana-Champaign
Gaby Neunzert	Colorado School of Mines
Su-Seng Pang	Louisiana State University
Frank Park	University of California-Irvine
William Predebon	Michigan Technological University
Dr. R. A. Raouf	U.S. Naval Academy

Mario Rivera   Union College
George Staab   Ohio State University
Major Curtis Thalken   U.S. Military Academy at West Point
Wallace Venable   West Virginia University
Z. U. A. Warsi   Mississippi State University
Charles White   Northeastern University
Dr. Junku Yuh   University of Hawaii-Manoa

We also want to thank the following students who took the time to complete a survey of the first edition. We used their input to help shape the second edition.

Sam Fortner, Bill Michel   California Maritime Academy
Russ Ahlberg, Jerry Allen,
    Patricia Johns, Shih-Hao Kuo   California State-Long Beach
Gregory J. Arserio, Paul Critchley,
    Mark Thomas   Clarkson
Philip J. Hruska   Cleveland State
Scott Francis   Colorado State
Annie Wing   Des Moines Area Community College
A. Marasco   Drexel
Matthew Ruminski   EEC North
Troy Brodstrom, Aaron Sanders,
    Dean Skinner   General Motors Institute: Engineering
                                          and Tech. Institute
Russ Ladley   Grand Rapids Community College
Matthew Greenlaw   Greenville Tech
Jenny Johannsen, Mitch Wagner,
    Yoke Theng Yoon   Iowa State
Charles Njendu, Jason Winebester   Mississippi State
Duane Bredel, Fred Jones,
    Tommy McBane, Ryan Pardee   Oklahoma State
David Boyd, Robert Harder,
    Roy Hedges, Thomas Smith   Oregon State
Jeff Weatheril   Pacific Lutheran
Albert Abramson, Tiffany M.
    Bonner, Budhi Sayopno,
    Amber L. Master, Parag Patel,
    Charlie Roth, Matt L. Walkeb,
    A. Zimmerman   Rensselaer Polytechnic Institute
Jason Gold   Roger Williams
Rob Allman   Ryerson
D. Jefferson   Spellman College
Valerie Mercer, Christine
    Rudakewycz   Stevens Institute of Technology
Wan Ci Lei   SUNY-Fredonia
Dawson Belobrajdic, Roger Canto,
    Christine Landry, Keli Musick,
    Eric Monkoe, Heath Penland,
    Adam Samuels, Julio Toro   Texas A&M
Renee Bailey, Althea F. Woodson   Tuskegee University
Julia Fernandez   Universidad de Turnabo
Kevin Bates, Charles Burgess,
    Tom Richardson   University of Alabama-Huntsville
Adrienne Cooper, Da-Yu Chang,
    Dioa A. Manly, Hoang Vu   University of California-Irvine

Salem Alshibani, R. Chapman,
  Jacqueline Long   University of Hartford
Eric Lee   University of Hawaii-Manoa
Manuel Bonilla, Kevin Boomsona,
  Danny H. Cruz, Anthony Duran,
  Brian Houston   University of Illinois-Chicago
Marcus Brewer   University of Kansas
Stacy Brown, James Knight   University of Maine
Stephen Ulhorn   University of Miami
Cuyler Larson, Ray Meyers,
  Jennifer L. Shryer   University of Missouri-Rolla
Ban Lin Lau   University of Oklahoma
Neit J. Nieves, Luis Ortiz,
  Javier Rodriguez, Vanessa Rosas,
  Hector Soto   University of Puerto Rico
Rainone Chin   University of Toledo
John Treybal   University of Vermont
Young Mee Cho   University of Virginia
Chad Helgerson, Bishwa Shrestha   University of Wyoming
Mike Capelli, Charles Guerrero,
  Kevin Wallace   U.S. Coast Guard Academy
Ken Nelson, Robert Ozanich,
  Brad Reisinger   U.S. Military Academy-West Point
Mark Elliott   Western Kentucky University
Peter Arangoudakis   Western New England College

William F. Riley
Leroy D. Sturges

A NOTE FROM THE PUBLISHER TO STUDENTS

We wish to acknowledge the extraordinary contributions of a number of engineering students to the development of the second edition of *Engineering Mechanics*. All of our textbooks depend on the experience with and understanding of students' needs that our authors bring to the work. This book benefited from the direct suggestions of students themselves.

Among the many important recommendations these students made were the following:

- They liked the idea of having homework problems grouped according to difficulty. This is a feature we added to this edition.
- They liked the idea of highlighting important equations. They found this to be a helpful study aid, especially when preparing for a test or reviewing the chapter. This is also a new feature to this edition.
- They liked the idea of helpful hints in the margins of worked-out example problems. They didn't want them at the end of the problem, but next to the specific step that they referred to. We added this feature to our text, and we made a concerted effort to place them at the appropriate place that would be the most helpful to students.
- The students wanted to see more photos. Photos make the concepts seem more realistic to them. When asked if they preferred an illustration to a photo on the chapter opener, they overwhelmingly preferred photos. Therefore, we are using photos for chapter openers.
- When the idea of a real-world conceptual example was explained to them, most of the students thought it was a good idea. We have about 2 of these qualitative examples per chapter. This is a new feature we have added to this edition.

These students told us that the software that accompanies the text should be an additional source of worked-out examples to provide further opportunity for learning, enhanced by the power of visualization and movement that software allows. The software that accompanies this book was developed with these suggestions in mind.

The following students participated in a focus group that helped guide the development of this book. We are grateful for their intelligent and thoughtful contributions.

Gavin Appel
Noel Aquino
Ney de Vasconcellos
Frank Fischer
Rosa Genao
Patrick Haran

Robert Ieraci
Paresh Rana
Anita Ristorucci
Anand Sarkar
Doug Stubbe

Charity Robey
Executive Editor

Madalyn Stone
Senior Development Editor

A GUIDE TO THE LEARNING FEATURES OF THIS BOOK

CONCEPTUAL EXAMPLES relate engineering principles to real-world situations using a minimum of math. They help students develop qualitative reasoning and problem-solving skills. We have created 16 Conceptual Examples for this edition, which are highlighted in the Table of Contents.

CONCEPTUAL EXAMPLE 13-1: THE GRAVITATIONAL CONSTANT-g

What is the difference between the gravitational constant g and the acceleration of gravity?

SOLUTION

According to Newton's Law of Gravitation, the earth exerts an attractive force

$$F_g = \frac{GM_e m}{r^2} \qquad (i)$$

on a body of mass m where G is the universal gravitational constant, M_e is the mass of the earth (constant), and r is the distance from the center of the earth to the body. If the body remains near the surface of the earth, then $r \cong r_e =$ constant and Eq. i can be written

$$F_g = mg \qquad (ii)$$

where

$$g = \frac{GM_e}{r^2_e} = \text{constant} \qquad (iii)$$

is called the gravitational constant. If the gravitational force is the only force acting on the body (for example, when the high diver jumps off of the platform), then the acceleration of the body is equal to the gravitational constant ($a = g$) and it acts in the direction of the gravitational force. Such motions are called projectile motions.

The gravitational constant defined by Eqs. ii and iii, however, is independent of the acceleration of the body. The gravitational force $F_g = mg$ acts on the high diver's body whether she is standing still on the platform preparing to dive or she is falling toward the water. In the first case, the gravitational force is opposed by the platform, the acceleration of the diver is zero, and the diver is in equilibrium. In the second case, the gravitational force is unopposed (assuming that air resistance is negligible) and the acceleration of the diver is $a = g$, downward.

Fig. CE13-1

CONCEPTUAL EXAMPLE 20-1: CONSERVATION OF ANGULAR MOMENTUM

How does a figure skater control her spinning by extending her arms or pulling her arms in close to her body?

SOLUTION

According to the angular impulse–momentum principle

$$(I\omega)_i + \int_i^f \textstyle\sum M \, dt = (I\omega)_f$$

the angular rotation rate of a body can be changed by a moment about the axis of rotation. However, the moments produced by air resistance and by ice friction are so small that they have little effect over short time periods such as 30 s to 60 s. Therefore, the skater's angular momentum is conserved

$$(I\omega)_i = (I\omega)_f$$

If the skater's body configuration does not change ($I_i = I_f$), then the skater's rotation rate will not change either ($\omega_i = \omega_f$). If the skater's body configuration does change, then her rotation rate must also change.

The figure skater starts her spin about a vertical axis with both arms and one leg outstretched. Although her arms and leg contribute only about 20 percent of her total body mass, the moment of inertia of a piece of mass dm is proportional to the square of its distance from the axis of rotation, $dI = r^2 \, dm$. By pulling her arms and leg in next to her body, the skater reduces her moment of inertia to about one-half of its initial value and nearly doubles her rate of rotation.

(a)   (b)

Fig. CE20-1

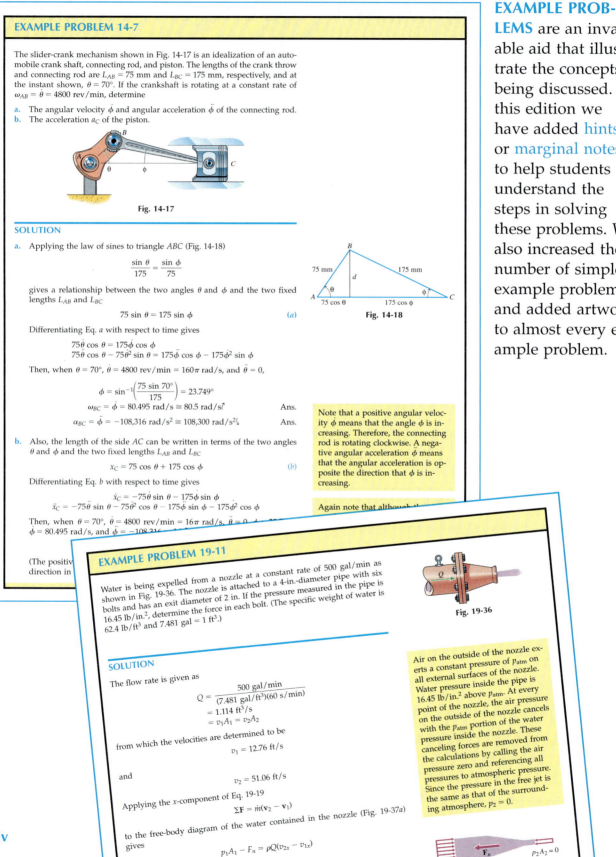

EXAMPLE PROBLEM 14-7

The slider-crank mechanism shown in Fig. 14-17 is an idealization of an automobile crank shaft, connecting rod, and piston. The lengths of the crank throw and connecting rod are $L_{AB} = 75$ mm and $L_{BC} = 175$ mm, respectively, and at the instant shown, $\theta = 70°$. If the crankshaft is rotating at a constant rate of $\omega_{AB} = \dot{\theta} = 4800$ rev/min, determine

a. The angular velocity $\dot{\phi}$ and angular acceleration $\ddot{\phi}$ of the connecting rod.
b. The acceleration a_C of the piston.

Fig. 14-17

SOLUTION

a. Applying the law of sines to triangle ABC (Fig. 14-18)

$$\frac{\sin \theta}{175} = \frac{\sin \phi}{75}$$

gives a relationship between the two angles θ and ϕ and the two fixed lengths L_{AB} and L_{BC}

$$75 \sin \theta = 175 \sin \phi \qquad (a)$$

Differentiating Eq. a with respect to time gives

$$75\dot{\theta} \cos \theta = 175\dot{\phi} \cos \phi$$
$$75\ddot{\theta} \cos \theta - 75\dot{\theta}^2 \sin \theta = 175\ddot{\phi} \cos \phi - 175\dot{\phi}^2 \sin \phi$$

Then, when $\theta = 70°$, $\dot{\theta} = 4800$ rev/min $= 160\pi$ rad/s, and $\ddot{\theta} = 0$,

$$\phi = \sin^{-1}\left(\frac{75 \sin 70°}{175}\right) = 23.749°$$

$$\omega_{BC} = \dot{\phi} = 80.495 \text{ rad/s} \cong 80.5 \text{ rad/s} \curvearrowright \qquad \text{Ans.}$$

$$\alpha_{BC} = \ddot{\phi} = -108,316 \text{ rad/s}^2 \cong 108,300 \text{ rad/s}^2 \curvearrowleft \qquad \text{Ans.}$$

b. Also, the length of the side AC can be written in terms of the two angles θ and ϕ and the two fixed lengths L_{AB} and L_{BC}

$$x_C = 75 \cos \theta + 175 \cos \phi \qquad (b)$$

Differentiating Eq. b with respect to time gives

$$\dot{x}_C = -75\dot{\theta} \sin \theta - 175\dot{\phi} \sin \phi$$
$$\ddot{x}_C = -75\ddot{\theta} \sin \theta - 75\dot{\theta}^2 \cos \theta - 175\ddot{\phi} \sin \phi - 175\dot{\phi}^2 \cos \phi$$

Then, when $\theta = 70°$, $\dot{\theta} = 4800$ rev/min $= 16\pi$ rad/s, $\ddot{\theta} = 0$
$\dot{\phi} = 80.495$ rad/s, and $\ddot{\phi} = -108\,316$

(The positiv...
direction in...

Fig. 14-18
75 mm   175 mm
d
θ ϕ
A 75 cos θ   175 cos φ   C

Note that a positive angular velocity $\dot{\phi}$ means that the angle ϕ is increasing. Therefore, the connecting rod is rotating clockwise. A negative angular acceleration $\ddot{\phi}$ means that the angular acceleration is opposite the direction that $\dot{\phi}$ is increasing.

Again note that although...

EXAMPLE PROBLEM 19-11

Water is being expelled from a nozzle at a constant rate of 500 gal/min as shown in Fig. 19-36. The nozzle is attached to a 4-in.-diameter pipe with six bolts and has an exit diameter of 2 in. If the pressure measured in the pipe is 16.45 lb/in.^{2}, determine the force in each bolt. (The specific weight of water is 62.4 lb/ft^3 and 7.481 gal = 1 ft^3.)

Fig. 19-36

SOLUTION

The flow rate is given as

$$Q = \frac{500 \text{ gal/min}}{(7.481 \text{ gal/ft}^3)(60 \text{ s/min})}$$
$$= 1.114 \text{ ft}^3/\text{s}$$
$$= v_1 A_1 = v_2 A_2$$

from which the velocities are determined to be

$$v_1 = 12.76 \text{ ft/s}$$

and

$$v_2 = 51.06 \text{ ft/s}$$

Applying the x-component of Eq. 19-19

$$\Sigma \mathbf{F} = \dot{m}(\mathbf{v}_2 - \mathbf{v}_1)$$

to the free-body diagram of the water contained in the nozzle (Fig. 19-37a) gives

$$p_1 A_1 - F_n = \rho Q(v_{2x} - v_{1x})$$

or

$$62.4 \text{ ...... } (51.06 - 12.76)$$

$p_1 A_1$ F_n $p_2 A_2 = 0$
y

Air on the outside of the nozzle exerts a constant pressure of p_{atm} on all external surfaces of the nozzle. Water pressure inside the pipe is 16.45 lb/in.^{2} above p_{atm}. At every point of the nozzle, the air pressure on the outside of the nozzle cancels with the p_{atm} portion of the water pressure inside the nozzle. These canceling forces are removed from the calculations by calling the air pressure zero and referencing all pressures to atmospheric pressure. Since the pressure in the free jet is the same as that of the surrounding atmosphere, $p_2 = 0$.

EXAMPLE PROBLEMS are an invaluable aid that illustrate the concepts being discussed. In this edition we have added hints or marginal notes to help students understand the steps in solving these problems. We also increased the number of simpler example problems, and added artwork to almost every example problem.

In this edition we have grouped the homework problems according to the degree of difficulty.

PROBLEMS

Introductory Problems

13-162 An eagle riding an updraft travels along an elliptical helical path (Fig. P13-162) described by the relations

$$\mathbf{r} = 15 \cos 0.2t\mathbf{i} + 10 \sin 0.2t\mathbf{j} + 0.8t\mathbf{k} \text{ m}$$

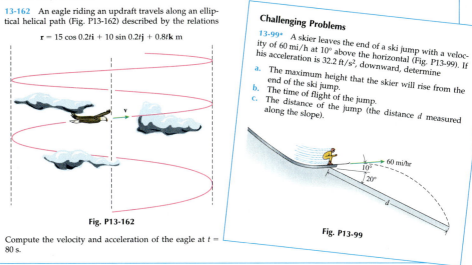

Fig. P13-162

Compute the velocity and acceleration of the eagle at $t = 80$ s.

Challenging Problems

13-99* A skier leaves the end of a ski jump with a velocity of 60 mi/h at 10° above the horizontal (Fig. P13-99). If his acceleration is 32.2 ft/s², downward, determine

a. The maximum height that the skier will rise from the end of the ski jump.
b. The time of flight of the jump.
c. The distance of the jump (the distance d measured along the slope).

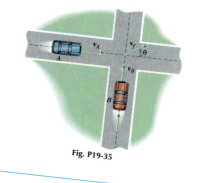

Fig. P13-99

At the end of every chapter, **REVIEW PROBLEMS** test students on all the concepts covered in the chapter. They often integrate topics and thus can employ realistic applications.

COMPUTER PROBLEMS allow students to tackle more realistic problems that would be difficult to solve without the use of a computer.

REVIEW PROBLEMS

21-135* A 60-lb child bounces up and down on a pair of elastic bands as shown in Fig. P21-135. If the amplitude of the oscillation is observed to decrease by 3 percent every 5 cycles and the 5 cycles take 6.5 s, determine the elastic modulus k and damping coefficient c of the elastic bands.

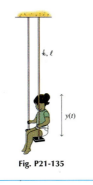

Fig. P21-135

Computer Problems

C19-35 Two cars collide at an intersection as shown in Fig. P19-35. Car A weighs 2200 lb and has an initial speed of $v_A = 30$ mi/h, whereas car B weighs 3500 lb. On impact, the wheels of both cars lock and both cars slide ($\mu_k = 0.3$). After impact, the cars become entangled and move as a single unit. Calculate and plot

a. The (common) speed v_f of the two cars immediately after the collision as a function of the initial speed of car B for 10 mi/h $< v_B <$ 50 mi/h.
b. The location (y versus x) where the cars will come to rest for 10 mi/h $< v_B <$ 50 mi/h. (Let the x-axis coincide with the initial path of car A and the y-axis coincide with the initial path of car B.)

Fig. P19-35

CONTENTS

*T*he batter has less than a half of a second to determine the trajectory of the ball and decide whether or not to swing the bat. The force of collision between the bat and the ball can be as high as 8000 lb, and the impact changes the velocity of the ball from 90 mi/h toward the plate to 110 mi/h toward the outfield fence.

Chapter 12

GENERAL PRINCIPLES

12-1 INTRODUCTION TO DYNAMICS

Mechanics has been defined as the branch of the physical sciences that deals with the response of bodies to the action of forces. For convenience, the study of mechanics is divided into three parts: namely, the mechanics of rigid bodies, the mechanics of deformable bodies, and the mechanics of fluids. Rigid-body mechanics can be further subdivided into **statics** (equilibrium of a rigid body) and **dynamics** (motion of a rigid body).

The portion of mechanics known as statics developed early in recorded history because the principles of statics are needed in building construction. The builders of the pyramids of Egypt understood and used such devices as the lever, the pulley, and the inclined plane.

The portion of mechanics known as dynamics developed much later, since the quantities involved (velocity and acceleration) depend on an accurate measurement of time. The experiments of Galileo Galilei (1564–1642) with falling bodies, pendulums, and rolling cylinders on an inclined plane started development of the field of dynamics. However, Galileo was handicapped in his work by a lack of adequate clocks for measuring the small time intervals involved in the experiments. Huygens (1629–1695) continued the work of Galileo and invented the pendulum clock. He also determined the acceleration of gravity and introduced theorems involving centrifugal force. Sir Isaac Newton (1642–1727) completed the formulation of the basic principles of mechanics with his discovery of the law of universal gravitation and his statement of the laws of motion. Newton's work on particles, based on geometry, was extended to rigid-body systems by Euler (1707–1793), who was also the first to use the term *moment of inertia*. D'Alembert

(1717–1783) introduced the concept of an inertia force. The sum of the forces on a body in motion is zero when an inertia force is included. Previous work in mechanics, which had been based largely on astronomical observations and geometrical concepts, was formalized by Lagrange (1736–1813), who derived the generalized equations of motion analytically by using energy concepts. The derivation of these equations, which are known as Lagrange's equations, represented a significant advancement in the development of classical mechanics. Another major advancement was made by Coriolis (1792–1843), who showed how the introduction of additional terms could validate Newton's laws when the reference frame is rotating.

The next major advances in mechanics were made by Max Planck (1858–1947), who formulated quantum mechanics, and Albert Einstein (1879–1955), who formulated the theory of relativity (1905). These new theories did not repudiate Newtonian mechanics; they were simply more general. Newtonian mechanics was and is applicable to the prediction of the motion of bodies where the speeds are small compared to the speed of light.

12-2 NEWTON'S LAWS

The foundations for studies in engineering mechanics are the laws formulated and published by Sir Isaac Newton in 1687. In a treatise called "The Principia," Newton stated the basic laws governing the motion of a particle as:[1]

Newton's Laws of Motion

Law 1. Every body perseveres in its state of rest or of uniform motion in a straight line, except in so far as it is compelled to change that state by impressed forces.

Law 2. Change of motion is proportional to the moving force impressed, and takes place in the direction of the straight line in which such force is impressed.

Law 3. Reaction is always equal and opposite to action; that is to say, the actions of two bodies upon each other are always equal and directly opposite.

These laws, which have come to be known as "Newton's Laws of Motion," are commonly expressed today as:

Law 1. A particle originally at rest will remain at rest. A particle originally moving with a constant velocity will continue to move with a constant velocity along a straight line, unless the particle is acted on by an unbalanced force.

Law 2. When a particle is acted on by an unbalanced force, the parti-

---

[1]Dr. Ernst Mach, "Die Mechanik in ihrer Entwickelung historisch-kritisch dargestellt," Professor an der Universitat zu Wien. Mit 257 Abbildungen. Leipzig, 1893. First translated from the German by Thomas J. McCormack in 1902. *The Science of Mechanics*, 9th ed. The Open Court Publishing Company, LaSalle, Ill., 1942.

cle will be accelerated in the direction of the force; the magnitude of the acceleration will be directly proportional to the force and inversely proportional to the mass of the particle.

Law 3. For every action there is an equal and opposite reaction. The forces of action and reaction between contacting bodies are equal in magnitude, opposite in direction, and collinear.

Newton's three laws were developed from a study of planetary motion (the motion of particles). During the eighteenth century, Leonhard Euler (1707–1783) extended Newton's work on particles to rigid-body systems.

The first law covers the case where a body is in equilibrium. Thus, the first law provides the foundation for the study of statics. The second law deals with accelerated motion of a body and provides the foundation for the study of dynamics.

The law that governs the mutual attraction between two isolated bodies was also formulated by Newton and is known as the **"Law of Gravitation."** The law of gravitation is very important in all studies involving the motion of planets, space craft, or artificial satellites.

12-3 FUNDAMENTAL QUANTITIES OF MECHANICS

The fundamental quantities of mechanics are space, time, mass, and force. Three of these quantities—space, time, and mass—are absolute quantities. This means that they are independent of each other and cannot be expressed in terms of the other quantities or in simpler terms. A force is not independent of the other three quantities but is related to the mass of the body and to the manner in which the velocity of the body varies with time. A brief description of these four quantities together with some other concepts of importance in dynamics follows:

Space is the geometric region commonly referred to as "the universe." The region extends without limit in all directions.

Time is the interval between two events. Measurement of this interval is made by making comparisons with some reproducible event such as the time required for the earth to rotate on its axis.

Matter is any substance that occupies space.

A **body** is matter bounded by a closed surface.

Inertia is the property of matter that causes resistance to a change in motion.

Mass is a quantitative measure of a body's resistance to a change in its motion.

A **force** is the action of one body on another body. Forces always occur in pairs, and the two forces have equal magnitude and opposite sense. The external effect of a force on a body is either development of resisting forces (reactions) on the body (statics problems) or accelerated motion of the body (dynamics problems).

A **particle** is an object whose size and shape can be ignored when studying its motion. Only the position of the mass center of the particle need be considered. The orientation of the object or rotation of the object plays no role in describing the motion of a particle. Particles can be very small or very large. Small does not always guarantee that an object can be modeled as a particle; large does not always prevent modeling an object as a particle. Whether a body is large or small relates to the length of path followed and/or the distance between bodies.

A **rigid body** is a collection of particles that remain at fixed distances from each other at all times and under all conditions of loading. The rigid-body concept represents an idealization of the true situation, since all real bodies will change shape to a certain extent when they are subjected to a system of forces. When it is assumed that the body is rigid (free of deformation), the material properties of the body are not required for the analysis of forces and their effects on the body. The bodies dealt with in this book, with the exception of deformable springs, will be considered to be rigid bodies.

The **position** of a point in space is specified by using linear and angular measurements with respect to a coordinate system whose origin is located at some reference point. The basic reference system used as an aid in solving mechanics problems is a primary inertial system, which is an imaginary set of rectangular axes that do not translate or rotate in space. Measurements made relative to this system are called absolute. The laws of Newtonian mechanics are valid for this reference system as long as any velocities involved are negligible with respect to the speed of light, which is 300,000 km/s (186,000 mi/s). A reference frame fixed to the surface of the earth moves with respect to the primary inertial system; however, corrections to account for the absolute motion of the earth are insignificant and may be neglected for most engineering problems involving machines and structures that remain on the surface of the earth.

12-4   UNITS OF MEASUREMENT

The physical quantities used to express the laws of mechanics are mass, length, force, time, velocity, acceleration, and so on. These quantities can be divided into fundamental quantities and derived quantities.

Fundamental quantities cannot be defined in terms of other physical quantities. The number of quantities regarded as fundamental is the minimum number needed to give a consistent and complete description of all of the physical quantities ordinarily encountered in the subject area. Examples of quantities viewed as fundamental in mechanics are length and time.

Derived quantities are those whose defining operations are based on measurements of other physical quantities. Examples of derived quantities in mechanics are area, volume, velocity, and acceleration.

Mass and force are examples of quantities that may be viewed as either fundamental or derived. In the SI system of units, mass is regarded as a fundamental quantity and force is a derived quantity. In the U.S. Customary System of units, force is regarded as a fundamental quantity and mass is a derived quantity.

Units of Length   The magnitude of each of the fundamental quantities is defined by an arbitrarily chosen unit or "standard." The familiar yard, foot, and inch, for example, come from the old practice of using the human arm, foot, and thumb as length standards. The first truly international standard of length was a bar of platinum-iridium alloy, called the standard meter,[2] which is kept at the International Bureau of Weights and Measures in Sèvres, France. The distance between two fine lines engraved on gold plugs near the ends of the bar is defined to be one meter. Historically, the meter was intended to be one ten-millionth of the distance from the pole to the equator along the meridian line through Paris. Accurate measurements made after the standard meter bar was constructed show that it differs from its intended value by approximately 0.023 percent.

In 1961 an atomic standard of length was adopted by international agreement. The wavelength in vacuum of the orange-red line from the spectrum of isotope krypton 86 was chosen. One meter (m) is now defined to be 1,650,763.73 wavelengths of this light. The choice of an atomic standard offers advantages other than increased precision in length measurements. Krypton 86 is readily available everywhere; the material is relatively inexpensive, and all atoms of the material are identical and emit light of the same wavelength. The particular wavelength chosen is uniquely characteristic of krypton 86 and is very sharply defined.

The definition of the yard, by international agreement, is 1 yard = 0.9144 m, exactly.[3] Thus, 1 inch = 25.4 mm, exactly; and 1 foot = 0.3048 m, exactly.

Units of Time   Similarly, time can be measured in a number of ways. Since early times, the length of the day has been an accepted standard of time measurement. The internationally accepted standard unit of time, the second (s), was defined in the past as 1/86,400 of a mean solar day or 1/31,557,700 of a mean solar year. Time defined in terms of the rotation of the earth was based on astronomical observations. Since these observations require at least several weeks, a good secondary terrestrial measure, calibrated by astronomical observations, was required for practical use. Quartz crystal clocks, based on the electrically sustained natural periodic vibrations of a quartz wafer, are used as secondary time standards for scientific work. The best of these quartz clocks keep time for a year with a maximum error of 0.02 s. To meet the need for an even better time standard, an atomic clock has been developed that uses the periodic atomic vibrations of isotope cesium 133. The second based on this cesium clock was adopted as the time standard by the Thirteenth General Conference on Weights and Measures in 1967. The second is defined as the duration of 9,192,631,770 cycles of vibration of isotope cesium 133. The cesium clock provides a significant improvement over the accuracy associated with other methods based on astronomical observations. Two cesium clocks will differ by no more than one second after running 3000 years.

Units of Mass and Weight   The standard unit of mass, the kilogram (kg), is defined by a cylinder of platinum-iridium alloy that is kept at the International Bureau of Weights and Measures in Sèvres, France.

---

[2]The United States has accepted the meter as a standard of length since 1893.
[3]*Guide for the Use of the International System of Units,* National Institute of Standards and Technology (NIST) Special Publication 811, September 1991.

In the SI System of Units, mass is considered to be a fundamental quantity and force is then a derived quantity. The unit of force, called a **newton** (N), is *the force required to give one kilogram of mass an acceleration of one meter per second squared.* Thus, $1 \text{ N} = 1\text{kg} \cdot \text{m/s}^2$. Since it is based on mass which is a measure of the inertia or resistance to change in motion, the SI System of Units is an **inertial** or **absolute system of units.**

In the U.S. Customary System of Units, the unit of force is called a pound (lb) and is defined as the weight at sea level and at a latitude of 45 degrees of a platinum standard, which is kept at the National Institute of Standards and Technology (NIST) in Washington, D.C. This platinum standard has a mass of 0.453,592,43 kg. The unit of mass in this system is derived and is called a slug. One slug is the mass that is accelerated one foot per second squared by a force of one pound, or 1 slug equals $1 \text{ lb} \cdot \text{s}^2/\text{ft}$. Since the weight of the platinum standard depends on the gravitational attraction of the earth, the U.S. Customary System is a **gravitational system** rather than an absolute system of units.

As an aid to interpreting the physical significance of answers in SI units for those more accustomed to the U.S. Customary System, some conversion factors for the quantities normally encountered in mechanics are provided in Table 12-1.

TABLE 12-1 CONVERSION FACTORS BETWEEN THE SI AND U.S. CUSTOMARY SYSTEMS

Quantity	U.S. Customary to SI	SI to U.S. Customary
Length	1 in. = 25.40 mm	1 m = 39.37 in.
	1 ft = 0.3048 km	1 m = 3.281 ft
	1 mi = 1.609 km	1 km = 0.6214 mi
Area	$1 \text{ in.}^2 = 645.2 \text{ mm}^2$	$1 \text{ m}^2 = 1550 \text{ in.}^2$
	$1 \text{ ft}^2 = 0.0929 \text{ m}^2$	$1 \text{ m}^2 = 10.76 \text{ ft}^2$
Volume	$1 \text{ in.}^3 = 16.39(10^3) \text{ mm}^3$	$1 \text{ mm}^3 = 61.02(10^{-6}) \text{ in.}^3$
	$1 \text{ ft}^3 = 0.02832 \text{ m}^3$	$1 \text{ m}^3 = 35.31 \text{ ft}^3$
	$1 \text{ gal} = 3.785 \text{ L}^a$	1 L = 0.2642 gal
Velocity	1 in./s = 0.0254 m/s	1 m/s = 39.37 in./s
	1 ft/s = 0.3048 m/s	1 m/s = 3.281 ft/s
	1 mi/h = 1.609 km/h	1 km/h = 0.6214 mi/h
Acceleration	$1 \text{ in./s}^2 = 0.0254 \text{ m/s}^2$	$1 \text{ m/s}^2 = 39.37 \text{ in./s}^2$
	$1 \text{ ft/s}^2 = 0.3048 \text{ m/s}^2$	$1 \text{ m/s}^2 = 3.281 \text{ ft/s}^2$
Mass	1 slug = 14.59 kg	1 kg = 0.06854 slug
Second moment of area	$1 \text{ in.}^4 = 0.4162(10^6) \text{ mm}^4$	$1 \text{ mm}^4 = 2.402(10^{-6}) \text{ in.}^4$
Force	1 lb = 4.448 N	1 N = 0.2248 lb
Distributed load	1 lb/ft = 14.59 N/m	1 kN/m = 68.54 lb/ft
Pressure or stress	1 psi = 6.895 kPa	1 kPa = 0.1450 psi
	1 ksi = 6.895 MPa	1 MPa = 145.0 psi
Bending moment or torque	$1 \text{ ft} \cdot \text{lb} = 1.356 \text{ N} \cdot \text{m}$	$1 \text{ N} \cdot \text{m} = 0.7376 \text{ ft} \cdot \text{lb}$
Work or energy	$1 \text{ ft} \cdot \text{lb} = 1.356 \text{ J}$	$1 \text{ J} = 0.7376 \text{ ft} \cdot \text{lb}$
Power	$1 \text{ ft} \cdot \text{lb/s} = 1.356 \text{ W}$	$1 \text{ W} = 0.7376 \text{ ft} \cdot \text{lb/s}$
	1 hp = 745.7 W	1 kW = 1.341 hp

^{a}Both L and l are accepted symbols for liter. Because "l" can easily be confused with the numeral "1," the symbol "L" is recommended for United States use by the National Institute of Standards and Technology (see NIST Special Publication 811, September 1991).

12-5 DIMENSIONAL CONSIDERATIONS

All the physical quantities encountered in engineering mechanics can be expressed dimensionally in terms of three fundamental or base quantities: mass, length, and time, denoted respectively by M, L, and T. The dimensions of other physical quantities follow from their definitions or from physical laws. For example, the dimension of velocity follows from the definition of velocity as the quotient of length and time (L/T). From Newton's second law, force is defined as the product of mass and acceleration; therefore, force has the dimension ML/T^2. The dimensions of other physical quantities commonly encountered in engineering mechanics are given in Table 12-2.

12-5.1 Dimensional Homogeneity

When an equation is used to describe a physical process, the equation is said to be dimensionally homogeneous if the form of the equation does not depend on the units of measurement. For example, the equation describing the distance h a body travels when released at rest in the earth's gravitational field is $h = gt^2/2$, where h is the distance traveled, t is the time since release, and g is the gravitational constant. This equation is valid whether length is measured in feet, meters, or inches and whether time is measured in hours, years, or seconds, provided g is measured in the same units of length and time as h and t. Dimen-

TABLE 12-2   DIMENSIONS OF THE PHYSICAL QUANTITIES OF MECHANICS

Physical Quantity	Dimension	Common Units	
		SI System	U.S. Customary System
Length	L	m, mm	in., ft
Area	L^2	m², mm²	in.², ft²
Volume	L^3	m³, mm³	in.³, ft³
Angle	$1\ (L/L)$	rad, degree	rad, degree
Time	T	s	s
Linear velocity	L/T	m/s	ft/s
Linear acceleration	L/T^2	m/s²	ft/s²
Angular velocity	$1/T$	rad/s	rad/s
Angular acceleration	$1/T^2$	rad/s²	rad/s²
Mass	M	kg	slug
Force	ML/T^2	N	lb
Moment of a force	ML^2/T^2	N · m	ft · lb
Pressure	M/LT^2	Pa, kPa	psi, ksi
Stress	M/LT^2	Pa, MPa	psi, ksi
Energy	ML^2/T^2	J	ft · lb
Work	ML^2/T^2	J	ft · lb
Power	ML^2/T^3	W	hp
Linear impulse	ML/T	N · s	lb · s
Momentum	ML/T	N · s	lb · s
Specific weight	M/L^2T^2	N/m³	lb/ft³
Density	M/L^3	kg/m³	slug/ft³
Second moment of area	L^4	m⁴, mm⁴	in.⁴, ft⁴
Moment of inertia	ML^2	kg · m²	slug · ft²

sionally homogeneous equations are usually preferred because of the potential confusion connected with the units of constants appearing in dimensionally inhomogeneous equations.

12-6   METHOD OF PROBLEM SOLVING

The principles of mechanics are few and relatively simple; however, the applications are infinite in their number, variety, and complexity. Success depends to a large degree on a well-disciplined method of problem solving. Problem solving typically consists of the following phases.

Three Phases of Professional Problem Solving

1. Problem definition and identification.
2. Model development and simplification.
3. Mathematical solution and result interpretation.

The problem-solving method outlined in this section will prove useful for this course, for the engineering courses that follow, and for most situations encountered later in engineering practice.

Problem Definition and Identification   Problems in engineering mechanics (statics, dynamics, and mechanics of deformable bodies) are concerned primarily with the external effects of a system of forces on a physical body. The approach usually followed in solving an engineering mechanics problem requires identification of all external forces acting on the body of interest. This can be accomplished by preparing a **free-body diagram** that shows the body of interest isolated from all other bodies and with all external forces applied. This simple tool provides a powerful means for distinguishing between cause (the external forces) and effect (motion or deformation of the body of interest) and helps focus attention on the principles and information required for solution of the problem.

Model Development and Simplification   Since relationships between cause and effect are stated in mathematical form, the true physical situation must be represented by a mathematical model in order to obtain the required solution. Often it is necessary to make simplifying assumptions or approximations in setting up this model in order to solve the problem. The most common approximation is to treat most of the bodies in statics and dynamics problems as rigid bodies. No real body is absolutely rigid; however, the changes in shape of a real body usually have a negligible effect on the acceleration produced by a force system or on the reactions required to maintain equilibrium of the body. Considerations of the changes in shape under these circumstances would be an unnecessary complication of the problem.

Mathematical Solution and Result Interpretation Usually, an actual physical problem cannot be solved exactly or completely. However, even in complicated problems, a simplified model can provide good qualitative results. Appropriate interpretation of such results can lead to approximate predictions of physical behavior or be used to verify the "reasonableness" of more sophisticated analytical or numerical results. The engineer must constantly be aware of the actual physical problem being considered and of any limitations associated with the mathematical model being used. Assumptions must be continually evaluated to ensure that the mathematical problem being solved provides an adequate representation of the physical process or device of interest.

The most effective way to learn the material contained in engineering mechanics courses is to solve a variety of problems. In order to become an effective engineer, the student must develop the ability to reduce complicated problems to simple parts that can be easily analyzed and to present results of the work in a clear, logical, and neat manner. This can be accomplished by using the following sequence of steps.

Steps for Analyzing and Solving Problems

1. Read the problem carefully.
2. Identify the result requested.
3. Identify the principles to be used to obtain the result.
4. Tabulate the information provided.
5. Draw the appropriate free-body diagrams.
6. Apply the appropriate principles and equations.
7. Report the answer with the appropriate number of significant figures and the appropriate units.
8. Study the answer and determine if it is reasonable.

Accepted engineering practice requires that all work be neat and orderly, since neatness is usually associated with clear and orderly thinking. Sloppy solutions that cannot be easily read and understood by others, because they contain superfluous or confusing details, have little or no value. Developing an ability to apply an orderly approach to problem solving is a significant part of an engineering education. Also note that the problem identification, model simplification, and result interpretation phases of engineering problem solving are often more important than the mathematical solution phase. No matter how elegantly or cleverly or exactly the mathematical solution is, it is of no value if it does not accurately reflect the physical problem that was to be solved.

12-7 SIGNIFICANCE OF NUMERICAL RESULTS

The speed and accuracy of pocket electronic calculators facilitates the numerical computations involved in the solution of engineering problems. However, the number of significant figures that can easily be obtained should not be taken as an indication of the accuracy of the solution. Rather, *final results should be reported with a number of significant figures that is indicative of the accuracy of the solution.*

The accuracy of solutions to real engineering problems depends on three factors: the accuracy of the known physical data, the accuracy of the physical model, and the accuracy of the computations performed.

Accuracy of the Known Physical Data   Calculated results should always be rounded off to the number of significant figures that will yield the same degree of accuracy as the data on which they are based. Engineering data is seldom known to an accuracy greater than 0.2 percent. A practical rule for rounding off the final numbers obtained in the computations involved in engineering analysis, one that provides answers to approximately this degree of accuracy, is to retain four significant figures for numbers beginning with the figure "1" and to retain three significant figures for numbers beginning with any figure from "2" through "9".

Accuracy of the Physical Model   In order to obtain a solution to a problem, the true physical situation must be represented by a mathematical model. Of course, the final result is only as accurate as the physical model used. Care must be exercised to ensure that the model and the associated mathematical problem being solved provide an adequate representation of the physical process or device that they represent.

Often it is necessary to make simplifying assumptions or approximations in setting up a mathematical model of a real physical situation. Common assumptions are that bodies are rigid and do not deform, that air resistance has a negligible effect on the motion of a projectile, and that gravity is constant and independent of latitude or altitude. Any assumption that is not obvious should be clearly stated at the beginning of the problem. Once the solution is completed, the appropriateness of these assumptions should be checked. If the results are not consistent with the assumptions used to produce the results, then the assumptions must be modified and the problem solved again.

Accuracy of the Computations Performed   Even if the physical model and the physical data are exact, computations almost always involve some round-off error. For example, calculators and computers round off fractions like "1/3" and values of functions like "sin 60°". The number of significant figures in the result depends on the exact machine used but is generally between 7 and 10 significant figures. Arithmetic operations on these approximate numbers reduces the number of significant figures even further. A common practice is to treat the result as having one less significant figure than either of the numbers used to produce it. However, when two approximately equal numbers are subtracted, the result may have only one significant digit.

Extreme accuracy is not the goal of every problem solution. A conceptual study may require only a crude simple model with approximate data and a final accuracy of 50 to 100 percent. A feasibility study would require a more accurate model with more representative data and a final accuracy of 5 to 20 percent. A final, detailed design would seek the most accurate model and the best data available. *It is the engineer's task to estimate how accurate the model and the data are and to re-*

port the results accordingly. Results reported with too many significant digits are not only misleading as to the real significance of the answer, they may also lead to an unexpected failure and tragedy.

In a book, of course, students have no way of knowing how accurate the given data is. When a length is given as 200 mm, there is no way of knowing whether the real measurement was 245 mm and rounded down to 200 mm (1 significant digit), or 204 mm and rounded down to 200 mm (2 significant digits), or 200.4 mm and rounded down to 200 mm (3 significant digits), and so on. Therefore, the example problems in this book are solved with (and the homework problems should be solved with) the assumption that the data provided are sufficiently accurate to justify at least three or four significant figures in the final answer. Intermediate calculations are (should be) worked out to one or two significant figures more than are reported in the final answer.

SUMMARY

Mechanics is the branch of the physical sciences that deals with the response of bodies to the action of forces. Mechanics is based on Newton's laws. Statics is the study of bodies in equilibrium and is based primarily on Newton's first and third laws. Dynamics is the study of bodies in motion and is based primarily on Newton's second and third laws.

Dynamics problems in which the motions are known are similar to statics problems, in that application of Newton's laws leads to equations that can be easily solved for the unknown forces. Dynamics problems in which some aspects of the motion are not known are more difficult. In these problems, application of Newton's second law normally results in a set of differential equations, which may be relatively easy or very difficult to solve.

Physical quantities used to express the laws of mechanics are of two types: fundamental quantities and derived quantities. The magnitude of each fundamental quantity is defined by an arbitrarily chosen unit or "standard." The fundamental units used in the SI system are the meter (m) for length, the kilogram (kg) for mass, and the second (s) for time. The unit of force is a derived unit called a newton (N). The SI system is an inertial or absolute system of units. In the U.S. Customary System of Units the fundamental units used are the foot (ft) for length, the pound (lb) for force, and the second (s) for time. The unit of mass is a derived unit called a slug. The U.S. Customary System is a gravitational system of units.

The terms of an equation used to describe a physical process should not depend on the units of measurement; equations should be dimensionally homogeneous. If an equation is dimensionally homogeneous, the equation is valid for use with any system of units provided all quantities in the equation are measured in the same system. Use of dimensionally homogeneous equations eliminates the need for unit conversion factors.

Success in engineering depends to a large degree on a well-disciplined method of problem solving. Professional problem solving consists of three phases:

1. Problem definition and identification.
2. Model development and simplification.
3. Mathematical solution and result interpretation.

The problem identification, model simplification, and result interpretation phases of engineering problem solving are often more important than the mathematical solution phase. No matter how elegantly or cleverly or exactly the mathematical solution is, it is of no value if it does not accurately reflect the physical problem that was to be solved.

Accepted engineering practice requires that all work be neat and orderly. Sloppy solutions that cannot be easily read and understood by others have little or no value. Final results should always be reported with a number of significant figures that is indicative of the accuracy of the solution.

*A*ir traffic control systems use principles of kinematics to keep track of airplanes. Radar measurements give the position of each plane in terms of radial and transverse coordinates. These measurements need to be converted to fixed rectangular coordinates to determine the relative positions and relative motions of the various planes.

Chapter 13

KINEMATICS OF PARTICLES

13-1 INTRODUCTION

 The study of dynamics consists of two related topics: **kinematics,** which is the study of how objects move, and **kinetics,** which is the study of the relationship between the motion and the forces that cause the motion. Kinematics describes how the velocity and acceleration of a body change with time and with changes in the position of the body. Not only is a sound understanding of kinematics a necessary background for the later study of kinetics, but kinematics is an important field of study in its own right. The design of many machine parts to create specific motions is based almost exclusively on kinematics. The study of the motion of projectiles, rocket ships, and space satellites is also based heavily on kinematics. Kinematics is the subject of this chapter and of Chapter 14.

A **particle** is an object whose size can be ignored when studying its motion. Only the position of the mass center of the particle need be considered. The orientation of the object or rotation of the object plays no role in describing the motion of a particle. Particles can be very small or very large. Small does not always guarantee that an object can be modeled as a particle; large does not always prevent modeling an object as a particle. Whether a body is large or small relates to the length of path followed, the distance between bodies, or both. The kinematics of particles is covered in this chapter. The kinematics of rigid bodies (bodies for which the orientation and rotation are important) is covered in Chapter 14.

13-2 POSITION, VELOCITY, AND ACCELERATION

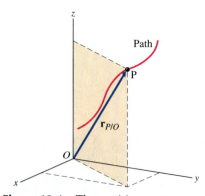

Figure 13-1 The position vector $\mathbf{r}_{P/O}$ measures the position of the particle P relative to the origin O of some coordinate system.

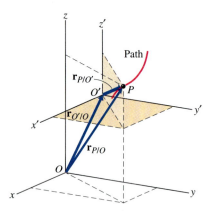

Figure 13-2 The position vector $\mathbf{r}_{P/O}$ depends on the location of the origin O as well as the particle P. Even if xyz and $x'y'z'$ are both fixed, Cartesian coordinate systems with the same orientation, the position vectors $\mathbf{r}_{P/O}$ and $\mathbf{r}_{P/O'}$ will generally have different magnitudes and different directions. However, the expressions for the velocity and acceleration of the particle will be the same in both coordinate systems.

Suppose a particle travels along a path as shown in Figure 13-1. At some instant of time t, the particle is at point P. In terms of the fixed, rectangular, Cartesian coordinate system indicated, the position of the particle is given by the position vector of P relative to the origin O, which can be written

$$\mathbf{r}_{P/O}(t) = x(t)\mathbf{i} + y(t)\mathbf{j} + z(t)\mathbf{k} \qquad (13\text{-}1)$$

where the basis vectors, $\mathbf{i}$, $\mathbf{j}$, and $\mathbf{k}$, are unit vectors in the x-, y-, and z-coordinate directions, respectively. The $/O$ on the subscript indicates that the components $x(t)$, $y(t)$, and $z(t)$ depend not only on the time t and the orientation of the coordinate system but also on the location of the origin O. In terms of a second fixed, rectangular, Cartesian, coordinate system oriented the same way as the first (having the same $\mathbf{i}$, $\mathbf{j}$, and $\mathbf{k}$) but having an origin at O' (Fig. 13-2), the position of the particle is written

$$\mathbf{r}_{P/O'}(t) = x'(t)\mathbf{i} + y'(t)\mathbf{j} + z'(t)\mathbf{k} \qquad (13\text{-}2)$$

By the triangle law of addition for vectors, these position vectors are related by

$$\mathbf{r}_{P/O} = \mathbf{r}_{O'/O} + \mathbf{r}_{P/O'} \qquad (13\text{-}3)$$

where $\mathbf{r}_{O'/O}$ is a constant vector.

The difference in position of the particle at two instants of time is called the **displacement** of the particle. If the particle that is at P at time t is at Q at time $t + \Delta t$ (Fig. 13-3), then the displacement $\delta\mathbf{r}$ is given by

$$\delta\mathbf{r} = \mathbf{r}_{Q/O} - \mathbf{r}_{P/O} \qquad (13\text{-}4)$$

Note that the displacement is independent of the location of the origin of the coordinate system since

$$\mathbf{r}_{Q/O'} - \mathbf{r}_{P/O'} = (\mathbf{r}_{Q/O} - \mathbf{r}_{O'/O}) - (\mathbf{r}_{P/O} - \mathbf{r}_{O'/O})$$
$$= \mathbf{r}_{Q/O} - \mathbf{r}_{P/O}$$

The **velocity** of a particle is defined to be the time rate of change of its position:[1]

$$\mathbf{v}_P(t) = \lim_{\delta t \to 0} \frac{\delta\mathbf{r}}{\delta t} = \frac{d\mathbf{r}_{P/O}}{dt} = \dot{\mathbf{r}}_{P/O} \qquad (13\text{-}5)$$

Since the displacement $\delta\mathbf{r}$ is independent of the location of the origin of the coordinate system, it is obvious that the velocity $\mathbf{v}_P$ is also independent of the location of the origin of the coordinate system. Furthermore, it is obvious that the direction of the velocity $\mathbf{v}_P$ is in the direction of the displacement $\delta\mathbf{r}$, or tangent to the path of the particle.

---

[1]Derivatives with respect to time occur so often in dynamics that a shorthand notation has been developed to express time derivatives. A dot over any symbol means the *time derivative* of that quantity; two dots over a symbol means the second time derivative, etc. Therefore, $\dot{s} \equiv \dfrac{ds}{dt}$; $\ddot{s} \equiv \dfrac{d^2s}{dt^2}$; etc.

Since the basis vectors are also constant, the velocity can be written in terms of components as

$$\mathbf{v}_P(t) = v_x(t)\mathbf{i} + v_y(t)\mathbf{j} + v_z(t)\mathbf{k} \qquad (13\text{-}6a)$$
$$= \dot{x}(t)\mathbf{i} + \dot{y}(t)\mathbf{j} + \dot{z}(t)\mathbf{k} \qquad (13\text{-}6b)$$

The **acceleration** of a particle is defined to be the time rate of change of its velocity

$$\mathbf{a}_P(t) = \frac{d\mathbf{v}_P}{dt} = \dot{\mathbf{v}}_P = \frac{d^2\mathbf{r}_{P/O}}{dt^2} = \ddot{\mathbf{r}}_{P/O} \qquad (13\text{-}7)$$

The acceleration is also independent of the location of the origin of the coordinate system. The acceleration can be written in terms of components as

$$\mathbf{a}_P(t) = a_x(t)\mathbf{i} + a_y(t)\mathbf{j} + a_z(t)\mathbf{k} \qquad (13\text{-}8a)$$
$$= \dot{v}_x(t)\mathbf{i} + \dot{v}_y(t)\mathbf{j} + \dot{v}_z(t)\mathbf{k} \qquad (13\text{-}8b)$$
$$= \ddot{x}(t)\mathbf{i} + \ddot{y}(t)\mathbf{j} + \ddot{z}(t)\mathbf{k} \qquad (13\text{-}8c)$$

If a coordinate system can be found such that the y- and z-components of the position, velocity, and acceleration are all zero for all time, the motion is called **rectilinear motion.** In this case the particle moves in a straight line (along the x-axis) with varying speed and acceleration. Rectilinear motion is covered in Sections 13-3 and 13-4.

If the particle is not moving in rectilinear motion but a coordinate system can be found such that the z-components of the position, velocity, and acceleration are all zero for all time, the motion is called **plane curvilinear motion.** The kinematics of plane curvilinear motion is covered in Sections 13-5 and 13-6.

Motion for which no coordinate system can be found that makes at least one component of the position, velocity, and acceleration zero for all time is called **general curvilinear motion** or **space curvilinear motion.** Kinematics of space curvilinear motion is covered in Section 13-7.

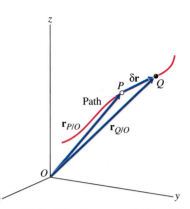

Figure 13-3 Displacement is the difference in the position of the particle at two instants of time. The displacement $\delta\mathbf{r}$ is independent of the location of the origin of the coordinate system.

13-3   RECTILINEAR MOTION

Motion in this case is along a straight line. The coordinate system will be oriented such that the x-axis coincides with the line of motion. For convenience this line will be assumed horizontal with positive to the right and negative to the left.

Since the position, velocity, and acceleration are completely prescribed by giving only their x-components, the vector notation will be dropped. The equations for position, displacement, velocity, and acceleration will be written simply as

$$\mathbf{r}_{P/O} = x \qquad\qquad \delta x = x_Q - x_P \qquad (13\text{-}9a,b)$$
$$v_P = v = \dot{x} \qquad\qquad a_P = a = \dot{v} = \ddot{x} \qquad (13\text{-}9c,d)$$

where the sign of the number indicates whether the vector is in the direction of the positive x-axis or the negative x-axis. Thus, if x is positive, the particle is to the right of the origin; if x is negative, it's to the left of the origin. When δx is positive, the particle's final position Q is to the right of its initial position, P; when δx is negative, Q is to the left of P. If v is positive, the particle is moving to the right; if v is negative, it's moving to the left. When the velocity and acceleration have the same sign, the velocity is increasing and the particle is said to be **accelerating.** When the velocity and acceleration have opposite signs, the velocity is decreasing and the particle is said to be **decelerating.**

It should be noted that the displacement of a particle is not the same thing as the **distance traveled by a particle.** The displacement is the (vector) difference in position between the beginning and end of a particle's path. If a particle begins and ends at the same point, its displacement is **0**. For example, the displacement of a train that travels from Baltimore to Washington, D.C., and back to Baltimore is **0**. The distance traveled, on the other hand, keeps track of what happens between the beginning and end of the path. It measures the total length of the path without regard to direction. The distance traveled is a scalar quantity—just a number. For the train that travels from Baltimore to Washington, D.C. (50 mi), and back (also 50 mi), the distance traveled is 100 mi. The distance traveled is always a positive number.

Equations 13-9 relate the four main variables of interest: position, velocity, acceleration, and time. In typical applications, a relationship between two of the variables is given and it is desired to find the other two variables. Some of the more common combinations will be considered in the following sections. Students are strongly encouraged not to try to memorize these formulas. Memorization invariably leads to using the formulas for situations in which they are not appropriate. These equations are basically just simple manipulations of the definitions of velocity and acceleration. It is these manipulations and not the resulting equations that are important.

13-3.1 Given $x(t)$; Postion as a Function of Time

A common requirement in the design of machinery is for various points to travel designated paths at specified times. As will be seen in Chapter 17, the power required to move a particle depends on its velocity which is found (by definition) by differentiating the position with respect to time (Eq. 13-9c)

$$v(t) = \frac{dx}{dt}$$

Also, in Chapter 15 it will be shown that the force required to produce the given motion is proportional to the acceleration of the particle which is found (by definition) by differentiating the velocity with respect to time (Eq. 13-9d)

$$a(t) = \frac{dv}{dt}$$

13-3.2 Given $v(t)$; Velocity as a Function of Time

Sometimes, the speed with which a machine operates is more important than where certain particles are at any instant of time. When the velocity of a particle is given as a function of time, the acceleration of the particle can be found by differentiating the velocity with respect to time as above (Eq. 13-9d). The position is found by integrating the definition of velocity (Eq. 13-9c)

$$v(t) = \frac{dx}{dt}$$

which gives

$$\int dx = \int v(t)\, dt \qquad (a)$$

or

$$x = \int v(t)\, dt \qquad (b)$$

The constant of integration must be chosen so that $x = x_0$ when $t = t_0$.

13-3.3 Given $a(t)$; Acceleration as a Function of Time

According to Newton's second law, the acceleration of a particle of mass m is directly proportional to the resultant force acting on the particle: $a = F/m$. Therefore, if the resultant force acting on the particle is known as a function of time, then the acceleration of the particle will also be known as a function of time. In this case, the velocity of the particle is then found by integrating the definition of acceleration (Eq. 13-9d)

$$a(t) = \frac{dv}{dt}$$

which gives

$$\int dv = \int a(t)\, dt \qquad (c)$$

or

$$v = \int a(t)\, dt \qquad (d)$$

The constant of integration must be chosen so that $v = v_0$ when $t = t_0$. The position is then found by integrating the definition of velocity as before (Eq. b).

13-3.4 Given $a(x)$; Acceleration as a Function of Position

In many situations, the force acting on a particle depends on the position of the particle rather than on time. Perhaps the most common example is a particle attached to an elastic spring. Since the acceleration of a particle is directly proportional to the resultant force acting on the particle, the acceleration of a particle attached to an elastic spring will be known as a function of position. In this case, the definition of the acceleration (Eq. 13-9d) must first be rewritten using the chain rule of differentiation

$$a(x) = \frac{dv}{dt} = \frac{dv}{dx}\frac{dx}{dt} = v\frac{dv}{dx} \qquad (e)$$

Then integration gives

$$\int v\, dv = \int a(x)\, dx \qquad (f)$$

or

$$\frac{v^2}{2} = \int a(x)\, dx \qquad (g)$$

where the constant of integration must be chosen so that $v = v_0$ when $x = x_0$. Once the velocity of the particle is known as a function of position, the position of the particle can be found as a function of time by integrating the definition of velocity (Eq. 13-9c)

$$v(x) = \frac{dx}{dt}$$

which gives

$$\int \frac{dx}{v(x)} = \int dt = t \tag{h}$$

where the constant of integration must be chosen so that $x = x_0$ when $t = t_0$. Note that before the integration can be carried out, all terms involving one variable (in this case x) must be taken to one side of the equation and all terms involving the other variable (in this case t) must be taken to the other side of the equation. Constants can appear on either side of the equation.

13-3.5   Given $a(v)$; Acceleration as a Function of Velocity

Particles moving rapidly through fluids such as air or water or sliding on a thin layer of fluid such as oil experience a force which depends on the velocity of the particle rather than on time or position. Since the acceleration of a particle is directly proportional to the resultant force acting on the particle, the acceleration of a particle moving through or sliding on a fluid will be known as a function of velocity. In this case, the velocity can be found as a function of time by integrating the definition of acceleration (Eq. 13-9d)

$$a(v) = \frac{dv}{dt}$$

which gives

$$\int \frac{dv}{a(v)} = \int dt = t \tag{i}$$

where the constant of integration must be chosen so that $v = v_0$ when $t = t_0$. Again note that before the integration can be carried out, all terms involving one variable (in this case v) must be taken to one side of the equation and all terms involving the other variable (in this case t) must be taken to the other side of the equation. Constants can appear on either side of the equation. Once the velocity is known as a function of time, the velocity can then be integrated as in Section 13-3.2 to get the position as a function of time.

Alternatively, the velocity can be found as a function of position by integrating Eq. e

$$a(v) = \frac{dv}{dt} = \frac{dv}{dx}\frac{dx}{dt} = v\frac{dv}{dx}$$

to get

$$\int \frac{v\,dv}{a(v)} = \int dx = x \tag{j}$$

13-3.6 Given a = constant

The case of constant acceleration (called **uniformly accelerated motion**) is included in the foregoing analysis because if the acceleration is constant, it can be treated as a function of time or of position or of velocity, whichever is convenient in the problem. However, the special case of uniformly accelerated motion (and the related special case of **uniform motion** in which $a = 0$) arises so often in mechanics that it is worth treating it as a separate case.

If the acceleration is constant, then the integrations of Section 13-3.3 can be performed immediately to get

$$v = a \int dt = v_0 + a(t - t_0) \qquad (k)$$

and

$$x = \int v \, dt = x_0 + v_0(t - t_0) + \frac{1}{2}a(t - t_0)^2 \qquad (l)$$

Similarly, the integration of Section 13-3.4 gives

$$v^2 - v_0^2 = 2a(x - x_0) \qquad (m)$$

The constants of integration in Eqs. k, l, and m have been chosen so that $x = x_0$ and $v = v_0$ when $t = t_0$.

It cannot be emphasized enough that Eqs. k, l, and m are valid only when the acceleration is a constant. A very common error is to use these equations when the acceleration is not constant.

13-3.7 Graphical Analysis

Typical graphs of the position, velocity, and acceleration versus time are drawn in Fig. 13-4. In these graphs the acceleration is the slope of the velocity graph, since

$$a(t) = \frac{dv}{dt} = \text{slope of the velocity graph at } t$$

If the value of the acceleration is positive, then the velocity is increasing; if the value of the acceleration is negative, the velocity is decreasing. The more positive or negative the acceleration is, the faster the velocity is increasing or decreasing. Furthermore, the change in the velocity from time t_0 to t_1 is equal to the (signed) area under the a–t graph between those times, since

$$v_1 - v_0 = \int_{t_0}^{t_1} a(t) \, dt = \text{area under the } a\text{–}t \text{ graph between } t_0 \text{ and } t_1$$

Similarly, the velocity is the slope of the position graph, since

$$v(t) = \frac{dx}{dt} = \text{slope of the position graph at } t$$

If the value of the velocity is positive, then the position is increasing; if the value of the velocity is negative, the position is decreasing. Also, the change in position from time t_0 to t_1 is equal to the (signed) area under the v–t graph between those times, since

$$x_1 - x_0 = \int_{t_0}^{t_1} v(t) \, dt = \text{area under the } v\text{–}t \text{ graph between } t_0 \text{ and } t_1$$

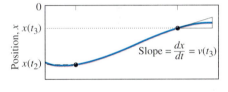

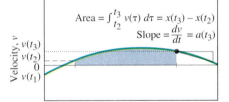

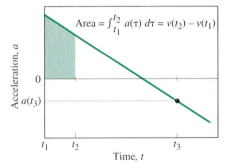

Figure 13-4 Since acceleration is the time derivative of velocity, it is the slope of the v–t graph. That is, when the acceleration is positive, the velocity is increasing, and when the acceleration is negative, the velocity is decreasing. Similarly, velocity is the time derivative of position and is therefore the slope of the x–t graph.

If the area is positive, the particle moves a distance $x_1 - x_0$ to the right (in the positive x-direction). If the area is negative, the particle moves a distance $|x_1 - x_0|$ to the left (in the negative x-direction). The displacement of the particle between time t_0 and t_1 is then the sum of all these positive and negative areas. The displacement may be positive or negative, depending on the relative amounts of positive and negative areas.

As defined previously, the distance traveled by a particle measures the total amount of motion that has taken place without regard to the direction of the motion. That is, it is the sum of the positive areas under the v–t graph (which measures the amount of motion in the positive coordinate direction) and the absolute value of the negative areas under the v–t graph (which measures the amount of motion in the negative coordinate direction). Mathematically, the distance traveled can be expressed as

$$\text{distance traveled} = \int_{t_0}^{t_1} |v(t)|\, dt \qquad (n)$$

The area–slope relations and related graphs presented in this section can be used to determine the position of a particle moving with a constant velocity or the velocity of a particle moving with constant acceleration. For more complicated motions, the computations involved in finding the appropriate areas and slopes are usually more difficult than the direct calculations using the definitions of velocity and acceleration. Even in these more complicated motions, however, the area–slope relations are often used to graphically depict the motion and to aid in understanding the motion.

EXAMPLE PROBLEM 13-1

A particle moves along the y-axis with an acceleration given by $a(t) = 5 \sin \omega t$ ft/s^2 where $\omega = 0.7$ rad/s. Initially (at $t = 0$) the particle is 2 ft above the origin and is moving downward with a speed of 5 ft/s.

a. Determine the velocity and position of the particle as functions of time.
b. Show the position, velocity, and acceleration on a graph.
c. Determine the displacement of the particle δy between $t = 0$ s and $t = 4$ s.
d. Determine the total distance traveled by the particle s between $t = 0$ s and $t = 4$ s.

SOLUTION

a. Since the acceleration is given as a function of time, the velocity and position can be obtained simply by integrating the definitions. First,

$$\frac{dv}{dt} = a(t) = 5 \sin (0.7\, t)$$

which can be integrated immediately to get

$$v(t) = -5 - \frac{5}{0.7}[\cos (0.7\,t) - 1] \text{ ft/s} \qquad \text{Ans.}$$

where the constant of integration has been chosen to satisfy the initial condition that $v = -5$ ft/s when $t = 0$ s. Then integrating

$$\frac{dy}{dt} = v(t) = -5 - \frac{5}{0.7}[\cos (0.7\,t) - 1]$$

gives

$$y(t) = 2 - 5t - \frac{5}{0.7}\left[\frac{\sin (0.7\,t)}{0.7} - t\right] \text{ ft} \qquad \text{Ans.}$$

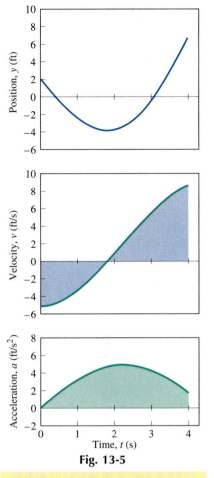

b. The position, velocity, and acceleration of the particle are sketched in Fig. 13-5. Notice that the value of the acceleration is positive for the entire first four seconds and hence the slope of the velocity graph is also positive for the entire first four seconds. Similarly, the value of the velocity is negative for about the first 1.8 s, during which time the slope of the position graph is also negative. After $t \cong 1.8$ s the value of the velocity is positive and the slope of the position graph is also positive. At $t \cong 1.8$ s the value of the velocity is zero, $v(t) = dy/dt = 0$, and the position takes on its minimum value there.

c. The displacement of the particle between $t = 0$ s and $t = 4$ s is just the difference in position at those two times

$$\delta y = y(4) - y(0) = 7.153 - 2 = 5.15 \text{ ft} \qquad \text{Ans.}$$

d. The distance traveled between $t = 0$ s and $t = 4$ s is greater than the displacement since the particle first moved below the origin and then came back up above the origin. The point where the particle turned around is found by determining when $dy/dt = 0$ (or equivalently where $v(t) = 0$)

$$v(t) = -5 - \frac{5}{0.7}[\cos (0.7\,t) - 1] = 0$$

which gives $t = 1.809$ s. Then

$$s = |y(1.809) - y(0)| + |y(4) - y(1.809)| \qquad \text{Ans.}$$
$$= 5.858 + 11.011 = 16.87 \text{ ft}$$

Fig. 13-5

During the first 1.8 s, the velocity of the particle is negative which means that the particle is moving in the negative y-direction. The speed of the particle, however, decreases from 5 ft/s when $t = 0$ s to 0 ft/s when $t = 1.8$ s. Since the particle is slowing down, it is said to be *decelerating*, even though its acceleration is positive.

EXAMPLE PROBLEM 13-2

A particle hanging from a spring moves with an acceleration that is proportional to its position and has the opposite sign. Suppose that $a(x) = -4x$ m/s^2 and that the velocity of the particle is 2 m/s upward when it passes through the origin.

a. Determine the velocity of the particle as a function of position.
b. If the particle is at the origin at $t = 1$ s, determine its position, velocity, and acceleration as a function of time.

SOLUTION

a. Since the acceleration is given as a function of position, the basic definition of acceleration needs to be rewritten using the chain rule

$$a(x) = \frac{dv}{dt} = \frac{dv}{dx}\frac{dx}{dt} = \frac{dv}{dx}v$$

Then the velocity is obtained by integrating this relationship

$$\int v\,dv = \int a(x)\,dx = \int (-4\,x)\,dx$$

which gives

$$\frac{v^2 - v_0^2}{2} = -2(x^2 - x_0^2)$$

Using the given conditions that $v = v_0 = 2$ m/s when $x = x_0 = 0$ and rearranging gives

$$v(x) = 2\sqrt{1 - x^2} \text{ m/s} \qquad\qquad \text{Ans.}$$

b. This last expression can now be integrated to get the position as a function of time. The definition gives

$$\frac{dx}{dt} = v(x) = 2\sqrt{1 - x^2} \qquad\qquad (a)$$

which can be rewritten

$$\frac{dx}{\sqrt{1 - x^2}} = 2\,dt$$

Integration of this equation gives

$$\sin^{-1} x = 2t + \text{constant} \qquad \text{or} \qquad x(t) = \sin(2t - 2) \text{ m} \qquad \text{Ans.}$$

where the constant of integration has been chosen to make $x = 0$ when $t = 1$ s. Substituting this expression into the given formula for the acceleration gives

$$a(t) = -4x = -4\sin(2t - 2) \text{ m/s}^2 \qquad \text{Ans.}$$

The equation for the velocity as a function of time can be obtained either by substitution into Eq. a

$$v(x) = 2\sqrt{1 - x^2} = 2\sqrt{1 - \sin^2(2t - 2)} = 2\cos(2t - 2) \text{ m/s}$$

or by direct differentiation of the position

$$v(t) = \frac{dx}{dt} = 2\cos(2t - 2) \text{ m/s} \qquad \text{Ans.}$$

These results are graphed in Fig. 13-6.

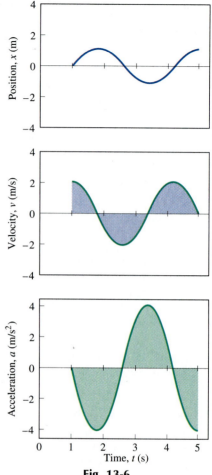

Fig. 13-6

The motion of the particle oscillates between $x = +1$ m and $x = -1$ m and it repeats every 3.14 s. The position is at its most positive value and the acceleration is at its most negative value when the velocity is zero. The velocity is at its most positive value when the position and acceleration are both zero. This behavior is typical of the vibratory motions studied in Chapter 21.

The acceleration of a ball falling in air is given by

$$\frac{W}{g}a = -C_D\frac{1}{2}\rho v^2 A + W$$

where g is the gravitational constant ($= 32.2$ ft/s^2), W is the weight of the ball, C_D is its drag coefficient, A is its cross-sectional area ($= \pi r^2$), v is its velocity, and ρ is the density of air. Given that $W = 1$ lb, $r = 2.5$ in., $C_D = 1.0$, $\rho = 0.0023$ slugs/ft^3, and that the ball starts from rest, determine the velocity of the ball as a function of its height.

SOLUTION

Since the acceleration is given as a function of velocity,

$$a(v) = 32.2 - 0.005049v^2$$

the basic definition of acceleration needs to be rewritten using the chain rule

$$a(v) = \frac{dv}{dt} = \frac{dv}{dy}\frac{dy}{dt} = v\frac{dv}{dy}$$

where a, v, and y are all positive in the same direction, downward. This relationship can be rearranged and integrated

$$\int \frac{v\,dv}{32.2 - 0.005049v^2} = \int dy$$

to get

$$-\frac{1}{0.01010}\ln(32.2 - 0.005049v^2) = y + C \qquad (a)$$

Using the initial condition that the ball starts from rest ($v = 0$ when $y = 0$), gives the constant of integration C

$$-\frac{1}{0.01010}\ln(32.2) = C \qquad (b)$$

Substituting the constant of integration (Eq. b) back into Eq. a and rearranging a little gives

$$-\frac{1}{0.01010}[\ln(32.2 - 0.005049v^2) - \ln(32.2)] = y$$

or

$$v = \left[\frac{32.2(1 - e^{-0.0101y})}{0.005049}\right]^{1/2} \text{ ft/s} \qquad \text{Ans.}$$

where y and v are both measured positive downward. This result is drawn in Fig. 13-7.

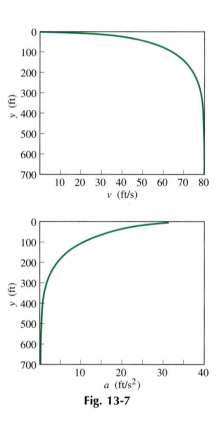

Fig. 13-7

Note that as the ball falls, the exponential term gets very small and the velocity approaches a constant value of about 79.9 ft/s. This limiting velocity is called the **terminal velocity** of the body.

The small electric car shown in Fig. 13-8 has a maximum constant acceleration of 1 m/s^2 and a maximum constant deceleration of 2 m/s^2. Determine the minimum distance it would take this car to accelerate to its top speed of 80 km/h and immediately decelerate back to rest.

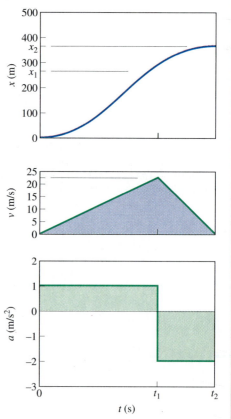

Fig. 13-8

SOLUTION

Analytical Solution

For $0 < t < t_1$ the car accelerates at 1 m/s^2 to its top speed of 80 km/h = 22.222 m/s. Therefore,

$$a = 1 \text{ m/s}^2 = \text{constant}$$
$$v = t + C_1 = t \text{ m/s} \qquad (a)$$
$$x = 0.5t^2 + C_2 = 0.5t^2 \text{ m} \qquad (b)$$

where the constants of integration $C_1 = C_2 = 0$ since the car starts from rest ($x = v = 0$ when $t = 0$). From Eq. a, the car reaches its top speed when

$$22.222 = t_1$$

or $t_1 = 22.222$ s, at which time the position of the car is (Eq. b)

$$x_1 = 0.5(22.222)^2 = 246.91 \text{ m}$$

For $t_1 < t < t_2$, the car decelerates at 2 m/s^2 back to zero speed. Therefore,

$$a = -2 \text{ m/s}^2 = \text{constant}$$
$$v = -2t + C_3 = 66.666 - 2t \text{ m/s} \qquad (c)$$
$$x = 66.666t - t^2 + C_4 = 66.666t - t^2 - 740.741 \text{ m} \qquad (d)$$

where the constants of integration $C_3 = 66.666$ m/s and $C_4 = -740.741$ m to match the conditions that $x = 246.91$ m and $v = 22.222$ m/s when $t = t_1 = 22.222$ s. From Eq. c, the car comes to a stop when

$$0 = 66.666 - 2t_2$$

or $t_2 = 33.333$ s, at which time the position of the car is (Eq. d)

$$x = 66.666(33.333) - (33.333)^2 - 740.741 \cong 370 \text{ m} \qquad \text{Ans.}$$

(The position, velocity, and acceleration of the car are graphed as functions of time in Fig. 13-9.)

Graphical Solution

The acceleration graph is constant at 1 m/s^2 for $0 < t < t_1$ and is constant at -2 m/s^2 for $t_1 < t < t_2$ as shown in the bottom graph of Fig. 13-9. Since the acceleration is the slope of the velocity ($a = dv/dt$), the velocity graph has a constant positive slope for $0 < t < t_1$, a constant negative slope for $t_1 < t < t_2$, and a peak value of 80 km/h = 22.222 m/s as shown in the middle graph of Fig. 13-9. Furthermore, the change in value on the velocity graph is equal to the area under the acceleration graph. Therefore, for the two rectangular areas,

$$1(t_1 - 0) = 22.222 = 2(t_2 - t_1)$$

or $t_1 = 22.222$ s and $t_2 = 33.333$ s.

Next, the velocity is the slope of the position graph ($v = dx/dt$). Therefore, the position graph starts with a zero slope (because the velocity is initially zero), the slope of the position graph increases until $t = t_1$, the slope of the position graph decreases from t_1 until t_2, and the slope of the position graph ends at zero (because the velocity ends at zero). Furthermore, the change in value

Fig. 13-9

on the position graph is equal to the area under the velocity graph. Therefore, for the two triangular areas,

$$x_1 - 0 = \frac{1}{2}(22.222)(22.222 - 0) = 246.91 \text{ m}$$

$$x_2 - x_1 = \frac{1}{2}(22.222)(33.333 - 22.222) = 123.46 \text{ m}$$

and the total distance traveled is

$$x_2 = 123.46 + 246.91 \cong 370 \text{ m} \qquad\qquad \text{Ans.}$$

as before.

PROBLEMS

Introductory Problems

13-1–13-4 The position of a particle moving along the *x*-axis is given as a function of time. In each problem

a. Determine the velocity of the particle as a function of time.
b. Determine the acceleration of the particle as a function of time.
c. Evaluate the position, velocity, and acceleration of the particle at $t = 5$ s.
d. Determine the total distance traveled by the particle between $t = 0$ and $t = 5$ s.
e. Sketch $x(t)$, $v(t)$, and $a(t)$; $0 \leq t \leq 8$ s.

13-1* $x(t) = 5t^2 - 30t + 20$ ft

13-2* $x(t) = 15 - 4t$ m

13-3* $x(t) = 3e^{-t/3}$ ft

13-4 $x(t) = 4 \sin \omega t$ m; $\omega = 3/2$ rad/s

13-5 Given the graph of position versus time shown in Fig. P13-5, construct the corresponding graphs of velocity versus time and acceleration versus time.

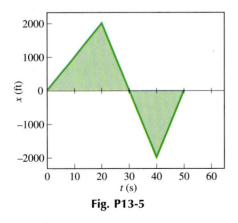

Fig. P13-5

13-6 Given the graph of position versus time shown in Fig. P13-6, construct the corresponding graphs of velocity versus time and acceleration versus time.

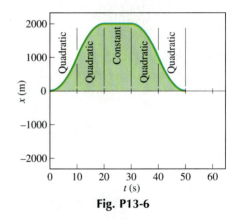

Fig. P13-6

13-7–13-10 The velocity of a particle moving along the *x*-axis is given as a function of time; the position of the particle is given at some instant. In each problem

a. Determine the position of the particle as a function of time.
b. Determine the acceleration of the particle as a function of time.
c. Evaluate the position, velocity, and acceleration of the particle at $t = 8$ s.
d. Determine the total distance traveled by the particle between $t = 5$ s and $t = 8$ s.
e. Sketch $x(t)$, $v(t)$, and $a(t)$; $0 \leq t \leq 10$ s.

13-7* $v(t) = 48 - 16t$ ft/s
$x(0) = 60$ ft

13-8* $v(t) = -4t^2 + 40t - 70$ m/s
$x(0) = 20$ m

13-9* $v(t) = 30e^{-t/3}$ ft/s
$x(3) = 20$ ft

13-10 $v(t) = 40 \cos \omega t$ m/s; $\omega = 3/2$ rad/s
$x(2) = 3$ m

13-11 Given the graph of velocity versus time shown in Fig. P13-11, construct the corresponding graphs of position versus time and acceleration versus time. The particle starts from $x = 0$ ft when $t = 0$.

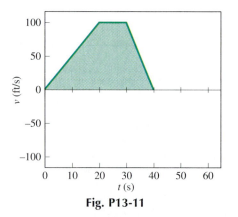

Fig. P13-11

13-12 Given the graph of velocity versus time shown in Fig. P13-12, construct the corresponding graphs of position versus time and acceleration versus time. The particle starts from $x = 0$ m when $t = 0$.

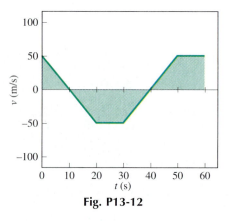

Fig. P13-12

13-13–13-16 The acceleration of a particle moving along the x-axis is given as a function of time; the position and velocity of the particle are given at some instant. In each problem

a. Determine the position of the particle as a function of time.
b. Determine the velocity of the particle as a function of time.
c. Evaluate the position, velocity, and acceleration of the particle at $t = 3$ s.
d. Determine the total distance traveled by the particle between $t = 3$ s and $t = 8$ s.
e. Sketch $x(t)$, $v(t)$, and $a(t)$, $0 \le t \le 10$ s.

13-13* $a(t) = 18 - 3t$ ft/s^2
$x(0) = -40$ ft; $v(0) = -30$ ft/s

13-14* $a(t) = -9.81$ m/s^2
$x(2) = 50$ m; $v(2) = 25$ m/s

13-15* $a(t) = 12e^{-t/6}$ ft/s^2
$x(8) = 5$ ft; $v(8) = -2$ ft/s

13-16 $a(t) = 20 \sin \omega t$ m/s^2; $\omega = 3/2$ rad/s
$x(10) = 40$ m; $v(10) = 5$ m/s

13-17 Given the graph of acceleration versus time shown in Fig. P13-17, construct the corresponding graphs of position versus time and velocity versus time. At $t = 0$ the particle is at $x = 0$ ft and has a velocity of $v = 0$ ft/s.

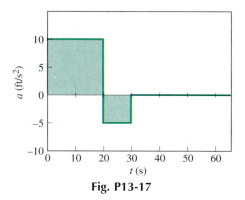

Fig. P13-17

13-18 Given the graph of acceleration versus time shown in Fig. P13-18, construct the corresponding graphs of position versus time and velocity versus time. At $t = 0$ the particle is at $x = 0$ m and has a velocity of $v = 50$ m/s.

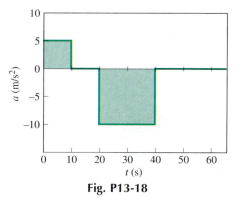

Fig. P13-18

Intermediate Problems

13-19* The small electric car shown in Fig. P13-19 accelerates uniformly from a speed of 20 mi/h to a speed of 45 mi/h in 5 s. Determine the acceleration of the car and how far the car will travel during the 5-s interval that it is accelerating.

Fig. P13-19

13-20 The automobile shown in Fig. P13-20 accelerates uniformly from a constant speed of 60 km/h to a speed of 80 km/h in 5 s. Determine the acceleration of the car and the distance traveled by the car during the 5-s interval that it is accelerating.

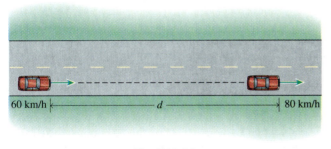

Fig. P13-20

13-21 The hockey puck shown in Fig. P13-21 decelerates uniformly from a speed of 60 ft/s to a speed of 40 ft/s in just 3 s. Determine the acceleration of the puck and the distance traveled by the puck during the 3-s interval that it is decelerating.

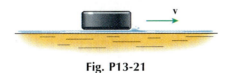

Fig. P13-21

13-22* The child's sled shown in Fig. P13-22 decelerates uniformly from a speed of 4 m/s to a speed of 2 m/s in 6 s. Determine the acceleration of the sled and the distance traveled by the sled during the 6-s interval that it is decelerating.

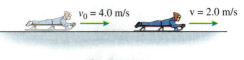

Fig. P13-22

13-23* The fully loaded Boeing 747 shown in Fig. P13-23 has a take-off acceleration of 10 ft/s^2. Determine the required length of the runway for a take-off speed of 140 mi/h.

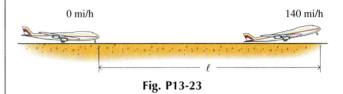

Fig. P13-23

13-24 A train is moving at 30 km/h when a coupling breaks and the last car separates from the train. As soon as the car separates, the brakes are automatically applied, locking all wheels of the runaway car and causing a uniform deceleration of 2.5 m/s^2. Determine the distance that the car will travel before coming to a stop.

13-25 The jet shown in Fig. P13-25 is catapulted from the deck of an aircraft carrier by a hydraulic ram. Determine the acceleration of the jet if it accelerates uniformly from rest to 160 mi/h in 300 ft.

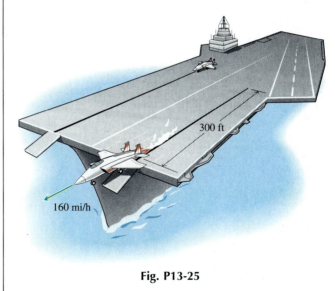

Fig. P13-25

13-26* A car is traveling at 90 km/h when a boulder rolls onto the highway 60 m in front of it (Fig. P13-26). Determine the minimum deceleration that the car must have in order to stop before hitting the boulder.

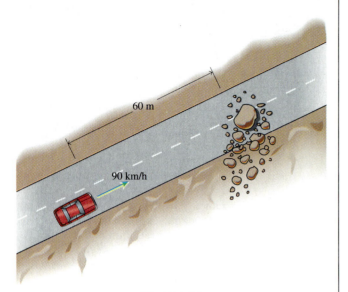

Fig. P13-26

13-27 The freeway on-ramp shown in Fig. P13-27 is 1200 ft long. A car starts up the ramp with a speed of 10 mi/h. Determine the minimum acceleration the car must have to merge smoothly with traffic moving at 60 mi/h at the end of the ramp.

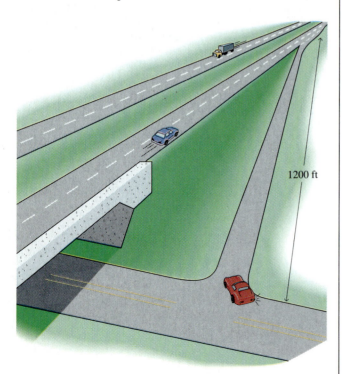

Fig. P13-27

13-28* A car traveling at a speed of 100 km/h exits from a freeway into a rest area. Determine the minimum deceleration that the car must have in order to slow to 15 km/h by the end of the exit ramp, which is only 300 m long.

13-29* A ball hanging from the end of an elastic cord (Fig. P13-29) has an acceleration proportional to its position but of opposite sign

$$a(y) = -3y \text{ ft/s}^2$$

Determine the velocity of the ball when $y = 1$ ft if the ball is released from rest when $y = -2$ ft.

Fig. P13-29

13-30 A ball hanging from the end of an elastic cord (Fig. P13-29) has an acceleration proportional to its position but of opposite sign

$$a(y) = -y \text{ m/s}^2$$

If the ball has a velocity of $v = -2$ m/s when $y = 0$ m and the ball passes through the point $y = 1$ m with a positive velocity when $t = 3$ s, determine the position, velocity, and acceleration of the ball when $t = 5$ s.

13-31* A cart attached to a spring (Fig. P13-31) moves with an acceleration proportional to its position but of opposite sign

$$a(x) = -2x \text{ ft/s}^2$$

Determine the velocity of the cart when $x = 10$ ft if the cart has a velocity of $v = 15$ ft/s when $x = 2$ ft.

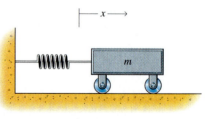

Fig. P13-31

13-32 A cart attached to a spring (Fig. P13-31) moves with an acceleration proportional to its position but of opposite sign

$$a(x) = -2x \text{ m/s}^2$$

If the cart has a velocity of $v = 5$ m/s when $x = 1$ m and passes through the point $x = 3$ m with a positive velocity when $t = 3$ s, determine the position, velocity, and acceleration of the cart when $t = 5$ s.

13-33 The acceleration of a rocket ship launched straight upward is given by (after the engines have stopped)

$$a = -g_0 \frac{R^2}{(R + h)^2}$$

where g_0 is the gravitational constant at the surface of the earth (32.2 ft/s^2), R is the radius of the earth (3960 mi), and h is the height of the rocket above the earth. Determine the maximum height attained by the rocket if the engines shut off at $h = 20$ mi and the velocity of the rocket at that point is 12,000 mi/h.

13-34* A cart is attached between two springs whose coils are very close together. The acceleration in this case is given by

$$a(x) = -x - 3x^2 \text{ m/s}^2$$

Determine the maximum position of the cart if it has a velocity of $v = 2$ m/s when $x = -1$ m.

13-35 A spherical ball falling through air (Fig. P13-35) has an acceleration

$$a(v) = 32.2 - 0.001v^2 \text{ ft/s}^2$$

where the velocity is in feet per second and the positive direction is downward. Determine the velocity of the ball as a function of height if the ball has a downward velocity of 10 ft/s when $y = 0$. Also, determine the terminal velocity of the ball.

Fig. P13-35

13-36* A spherical ball thrown upward through air has an acceleration

$$a(v) = -9.81 - 0.003v^2 \text{ m/s}^2$$

where the velocity is in meters per second and the positive direction is upward. Determine the velocity of the ball as

a function of height if the ball is thrown upward with an initial velocity of 30 m/s. Also, determine the maximum height reached by the ball.

13-37* A hockey puck sliding on a thin film of water on a horizontal surface (Fig. P13-37) has a deceleration directly proportional to its speed

$$a(v) = -0.50v \text{ ft/s}^2$$

where the velocity ($v > 0$) is in feet per second. If the puck has a velocity of 50 ft/s when $x = 0$, determine its velocity as a function of distance and calculate the velocity of the puck when $x = 75$ ft.

Fig. P13-37

13-38 A hockey puck sliding on a thin film of water on a horizontal surface (Fig. P13-37) has a deceleration directly proportional to its speed

$$a(v) = -0.50v \text{ m/s}^2$$

where the velocity ($v > 0$) is in meters per second. If the puck has a velocity of 15 m/s when $x = t = 0$, determine its velocity and position as functions of time. Also, calculate how long it takes the puck to slow to 0.1 m/s and its position at this time.

13-39* Because the drag on objects moving through air increases as the square of the velocity, the acceleration of a bicyclist coasting down a slight hill (Fig. P13-39) is

$$a(v) = 0.4 - 0.0002v^2 \text{ ft/s}^2$$

where the velocity is in feet per second. Determine the velocity of the bicyclist as a function of distance if the velocity is zero when $x = 0$. Also, determine the maximum velocity that the cyclist could attain coasting down this hill.

Fig. P13-39

13-40 Because the drag on objects moving through air increases as the square of the velocity, the acceleration of a bicyclist coasting down a slight hill (Fig. P13-39) is

$$a(v) = 0.12 - 0.0006v^2 \text{ m/s}^2$$

where the velocity is in meters per second. If the cyclist starts from rest when $t = 0$, determine the velocity as a function of time and calculate how long it takes the cyclist to reach a velocity of 6 m/s.

Challenging Problems

13-41* A hockey puck sliding on a thin film of water on a horizontal surface (Fig. P13-41) has a deceleration directly proportional to its speed

$$a(v) = -0.50v \text{ ft/s}^2$$

where the velocity ($v > 0$) is in feet per second. If the puck starts with a velocity of 50 ft/s when $x = t = 0$, determine

a. The velocity of the puck when $x = 60$ ft.
b. The elapsed time for the puck to slide the 60 ft.

Fig. P13-41

13-42* Because the drag on objects moving through air increases as the square of the velocity, the acceleration of a bicyclist coasting down a slight hill (Fig. P13-42) is

$$a(v) = 0.12 - 0.0006v^2 \text{ m/s}^2$$

where the velocity is in meters per second. If the cyclist starts from rest at the top of a 400-m long hill, determine

a. The velocity of the cyclist at the bottom of the hill.
b. The elapsed time for the cyclist to reach the bottom of the hill.

Fig. P13-42

13-43 A truck is traveling at 65 mi/h when the driver notices a moose on the road 200 ft ahead (Fig. P13-43). Determine the minimum constant deceleration required to stop the truck if

a. The driver immediately steps on the brake and begins to slow down.
b. The driver hesitates for 0.3 s before stepping on the brake and slowing down.

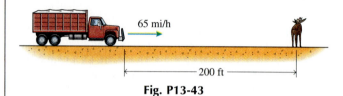

Fig. P13-43

13-44* A car is traveling at 70 km/h when the driver notices the traffic light turning red 90 m ahead. Determine the minimum constant deceleration required to stop the car if

a. The driver immediately steps on the brake and begins to slow down.
b. The driver hesitates for 1 s before stepping on the brake and slowing down.

13-45 A particle has the acceleration history shown in Fig. P13-45. If the particle starts from rest when $x = t = 0$, determine the position and velocity of the particle

a. When $t = 3$ s.
b. When $t = 8$ s.

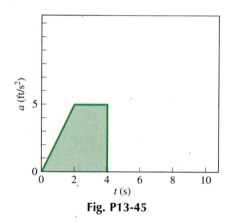

Fig. P13-45

13-46* A particle has the acceleration history shown in Fig. P13-46. If the particle starts from rest when $x = t = 0$, determine the position and velocity of the particle

a. When $t = 4$ s.
b. When $t = 8$ s.

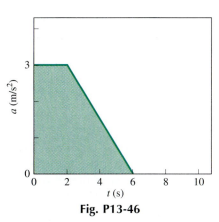

Fig. P13-46

13-47* A car has a maximum constant acceleration of 10 ft/s² and a maximum constant deceleration of 15 ft/s². Determine the minimum amount of time it would take to drive one mile assuming that the car starts and ends at rest and never exceeds the speed limit (55 mi/h).

13-48 A small, electric car (Fig. P13-48) has a maximum constant acceleration of 1 m/s², a maximum constant deceleration of 2 m/s², and a top speed of 80 km/h. Determine the amount of time it would take to drive this car 1 km—starting from and finishing at rest.

Fig. P13-48

13-49 The city bus shown in Fig. P13-49 has a constant acceleration of 3 ft/s² and a constant deceleration of 5 ft/s². The bus stops every 660 ft to take on passengers. If the bus departs from the first stop at 1:00 P.M. and the duration of each stop is 30 s, determine

a. The time of arrival at the next three stops.
b. The maximum velocity attained by the bus between stops.

Fig. P13-49

13-50* A commuter train has a constant acceleration of 1.5 m/s², a constant deceleration of 2 m/s², and a top speed of 100 km/h. The train stops every 5 km to take on passengers. If the train departs from the station at 7:00 A.M. and the duration of each stop is 2 min, determine the time of arrival at the next three stops.

13-4 RELATIVE MOTION ALONG A LINE

When two or more particles move in rectilinear motion, separate equations may be written to describe their motion. The particles may be moving along the same line or along separate lines. If the n particles are described by their various n coordinates but only m of the coordinates may be changed independently, then the system has m **degrees of freedom** (DOF). If $m = n$, then each particle can move independently and the particles are said to be in **independent relative motion.** If $m < n$, then the motion of one or more of the particles is completely determined by the motion of the other particles and the particles are said to be in **dependent relative motion.**

Whether the particles are in independent or dependent relative motion, the motion of any particular particle can be written relative to the motion of one or more of the other particles. The need for relative description of motion arises often in engineering. For example, in structural applications, it is the relative position of two particles rather than the absolute position that describes how severely deformed a structure is and whether or not it is likely to break or collapse. In vehicle crashes it is the relative velocity and not the absolute velocity that determines the severity of the crash. When police use radar to measure the speed of cars, it is the speed of the cars relative to the speed of the police car and not the absolute speed that is being measured. In each case, the observed position or speed must be converted into the desired values—either relative or absolute.

The first part of this section describes relative motion in general and independent relative motion in particular. The second part applies the principles of relative motion to the case of dependent relative motion.

13-4.1   Independent Relative Motion

Let A and B be two particles moving along the same straight line as shown in Fig. 13-10. The positions x_A and x_B are both measured relative to the fixed origin O and are called the **absolute positions** of the particles. The position of particle B as measured from the moving particle A is denoted $x_{B/A}$ and is called the **relative position** of B measured with respect to A or more simply just the position of B relative to A. These positions are related simply by

$$x_B = x_A + x_{B/A} \qquad (13\text{-}10)$$

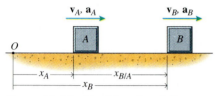

Figure 13-10   Particles A and B exhibit independent relative motion along a straight line.

(see Fig. 13-10). Differentiating Eq. 13-10 with respect to time gives

$$v_B = v_A + v_{B/A} \qquad (13\text{-}11)$$
$$a_B = a_A + a_{B/A} \qquad (13\text{-}12)$$

That is, the velocity of particle B measured relative to particle A is the difference in the absolute velocities (velocities measured relative to a fixed coordinate system) of particles A and B. Similarly, the acceleration of particle B measured relative to particle A is the difference in the absolute accelerations of particles A and B.

13-4.2   Dependent Relative Motion

In many engineering situations, two particles are not able to move independently. The motion of one particle will depend somehow on the motion of the other. A common dependency, or **constraint,** is that the particles are connected by a cord of fixed length. In this case, an equation representing the constraint replaces Eq. 13-10.

Although both particles are moving in rectilinear motion, they need not be moving along the same straight line. Both particles must be measured relative to a fixed origin, but it is often convenient to use a different origin for each particle. Even when the two particles are moving along the same straight line and are measured relative to the same fixed origin, it is often convenient to set the positive direction for each particle separately.

A constraint equation is then written in terms of the coordinates of the individual particles. This constraint equation is differentiated to get the relationship between the absolute velocities and accelerations of the particles. Care must be taken to interpret the directions of positive velocity and positive acceleration in accordance with the assumed positive coordinate directions.

For example, the cars and wreckers of Fig. 13-11 are connected by a cable of length L. In Fig. 13-11a, both vehicles move along the same straight line and their positions are measured from the same origin. In

Fig. 13-11b, the vehicles move along different straight lines, but their positions are both measured from the same origin. In both cases, the positions of the two vehicles (s_A and s_B) are both measured as distances from the fixed origin O. The constraint equation in both cases is

$$L = 3s_A + s_B = \text{constant} \qquad \textbf{(a)}$$

Taking the time derivatives of the constraint equation gives relationships between the velocities and accelerations of the vehicles

$$\dot{L} = 3\dot{s}_A + \dot{s}_B = 0 \qquad \textbf{(b)}$$
$$\ddot{L} = 3\ddot{s}_A + \ddot{s}_B = 0 \qquad \textbf{(c)}$$

If $\dot{s}_A$ is positive, then s_A is increasing and the velocity of the car is to the left. If $\dot{s}_A$ is negative, then s_A is decreasing and the velocity of the car is to the right. Similar interpretations apply to $\dot{s}_B$, $\ddot{s}_A$, and $\ddot{s}_B$.

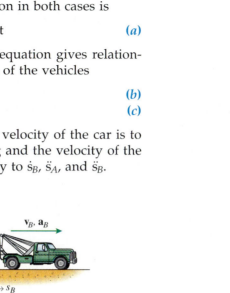

(a)

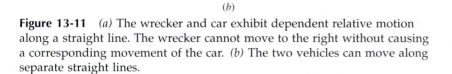

(b)

Figure 13-11 (a) The wrecker and car exhibit dependent relative motion along a straight line. The wrecker cannot move to the right without causing a corresponding movement of the car. (b) The two vehicles can move along separate straight lines.

Two race cars start from rest at the same position (Fig. 13-12). The acceleration of car A is

$$a_A = 30e^{-t/5} \text{ m/s}^2$$

and the acceleration of car B is

$$a_B = 20e^{-t/10} \text{ m/s}^2$$

Determine the distance at which car B overtakes car A and their relative velocity at that point.

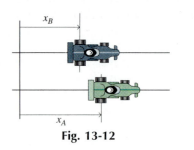

Fig. 13-12

SOLUTION

Integrating the given accelerations with respect to time gives the velocities of the two cars

$$v_A = 150(1 - e^{-t/5}) \text{ m/s} \qquad (a)$$
$$v_B = 200(1 - e^{-t/10}) \text{ m/s} \qquad (b)$$

Integrating Eqs. a and b again with respect to time gives the positions of the cars (Fig. 13-13).

$$x_A = 150[t + 5(e^{-t/5} - 1)] \text{ m} \qquad (c)$$
$$x_B = 200[t + 10(e^{-t/10} - 1)] \text{ m} \qquad (d)$$

Car B will overtake car A when the positions of the two cars are equal (and the relative position $x_{B/A} = x_B - x_A = 0$). Setting Eqs. c and d equal gives

$$150[t + 5(e^{-t/5} - 1)] = 200[t + 10(e^{-t/10} - 1)]$$

or

$$50t + 2000e^{-t/10} - 750e^{-t/5} - 1250 = 0 \qquad (e)$$

Solving Eq. e (using trial and error, the Newton–Raphson method, or Math-CAD) gives the time at which car B overtakes car A as $t = 19.727$ s. Substituting this time into the position equations (Eqs. c and d) gives the position at which car B overtakes car A as

$$x_A = x_B = 2223.5 \text{ m} \cong 2220 \text{ m} \qquad \text{Ans.}$$

The velocity of the cars at the time car B overtakes car A is found by substituting $t = 19.727$ s into the velocity equations (Eqs. a and b), which gives

$$v_A = 147.1 \text{ m/s} \qquad v_B = 172.2 \text{ m/s}$$

The relative velocity is then (Eq. 13-11)

$$v_{B/A} = v_B - v_A = 25.1 \text{ m/s} \qquad \text{Ans.}$$

> Car B has a smaller initial acceleration than car A (20 m/s^2 compared to 30 m/s^2) and will initially fall behind car A. However, car B has a greater top speed than car A (200 m/s compared to 150 m/s) and will eventually overtake and pass car A.

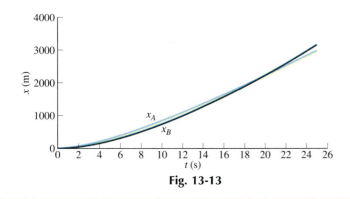

Fig. 13-13

EXAMPLE PROBLEM 13-6

Car A of Fig. 13-14 is stuck and is being pulled out by the wrecker truck B. If the truck B is moving to the right with a speed of 18 ft/s, determine the speed of the car A. Also, if the speed of the truck B is decreasing at a rate of 3 ft/s^2, determine the acceleration of the car A.

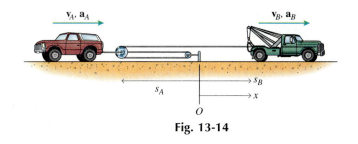

Fig. 13-14

SOLUTION

Both vehicles move along the same straight line. Their positions will both be measured relative to a fixed origin at the fixed pulley. The car is a distance s_A to the left of the origin, and its position is given by $x_A = -s_A$. The wrecker is a distance s_B to the right of the origin, and its position is given by $x_B = s_B$. The positions of the two vehicles are related by the length of the cable, which is a constant,

$$L = 3s_A + s_B + C \qquad (a)$$

where C is a constant to account for the length of cable that is wound around the pulleys (which is constant) and the distance between the centers of the pulleys and the vehicles (which is also constant). Differentiating Eq. a with respect to time gives

$$\dot{L} = 0 = 3\dot{s}_A + \dot{s}_B$$

or

$$3\dot{s}_A = -\dot{s}_B \qquad (b)$$

where $v_B = \dot{x}_B = \dot{s}_B = 18$ ft/s is positive to the right (the same direction as x_B and s_B are measured), $v_A = \dot{x}_A$ is positive to the right (the same direction as x_A is measured), and $\dot{s}_A = -\dot{x}_A$ is positive to the left (the same direction as s_A is measured). Therefore,

$$\dot{s}_A = -\frac{1}{3}\dot{s}_B = -6 \text{ ft/s}$$

and

$$v_A = \dot{x}_A = -\dot{s}_A = 6 \text{ ft/s} \rightarrow \qquad \text{Ans.}$$

Differentiating Eq. b again with respect to time gives

$$3\ddot{s}_A = -\ddot{s}_B$$

where $a_B = \ddot{x}_B = \ddot{s}_B = -3$ ft/s^2 (since the wrecker is slowing down). Therefore,

$$\ddot{s}_A = -\frac{1}{3}\ddot{s}_B = -\frac{1}{3}(-3) = 1 \text{ ft/s}^2$$

and

$$a_A = \ddot{x}_A = -\ddot{s}_A = -1 \text{ ft/s}^2 = 1 \text{ ft/s}^2 \leftarrow \qquad \text{Ans.}$$

When the wrecker moves to the right, the car must also move to the right because the length of the cable is constant. The exact relationship between the motion of the wrecker and the motion of the car depends on the configuration of the pulleys. Although the motions of both vehicles could be expressed in terms of their positions along the x-axis, relating the positions and motions of the vehicles to the length of the cable is made easier by measuring the distances s_A and s_B from the fixed post to the two vehicles. Once the relationship between these distances is known, it is easy to relate the distances s_A and s_B and their derivatives to the position and motion of the vehicles along the x-axis.

Car A of Fig. 13-15 is stuck and is being pulled out by the wrecker truck B. If the truck B is moving south with a speed of 6 m/s, determine the speed of the car A. Also, if the speed of the truck B is decreasing at a rate of 0.9 m/s^2, determine the acceleration of the car A.

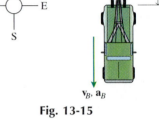

Fig. 13-15

SOLUTION

The car A moves along an east-west straight line while the wrecker truck B moves along a north-south straight line. Both vehicles positions will be measured relative to a fixed origin at the fixed pulley. The car is a distance s_A west of the origin and the wrecker is a distance s_B south of the origin. The positions of the two vehicles are related by the length of the cable which is a constant

$$L = 3s_A + s_B + C \qquad (a)$$

where C is a constant to account for the length of cable that is wound around the pulleys (which is constant) and the distance between the centers of the pulleys and the vehicles (which is also constant). Differentiating Eq. a with respect to time gives

$$\dot{L} = 0 = 3\dot{s}_A + \dot{s}_B$$

or

$$3\dot{s}_A = -\dot{s}_B \qquad (b)$$

where $v_B = \dot{s}_B = 6$ m/s is positive to the south (the same direction as s_B is measured), v_A is positive to the east, and $\dot{s}_A$ is positive to the west (the same direction as s_A is measured). Therefore,

$$\dot{s}_A = -\frac{1}{3}\dot{s}_B = -2 \text{ m/s}$$

and

$$v_A = -\dot{s}_A = 2 \text{ m/s} \qquad \text{(east)} \qquad\qquad \text{Ans.}$$

Differentiating Eq. b again with respect to time gives

$$3\ddot{s}_A = -\ddot{s}_B$$

where $a_B = \ddot{s}_B = -0.9$ m/s^2 (since the wrecker is slowing down). Therefore,

$$\ddot{s}_A = -\frac{1}{3}\ddot{s}_B = -\frac{1}{3}(-0.9) = 0.3 \text{ m/s}^2$$

and

$$a_A = -\ddot{s}_A = -0.3 \text{ m/s}^2 = 0.3 \text{ m/s}^2 \qquad \text{(west)} \qquad\qquad \text{Ans.}$$

PROBLEMS

Introductory Problems

13-51* Train A is traveling eastward at 80 mi/h while train B is traveling westward at 60 mi/h (Fig. P13-51). Determine

a. The velocity of train A relative to train B.
b. The velocity of train B relative to train A.

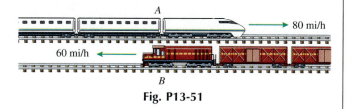

Fig. P13-51

13-52* Boat A travels down a straight river at 20 m/s while boat B travels up the river at 15 m/s (Fig. P13-52). Determine

a. The velocity of boat A relative to boat B.
b. The velocity of boat B relative to boat A.

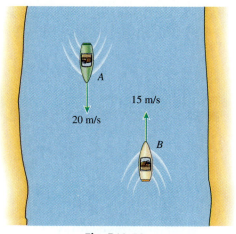

Fig. P13-52

13-53 A straight river flows at a speed of 5 mi/h. If a boat can maintain a speed of 30 mi/h relative to the current, determine

a. How long it will take the boat to travel from town A to town B, which is 30 miles down the river from town A.
b. How long it will take the boat to make the return trip against the current from town B to town A.

13-54* The jet stream flows from west to east at 180 km/h. If an airplane can maintain an indicated air speed (relative to the jet stream) of 540 km/h, determine

a. How long it will take the airplane to travel from city A to city B which is 800 km to the east of city A.
b. How long it will take the airplane to make the return trip against the jet stream from city B to city A.

13-55 A straight river flows at a speed of 5 mi/h. Boat A travels down river at 30 mi/h relative to the current while boat B travels up the river at 30 mi/h relative to the current. The boats are traveling between two towns which are 30 miles apart. If the boats both start from their respective towns at 12:00 noon, determine when and where the boats will meet.

13-56* The jet stream flows from west to east at 180 km/h. Airplane A is flying from west to east at an indicated air speed (relative to the jet stream) of 540 km/h. Airplane B is flying from east to west at an indicated air speed of 540 km/h. The airplanes are traveling between two cities 800 km apart. If the airplanes both start from their respective cities at 8:00 P.M., determine when and where the airplanes will meet.

13-57* A barge breaks away from its moorings and floats down river on the current at 10 ft/s. A tugboat goes after the runaway barge at a speed of 15 ft/s relative to the current. If the tugboat starts at a distance of 1500 feet behind the barge, determine the time it will take the tugboat to catch the barge and the total distance traveled by the tugboat in that time.

13-58 Spheres falling through still water fall at a steady speed that is inversely proportional to their diameters. Sphere A is falling at a speed of 5 m/s. Sphere B is half as big and is falling at 10 m/s. If at some instant of time sphere A is ahead of sphere B by 20 m, determine the amount of time it will take sphere B to overtake sphere A and the total distance travelled by sphere B in that time.

Intermediate Problems

13-59* Two cars travel between towns 50 miles apart. Both cars start at 3:00 P.M. from the same town, but the first car travels at 50 mi/h while the second car travels at 30 mi/h. If the first car stops in the second town for 5 min and then returns (still at 50 mi/h), determine when and where the two cars will meet.

13-60* Two bicyclists start riding toward each other at 1:00 P.M. from towns 20 km apart. The first cyclist is riding with the wind and maintains a speed of 7 m/s. The second cyclist is riding against the wind at a speed of 5 m/s and stops to rest for 5 min every 4 km. Determine when and where the cyclists will meet.

13-61 Two cars are separated by 60 ft and traveling in the same direction at 50 mi/h when car A (the front car) suddenly begins to brake at 12 ft/s². One second later the driver of the car B (the back car) begins to brake at 15 ft/s². Determine the separation distance between the cars when they are both stopped.

13-62* Two cars are traveling in the same direction at 80 km/h when car A (the front car) suddenly begins to brake at 4 m/s². If the reaction time of the driver of car B (the back car) is 1 s and car B also brakes at 4 m/s², determine the safe following distance—the distance between the two cars such that car B will stop before hitting car A.

13-63 A motorcycle is stopped by the side of the road when a car passes at 50 mi/h. Twenty seconds later the motorcycle starts chasing the car. Assume that the motorcycle accelerates at 8 ft/s² until it reaches 60 mi/h and then travels at a constant speed. Find the amount of time it will take the motorcycle to overtake the car and the total distance traveled by the motorcycle in that time.

13-64 Two fighter airplanes are flying in the same direction at 1100 km/h and are separated by 3 km when the back airplane fires a missile at the front airplane. Determine

a. The constant acceleration that the missile must have to catch the front airplane is 5 s.
b. The velocity of the missile relative to the front airplane at the time of impact.

13-65* A car is traveling at 60 mi/h on a road that is parallel to a railroad track when it comes upon a train. If the train is one-half mile long and moving at 40 mi/h, determine the length of time it will take the car to pass the train if they are traveling

a. In the same direction.
b. In opposite directions.

13-66 Two cars are approaching each other on a narrow straight road. Car A has an initial speed of 60 km/h and car B has an initial speed of 30 km/h. If both drivers apply their brakes when they are 45 m apart and both cars slow down at a rate of 3 m/s², determine

a. If the two cars will collide.
b. The relative speed of the two cars when they collide (if they do in fact collide).

13-67* The car and truck shown in Fig. P13-67a are both traveling at 50 mi/h when the car decides to pass the truck. If the car accelerates at 4 ft/s², passes the truck, and returns to the right-hand lane when it is 35 ft ahead of the truck (Fig. P13-67b), determine

a. The distance that the car travels while passing the truck.
b. The speed of the car when it pulls back into the right-hand lane.

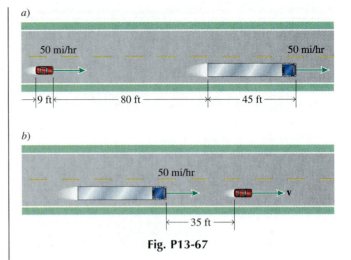

Fig. P13-67

13-68 A police car is parked in a school zone when a car speeds by. The police car starts from rest just as the car passes it, accelerates at 2 m/s² until it reaches a speed of 80 km/h, and then maintains a constant speed. If the speed of the first car is constant at 50 km/h, determine how far the police car will have to chase the speeder before catching him.

13-69 In Fig. P13-69 the elevator E is moving downward at a speed of 3 ft/s and its speed is increasing at the rate of 0.3 ft/s². Determine the velocity and acceleration of the counterweight C.

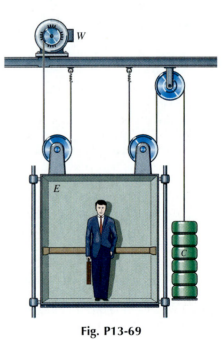

Fig. P13-69

13-70* In Fig. P13-70 block *A* is moving to the left with a speed of 2 m/s and its speed is increasing at the rate of 0.2 m/s². Determine the velocity and acceleration of block *B*.

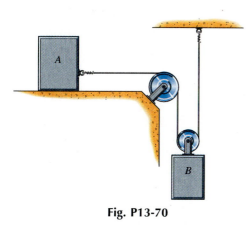

Fig. P13-70

13-71 A tow truck is pulling a car up a 25° incline using pulleys as shown in Fig. P13-71. If the tow truck accelerates at a constant rate of 4 ft/s², determine the speed and acceleration of the car 5 s after the tow truck starts from rest.

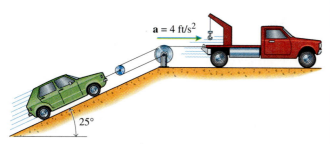

Fig. P13-71

13-72* In Fig. P13-72 block *B* has a constant downward acceleration of 0.8 m/s². Determine the speed and acceleration of block *A* 5 s after the system starts from rest.

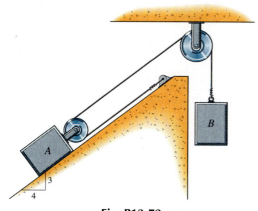

Fig. P13-72

13-73* In Fig. P13-73 the elevator *E* is moving upward at a speed of 6 ft/s and its speed is decreasing at the rate of 0.5 ft/s². Determine the velocity and acceleration of the counterweight *C*.

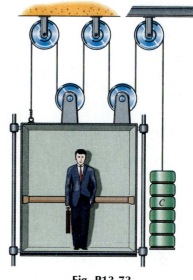

Fig. P13-73

13-74 In Fig. P13-74 block *A* is moving to the right with a speed of 5 m/s and its speed is decreasing at the rate of 0.2 m/s². Determine the velocity and acceleration of block *B*.

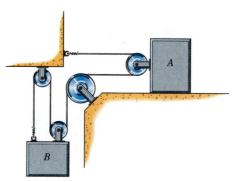

Fig. P13-74

13-75* In Fig. P13-75 block *B* moves to the right with a speed of 10 ft/s and its speed is decreasing at the rate of 1 ft/s². Determine

a. The velocity of block *C* if block *A* is fixed and does not move.
b. The velocity of block *A* if block *C* is fixed and does not move.

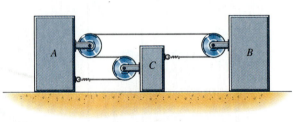

Fig. P13-75

13-76 In Fig. P13-76 block *B* moves to the right with a speed of 2 m/s and its speed is increasing at the rate of 0.3 m/s². Determine

a. The velocity of block *C* if block *A* is fixed and does not move.
b. The velocity of block *A* if block *C* is fixed and does not move.

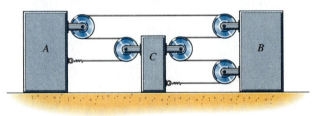

Fig. P13-76

Challenging Problems

13-77* A railroad car is loose on a siding and rolling at a constant speed of 8 mi/h (Fig. P13-77). A switch engine dispatched to catch the runaway car has a maximum acceleration of 3 ft/s², a maximum deceleration of 5 ft/s², and a maximum speed of 45 mi/h. Determine the minimum distance required to catch the runaway car. (Assume that the switch engine starts from rest when the runaway car is 500 ft down the track and that the relative velocity when the engine catches the car must be less than 3 mi/h.)

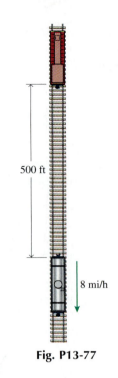

Fig. P13-77

13-78 A high-speed train travels from Washington D.C. to Philadelphia (a distance of 80 km) in just 35 min. The top speed of the train is 225 km/h. If the train decelerates twice as quickly as it accelerates, determine

a. The deceleration of the train.
b. The distance that the train travels at its maximum speed.

13-79 In Fig. P13-79 the winch *W* reels in cable at the constant rate of 5 ft/s. If the block that the winch is mounted on is stationary, determine the velocity of block *A*.

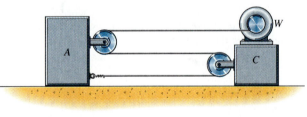

Fig. P13-79

13-80* In Fig. P13-80 the winch *W* is drawing in cable at a constant rate of 2 m/s. Determine the velocity of the counterweight *C* relative to the elevator *E*.

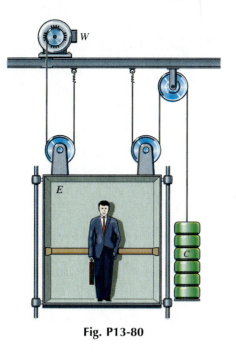

Fig. P13-80

13-81 In Fig. P13-81 the winch W reels in cable at the constant rate of 5 ft/s. If the block that the winch is mounted on C is moving to the left with a speed of 1 ft/s and the speed of C is increasing at the rate of 0.5 ft/s², determine

a. The velocity of block A.
b. The velocity of acceleration of A relative to C.

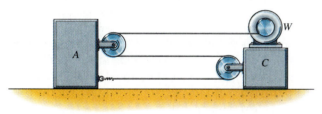

Fig. P13-81

13-82* In Fig. P13-82 the free end of the rope is being pulled to right at a constant speed of 3 m/s. Determine the speed and acceleration of slider on the vertical pole when $d = 2$ m.

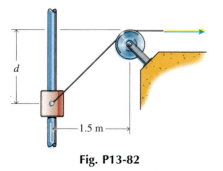

Fig. P13-82

13-83* In Fig. P13-83 cart A is being pulled to the left with a constant speed of 5 ft/s. Determine the speed and acceleration of cart B when $x = 4$ ft.

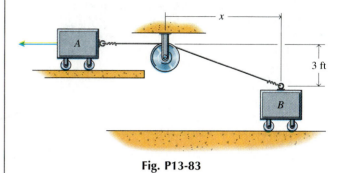

Fig. P13-83

13-84 At the instant shown in Fig. P13-84, block A is falling with a speed of 2 m/s, its speed is increasing at a rate of 0.8 m/s², and $y = 3$ m. Determine the speed and acceleration of block B at this instant.

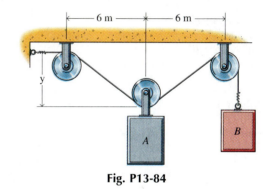

Fig. P13-84

Computer Problems

C13-85 In Fig. P13-85 the free end of the rope is being pulled to the left with a constant speed of 5 ft/s. Calculate and plot

a. The speed v and acceleration a of the cart as functions of the distance x, $0.5 \leq x \leq 15$ ft.
b. The position x, the speed v, and the acceleration a of the cart as functions of time t as the cart moves from $x = 15$ ft to $x = 0.5$ ft.

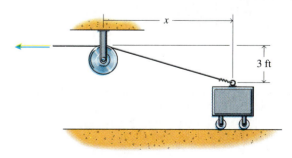

Fig. P13-85

C13-86 In Fig. P13-86 the free end of the rope is being pulled downward at a constant speed of 2 m/s. Calculate and plot

a. The speed v and acceleration a of the block as functions of the distance y, $0.2 \leq y \leq 12$ m.

b. The position y, the speed v, and the acceleration a of the block as functions of time t as the block moves from $y = 12$ m to $y = 0.2$ m.

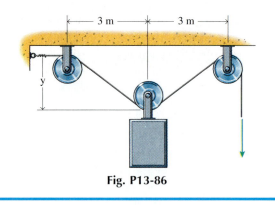

Fig. P13-86

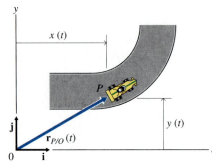

Figure 13-16 In a rectangular coordinate system, the position of a particle is described by giving its distance from two fixed orthogonal lines.

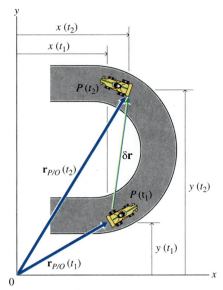

Figure 13-17 In a rectangular coordinate system, the x-component of the displacement is just the difference in the x-coordinates of the particle at times t_2 and t_1. Similarly, the y-component of the displacement is just the difference in the y-coordinates of the particle at times t_2 and t_1.

13-5 PLANE CURVILINEAR MOTION

When motion occurs in a single plane, two coordinates are required to describe the motion. The choice of which coordinates to use in a particular problem will depend on the geometry of the problem, on the way in which data are given for the problem, and on the type of solution that is desired. Three of the more commonly used coordinate systems used to represent the motion are rectangular coordinates, polar coordinates, and normal/tangential coordinates. These will be discussed in turn in the next three sections.

13-5.1 Rectangular Coordinates

In a rectangular coordinate system (in a plane), the position of a particle is described by giving its distance from two fixed orthogonal lines (Fig. 13-16). These two lines are called the x- and y-axes, and the coordinates are called the x- and y-components of the position. Unit vectors along the x- and y-axes will be denoted by **i** and **j**, respectively. Although the x- and y-coordinate directions (**i** and **j**) need not be horizontal and vertical, once they are chosen, they must remain fixed.

The position of a particle P with respect to the origin O of the fixed coordinate system is given by (Fig. 13-16)

$$\mathbf{r}_{P/O}(t) = x(t)\mathbf{i} + y(t)\mathbf{j} \qquad \textbf{(13-13)}$$

where $x(t)$ is the (time-dependent) x-component of the position and $y(t)$ is the (time-dependent) y-component of the position. The displacement of the particle between times t_1 and $t_2 > t_1$ (Fig. 13-17)

$$\delta\mathbf{r} = \mathbf{r}_{P/O}(t_2) - \mathbf{r}_{P/O}(t_1)$$
$$= [x(t_2) - x(t_1)]\mathbf{i} + [y(t_2) - y(t_1)]\mathbf{j} \qquad \textbf{(a)}$$

Since the directions as well as the magnitudes of the unit vectors **i** and **j** are fixed, their derivatives are zero. Then the velocity and acceleration of the particle are

$$\mathbf{v}_P(t) = v_x(t)\mathbf{i} + v_y(t)\mathbf{j}$$
$$= \dot{\mathbf{r}}_{P/O}(t) = \dot{x}(t)\mathbf{i} + \dot{y}(t)\mathbf{j} \qquad \textbf{(13-14)}$$

and

$$\mathbf{a}_P(t) = a_x(t)\mathbf{i} + a_y(t)\mathbf{j}$$
$$= \dot{\mathbf{v}}_P(t) = \dot{v}_x(t)\mathbf{i} + \dot{v}_y(t)\mathbf{j}$$
$$= \ddot{\mathbf{r}}_P(t) = \ddot{x}(t)\mathbf{i} + \ddot{y}(t)\mathbf{j} \qquad \textbf{(13-15)}$$

respectively.

The rectangular coordinate system is usually the most convenient one to use when the x- and y-components of the motion are specified separately from each other, do not depend on each other, or both. Typical examples are motion plotted on map grids (where x may be the longitude and y may be the latitude of the particle) and trajectory motion (where x may be the distance measured along the ground and y may be the height of the particle above the ground).

EXAMPLE PROBLEM 13-8

A basketball player shoots the ball at the basket which is 25 ft away, as shown in Fig. 13-18. Air resistance is negligible, and the acceleration of the ball during flight is 32.2 ft/s² downward. If the ball passes through the center of the basket, determine

a. The required initial speed v_0.
b. The angle ϕ that the trajectory of the ball will make with the vertical when the ball goes through the basket.

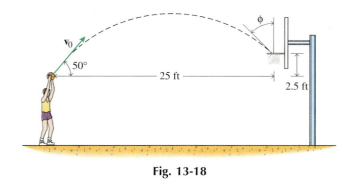

Fig. 13-18

SOLUTION

The velocity and position of the ball are obtained by integrating the acceleration

$$\mathbf{a} = \frac{d\mathbf{v}}{dt} = -32.2\mathbf{j} \text{ ft/s}^2$$

with respect to time to get

$$\mathbf{v} = v_0 \cos 50°\mathbf{i} + (v_0 \sin 50° - 32.2t)\mathbf{j} \text{ ft/s} \qquad (a)$$
$$\mathbf{r} = v_0 t \cos 50°\mathbf{i} + (v_0 t \sin 50° - 16.1t^2)\mathbf{j} \text{ ft} \qquad (b)$$

where the constants of integration have been chosen to satisfy the initial velocity ($\mathbf{v} = v_0 \cos 50°\mathbf{i} + v_0 \sin 50°\mathbf{j}$ ft/s) and initial position ($\mathbf{r} = \mathbf{0}$) when $t = 0$ s. The ball goes through the basket when

$$x = v_0 t \cos 50° = 25 \text{ ft} \qquad (c)$$
$$y = v_0 t \sin 50° - 16.1t^2 = 2.5 \text{ ft} \qquad (d)$$

$$10 = \frac{\cos 50}{\sin 50} - 16t^2$$

Solving Eqs. *c* and *d* gives $t = 1.3020$ s and

$$v_0 = 29.87 \cong 29.9 \text{ ft/s} \qquad \text{Ans.}$$

The velocity of the ball when it goes through the basket is found by substituting $t = 1.3020$ s and $v_0 = 29.87$ ft/s into Eq. *a*, which gives

$$\mathbf{v} = 29.87 \cos 50°\mathbf{i} + [29.87 \sin 50° - 32.2(1.3020)]\mathbf{j}$$
$$= 19.20\mathbf{i} - 19.04\mathbf{j} = 27.04 \text{ ft/s} \searrow 44.76°$$
$$\phi = 90° - 44.76° = 45.24° \qquad \text{Ans.}$$

Since the velocity vector is always tangent to the motion, the direction of the velocity vector can be used to determine the angle that the trajectory makes with the horizontal at any time of interest.

CONCEPTUAL EXAMPLE 13-1: THE GRAVITATIONAL CONSTANT-g

*What is the difference between the gravitational constant **g** and the acceleration of gravity?*

SOLUTION

According to Newton's Law of Gravitation, the earth exerts an attractive force

$$F_g = \frac{GM_e m}{r^2} \qquad (i)$$

on a body of mass *m* where *G* is the universal gravitational constant, M_e is the mass of the earth (constant), and *r* is the distance from the center of the earth to the body. If the body remains near the surface of the earth, then $r \cong r_e =$ constant and Eq. *i* can be written

$$F_g = mg \qquad (ii)$$

where

$$g = \frac{GM_e}{r_e^2} = \text{constant} \qquad (iii)$$

is called the gravitational constant. If the gravitational force is the only force acting on the body (for example, when the high diver jumps off of the platform), then the acceleration of the body is equal to the gravitational constant ($a = g$) and it acts in the direction of the gravitational force. Such motions are called projectile motions.

The gravitational constant defined by Eqs. *ii* and *iii*, however, is independent of the acceleration of the body. The gravitational force $F_g = mg$ acts on the high diver's body whether she is standing still on the platform preparing to dive or she is falling toward the water. In the first case, the gravitational force is opposed by the platform, the acceleration of the diver is zero, and the diver is in equilibrium. In the second case, the gravitational force is unopposed (assuming that air resistance is negligible) and the acceleration of the diver is $a = g$, downward.

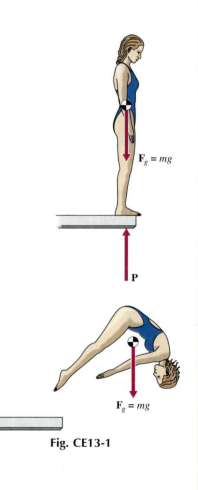

Fig. CE13-1

48

EXAMPLE PROBLEM 13-9

When air resistance is neglected, a cannonball fired through the air has a downward acceleration of g. If the cannonball has an initial velocity of v_0 with an angle of θ above the horizontal (Fig. 13-19), determine the angle θ_r that will give the maximum range.

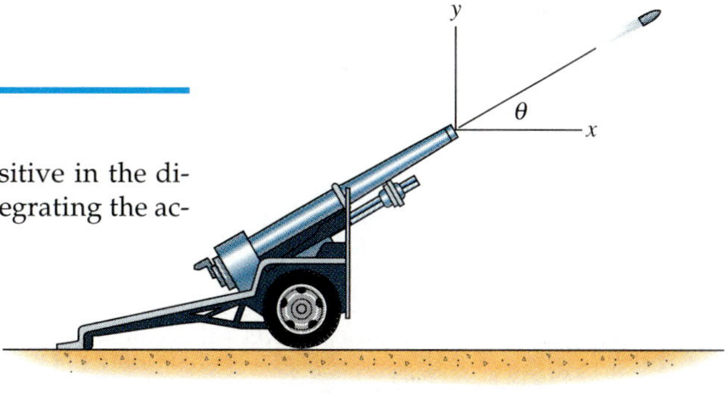

Fig. 13-19

SOLUTION

Choose rectangular coordinates with the x-axis horizontal (positive in the direction of motion) and the y-axis vertical (positive upward). Integrating the acceleration

$$\mathbf{a}(t) = -g\mathbf{j} = \dot{v}_x(t)\mathbf{i} + \dot{v}_y(t)\mathbf{j}$$

and using the given initial velocity

$$\mathbf{v}_0 = v_0 \cos\theta\mathbf{i} + v_0 \sin\theta\mathbf{j}$$

gives

$$\begin{aligned}\mathbf{v}(t) &= v_0 \cos\theta\mathbf{i} + (v_0 \sin\theta - gt)\mathbf{j} \\ &= \dot{x}(t)\mathbf{i} + \dot{y}(t)\mathbf{j}\end{aligned}$$

Integrating again gives

$$x(t) = v_0 t \cos\theta \qquad (a)$$

$$y(t) = \left(v_0 t \sin\theta - \frac{1}{2}gt^2\right) \qquad (b)$$

To find the time t_r when the cannonball again reaches the ground, the height is set to zero

$$y(t) = \left(v_0 t_r \sin\theta - \frac{1}{2}gt_r^2\right) = 0$$

which gives

$$t_r = \frac{2v_0 \sin\theta}{g}$$

The x-position of the particle at this time is

$$\begin{aligned}x(t_r) = x_{\text{range}} &= \frac{2v_0^2 \sin\theta\cos\theta}{g} \\ &= \frac{v_0^2 \sin 2\theta}{g}\end{aligned} \qquad (c)$$

The angle that gives the maximum x_{range} is then found by differentiating Eq. c with respect to θ and setting the derivative equal to zero

$$\frac{dx_{\text{range}}}{d\theta} = \frac{2v_0^2 \cos 2\theta_r}{g} = 0$$

which gives

$$\theta_r = 45° \qquad \text{Ans.}$$

Note that the result derived here is independent of the initial velocity v_0 and the gravitational constant g. However, the result does require that the initial and final heights be the same.

PROBLEMS

Introductory Problems

13-87* A soldier is firing a mortar which has a muzzle velocity of 600 ft/s at a hilltop target (Fig. P13-87). If the acceleration of the mortar shell is 32.2 ft/s², downward, and the horizontal and vertical distances to the target are 0.5 mi and 0.25 mi, respectively, determine the angle θ at which the mortar should be fired.

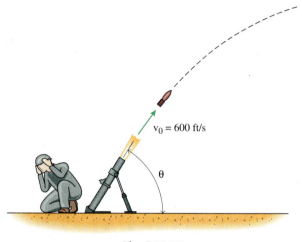

Fig. P13-87

13-88* An airplane flying horizontally at 600 km/h drops a bomb from an altitude of 2 km (Fig. P13-88). The acceleration of the bomb is 9.81 m/s², downward. Determine the horizontal distance d traveled by the bomb before it hits the target.

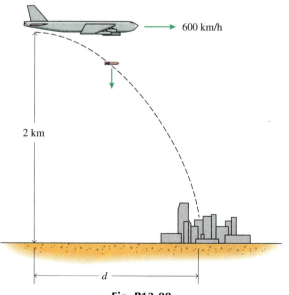

Fig. P13-88

13-89 The water skier shown in Fig. P13-89 leaves the end of the ski ramp with a speed of 20 mi/h and an angle of 25°. If he lets go of the tow rope immediately as he leaves the end of the ramp, his acceleration will be 32.2 ft/s², downward. Determine

a. The maximum height h attained by the skier.
b. The distance R from the end of the ramp to his point of landing.

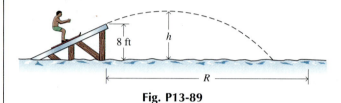

Fig. P13-89

13-90* For a circus act a human cannonball is to be shot over three tall Ferris wheels as shown in Fig. P13-90. If the launch angle is $\theta_0 = 53°$ and the acceleration is 9.81 m/s², downward, determine

a. The minimum initial velocity v_0 required to clear the three Ferris wheels.
b. The location R at which the net must be placed to catch the human cannonball.

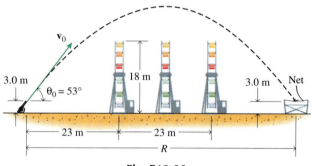

Fig. P13-90

13-91 A cyclist plans to jump her motorcycle over school buses as shown in Fig. P13-91. If the initial speed of the cyclist is $v_0 = 60$ mi/h and the acceleration is 32.2 ft/s², downward, determine

a. The maximum number of buses that the cyclist can jump over.
b. The location R at which the second ramp must be placed to complete the jump.

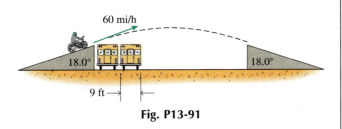

60 mi/h

18.0°   18.0°

9 ft

Fig. P13-91

13-92 A shore cannon is firing at a pirate ship 560 m offshore (Fig. P13-92). If the initial velocity of the cannon ball is $v_0 = 80$ m/s and its acceleration is 9.81 m/s², downward, determine

a. The initial angle θ_0 at which the cannon must be fired.
b. The time T that the pirate ship will have to evade the cannon ball.

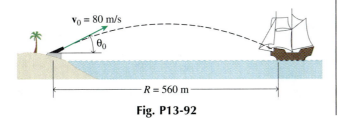

$v_0 = 80$ m/s

θ_0

$R = 560$ m

Fig. P13-92

Intermediate Problems

13-93* A cat burglar makes her escape by jumping from rooftop to rooftop as shown in Fig. P13-93. If the acceleration of the cat burglar is 32.2 ft/s², downward, determine

a. The initial speed v_0 required to complete the jump.
b. The horizontal and vertical components of the cat burglar's velocity when she lands on the lower roof.

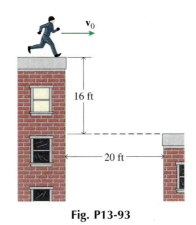

v_0

16 ft

20 ft

Fig. P13-93

13-94* A daredevil jumps his motorcycle across a canyon as shown in Fig. P13-94. If the acceleration of the daredevil is 9.81 m/s², downward, determine

a. The initial speed v_0 required to complete the jump.
b. The landing speed v_f.
c. The surface angle θ required at the end of the jump to ensure a smooth landing.

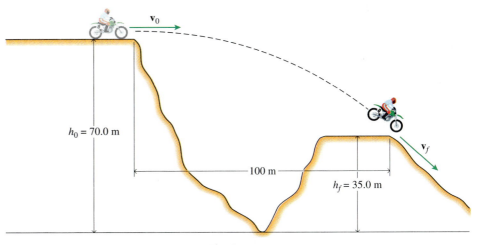

v_0

$h_0 = 70.0$ m

100 m

$h_f = 35.0$ m

v_f

Fig. P13-94

13-95 A 1/2-in.-diameter marble rolls off a horizontal step 8 ft high with an initial speed of $v_0 = 5 \pm 0.5$ ft/s (Fig. P13-95). Determine the minimum size hole D that this marble will be certain to fall through. (The acceleration of the marble is 32.2 ft/s^2, downward.)

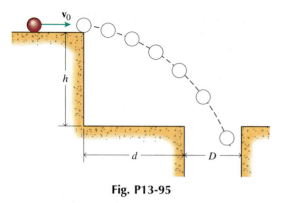

Fig. P13-95

13-96 A boy standing 5 m away from the bottom of a building is trying to throw a small ball through a 1-m-high window 7 m up (Fig. P13-96). If the ball is to pass through the window at the peak of its trajectory and its acceleration is 9.81 m/s^2, downward, determine

a. The initial velocity (magnitude and direction) required to just clear the bottom of the window.
b. The initial velocity (magnitude and direction) required to just clear the top of the window.

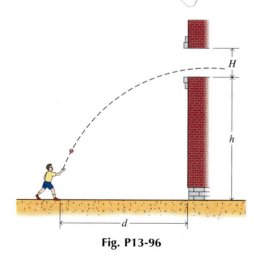

Fig. P13-96

13-97* In baseball the distance between the pitcher and the plate is 60 ft (Fig. P13-97). If the pitcher throws a fastball with an initial speed of 96 mi/h and its acceleration is 32.2 ft/s^2, downward, determine

a. The distance a that the ball will drop if the ball is pitched horizontally ($\theta_0 = 0$).
b. The initial angle θ_0 for which the baseball will reach the catcher at its initial level ($a = 0$).

c. The maximum height that the ball will attain if thrown at the angle θ_0 of part b.

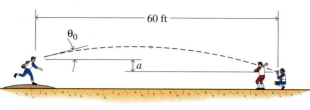

Fig. P13-97

13-98 A shotputter tosses a shot at 40° to the horizontal from a height of 1.8 m (Fig. P13-98). If the shot lands 15 m away and its acceleration is 9.81 m/s^2, downward, determine

a. The initial speed v_0 of the shot.
b. The maximum height h attained by the shot.
c. The distance d at which the maximum height occurs.

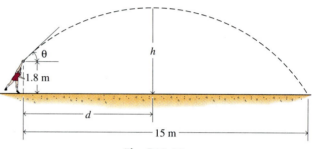

Fig. P13-98

Challenging Problems

13-99* A skier leaves the end of a ski jump with a velocity of 60 mi/h at 10° above the horizontal (Fig. P13-99). If his acceleration is 32.2 ft/s^2, downward, determine

a. The maximum height that the skier will rise from the end of the ski jump.
b. The time of flight of the jump.
c. The distance of the jump (the distance d measured along the slope).

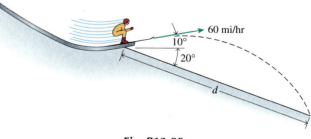

Fig. P13-99

13-100 In Example Problem 13-9 it was shown that the initial angle θ_0 that gives the maximum range of a projectile is 45°. Although this result holds for any given initial speed, it does require that the initial and final heights are the same. Determine the initial angle θ_0 that gives the maximum range of a projectile if the final height is above the initial height as shown in Fig. P13-100.

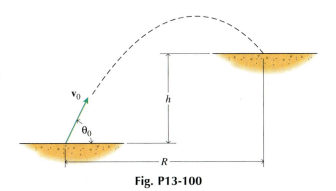

Fig. P13-100

Computer Problems

C13-101 Pin P of Fig. P13-101 slides in the horizontal and vertical grooves attached to collars A and B. Collar A slides in a horizontal plane with a position given by $x(t) = 10 \cos at$ while collar B slides in a vertical plane with a position given by $y(t) = 10 \sin bt$, where $a = 3$ rad/s, $b = 4$ rad/s, t is in seconds, and x and y are both in inches.

a. Calculate and plot the position of the pin $y(t)$ versus $x(t)$ for $0 < t < 6.5$ s.
b. Evaluate the velocity $\mathbf{v}_P(t)$ and the acceleration $\mathbf{a}_P(t)$ of the pin at $t = 1.0$ s and show them on the plot of part a.

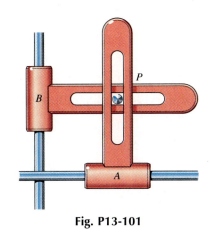

Fig. P13-101

C13-102 A basketball player shoots the ball at the basket with an initial speed v_0 and initial angle θ_0 as shown in Fig. P13-102. For the ball to go through the hoop, it must be coming down at an angle θ less than 70°. A further constraint is that the maximum arch of the ball should not be greater than $h = 3$ m.

a. Calculate and plot the initial velocity v_0 versus initial angle θ_0 ($30° \le \theta_0 \le 70°$) for successful shots. (Don't count shots that bounce off the backboard and into the hoop!)
b. Since the hoop is a little larger than the ball, a successful shot need not hit the center of the basket. However, shots that enter the basket at a low angle θ must be more precise than shots that enter the basket at a high angle. Assume that the ball can be short or long by

10 mm	for	$65° \le \theta \le 70°$
25 mm	for	$60° \le \theta < 65°$
50 mm	for	$50° \le \theta < 60°$
100 mm	for	$40° \le \theta < 50°$
125 mm	for	$30° \le \theta < 40°$
150 mm	for	$15° \le \theta < 30°$
175 mm	for	$0° \le \theta < 15°$

and plot the range of initial velocities that are acceptable for each angle θ_0 ($30° \le \theta_0 \le 70°$).

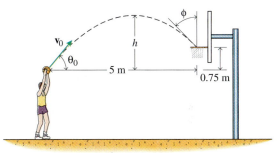

Fig. P13-102

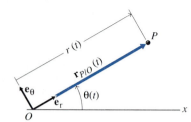

Figure 13-20 In a polar coordinate system, the position of a particle is described by giving the distance from a fixed point and the angular displacement relative to a fixed line.

13-5.2   Polar Coordinates (Radial/Transverse Coordinates)

In a polar coordinate system, the position of a particle is described by giving the distance from a fixed point and the angular displacement relative to a fixed line (Fig. 13-20). The coordinate directions ($\mathbf{e}_r$ and $\mathbf{e}_\theta$) are taken to be radially outward from the fixed point and perpendicular to the radial line in the direction of increasing θ.

In polar coordinates, the position of particle P with respect to the origin O is given by (Fig. 13-20)

$$\mathbf{r}_{P/O}(t) = r(t)\mathbf{e}_r \qquad (13\text{-}16)$$

where $r(t)$ is the (time-dependent) r-component of the position. The dependence of the position vector on the angle $\theta(t)$ is hidden in the unit vector $\mathbf{e}_r$, which depends on θ (which may depend on time).

Since the directions of the unit vectors $\mathbf{e}_r$ and $\mathbf{e}_\theta$ are not necessarily fixed, their changes must be considered when the position vector, Eq. 13-16, is differentiated. The derivative of $\mathbf{e}_r$ with respect to time is calculated using the chain rule of differentiation

$$\dot{\mathbf{e}}_r = \frac{d\mathbf{e}_r}{dt} = \frac{d\mathbf{e}_r}{d\theta}\frac{d\theta}{dt} \qquad (b)$$

where

$$\frac{d\mathbf{e}_r}{d\theta} = \lim_{\Delta\theta\to 0} \frac{\mathbf{e}_r(\theta + \Delta\theta) - \mathbf{e}_r(\theta)}{\Delta\theta} \qquad (c)$$

But in the limit as $\Delta\theta \to 0$, the distance $|\mathbf{e}_r(\theta + \Delta\theta) - \mathbf{e}_r(\theta)|$ tends to the arc length along a unit circle $\Delta s = 1\,\Delta\theta$ and the angle α tends to 90° (see Fig. 13-21). Therefore, the vector $\mathbf{e}_r(\theta + \Delta\theta) - \mathbf{e}_r(\theta)$ has magnitude $\Delta\theta$ and points in the $\mathbf{e}_\theta$-direction and

$$\dot{\mathbf{e}}_r = \dot{\theta} \lim_{\Delta\theta\to 0} \frac{\Delta\theta\mathbf{e}_\theta}{\Delta\theta}$$

$$= \dot{\theta}\mathbf{e}_\theta \qquad (13\text{-}17)$$

where $\dot{\theta} = d\theta/dt$.

Similarly, the derivative of $\mathbf{e}_\theta$ with respect to time can be calculated:

$$\dot{\mathbf{e}}_\theta = \frac{d\mathbf{e}_\theta}{dt} = \frac{d\mathbf{e}_\theta}{d\theta}\frac{d\theta}{dt} \qquad (d)$$

where

$$\frac{d\mathbf{e}_\theta}{d\theta} = \lim_{\Delta\theta\to 0} \frac{\mathbf{e}_\theta(\theta + \Delta\theta) - \mathbf{e}_\theta(\theta)}{\Delta\theta} \qquad (e)$$

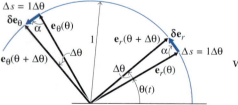

Figure 13-21 In the limit as $\Delta t \to 0$, the angle $\Delta\theta \to 0$, the displacement $\delta\mathbf{e}_r$ points in the direction of $\mathbf{e}_\theta$, the displacement $\delta\mathbf{e}_\theta$ points in the direction of $-\mathbf{e}_r$, and the lengths of the displacement vectors $|\delta\mathbf{e}_r| = |\delta\mathbf{e}_\theta| \to \Delta s = 1\,\Delta\theta$.

But in the limit as $\Delta\theta \to 0$, the distance $|\mathbf{e}_\theta(\theta + \Delta\theta) - \mathbf{e}_\theta(\theta)|$ again tends to the arc length along a unit circle $\Delta s = 1\,\Delta\theta$ and the angle α again tends to 90° (see Fig. 13-21). Therefore, the vector $\mathbf{e}_\theta(\theta + \Delta\theta) - \mathbf{e}_\theta(\theta)$ has magnitude $\Delta\theta$ and points in the negative $\mathbf{e}_r$-direction and

$$\dot{\mathbf{e}}_\theta = \dot{\theta} \lim_{\Delta\theta\to 0} \frac{-\Delta\theta\mathbf{e}_r}{\Delta\theta} = -\dot{\theta}\mathbf{e}_r \qquad (13\text{-}18)$$

An alternative way to evaluate the derivatives of $\mathbf{e}_r$ and $\mathbf{e}_\theta$ with respect to θ, one that may be easier for the student to understand and remember, is to write $\mathbf{e}_r$ and $\mathbf{e}_\theta$ in terms of their rectangular components and then take the derivatives. With reference to Fig. 13-22,

$$\mathbf{e}_r = \cos\theta\,\mathbf{i} + \sin\theta\,\mathbf{j} \tag{f}$$
$$\mathbf{e}_\theta = -\sin\theta\,\mathbf{i} + \cos\theta\,\mathbf{j} \tag{g}$$

The derivatives are then

$$d\mathbf{e}_r/d\theta = -\sin\theta\,\mathbf{i} + \cos\theta\,\mathbf{j} = \mathbf{e}_\theta \tag{h}$$
$$d\mathbf{e}_\theta/d\theta = -\cos\theta\,\mathbf{i} - \sin\theta\,\mathbf{j} = -\mathbf{e}_r \tag{i}$$

which is the same as above.

The velocity of the particle can now be computed:

$$\mathbf{v}_P(t) = v_r\mathbf{e}_r + v_\theta\mathbf{e}_\theta = \dot{\mathbf{r}}_{P/O}(t)$$
$$= \dot{r}\mathbf{e}_r + r\dot{\mathbf{e}}_r$$
$$= \dot{r}\mathbf{e}_r + r\dot\theta\mathbf{e}_\theta \tag{13-19}$$

Finally, the acceleration is computed:

$$\mathbf{a}_P(t) = a_r\mathbf{e}_r + a_\theta\mathbf{e}_\theta = \dot{\mathbf{v}}_P(t)$$
$$= \ddot{r}\mathbf{e}_r + \dot{r}\dot{\mathbf{e}}_r + (\dot{r}\dot\theta + r\ddot\theta)\mathbf{e}_\theta + r\dot\theta\dot{\mathbf{e}}_\theta$$
$$= \ddot{r}\mathbf{e}_r + \dot{r}(\dot\theta\mathbf{e}_\theta) + (\dot{r}\dot\theta + r\ddot\theta)\mathbf{e}_\theta - r\dot\theta(\dot\theta\mathbf{e}_r)$$
$$= (\ddot{r} - r\dot\theta^2)\mathbf{e}_r + (r\ddot\theta + 2\dot{r}\dot\theta)\mathbf{e}_\theta \tag{13-20}$$

For the special case of a particle in circular motion, $r =$ constant, Eqs. 13-19 and 13-20 reduce to

$$\mathbf{v}_P(t) = r\dot\theta\mathbf{e}_\theta \tag{13-21}$$
$$\mathbf{a}_P(t) = -r\dot\theta^2\mathbf{e}_r + r\ddot\theta\mathbf{e}_\theta \tag{13-22}$$

The polar coordinate system is usually the most convenient one to use when the distance and direction of a particle are measured relative to a fixed point (such as a radar tracking an airplane in a plane) or when a particle is fixed on or moves along a rotating arm.

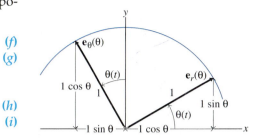

Figure 13-22 If the angle θ is measured relative to the x-axis of a rectangular Cartesian coordinate system, then the rectangular components of the polar coordinate unit vectors are $\mathbf{e}_r = 1\cos\theta\mathbf{i} + 1\sin\theta\mathbf{j}$ and $\mathbf{e}_\theta = -1\sin\theta\mathbf{i} + 1\cos\theta\mathbf{j}$.

EXAMPLE PROBLEM 13-10

A radar tracking an airplane gives the coordinates of the plane as $r(t)$ and $\theta(t)$ (Fig. 13-23). At some instant of time, $\theta = 40°$ and $r = 6400$ ft. From successive measurements of r and θ, the derivatives at this instant are estimated as $\dot{r} = 312$ ft/s, $\dot\theta = -0.039$ rad/s, $\ddot{r} = 9.751$ ft/s², and $\ddot\theta = 0.003807$ rad/s². Calculate the velocity and acceleration of the airplane at this instant.

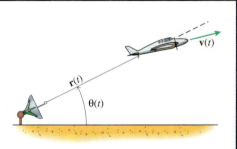

SOLUTION

Choosing polar coordinates centered at the radar as indicated (Fig. 13-23), the radial component of velocity is

$$v_r = \dot{r} = 312 \text{ ft/s}$$

and the θ-component is

$$v_\theta = r\dot\theta = (6400)(-0.039) = -250 \text{ ft/s}$$

The resultant then has magnitude

$$v = \sqrt{312^2 + 250^2} = 400 \text{ ft/s} \qquad\qquad \text{Ans.}$$

Fig. 13-23

and acts at an angle

$$\phi_v = \tan^{-1}\frac{250}{312} = 38.7°$$ Ans.

measured clockwise from the radial direction (Fig. 13-24a).
The radial component of acceleration is

$$a_r = \ddot{r} - r\dot{\theta}^2 = (9.751) - (6400)(-0.039)^2 = 0.017 \text{ ft/s}^2$$

and the θ-component is

$$a_\theta = r\ddot{\theta} + 2\dot{r}\dot{\theta} = (6400)(0.003807) + 2(312)(-0.039)$$
$$= 0.029 \text{ ft/s}^2$$

The resultant then has magnitude

$$a = \sqrt{0.017^2 + 0.029^2} = 0.034 \text{ ft/s}^2$$ Ans.

and acts at an angle

$$\phi_a = \tan^{-1}\frac{0.029}{0.017} = 59.6°$$ Ans.

measured clockwise from the radial direction (Fig. 13-24b).

Note that the numbers obtained for the acceleration terms are probably not very accurate. They resulted from subtracting two numbers that were accurate to the second decimal place at best. Therefore, the answer probably has no more than one significant figure.

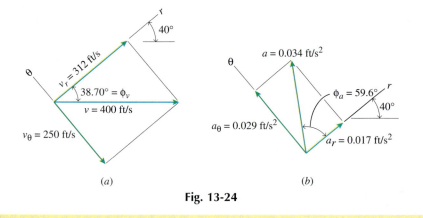

(a)   (b)

Fig. 13-24

A cam lobe has a shape given by $r = 20 + 15 \cos\theta$ mm (Fig. 13-25). Pin P slides in a slot along arm AB and is held against the cam by a spring. The arm AB rotates counterclockwise about A at a rate of 30 rev/min. Given that $\theta = 0$ at $t = 0$:

a. Determine the velocity and acceleration of the pin.
b. Evaluate the expressions of part a for the velocity and acceleration at $t = 0.75$ s.
c. Show the velocity and acceleration of part b on a suitable sketch.

(Assume that the pin is very small so that the center of the pin follows the contour of the cam lobe.)

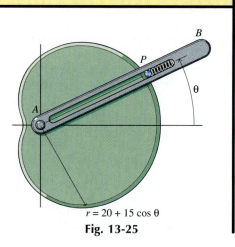

$r = 20 + 15 \cos\theta$
Fig. 13-25

SOLUTION

a. First, integrating the given angular velocity

$$\dot{\theta} = \frac{(30 \text{ rev/min})}{(60 \text{ sec/min})}(2\pi \text{ rad/rev}) = \pi \text{ rad/s}$$

gives

$$\theta = \pi t \text{ rad}$$

where the constant of integration is zero, since $\theta = 0$ at $t = 0$. Also, since the rotation rate is a constant,

$$\ddot{\theta} = 0$$

Next, differentiating the radial function

$$r = 20 + 15 \cos \theta = 20 + 15 \cos \pi t$$

gives

$$\dot{r} = -15\pi \sin \pi t$$

and

$$\ddot{r} = -15\pi^2 \cos \pi t$$

Now, the velocity components can be computed

$$v_r = \dot{r} = -15\pi \sin \pi t \qquad \text{Ans.}$$
$$v_\theta = r\dot{\theta} = (20 + 15 \cos \pi t)(\pi)$$
$$= 20\pi + 15\pi \cos \pi t \qquad \text{Ans.}$$

Similarly, the acceleration components are

$$a_r = \ddot{r} - r\dot{\theta}^2 = (-15\pi^2 \cos \pi t) - (20 + 15 \cos \pi t)(\pi)^2$$
$$= -\pi^2(20 + 30 \cos \pi t) \qquad \text{Ans.}$$
$$a_\theta = r\ddot{\theta} + 2\dot{r}\dot{\theta} = (20 + 15 \cos \pi t)(0) + 2(-15\pi \sin \pi t)(\pi)$$
$$= -30\pi^2 \sin \pi t \qquad \text{Ans.}$$

> The radial function can also be differentiated implicitly with respect to time. That is, $\dot{r} = \dfrac{dr}{dt} = \dfrac{dr}{d\theta}\dfrac{d\theta}{dt}$ where $\dfrac{dr}{d\theta} = -15 \sin \theta$ and $\dfrac{d\theta}{dt} = \dot{\theta} = \pi$ rad/s. Therefore, $\dot{r} = -15\pi \sin \theta$ and when $\theta = 3\pi/4$ rad $= 135°$, $v_r = \dot{r} = -33.3$ m/s.

b. At $t = 0.75$ s, $\theta = 3\pi/4$ rad $= 135°$. The velocity and acceleration components are

$$v_r = -15\pi \sin 3\pi/4 = -33.3 \text{ mm/s} \qquad \text{Ans.}$$
$$v_\theta = 20\pi + 15\pi \cos 3\pi/4 = 29.5 \text{ mm/s} \qquad \text{Ans.}$$
$$a_r = -\pi^2(20 + 30 \cos 3\pi/4)$$
$$= 11.97 \text{ mm/s}^2 \qquad \text{Ans.}$$
$$a_\theta = -30\pi^2 \sin 3\pi/4 = -209.4 \text{ mm/s}^2 \qquad \text{Ans.}$$

c. The magnitude of the velocity is

$$v = \sqrt{33.3^2 + 29.5^2} = 44.5 \text{ mm/s}$$

and the direction of the velocity is

$$\phi_v = \tan^{-1}\frac{29.5}{33.3} = 41.5°$$

measured clockwise from the negative r-direction. The magnitude of the acceleration is

$$a = \sqrt{11.97^2 + 209.4^2} = 209.7 \text{ mm/s}^2$$

and the direction of the acceleration is

$$\phi_a = \tan^{-1}\frac{209.4}{11.97} = 86.7°$$

measured clockwise from the positive r-direction. These values are shown in Fig. 13-26.

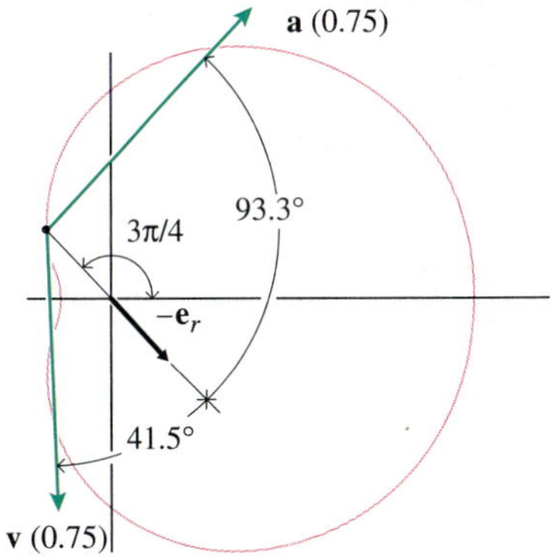

Fig. 13-26

PROBLEMS

Introductory Problems

13-103* A radar is tracking a rocket (Fig. P13-103). At some instant of time, the distance, r, and angle, θ, are measured as 10 mi and 30°, respectively. From successive measurements, the derivatives, $\dot{r}$, $\ddot{r}$, $\dot{\theta}$, and $\ddot{\theta}$ are estimated to be 650 ft/s, 165 ft/s², 0.031 rad/s, and 0.005 rad/s², respectively. Determine the velocity and acceleration of the rocket.

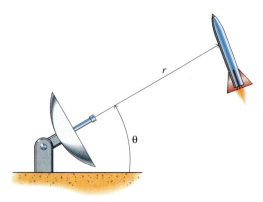

Fig. P13-103

13-104* A particle is following a spiral path given by $r(t) = 5\theta/3$ where $\theta(t)$ is in radians and r is in millimeters. Given that $\dot{\theta} = 10/t$ rad/s and that $\theta = 0$ when $t = 1$ s, determine the velocity and acceleration of the particle when $\theta = 240°$.

13-105 A particle is following a path given by $r(t) = 5 \sin \theta \cos^2 \theta$ where $\theta(t)$ is in radians and r is in inches. Given that $\dot{\theta} = 2$ rad/s (constant) and that $\theta = 0$ when $t = 0$, determine the velocity and acceleration of the particle when $t = 2$ s.

13-106* A particle is following a path given by $r(t) = 50 \cos 3\theta$, where $\theta(t)$ is in radians and r is in millimeters. Given that $\dot{\theta} = 2.5$ rad/s (constant) and that $\theta = 0$ when $t = 0$, determine the velocity and acceleration of the particle when $t = 2$ s.

Intermediate Problems

13-107* A radar tracking an airplane gives the coordinates of the plane as $r(t)$ and $\theta(t)$ (Fig. P13-107). During some segment of the tracking, the radar recorded the following values:

t	$r(t)$	$\theta(t)$
0 s	8,020 ft	29.9°
2 s	8,590 ft	28.7°
4 s	9,285 ft	27.6°
6 s	10,110 ft	26.5°

For this segment of the flight,

a. Estimate the velocity components v_r and v_θ.
b. Estimate the acceleration components a_r and a_θ.

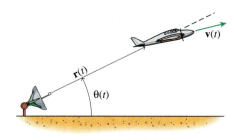

Fig. P13-107

13-108 A radar tracking a rocket gives the coordinates of the rocket as $r(t)$ and $\theta(t)$ (Fig. P13-108). During some segment of the tracking, the radar recorded the following values:

t	$r(t)$	$\theta(t)$
0 s	5000 m	60.0°
2 s	6000 m	64.6°
4 s	7145 m	68.1°
6 s	8422 m	70.9°

For this segment of the flight,

a. Estimate the velocity components v_r and v_θ.
b. Estimate the acceleration components a_r and a_θ.

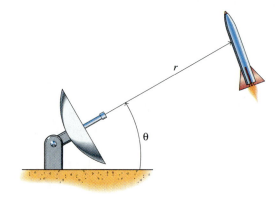

Fig. P13-108

13-109 A collar that slides along a horizontal rod has a pin that is constrained to move in the slot of arm AB (Fig. P13-109). The arm oscillates with angular position given by $\theta(t) = 90 - 30 \cos \omega t$ where $\omega = 1.5$ rad/s, t is in seconds, and θ is in degrees. For $t = 5$ s,

a. Determine the radial distance r from the pivot A to the pin B.
b. Determine the velocity components v_r and v_θ of the collar.

c. Determine the acceleration components a_r and a_θ of the collar.

d. Verify that the velocity vector **v** and the acceleration vector **a** are both directed along the horizontal rod.

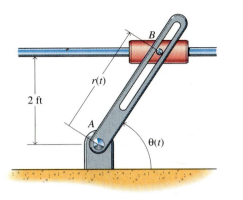

Fig. P13-109

13-110* A collar that slides around a circular wire has a pin that is constrained to move in the slot of arm AB (Fig. P13-110). The arm rotates counterclockwise at a constant angular speed of $\omega = 2$ rad/s. When the arm is 30° above the horizontal,

a. Determine the radial distance r from the pivot A to the pin B.

b. Determine the velocity components v_r and v_θ of the collar.

c. Determine the acceleration components a_r and a_θ of the collar.

d. Verify that the velocity vector **v** is directed along the wire.

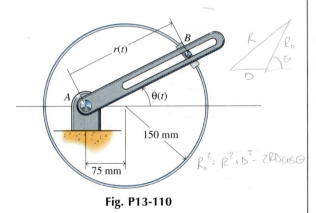

Fig. P13-110

13-111* Arm AC of the cam follower mechanism shown in Fig. P13-111 is rotating at a constant angular speed of $\omega = 150$ rev/min. A spring holds the pin B against the cam lobes. The equation that describes the shape of the cam lobes is

$$R = 2.5 + 0.5 \cos 3\theta$$

where R is in inches. When $\theta = 10°$,

a. Determine the velocity components v_r and v_θ of the pin B.

b. Determine the acceleration components a_r and a_θ of the pin B.

c. Verify that the velocity vector **v** is directed along the cam surface.

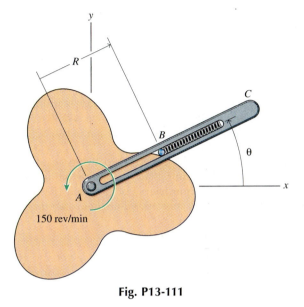

Fig. P13-111

13-112 Arm AB of the cam follower mechanism shown in Fig. P13-112 is rotating at a constant angular speed of $\omega = 60$ rev/min. A spring holds the pin P against the cam lobes. The equation that describes the shape of the cam lobes is

$$r = 20 + 15 \cos \theta$$

where r is in millimeters. When $\theta = 75°$,

a. Determine the velocity components v_r and v_θ of the pin P.

b. Determine the acceleration components a_r and a_θ of the pin P.

c. Verify that the velocity vector **v** is directed along the cam surface.

ANSWERS $v = -91.036e_r + 150 \cdot e_\theta$

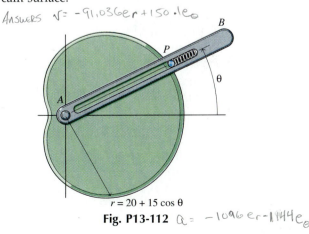

$r = 20 + 15 \cos \theta$

Fig. P13-112 $a = -1096 e_r - 1144 e_\theta$

59

Challenging Problems

13-113* A particle is following a path given by $r(t) = 5 \sin \theta \cos^2 \theta$ where $\theta(t)$ is in radians and r is in inches. If the speed of the particle is not to exceed 10 ft/s, determine the maximum constant value that $\dot{\theta}$ can be.

13-114 A particle is following a path given by $r(t) = 50 \cos 3\theta$, where $\theta(t)$ is in radians and r is in millimeters. If the speed of the particle is not to exceed 5 m/s, determine the maximum constant value that $\dot{\theta}$ can be.

13-115 A particle is following a path given by $r(t) = 5 \sin \theta \cos^2 \theta$ where $\theta(t)$ is in radians and r is in inches. If the acceleration of the particle is not to exceed 30 ft/s², determine the maximum constant value that $\dot{\theta}$ can be.

13-116* A particle is following a path given by $r(t) = 50 \cos 3\theta$, where $\theta(t)$ is in radians and r is in millimeters. If the acceleration of the particle is not to exceed 15 m/s², determine the maximum constant value that $\dot{\theta}$ can be.

Computer Problems

13-117 A particle is following a path given by $r(t) = 5 \sin \theta \cos^2 \theta$ where $\theta(t)$ is in radians and r is in inches. Given that $\dot{\theta} = 2$ rad/s (constant) and that $\theta = 0$ when $t = 0$,

a. Compute and plot the position of the particle for $0 < t < 2$ s.
b. Compute the velocity $\mathbf{v}(t)$ and the acceleration of the particle $\mathbf{a}(t)$. Evaluate $\mathbf{v}(t)$ and $\mathbf{a}(t)$ when $\theta = 150°$ and show them on the plot of part *a*.

13-118 Arm AC of the cam follower mechanism shown in Fig. P13-118 is rotating at a constant angular speed of $\omega = 150$ rev/min. A spring holds the pin B against the cam lobes. If the equation that describes the shape of the cam lobes is

$$R = 125 + 50 \cos 3\theta$$

where R is in millimeters,

a. Calculate and plot the magnitude of the velocity v_B and the acceleration a_B of the pin B as functions of θ for $0 < \theta < 180°$.
b. Will the shape of the curves change if the angular speed ω is doubled?

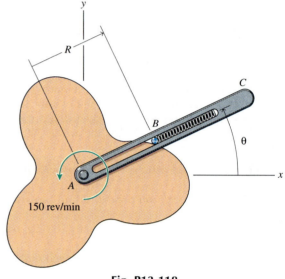

Fig. P13-118

13-5.3   Normal and Tangential Coordinates

In some problems the motion is specified by giving the path that the particle is moving along and the speed of the particle at each point along the path. Coordinates are chosen at each point along the path with unit vectors $\mathbf{e}_t$ tangential to the path and pointing in the direction of motion and $\mathbf{e}_n$ normal to the path and pointing toward the center of curvature (Fig. 13-27).

The velocity of the particle has direction $\mathbf{e}_t$ and a magnitude equal to the rate at which the particle moves along the path. To see that this is so, draw the position of the particle at two instants of time (Fig. 13-28). For Δt small, the magnitude of the displacement is nearly the same as the distance along the curve Δs and the direction of the displacement tends to the direction of the unit tangent vector $\mathbf{e}_t$. The velocity is then

$$\mathbf{v}_P(t) = \lim_{\Delta t \to 0} \frac{\delta \mathbf{r}(t)}{\Delta t} = \lim_{\Delta t \to 0} \frac{\Delta s(t)}{\Delta t} \mathbf{e}_t$$

$$= \dot{s} \mathbf{e}_t = v \mathbf{e}_t \tag{13-23}$$

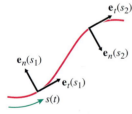

Figure 13-27 In the normal and tangential coordinate system, the position of the particle is specified by giving the distance it has moved along its path. The tangential unit vector always points in the direction of motion. The normal unit vector is perpendicular to the motion and points in the direction that the particle is turning.

where $v = \dot{s}$ is the magnitude of the velocity and the direction of the unit tangent vector $\mathbf{e}_t$ varies with position (which varies with time).

Since the direction of the unit vector $\mathbf{e}_t$ is not fixed, its change must also be considered when differentiating the velocity to find the acceleration. Using the chain rule of differentiation, the derivative of $\mathbf{e}_t$ with respect to time is calculated by

$$\frac{d\mathbf{e}_t}{dt} = \frac{d\mathbf{e}_t}{ds}\frac{ds}{dt} \qquad (j)$$

To evaluate the derivative of the unit tangent vector $\mathbf{e}_t$ with respect to s, let s be the position of the particle at time t and $s + \Delta s$ be its position at time $t + \Delta t$ (Fig. 13-29). Draw a circle having its center at the intersection of $\mathbf{e}_n(s)$ and $\mathbf{e}_n(s + \Delta s)$ that passes through the points s and $s + \Delta s$. The relationship between the unit tangent and unit normal vectors at s and $s + \Delta s$ is shown in Figure 13-30. Then

$$\frac{d\mathbf{e}_t}{ds} = \lim_{\Delta s \to 0} \frac{\mathbf{e}_t(s + \Delta s) - \mathbf{e}_t(s)}{\Delta s} \qquad (k)$$

But in the limit as $\Delta s \to 0$, the distance $|\mathbf{e}_t(s + \Delta s) - \mathbf{e}_t(s)|$ tends to the arc length along a unit circle $1\,\Delta\phi$ and the angle α tends to 90°. Therefore the vector $\mathbf{e}_t(s + \Delta s) - \mathbf{e}_t(s)$ has magnitude $\Delta\phi$ and points in the $\mathbf{e}_n$-direction and

$$\frac{d\mathbf{e}_t}{ds} = \lim_{\Delta s \to 0} \frac{\Delta\phi}{\Delta s}\mathbf{e}_n(s) \qquad (l)$$

But from Fig. 13-29 $\Delta s = \rho\,\Delta\phi$, so finally

$$\dot{\mathbf{e}}_t = \dot{s} \lim_{\Delta s \to 0} \frac{\Delta s}{\rho\,\Delta s}\mathbf{e}_n(s) = \frac{\dot{s}}{\rho}\mathbf{e}_n(s) \qquad (13\text{-}24a)$$

where

$$v = \dot{s} = \lim_{\Delta t \to 0} \frac{\Delta s}{\Delta t} = \lim_{\Delta t \to 0} \frac{\rho\,\Delta\phi}{\Delta t} = \rho\dot{\phi} \qquad (13\text{-}24b)$$

The acceleration of a particle in normal and tangential coordinates then is given by

$$\mathbf{a}_P(t) = a_t\mathbf{e}_t + a_n\mathbf{e}_n = \dot{\mathbf{v}}_P(t) = \dot{v}\mathbf{e}_t + v\dot{\mathbf{e}}_t$$

$$= \dot{v}\mathbf{e}_t + v\frac{\dot{s}}{\rho}\mathbf{e}_n = \dot{v}\mathbf{e}_t + \frac{v^2}{\rho}\mathbf{e}_n \qquad (13\text{-}25)$$

For the special case of a particle in circular motion, $\rho = r = $ constant, $\mathbf{e}_r = -\mathbf{e}_n$ (since $\mathbf{e}_r$ points outward from the center of the circle and $\mathbf{e}_n$ points toward the center of curvature), and $\mathbf{e}_t = \mathbf{e}_\theta$. Then

$$\mathbf{v}_P(t) = v_t\mathbf{e}_t = r\dot{\theta}\mathbf{e}_\theta \qquad (13\text{-}26)$$

and

$$\mathbf{a}_P(t) = \dot{v}\mathbf{e}_\theta + \frac{v^2}{\rho}(-\mathbf{e}_r) = r\ddot{\theta}\mathbf{e}_\theta - \frac{(r\dot{\theta})^2}{r}\mathbf{e}_r$$

$$= -r\dot{\theta}^2\mathbf{e}_r + r\ddot{\theta}\mathbf{e}_\theta \qquad (13\text{-}27)$$

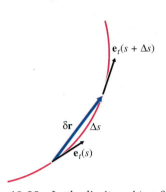

Figure 13-28 In the limit as $\Delta t \to 0$, the magnitude of the displacement $|\delta \mathbf{r}| \to \Delta s$, and the direction of the displacement approaches the tangential direction.

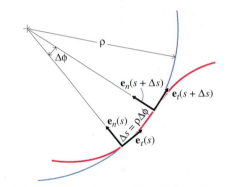

Figure 13-29 The radius of curvature ρ is the radius of the circle that best fits the path at the point of interest. In the limit as $\Delta t \to 0$, the angle $\Delta\phi \to 0$ and the magnitude of the displacement $|\delta\mathbf{r}| \to \Delta s = \rho\,\Delta\phi$.

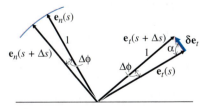

Figure 13-30 In the limit as $\Delta t \to 0$, the angle $\Delta\phi \to 0$, the displacement $\delta\mathbf{e}_t$ points in the direction of $\mathbf{e}_n$, and the length of the displacement vector $|\delta\mathbf{e}_t| \to 1\,\Delta\phi$.

But Eqs. 13-26 and 13-27 are the same as Eqs. 13-21 and 13-22, which were derived for the velocity and acceleration in polar coordinates for the case where r is a constant.

Normal and tangential coordinates are the most convenient to use when particles move along a surface of known shape. In such cases, the normal accelerations are required to determine the contact force between the particle and the surface. When the contact force becomes negative, as in the design of roller coasters, special tracks must be used to keep the particle following the curve. Also, the normal contact force is often required to compute the tangential (friction) force and thereby determine the tangential acceleration and velocity.

CONCEPTUAL EXAMPLE 13-2: SMOOTHING HIGHWAY TURNS

Why do arcs of circles not make good curves for highways? Why are transition curves required to blend between a circular curve and a straight segment of highway?

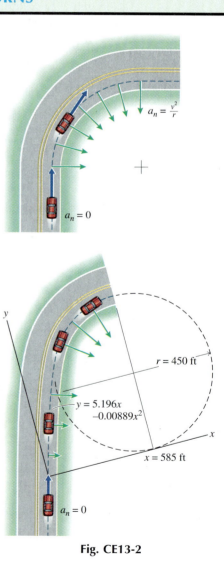

Fig. CE13-2

SOLUTION

When vehicles travel at a constant speed, their tangential component of acceleration is always zero: $a_t = \dot{v} = 0$. However, the normal component of acceleration is directly proportional to the square of the speed v^2 of the vehicle and inversely proportional to the radius of curvature ρ of the path the vehicle is traveling, $a_n = v^2/\rho$. Therefore, for a given speed v, the sharper the curve (the smaller the radius of curvature ρ), the greater will be the normal component of the acceleration. Also, for a circular arc ($\rho = r = $ constant), the normal component of acceleration will be constant from the beginning of the curve to the end of the curve.

Therefore, when a car changes from moving in a straight line to traveling around a circular arc, the car (and its occupants) will suddenly experience an acceleration of magnitude $a_n = v^2/\rho$ directed toward the center of the circular curve. For a car traveling at 60 mi/h = 88 ft/s and a curve of radius 450 ft, the normal acceleration is $a_n = 17.2$ ft/s$^2 \cong 0.5g$. Since objects tend to move with a constant speed in a straight line unless acted on by an outside force (Newton's second law of motion), passengers in the car will feel that they are being thrown abruptly against the door of the car on the outside of the turn. Of course, when the car again starts going straight at the end of the curve, the normal component of acceleration will vanish and passengers in the car will feel a jerk in the other direction.

A better curve in a highway will have a transition section between the straight section of the roadway and the circular curve. For the simple quadratic polynomial shown, the curvature varies smoothly from $\rho = 16,667$ ft at the beginning of the curve (giving $a_n = 88^2/16,667 = 0.5$ ft/s^2) to $\rho = r = 450$ ft (giving $a_n = 88^2/450 = 17.2$ ft/s^2) where it meets the circular curve. Passengers will still feel a large normal component of acceleration while traveling around the circular portion of the curve, but at least there is no sudden jerk when the curve begins. A similar transition would be used at the end of the circular curve to reduce the normal acceleration smoothly back to zero.

EXAMPLE PROBLEM 13-12

A turn in a country highway has a radius of curvature that varies from infinity at the beginning and end of the turn to ρ_{min} at the middle of the turn (Fig. 13-31). If the tires of a car going around the turn begin to slide when the normal acceleration reaches 12 ft/s², determine

a. The maximum constant speed at which the car can go around the turn for $\rho_{min} = 500$ ft.
b. The smallest ρ_{min} for which the car can go around the turn at 60 mi/h.

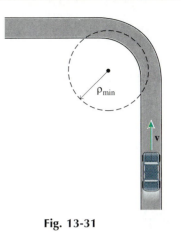

Fig. 13-31

SOLUTION

a. The normal component of acceleration is given by

$$a_n = v^2/\rho$$

Using $\rho = \rho_{min}$ and solving for v gives the maximum speed the car can have as

$$v = \sqrt{(\rho_{min}a_n)} = \sqrt{(500)(12)}$$
$$= 77.5 \text{ ft/s} \qquad \text{Ans.}$$

b. Solving for ρ and using $v = 60$ mi/h $= 88$ ft/s gives the smallest ρ_{min} that the turn can have as

$$\rho_{min} = \frac{v^2}{a_n} = \frac{88^2}{12} = 645 \text{ ft} \qquad \text{Ans.}$$

EXAMPLE PROBLEM 13-13

A box slides down a chute, which is bent in the shape of a hyperbola (Fig. 13-32). When the box reaches the point $x = 5$ m, it has a speed of 5 m/s, and the speed is decreasing at the rate of 0.5 m/s². Determine the normal and tangential components of the acceleration of the box.

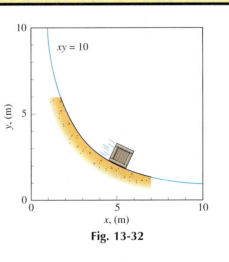

Fig. 13-32

SOLUTION

The tangential direction is found by computing the slope of the curve

$$\frac{dy}{dx} = \frac{d}{dx}\left(\frac{10}{x}\right) = -\frac{10}{x^2} = \tan \phi$$

So, at $x = 5$ m

$$\phi = \tan^{-1}\left(-\frac{10}{5^2}\right) = -21.80°$$

(below the horizontal). The tangential component of the acceleration is then

$$\mathbf{a}_t = \dot{v}\mathbf{e}_t = (-0.5)\mathbf{e}_t \text{ m/s}^2 = 0.5 \text{ m/s}^2 \searrow 21.80°$$

The normal component of the acceleration is

$$a_n = v^2/\rho$$

63

where the radius of curvature is given by (see any elementary calculus book)

$$\frac{1}{\rho} = \frac{\left|\dfrac{d^2y}{dx^2}\right|}{\left[1 + \left(\dfrac{dy}{dx}\right)^2\right]^{3/2}}$$

and the absolute value is to guarantee that ρ is positive. Calculating the second derivative gives

$$\frac{d^2y}{dx^2} = \frac{20}{x^3}$$

so that at $x = 5$ m

$$\frac{1}{\rho} = \frac{\left|\dfrac{20}{5^3}\right|}{\left[1 + \left(-\dfrac{10}{5^2}\right)^2\right]^{3/2}} = 0.1281 \ \text{m}^{-1}$$

Finally, the normal component of the acceleration is

$$\mathbf{a}_n = (5)^2(0.1281)\mathbf{e}_n$$
$$= 3.20 \ \text{m/s}^2 \ \measuredangle \ 68.2° \qquad\qquad \text{Ans.}$$

PROBLEMS

Introductory Problems

13-119* A car drives over the top of a hill that has a radius of curvature of 110 ft (Fig. P13-119). If the normal component of acceleration necessary to keep the car on the road becomes greater than that provided by gravity, the car will become airborne. Determine the maximum constant speed v at which the car can go over the hill.

13-120* The car of Fig. P13-120 has a speed of $v = 100$ km/h and its speed is increasing at the rate of $\dot{v} = 5$ m/s^2 at the instant shown. If the radius of curvature at the bottom of the hill is 80 m, determine the acceleration (magnitude and direction) of the car.

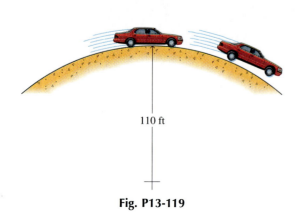

110 ft

Fig. P13-119

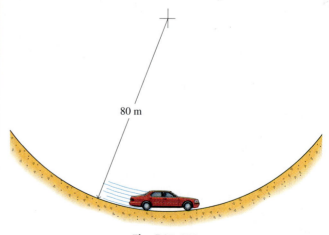

80 m

Fig. P13-120

13-121 A car is traveling around a curve as shown in Fig. P13-121. At some instant of time, the car has a speed of $v = 45$ mi/h in a direction 30° north of east, its speed is increasing at a rate of $\dot{v} = 5$ ft/s², and the radius of curvature is 450 ft. Determine the acceleration (magnitude and direction) of the car.

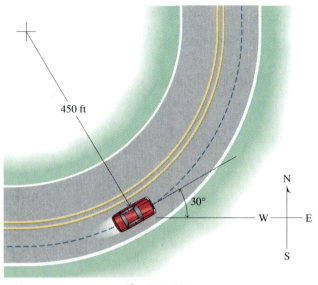

Fig. P13-121

13-122 The roller coaster cars shown in Fig. P13-122 are traveling at a speed of $v = 50$ km/h (and $\dot{v} = 0$) when they pass over the top of the hill. If the radius of curvature is $\rho = 21$ m, determine the acceleration of the cars as they pass over the top of the hill.

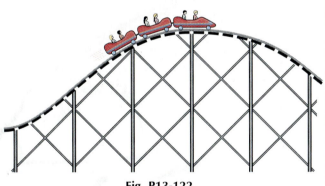

Fig. P13-122

Intermediate Problems

13-123* When the total acceleration of a car going around a curve exceeds one-third of the gravitational acceleration ($g = 32.2$ ft/s²), the tires of the car will begin to slide. For a car increasing in speed at $\dot{v} = 5$ ft/s² around a corner having a radius of curvature of 200 ft, determine the speed v at which the tires will begin to slide.

13-124* The bead shown in Fig. P13-124 is sliding around a circular ring 3 m in diameter. At the instant shown ($\theta = 30°$), the speed of the bead is $v = 4$ m/s and the speed is increasing at a rate of $\dot{v} = 5$ m/s². Determine the acceleration (magnitude and direction) of the bead.

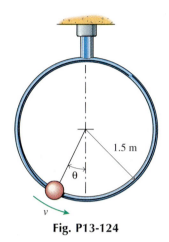

Fig. P13-124

13-125 As the car shown in Fig. P13-125 moves along the circular road, its speed increases uniformly from $v_A = 30$ mi/h at A to $v_B = 60$ mi/h at B. Determine the magnitude of the car's acceleration when $s = 300$ ft.

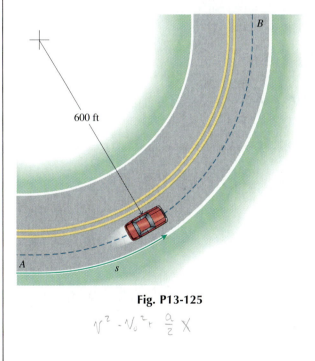

Fig. P13-125

$v^2 - v_0^2 + \frac{a}{2} x$

65

13-126 The bead shown in Fig. P13-126 moves along a circular path in a horizontal plane according to the relation $s = t^3 + 6t$ where s is measured in meters and t is the time in seconds. If the magnitude of the acceleration of the bead is 20 m/s^2 when $t = 2$ s, determine the radius of the circle.

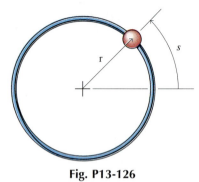

Fig. P13-126

13-127 A girl is practicing on a quarter-mile track composed of two straight parallel sides connected by two semicircles, each of radius 150 ft (Fig. P13-127). If the girl runs at a constant speed and completes one lap in 60 s, determine her acceleration on both the straight and curved portions of the track.

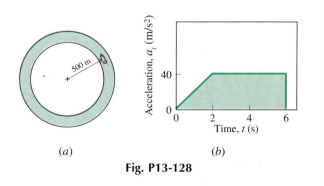

Fig. P13-127

13-128* An unmanned rocket sled is moving along the circular track shown in Fig. P13-128a. If the tangential component of the acceleration of the sled is as shown in Fig. P13-128b, determine

a. The maximum speed attained by the sled.
b. The magnitude of the maximum acceleration of the sled.

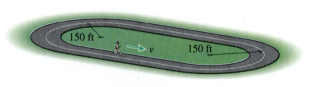

(a)   (b)

Fig. P13-128

Challenging Problems

13-129* A sphere slides along a rod that is bent in a vertical plane into a shape that can be described by the equation $x^2 = 8y$, where x and y are both measured in feet. When the sphere is at the point $x = -8$ ft, $y = 8$ ft, as shown in Fig. P13-129, it is moving along the rod (down and to the right) at a speed of $v = 15$ ft/s and is slowing down at a rate of 3 ft/s^2. Determine the acceleration (magnitude and direction) of the sphere at this time.

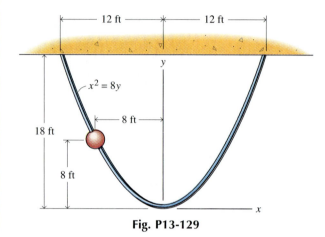

Fig. P13-129

13-130 A sphere slides along a rod that is bent in a vertical plane into a shape that can be described by the equation $y = \sqrt{x}$, where x and y are both measured in meters. When the sphere is at point A ($x = 2$ m), as shown in Fig. P13-130, it is moving along the rod (up and to the right) at a speed of $v = 7$ m/s and is slowing down at a rate of 2 m/s^2. Determine the acceleration (magnitude and direction) of the sphere at this time.

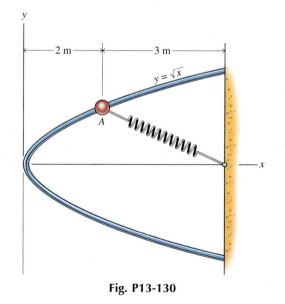

Fig. P13-130

66

13-131 The chute of Fig. P13-131 is elliptic in shape; that is,

$$y = f(x) = 1 - 0.5\sqrt{4 - x^2}$$

where x and y are both in feet. A marble rolling down the chute passes point $A(x_0 = 1.5$ ft$)$ with a speed $v = 12$ ft/s, and its speed is increasing at the rate of $\dot{v} = 8$ ft/s^2. Determine the acceleration (magnitude and direction) of the marble as it passes point A.

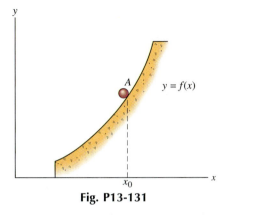

Fig. P13-131

13-132* A ski slope has a shape given by

$$y = \frac{400}{x + 15}$$

where x and y are both in meters. When a skier is at $x = 20$ m, she has a speed of $v = 15$ m/s and her speed is increasing at the rate of $\dot{v} = 2$ m/s^2. Determine the acceleration (magnitude and direction) of the skier at this point.

13-5.4   Combination of Coordinate Systems

The choice of which coordinate system to use to solve any particular problem depends on how the data is given and the type of information desired. The rectangular coordinate system is usually the most convenient system to use when the x- and y-components of the motion are specified separately from each other and/or do not depend on each other. Similarly, the polar coordinate system is usually the most convenient system to use when the distance and direction of a particle are measured relative to a fixed point or when a particle is fixed on or moves along a rotating arm. Normal and tangential coordinates are the most convenient coordinates to use when particles move along a surface of known shape.

Sometimes, however, some components of velocity or acceleration will be given in one coordinate system while other components of velocity or acceleration are given in a different coordinate system. For example, the speed of a car traveling along a straight road is to be determined from the rotation rate of a fixed camera tracking the car. Since the road is straight, the motion of the car is conveniently described in terms of a rectangular coordinate system. Relative to the fixed camera, however, the motion is more conveniently described in terms of a polar coordinate system. The velocity and acceleration must be written in terms of both coordinate systems and then the two coordinate systems must be related (see Example Problem 13-14).

Sometimes the velocity or acceleration will be given in one system while one or more components of velocity or acceleration are needed in a different coordinate system. For example, the normal acceleration of a particle moving along a surface is required to determine the contact force between the particle and the surface. Also, the normal contact force is often required to compute the tangential (friction) force and thereby determine the tangential acceleration and velocity. Regardless of the coordinate system used to compute the acceleration of the particle, the acceleration must be converted into normal and tangential components (see Example Problem 13-15).

For convenience, the major equations of plane curvilinear motion are summarized as follows:

Rectangular Coordinates

$$\mathbf{r}_{P/O}(t) = x(t)\mathbf{i} + y(t)\mathbf{j}$$
$$\mathbf{v}_P(t) = \dot{x}(t)\mathbf{i} + \dot{y}(t)\mathbf{j}$$
$$= v_x(t)\mathbf{i} + v_y(t)\mathbf{j}$$
$$\mathbf{a}_P(t) = \ddot{x}(t)\mathbf{i} + \ddot{y}(t)\mathbf{j}$$
$$= \dot{v}_x(t)\mathbf{i} + \dot{v}_y(t)\mathbf{j}$$
$$= a_x(t)\mathbf{i} + a_y(t)\mathbf{j}$$

Polar Coordinates

$$\mathbf{r}_{P/O}(t) = r(t)\mathbf{e}_r$$
$$\mathbf{v}_P(t) = \dot{r}\mathbf{e}_r + r\dot{\theta}\mathbf{e}_\theta$$
$$= v_r\mathbf{e}_r + v_\theta\mathbf{e}_\theta$$
$$\mathbf{a}_P(t) = (\ddot{r} - r\dot{\theta}^2)\mathbf{e}_r + (r\ddot{\theta} + 2\dot{r}\dot{\theta})\mathbf{e}_\theta$$
$$= a_r\mathbf{e}_r + a_\theta\mathbf{e}_\theta$$

Normal and Tangential Coordinates

$$\mathbf{v}_P(t) = \dot{s}(t)\mathbf{e}_t = v_t\mathbf{e}_t$$

$$\mathbf{a}_P(t) = \ddot{s}\mathbf{e}_t + \frac{\dot{s}^2}{\rho}\mathbf{e}_n$$

$$= a_t\mathbf{e}_t + a_n\mathbf{e}_n$$

Circular Motion

$$\mathbf{v}_P(t) = r\dot{\theta}\mathbf{e}_\theta = v\mathbf{e}_t$$

$$\mathbf{a}_P(t) = -r\dot{\theta}^2\mathbf{e}_r + r\ddot{\theta}\mathbf{e}_\theta = r\ddot{\theta}\mathbf{e}_t + \frac{v^2}{r}\mathbf{e}_n$$

EXAMPLE PROBLEM 13-14

A camera is filming a car that is traveling at a constant speed along a straight road as shown in Fig. 13-33. At the instant shown, the angle θ is 60° and the camera must rotate at the rate of 0.100 rad/s to follow the car. Determine the speed of the car and the rate at which $\dot{\theta}$ is changing.

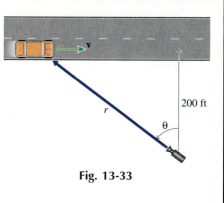

Fig. 13-33

SOLUTION

In terms of a rectangular coordinate system with the x-axis to the right and the y-axis up (Fig. 13-34), the velocity of the car can be written

$$\mathbf{v} = v\mathbf{i} \text{ ft/s}$$

However, in order to relate this velocity to the rate at which the camera is turning, the velocity must be written in terms of polar coordinates. In terms of the polar coordinate system shown, the velocity components are

$$v_r = \dot{r} = -v \sin 60° \qquad (a)$$
$$v_\theta = r\dot{\theta} = -v \cos 60° \qquad (b)$$

where $r = 200/\cos 60° = 400$ ft and $\dot{\theta} = -0.100$ rad/s. Then, from Eq. b

$$v = \frac{-(400)(-0.100)}{\cos 60°} = 80 \text{ ft/s} \cong 54.5 \text{ mi/h} \qquad \text{Ans.}$$

while from Eq. a

$$\dot{r} = -(80) \sin 60° = -69.28 \text{ ft/s}$$

Since the velocity is constant in both magnitude and direction, $\mathbf{a} = \mathbf{0}$ ft/s² in every coordinate system including the rectangular coordinate system and the polar coordinate system. In terms of the polar coordinate system, the acceleration components are

$$a_r = \ddot{r} - r\dot{\theta}^2 = 0 \qquad (c)$$
$$a_\theta = r\ddot{\theta} + 2\dot{r}\dot{\theta} = 0 \qquad (d)$$

Equation d gives

$$\ddot{\theta} = \frac{d\dot{\theta}}{dt} = \frac{-2(-69.28)(-0.100)}{400} = -0.0346 \text{ rad/s}^2 \qquad \text{Ans.}$$

> Since r is the distance from the camera to the car and θ is the direction that the camera is pointed, the polar coordinate system is centered at the camera. The unit vectors $\mathbf{e}_r$ and $\mathbf{e}_\theta$ point in the directions that r and θ are increasing (outward and counterclockwise). The car's velocity points in the negative r and negative θ directions and has components $v_r = -v \sin \theta$ and $v_\theta = -v \cos \theta$.

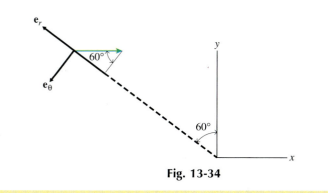

Fig. 13-34

A cam lobe has a shape given by $r = 20 + 15 \cos \theta$ mm (Fig. 13-35). Pin P slides in a slot along arm AB and is held against the cam by a spring. The arm AB rotates counterclockwise about A at a constant rate of 30 rev/min. When $\theta = 40°$, determine

a. The velocity of the pin P.
b. The tangential component of the acceleration of the pin P.
c. The normal component of the acceleration of the pin P.

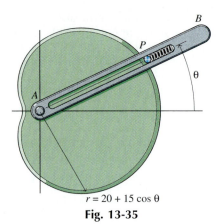

$r = 20 + 15 \cos \theta$

Fig. 13-35

SOLUTION

Since the position of the pin is given in terms of distance and direction, polar coordinates will be used compute its velocity and acceleration. The rotation rate of the arm is constant

$$\dot{\theta} = 30 \text{ rev/min} = \pi \text{ rad/s} = \text{constant}$$
$$\ddot{\theta} = 0 \text{ rad/s}^2$$

and taking the time derivatives of the radial position function gives

$$r = 20 + 15 \cos \theta \text{ mm}$$
$$\dot{r} = -15\dot{\theta} \sin \theta \text{ mm/s}$$
$$\ddot{r} = -15\ddot{\theta} \sin \theta - 15\dot{\theta}^2 \cos \theta \text{ mm/s}^2$$

Then, when $\theta = 40°$,

$$r = 31.491 \text{ mm} \qquad \dot{r} = -30.291 \text{ mm/s} \qquad \ddot{r} = -113.408 \text{ mm/s}^2$$

Substituting these values into the expressions for velocity components in polar coordinates (Eq. 13-19) gives

$$v_r = \dot{r} = -30.291 \text{ mm/s}$$
$$v_\theta = r\dot{\theta} = (31.491)\pi = 98.932 \text{ mm/s}$$
$$\mathbf{v} = -30.3\mathbf{e}_r + 98.9\mathbf{e}_\theta \text{ mm/s} \qquad\qquad \text{Ans.}$$

That is, the velocity has a magnitude

$$v = \sqrt{30.291^2 + 98.932^2} = 103.465 \text{ mm/s}$$

and its direction is (Fig. 13-36a)

$$\phi_v = \tan^{-1} \frac{30.291}{98.932} = 17.024°$$

relative to the positive θ-direction. But, in terms of normal/tangential coordinates, the velocity is

$$\mathbf{v} = v\mathbf{e}_t = 103.465\mathbf{e}_t \text{ mm/s}$$

Therefore, the unit vector in the tangential direction has the polar coordinate components

$$\mathbf{e}_t = \frac{\mathbf{v}}{v} = -0.299277\mathbf{e}_r + 0.95619\mathbf{e}_\theta$$

Similarly, the acceleration components in polar coordinates (Eq. 13-20) are

$$a_r = \ddot{r} - r\dot{\theta}^2 = (-113.408) - (31.491)(\pi)^2 = -424.212 \text{ mm/s}^2$$
$$a_\theta = r\ddot{\theta} + 2\dot{r}\dot{\theta} = 0 + 2(-30.291)(\pi) = -190.324 \text{ mm/s}^2$$
$$\mathbf{a} = -424.212\mathbf{e}_r - 190.324\mathbf{e}_\theta \text{ mm/s}^2$$

Because of the way that the motion of the pin is given, the velocity and acceleration of the pin are easy to write in terms of polar coordinates. The velocity vector always points in the tangential direction, however, so dividing the velocity vector by its magnitude gives the r- and θ-components of a unit vector in the tangential direction. Then, the scalar product of the acceleration (in r- and θ-coordinates) with the unit vector $\mathbf{e}_t$ (also expressed in r- and θ-coordinates) gives the magnitude of the tangential component of the acceleration.

That is, the acceleration has a magnitude

$$a = \sqrt{424.212^2 + 190.324^2} = 464.950 \text{ mm/s}^2$$

and its direction is (Fig. 13-36b)

$$\phi_a = \tan^{-1}\frac{190.324}{424.212} = 24.164°$$

relative to the negative r-direction. The tangential component of the acceleration can be determined using the scalar product

$$\mathbf{a}_t = \mathbf{a} \cdot \mathbf{e}_t = (-424.212)(-0.29277) + (-190.324)(0.95619)$$
$$= -57.789 \text{ mm/s}^2 \cong -57.8 \text{ mm/s}^2 \qquad \text{Ans.}$$

Finally, the normal component of acceleration is found from

$$a = \sqrt{a_t^2 + a_n^2} = \sqrt{a_r^2 + a_\theta^2} = 464.950 \text{ mm/s}^2$$

or

$$a_n = \sqrt{464.950^2 - 57.789^2} = 461 \text{ mm/s}^2 \qquad \text{Ans.}$$

Alternatively, the normal component of acceleration can be found directly from the acceleration vector. The acceleration vector makes an angle of 24.164° with the negative r-direction while the normal direction makes an angle of 17.024° with the negative r-direction (Fig. 13-36c). Therefore, the acceleration components are

$$a_n = 464.950 \cos(24.164° - 17.024°) = 461 \text{ mm/s}^2$$
$$a_t = -464.950 \sin(24.164° - 17.024°) = -57.8 \text{ mm/s}^2 \qquad \text{Ans.}$$

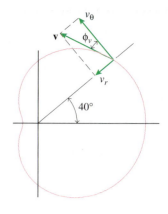

(a)

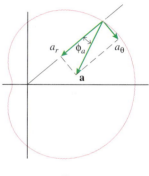

(b)

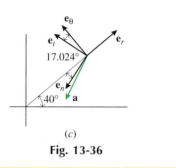

(c)

Fig. 13-36

71

PROBLEMS

Intermediate Problems

13-133* A small bead moves along the path shown in Fig. P13-133. The position of the bead is located by the equations $x = 2t^2$ and $y = 4t$, where x and y are measured in feet and t is the time measured in seconds.

a. Show that the path followed by the bead is a parabola.
b. Determine the x- and y-components of the velocity and acceleration of the bead when $t = 2$ s.
c. Determine the normal and tangential components of the velocity and acceleration of the bead when $t = 2$ s.
d. Determine the radius of curvature of the path when $t = 2$ s.

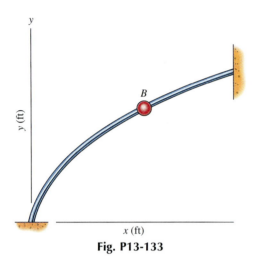

Fig. P13-133

13-134* The pin P is free to slide in the slots of both members shown in Fig. P13-134. The curved slot has the shape of a parabola, $y = 0.1x^2$, where both x and y are measured in millimeters. If the position of the pin is given by $x = 30t$ where t is the time in seconds, determine

a. The magnitude and direction of the velocity of the pin when $t = 2$ s. Verify that the direction of the velocity is tangent to the slot.
b. The x- and y-components of the acceleration of the pin when $t = 2$ s.
c. The normal and tangential components of the acceleration of the pin when $t = 2$ s.
d. Verify that the magnitude of the acceleration computed from the x- and y-components is the same as the magnitude computed from the normal and tangential components.

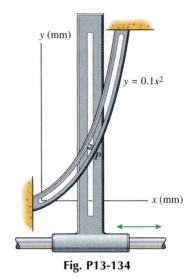

Fig. P13-134

13-135 An inextensible cord connects the mass m and the end of a 12-in.-long arm that rotates in a vertical plane with a constant clockwise angular velocity $\dot{\theta} = 0.2$ rad/s (Fig. P13-135). Determine the velocity of the mass when $\theta = 45°$ and when $\theta = 315°$. Neglect the size of the pulley at P.

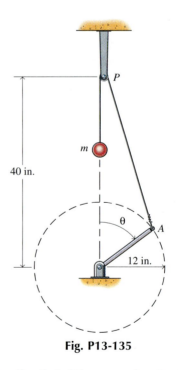

Fig. P13-135

13-136 A collar that slides around a circular wire has a pin that is constrained to move in the slot of arm AB (Fig. P13-136). The arm rotates counterclockwise at a constant angular speed of $\dot{\theta} = 2$ rad/s. When the arm is 30° above the horizontal, determine

a. The radial distance $r(t)$.
b. The magnitude and direction of the velocity of the collar when $t = 2$ s. Verify that the direction of the velocity is tangent to the wire.
c. The r- and θ-components of the acceleration of the collar when $t = 2$ s.
d. The normal and tangential components of the acceleration of the collar when $t = 2$ s. Verify that the magnitude of the acceleration computed from the r- and θ-components is the same as the magnitude computed from the normal and tangential components.

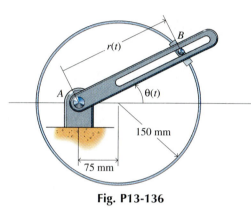

Fig. P13-136

13-137* A radar tracking a rocket gives the coordinates of the rocket as $r(t)$ and $\theta(t)$ (Fig. P13-137). At some instant of time, $\theta = 30°$ and $r = 10$ mi. From successive measurements, the derivatives $\dot{r}$, $\ddot{r}$, $\dot{\theta}$, and $\ddot{\theta}$ are estimated to be 650 ft/s, 165 ft/s², 0.031 rad/s, and 0.005 rad/s², respectively. Determine

a. The r- and θ-components of the rocket's velocity and acceleration.
b. The magnitude and direction of the rocket's velocity and the magnitude and direction of the rocket's acceleration.
c. The normal and tangential components of the rocket's velocity and acceleration.

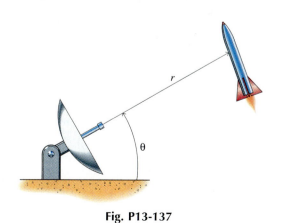

Fig. P13-137

13-138 A radar tracking an airplane gives the coordinates of the plane as $r(t)$ and $\theta(t)$ (Fig. P13-138). At some instant of time, $\theta = 40°$ and $r = 8$ km. From successive measurements, the derivatives $\dot{r}$, $\ddot{r}$, $\dot{\theta}$, and $\ddot{\theta}$ are estimated to be 173 m/s, 8 m/s², −0.0125 rad/s, and 0.00084 rad/s², respectively. Determine

a. The r- and θ-components of the plane's velocity and acceleration.
b. The magnitude and direction of the plane's velocity and the magnitude and direction of the plane's acceleration.
c. The normal and tangential components of the plane's velocity and acceleration.

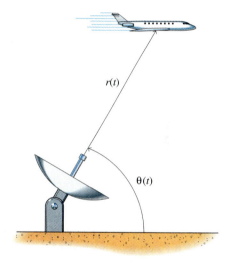

Fig. P13-138

Challenging Problems

13-139 A ski slope has a shape given by

$$y = 0.003(x - 150)^2$$

where x and y are both in feet. When a skier is at $x = 100$ ft, he has a speed of 30 ft/s and his speed is increasing at the rate of 4 ft/s². Determine

a. The x- and y-components of the velocity of the skier at this point.
b. The normal and tangential components a_n and a_t of the acceleration of the skier at this point.
c. The angle between the velocity vector and the acceleration vector of the skier at this point.

13-140* The chute of Fig. P13-140 is parabolic in shape; that is,

$$y = f(x) = x^2 - 6x + 9$$

where x and y are both in meters. A marble rolling down the chute passes point $A(x_0 = 5 \text{ m})$ with a speed of 3 m/s, and its speed is increasing at the rate of 5 m/s^2. Determine

a. The x- and y-components of the velocity of the marble as it passes point A.
b. The normal and tangential components a_n and a_t of the acceleration of the marble as it passes point A.
c. The angle between the velocity vector and the acceleration vector at point A.

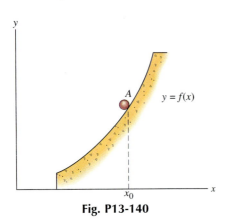

Fig. P13-140

13-141* Arm AC of the cam follower mechanism shown in Fig. P13-141 is rotating at a constant angular speed of $\omega = 150$ rev/min. A spring holds the pin B against the cam lobes. The equation that describes the shape of the cam lobes is

$$R = 2.5 + 0.5 \cos 3\theta$$

where R is in inches. When $\theta = 10°$, determine

a. The r- and θ-components of the velocity of the pin B.
b. The normal and tangential components a_n and a_t of the acceleration of the pin B.
c. The angle between the velocity vector and the acceleration vector.

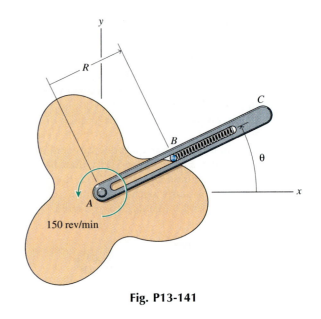

Fig. P13-141

13-142 Arm AB of the cam follower mechanism shown in Fig. P13-142 is rotating at a constant angular speed of $\omega = 60$ rev/min. A spring holds the pin P against the cam lobes. The equation that describes the shape of the cam lobes is

$$r = 20 + 15 \cos \theta$$

where r is in millimeters. When $\theta = 75°$, determine

a. The r- and θ-components of the velocity of the pin P.
b. The normal and tangential components a_n and a_t of the acceleration of the pin P.
c. The angle between the velocity vector and the acceleration vector.

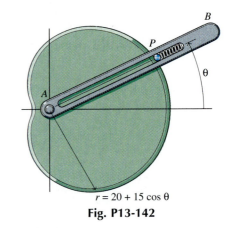

$r = 20 + 15 \cos \theta$

Fig. P13-142

13-6 RELATIVE MOTION IN A PLANE

The motion of two separate particles moving in plane curvilinear motion can be related just as was the motion of two particles moving in rectilinear motion in Section 13-4. The difference, of course, is that now the relative motion, like the individual motions, must be described with vectors.

The relationship between the positions of the individual particles and the relative position is obtained from the vector law of addition (Fig. 13-37)

$$\mathbf{r}_{B/O} = \mathbf{r}_{A/O} + \mathbf{r}_{B/A} \tag{13-28}$$

where $\mathbf{r}_{B/A}$ is the position of particle B relative to the position of particle A. Differentiating Eq. 13-28 with respect to time gives

$$\mathbf{v}_B = \mathbf{v}_A + \mathbf{v}_{B/A} \tag{13-29}$$
$$\mathbf{a}_B = \mathbf{a}_A + \mathbf{a}_{B/A} \tag{13-30}$$

That is, the velocity of particle B measured relative to particle A is the difference in the absolute velocities (velocities measured relative to a fixed coordinate system) of particles B and A. Similarly, the acceleration of particle B measured relative to particle A is the difference in the absolute accelerations of particles B and A.

The individual terms of these equations can be written in any convenient coordinate system: rectangular, polar, normal/tangential. However, all components must be converted to a common coordinate system (usually the rectangular coordinate system) before they can be added.

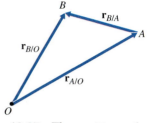

Figure 13-37 The positions of particles A and B relative to the fixed origin O and the position of particle B relative to particle A are related by the vector law of addition.

EXAMPLE PROBLEM 13-16

An airplane is trying to fly straight north (Fig. 13-38). However, a 20 m/s wind blowing due east carries the airplane off course unless the airplane flies at an angle to the desired direction. If the speed of the airplane is 250 km/h, determine:

a. The direction in which the airplane must fly so that it travels due north.
b. The time required for the airplane to fly 250 km in the northerly direction.

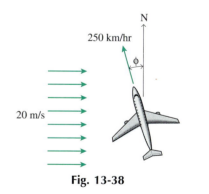

Fig. 13-38

SOLUTION

a. In a coordinate system in which east corresponds to the x-direction and north corresponds to the y-direction, the wind velocity is

$$\mathbf{v}_w = 20\mathbf{i} \text{ m/s}$$

and the desired velocity of the airplane is

$$\mathbf{v}_a = V\mathbf{j} \text{ m/s}$$

The velocity of the airplane relative to the wind is

$$\mathbf{v}_{a/w} = -250 \sin \phi \, \mathbf{i} + 250 \cos \phi \, \mathbf{j} \text{ km/h}$$
$$= -69.44 \sin \phi \, \mathbf{i} + 69.44 \cos \phi \, \mathbf{j} \text{ m/s}$$

where ϕ is the heading (west of north) at which the airplane must fly so that its absolute velocity will be due north.

Putting these all together gives

$$\mathbf{v}_a = \mathbf{v}_w + \mathbf{v}_{a/w}$$

or

$$V\mathbf{j} = 20\mathbf{i} + (-69.44 \sin \phi\,\mathbf{i} + 69.44 \cos \phi\,\mathbf{j})$$

Then the x-component of this equation gives the heading

$$\phi = \sin^{-1}\frac{20}{69.44}$$
$$= 16.74° \text{ (west of north)} \qquad \text{Ans.}$$

and the y-component gives the absolute speed of the airplane in the northerly direction

$$V = 69.44 \cos 16.74° = 66.50 \text{ m/s} = 239.4 \text{ km/h}$$

b. The time required to fly 250 km to the north is then

$$t = \frac{250}{239.4} = 1.044 \text{ h} \qquad \text{Ans.}$$

Two bicyclists are riding around a circular track (Fig. 13-39). Cyclist 1 rides around the inside of the track where the radius is 200 ft, while cyclist 2 rides around the outside of the track where the radius is 210 ft. Both start at $\theta = 0$ and $v = 0$ at $t = 0$. Both accelerate at a constant rate of 2 ft/s² until they reach a speed of 20 ft/s after which they maintain a constant speed. When the first cyclist reaches B determine:

a. The angular position of cyclist 2, θ_2.
b. The relative position $\mathbf{r}_{2/1}$.
c. The relative velocity $\mathbf{v}_{2/1}$.
d. The relative acceleration $\mathbf{a}_{2/1}$.

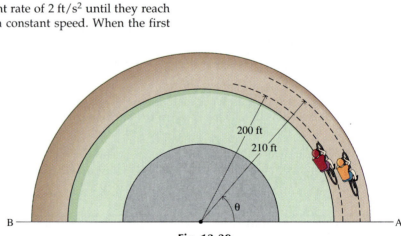

Fig. 13-39

SOLUTION

a. For cyclist 1, $r_1 = 200$ ft = constant, $\dot{r}_1 = 0$, $\ddot{r}_1 = 0$. Therefore, the velocity and acceleration are given by

$$\mathbf{v}_1 = \dot{r}_1\mathbf{e}_r + r_1\dot{\theta}_1\mathbf{e}_\theta = r_1\dot{\theta}_1\mathbf{e}_\theta$$
$$\mathbf{a}_1 = (\ddot{r}_1 - r_1\dot{\theta}_1^2)\mathbf{e}_r + (r_1\ddot{\theta}_1 + 2\dot{r}_1\dot{\theta}_1)\mathbf{e}_\theta$$
$$= -r_1\dot{\theta}_1^2\mathbf{e}_r + r_1\ddot{\theta}_1\mathbf{e}_\theta$$

76

Initially, the acceleration along the track (θ-component) is 2 ft/s² = constant, so

$$\ddot{\theta}_1 = \frac{2}{200} = 0.010 \text{ rad/s}^2 \qquad (a)$$

Integrating Eq. *a* gives

$$\dot{\theta}_1 = 0.010t \text{ rad/s}$$

and

$$\theta_1 = 0.005t^2 \text{ rad}$$

Cyclist 1 accelerates until his speed reaches 20 ft/s or until

$$20 = (200)(0.010t)$$

which gives $t = 10$ s. The angular speed and position at this time are

$$\dot{\theta}_1 = 0.10 \text{ rad/s}$$
$$\theta_1 = 0.50 \text{ rad } (= 28.6°)$$

After $t = 10$ s, the speed (and therefore the angular speed) of cyclist 1 remains constant

$$\dot{\theta}_1 = 0.10 \text{ rad/s} = \text{constant}$$

Integrating to get the angular position as a function of time gives

$$\theta_1 = 0.10t - 0.50 \text{ rad}$$

where the constant of integration has been chosen so that $\theta_1 = 0.50$ rad when $t = 10$ s. Finally, the time at which cyclist 1 is at B ($\theta_1 = 180° = \pi$ rad) is found

$$t = \frac{\pi + 0.50}{0.10} = 36.42 \text{ s}$$

at which time his position, velocity, and acceleration are

$$\mathbf{r}_1 = 200\mathbf{e}_r \text{ ft} = -200\mathbf{i} \text{ ft}$$
$$\mathbf{v}_1 = (200)(0.10)\mathbf{e}_\theta = 20\mathbf{e}_\theta \text{ ft/s} = -20\mathbf{j} \text{ ft/s}$$
$$\mathbf{a}_1 = -(200)(0.10)^2\mathbf{e}_r = -2\mathbf{e}_r \text{ ft/s}^2 = 2\mathbf{i} \text{ ft/s}^2$$

Similarly for cyclist 2, $r_2 = 210$ ft = constant, $\dot{r}_2 = 0$, and $\ddot{r}_2 = 0$. Initially,

$$\ddot{\theta} = (2)/(210) = 0.00952 \text{ rad/s}^2$$
$$\dot{\theta}_2 = 0.00952t \text{ rad/s}$$
$$\theta_2 = 0.00476t^2 \text{ rad}$$

Cyclist 2 also reaches a speed of 20 ft/s at

$$t = (20)/(210)(0.00952) = 10 \text{ s}$$

at which time his angular speed and angular position are

$$\dot{\theta}_2 = 0.0952 \text{ rad/s}$$
$$\theta_2 = 0.476 \text{ rad } (= 27.3°)$$

After $t = 10$ s, the speed (and therefore the angular speed) of cyclist 2 remains constant so that

$$\dot{\theta}_2 = 0.0952 \text{ rad/s}$$

and integrating with respect to time gives

$$\theta_2 = 0.0952t - 0.476 \text{ rad}$$

where the constant of integration has been chosen such that $\theta_2 = 0.476$ rad when $t = 10$ s. Then the position, velocity, and acceleration of cyclist 2 at $t = 36.42$ s (the time when cyclist 1 is at B) is

$$\theta_2 = 2.991 \text{ rad } (= 171.4°) \qquad \text{Ans.}$$
$$\mathbf{r}_2 = 210\mathbf{e}_r \text{ ft} = -207.6\mathbf{i} + 31.55\mathbf{j} \text{ ft}$$
$$\mathbf{v}_2 = 20\mathbf{e}_\theta \text{ ft/s} = -3.005\mathbf{i} - 19.77\mathbf{j} \text{ ft/s}$$
$$\mathbf{a}_2 = -1.903\mathbf{e}_r \text{ ft/s}^2 = 1.881\mathbf{i} - 0.286\mathbf{j} \text{ ft/s}^2$$

b. Now the relative position can be computed

$$\mathbf{r}_{2/1} = \mathbf{r}_2 - \mathbf{r}_1 = -7.6\mathbf{i} + 31.55\mathbf{j} \text{ ft} \qquad \text{Ans.}$$

c. Similarly, the relative velocity is

$$\mathbf{v}_{2/1} = \mathbf{v}_2 - \mathbf{v}_1 = -3.00\mathbf{i} + 0.23\mathbf{j} \text{ ft/s} \qquad \text{Ans.}$$

d. Finally, the relative acceleration is

$$\mathbf{a}_{2/1} = \mathbf{a}_2 - \mathbf{a}_1$$
$$= -0.119\mathbf{i} - 0.286\mathbf{j} \text{ ft/s}^2 \qquad \text{Ans.}$$

Because the cyclists are traveling in circular paths, it is easy to compute the position, velocity, and acceleration in polar coordinates. The polar coordinates, however, cannot be used directly to compute the relative position $\mathbf{r}_{2/1}$, relative velocity $\mathbf{v}_{2/1}$, and relative acceleration $\mathbf{a}_{2/1}$. For example, $\mathbf{r}_{2/1} = \mathbf{r}_2 - \mathbf{r}_1 \neq 210 \, \mathbf{e}_r - 200 \, \mathbf{e}_r$ because the $\mathbf{e}_r$ unit vectors in the two terms are not the same. For cyclist 1 $\mathbf{e}_r$ points in the negative x-direction ($\theta_1 = 180°$), while for cyclist 2 $\mathbf{e}_r$ points ⬈ 8.6° ($\theta_2 = 171.4°$).

PROBLEMS

Introductory Problems

13-143* A boat is trying to travel straight across a river as shown in Fig. P13-143. The river is 2000 ft wide and has a current of 5 mi/h. If the boat is traveling at 15 mi/h, determine

a. The time T required to travel straight from A to B.
b. The angle ϕ at which the boat must head to travel straight from A to B.

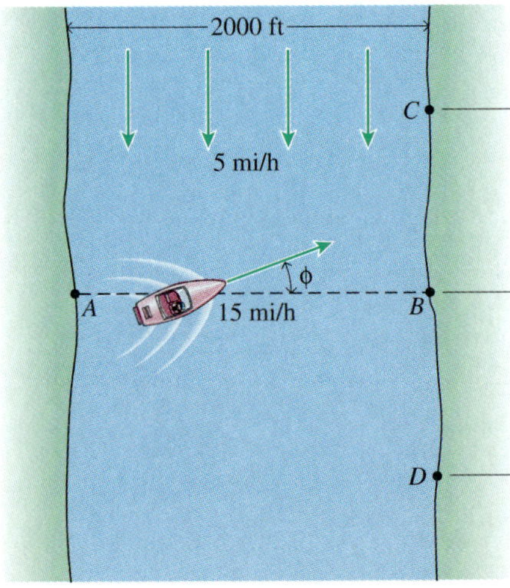

13-144* A man distributing newspapers by car tosses a bundle of papers from the car as shown in Fig. P13-144. If the car is traveling at 15 km/h and the papers are tossed with a velocity of 5 m/s relative to the car and perpendicular to the motion of the car, determine

a. The velocity $\mathbf{v}_P$ of the papers relative to the sidewalk.
b. The angle ϕ between the velocities $\mathbf{v}_P$ and $\mathbf{v}_c$.

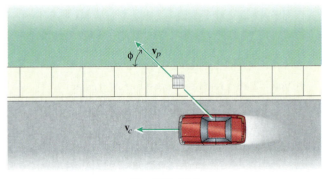

Fig. P13-144

Fig. P13-143

13-145 The water in the river of Fig. P13-145 has an average speed of 3 mi/h. A boat leaves from position A, and the pilot keeps the boat headed in a direction that is always perpendicular to the banks of the river. If the constant speed of the boat relative to the water is 5 mi/h, determine

a. The distance traveled by the boat when it lands at B.
b. The time to travel from A to B.

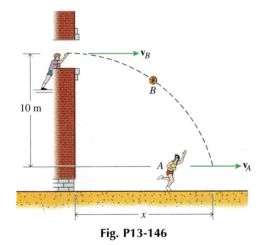

Fig. P13-145

13-146 A boy throws a ball from a window 10 m above the ground as shown in Fig. P13-146. The initial speed of the ball is 10 m/s, and the ball has a constant downward acceleration of 9.81 m/s^2. A second boy runs along the ground at a speed of 5 m/s and catches the ball on the run. Determine

a. The distance x at which the boy will catch the ball.
b. The relative velocity $\mathbf{v}_{B/A}$ of the ball with respect to the boy when he catches the ball.

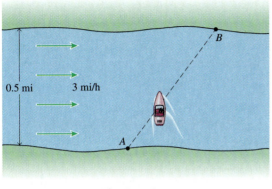

Fig. P13-146

13-147* A football player throws a pass to a receiver as shown in Fig. P13-147. The initial velocity of the football is 35 ft/s at $\theta = 30°$, and the ball has a constant downward acceleration of 32.2 ft/s^2. If the receiver runs at a constant speed of 15 ft/s, determine

a. The distance x at which the receiver will catch the football.
b. The relative velocity $\mathbf{v}_{B/A}$ of the ball with respect to the receiver when he catches the ball.

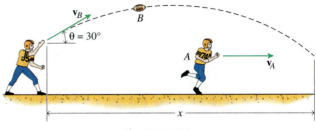

Fig. P13-147

13-148 Car A is traveling due east at a constant speed of 80 km/h while car B is traveling due north at a speed of 35 km/h. At the instant shown in Fig. P13-148, the speed of car B is increasing at a constant rate of 2 m/s^2. Determine the relative velocity $\mathbf{v}_{B/A}$ and the relative acceleration $\mathbf{a}_{B/A}$ of car B with respect to car A at this instant.

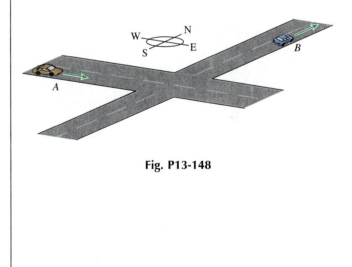

Fig. P13-148

Intermediate Problems

13-149* Two boats leave a dock at the same time ($t = 0$) as shown in Fig. P13-149. Boat A travels at a steady speed of 15 mi/h while boat B travels at a steady speed of 20 mi/h. For $t = 30$ s, determine

a. The distance d between the boats.
b. The rate $\dot{d}$ at which the boats are separating.

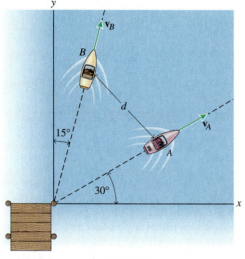

Fig. P13-149

13-150* Rain is falling with a speed of 30 m/s and an angle of 20° to the vertical. For a car traveling into the rain (Fig. P13-150), determine

a. The angle ϕ at which the rain appears to strike the windshield if the car is traveling at 60 km/h.
b. The speed of the car for which $\phi = 90°$.

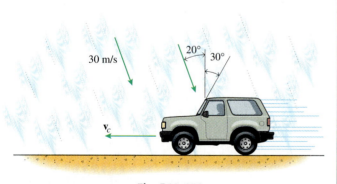

Fig. P13-150

13-151 Rain falls (vertical component of velocity) at 90 ft/s and is blown sideways (horizontal component of velocity) by the wind at 15 ft/s (Fig. P13-151). For a man walking briskly at 6 ft/s, determine the angle ϕ at which the man should hold the umbrella (the angle of the relative velocity) if the man is walking

a. With the wind.
b. Into the wind.

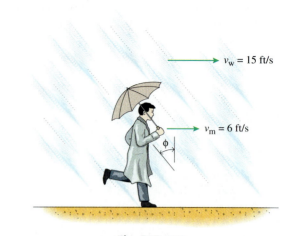

Fig. P13-151

13-152* Two airplanes are flying straight and level at the same altitude as shown in Fig. P13-152. At $t = 0$ s the distances AC and BC are 20 km and 30 km, respectively. If the planes maintain constant speeds ($v_A = 300$ km/h and $v_B = 400$ km/h), determine

a. The relative position of the planes $\mathbf{r}_{B/A}$ at $t = 3$ min.
b. The relative velocity of the planes $\mathbf{v}_{B/A}$ at $t = 3$ min.
c. The distance d separating the planes at $t = 3$ min.
d. The time T at which the separation distance d is a minimum.

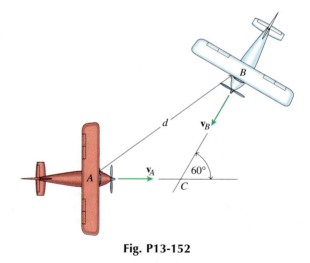

Fig. P13-152

13-153 A small ball moving to the right at a constant speed of 30 ft/s hits the curved track shown in Fig. P13-153. The curved track has a constant speed to the right of 10 ft/s. If the speed of the ball with respect to the track is constant, determine the velocity of the ball as it leaves the track.

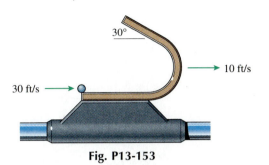

Fig. P13-153

13-154 Track A of Fig. P13-154 moves to the right with a constant speed of $v_A = 1$ m/s causing mass B to move. Mass B is attached to the end of the 2-m-long rod OB which is pinned at O. For $\theta = 30°$ determine

a. The velocity of B with respect to the track.
b. The angular velocity $\dot{\theta}$ of the rod OB.

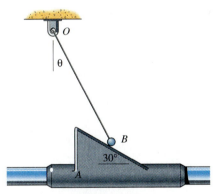

Fig. P13-154

Challenging Problems

13-155* Two boys are playing catch on a hill as shown in Fig. P13-155. The first boy throws the ball with an initial speed of v_B in a horizontal direction and the ball has a constant downward acceleration of 32.2 ft/s². If the second boy starts at $s = 20$ ft and runs at a constant speed of 15 ft/s, determine

a. The initial speed of the ball v_B such that the second boy can catch it on the run.
b. The distance s at which the second boy will catch the ball.
c. The relative velocity $\mathbf{v}_{B/A}$ of the ball with respect to the boy when he catches the ball.

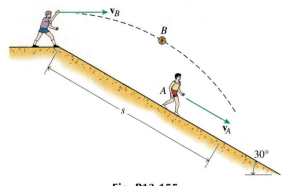

Fig. P13-155

13-156 An airplane towing a glider (Fig. P13-156) is flying straight and level at a constant speed of 70 m/s. The tow rope is 50 m long and makes an angle θ of 10° with the horizontal. If $\dot{\theta} = 0.40$ rad/s and $\ddot{\theta} = -0.25$ rad/s², determine

a. The rate of climb of the glider v_{By}.
b. The acceleration of the glider $\mathbf{a}_B$.

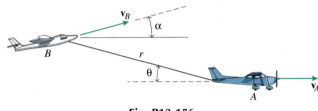

Fig. P13-156

Computer Problems

13-157 Rain is falling with a speed of 90 ft/s and an angle of 20° to the vertical (Fig. P13-157). Plot the angle ϕ at which the rain appears to strike the car's windshield as a function of the car's speed v_c $(0 < v_c < 80$ mi/h$)$

a. For a car traveling into the rain.
b. For a car traveling with the rain.

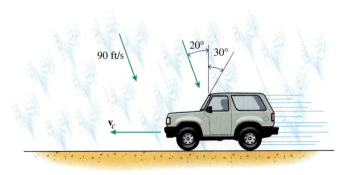

Fig. P13-157

13-158 A fox starts chasing a rabbit as shown in Fig. P13-158a. The rabbit runs at a constant speed of 6 m/s around a circular path of radius 12 m, and the fox runs at a constant speed of 7 m/s. The fox follows a path that is always directly toward the rabbit's current position.

a. Use numerical integration methods to compute the position of the fox as a function of time until it catches the rabbit (assume that the fox has caught the rabbit when the distance between them is less than 0.1 m). Plot the position of both animals at 0.5 s intervals until the fox catches the rabbit.

   Instead of always running toward the rabbit, where should the fox run to catch the rabbit in the shortest amount of time?

b. Repeat part *a* for the case where the rabbit runs around a circle away from the fox (Fig. P13-158b) and for the case where the rabbit runs a zigzag path away from the fox (Fig. P13-158c).

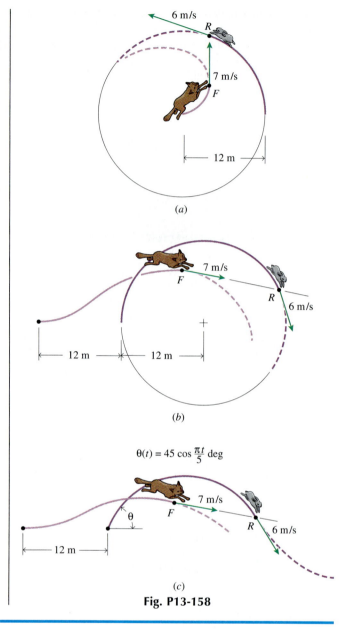

(a)

(b)

$$\theta(t) = 45 \cos \frac{\pi t}{5} \text{ deg}$$

(c)

Fig. P13-158

13-7   SPACE CURVILINEAR MOTION

Three coordinates are required to describe motion along a curve in three-dimensional space. The most commonly used coordinate systems are the rectangular coordinate system and the cylindrical coordinate system, which are described in detail. A less commonly used coordinate system—the spherical coordinate system—is also described briefly. Although a modified version of the normal and tangential coordinate system can be made for general three-dimensional motion, it is not of general interest in an elementary course, since the plane of motion (called the osculating plane and defined by the tangent and the principal normal directions) varies from point to point along the curve and from instant to instant in time.

13-7.1 Rectangular Coordinates

The three-dimensional rectangular coordinate system starts with the rectangular coordinates x and y (Section 13-5.1) and then adds a z-coordinate—the distance from the xy plane (Fig. 13-40). The unit vector in the z-direction is called $\mathbf{k}$ and the position of a particle P relative to the coordinate origin O is

$$\mathbf{r}_{P/O}(t) = x(t)\mathbf{i} + y(t)\mathbf{j} + z(t)\mathbf{k} \qquad (13\text{-}31)$$

Like $\mathbf{i}$ and $\mathbf{j}$ the unit vector $\mathbf{k}$ is constant in both magnitude and direction so that the derivatives of the position are

$$\mathbf{v}_P(t) = \dot{\mathbf{r}}_{P/O}(t) = \dot{x}(t)\mathbf{i} + \dot{y}(t)\mathbf{j} + \dot{z}(t)\mathbf{k} \qquad (13\text{-}32)$$
$$\mathbf{a}_P(t) = \dot{\mathbf{v}}_P(t) = \ddot{\mathbf{r}}_{P/O}(t) = \ddot{x}(t)\mathbf{i} + \ddot{y}(t)\mathbf{j} + \ddot{z}(t)\mathbf{k} \qquad (13\text{-}33)$$

Again, the (x,y)-coordinate directions need not lie in a horizontal or vertical plane. The coordinate directions can be chosen arbitrarily so long as they are orthogonal. Once chosen, however, they must remain fixed.

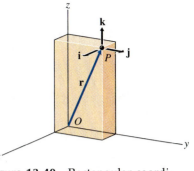

Figure 13-40 Rectangular coordinates in three dimensions describes the position of a particle by giving its distance from three orthogonal planes—x, the distance from the yz-plane; y, the distance from the xz-plane; and z, the distance from the xy-plane.

13-7.2 Cylindrical Coordinates

The cylindrical coordinate system starts with the two-dimensional polar coordinate system of Section 13-5.2 and adds a z-coordinate—the distance from the $r\theta$ plane (Fig. 13-41). The unit vector in the z-direction is again called $\mathbf{k}$ and is again constant in both magnitude and direction. Since $\mathbf{k}$ is fixed and $\mathbf{e}_r$ and $\mathbf{e}_\theta$ are independent of the z-coordinate, the r- and θ-components of the position, velocity, and acceleration are the same as for polar coordinates, and the z-components are the same as for rectangular coordinates:

$$\mathbf{r}_{P/O}(t) = r(t)\mathbf{e}_r + z(t)\mathbf{k} \qquad (13\text{-}34)$$
$$\mathbf{v}_P(t) = \dot{\mathbf{r}}_{P/O}(t) = \dot{r}\mathbf{e}_r + r\dot{\theta}\mathbf{e}_\theta + \dot{z}\mathbf{k} \qquad (13\text{-}35)$$
$$\mathbf{a}_P(t) = \dot{\mathbf{v}}_P(t) = \ddot{\mathbf{r}}_{P/O}(t) = (\ddot{r} - r\dot{\theta}^2)\mathbf{e}_r + (r\ddot{\theta} + 2\dot{r}\dot{\theta})\mathbf{e}_\theta + \ddot{z}\mathbf{k} \qquad (13\text{-}36)$$

As with rectangular coordinates, the (r,θ)-coordinate directions need not lie in a horizontal or vertical plane. The cylindrical coordinate system is usually used for a body rotating about an axis. The z-axis is usually chosen along the axis of rotation.

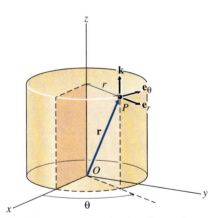

Figure 13-41 In cylindrical coordinates, the position of a particle is described by giving its position in a plane (using r and θ) as in polar coordinates and by giving its distance (z) from the plane.

13-7.3 Spherical Coordinates

The spherical coordinate system describes the position of a particle in terms of a radial distance and two angles, as in Fig. 13-42. The θ-coordinate is measured in a plane as in polar and cylindrical coordinates. However, the distance from the θ-plane is given by ϕ, the angle between the position vector and the normal to the θ-plane. The three unit

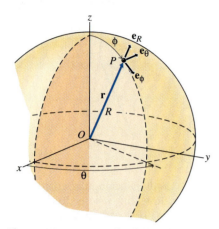

Figure 13-42 In spherical coordinates, the position of a particle is described by giving its distance R from an origin O, the angle θ from the xz-plane to the plane containing the position vector $\mathbf{r}$ and z-axis, and the angle ϕ from the z-axis to the position vector $\mathbf{r}$.

vectors $\mathbf{e}_R$, $\mathbf{e}_\theta$, and $\mathbf{e}_\phi$ are perpendicular to each other and point in the sense that their coordinates increase. Clearly, the directions of all three unit vectors depend on both angles θ and ϕ, which in turn depend on time, and any time derivatives must consider variations in all three unit vectors with respect to both coordinates.

Therefore, although the expression for position in spherical coordinates is very simple

$$\mathbf{r}_{P/O}(t) = R(t)\mathbf{e}_R \qquad (13\text{-}37)$$

derivation of the expressions for the velocity and the acceleration in spherical coordinates is not. Since the derivation is not central to understanding the kinematics, it will not be given. However, the results are included here for completeness:

$$\mathbf{v}_P(t) = \dot{\mathbf{r}}_{P/O}(t) = \dot{R}\mathbf{e}_R + R\dot{\theta}\sin\phi\,\mathbf{e}_\theta + R\dot{\phi}\,\mathbf{e}_\phi \qquad (13\text{-}38)$$

$$\begin{aligned}\mathbf{a}_P(t) = \dot{\mathbf{v}}_P(t) = \ddot{\mathbf{r}}_{P/O}(t) \\ = (\ddot{R} - R\dot{\phi}^2 - R\dot{\theta}^2\sin^2\phi)\mathbf{e}_R \\ + (R\ddot{\theta}\sin\phi + 2\dot{R}\dot{\theta}\sin\phi + 2R\dot{\theta}\dot{\phi}\cos\phi)\mathbf{e}_\theta \\ + (R\ddot{\phi} + 2\dot{R}\dot{\phi} - R\dot{\theta}^2\sin\phi\cos\phi)\mathbf{e}_\phi \qquad (13\text{-}39)\end{aligned}$$

The spherical coordinate system is most often used in radar observations of aircraft or spacecraft positions and in describing the position and motion of robotic arms.

EXAMPLE PROBLEM 13-18

A particle following a space curve has a velocity given by

$$\mathbf{v}(t) = 12t^2\mathbf{i} + 16t^3\mathbf{j} + \sin\pi t\,\mathbf{k}\ \text{ft/s}$$

If at $t = 0$ the particle has the position $\mathbf{r}_0 = 4\mathbf{j} + 3\mathbf{k}$ ft, find

a.   The acceleration of the particle $\mathbf{a}(t)$.
b.   The position of the particle $\mathbf{r}(t)$.

SOLUTION

a.   The acceleration is obtained by simply differentiating the velocity to get

$$\mathbf{a}(t) = 24t\mathbf{i} + 48t^2\mathbf{j} + \pi\cos\pi t\,\mathbf{k}\ \text{ft/s}^2 \qquad \text{Ans.}$$

b.   The position of the particle is obtained by integrating the velocity, giving

$$\mathbf{r}(t) = 4t^3\mathbf{i} + 4t^4\mathbf{j} - (1/\pi)\cos\pi t\,\mathbf{k} + \mathbf{C}$$

where $\mathbf{C}$ is a constant of integration to be determined using the initial condition. At $t = 0$

$$\mathbf{r}(0) = -(1/\pi)\mathbf{k} + \mathbf{C} = \mathbf{r}_0 = 4\mathbf{j} + 3\mathbf{k}\ \text{ft}$$

Therefore

$$\mathbf{C} = 4\mathbf{j} + (3 + 1/\pi)\mathbf{k}\ \text{ft}$$

and

$$\begin{aligned}\mathbf{r}(t) = 4t^3\mathbf{i} + 4(1 + t^4)\mathbf{j} \\ + [3 + 1/\pi - (1/\pi)\cos\pi t]\mathbf{k}\ \text{ft} \qquad \text{Ans.}\end{aligned}$$

The exit ramp of a parking garage is in the shape of a helix

$$r(\theta) = 15 + 3 \sin \theta \text{ m}$$

which drops 6 m for each complete revolution. For a car traveling down the ramp such that $\dot{\theta} = 0.3 \text{ rad/s} = \text{constant}$,

a. Determine the velocity and acceleration when $\theta = 0°$.
b. Determine the velocity and acceleration when $\theta = 90°$.
c. Show that the velocity and acceleration are perpendicular when $\theta = 90°$.

SOLUTION

a. The position vector in cylindrical coordinates is

$$\mathbf{r}(t) = r(t)\mathbf{e}_r + z(t)\mathbf{k}$$

where

$$r(t) = 15 + 3 \sin \theta \text{ m}$$
$$z(t) = A - (6\theta/2\pi) \text{ m}$$
$$\theta(t) = B + 0.3t \text{ rad}$$

and A and B constants. The velocity and acceleration are given by

$$\mathbf{v}(t) = \dot{r}\mathbf{e}_r + r\dot{\theta}\mathbf{e}_\theta + \dot{z}\mathbf{k}$$
$$\mathbf{a}(t) = (\ddot{r} - r\dot{\theta}^2)\mathbf{e}_r + (r\ddot{\theta} + 2\dot{r}\dot{\theta})\mathbf{e}_\theta + \ddot{z}\mathbf{k}$$

where

$$\dot{r} = 3\dot{\theta}\cos\theta \text{ m/s} \qquad\qquad \ddot{r} = 3\ddot{\theta}\cos\theta - 3\dot{\theta}^2\sin\theta \text{ m/s}^2$$
$$\dot{\theta} = 0.3 \text{ rad/s} \qquad\qquad \ddot{\theta} = 0 \text{ rad/s}^2$$

$$\dot{z} = -\frac{6\dot{\theta}}{2\pi} \text{ m/s} = -0.286 \text{ m/s} \qquad \ddot{z} = 0 \text{ m/s}^2$$

Then when $\theta = 0°$

$$r = 15 \text{ m} \qquad \dot{r} = 0.9 \text{ m/s} \qquad \ddot{r} = 0 \text{ m/s}^2$$

and

$$\mathbf{v} = 0.900\mathbf{e}_r + 4.500\mathbf{e}_\theta - 0.286\mathbf{k} \text{ m/s} \qquad \text{Ans.}$$
$$\mathbf{a} = -1.350\mathbf{e}_r + 0.540\mathbf{e}_\theta \text{ m/s}^2 \qquad \text{Ans.}$$

b. When $\theta = 90°$

$$r = 18 \text{ m} \qquad \dot{r} = 0 \text{ m/s} \qquad \ddot{r} = -0.270 \text{ m/s}^2$$

and

$$\mathbf{v} = 5.400\mathbf{e}_\theta - 0.286\mathbf{k} \text{ m/s} \qquad \text{Ans.}$$
$$\mathbf{a} = -1.890\mathbf{e}_r \text{ m/s}^2$$

c. Checking the scalar product of the velocity and acceleration vectors of part b gives

$$\mathbf{v} \cdot \mathbf{a} = 0$$

which shows that the velocity and acceleration vectors are indeed perpendicular. Ans.

The radar of Fig. 13-43 is tracking an airplane. At the instant shown, the position of the airplane is given by $R = 65{,}000$ ft, $\theta = 110°$, and $\phi = 60°$. Comparison with previous positions allows the estimation of the derivatives $\dot{R} = -285$ ft/s, $\ddot{R} = 15$ ft/s^2, $\dot{\theta} = 9.0(10^{-3})$ rad/s, $\ddot{\theta} = 20(10^{-6})$ rad/s^2, $\dot{\phi} = 2.5(10^{-3})$ rad/s, and $\ddot{\phi} = 80(10^{-6})$ rad/s^2. For this instant, determine

a. The velocity and acceleration of the airplane in spherical coordinates (R, ϕ, θ).
b. The velocity and acceleration of the airplane in rectangular coordinates in which the z-axis corresponds to the axis $\phi = 0°$ and the x-axis corresponds to the axis $\phi = 90°$ and $\theta = 0°$.
c. The magnitudes of the velocity and acceleration of the airplane.

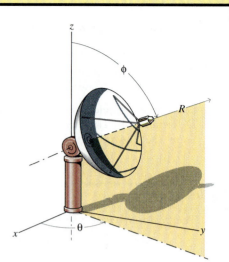

Fig. 13-43

SOLUTION

a. The velocity of the airplane is given in spherical coordinates by Eq. 13-38

$$\mathbf{v}(t) = \dot{R}\mathbf{e}_R + R\dot{\theta}\sin\phi\,\mathbf{e}_\theta + R\dot{\phi}\mathbf{e}_\phi$$
$$= -285\mathbf{e}_R + 506.6\mathbf{e}_\theta + 162.5\mathbf{e}_\phi \text{ ft/s} \qquad \text{Ans.}$$

and the acceleration is given in spherical coordinates by Eq. 13-39

$$\mathbf{a}(t) = (\ddot{R} - R\dot{\phi}^2 - R\dot{\theta}^2\sin^2\phi)\mathbf{e}_R$$
$$+ (R\ddot{\theta}\sin\phi + 2\dot{R}\dot{\theta}\sin\phi + 2R\dot{\theta}\dot{\phi}\cos\phi)\mathbf{e}_\theta$$
$$+ (R\ddot{\phi} + 2\dot{R}\dot{\phi} - R\dot{\theta}^2\sin\phi\cos\phi)\mathbf{e}_\phi$$
$$= 10.65\mathbf{e}_R - 1.854\mathbf{e}_\theta + 1.495\mathbf{e}_\phi \text{ ft/s}^2 \qquad \text{Ans.}$$

b. With reference to Fig. 13-42, the spherical coordinate unit vectors $\mathbf{e}_R$, $\mathbf{e}_\theta$, and $\mathbf{e}_\phi$ can be related to the rectangular unit vectors $\mathbf{i}$, $\mathbf{j}$, and $\mathbf{k}$ by

$$\mathbf{e}_R = \sin\phi\cos\theta\,\mathbf{i} + \sin\phi\sin\theta\,\mathbf{j} + \cos\phi\,\mathbf{k}$$
$$\mathbf{e}_\theta = -\sin\theta\,\mathbf{i} + \cos\theta\,\mathbf{j}$$

and (since R-ϕ-θ form a right-handed orthogonal coordinate system)

$$\mathbf{e}_\phi = \mathbf{e}_\theta \times \mathbf{e}_R$$

Then when $\theta = 110°$ and $\phi = 60°$

$$\mathbf{e}_R = -0.2962\mathbf{i} + 0.8138\mathbf{j} + 0.5000\mathbf{k}$$
$$\mathbf{e}_\theta = -0.9397\mathbf{i} - 0.3420\mathbf{j}$$
$$\mathbf{e}_\phi = -0.1710\mathbf{i} + 0.4699\mathbf{j} - 0.8660\mathbf{k}$$

and

$$\mathbf{v}(t) = -285\,(-0.2962\mathbf{i} + 0.8138\mathbf{j} + 0.5000\mathbf{k})$$
$$+506.6\,(-0.9397\mathbf{i} - 0.3420\mathbf{j})$$
$$+162.5\,(-0.1710\mathbf{i} + 0.4699\mathbf{j} - 0.8660\mathbf{k})$$
$$= -419.4\mathbf{i} - 328.9\mathbf{j} - 283.2\mathbf{k} \text{ ft/s} \qquad \text{Ans.}$$
$$\mathbf{a}(t) = 10.65\,(-0.2962\mathbf{i} + 0.8138\mathbf{j} + 0.5000\mathbf{k})$$
$$-1.854\,(-0.9397\mathbf{i} - 0.3420\mathbf{j})$$
$$+1.495\,(-0.1710\mathbf{i} + 0.4699\mathbf{j} - 0.8660\mathbf{k})$$
$$= -1.666\mathbf{i} + 10.00\mathbf{j} + 4.03\mathbf{k} \text{ ft/s}^2 \qquad \text{Ans.}$$

c. The magnitudes of the velocity and acceleration can be computed from the components in either the spherical or rectangular coordinate system. Using the spherical coordinate components

$$v = \sqrt{285^2 + 506.6^2 + 162.5^2} = 604 \text{ ft/s} \qquad \text{Ans.}$$
$$a = \sqrt{10.65^2 + 1.854^2 + 1.495^2} = 10.91 \text{ ft/s}^2 \qquad \text{Ans.}$$

PROBLEMS

Introductory Problems

13-159* The three-dimensional motion of a particle is described by the relation

$$\mathbf{r} = 5t^2\mathbf{i} + 3t\mathbf{j} + 15t^3\mathbf{k} \text{ ft}$$

Compute the velocity and acceleration of the particle.

13-160* The three-dimensional motion of a particle is described by the relations

$$\mathbf{r} = 6\sin 6t\mathbf{i} + 3\sqrt{3}\cos 6t\mathbf{j} + 3\cos 6t\mathbf{k} \text{ m}$$

Compute the acceleration of the particle, and show that it has constant magnitude.

13-161 The motion of a particle moving down a helical wire (Fig. P13-161) is described by the relation

$$\mathbf{r} = 2\cos 0.3t\mathbf{i} + 2\sin 0.3t\mathbf{j} - 0.24t\mathbf{k} \text{ ft}$$

Determine the magnitudes of the velocity and acceleration of the particle when $t = 5$ s.

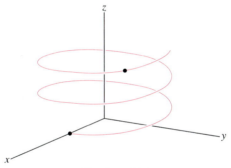

Fig. P13-161

13-162 An eagle riding an updraft travels along an elliptical helical path (Fig. P13-162) described by the relations

$$\mathbf{r} = 15\cos 0.2t\mathbf{i} + 10\sin 0.2t\mathbf{j} + 0.8t\mathbf{k} \text{ m}$$

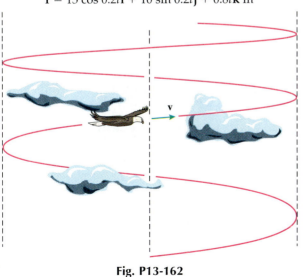

Fig. P13-162

Compute the velocity and acceleration of the eagle at $t = 80$ s.

Intermediate Problems

13-163* A projectile is fired with an initial velocity of 1200 ft/s in the xy-plane as shown in Fig. P13-163. A strong wind is blowing in the z-direction, causing a z-component of velocity given by $v_z = 7t^2$ where v_z is in ft/s and t is the time in seconds. Neglect air resistance in the x- and y-directions, and assume that the acceleration due to gravity in the y-direction is constant and equal to $g = 32.2$ ft/s^2. Determine the equations for the x-, y-, and z-coordinates of the projectile at any time t.

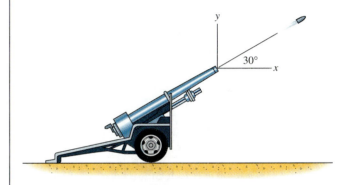

Fig. P13-163

13-164* The three-dimensional motion of a particle on the surface of a right circular cylinder is described by the relations

$$r = 2 \text{ m} \qquad \theta = \pi t \text{ rad} \qquad z = \sin 6\theta \text{ m}$$

Compute the velocity and acceleration of the particle at $t = 3$ s.

13-165 The three-dimensional motion of a particle on the surface of a right circular cone 3 ft tall is described by the relations

$$r = z\tan\beta \text{ ft} \qquad \theta = 2\pi t \text{ rad} \qquad z = \frac{h\theta}{2\pi} \text{ ft}$$

where $\beta = 20°$ is the apex angle of the cone and $h = 0.5$ ft is the distance the particle rises in one trip around the cone. Calculate the velocity and acceleration of the particle at

a. The apex of the cone.
b. The top of the cone.

13-166 The three-dimensional motion of a particle is described by the relations

$$r = 5(1 - e^{-t}) \text{ m} \qquad \theta = 2\pi t \text{ rad} \qquad z = 3\sin 3\theta \text{ m}$$

Compute the magnitude of the velocity and the magnitude of the acceleration of the particle for

a. $t = 1$ s.
b. $t = 4$ s.
c. $t = 20$ s.

13-167* A car is traveling down the exit ramp of a parking garage at a constant speed of 10 mi/h (Fig. P13-167). The ramp is a circle of diameter 120 ft and drops 20 ft for every full revolution ($\theta = 2\pi$). Determine the magnitude of the car's acceleration as it moves down the ramp.

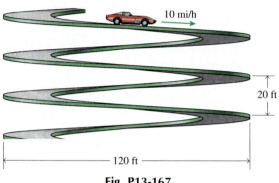

Fig. P13-167

13-168* An airplane is descending in a circular pattern that has a constant radius of 250 m (Fig. P13-168). If the airplane has a horizontal speed of 75 m/s (constant) and a downward speed of 5 m/s (which is increasing at a rate of 2 m/s²), determine the acceleration of the airplane.

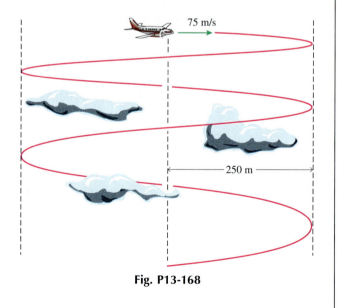

Fig. P13-168

Challenging Problems

13-169 A circular hoop of 2.5 ft radius (Fig. P13-169) is rotating about a vertical axis with an angular velocity given by $\dot{\theta} = t^2 + 1$ rad/s, where t is the time in seconds. A small bead starts from rest at the top of the hoop and moves along the hoop according to $\dot{\phi} = 2t - 1$ rad/s. Determine the magnitude of the velocity of the bead when $t = 2$ s.

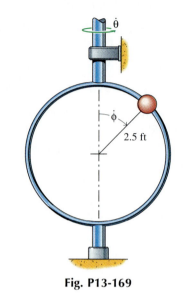

Fig. P13-169

13-170* The crane of Fig. P13-170 rotates about the axis CD at a constant rate ω. At the same time the 20-m-long boom AB is being lowered at a constant rate of 3 rad/min. Determine the maximum rotation rate ω for which the acceleration of point B does not exceed 0.25 m/s² when $\phi = 30°$.

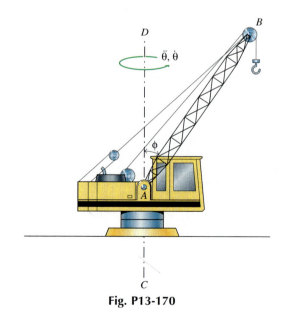

Fig. P13-170

13-171 An airplane is flying due west at a constant speed of 300 mi/h and a constant altitude of 5000 ft. The ground track of the airplane passes 3000 ft to the north of a radar tracking site (Fig. P13-171). Determine the rates θ, $\dot{\theta}$, ϕ, and $\dot{\phi}$ at which the radar dish must be turned to track the airplane when it is 6000 ft east of the radar site.

13-172 An airplane is flying due west at a constant speed of 100 m/s and a constant altitude of 1500 m. The ground track of the airplane passes 2 km to the north of a radar tracking site (Fig. P13-172). Determine the rates θ, $\dot{\theta}$, ϕ, and $\dot{\phi}$ at which the radar dish must be turned to track the airplane when it is directly north of the radar site.

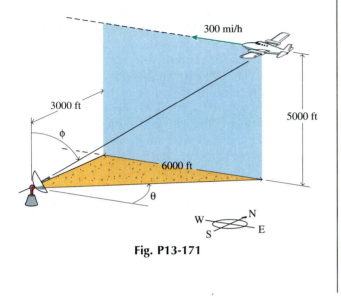

Fig. P13-171

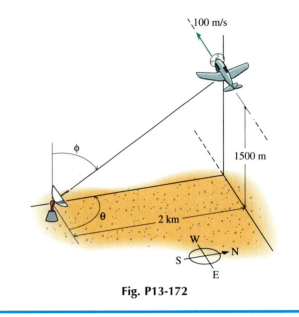

Fig. P13-172

SUMMARY

Kinematics is the study of how particles move. It describes how the velocity and acceleration of a body change with time and with changes in the position of the body. A sound understanding of kinematics is a necessary background for the study of kinetics, which relates the motion to the forces that cause it.

A particle is an object whose size can be ignored when studying its motion. Only the position of the mass center of a particle need be considered. The orientation of the object or rotation of the object plays no role in describing the motion of a particle. Particles can be very small or very large. Small does not always guarantee that an object can be modeled as a particle; large does not always prevent modeling an object as a particle. Whether a body is large or small relates to the length of path followed, the distance between bodies, or both.

The kinematic quantities used to describe the motion of a particle are time, position (including displacement and total distance traveled), velocity

$$\mathbf{v}_P(t) = \frac{d\mathbf{r}_{P/O}}{dt} = \dot{\mathbf{r}}_{P/O} \qquad \text{(13-5)}$$

and acceleration

$$\mathbf{a}_P(t) = \frac{d\mathbf{v}_P}{dt} = \dot{\mathbf{v}}_P = \frac{d^2\mathbf{r}_{P/O}}{dt^2} = \ddot{\mathbf{r}}_{P/O} \qquad \text{(13-7)}$$

Unlike the position vector $\mathbf{r}_{P/O}$, the velocity $\mathbf{v}_P$ and the acceleration $\mathbf{a}_P$ of a particle P are independent of the location of the origin of the coordinate system. The direction of the velocity vector $\mathbf{v}_P$ is always tangent to the path of the particle. The acceleration, in general, has components both tangent and normal to the path of the particle.

The problems of kinematics consist of determining one or more of the preceding quantities from the given data. In most problems, the acceleration of a particle is derived from the forces that act on the particle. Depending on the nature of the forces, the acceleration may be known as a function of time, as a function of velocity, or as a function of the position of the particle. The velocity and position of the particle are obtained by integrating the definitions of acceleration and velocity.

Rectilinear motion is motion along a straight line. If the coordinate system is oriented such that the x-axis coincides with the line of motion, then the position, velocity, and acceleration are completely prescribed by giving only their x-components. That is, the position vector, the velocity vector, and the acceleration vector may be specified by giving their "signed magnitudes" x, $v = \dot{x}$, and $a = \ddot{x}$, respectively. Positive values for x, v, and a indicate that the *vectors* are in the positive coordinate direction; negative values indicates that the *vectors* are in the negative coordinate direction.

When two or more particles move in rectilinear motion, separate equations may be written to describe their motion. The particles may be moving along the same line or along separate lines. If the position of one particle depends on the position of another or several other particles, a constraint equation can be written relating the positions of the particles. The constraint equation is then differentiated to get the relative velocity and relative acceleration equations.

The choice of which coordinate system to use in a particular problem will depend on the geometry of the problem, on the way in which data are given for the problem, and on the type of solution desired. Rectangular coordinates are usually the most convenient coordinates to use when the x- and y-components of the motion are specified independently and do not depend on each other. Polar coordinates are usually the most convenient coordinates to use when the distance and direction of a particle are measured relative to a fixed point or when the particle is moving along a rotating arm. Normal/tangential coordinates are usually the most convenient coordinates to use when particles move along a surface of known shape.

REVIEW PROBLEMS

13-173* A ball is suspended between two elastic bands that are both stretched near their elastic limit. The acceleration in this case is not linear but is given by

$$a(x) = -3x - 5x^3 \text{ ft/s}^2$$

Determine the maximum velocity of the ball if it has a velocity $v = -4$ ft/s when $x = 1$ ft.

13-174* The rocket sled shown in Fig. P13-174 is released from rest and accelerates according to the relation

$$a(t) = 12 + 1.8t - 0.12t^2$$

where $a(t)$ is the acceleration in m/s^2 and t is the time in seconds. Determine

a. The velocity of the sled when $t = 5$ s.

b. The distance traveled by the sled during the 5 s.
c. The maximum acceleration attained by the sled during the first 5 s.

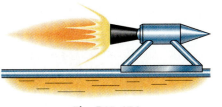

Fig. P13-174

13-175 A car and a truck are traveling along a straight road as shown in Fig. P13-175. The car has a constant speed of 65 mi/h, and the truck has a constant speed of 55 mi/h. If the car is initially 500 ft behind the truck, determine the time, in seconds, for the car to overtake the truck.

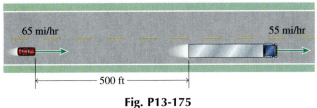

Fig. P13-175

13-176 A bicyclist starts from rest at A and moves along a circular track, half of which is shown in Fig. P13-176a. If the cyclist moves according to the speed–time relationship shown in Fig P13-176b. Determine

a. The magnitude of the acceleration of the cyclist when $t = 20$ s.
b. The distance, measured along the circular track, traveled by the cyclist from $t = 0$ to $t = 20$ s.
c. The time for the cyclist to complete one lap around the track.

13-177 In Fig. P13-177 block B is moving to the right with a speed of 10 ft/s; its speed is decreasing at the rate of 1 ft/s^2. Block C is also moving to the right with a speed of 2 ft/s, and its speed is increasing at the rate of 0.5 ft/s^2. Determine the velocity and acceleration of block A.

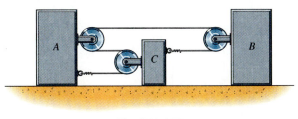

Fig. P13-177

13-178 In Fig. P13-178 block A is moving to the left with a speed of 4 m/s; the speed is decreasing at the rate of 0.15 m/s^2. At the instant shown, $d_A = 8$ m and $d_B = 6$ m. Determine the relative velocity $\mathbf{v}_{B/A}$ and the relative acceleration $\mathbf{a}_{B/A}$.

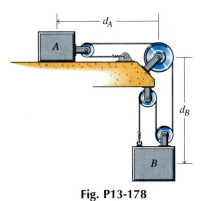

Fig. P13-178

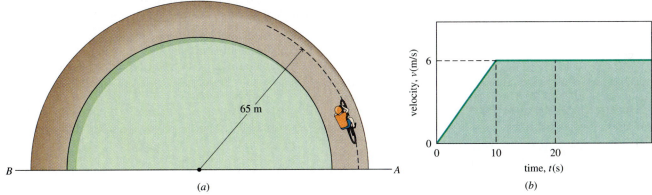

(a)   (b)

Fig. P13-176

13-179* Water is leaking from a tank through a small hole in the tank wall as shown in Fig. P13-179. The horizontal velocity of the water as it leaves the tank is given by $v = \sqrt{2gh}$, where h is the vertical distance from the hole to the free surface of the water. Determine the relationship between d, h, and a.

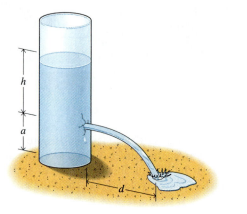

Fig. P13-179

13-180* A 10-mm-diameter marble rolls off a horizontal step 4 m high as shown in Fig. P13-180. Determine the minimum and maximum speed v_0 that the marble can have if it is to pass through a 200-mm-wide hole 2 m away from the bottom of the step. (The acceleration of the marble is 9.81 m/s², downward.)

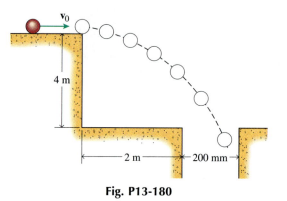

Fig. P13-180

13-181 A particle is sliding along a flat, horizontal, frictionless surface at $v_i = 10$ ft/s as shown in Fig. P13-181. When the particle is 20 ft from the wall, it explodes and splits into two pieces. If $\theta_A = 14°$, $\theta_B = 26°$, $v_A = 12$ ft/s, and $v_B = 7$ ft/s, determine

a. Where each particle will strike the wall, y_A and y_B.
b. The time difference Δt between when particle A strikes the wall and when particle B strikes the wall.

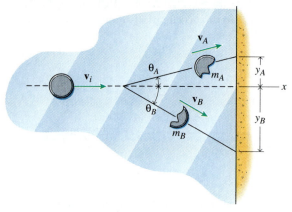

Fig. P13-181

13-182 A ball falls on a hard surface and bounces over a vertical wall as shown in Fig. P13-182. If the velocity of the ball as it leaves the inclined surface is $v_i = 3$ m/s and the ball just clears the wall at the peak of its bounce, determine the distances b, c, and d in the figure.

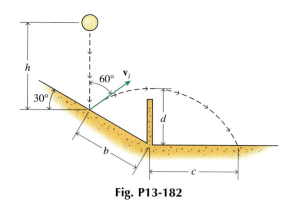

Fig. P13-182

13-183 The rod shown in Fig. P13-183 is rotating in a vertical plane with a constant angular velocity of 20 rev/min. The small bead slides along the rod with a constant velocity relative to the rod of 4 ft/s. When the bead is 3 ft from the axis of rotation, determine

a. The radial and transverse components of the velocity and acceleration of the bead.
b. The magnitude of the velocity and acceleration of the bead.

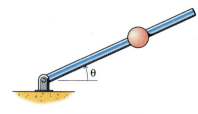

Fig. P13-183

13-184* One collar shown in Fig. P13-184 is free to slide in a circular slot while the other collar is free to slide along the rod that rotates about the pin at A. The two collars are pin-connected at B. When $\theta = 30°$ the speed of the collar in the circular slot is 125 mm/s and is increasing at a rate of 20 mm/s². Determine

a. The component of the velocity of the collar attached to the rod in the direction of the rod.
b. The component of the acceleration of the collar attached to the rod in the direction of the rod.

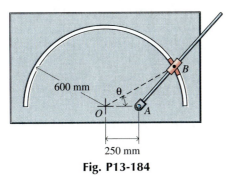

Fig. P13-184

13-185* A boat is trying to travel across a 2000-ft wide river as shown in Fig. P13-185. If the river has a current of 5 mi/h and the boat is traveling at 15 mi/h, determine the time T and the angle θ required to travel directly

a. From A to C.
b. From A to D.

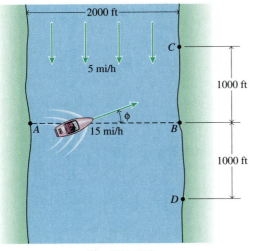

Fig. P13-185

13-186* Rain is falling with a speed of 30 m/s and an angle of 20° to the vertical (Fig. P13-186). For a car traveling with the rain, determine

a. The angle ϕ at which the rain appears to strike the windshield if the car is traveling at 60 km/h.
b. The speed of the car for which $\phi = 90°$.

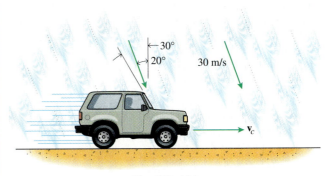

Fig. P13-186

13-187* The boom AB of the crane shown in Fig. P13-187 is 75 ft long. When $\phi = 30°$ the crane is rotating about the axis CD with $\dot\theta = 3$ rad/min, $\ddot\theta = -1$ rad/min², $\dot\phi = -5$ rad/min, and $\ddot\phi = 2$ rad/min². Calculate the acceleration of point B.

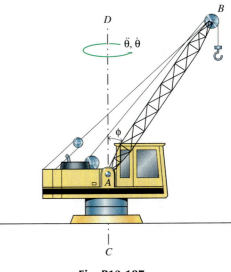

Fig. P13-187

13-188 A cannonball is fired with an initial velocity of 125 m/s and $\theta_o = 75°$ as shown in Fig. P13-188. At the peak of its trajectory, the ball explodes and splits into two pieces. If the velocities of the two pieces immediately after the explosion are $\mathbf{v}_1 = 4.0\mathbf{i} - 3.8\mathbf{j} + 3.8\mathbf{k}$ m/s and $\mathbf{v}_2 = -2.6\mathbf{i} + 56.0\mathbf{j} - 2.5\mathbf{k}$ m/s, determine

a. The locations $\mathbf{r}_1 = x_1\mathbf{i} + y_1\mathbf{j}$ and $\mathbf{r}_2 = x_2\mathbf{i} + y_2\mathbf{j}$ where the two pieces hit the ground.
b. The time Δt between when piece 1 hits the ground and when piece 2 hits the ground.

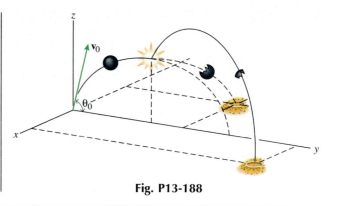

Fig. P13-188

Although the principal motion exhibited by a Ferris wheel is a fixed-axis rotation, the motion of the chairs and the riders is plane curvilinear translation. That is, the chairs and riders remain upright at all times. If the chairs were rigidly attached to the Ferris wheel and rotated with it, the passengers would be dumped out of the chairs at the top of the wheel.

Chapter 14

KINEMATICS OF RIGID BODIES

14-1   INTRODUCTION

The kinematics of particles was analyzed in the preceding chapter. In that analysis, the location of a particle at all instants of time was all that was necessary to fully describe the motion of the particle. For solid bodies, however, a full description of the motion requires giving both the location and the orientation of the body. The kinematics of solid bodies involves both linear and angular quantities.

All solid bodies considered in this chapter and in the rest of the book will be considered to be rigid. In a rigid body, the distance between all pairs of particles is fixed and independent of time (Fig. 14-1). Clearly, if the distances between all pairs of particles are fixed, then so are the angles between triples of points (Fig. 14-1).

No real body is absolutely rigid, of course. However, for most engineering applications, the deformations due to applied forces are relatively small, and the changes in the shape of the body due to applied forces have a negligible effect on the acceleration produced by a force system or on the forces required to produce a given motion. After the kinetic analysis is completed, the deformations should be computed. If the deformations are large, the kinematic and kinetic analyses may have to be repeated taking into account the deformation.

There are five general types of rigid body motion to be considered:

1. **Translation**   In the translation of a rigid body, the orientation of every straight line in the body is fixed. That is, horizontal lines remain horizontal, vertical lines remain vertical, and so on. A motion for which one line is always aligned with the velocity, such as

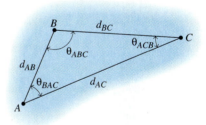

Figure 14-1   For a perfectly rigid body, the distance between any pair of points is fixed. If the distances are all fixed, however, then the angle between any two lines in the body must also be fixed.

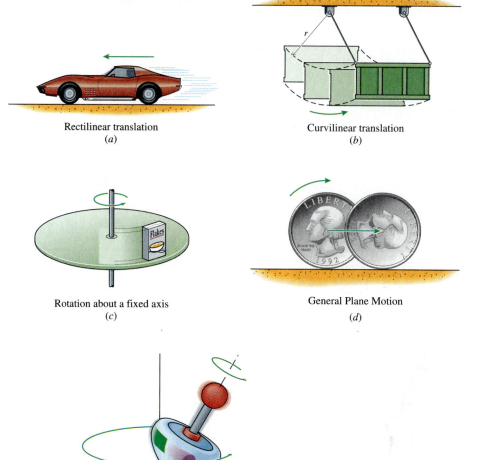

Rectilinear translation
(a)

Curvilinear translation
(b)

Rotation about a fixed axis
(c)

General Plane Motion
(d)

Rotation about a fixed point
(e)

Figure 14-2 *(a)* A car traveling along a straight road exhibits rectilinear translation. A line in the car that is initially aligned with the motion will always be aligned with the motion.

(b) The swinging crate exhibits plane curvilinear translation. If the wires are parallel and have the same length, then the top of the crate will remain parallel with the ceiling as the crate swings back and forth in a vertical plane. All points in the crate travel along parallel circular paths, but all horizontal lines remain horizontal.

(c) Both the horizontal turntable and the cereal box exhibit fixed axis rotation. Every point on the turntable rotates about the axis of the shaft fixed to its center, whereas points along the axis of the shaft do not move. Every point of the cereal box also rotates in a circular path about the axis of the shaft, even though the axis does not intersect the cereal box.

(d) The rolling quarter exhibits general plane motion. Although the quarter rotates about an axis through its center, the center is not fixed.

(e) The child's spinning top exhibits rotation about a fixed point (as long as the point of the top is sharp and does not move along the surface). Although the top always spins about its axis of symmetry, the orientation of that axis may (and usually does) change from instant to instant.

horizontal lines on the body of a car driving on a straight level road (Fig. 14-2a), is called **rectilinear translation.** In rectilinear translation, every particle in the body follows a straight-line path in the direction of the motion. In **curvilinear translation** the orientation of every straight line is still fixed, but individual particles do not follow straight-line paths (Fig. 14-2b). In **coplanar translation,** the path of each particle—whether straight or curved—remains in a single plane.

2. **Rotation about a Fixed Axis**   In rotation about a fixed axis, one straight line in the body, the axis of rotation, is fixed. Particles that are not on the axis travel in circular paths centered on the axis (Fig. 14-2c). If the axis of rotation does not intersect the body, the body may be imagined to be extended to include the axis of rotation. That is, for the purpose of kinematics, the motion of the body is the same as it would be if it were part of a larger rigid body that includes the axis of rotation. Since each circular path is contained in a single plane, the rotation of a body about a fixed axis is a plane motion.

3. **General Plane Motion**   In a plane motion, each particle in the body remains in a single plane. Coplanar translation and rotation about a fixed axis are specific types of plane motion in which lines in the body are fixed in special ways. Any other plane motion is called general plane motion (Fig. 14-2d).

4. **Rotation about a Fixed Point**   In rotation about a fixed point, one point in the body is fixed (Fig. 14-2e). Particles move along paths on the surface of a sphere centered at the point.

5. **General Motion**   All other motions are called general motion.

14-2   TRANSLATION

In the translation of a rigid body, the orientation of every straight line in the body is fixed. That is, horizontal lines remain horizontal, vertical lines remain vertical, and so on. If A and B are two arbitrary points in the body, the position of the points are related by the triangle law of addition for vectors (Fig. 14-3):

$$\mathbf{r}_B = \mathbf{r}_A + \mathbf{r}_{B/A} \qquad (14\text{-}1)$$

where $\mathbf{r}_A$ and $\mathbf{r}_B$ are the absolute positions of the particles A and B (the positions relative to a fixed origin O), respectively, and $\mathbf{r}_{B/A}$ is the position of B relative to the moving point A. Since the relative position $\mathbf{r}_{B/A}$ is constant in both magnitude (since the body is rigid) and direction (since the body is translating), its derivative is zero and the time derivative of Eq. 14-1 gives simply

$$\mathbf{v}_B = \mathbf{v}_A \qquad (14\text{-}2)$$

where $\mathbf{v}_A$ and $\mathbf{v}_B$ are the absolute velocities (relative to a fixed coordinate system) of particles A and B, respectively. That is, *the velocity of every particle in a translating body is the same as the velocity of every other particle in the body.*

Equation 14-2 can be differentiated with respect to time again to get

$$\mathbf{a}_B = \mathbf{a}_A \qquad (14\text{-}3)$$

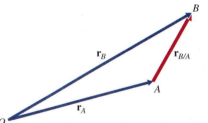

Figure 14-3   The positions of particles A and B relative to the fixed origin O and the position of particle B relative to particle A are related by the vector law of addition.

where $\mathbf{a}_A$ and $\mathbf{a}_B$ are the absolute accelerations (relative to a fixed coordinate system) of particles A and B, respectively. Equation 14-3 says that *the acceleration of every particle in a translating body is also the same as the acceleration of every other particle in the body.*

Since the motion of every point is the same as every other point, no distinction need be made between the motion of particle A and the motion of particle B. It is called simply the motion of the body. Since the size, shape, and orientation of the body are not important in describing the motion of the body, the kinematics of the particles that make up a rigid body undergoing translation is identical to the kinematics of particle motion discussed in Chapter 13. All the results of Chapter 13 apply, and no further discussion of a rigid body in translation will be made here.

14-3 PLANAR MOTION

In a plane motion, each particle in the body remains in a single plane. Since all points along lines perpendicular to the plane have the same motion, only the motion in a single plane need be considered. In the discussion that follows, the plane that contains the mass center, called **the plane of motion,** will be used.

Since particles cannot move out of the plane of motion, the position of a rigid body in plane motion is completely determined by giving the location of one point and the orientation of one line in the plane of motion (Fig. 14-4). The orientation of the line may be given either by giving the angle it makes with a fixed direction (Fig. 14-4*a*) or by giving the location of any two points on the line (Fig. 14-4*b*). The motion of the entire body can be determined from the motion of this one point and the motion of the line.

It is important to note that the angular motion of lines in the plane of motion is the same for every straight line in a rigid body. For example, consider the body of Fig. 14-5, on which have been drawn two lines separated by a fixed angle β. Both lines are in the plane of motion, and the angles between these lines and a fixed reference direction are θ_{AB} and θ_{CD}, respectively, as shown. From Fig. 14-5 these angles are related by

$$\theta_{CD} = \theta_{AB} + \beta \qquad (a)$$

As the body moves, the angles θ_{AB} and θ_{CD} will change. Since the body is rigid, however, the angle β is fixed and differentiation of Eq. *a* with respect to time gives

$$\omega_{CD} = \dot{\theta}_{CD} = \dot{\theta}_{AB} = \omega_{AB} \qquad (b)$$

where ω, the time rate of change of angular position, is called the **angular velocity.** Equation *b* gives that the angular velocity of every line in the body is the same. Therefore, $\omega_{AB} = \omega_{CD}$ will be called simply the *angular velocity of the body.*

Equation *b* can be differentiated with respect to time again to get

$$\alpha_{CD} = \dot{\omega}_{CD} = \ddot{\theta}_{CD} = \ddot{\theta}_{AB} = \dot{\omega}_{AB} = \alpha_{AB} \qquad (c)$$

where α, the time rate of change of angular velocity, is called the **angular acceleration.** Like the angular velocity, the angular acceleration

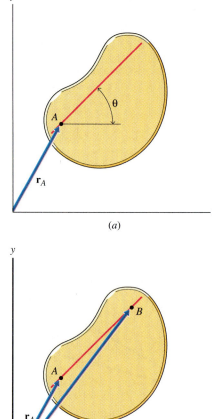

(a)

(b)

Figure 14-4   The position of a rigid body in plane motion is completely specified by giving the location of any one point in the body and the orientation of any line in the plane of motion. The orientation of the line may be given either by giving the angle it makes with a fixed direction or by giving the location of any two points on the line.

of every line in the body is the same, and $\alpha_{AB} = \alpha_{CD}$ will be called simply the *angular acceleration of the body.*

14-4   ROTATION ABOUT A FIXED AXIS

It has been noted that the position of a rigid body in plane motion is completely determined by giving the location of one point and the orientation of one line in the plane of motion. The motion of the entire body can be determined from the motion of this one point and the motion of the line. For rotation about a fixed axis, however, the point on the axis remains on the axis—always. Therefore, the motion of the entire body can be determined from the motion of a line.

14-4.1   Motion of a Line in Fixed Axis Rotation

In fixed axis rotation the position of the body is completely determined by giving the angular position θ of any line in the plane of motion. The time derivatives of the angular position give the angular velocity $\omega(t)$

$$\omega(t) = \frac{d\theta}{dt} \tag{14-4}$$

and the angular acceleration $\alpha(t)$

$$\alpha(t) = \frac{d\omega}{dt} = \frac{d^2\theta}{dt^2} \tag{14-5}$$

of the rigid body.

Equations 14-4 and 14-5 relating the angular position, angular velocity, and angular acceleration of a rigid body are analogous to Eqs. 13-9*c* and 13-9*d*, which relate the position, velocity, and acceleration of a particle in rectilinear motion. Just as Eqs. 13-9 were integrated in Section 13-3 to get general relationships between the position, velocity, and acceleration of a particle in rectilinear motion, Eqs. 14-4 and 14-5 can be integrated to get general relationships between the angular position, angular velocity, and angular acceleration of a rigid body.

In particular, if the angular acceleration is known as a function of time, then it can be integrated to get the angular velocity

$$\omega(t) = \int \alpha(t)\, dt + C_1 \tag{d}$$

and the angular position

$$\theta(t) = \int \omega(t)\, dt + C_2 \tag{e}$$

as functions of time. As in Chapter 13, the constants of integration C_1 and C_2 are determined from knowing the angular velocity and angular position at some instant. For the special case in which the angular acceleration is constant, these integrals may be evaluated immediately to get

$$\omega(t) = \omega_0 + \alpha t \tag{f}$$

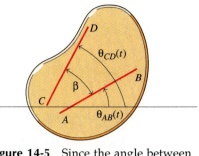

Figure 14-5   Since the angle between any two lines in a rigid body is fixed, the lines must rotate the same amount in any interval of time. That is, all lines in the body must rotate at the same rate. Therefore, $\omega = \dot{\theta}$ and $\alpha = \dot{\omega}$ are the angular velocity and angular acceleration *of the body* and not just of some particular line in the body.

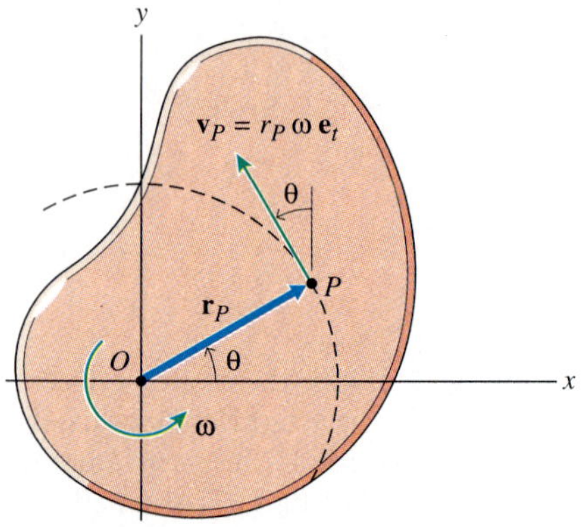

and

$$\theta(t) = \theta_0 + \omega_0 t + \frac{1}{2}\alpha t^2 \qquad (g)$$

where the constants of integration have been chosen so that $\theta = \theta_0$ and $\omega = \omega_0$ when $t = 0$.

If the angular acceleration is known as a function of angular position instead of time, then using the chain rule of differentiation gives

$$\alpha(\theta) = \frac{d\omega}{dt} = \frac{d\omega}{d\theta}\frac{d\theta}{dt} = \omega\frac{d\omega}{d\theta} \qquad (h)$$

which can be integrated to get the angular velocity as a function of angular position

$$\frac{\omega_2^2}{2} - \frac{\omega_1^2}{2} = \int_{\theta_1}^{\theta_2}\alpha(\theta)d\theta \qquad (i)$$

Equations d through i are directly analogous to the formulas developed in Section 13-3 for a particle in rectilinear motion. All the results derived there also apply to the rotation of a rigid body about a fixed axis by simply replacing x with θ, v with ω, and a with α.

14-4.2   Motion of a Point in Fixed Axis Rotation

In fixed axis rotation, particles that are not on the axis travel in circular paths centered on the axis. If $\mathbf{r}_P$ is the position vector of point P measured relative to the axis of rotation (Fig. 14-6), then the velocity of point P can be written in nt coordinates[1] (Eq. 13-26)

$$\mathbf{v}_P = r_P\omega\mathbf{e}_t \qquad (14\text{-}6a)$$

where $\mathbf{e}_t$ is a unit vector tangent to the circular path at P (perpendicular to $\mathbf{r}_P$). The x- and y-components of this velocity are easily found to be

$$\mathbf{v}_P = -r_P\omega\sin\theta\,\mathbf{i} + r_P\omega\cos\theta\,\mathbf{j} \qquad (j)$$

where θ is the angle between the position vector $\mathbf{r}_P$ and the x-axis.

The velocity of P can also be written in terms of a vector angular velocity $\boldsymbol{\omega}$ defined by

$$\boldsymbol{\omega} = \omega\mathbf{k}$$

(Fig. 14-7). The direction of this vector represents the axis about which the body is rotating. The sense of rotation is according to the *right-hand rule*. That is, if you hold your right hand with your thumb pointing in the direction of the vector, your fingers curl in the sense of the rotation (in this case, counterclockwise when the plane of motion is viewed downward along the z-axis). Then the vector product

$$\boldsymbol{\omega} \times \mathbf{r}_P$$

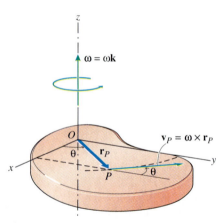

Figure 14-6   In fixed axis rotation, particles that are not on the axis of rotation will travel in circular paths centered on the axis. The velocity of such particles will be tangent to the circular path (perpendicular to the radius or $\mathbf{r}$ vector) with a magnitude of $v_P = r_P\omega$.

Figure 14-7   The angular velocity vector $\boldsymbol{\omega} = \omega\mathbf{k}$ is defined such that its direction is parallel to the axis about which the rotation occurs and its magnitude is equal to the rate of rotation. Then the velocity of any point in the body can be determined using the vector product $\mathbf{v}_P = \boldsymbol{\omega} \times \mathbf{r}_P$.

[1]Equations 14-6a and 14-7a could just as easily have been written in terms of $r\theta$ coordinates. As shown in Section 13-5, when a particle travels around a circular path of constant radius, the expressions for the velocity in the nt coordinate system (Eq. 13-26) and in the $r\theta$ coordinate system (Eq. 13-21) are the same. Similarly, when a particle travels around a circular path of constant radius, the expressions for the acceleration in the nt coordinate system (Eq. 13-27) and in the $r\theta$ coordinate system (Eq. 13-22) are the same.

gives a vector **b**, which is perpendicular to **ω** and perpendicular to **r**$_P$. Hence, the vector **b** lies in the plane of motion and points in the direction of the velocity **v**$_P$. Furthermore, since **ω** and **r**$_P$ are perpendicular, the vector **b** has magnitude $r_P\omega \sin 90° = r_P\omega$. Therefore, the vector **b** is the velocity of point P,

$$\mathbf{v}_P = \boldsymbol{\omega} \times \mathbf{r}_P = r_P\omega \mathbf{e}_t \qquad (14\text{-}6b)$$

Evaluating the vector product of Eq. 14-6b in terms of xy coordinates gives

$$\begin{aligned}\mathbf{v}_P &= (\omega\mathbf{k}) \times (r_P \cos \theta \,\mathbf{i} + r_P \sin \theta \,\mathbf{j}) \\ &= -r_P\omega \sin \theta \,\mathbf{i} + r_P\omega \cos \theta \,\mathbf{j}\end{aligned} \qquad (k)$$

which is the same as Eq. j.

The acceleration of point P is obtained by taking the time derivative of the velocity. Using Eq. 14-6a, this gives

$$\begin{aligned}\mathbf{a}_P &= \frac{d}{dt}(r_P\omega\mathbf{e}_t) = r_P\dot{\omega}\mathbf{e}_t + r_P\omega\dot{\mathbf{e}}_t = r_P\alpha\mathbf{e}_t + r_P\omega(\omega\mathbf{e}_n) \\ &= r_P\alpha\mathbf{e}_t + r_P\omega^2\mathbf{e}_n\end{aligned} \qquad (14\text{-}7a)$$

Note that Eq. 14-7a is the same as Eq. 13-27 for the acceleration of a particle in circular motion. That is, the point P is moving along a circular path about the axis of rotation and it has components $(\mathbf{a}_P)_t = r_P\alpha\mathbf{e}_t$, which is tangent to the circular path, and $(\mathbf{a}_P)_n = r_P\omega^2\mathbf{e}_n$, which is normal to the circular path, as shown in Fig. 14-8. The x- and y-components of the acceleration are

$$\mathbf{a}_P = -r_P\alpha \sin \theta \,\mathbf{i} + r_P\alpha \cos \theta \,\mathbf{j} - r_P\omega^2 \cos \theta \,\mathbf{i} - r_P\omega^2 \sin \theta \,\mathbf{j} \qquad (l)$$

Alternatively, differentiating Eq. 14-6b to get the acceleration of the point P gives

$$\begin{aligned}\mathbf{a}_P &= \frac{d}{dt}(\boldsymbol{\omega} \times \mathbf{r}_P) = \dot{\boldsymbol{\omega}} \times \mathbf{r}_P + \boldsymbol{\omega} \times \dot{\mathbf{r}}_P = \boldsymbol{\alpha} \times \mathbf{r}_P + \boldsymbol{\omega} \times \mathbf{v}_P \\ &= \boldsymbol{\alpha} \times \mathbf{r}_P + \boldsymbol{\omega} \times (\boldsymbol{\omega} \times \mathbf{r}_P)\end{aligned} \qquad (14\text{-}7b)$$

where $\boldsymbol{\alpha} = \dfrac{d\boldsymbol{\omega}}{dt} = \dfrac{d(\omega\mathbf{k})}{dt} = \dot{\omega}\mathbf{k} = \alpha\mathbf{k}$. That is, the angular acceleration vector also points along the axis of rotation and its magnitude is the rate of change of the angular velocity (Fig. 14-9). The first term of Eq. 14-7b is the tangential component of the acceleration (Fig. 14-9)

$$(\alpha\mathbf{k}) \times [r_P(-\mathbf{e}_n)] = \alpha r_P\mathbf{e}_t = (\mathbf{a}_P)_t$$

and the second term is the normal component of the acceleration (Fig. 14-9)

$$\boldsymbol{\omega} \times \mathbf{v}_P = (\omega\mathbf{k}) \times (\omega r_P\mathbf{e}_t) = r_P\omega^2\mathbf{e}_n = (\mathbf{a}_P)_n$$

(Note that the parentheses are needed in the last term of Eq. 14-7b because

$$\boldsymbol{\omega} \times (\boldsymbol{\omega} \times \mathbf{r}_P) \neq (\boldsymbol{\omega} \times \boldsymbol{\omega}) \times \mathbf{r}_P = 0$$

and the order in which the cross products are performed is important.)

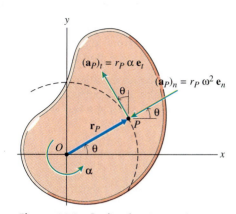

Figure 14-8 In fixed axis rotation, particles that are not on the axis of rotation will travel in circular paths centered on the axis. The acceleration of such particles will have components both tangential to the circular path (perpendicular to the radius or **r** vector) with a magnitude of $a_{Pt} = r_P\alpha$ and normal to the circular path (toward the center of the circle) with a magnitude of $a_{Pn} = r_P\omega^2$.

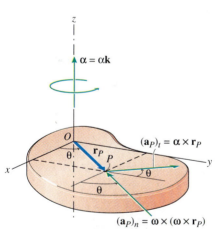

Figure 14-9 The angular acceleration vector $\boldsymbol{\alpha} = \alpha\mathbf{k} = \dot{\omega}\mathbf{k}$ is the time derivative of the angular velocity vector. Its direction is also parallel to the axis about which the rotation occurs and its magnitude is equal to the rate of change of the rotation rate. The acceleration of any point in the body is given by the vector product $\mathbf{a}_P = \boldsymbol{\alpha} \times \mathbf{r}_P + \boldsymbol{\omega} \times \mathbf{v}_P$.

The turntable of the record player shown in Fig. 14-10a attains its operating speed of $33\frac{1}{3}$ rev/min in 5 revolutions after being turned on. Determine the initial angular acceleration α_0 of the turntable if

a. The angular acceleration is constant ($\alpha = \alpha_0 =$ constant).

b. The angular acceleration decreases linearly with angular velocity from α_0 when $\omega = 0$ to $\alpha_0/4$ when $\omega = 33\frac{1}{3}$ rev/min (Fig. 14-10b).

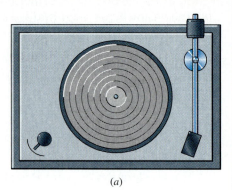

(a)

SOLUTION

a. First, convert the angular velocity from rev/min to rad/s

$$\frac{(33\frac{1}{3} \text{ rev/min})(2\pi \text{ rad/rev})}{60 \text{ s/min}} = 3.491 \text{ rad/s}$$

and the angular displacement to radians

$$(5 \text{ rev})(2\pi \text{ rad/rev}) = 10\pi \text{ rad}$$

Then, since a relationship between the angular displacement and angular velocity is desired, use the chain rule of differentiation to rewrite the definition of angular acceleration

$$\alpha = \frac{d\omega}{dt} = \frac{d\omega}{d\theta}\frac{d\theta}{dt} = \omega\frac{d\omega}{d\theta} = \alpha_0$$

and integrate

$$\int_0^{3.491} \omega \, d\omega = \alpha_0 \int_0^{10\pi} d\theta \qquad (a)$$

to get

$$\frac{3.491^2}{2} = \alpha_0(10\pi)$$

or

$$\alpha_0 = 0.1939 \text{ rad/s}^2 \qquad \text{Ans.}$$

b. This time the angular acceleration is to decrease linearly with angular velocity, so

$$\alpha(\omega) = \alpha_0(1 - 0.2148\omega)$$

and Eq. a becomes

$$\int_0^{3.491} \frac{\omega \, d\omega}{1 - 0.2148\omega} = \alpha_0 \int_0^{10\pi} d\theta$$

which gives

$$\left[\frac{1 - 0.2148\omega - \ln(1 - 0.2148\omega)}{0.2148^2}\right]_0^{3.491} = \alpha_0\left[\theta\right]_0^{10\pi}$$

$$13.78214 = \alpha_0(10\pi)$$

or

$$\alpha_0 = 0.439 \text{ rad/s}^2 \qquad \text{Ans.}$$

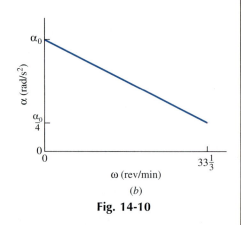

α_0

α (rad/s^2)

$\dfrac{\alpha_0}{4}$

0

0

$33\frac{1}{3}$

ω (rev/min)

(b)

Fig. 14-10

EXAMPLE PROBLEM 14-2

A gear is rotating at a rate of 20 rad/s when a torque is applied to the gear giving it the angular acceleration shown in Fig. 14-11. Determine

a. The angular velocity of the gear for $t > 9$ s.
b. The number of revolutions that the gear turns while it is accelerating.

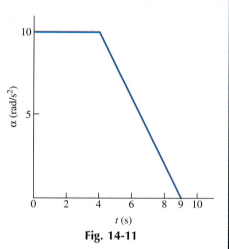

Fig. 14-11

SOLUTION

The angular acceleration changes character at $t = 4$ s, so the problem will have to be solved in two parts. First, for $0 < t < 4$ s,

$$\alpha = \frac{d\omega}{dt} = 10 \text{ rad/s}^2 = \text{constant}$$

Integrating with respect to time gives the angular velocity for $0 < t < 4$ s

$$\omega = 10t + c_1 + 20 \text{ rad/s} \qquad (a)$$

where the constant of integration ($c_1 = 20$ rad/s) has been chosen so that $\omega = \omega_0 = 20$ rad/s when $t = 0$. Then, integrating the angular velocity ($\omega = d\theta/dt$) with respect to time gives the angular position for $0 < t < 4$ s

$$\theta = 5t^2 + 20t + c_2 = 5t^2 + 20t \qquad (b)$$

where the constant of integration ($c_2 = 0$ rad) has been chosen so that $\theta = \theta_0 = 0$ when $t = 0$. Evaluating Eqs. a and b at $t = 4$ s gives

$$\omega = 10(4) + 20 = 60 \text{ rad/s} \qquad \text{at } t = 4 \text{ s}$$
$$\theta = 5(4^2) + 20(4) = 160 \text{ rad} \qquad \text{at } t = 4 \text{ s}$$

Next, when $4 < t < 9$ s the angular acceleration decreases linearly with respect to time

$$\alpha = \frac{d\omega}{dt} = 18 - 2t \text{ rad/s}^2$$

and integrating with respect to time gives the angular velocity for $4 < t < 9$ s

$$\omega = 18t - t^2 + c_3 = 18t - t^2 + 4 \text{ rad/s} \qquad (c)$$

where the constant of integration ($c_3 = 4$ rad/s) has been chosen so that $\omega = 60$ rad/s when $t = 4$ s. Finally, integrating Eq. c with respect to time gives the angular position for $4 < t < 9$ s

$$\theta = 9t^2 - \frac{1}{3}t^3 + 4t + c_4 = 9t^2 - \frac{1}{3}t^3 + 4t + 21.333 \text{ rad} \qquad (d)$$

where the constant of integration ($c_4 = 21.333$ rad) has been chosen so that $\theta = 160$ rad when $t = 4$ s. Therefore, the angular velocity and angular position for $t = 9$ s are

$$\omega = 18(9) - (9^2) + 4 = 85.0 \text{ rad/s} \qquad \text{Ans.}$$
$$\theta = 9(9^2) - \frac{1}{3}(9^3) + 4(9) + 21.333$$
$$= 543.33 \text{ rad} \cong 86.5 \text{ rev} \qquad \text{Ans.}$$

EXAMPLE PROBLEM 14-3

An 80-mm-diameter gear rotates about an axle through its center O (Fig. 14-12). At some instant of time, the angular velocity of the gear is 2 rad/s counterclockwise and is increasing at the rate of 1 rad/s². Determine the acceleration (magnitude and direction) of tooth A at this instant.

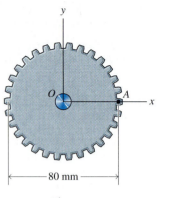

Fig. 14-12

SOLUTION

Every tooth on the gear travels along a circular path around the fixed axis of rotation through O. The acceleration of a point on a rigid body in fixed axis rotation is (Eq. 14-7a)

$$\mathbf{a}_A = r\alpha\mathbf{e}_t + r\omega^2\mathbf{e}_n = (40)(1)\mathbf{j} + (40)(2)^2(-\mathbf{i})$$
$$= -160\mathbf{i} + 40\mathbf{j} \text{ mm/s}^2$$
$$= 164.9 \text{ mm/s}^2 \text{ ⦣ } 14.04° \qquad \text{Ans.}$$

Alternatively, using the vector product form for the acceleration (Eq. 14-7b)

$$\mathbf{a}_A = \boldsymbol{\alpha} \times \mathbf{r} + \boldsymbol{\omega} \times (\boldsymbol{\omega} \times \mathbf{r})$$
$$= (1\mathbf{k}) \times (40\mathbf{i}) + (2\mathbf{k}) \times [(2\mathbf{k}) \times (40\mathbf{i})]$$
$$= 40\mathbf{j} + (2\mathbf{k}) \times (80\mathbf{j}) = 40\mathbf{j} + 160(-\mathbf{i})$$
$$= -160\mathbf{i} + 40\mathbf{j} \text{ mm/s}^2$$
$$= 164.9 \text{ mm/s}^2 \text{ ⦣ } 14.04° \qquad \text{Ans.}$$

as before.

EXAMPLE PROBLEM 14-4

A 16-in.-diameter wheel is at rest when it is brought into contact with a belt moving at a speed of 30 ft/s (Fig. 14-13). If friction between the belt and the wheel gives the wheel a constant angular acceleration of 10 rad/s², determine

a. The amount of time that the wheel slips on the belt.
b. The number of revolutions the wheel turns before it stops slipping.

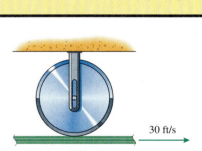

Fig. 14-13

SOLUTION

a. The wheel will stop slipping when the point on the wheel that is in contact with the belt has the same speed as the belt. Since the speed of a point rotating about a fixed axis is given by (Eq. 14-6)

$$v = r\omega \qquad (a)$$

the angular velocity must first be found by integrating the angular acceleration

$$\alpha = \frac{d\omega}{dt} = 10 \text{ rad/s}^2 = \text{constant}$$

This gives

$$\omega = 10t + c_1 = 10t \qquad (b)$$

where the constant of integration ($c_1 = 0$ rad/s) has been chosen so that $\omega = \omega_0 = 0$ rad/s when $t = 0$. Then, the velocity of the contact point will be the same as the belt when (Eq. a)

$$30 = \frac{8}{12}(10t)$$

which gives

$$t = 4.5 \text{ s} \qquad\qquad \text{Ans.}$$

b. Integrating the angular velocity (Eq. b)

$$\omega = d\theta/dt = 10t$$

with respect to time gives the angular position

$$\theta = 5t^2 + c_2 = 5t^2 \qquad\qquad (c)$$

where the constant of integration ($c_2 = 0$ rad) has been chosen so that $\theta = \theta_0 = 0$ when $t = 0$. Evaluating Eq. c when $t = 4.5$ s gives

$$\theta = 5(4.5)^2 = 101.25 \text{ rad} \cong 16.11 \text{ rev} \qquad \text{Ans.}$$

PROBLEMS

Introductory Problems

14-1* The electric motor shown in Fig. P14-1 gives the grinding wheel a constant angular acceleration when it is turned on. If the motor attains its operating speed of 3600 rev/min within 3 s after it is turned on, determine the angular acceleration of the grinding wheel.

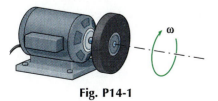

Fig. P14-1

14-2* The drive motor of the computer disk drive shown in Fig. P14-2 gives the disk a constant angular acceleration of 150 rad/s^2 when it is turned on. Determine

a. The time required for the disk drive to achieve its operating speed of 3600 rev/min after it is turned on.
b. The number of revolutions the disk turns before disk drive attains its operating speed.

Fig. P14-2

14-3 The turntable of Fig. P14-3 has an operating speed of $33\frac{1}{3}$ rev/min. When it is turned off, friction in the bearings causes the platter to coast to rest in 2 min with a constant angular deceleration. Determine the angular deceleration of the platter caused by friction.

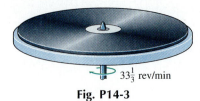

Fig. P14-3

14-4* The gear shown in Fig. P14-4 is rotating with an angular speed of 3600 rev/min when its motor is shut off. If friction in the bearing of the drive shaft causes a constant angular deceleration of 3 rad/s², determine

a. The time required for the gear to coast to rest from the operating speed of 3600 rev/min.
b. The number of revolutions that the gear turns before it comes to rest.

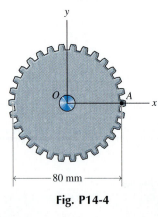

Fig. P14-4

14-5 The angular position of a gear is given by

$$\theta = 5e^{-t} \sin 2t$$

where θ is in radians and t is in seconds. Determine the angular velocity and the angular acceleration of the gear when $t = 2$ s.

14-6* The angular velocity of a gear is given by

$$4\omega^2 + 9\theta^2 = 25$$

where $\omega = \dot{\theta}$ and $\theta = \theta(t)$. Determine

a. The angular acceleration of the gear as a function of the angular position.
b. The angular position of the gear as a function of time.

Intermediate Problems

14-7* A variable torque drive motor gives a circular disk an angular acceleration that varies linearly with angular position as shown in Fig. P14-7. If the angular velocity of the disk is 10 rad/s when $\theta = 0$, determine the angular velocity of the disk after 50 revolutions.

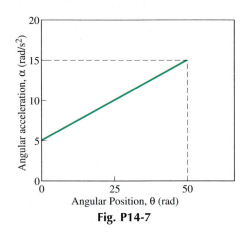

Fig. P14-7

14-8 A variable torque drive motor gives a circular disk an angular acceleration that is inversely proportional to its angular velocity

$$\alpha = \frac{k}{\omega}$$

where ω is in rad/s, α is in rad/s², k is a constant, and $\omega = 0$ rad/s when $\theta = 0$ rad. If the angular velocity of the disk is 40 rad/s after 25 revolutions, determine the angular velocity of the disk after 50 revolutions.

14-9* A variable torque drive motor gives a circular disk an angular acceleration of

$$\alpha = \left(\frac{\omega}{16} - 8\right)^2$$

where ω is in rad/s, α is in rad/s², and $\omega = \theta = 0$ at $t = 0$. Determine

a. The time it takes the motor to rotate the disk 50 revolutions.
b. The angular velocity of the disk after 50 revolutions.

14-10 A variable torque drive motor gives a circular disk an angular acceleration of

$$\alpha = 8 - 0.5\omega$$

where ω is in rad/s, α is in rad/s², and $\omega = \theta = 0$ at $t = 0$. Determine

a. The time it takes the motor to rotate the disk 50 revolutions.
b. The angular velocity of the disk after 50 revolutions.

14-11 A small block B rotates with the horizontal turntable A of Fig. P14-11. The distance between the block and the axis of rotation is $r = 3$ in., the angular acceleration of the turntable is $\alpha = 5$ rad/s² = constant, and the initial angular velocity is zero. Determine

a. The time t_1 when the normal and tangential components of the acceleration of the block are equal.
b. The number of revolutions N of the turntable between $t = 0$ and $t = t_1$.
c. The angular velocity of the turntable at $t = t_1$.

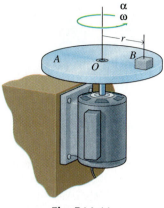

Fig. P14-11

14-12* A small block B rotates with the horizontal turntable A of Fig. P14-11. The distance between the block and the axis of rotation is $r = 80$ mm; the angular acceleration of the turntable is $\alpha = -3$ rad/s^2 = constant, and the initial angular velocity is $\omega_0 = 15$ rad/s. Determine the time t_1 when the angle ϕ between the acceleration of the block $\mathbf{a}_B$ and the radial line OB is 30°.

14-13* The 6-in.-diameter sprocket drive of Fig. P14-13 is driven by an electric motor that gives it a constant angular acceleration α. If link A on the chain has a velocity of $v_A = 20$ ft/s just 5 s after the motor is started, determine

a. The angular acceleration α of the sprocket.
b. The angular velocity ω of the sprocket at the instant when $v_A = 20$ ft/s.
c. The acceleration $\mathbf{a}_B$ of tooth B of the sprocket when $v_A = 20$ ft/s.

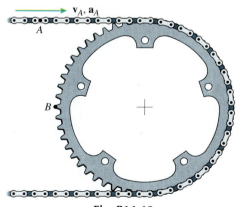

Fig. P14-13

14-14 The space station shown in Fig. P14-14 rotates about its axis of symmetry. The radius of the outer ring is $r_A = 900$ m, and the radius of the inner ring is $r_B = 400$ m.

a. With what constant angular velocity ω must the station rotate such that the acceleration of a person in the outer ring of the station will be $a_A = g = 9.81$ m/s^2?
b. If the acceleration of a person in the outer ring of the station is $a_A = g$, what will be the acceleration a_B of a person in the inner ring of the station.

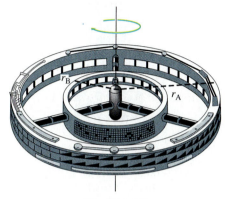

Fig. P14-14

14-15 The belt drive shown in Fig. P14-15 is used to drive the turntable of a record player. The diameters of the motor pulley and the turntable pulley are 0.5 in. and 6 in., respectively. If the angular acceleration of the motor is $\alpha = 10$ rad/s^2 = constant, determine

a. The time t_1 required for the turntable to achieve its operating speed of $33\frac{1}{3}$ rev/min starting from rest.
b. The number of revolutions N that the turntable will turn while it is spinning up.

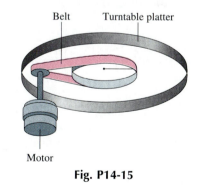

Belt   Turntable platter

Motor

Fig. P14-15

14-16* Weights A and B are supported by cords wrapped around a stepped drum as shown in Fig. P14-16. At the instant shown, weight A has a downward velocity of 2 m/s and its speed is decreasing at a rate of 1.5 m/s². For this instant, determine

a. The acceleration of weight B.
b. The acceleration of point D on the rim of the wheel.

Fig. P14-16

14-17* The rock tumbler C shown in Fig. P14-17 is rotating on two small wheels A and B. When the motor is started, it rotates wheel A with a constant angular acceleration α_A. If the tumbler reaches its operating speed of 20 rev/min in 3 s after being started, determine

a. The angular acceleration of wheel A.
b. The number of revolutions the tumbler rotates in the first 3 seconds.

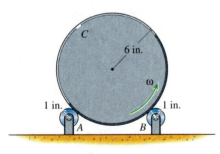

Fig. P14-17

14-18 The gear system shown in Fig. P14-18 is used to raise the load L. If the system is initially at rest and a constant torque motor attached to the smaller gear A gives it a constant angular acceleration of $\alpha_A = 2.5$ rad/s², determine

a. The amount of time required to raise the load L a distance of 5 m.
b. The number of revolutions N_B that the larger gear turns while lifting the load L a distance of 5 m.
c. The velocity of load L when it has risen 5 m.

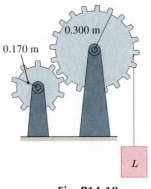

Fig. P14-18

Challenging Problems

14-19* Initially, pulley C of Fig. P14-19 is at rest and pulley A is rotating at 600 rev/min. When the idler pulley B removes the slack from the belt, the belt slips on pulley A for a period of 10 s, during which time the angular acceleration of each pulley is constant. At the end of the 10 s, the belt is no longer slipping and pulley A has slowed to a final angular velocity of 250 rev/min. Determine the angular acceleration of each pulley and the final angular velocity of pulley C.

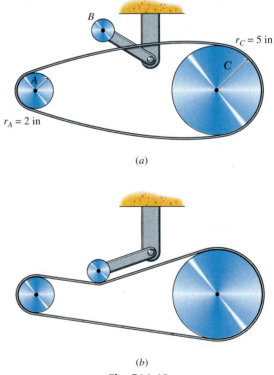

Fig. P14-19

14-20* It is desired to bring the two disks of Fig. P14-20 together without slipping. Initially, disk B is at rest and disk A is rotating at 600 rev/min. If disk A is given a constant angular deceleration of 3 rad/s² and disk B is given a constant angular acceleration of 5 rad/s², determine the time at which the two disks can be brought together without slipping and the angular velocity of each disk at that time.

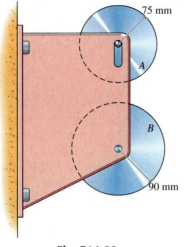

Fig. P14-20

14-21 It is desired to bring the two disks of Fig. P14-21 together without slipping. Initially, disk B is at rest and disk A is rotating at 750 rev/min. If disk A is given a constant angular deceleration of 5 rad/s² and disk B is given a constant angular acceleration of 8 rad/s², determine the time at which the two disks can be brought together without slipping and the angular velocity of each disk at that time.

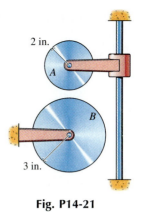

Fig. P14-21

14-22 A small block B rotates with the horizontal turntable A of Fig. P14-22. The block will begin to slip when its acceleration exceeds 0.6g. The turntable starts from rest and accelerates to 30 rev/min in 1 revolution, after which the angular velocity is constant. Determine the maximum distance r for which the block will not slip if

a. The angular acceleration α = constant.
b. The angular acceleration α decreases linearly with respect to θ from $\alpha = \alpha_0$ when $\omega = 0$ to $\alpha = 0$ after one revolution.

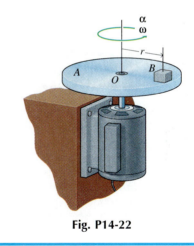

Fig. P14-22

14-5 GENERAL PLANE MOTION

A general plane motion is any plane motion for which lines in the body rotate but no point in the body is fixed. It will be seen that general plane motions are a superposition of translation and fixed axis rotation.

There are two general approaches to solving general plane motion problems: absolute motion analysis and relative motion analysis. In the first approach, geometric relationships are written that describe the constraints that act on the body and its interaction with other bodies. These relationships are then used to describe the location and motion of other points on the body. The second approach uses the concept of the relative motion of particles developed in Section 13-6. Since the distance between two points in a rigid body is fixed, the expressions for the relative velocity and relative acceleration take particularly simple forms that depend only on the angular velocity and angular acceleration of the body.

Either approach may be used to solve any particular problem. Some problems are easily described geometrically and are easily handled by the absolute motion approach. Problems that are not easily described geometrically are usually solved using the relative motion approach. For many problems the choice is a matter of personal preference.

14-5.1   Absolute Motion Analysis

Equations relating the angular motion of the rigid body and the motion of some point of the rigid body can be obtained from careful geometric analysis of the relationship between points and lines on a rigid body. The location of some point on the body is first obtained as a function of the angular orientation of the body. Then the time derivatives of this relationship give the velocity and acceleration of the point as functions of the angular orientation, the angular velocity, and the angular acceleration of the body.

Since the absolute motion approach relies totally on the geometric description of the body or bodies in a problem, no general formulas can be derived. Specific formulas must be derived for each specific problem, as illustrated in the following examples.

CONCEPTUAL EXAMPLE 14-1: ROTARY START-STOP MOTION

The number wheels in an odometer cannot be connected by simple gears. If they did, then they would all rotate continuously and would seldom have their numbers centered in the display windows. The tens digit, however, reads a constant value until the units digit reaches nine. Then the tens digit changes smoothly while the units digit rotates from nine to zero. So how does the odometer accomplish this start-stop motion?

SOLUTION

A sequence of Geneva wheels can create exactly the type of motion described. Each of the number wheels consists of three disks that are attached together and rotate as a single unit. The center disk is inscribed with the numbers that are seen through the windows of the odometer face. The disk on one side of each of the number wheels has ten slots, and the disk on the other side is a gear the same size as the Geneva wheel. Every time the number wheel rotates one full turn, the Geneva wheel also rotates one full turn. As the Geneva wheel completes its rotation, its pin enters one of the slots on the back side of the next number wheel and advances that wheel 1/10 turn. Of course, that number wheel is also connected to another Geneva wheel. After ten rotations of the first number wheel, the second number wheel will complete one rotation. The pin of the Geneva wheel attached to the second number wheel will advance the third number wheel 1/10 turn, and so on. The process continues for however many number wheels there are in the display.

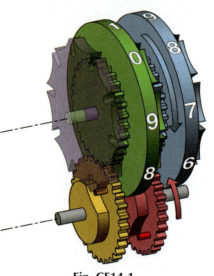

Fig. CE14-1

Derive an expression relating the position of a point on the rim of a wheel and the rotation of the wheel as it rolls without slipping on a stationary horizontal surface. Use the expression to

a. Give the velocity of the point as a function of θ and ω.
b. Show that the velocity of the point of contact between the wheel and the surface is instantaneously zero.
c. Give the acceleration of the point as a function of θ, ω, and α.
d. Show that the acceleration of the point of contact with the surface is normal to the surface and is not zero.

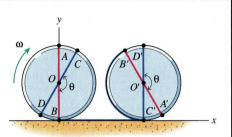

Fig. 14-14

SOLUTION

a. Let A, B, C, and D be points on the rim of the wheel as shown in Fig. 14-14. When the wheel rotates through an angle θ, the center of the wheel will move from O to O' and point C will rotate down to contact the surface at C'. The location of point A can be written in terms of the angle θ

$$\mathbf{r}_A = (x + r \sin \theta)\mathbf{i} + (r + r \cos \theta)\mathbf{j}$$

where $x = OO' = BC'$ is the distance traveled by the center of the wheel as it rotates. Since the wheel does not slip as it rolls, the distance BC' must equal the arc length $BC = r\theta$ and therefore

$$\mathbf{r}_A = (r\theta + r \sin \theta)\mathbf{i} + (r + r \cos \theta)\mathbf{j} \qquad \text{Ans.}$$

(The path described by this equation is called a cycloid and is drawn in Fig. 14-15a).

The velocity of point A is given by the time derivative of the position

$$\mathbf{v}_A = \dot{\mathbf{r}}_A = \frac{d\mathbf{r}_A}{d\theta} \frac{d\theta}{dt}$$
$$= r\omega[(1 + \cos \theta)\mathbf{i} - \sin \theta\, \mathbf{j}] \qquad \text{Ans.}$$

The x- and y-components of the velocity are drawn in Fig. 14-15b.

b. Evaluating the velocity of A when $\theta = 180°$ (and A is in contact with the surface) gives

$$\mathbf{v}_A(180°) = r\omega[(1 - 1)\mathbf{i} - 0\mathbf{j}] = \mathbf{0} \qquad \text{Ans.}$$

Therefore the contact point is instantaneously at rest: it is an instantaneous center of zero velocity.

c. The acceleration of A is given by the time derivative of the velocity

$$\mathbf{a}_A = \dot{\mathbf{v}}_A = r[\alpha(1 + \cos \theta) - \omega^2 \sin \theta]\mathbf{i}$$
$$- r[\alpha \sin \theta + \omega^2 \cos \theta]\mathbf{j} \qquad \text{Ans.}$$

d. Evaluating the acceleration of A when $\theta = 180°$ (and A is in contact with the surface) gives

$$\mathbf{a}_A = r\omega^2\mathbf{j} \qquad \text{Ans.}$$

Therefore, the contact point is not an instantaneous center of zero acceleration. The acceleration of the contact point is toward the center of the wheel and is perpendicular to the surface on which the wheel rolls.

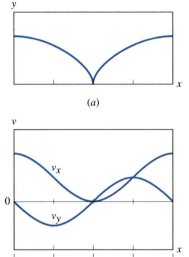

(a)

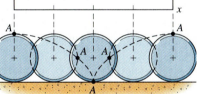

(b)

Fig. 14-15

As the point A comes in contact with the surface, both its x- and y-components of velocity become zero and the contact point is instantaneously at rest. Although the x-component of acceleration is also zero (that is the slope of v_x is zero), the y-component of acceleration is not zero (the slope of v_y is not zero).

EXAMPLE PROBLEM 14-6

The 3-m-long ladder AB slides along a corner as shown in Fig. 14-16. At some instant of time, the lower end of the ladder is 1.2 m from the corner and is moving to the right at a constant speed of 0.5 m/s. For this instant, determine

a. The angular velocity and angular acceleration of the ladder.
b. The acceleration $\mathbf{a}_G$ of the mass center of the ladder.

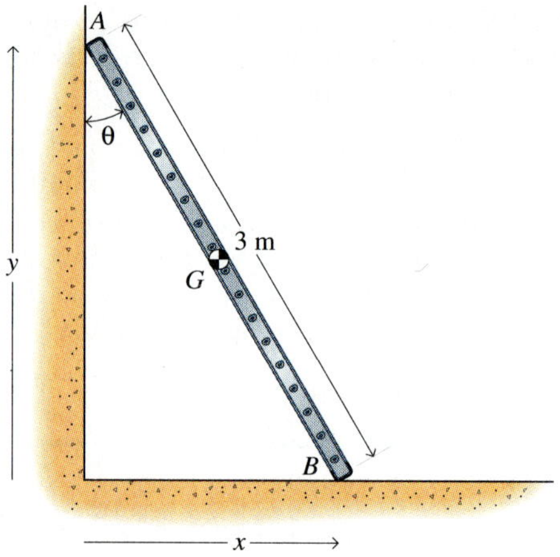

SOLUTION

a. There are four fairly obvious geometrical relationships that could be used in this problem:

$$x^2 + y^2 = 3^2 \qquad (a)$$
$$\sin \theta = x/3 \qquad (b)$$
$$\cos \theta = y/3 \qquad (c)$$
$$\tan \theta = x/y \qquad (d)$$

Equation b will be used because it relates the position of end B of the ladder (for which information is given) and the angular position of the ladder (for which information is desired). Rewriting Eq. b

$$x = 3 \sin \theta$$

and differentiating with respect to time gives

$$\dot{x} = 3\dot{\theta} \cos \theta$$
$$\ddot{x} = 3\ddot{\theta} \cos \theta + 3\dot{\theta}(-\dot{\theta} \sin \theta) = 3(\ddot{\theta} \cos \theta - \dot{\theta}^2 \sin \theta)$$

Then, when $x = 1.2$ m, $\dot{x} = 0.5$ m/s, and $\ddot{x} = 0$ m/s²,

$$\theta = \sin^{-1}(1.2/3) = 23.578°$$
$$\omega = \dot{\theta} = 0.18185 \text{ rad/s} \cong 0.1819 \text{ rad/s} \text{↰} \qquad \text{Ans.}$$
$$\alpha = \ddot{\theta} = 0.01443 \text{ rad/s}^2 \text{↰} \qquad \text{Ans.}$$

b. By similar triangles, the location of the mass center is

$$x_G = x/2 \qquad y_G = y/2$$

Differentiating these equations with respect to time gives

$$\dot{x}_G = \dot{x}/2 \qquad \dot{y}_G = \dot{y}/2$$
$$\ddot{x}_G = \ddot{x}/2 \qquad \ddot{y}_G = \ddot{y}/2$$

Now that x, $\dot{x}$, $\ddot{x}$, θ, $\dot{\theta}$, and $\ddot{\theta}$ are all known, any of Eqs. a, c, or d can be used to find $\dot{y}$ and $\ddot{y}$. Differentiating Eq. a with respect to time gives

$$2x\dot{x} + 2y\dot{y} = 0$$
$$x\ddot{x} + \dot{x}^2 + y\ddot{y} + \dot{y}^2 = 0$$

Then, when $x = 1.2$ m, $\dot{x} = 0.5$ m/s, and $\ddot{x} = 0$ m/s²,

$$y = \sqrt{3^2 - 1.2^2} = 2.74955 \text{ m}$$
$$\dot{y} = -0.21822 \text{ m/s} \cong 0.218 \text{ m/s}↓$$
$$\ddot{y} = -0.10824 \text{ m/s}^2 \cong 0.1082 \text{ m/s}^2↓$$

Finally, the acceleration of the mass center is

$$\mathbf{a}_G = \ddot{x}_G \mathbf{i} + \ddot{y}_G \mathbf{j} = (\ddot{x}/2)\mathbf{i} + (\ddot{y}/2)\mathbf{j}$$
$$= -0.05412\mathbf{j} \text{ m/s}^2 \cong 0.0541 \text{ m/s}^2↓ \qquad \text{Ans.}$$

Fig. 14-16

Note that a positive angular velocity $\dot{\theta}$ means that the angle θ is increasing. Therefore, the ladder is rotating counterclockwise.

It is interesting to note that although $\ddot{x} = 0$, neither the angular acceleration $\ddot{\theta}$ nor the accelerations $\ddot{y}$ or $\mathbf{a}_G$ are zero.

EXAMPLE PROBLEM 14-7

The slider-crank mechanism shown in Fig. 14-17 is an idealization of an automobile crank shaft, connecting rod, and piston. The lengths of the crank throw and connecting rod are $L_{AB} = 75$ mm and $L_{BC} = 175$ mm, respectively, and at the instant shown, $\theta = 70°$. If the crankshaft is rotating at a constant rate of $\omega_{AB} = \dot{\theta} = 4800$ rev/min, determine

a. The angular velocity $\dot{\phi}$ and angular acceleration $\ddot{\phi}$ of the connecting rod.
b. The acceleration a_C of the piston.

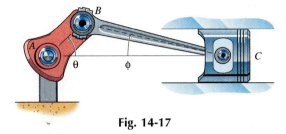

Fig. 14-17

SOLUTION

a. Applying the law of sines to triangle ABC (Fig. 14-18)

$$\frac{\sin \theta}{175} = \frac{\sin \phi}{75}$$

gives a relationship between the two angles θ and ϕ and the two fixed lengths L_{AB} and L_{BC}

$$75 \sin \theta = 175 \sin \phi \tag{a}$$

Differentiating Eq. a with respect to time gives

$$75\dot{\theta} \cos \theta = 175\dot{\phi} \cos \phi$$
$$75\ddot{\theta} \cos \theta - 75\dot{\theta}^2 \sin \theta = 175\ddot{\phi} \cos \phi - 175\dot{\phi}^2 \sin \phi$$

Then, when $\theta = 70°$, $\dot{\theta} = 4800$ rev/min $= 160\pi$ rad/s, and $\ddot{\theta} = 0$,

$$\phi = \sin^{-1}\left(\frac{75 \sin 70°}{175}\right) = 23.749°$$
$$\omega_{BC} = \dot{\phi} = 80.495 \text{ rad/s} \cong 80.5 \text{ rad/s} \nearrow \qquad \text{Ans.}$$
$$\alpha_{BC} = \ddot{\phi} = -108,316 \text{ rad/s}^2 \cong 108,300 \text{ rad/s}^2 \searrow \qquad \text{Ans.}$$

b. Also, the length of the side AC can be written in terms of the two angles θ and ϕ and the two fixed lengths L_{AB} and L_{BC}

$$x_C = 75 \cos \theta + 175 \cos \phi \tag{b}$$

Differentiating Eq. b with respect to time gives

$$\dot{x}_C = -75\dot{\theta} \sin \theta - 175\dot{\phi} \sin \phi$$
$$\ddot{x}_C = -75\ddot{\theta} \sin \theta - 75\dot{\theta}^2 \cos \theta - 175\ddot{\phi} \sin \phi - 175\dot{\phi}^2 \cos \phi$$

Then, when $\theta = 70°$, $\dot{\theta} = 4800$ rev/min $= 16\pi$ rad/s, $\ddot{\theta} = 0$, $\phi = 23.749°$, $\dot{\phi} = 80.495$ rad/s, and $\ddot{\phi} = -108,316$ rad/s^2,

$$a_C = \ddot{x}_C = 114,726 \text{ mm/s}^2 \cong 114.7 \text{ m/s}^2 \rightarrow \qquad \text{Ans.}$$

(The positive directions for velocity $\dot{x}_C$ acceleration $\ddot{x}_C$ are the same as the direction in which x_C increases.)

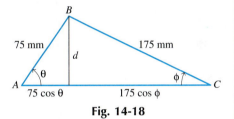

Fig. 14-18

Note that a positive angular velocity $\dot{\phi}$ means that the angle ϕ is increasing. Therefore, the connecting rod is rotating clockwise. A negative angular acceleration $\ddot{\phi}$ means that the angular acceleration is opposite the direction that ϕ is increasing.

Again note that although the crankshaft rotates at a constant angular rate $\dot{\theta} = 160\pi$ rad/s = constant, neither the angular acceleration of the connecting rod $\ddot{\phi}$ nor the acceleration of the piston a_C are zero.

PROBLEMS

Introductory Problems

14-23* Rod CD of Fig. P14-23 moves in the horizontal direction, causing lever AB to rotate. The circular disk has a radius $r = 2$ in., and at the instant shown $x = 5$ in. and the velocity of the plunger is 15 ft/s to the right. Determine the angular velocity ω_{AB} of the lever AB.

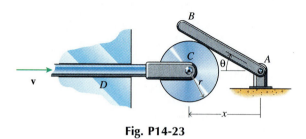

Fig. P14-23

14-24* Rotation of the circular cam of Fig. P14-24 causes the plunger to move up and down. The radius of the cam is $r = 50$ mm, and it is mounted on a shaft $b = 35$ mm from its center. At the instant shown, the cam is rotating at a constant angular speed of $\omega = 15$ rad/s and $\theta = 60°$. Determine the velocity and acceleration of the plunger at this instant.

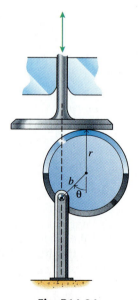

Fig. P14-24

14-25 The mechanism shown in Fig. P14-25 is used to convert the rotary motion of the arm AB into translational motion of the plunger CD. Express the velocity and acceleration of the plunger in terms of θ, ω, α, and b, the angular position, angular velocity, angular acceleration, and length of the rod AB.

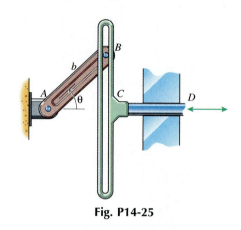

Fig. P14-25

14-26 Sliders A and B of Fig. P14-26 are constrained to move in vertical and horizontal slots, respectively, and are connected by a rigid rod 800 mm long. At the instant shown $\theta = 75°$ and slider B is moving to the right with a constant speed of 25 mm/s. Determine the angular velocity ω_{AB} and angular acceleration α_{AB} of the bar AB at this instant.

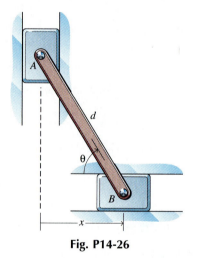

Fig. P14-26

14-27* The 7-ft-long rod AB of Fig. P14-27 slides on a 4-ft-high step. If end B of the rod is made to move to the right at a constant speed of 0.5 ft/s, determine the angular velocity ω_{AB} of the rod at the instant when $x = 2$ ft.

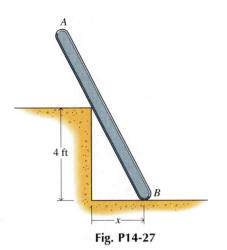

Fig. P14-27

14-28* As the circular cam of Fig. P14-28 rotates, the follower rod moves up and down. The radius of the cam is $r_1 = 40$ mm, and is mounted on a shaft $b = 25$ mm from its center. The radius of the smaller disk is $r_2 = 30$ mm. At the instant shown, the cam is rotating at a constant angular speed of $\omega = 10$ rad/s and $\theta = 30°$. Determine the velocity and acceleration of the follower rod at this instant.

Fig. P14-28

Intermediate Problems

14-29* A control rod BC is attached to a rotating crank AB as shown in Fig. P14-29. At the instant shown, the control rod BC is sliding through the pivoted guide D at a rate of 15 in./s. Determine the angular velocity ω_{AB} of the crank at this instant.

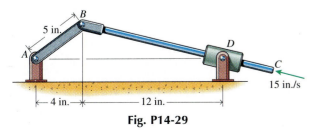

Fig. P14-29

14-30 In the mechanism of Fig. P14-30, arm AB is rotating counterclockwise with an angular velocity of 2 rad/s at the instant shown. Determine the angular velocity ω_{CD} of arm CD and the speed v of the slider along arm AB at this instant.

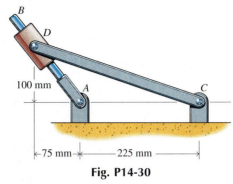

Fig. P14-30

14-31 As the 6-in.-long arm AB of the mechanism shown in Fig. P14-31 oscillates, the pin B slides up and down in the slot of arm CD. Given that $\theta = \cos \pi t$ rad where t is in seconds, determine the angular velocity ω_{CD} of the arm CD when $t = \frac{1}{2}$ s.

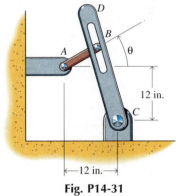

Fig. P14-31

14-32 The wheel of Fig. P14-32 is rotating clockwise at a constant rate of 120 rev/min. Pin D is fastened to the wheel at a point 125 mm from its center and slides in the slot of arm AB. Determine the angular velocity ω_{AB} of arm AB at the instant shown.

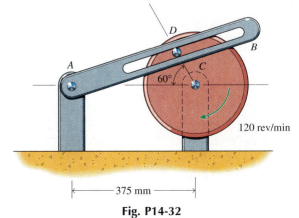

Fig. P14-32

14-33* The crank AB of Fig. P14-33 has a constant counterclockwise angular velocity of 60 rev/min. If at the instant shown $\theta = 40°$, determine the angular velocity ω_{BC} of member BC and the rate of slip at the contact point D (the relative velocity between the point on the arm BC and the stationary pivot point).

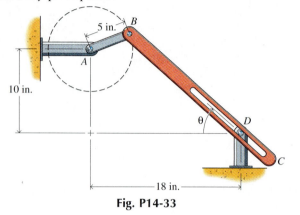

Fig. P14-33

14-34 The Geneva mechanism of Fig. P14-34 is used to produce intermittent motion. Both wheels rotate on fixed axles through their centers. If wheel A has a constant clockwise angular velocity of 30 rev/min, determine the angular velocity of wheel B when $\theta = 30°$.

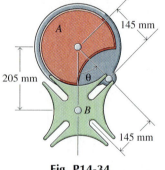

Fig. P14-34

14-35 Movement of the solenoid plunger shown in Fig. P14-35 causes a gear to rotate. If at the instant shown the angular velocity of the gear is $\omega_0 = 4$ rad/s counterclockwise, determine the angular velocity ω_{AB} of rod AB and the velocity $\mathbf{v}_A$ of the plunger.

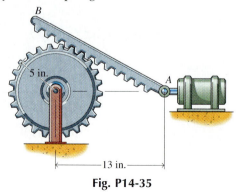

Fig. P14-35

14-36* As the 400-mm-long arm BC of the mechanism shown in Fig. P14-36 oscillates, the collar C slides up and down the arm AD. Given that $\phi = 1.5 \sin \pi t$ rad where t is in seconds, determine the rotation rate ω_{AD} of arm AD and the speed v of the slider along arm AD when $t = \frac{1}{3}$ s.

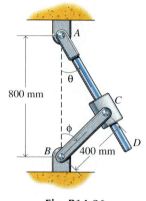

Fig. P14-36

Challenging Problems

14-37* In the mechanism of Fig. P14-37, arm AB is rotating counterclockwise with a constant angular velocity of 2 rad/s at the instant shown. Determine the angular acceleration α_{CD} of arm CD and the acceleration a of the slider along arm CD at this instant.

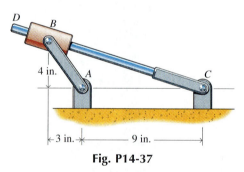

Fig. P14-37

14-38* As the 75-mm-long arm AB of the mechanism shown in Fig. P14-38 oscillates, the pin B slides up and down in the slot of arm CD. Given that $\theta = 3 \sin 2\pi t$ rad where t is in seconds, determine the angular velocity ω_{CD} and the angular acceleration α_{CD} of the arm CD when $t = 0.1$ s.

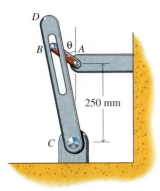

Fig. P14-38

14-39 The wheel of Fig. P14-39 is rotating clockwise at a constant rate of 120 rev/min. Pin D is fastened to the wheel at a point 5 in. from its center and slides in the slot of arm AB. Determine the angular acceleration α_{AB} of arm AB at the instant shown.

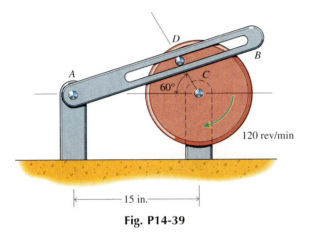

Fig. P14-39

14-40 The Geneva mechanism of Fig. P14-40 is used to produce intermittent motion. Both wheels rotate on fixed axles through their centers. If wheel A has a constant clockwise angular velocity of 30 rev/min, determine the angular acceleration of wheel B when $\theta = 30°$.

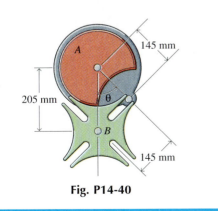

Fig. P14-40

Computer Problems

C14-41 A Geneva wheel (Fig. P14-41) is a mechanism often used to create intermittent motion. The size and location of the input wheel A is such that the pin P enters and exits the slots smoothly. If the input wheel rotates with a constant angular speed of $\omega_A = 5$ rad/s

a. Compute and plot the angular position θ_B of the Geneva wheel versus θ_A, the angular position of the input wheel, for one complete revolution of the input wheel (that is, for $0° < \theta_A < 360°$).

b. Compute and plot the angular velocity ω_B of the Geneva wheel versus θ_A for one complete revolution of the input wheel.

c. Compute and plot the angular acceleration α_B of the Geneva wheel versus θ_A for one complete revolution of the input wheel.

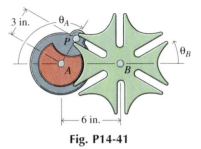

Fig. P14-41

C14-42 The slider-crank mechanism of Fig. P14-42 is an idealization of an automobile crank shaft, connecting rod, and piston. If the crankshaft is rotating at a constant rate $\dot{\theta}$,

a. Write expressions for a_C, the acceleration of the piston; $\dot{\phi}$, the angular velocity of the connecting rod; and $\ddot{\phi}$, the angular acceleration of the connecting rod in terms of θ, $\dot{\theta}$, ℓ_{AB}, and ℓ_{BC}.

b. Using $\ell_{AB} = 50$ mm, $\ell_{BC} = 650$ mm, and $\dot{\theta} = 4800$ rev/min, plot a_C, $\dot{\phi}$, and $\ddot{\phi}$ as functions of θ ($0 < \theta < 360°$).

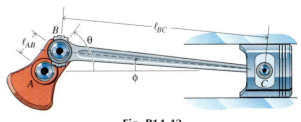

Fig. P14-42

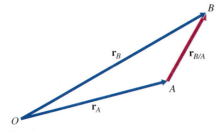

Figure 14-19 The positions of particles A and B relative to the fixed origin O and the position of particle B relative to particle A are related by the vector law of addition.

14-5.2 Relative Velocity

If A and B are any two particles, their positions are related by the triangle law of addition for vectors (Fig. 14-19)

$$\mathbf{r}_B = \mathbf{r}_A + \mathbf{r}_{B/A} \qquad (14\text{-}8)$$

where $\mathbf{r}_A$ and $\mathbf{r}_B$ are the absolute positions of the particles A and B, respectively, and $\mathbf{r}_{B/A}$ is the position of B relative to A. The time derivative of Eq. 14-8 gives

$$\mathbf{v}_B = \mathbf{v}_A + \mathbf{v}_{B/A} \qquad (14\text{-}9a)$$

where $\mathbf{v}_A$ is the absolute velocity (measured relative to a fixed coordinate system) of particle A, $\mathbf{v}_B$ is the absolute velocity of particle B, and $\mathbf{v}_{B/A}$ is the relative velocity of particle B (measured relative to particle A). Equations 14-8 and 14-9a apply to any two particles—whether they are part of a rigid body or not.

If particles A and B are two particles in a rigid body, however, then the distance between them is constant and particle B appears to travel a circular path around particle A. Therefore, the relative velocity $\mathbf{v}_{B/A}$ is given by (Eqs. 14-6)

$$\mathbf{v}_{B/A} = r_{B/A}\omega\mathbf{e}_t = \omega\mathbf{k} \times \mathbf{r}_{B/A} \qquad (14\text{-}9b)$$

where $\mathbf{r}_{B/A} = \mathbf{r}_B - \mathbf{r}_A$, $\mathbf{e}_t$ is a unit vector tangent to the relative motion (tangent to the circle centered at A), and ω is the angular velocity of the body. Then

$$\mathbf{v}_B = \mathbf{v}_A + \mathbf{v}_{B/A} = \mathbf{v}_A + r_{B/A}\omega\mathbf{e}_t$$
$$= \mathbf{v}_A + \omega\mathbf{k} \times \mathbf{r}_{B/A} \qquad (14\text{-}9c)$$

Therefore, the velocity of particle B consists of the sum of two parts: $\mathbf{v}_A$, which represents a translation of the entire body with particle A, and $r_{B/A}\omega\mathbf{e}_t$, which represents fixed axis rotation of the body about an axis through A (Fig. 14-20). In words, Eq. 14-9c says: *The velocity of any point B of a rigid body consists of a translation of the entire rigid body with point A plus a rotation of the rigid body about an axis through A.*

The relative velocity equation (Eq. 14-9c) is a vector equation that, for planar motion, has two independent scalar components. Therefore, the relative velocity equation can be used to find the two components

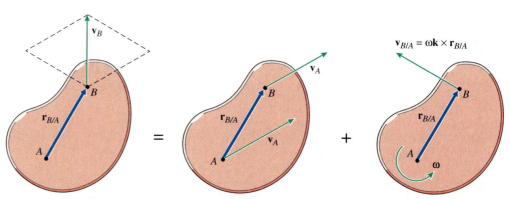

Figure 14-20 The velocity of any point B of a rigid body consists of a translation of the entire body with point A plus a rotation of the rigid body about an axis through A.

of the velocity of some point $\mathbf{v}_B$ when the angular velocity of the body ω and the velocity of some other point in the body $\mathbf{v}_A$ are known. This equation can also be solved when the directions of the velocities $\mathbf{v}_A$ and $\mathbf{v}_B$ are known (for example, if they slide along fixed guides) and one of the three magnitudes v_A, v_B, or ω is given.

When two or more rigid bodies are pinned together as in Fig. 14-21, relative velocity equations can be written for each of the bodies separately. One of the points used in each of the equations should be the common point (point B of Fig. 14-21) connecting the two bodies; its velocity will be the same for each body. The other point in each equation should be some other point (A or C) whose velocity is known or is to be found. Then the velocities and angular velocities of the bodies can be related by equating the two expressions for the velocity of the common point

$$\mathbf{v}_B = \mathbf{v}_A + \mathbf{v}_{B/A} = \mathbf{v}_C + \mathbf{v}_{B/C} \qquad (a)$$

or

$$\mathbf{v}_A + \omega_{AB}\mathbf{k} \times \mathbf{r}_{B/A} = \mathbf{v}_C + \omega_{BC}\mathbf{k} \times \mathbf{r}_{B/C} \qquad (b)$$

Equation b can be solved for any two unknowns if the other quantities are given.

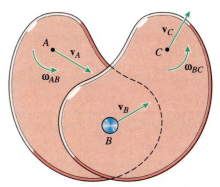

Figure 14-21 When two bodies are connected at some point B, the velocity of that common point can be used in the relative velocity equations for both bodies (and it will have the same value in both equations).

CONCEPTUAL EXAMPLE 14-2: LINEAR START-STOP MOTION

The film advance mechanism in a movie projector cannot simply pull the film through the projector at a constant rate. That would create a blurred image on the movie screen. The film advance mechanism must quickly pull the film from one frame to the next but must then allow the film to stay in one place for a little while to show the picture before advancing the film to the next picture. So how does the film advance mechanism accomplish this start-stop motion?

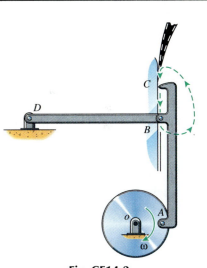

Fig. CE14-2

SOLUTION

A four-bar linkage such as the control wheel OA, the advance arm ABC and the rocker arm BD can create exactly the type of motion described. As the control wheel rotates from about 40° before the position shown to about 40° past the position shown, point A moves nearly straight downward and the pawl C of the advance arm ABC pulls the film down one frame. As the point A moves around to the back side of the control wheel, the pawl pivots away from the film and recycles up to hook and pull the film again.

Although the film is stationary for about 75 percent of each control wheel cycle, the film is still moving and the projected image is changing during about 25 percent of each control wheel cycle. This movement of the film is concealed by a shutter which covers the light source and hence prevents any image from being projected while the film is moving. This sequence of still images separated by brief blackouts is readily visible in some slow-motion projectors. For a normal projection rate of 30 frames per second, however, this blackout lasts only about 1/125 second per frame and is not normally perceived by film patrons.

The 3-m-long ladder AB slides along a corner as shown in Fig. 14-22. When the angle $\theta = 30°$, the bottom end of the ladder is moving to the right with a constant speed of 2.0 m/s. Determine the velocity of the top end of the ladder and the angular velocity of the ladder at this instant.

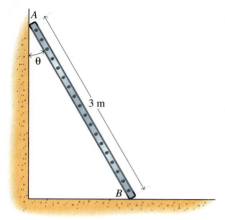

Fig. 14-22

SOLUTION

A coordinate system is chosen with x to the right and y up. It will be assumed that ends A and B of the ladder remain in contact with the vertical and horizontal surfaces, respectively. Then end B of the ladder must slide along the horizontal surface and its velocity can have no y-component. Its velocity is given as $\mathbf{v}_B = 2\mathbf{i}$ m/s. Similarly, end A of the ladder must slide along the vertical surface and its velocity can have no x-component. Although the speed of A is unknown, the velocity of A must have the form $\mathbf{v}_A = v_A\mathbf{j}$. Finally, the relative velocity equation

$$\mathbf{v}_A = \mathbf{v}_B + \mathbf{v}_{A/B}$$

gives

$$v_A\mathbf{j} = 2\mathbf{i} + \mathbf{v}_{A/B} \tag{a}$$

But from Fig. 14-23 the relative velocity is given by

$$\mathbf{v}_{A/B} = 3\omega\mathbf{e}_t = 3\omega(-\cos 30°\mathbf{i} - \sin 30°\mathbf{j}) \tag{b}$$

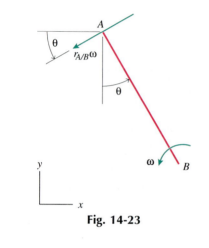

Fig. 14-23

Then the x-component of Eq. a gives

$$0 = 2 - 3\omega \cos 30°$$

or

$$\omega = 0.770 \text{ rad/s} \circlearrowleft \qquad \text{Ans.}$$

and the y-component of Eq. a gives

$$v_A = -(3)(0.770) \sin 30° = -1.155 \text{ m/s}$$

or

$$\mathbf{v}_A = 1.155 \text{ m/s} \downarrow \qquad \text{Ans.}$$

   Alternatively, the relative velocity term can be computed using the vector cross product

$$\mathbf{v}_{A/B} = \boldsymbol{\omega} \times \mathbf{r}_{A/B} = \omega\mathbf{k} \times (-3 \sin 30°\mathbf{i} + 3 \cos 30°\mathbf{j})$$
$$= -3\omega \sin 30°\mathbf{j} - 3\omega \cos 30°\mathbf{i}$$

which is identical to Eq. b.

The relative velocity $\mathbf{v}_{A/B}$ has magnitude $v_{A/B} = r_{A/B}\omega$, is perpendicular to the relative position vector $\mathbf{r}_{A/B}$, and points in the direction that ω rotates the body.

EXAMPLE PROBLEM 14-9

The wheel of the slider-crank mechanism shown in Fig. 14-24 is rotating counterclockwise at a constant rate of 10 rad/s. Determine the velocity of the slider $\mathbf{v}_B$ and the angular velocity of the crank arm ω_{AB} when $\theta = 60°$.

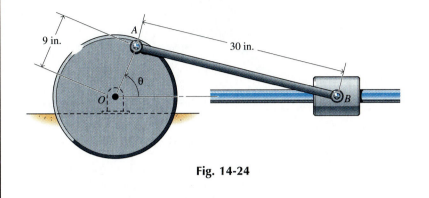

Fig. 14-24

SOLUTION

The wheel and the crank arm AB are separate rigid bodies connected at point A. The velocity of O, the axle of the wheel, is zero, and the velocity of B, the slider, has only a horizontal component. The velocity of the common point is written relative to the velocities of these points and the angular velocities of each of the bodies.

$$\mathbf{v}_A = \mathbf{v}_B + \mathbf{v}_{A/B} = \mathbf{v}_O + \mathbf{v}_{A/O} \qquad (a)$$

Using a coordinate system with x to the right and y up and referring to Fig. 14-25, gives the relative velocity terms as

$$\mathbf{v}_{A/B} = r_{A/B}\,\omega_{AB}\mathbf{e}_t = 30\omega_{AB}(-\sin\phi\,\mathbf{i} - \cos\phi\,\mathbf{j}) \text{ in./s}$$
$$\mathbf{v}_{A/O} = r_{A/O}\omega_{OA}\mathbf{e}_t = 9\omega_{OA}(-\sin 60°\,\mathbf{i} + \cos 60°\,\mathbf{j}) \text{ in./s}$$

where the angle ϕ is determined using the law of sines (Fig. 14-25c)

$$\frac{\sin\phi}{9} = \frac{\sin 60°}{30}$$

or

$$\phi = 15.06°$$

Then the x- and y-components of Eq. a

$$v_B - 30\omega_{AB}\sin 15.06° = -(9)(10)\sin 60°$$
$$-30\omega_{AB}\cos 15.06° = (9)(10)\cos 60°$$

are solved simultaneously to get

$$\omega_{AB} = -1.553 \text{ rad/s}$$
$$= 1.553 \text{ rad/s} \curvearrowright \qquad \text{Ans.}$$

and

$$v_B = -90.0 \text{ in./s}$$

or

$$\mathbf{v}_B = 90.0 \text{ in./s} \leftarrow \qquad \text{Ans.}$$

Fig. 14-25

> The relative velocity $\mathbf{v}_{A/B}$ (and $\mathbf{v}_{A/O}$) has magnitude $v_{A/B} = r_{A/B}\omega_{AB}$ (and $v_{A/O} = r_{A/O}\omega_{OA}$), is perpendicular to the relative position vector $\mathbf{r}_{A/B}$ (and $\mathbf{r}_{A/O}$), and points in the direction that ω_{AB} (and ω_{OA}) rotates the body.

PROBLEMS

Introductory Problems

14-43* Sliders *A* and *B* of Fig. P14-43 are constrained to move in vertical and horizontal slots, respectively, and are connected by a 32-in.-long rigid rod. At the instant shown, $\theta = 75°$ and slider *B* is moving to the right with a constant speed of 2 in./s. Determine the velocity $\mathbf{v}_A$ of slider *A* and the angular velocity ω_{AB} of bar *AB* at this instant.

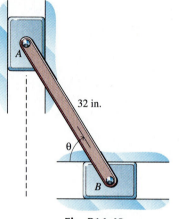

Fig. P14-43

14-44* One end of the 2-m-long bar *AB* shown in Fig. P14-44 is attached to a collar which slides on a smooth vertical shaft. The other end of the bar slides on a smooth horizontal surface. At the instant shown, $\theta = 30°$ and collar *A* is moving downward with a speed of 0.4 m/s. Determine the velocity $\mathbf{v}_B$ of end *B* and the angular velocity ω_{AB} of bar *AB* at this instant.

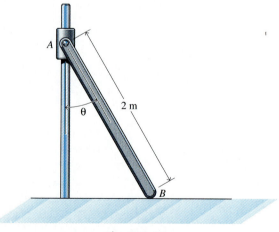

Fig. P14-44

14-45 One end of the 24-in.-long bar *AB* shown in Fig. P14-45 is attached to a collar that slides on a smooth shaft. The other end of the bar is attached to a slider that moves in a smooth horizontal slot. At the instant shown, the slider *B* is moving to the left with a speed of 1.5 in./s. Determine the velocity $\mathbf{v}_A$ of the collar *A* and the angular velocity ω_{AB} of bar *AB* at this instant.

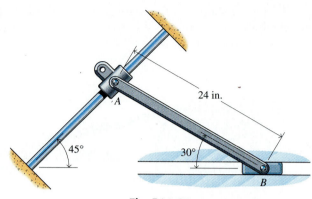

Fig. P14-45

14-46 The slender bar *AB* shown in Fig. P14-46 is attached to small wheels that roll on the horizontal and inclined surfaces. At the instant shown, the angular velocity of the bar is $\omega_{AB} = 0.2$ rad/s, clockwise. Determine the velocities of the pins *A* and *B* at this instant.

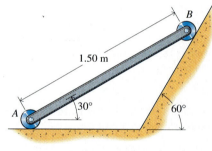

Fig. P14-46

14-47 One end of the 18-in.-long bar *AB* shown in Fig. P14-47 is attached to a slider which moves in a vertical slot. The other end is attached to a wheel that rolls on a horizontal surface. At the instant shown, the center of the wheel is moving to the right with a speed of 5 in./s. Determine the angular velocity ω_A of the wheel and the velocity $\mathbf{v}_B$ of the slider *B* at this instant.

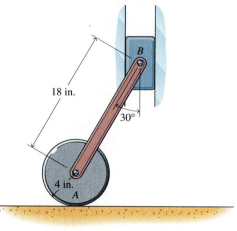

Fig. P14-47

14-48* The piston of Fig. P14-48 is connected to the crankshaft by a 650-mm-long connecting rod. At the instant shown, the angular velocity of the crankshaft is 360 rev/min clockwise. Determine the angular velocity ω_{BC} of the connecting rod and the velocity $\mathbf{v}_C$ of the piston at this instant.

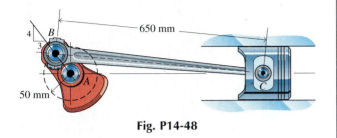

Fig. P14-48

Intermediate Problems

14-49* Movement of the solenoid plunger shown in Fig. P14-49 causes a gear to rotate. At the instant shown, the angular velocity of the gear is $\omega_O = 4$ rad/s counterclockwise. Determine the angular velocity ω_{AB} of the rod AB and the velocity $\mathbf{v}_A$ of the plunger at this instant.

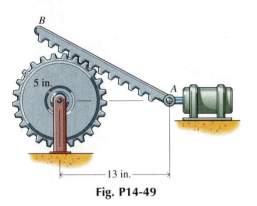

Fig. P14-49

14-50* The arm BC of Fig. P14-50 is attached to the sliding block and is made to rotate clockwise at a constant rate of 2 rev/min. Determine the velocity of the block $\mathbf{v}_C$ and the angular velocity ω_{AB} of the link AB at the instant shown when $\theta = 75°$.

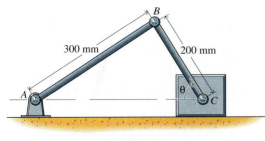

Fig. P14-50

14-51 The mechanism shown in Fig. P14-51 is a simplified sketch of a printing press. As the crank arm AB rotates ($\dot{\theta} = 5$ rev/min = constant) the printing drum C moves back and forth across the paper. For the instant shown ($\theta = 50°$), determine the angular velocity ω_{BC} of arm BC and the velocity $\mathbf{v}_C$ of the printing drum.

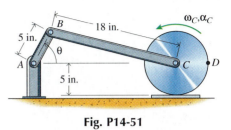

Fig. P14-51

14-52 At the instant shown, roller C of Fig. P14-52 has a velocity of 250 mm/s up the channel. Determine the angular velocities of both bars and the velocity of pin B at this instant.

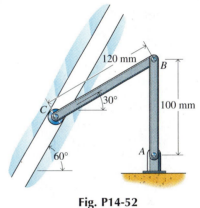

Fig. P14-52

14-53 A control rod BC is attached to a rotating crank AB as shown in Fig. P14-53. At the instant shown, the control rod BC is sliding through the pivoted guide D at a rate of 15 in./s. Determine the angular velocity ω_{AB} of the crank AB and the velocity $\mathbf{v}_B$ of pin B at this instant.

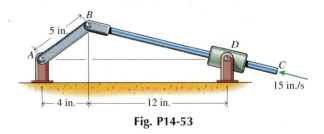

Fig. P14-53

14-54* The 400-mm-diameter wheel shown in Fig. P14-54 rolls without slipping on the horizontal surface. Bar AB is 750 mm long and is attached to the wheel by a smooth pin 150 mm from the center. At the instant shown, the center of the wheel has a velocity of 1.5 m/s to the left. Determine the angular velocity ω_{AB} of the bar AB and the velocity $\mathbf{v}_A$ of pin A at this instant.

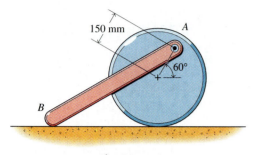

Fig. P14-54

14-55* The wheel shown in Fig. P14-55 rolls without slipping on the horizontal surface. At the instant shown, the wheel has an angular velocity of 6 rad/s clockwise. Determine the angular velocity of both links at this instant.

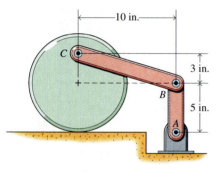

Fig. P14-55

14-56 The height of the platform CD shown in Fig. P14-56 is adjusted remotely using the hydraulic cylinder attached at A. At the instant shown, $d = 200$ mm and A has a velocity to the left of 50 mm/s. Determine the angular velocity ω_{AC} of the arm ABC and the velocity $\mathbf{v}_C$ of the platform CD at this instant.

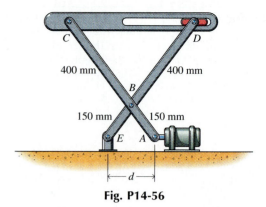

Fig. P14-56

14-57 The handle of the mechanism shown in Fig. P14-57 is controlled using the hydraulic cylinder attached at B. At the instant shown, B has a downward velocity of 5 in./s. Determine the angular velocity of bar AB and the velocity of point A at this instant.

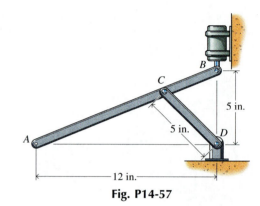

Fig. P14-57

14-58* In the four-bar linkage shown in Fig. P14-58, control link AB has a constant counterclockwise angular velocity of 100 rev/min. Bar BC is horizontal and bar CD is vertical when $\theta = 90°$. Determine the angular velocity ω_{BC} of bar BC and the velocity $\mathbf{v}_E$ of point E when $\theta = 90°$.

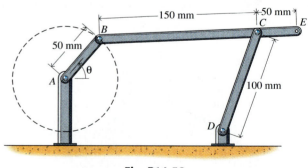

Fig. P14-58

Challenging Problems

14-59* The centers of the two gears shown in Fig. P14-59 are connected by smooth pins to the arm *ABC*, which rotates at a constant angular velocity of $\omega_{ABC} = 5$ rad/s clockwise about the fixed point *A*. The larger gear rolls on the inside of a fixed toothed drum. Determine the angular velocity of each of the gears and the velocity of tooth *D* for the position shown.

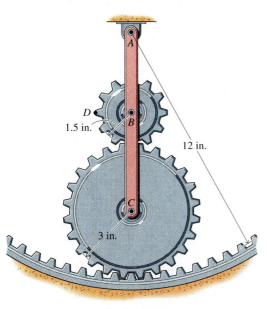

Fig. P14-59

14-60* The gears of an automatic transmission consist of a ring gear *R*, a sun gear *S*, three (or more) equal-sized planet gears *P*, and a planet carrier *C*, which is pin connected to the center of the planet gears (Fig. P14-60). Various gear ratios (ratio to the output angular velocity ω_{out} to the input angular velocity ω_{in}) are achieved by holding one of the parts fixed and driving the others. Determine the gear ratio of the transmission shown if the ring gear is held fixed, the sun gear is the input ($\omega_{in} = \omega_S$), and the planet carrier is the output ($\omega_{out} = \omega_C$).

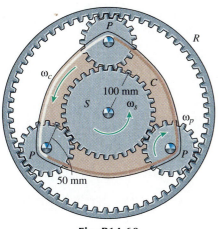

Fig. P14-60

14-61 The centers of the three gears shown in Fig. P14-61 are connected by smooth pins to the arm *ABC*, which rotates with an angular velocity of $\omega_{ABC} = 5$ rad/s counterclockwise. If the larger gear is fixed and does not rotate, determine the angular velocity of the smaller gears and the velocity of tooth *D* for the position shown.

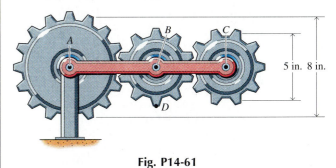

Fig. P14-61

14-62 In the four-bar linkage shown in Fig. P14-62, control link *AB* has a constant counterclockwise angular velocity of 100 rev/min. Bar *BC* is horizontal and bar *CD* is vertical when $\theta = 90°$. Determine the angular velocity ω_{BC} of bar *BC* and the velocity $\mathbf{v}_E$ of point *E* when $\theta = 0°$.

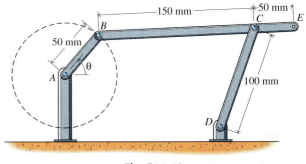

Fig. P14-62

14-63* The Geneva wheel mechanism of Fig. P14-63 is used to produce intermittent motion. Both wheels rotate on fixed axles through their centers. If wheel *A* has a constant clockwise angular velocity of $\omega_A = 30$ rev/min, determine the angular velocity ω_B of wheel *B* when $\theta = 30°$.

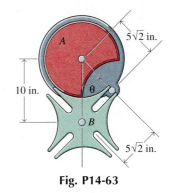

Fig. P14-63

14-64* Arm D of the mechanism shown in Fig. P14-64 has a sleeve that slides freely on the rod BC. The control link AB has an angular velocity of $\omega_{AB} = 60$ rev/min counterclockwise when in the position shown. For this position, determine

a. The angular velocity ω_D of the sleeve body D.
b. The velocity $\mathbf{v}_C$ of point C.

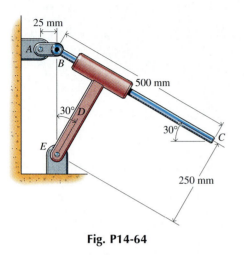

Fig. P14-64

14-65 The triangular plate of Fig. P14-65 oscillates as the control link AB rotates. Determine the maximum angular velocity ω_{AB} of the control link AB for which the velocity of point D is less than 6 in./s and the angular velocity of the plate is less than 1 rad/s at the instant shown.

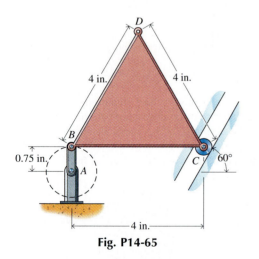

Fig. P14-65

14-66 In the four-bar linkage shown in Fig. P14-66, bar BC is horizontal and bar CD is vertical when $\theta = 90°$. Determine the maximum angular velocity ω_{AB} of the control link AB for which the velocity of point E is less than 1 m/s and the angular velocity of bar CD is less than 1.5 rad/s when $\theta = 0°$.

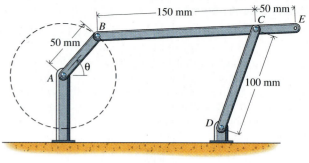

Fig. P14-66

Computer Problems

C14-67 The wheel of the slider-crank mechanism shown in Fig. P14-67 is rotating counterclockwise at a constant rate of 10 rad/s. Calculate and plot

a. The angular velocity ω_{AB} of the connecting rod AB as a function of θ for $0° < \theta < 360°$.
b. The velocity $\mathbf{v}_B$ of the slider B as a function of θ for $0° < \theta < 360°$.

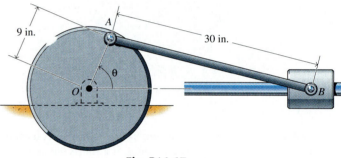

Fig. P14-67

C14-68 A 600-mm-diameter wheel rolls without slipping on a horizontal surface as shown in Fig. P14-68. The 1-m-long rod AB is attached to the wheel at a point 250 mm from its center, and end A slides freely along the horizontal surface. If the center of the wheel has a constant speed of 1.2 m/s to the right and $\theta = 0°$ when $t = 0$, calculate and plot

a. The angular velocity ω_{AB} of the rod AB as a function of time t for $0 < t < 3$ s.
b. The velocity $\mathbf{v}_A$ of end A of the rod as a function of time $(0 < t < 3$ s$)$.

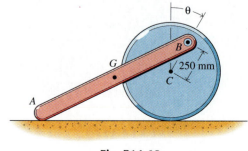

Fig. P14-68

C14-69 The Geneva wheel mechanism shown in Fig. P14-69 is used to produce intermittent motion. Both wheels rotate on fixed axles through their centers. The size and location of the input wheel A is such that the pin P enters and exits the slots smoothly. If the input wheel rotates with a constant angular speed of $\omega_A = 5$ rad/s, compute and plot

a. The angular position θ_B of the Geneva wheel versus θ_A, the angular position of the input wheel, for one complete revolution of the input wheel (that is, $-180° < \theta_A < 180°$).

b. The angular velocity ω_B of the Geneva wheel versus θ_A for one complete revolution of the input wheel.

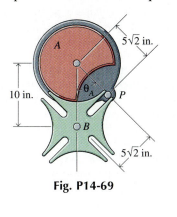

Fig. P14-69

C14-70 The mechanism shown in Fig. P14-70 is used to advance the film in a movie projector. As the input link AB rotates, the hook at P alternately engages the film and pulls it to the left, then lifts away from the film and moves back to the right. If link AB rotates at a constant angular speed of $\omega_{AB} = 900$ rev/min, compute and plot

a. The motion of the film pawl P (the position y_P versus x_P) for one complete revolution of the link AB.

b. The magnitude of the velocity of the film pawl P as a function of time for one complete revolution of the link AB. Let $t = 0$ when AB and CD are both vertical as shown.

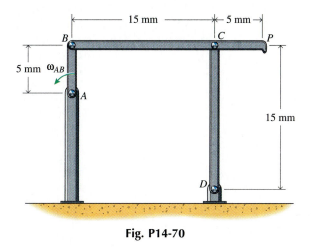

Fig. P14-70

14-5.3 Instantaneous Center of Zero Velocity

In the general plane motion of a rigid body, no point is fixed for all time. However, at any instant of time, it is always possible to find a point on the body (or the body extended) that has zero velocity.[2] This point is called the **instantaneous center of zero velocity** or simply the **instantaneous center.**

It is important to recognize that the instantaneous center of zero velocity for a rigid body in general plane motion is not fixed. The acceleration of the point that is the instantaneous center is usually not zero. Therefore, different points of the rigid body will be instantaneous centers at different instants of time and the location of the instantaneous center of zero velocity will move with respect to time.

In order to locate the instantaneous center, suppose that A and B are any two points in the rigid body whose velocity is known and that point C is an instantaneous center of zero velocity. Point C may lie in the body or in the body extended. Then, since $\mathbf{v}_C = \mathbf{0}$, the relative velocity equation (Eqs. 14-9) gives

$$\mathbf{v}_A = \mathbf{v}_{A/C} = \omega \mathbf{k} \times \mathbf{r}_{A/C} \qquad (c)$$

and point C must lie on the line through A which is perpendicular to $\mathbf{v}_A$. Similarly,

$$\mathbf{v}_B = \mathbf{v}_{B/C} = \omega \mathbf{k} \times \mathbf{r}_{B/C} \qquad (d)$$

---

[2]The case of ω being instantaneously zero requires that the point be at infinity.

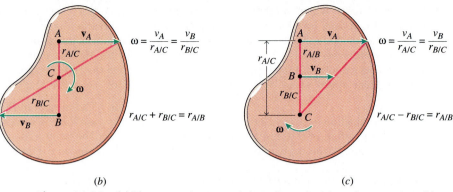

(a)

(b)

(c)

Figure 14-26   (a) If at some instant of time the velocities of two points (A and B) in a body are not parallel, then there exists a point C called the *instantaneous center of zero velocity* that is at rest at that instant. The instantaneous center of zero velocity is at the intersection of the lines through A and B that are perpendicular to the velocities of those points.
(b) If at some instant of time the velocities of two points (A and B) in a body are parallel (and perpendicular to the line joining the points) and in opposite directions, then the instantaneous center of zero velocity is between points A and B along the line that joins the two points.
(c) If at some instant of time the velocities of two points (A and B) in a body are parallel (and perpendicular to the line joining the points) and in the same direction, then the instantaneous center of zero velocity is along the line that joins A and B with both points on the same side of the instantaneous center.

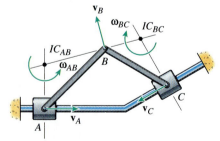

Figure 14-27   When two bodies are joined at some point B, each body will have an instantaneous center of zero velocity. The two instantaneous centers of zero velocity will generally not be the same point but will always lie on the same straight line through the common point joining the two bodies.

and point C must lie on the line through B which is perpendicular to $\mathbf{v}_B$. Assuming that $\mathbf{v}_A$ and $\mathbf{v}_B$ are not parallel, these two lines will intersect, and the point of intersection is point C (Fig. 14-26a).

If the velocities of points A and B are parallel, then the instantaneous center must lie on the line joining them. Since the magnitude of the relative velocity is ωr, the instantaneous center will be a distance $r_{A/C} = v_A/\omega$ from point A and a distance $r_{B/C} = v_B/\omega$ from point B, and its location can be found by similar triangles as shown in Figs. 14-26b and 14-26c.

If the velocities of points A and B are the same at any instant, then the body is instantaneously in translation and $\omega = 0$. This case can be included in the foregoing analysis if the instantaneous center is considered to be at infinity.

Once the instantaneous center is located, the velocity of any other point in the body is found using the relative velocity equation (Eq. 14-9)

$$\mathbf{v}_D = \mathbf{v}_C + \mathbf{v}_{D/C} = \omega\mathbf{k} \times \mathbf{r}_{D/C} \qquad (e)$$

When two or more bodies are pinned together, an instantaneous center of zero velocity can be found for each body. In general, the location of these separate instantaneous centers will not coincide. The location of each of the instantaneous centers can be found as before. Since the velocity of the point joining two bodies is the same for each body, the instantaneous centers of both bodies must lie on a single line through the common point of the two bodies (Fig. 14-27).

Use of the instantaneous center is not required to solve any particular problem. It is just another way of expressing the relative velocity equation.

CONCEPTUAL EXAMPLE 14-3: INSTANTANEOUS CENTER OF ZERO VELOCITY

In the photograph of a rolling bicycle wheel, spokes near the top of the wheel look blurred while spokes near the bottom of the wheel do not. Also, some spokes near the center of the wheel look blurred while other spokes near the center do not. Why do some of the spokes in the photograph look blurred while others do not?

SOLUTION

When a wheel rolls without slipping on a stationary surface, the contact point is an instantaneous center of zero velocity for the wheel. Then, the velocity of every other point of the wheel is directly proportional to its distance from the instantaneous center of zero velocity. The more distant the point, the faster it is moving. The faster the point is moving, the farther the point will move during the time of the exposure of the photograph and the more blurred the point will appear to be.

For example, if the center of a 26-in.-diameter bicycle wheel is moving at just 6 ft/s (about 4 mi/h), then during a 1/125 sec photographic exposure, the center of the wheel will move horizontally about 0.58 in. In the same 1/125 sec, the top of the wheel will move twice as far, about 1.15 in., since it is twice as far away from the instantaneous center of zero velocity. Points on the front and back edge of the wheel are $13\sqrt{2}$ in. from the instantaneous center of zero velocity and will travel about 0.8 in. in the 1/125 sec. All of these points leave a blurred image on the photograph.

Although all of the points near the center of the wheel move approximately the same distance (about 0.58 in.), some of the points move along a spoke whereas other points move perpendicularly to a spoke. When points move along a spoke they reinforce the image of the spoke rather than causing a blurred image.

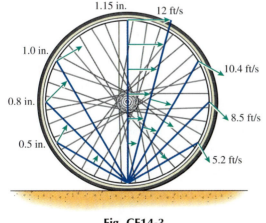

Fig. CE14-3

At the instant shown in Fig. 14-28, slider A is moving to the right with a speed of 3 m/s. Find the location of the instantaneous center of zero velocity and use it to find the angular velocity of the arm ω and the velocity of the slider $\mathbf{v}_B$.

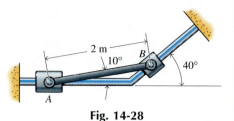

Fig. 14-28

SOLUTION

Since slider A moves in the horizontal direction, the instantaneous center of zero velocity must lie on the vertical line through A (Fig. 14-29). Similarly, the instantaneous center must lie on the line through B perpendicular to the right guide bar. The intersection of these two lines, point C, is the desired point. By the law of sines,

$$\frac{r_{A/C}}{\sin 60°} = \frac{2}{\sin 40°}$$
$$r_{A/C} = 2.69 \text{ m}$$

Therefore, the instantaneous center of zero velocity is located 2.69 m vertically above A.   Ans.

Since C is an instantaneous center of zero velocity,

$$\mathbf{v}_A = \mathbf{v}_C + \mathbf{v}_{A/C} = 0 + \mathbf{v}_{A/C}$$

and the velocity of point A is given by

$$v_A = r_{A/C}\omega$$

so

$$\omega = 1.113 \text{ rad/s} \curvearrowright$$ Ans.

Also, since C is an instantaneous center of zero velocity,

$$\mathbf{v}_B = \mathbf{v}_C + \mathbf{v}_{B/C} = 0 + \mathbf{v}_{B/C}$$

and the velocity of point B is given by

$$v_B = r_{B/C}\omega$$

Again using the law of sines to find the distance $r_{B/C}$ gives

$$\frac{r_{B/C}}{\sin 80°} = \frac{1}{\sin 40°}$$
$$r_{B/C} = 3.06 \text{ m}$$

Therefore, the velocity of slider B is

$$v_B = r_{B/C}\omega = 3.41 \text{ m/s} \measuredangle 40°$$ Ans.

Of course, the angular velocity of the arm ω_{AB} and the velocity of the slider $\mathbf{v}_B$ could just as easily have been solved using the relative velocity principles directly. In terms of a coordinate system with x to the right and y up, the velocities of the two sliders are

$$\mathbf{v}_A = v_A\mathbf{i} = 3\mathbf{i}$$
$$\mathbf{v}_B = v_B \cos 40°\mathbf{i} + v_B \sin 40°\mathbf{j}$$

and the velocity of B relative to A is (Fig. 14-30)

$$\mathbf{v}_{B/A} = r_{B/A}\omega\mathbf{e}_t = 2\omega(-\sin 10°\mathbf{i} + \cos 10°\mathbf{j})$$

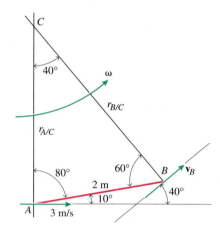

Fig. 14-29

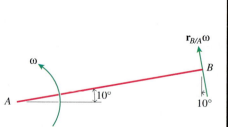

Fig. 14-30

Therefore, the relative velocity equation

$$\mathbf{v}_B = \mathbf{v}_A + \mathbf{v}_{B/A}$$

gives

$$v_B(\cos 40°\mathbf{i} + \sin 40°\mathbf{j}) = 3\mathbf{i} + 2\omega(-\sin 10°\mathbf{i} + \cos 10°\mathbf{j})$$

The x- and y-components of this equation

$$v_B \cos 40° = 3 - 2\omega \sin 10°$$
$$v_B \sin 40° = 2\omega \cos 10°$$

are solved simultaneously to get

$$\omega = 1.11334 \text{ rad/s} \cong 1.113 \text{ rad/s} \curvearrowleft$$
$$v_B = 3.4115 \text{ m/s} \cong 3.41 \text{ m/s} \measuredangle 40°$$

as before.

> The instantaneous center method trades geometric complexity for algebraic complexity. As a general rule, if the geometry is simple, then the instantaneous center method will be the easiest to use. If the geometry is complex, then the algebraic, relative velocity equations will be the easiest to use.

EXAMPLE PROBLEM 14-11

Three sliders are connected by rigid rods as shown in Fig. 14-31. At the instant shown, slider B is moving upward at a rate of 5 ft/s. Determine the angular velocity of both rods and the velocity of sliders A and C.

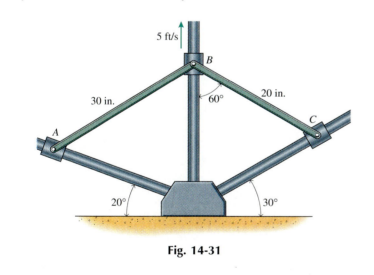

Fig. 14-31

SOLUTION

Slider B moves in a vertical direction. Therefore, the instantaneous center of zero velocity of any body connected to slider B must lie on a horizontal line through B. Slider A moves along the rod AD, so the instantaneous center of zero velocity of rod AB must lie on the line through A that is perpendicular to AD. Similarly, slider C moves along the rod CD and the instantaneous center of zero velocity of rod BC must lie on the line through C that is perpendicular to CD. The instantaneous centers of zero velocity for the two rods are shown on Fig. 14-32a.

The distances from the instantaneous centers of zero velocity to the three sliders are determined with the help of Fig. 14-32b. Triangle BCD is an equi-

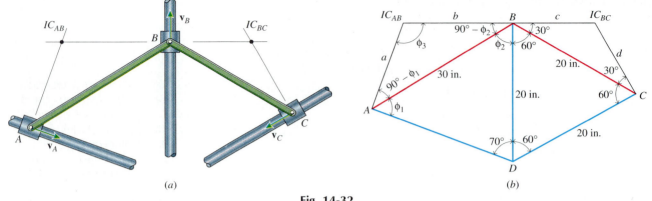

Fig. 14-32

lateral triangle and triangle $BC(IC)_{BC}$ is an isosceles triangle with

$$c = d = \frac{10}{\cos 30°} = 11.5470 \text{ in.}$$

Using the law of sines to find angle ϕ_1

$$\frac{\sin \phi_1}{20} = \frac{\sin 70°}{30}$$

gives $\phi_1 = 38.790°$. Then

$$\phi_2 = 180° - 70° - \phi_1 = 72.210°$$
$$\phi_3 = 180 - (90° - \phi_1) - (90° - \phi_2) = 110°$$

Finally, using the law of sines again to find the distances a and b

$$\frac{a}{\sin (90° - \phi_2)} = \frac{30}{\sin \phi_3}$$
$$a = 10.2829 \text{ in.}$$
$$\frac{b}{\sin (90° - \phi_1)} = \frac{30}{\sin \phi_3}$$
$$b = 24.8843 \text{ in.}$$

Now, using the relative velocity equation on rod AB

$$\mathbf{v}_B = \mathbf{v}_{IC} + \mathbf{v}_{B/IC} = 0 + \mathbf{v}_{B/IC}$$

where $\mathbf{v}_B = 5$ ft/s↑ and $\mathbf{v}_{B/IC} = b\omega_{AB}$↑. Therefore,

$$5 = b\omega_{AB} = \frac{24.8843}{12}\omega_{AB}$$
$$\omega_{AB} = 2.41116 \text{ rad/s} \cong 2.41 \text{ rad/s} \downdownarrows \qquad \text{Ans.}$$

But also for rod AB

$$\mathbf{v}_A = \mathbf{v}_{IC} + \mathbf{v}_{A/IC} = 0 + \mathbf{v}_{A/IC}$$

where $\mathbf{v}_A = v_A \searrow 20°$ and $\mathbf{v}_{A/IC} = a\omega_{AB} \searrow 20°$. Therefore,

$$v_A = a\omega_{AB} = \frac{10.2829}{12} (2.41116)$$
$$= 2.07 \text{ ft/s} \searrow 20° \qquad \text{Ans.}$$

Similarly, using the relative velocity equation on rod BC

$$\mathbf{v}_B = \mathbf{v}_{IC} + \mathbf{v}_{B/IC} = 0 + \mathbf{v}_{B/IC}$$

where $\mathbf{v}_B = 5 \text{ ft/s} \uparrow$ and $\mathbf{v}_{B/IC} = c\omega_{BC} \uparrow$. Therefore,

$$5 = c\omega_{BC} = \frac{11.5470}{12}\omega_{BC}$$
$$\omega_{BC} = 5.19615 \text{ rad/s} \cong 5.20 \text{ rad/s}\downarrow \qquad \text{Ans.}$$

And finally,

$$\mathbf{v}_C = \mathbf{v}_{IC} + \mathbf{v}_{C/IC} = \mathbf{0} + \mathbf{v}_{C/IC}$$

where $\mathbf{v}_C = v_C \nearrow 30°$ and $\mathbf{v}_{C/IC} = d\omega_{BC} \nearrow 30°$. Therefore,

$$v_C = d\omega_{BC} = \frac{11.5470}{12}(5.19615)$$
$$= 5.00 \text{ ft/s} \nearrow 30° \qquad \text{Ans.}$$

PROBLEMS

Introductory Problems

14-71* Sliders A and B of Fig. P14-71 are constrained to move in vertical and horizontal slots, respectively, and are connected by a 32-in.-long rigid rod. At the instant shown, $\theta = 75°$ and slider B is moving to the right with a constant speed of 2 in./s. For this instant,

a. Locate C, the instantaneous center of zero velocity of the bar AB.
b. Determine the angular velocity ω_{AB} of the bar AB.
c. Determine the velocity $\mathbf{v}_A$ of the slider A.

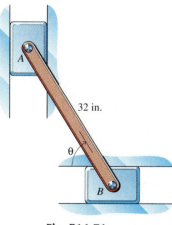

Fig. P14-71

14-72* One end of the 2-m-long bar AB shown in Fig. P14-72 is attached to a collar that slides on a smooth vertical shaft. The other end of the bar slides on a smooth horizontal surface. At the instant shown, $\theta = 30°$ and collar A is moving downward with a speed of 0.4 m/s. For this instant,

a. Locate C, the instantaneous center of zero velocity of the bar AB.
b. Determine the angular velocity ω_{AB} of the bar AB.
c. Determine the velocity $\mathbf{v}_B$ of the end B.

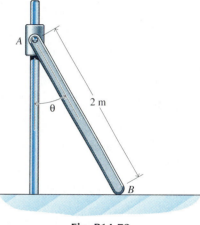

Fig. P14-72

14-73 One end of the 24-in.-long bar AB shown in Fig. P14-73 is attached to a collar that slides on a smooth shaft. The other end of the bar is attached to a slider that moves in a smooth horizontal slot. At the instant shown, the slider B is moving to the left with a speed of 1.5 in./s. For this instant,

a. Locate C, the instantaneous center of zero velocity of the bar AB.
b. Determine the angular velocity ω_{AB} of the bar AB.
c. Determine the velocity $\mathbf{v}_A$ of the collar A.

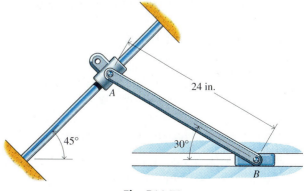

Fig. P14-73

14-74 The slender bar *AB* shown in Fig. P14-74 is attached to small wheels that roll on the horizontal and inclined surfaces. At the instant shown, the angular velocity of the bar is $\omega_{AB} = 0.2$ rad/s, clockwise. For this instant,

a. Locate *C*, the instantaneous center of zero velocity of the bar *AB*.
b. Determine the velocities of the pins *A* and *B*.

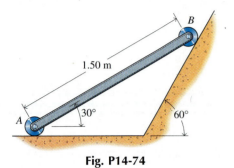

Fig. P14-74

14-75* One end of the 18-in.-long bar *AB* shown in Fig. P14-75 is attached to a slider that moves in a vertical slot. The other end is attached to a wheel which rolls on a horizontal surface. At the instant shown, the center of the wheel is moving to the right with a speed of 5 in./s. For this instant,

a. Locate *C*, the instantaneous center of zero velocity of the bar *AB*.
b. Determine the angular velocity ω_{AB} of the bar *AB*.
c. Determine the velocity $\mathbf{v}_A$ of the collar *A*.

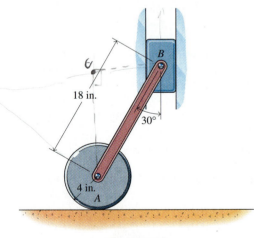

Fig. P14-75

14-76 The square plate of Fig. P14-76 is attached to a short link at *A* and a roller at *C*. At the instant shown, link *OA* has a counterclockwise angular velocity of 4 rad/s. For this instant,

a. Locate the instantaneous center of zero velocity of the plate *ABCD*.
b. Determine the angular velocity ω of the plate *ABCD*.
c. Determine the velocity $\mathbf{v}_C$ of the roller *C*.

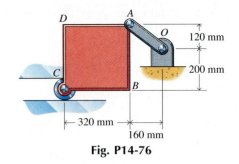

Fig. P14-76

Intermediate Problems

14-77* The triangular plate of Fig. P14-77 oscillates as the link *AB* rotates. At the instant shown, the control link *AB* is rotating counterclockwise at 60 rev/min. For this instant,

a. Locate the instantaneous center of zero velocity of the plate *BCD*.
b. Determine the angular velocity ω_{BC} of the plate *BCD*.
c. Determine the velocity $\mathbf{v}_D$ of point *D*.

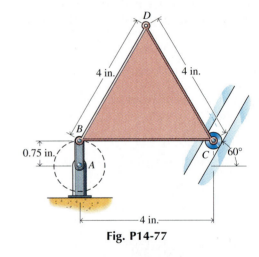

Fig. P14-77

14-78* The crank wheel *OA* of Fig. P14-78 is rotating counterclockwise at a steady angular velocity of 180 rev/min. For the position shown,

a. Locate *C*, the instantaneous center of zero velocity of the bar *AB*.
b. Determine the angular velocity ω_{AB} of the bar *AB*.
c. Determine the velocity $\mathbf{v}_B$ of the slider *B*.

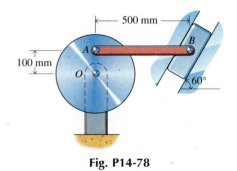

Fig. P14-78

14-79 For some short period of its motion, slider D of Fig. P14-79 has a velocity of 3 ft/s up the channel. For the instant shown,

a. Locate the instantaneous center of zero velocity of the bar BCD.
b. Determine the angular velocities ω_{AB} and ω_{CD} of both bars.
c. Determine the velocity $\mathbf{v}_C$ of the pin C.

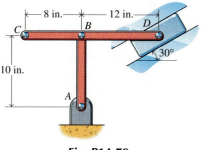

Fig. P14-79

14-80 At the instant shown, roller C of Fig. P14-80 has a velocity of 250 mm/s up the channel. For this instant,

a. Locate the instantaneous center of zero velocity of the bar BC.
b. Determine the angular velocities ω_{AB} and ω_{BC} of both bars.
c. Determine the velocity $\mathbf{v}_B$ of the pin B.

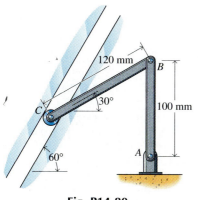

Fig. P14-80

14-81* Movement of the solenoid plunger shown in Fig. P14-81 gives pin A of the mechanism a velocity of 5 in./s horizontally to the right for some short interval of its motion. For this instant

a. Locate the instantaneous center of zero velocity of the bar AB.
b. Determine the angular velocities ω_{AB} and ω_{CD} of both bars.
c. Determine the velocity $\mathbf{v}_B$ of the pin B.

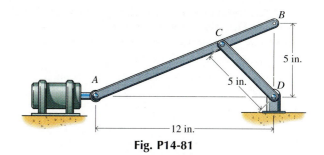

Fig. P14-81

14-82 The 400-mm-diameter wheel shown in Fig. P14-82 rolls without slipping on the horizontal surface. Bar AB is 750 mm long and is attached to the wheel by a smooth pin 150 mm from the center. At the instant shown, the center of the wheel has a velocity of 1.5 m/s to the left. For this instant,

a. Locate the instantaneous center of zero velocity of the bar AB.
b. Determine the angular velocity ω_{AB} of the bar AB.
c. Determine the velocity $\mathbf{v}_A$ of the pin A.

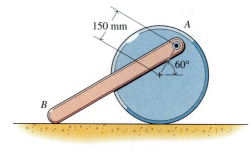

Fig. P14-82

14-83 The wheel shown in Fig. P14-83 rolls without slipping on the horizontal surface. At the instant shown, the wheel has an angular velocity of 6 rad/s clockwise. For this instant,

a. Locate the instantaneous center of zero velocity of the bar BC.
b. Determine the angular velocities ω_{AB} and ω_{BC} of both bars.
c. Determine the velocity $\mathbf{v}_C$ of the pin C.

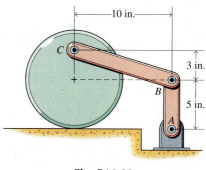

Fig. P14-83

14-84* A stepped wheel rolls without slipping on its hub as shown in Fig. P14-84. If the velocity of the center C is 4.5 m/s to the right at the instant shown, determine

a. The angular velocity of the wheel ω.
b. The velocities of points A, D, and E.

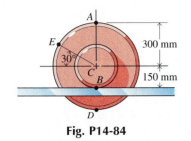

Fig. P14-84

14-85* The 1-ft-diameter wheels shown in Fig. P14-85 roll along a horizontal plane without slipping and are connected by a 3-ft-long rod AB. Pins A and B are each 4 in. from the centers of the wheels. At the instant shown, the velocity of point P is 2 ft/s to the right. Determine the velocity of point Q and the angular velocity of bar AB at this instant.

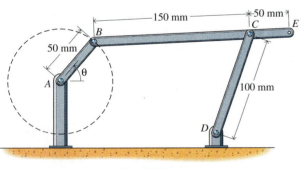

Fig. P14-85

14-86 In the four-bar linkage shown in Fig. P14-86 the control link AB has a constant counterclockwise angular velocity of 100 rev/min. Bar BC is horizontal and bar CD is vertical when $\theta = 90°$. Determine the angular velocity of bar BC and the velocity of point E when $\theta = 90°$.

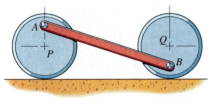

Fig. P14-86

Challenging Problems

14-87* The 1-ft-diameter wheels shown in Fig. P14-87 roll along a horizontal plane without slipping and are connected by a 3-ft-long rod AB. The pins A and B are each 4 in. from the centers of the wheels. At the instant shown, the velocity of point P is 2 ft/s to the right. For this instant,

a. Locate the instantaneous center of zero velocity of the bar AB.
b. Determine the angular velocity ω_{AB} of the bar AB.
c. Determine the velocity $\mathbf{v}_Q$ of the point Q.

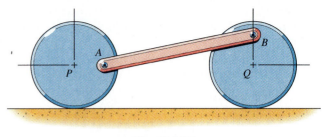

Fig. P14-87

14-88* In the four-bar linkage shown in Fig. P14-88 the control link AB has a constant counterclockwise angular velocity of 100 rev/min. Bar BC is horizontal and bar CD is vertical when $\theta = 90°$. For the instant when $\theta = 0°$,

a. Locate the instantaneous center of zero velocity of the bar BC.
b. Determine the angular velocity ω_{BC} of the bar BC.
c. Determine the velocity $\mathbf{v}_E$ of point E.

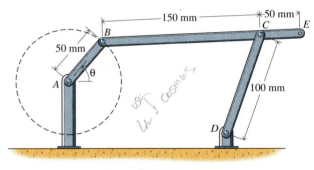

Fig. P14-88

14-89 The centers of the three gears shown in Fig. P14-89 are connected by smooth pins to the arm ABC, which rotates with a constant angular velocity of 5 rad/s counterclockwise about the fixed point A. If the larger gear is fixed and does not rotate, determine the angular velocity of the smaller gears and the velocity of tooth D using instantaneous center principles.

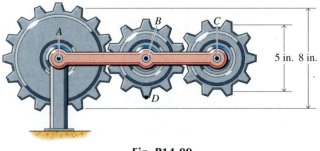

Fig. P14-89

138

14-90 In the four-bar linkage shown in Fig. P14-90 the control link *CD* has a constant clockwise angular velocity of 120 rev/min. For the instant shown, in which *B* is directly above *D*,

a. Locate the instantaneous center of zero velocity of the bar *BC*.

b. Determine the angular velocities ω_{AB} and ω_{BC} of both bars.

c. Determine the velocity $\mathbf{v}_B$ of the pin *B*.

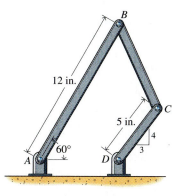

Fig. P14-90

14-91 The centers of the two gears shown in Fig. P14-91 are connected by smooth pins to the arm *ABC*, which rotates with a constant angular velocity of 5 rad/s clockwise about the fixed point *A*. The larger gear rolls on the inside of a fixed toothed drum. Determine the angular velocity of each of the gears and the velocity of tooth *D* using instantaneous center principles.

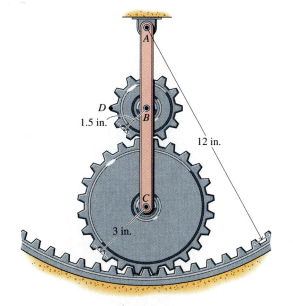

Fig. P14-91

14-92* A toy jeep is being driven on a treadmill as shown in Fig. P14-92. The treadmill is moving at 2 m/s, and the absolute velocity of the jeep is 0.3 m/s. If the diameter of the wheels is 50 mm, determine the angular velocity of the wheels and the velocity $\mathbf{v}_P$ of the point on the front of a wheel.

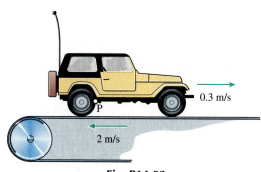

Fig. P14-92

Computer Problems

C14-93 The crank wheel *OA* of Fig. P14-93 is rotating counterclockwise at a steady angular velocity of 180 rev/min. Calculate and plot the location of *C* the instantaneous center of zero velocity of the bar *AB* as the crank wheel rotates through a full turn. That is, plot y_C versus x_C for $0° < \theta < 360°$ where x_C and y_C are the coordinates of the instantaneous center of zero velocity and $\theta = 0°$ when *A* is directly to the right of *O*.

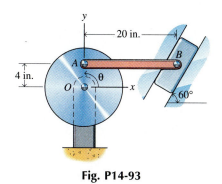

Fig. P14-93

C14-94 The square plate of Fig. P14-94 is attached to a short link at A and a roller at C. The 200-mm-long control link OA has a constant counterclockwise angular velocity of 4 rad/s. Calculate and plot the location of E, the instantaneous center of zero velocity of the plate $ABCD$ as the crank wheel rotates through a full turn. That is, plot y_E versus x_E for $0° < \theta < 360°$, where x_E and y_E are the coordinates of the instantaneous center of zero velocity and $\theta = 0°$ when A is directly to the right of O.

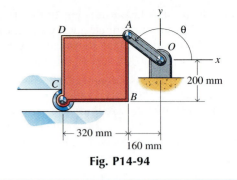

Fig. P14-94

14-5.4 Relative Acceleration

Taking two time derivatives of the relative position equation (Eq. 14-8) gives the relationship between the accelerations of particles A and B

$$\mathbf{a}_B = \mathbf{a}_A + \mathbf{a}_{B/A} \tag{14-10a}$$

where $\mathbf{a}_A$ is the absolute acceleration of particle A, $\mathbf{a}_B$ is the absolute acceleration of particle B, and $\mathbf{a}_{B/A}$ is the relative acceleration of particle B (measured relative to particle A). Equation 14-10a applies to any two particles—whether they are part of a rigid body or not.

If particles A and B are two particles in a rigid body, however, then the distance between them is constant and particle B appears to travel a circular path around particle A. Therefore, the relative acceleration $\mathbf{a}_{B/A}$ is given by (Eq. 14-7)

$$\mathbf{a}_{B/A} = (\mathbf{a}_{B/A})_t + (\mathbf{a}_{B/A})_n = \alpha\mathbf{k} \times \mathbf{r}_{B/A} + \omega\mathbf{k} \times (\omega\mathbf{k} \times \mathbf{r}_{B/A})$$
$$= r_{B/A}\alpha\mathbf{e}_t + r_{B/A}\omega^2\mathbf{e}_n \tag{14-10b}$$

and

$$\mathbf{a}_B = \mathbf{a}_A + \mathbf{a}_{B/A} = \mathbf{a}_A + \alpha\mathbf{k} \times \mathbf{r}_{B/A} + \omega\mathbf{k} \times (\omega\mathbf{k} \times \mathbf{r}_{B/A})$$
$$= \mathbf{a}_A + r_{B/A}\alpha\mathbf{e}_t + r_{B/A}\omega^2\mathbf{e}_n \tag{14-10c}$$

Therefore, the acceleration of particle B consists of two parts: $\mathbf{a}_A$, which represents a translation of the entire body with particle A; and $r_{B/A}\alpha\mathbf{e}_t + r_{B/A}\omega^2\mathbf{e}_n$, which represents fixed axis rotation of the body about an axis through A (Fig. 14-33).

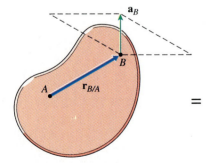
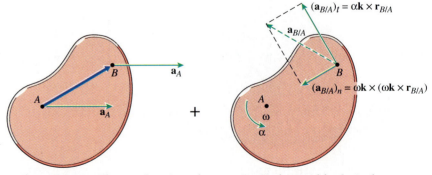

Figure 14-33   The acceleration of any point B of a rigid body is the same as a translation of the entire body with point A plus a rotation of rigid body about an axis through A.

The use of the relative acceleration equation (Eqs. 14-10) in solving problems involving a single body or multiple connected bodies is analogous to the use of the relative velocity equation. Since the normal component of the relative acceleration equation involves the angular velocity ω, the relative velocity problem usually must be solved before the relative acceleration problem can be solved. For multiple connected bodies, the acceleration of the common point of the two rigid bodies is written in terms of the acceleration of some other point (whose acceleration is known or desired) on each of the rigid bodies.

The instantaneous center of zero velocity for a rigid body in general plane motion is not fixed. The acceleration of the point that is the instantaneous center is usually not zero. Therefore, this point must not be used to compute accelerations. An analysis similar to that used to locate the instantaneous center of zero velocity can also be used to locate an instantaneous center of zero acceleration. However, the instantaneous center of zero acceleration is not usually useful in the solution of simple problems.

EXAMPLE PROBLEM 14-12

The 3-m-long ladder AB slides along a corner as shown in Fig. 14-34. When angle $\theta = 30°$, the bottom end of the ladder is moving to the right with a constant speed of 2.0 m/s. Determine the angular acceleration of the ladder and the acceleration of the top end of the ladder.

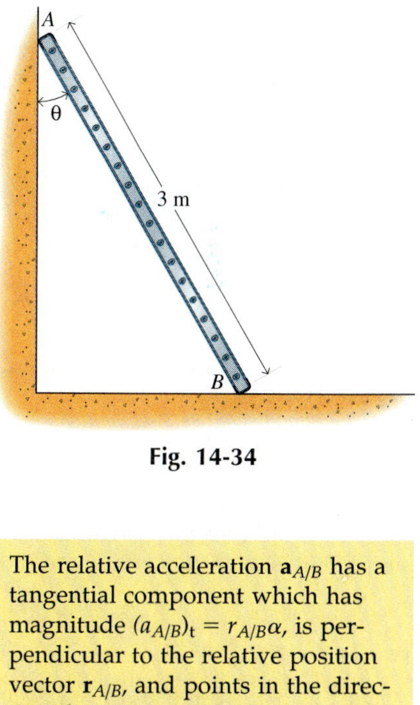

Fig. 14-34

SOLUTION

The acceleration of end B of the ladder is given as $\mathbf{a}_B = \mathbf{0}$, and the acceleration of end A of the ladder must be along the wall $\mathbf{a}_A = a_A\mathbf{j}$. Then the relative acceleration equation gives

$$a_A\mathbf{j} = \mathbf{a}_B + \mathbf{a}_{A/B} = \mathbf{0} + (\mathbf{a}_{A/B})_t + (\mathbf{a}_{A/B})_n \qquad (a)$$

But from Fig. 14-35 the relative acceleration terms are given by

$$(\mathbf{a}_{A/B})_t = 3\alpha\mathbf{e}_t = 3\alpha(-\cos 30°\mathbf{i} - \sin 30°\mathbf{j}) \qquad (b)$$

and

$$(\mathbf{a}_{A/B})_n = 3\omega^2\mathbf{e}_n = 3\omega^2(\sin 30°\mathbf{i} - \cos 30°\mathbf{j}) \qquad (c)$$

where from Example Problem 14-8, $\omega = 0.770$ rad/s. Then the x-component of Eq. a gives

$$0 = -3\alpha \cos 30° + 3(0.770)^2 \sin 30°$$
$$\alpha = 0.342 \text{ rad/s}^2 \curvearrowright \qquad \text{Ans.}$$

and the y-component gives

$$a_A = -(3)(0.342) \sin 30° - (3)(0.770)^2 \cos 30°$$
$$= -2.053 \text{ m/s}^2$$

or

$$\mathbf{a}_A = 2.053 \text{ m/s}^2 \downarrow \qquad \text{Ans.}$$

The relative acceleration $\mathbf{a}_{A/B}$ has a tangential component which has magnitude $(a_{A/B})_t = r_{A/B}\alpha$, is perpendicular to the relative position vector $\mathbf{r}_{A/B}$, and points in the direction that α rotates the body. It also has a normal component which has magnitude $(a_{A/B})_n = r_{A/B}\omega^2$, is along the relative position vector $\mathbf{r}_{A/B}$ and points toward B (the point about which the relative motion occurs).

Alternatively, the relative acceleration terms can be computed using the vector cross product

$$(\mathbf{a}_{A/B})_t = \alpha\mathbf{k} \times (-3 \sin 30°\mathbf{i} + 3 \cos 30°\mathbf{j})$$
$$= -3\alpha \sin 30°\mathbf{j} - 3\alpha \cos 30° \mathbf{i}$$
$$(\mathbf{a}_{A/B})_n = \omega\mathbf{k} \times [\omega\mathbf{k} \times (-3 \sin 30°\mathbf{i} + 3 \cos 30°\mathbf{j})]$$
$$= \omega\mathbf{k} \times [-3\omega \sin 30°\mathbf{j} - 3\omega \cos 30°\mathbf{i}]$$
$$= 3\omega^2 \sin 30°\mathbf{i} - 3\omega^2 \cos 30°\mathbf{j}$$

which are identical to Eqs. *b* and *c*.

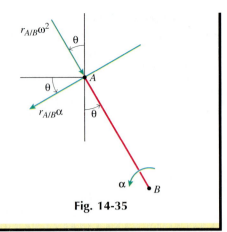

Fig. 14-35

EXAMPLE PROBLEM 14-13

The wheel of the slider-crank mechanism shown in Fig. 14-36 is rotating counterclockwise at a constant rate of 10 rad/s. Determine the acceleration of the slider $\mathbf{a}_B$ and the angular acceleration of the crank arm α_{AB} when $\theta = 60°$.

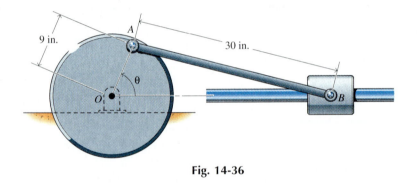

Fig. 14-36

SOLUTION

The acceleration of O, the axle of the wheel, is zero, and the acceleration of B, the slider, has only a horizontal component. Also, from Example Problem 14-9, $\omega_{OA} = 10$ rad/s, $\alpha_{OA} = 0$ rad/s², $\omega_{AB} = 1.533$ rad/s, and the angle between bar AB and the horizontal is $\phi = 15.06°$. The acceleration of the common point A is written relative to the accelerations of points O and B

$$\mathbf{a}_A = \mathbf{a}_B + \mathbf{a}_{A/B} = \mathbf{a}_O + \mathbf{a}_{A/O}$$

or

$$a_B\mathbf{i} + (\mathbf{a}_{A/B})_t + (\mathbf{a}_{A/B})_n = 0 + (\mathbf{a}_{A/O})_t + (\mathbf{a}_{A/O})_n \qquad (a)$$

But from Fig. 14-37 the relative acceleration terms are given by

$$(\mathbf{a}_{A/B})_t = 30\alpha_{AB}(-\sin \phi\mathbf{i} - \cos \phi\mathbf{j})$$
$$= -7.795\alpha_{AB}\mathbf{i} - 28.97\alpha_{AB}\mathbf{j} \text{ in./s}^2 \qquad (b)$$
$$(\mathbf{a}_{A/B})_n = 30\omega_{AB}^2(\cos \phi\mathbf{i} - \sin \phi\mathbf{j})$$
$$= 68.08\mathbf{i} - 18.32\mathbf{j} \text{ in./s}^2 \qquad (c)$$
$$(\mathbf{a}_{A/O})_t = 9\alpha_{OA}(-\sin 60°\mathbf{i} + \cos 60°\mathbf{j}) = 0 \text{ in./s}^2 \qquad (d)$$

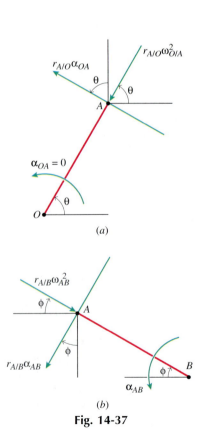

(a)

(b)

Fig. 14-37

and

$$(\mathbf{a}_{A/O})_n = 9\omega_{OA}^2(-\cos 60°\mathbf{i} - \sin 60°\mathbf{j})$$
$$= -450.0\mathbf{i} - 779.4\mathbf{j} \text{ in./s}^2 \qquad (e)$$

Then the y-component of Eq. e gives

$$-28.97\alpha_{AB} - 18.32 = -779.4$$

or

$$\alpha_{AB} = 26.3 \text{ rad/s}^2 \curvearrowright \qquad \text{Ans.}$$

and the x-component gives

$$a_B - 7.795\alpha_{AB} + 68.08 = -450.0$$
$$a_B = -313 \text{ in./s}^2$$
$$\mathbf{a}_B = 313 \text{ in./s}^2 \leftarrow \qquad \text{Ans.}$$

> The relative accelerations $\mathbf{a}_{A/B}$ and $\mathbf{a}_{A/O}$ each have components in both the tangential and normal directions. The magnitudes of the tangential components are $r_{A/B}\alpha_{AB}$ and $r_{A/O}\alpha_{OA}$. The magnitudes of the normal components are $r_{A/B}\omega_{AB}^2$ and $r_{A/O}\omega_{OA}^2$.

EXAMPLE PROBLEM 14-14

An 18-in.-long bar connects a wheel A, which rolls without slipping on a horizontal surface, and a slider B, which moves in a vertical guide (Fig. 14-38). At the instant shown the center of the wheel is moving to the right at a constant speed of 2.5 ft/s. Determine the acceleration $\mathbf{a}_G$ of the middle of the rod AB at this instant.

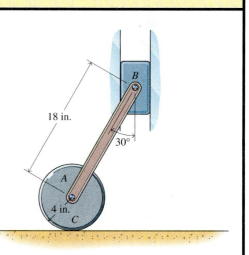

Fig. 14-38

SOLUTION

Even though the velocity and acceleration of the center of the wheel are given ($\mathbf{v}_A = 2.5\mathbf{i}$ ft/s, $\mathbf{a}_A = \mathbf{0}$), the relative acceleration equation relating the motion of A and G contains too many unknowns to solve directly. The angular velocities and angular accelerations must first be found using the known information about the motion of the slider and the contact point between the wheel and the horizontal surface.

The slider B has no horizontal motion, so its velocity and acceleration must have the form $\mathbf{v}_B = v_B\mathbf{j}$, $\mathbf{a}_B = a_B\mathbf{j}$. Since the wheel rolls without slipping on a stationary surface, the contact point between the wheel and the surface is an instantaneous center of zero velocity and $\mathbf{v}_C = \mathbf{0}$. Also as a consequence of rolling without slipping on a stationary surface, the component of acceleration along the surface is zero. The component of acceleration normal to the surface is not restricted, however, and the acceleration of the contact point of the wheel is $\mathbf{a}_C = a_C\mathbf{j}$.

The wheel and the rod are separate, connected rigid bodies. Therefore, separate relative velocity and relative acceleration equations must be written for them. Since the angular velocities will be required in the relative acceleration terms, the relative velocity problem must be solved first. For the wheel, the relative velocity equation is

$$\mathbf{v}_A = \mathbf{v}_C + \mathbf{v}_{A/C}$$

where $\mathbf{v}_{A/C} = (4/12)\omega_A\mathbf{i}$ (Fig. 14-39a). Therefore

$$2.5\mathbf{i} = 0 + \frac{4}{12}\omega_A\mathbf{i}$$
$$\omega_A = 7.50 \text{ rad/s} \curvearrowright$$

For the bar AB, the relative velocity equation is

$$\mathbf{v}_B = \mathbf{v}_A + \mathbf{v}_{B/A}$$

where $\mathbf{v}_{B/A} = (18/12)\omega_{AB}(-\cos 30°\mathbf{i} + \sin 30°\mathbf{j})$ (Fig. 14-39b). Therefore,

$$v_B\mathbf{j} = 2.5\mathbf{i} + \frac{18}{12}\omega_{AB}(-\cos 30°\mathbf{i} + \sin 30°\mathbf{j})$$

The x- and y-components of this equation

$$0 = 2.5 - \frac{18}{12}\omega_{AB}\cos 30°$$

$$v_B = \frac{18}{12}\omega_{AB}\sin 30°$$

give

$$\omega_{AB} = 1.92450 \text{ rad/s}\circlearrowleft$$
$$v_B = 1.44338 \text{ ft/s}\uparrow$$

For the wheel, the relative acceleration equation is

$$\mathbf{a}_A = \mathbf{a}_C + \mathbf{a}_{A/C}$$

where $\mathbf{a}_{A/C} = (4/12)\alpha_A\mathbf{i} - (4/12)\omega_A^2\mathbf{j}$ (Fig. 14-39c). Therefore,

$$\mathbf{0} = a_C\mathbf{j} + \left(\frac{4}{12}\alpha_A\mathbf{i} - \frac{4}{12}(7.5)^2\mathbf{j}\right)$$

and the x- and y-components of this equation

$$0 = \frac{4}{12}\alpha_A$$

$$0 = a_C - \frac{4}{12}(7.5)^2$$

give

$$\alpha_A = 0 \text{ rad/s}^2$$
$$a_C = 18.7500 \text{ ft/s}^2\uparrow$$

For the bar AB, the relative acceleration equation is

$$\mathbf{a}_B = \mathbf{a}_A + \mathbf{a}_{B/A}$$

where $\mathbf{a}_{B/A} = (18/12)\alpha_{AB}(-\cos 30°\mathbf{i} + \sin 30°\mathbf{j}) + (18/12)\omega_{AB}^2(-\sin 30°\mathbf{i} + \cos 30°\mathbf{j})$ (Fig. 14-39d). Therefore,

$$a_B\mathbf{j} = \mathbf{0} + \frac{18}{12}\alpha_{AB}(-\cos 30°\mathbf{i} + \sin 30°\mathbf{j})$$

$$+ \frac{18}{12}(1.92450)^2(-\sin 30°\mathbf{i} + \cos 30°\mathbf{j})$$

The x- and y-components of this equation

$$0 = -\frac{18}{12}\alpha_{AB}\cos 30° - \frac{18}{12}(1.92450)^2 \sin 30°$$

$$a_B = \frac{18}{12}\alpha_{AB}\sin 30° + \frac{18}{12}(1.92450)^2\cos 30°$$

give

$$\alpha_{AB} = -2.13833 \text{ rad/s}^2 = 2.13833 \text{ rad/s}^2\circlearrowright$$
$$a_B = -6.41500 \text{ ft/s}^2 = 6.41500 \text{ ft/s}^2\downarrow$$

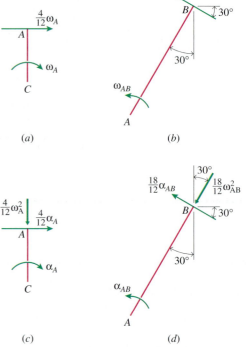

(a) (b)

(c) (d)

Fig. 14-39

Finally, the relative acceleration equation relating points A and G (which are both on the bar AB)

$$\mathbf{a}_G = \mathbf{a}_A + \mathbf{a}_{G/A}$$

$$\mathbf{a}_G = 0 + \frac{9}{12}\alpha_{AB}(-\cos 30°\mathbf{i} + \sin 30°\mathbf{j})$$

$$+ \frac{9}{12}(1.92450)^2(-\sin 30°\mathbf{i} + \cos 30°\mathbf{j})$$

gives

$$\mathbf{a}_G = -3.20750\mathbf{j} \text{ ft/s}^2 \cong 3.21 \text{ ft/s}^2\downarrow \qquad \text{Ans.}$$

PROBLEMS

Introductory Problems

14-95* Sliders A and B of Fig. P14-95 are constrained to move in vertical and horizontal slots, respectively, and are connected by a 32-in.-long rigid rod. At the instant shown, $\theta = 75°$ and slider B is moving to the right with a constant speed of 2 in./s. Determine the acceleration $\mathbf{a}_A$ of slider A and the angular acceleration α_{AB} of bar AB at this instant.

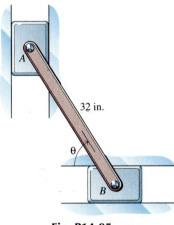

32 in.

Fig. P14-95

14-96* One end of the 2-m-long bar AB shown in Fig. P14-96 is attached to a collar, which slides on a smooth vertical shaft. The other end of the bar slides on a smooth horizontal surface. At the instant shown, $\theta = 30°$ and collar A is moving downward with a constant speed of 0.4 m/s. Determine the acceleration $\mathbf{a}_B$ of end B and the angular acceleration α_{AB} of bar AB at this instant.

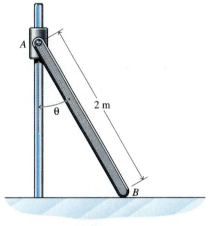

2 m

Fig. P14-96

14-97 One end of the 24-in.-long bar AB shown in Fig. P14-97 is attached to a collar, which slides on a smooth shaft. The other end of the bar is attached to a slider, which moves in a smooth horizontal slot. At the instant shown, the slider B is moving to the left with a speed of 1.5 in./s and the speed is decreasing at a rate of 0.2 in./s². Determine the acceleration $\mathbf{a}_A$ of the collar A and the angular acceleration α_{AB} of bar AB at this instant.

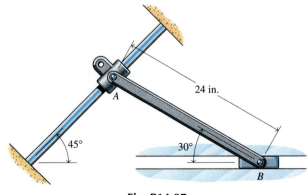

24 in.

45°   30°

Fig. P14-97

145

14-98* The slender bar AB shown in Fig. P14-98 is attached to small wheels, which roll on the horizontal and inclined surfaces. At the instant shown, the angular velocity and angular acceleration of the bar are $\omega_{AB} = 0.2$ rad/s and $\alpha_{AB} = 0.1$ rad/s², both clockwise. Determine the accelerations of the pins A and B at this instant.

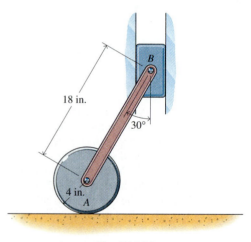

Fig. P14-98

14-99 One end of the 18-in.-long bar AB shown in Fig. P14-99 is attached to a slider, which moves in a vertical slot. The other end is attached to a wheel that rolls on a horizontal surface. At the instant shown, the center of the wheel is moving to the right with a speed of 5 in./s and the speed is increasing at a rate of 2 in./s². Determine the angular acceleration α_A of the wheel and the acceleration $\mathbf{a}_B$ of the slider B at this instant.

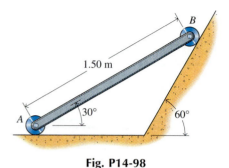

Fig. P14-99

14-100 The arm BC of Fig. P14-100 is attached to the sliding block and is made to rotate clockwise at a constant rate of 2 rev/min. Determine the acceleration $\mathbf{a}_C$ of the block C and the angular acceleration α_{AB} of the link AB at the instant shown when $\theta = 75°$.

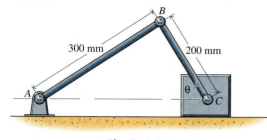

Fig. P14-100

Intermediate Problems

14-101* Movement of the solenoid plunger shown in Fig. P14-101 causes a gear to rotate. At the instant shown, the angular velocity of the gear is $\omega_O = 4$ rad/s counterclockwise and the speed of the plunger is decreasing at a rate of 0.25 ft/s². Determine the angular acceleration α_O of the gear and the angular acceleration α_{AB} of the rod AB at this instant.

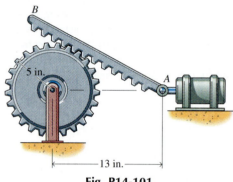

Fig. P14-101

14-102* The square plate of Fig. P14-102 is attached to a short link at A and a roller at C. At the instant shown, the control link OA has a counterclockwise angular velocity of 4 rad/s. If the speed of point C is slowing down at the rate of 50 mm/s², determine the acceleration of points A and B at this instant.

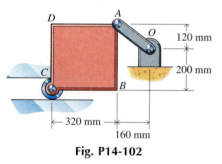

Fig. P14-102

14-103 The triangular plate of Fig. P14-103 oscillates as the control link AB rotates. At the instant shown, link AB has an angular velocity of 60 rev/min and an angular acceleration of 10 rad/s², both counterclockwise. Determine the angular acceleration α_{BC} of the plate and the acceleration $\mathbf{a}_D$ of point D at this instant.

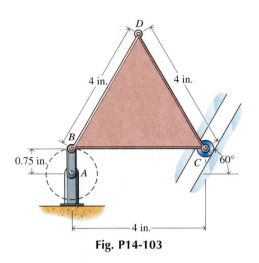

Fig. P14-103

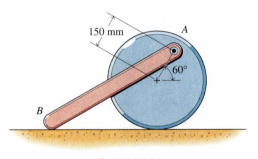

Fig. P14-104

14-104 The 400-mm-diameter wheel shown in Fig. P14-104 rolls without slipping on the horizontal surface. Bar AB is 750 mm long and is attached to the wheel by a smooth pin 150 mm from the center. At the instant shown, the center of the wheel has a velocity of 1.5 m/s to the left and an acceleration of 0.80 m/s² to the right. Determine the angular acceleration of the wheel α_O and the angular acceleration of the bar α_{AB} at this instant.

14-105* The wheel shown in Fig. P14-105 rolls without slipping on the horizontal surface. At the instant shown, the wheel has an angular velocity of 6 rad/s and an angular acceleration of 2 rad/s², both clockwise. Determine the angular acceleration of both links at this instant.

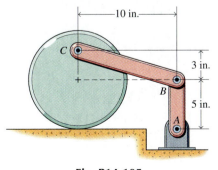

Fig. P14-105

14-106 At the instant shown in Fig. P14-106, the crank wheel OA is rotating counterclockwise at 180 rev/min and is slowing down at a rate of 0.3 rad/s². Determine the angular acceleration α_{AB} of the bar AB and the acceleration $\mathbf{a}_B$ of the slider B at this instant.

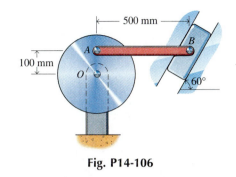

Fig. P14-106

14-107 At the instant shown in Fig. P14-107, the slider D has a velocity of 3 ft/s up the channel and its speed is decreasing at a rate of 0.5 ft/s². Determine the angular acceleration of both bars and the acceleration of point C at this instant.

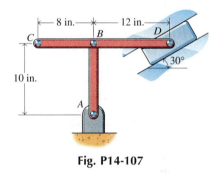

Fig. P14-107

14-108* In the four-bar linkage shown in Fig. P14-108 bar BC is horizontal and bar CD is vertical when $\theta = 90°$. If the control link AB has an angular velocity of 100 rev/min and an angular acceleration of 12 rad/s², both counterclockwise, at the instant when $\theta = 90°$, determine the angular acceleration α_{BC} of bar BC and the acceleration $\mathbf{a}_E$ of point E at this instant.

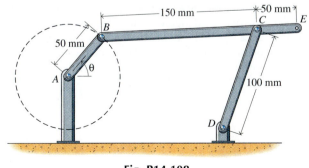

Fig. P14-108

Challenging Problems

14-109* In the four-bar linkage shown in Fig. P14-109, the control link *CD* has a constant clockwise angular velocity of 120 rev/min. Determine the angular acceleration of bars *AB* and *BC* at the instant shown, in which *B* is directly above *D*.

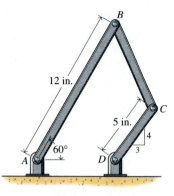

Fig. P14-109

14-110 In the four-bar linkage shown in Fig. P14-110, bar *BC* is horizontal and bar *CD* is vertical when $\theta = 90°$. When $\theta = 0°$, the control link *AB* has an angular velocity of 100 rev/min and an angular acceleration of 12 rad/s², both counterclockwise. Determine the angular acceleration α_{BC} of bar *BC* and the acceleration $\mathbf{a}_E$ of point *E* at this instant.

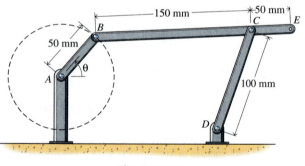

Fig. P14-110

14-111* The centers of the two gears shown in Fig. P14-111 are connected by smooth pins to the arm *ABC*, which rotates about the fixed point *A*. The larger gear rolls on the inside of a fixed toothed drum. At the instant shown, the arm has a clockwise angular velocity of 5 rad/s and a counterclockwise angular acceleration of 2 rad/s². Determine the angular acceleration of each of the gears and the acceleration of tooth *D* at this instant.

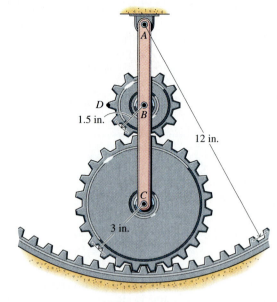

Fig. P14-111

14-112 The gears of an automatic transmission consist of a ring gear *R*, a sun gear *S*, three (or more) equal-sized planet gears *P*, and a planet carrier *C*, which is pin connected to the center of the planet gears (Fig. P14-112). Various gear ratios (ratio of the output angular velocity ω_{out} to the input angular velocity ω_{in}) are achieved by holding one of the parts fixed and driving the others. At instant shown, the ring gear is held fixed and the sun gear is rotating with an angular velocity of 3000 rev/min and an angular acceleration of 5 rad/s². For this instant determine

a. The angular accelerations α_C of the planet carrier and α_P of a planet gear.

b. The acceleration $\mathbf{a}_P$ of the center of the upper planet gear.

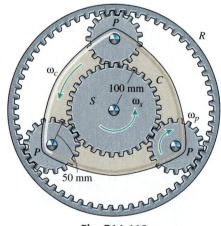

Fig. P14-112

14-113 The centers of the three gears shown in Fig. P14-113 are connected by smooth pins to the arm ABC, which rotates about the fixed point A. At the instant shown, the arm is rotating with an angular velocity of 5 rad/s, counterclockwise, and an angular acceleration of 2 rad/s^2, clockwise. If the larger gear is fixed and does not rotate, determine the angular acceleration of the smaller gears and the acceleration of tooth D.

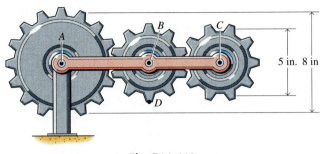

Fig. P14-113

14-114* The 0.5-m-diameter wheel of Fig. P14-114 rolls without slipping inside a 2.0-m-diameter fixed drum. When the wheel passes the bottom of the drum, the center of the wheel has a velocity of 1.5 m/s to the right and the speed is decreasing at a rate of 0.5 m/s^2. Determine the angular acceleration of the wheel and the acceleration of point C (the point of the wheel in contact with the drum) at that instant. (Hint: Pretend that the center of the drum A and the center of the wheel B are connected by a rigid rod. Relate the angular motions of the wheel and the imaginary rod through the given motion of the center of the wheel.)

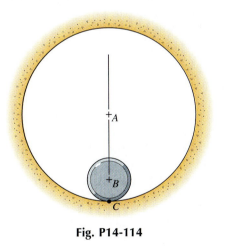

Fig. P14-114

Computer Problems

C14-115 The slider-crank mechanism of Fig. P14-115 is an idealization of an automobile crank shaft ($L_{AB} = 3$ in.), connecting rod ($L_{BC} = 7$ in.), and piston. If the crankshaft is rotating a constant rate $\dot{\theta} = 4800$ rev/min, calculate and plot

a. The velocity v_C and acceleration a_C of the piston as functions of θ ($0 < \theta < 360°$).
b. The angular velocity ω_{BC} and the angular acceleration α_{BC} of the connecting rod as functions of θ ($0 < \theta < 360°$).

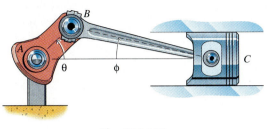

Fig. P14-115

C14-116 The mechanism shown in Fig. P14-116 is used to advance the film in a movie projector. As the input link AB rotates, the hook at P alternately engages the film and pulls it to the left, then lifts away from the film and moves back to the right. If link AB rotates at a constant angular speed of $\omega_{AB} = 900$ rev/min, compute and plot

a. The motion of the film pawl P (the position y_P versus x_P) for one complete revolution of the link AB.
b. The magnitude of the velocity of the film pawl P as a function of time for one complete revolution of the link AB. Let $t = 0$ when AB and CD are both vertical as shown.
c. The magnitude of the acceleration of the film pawl P as a function of time for one complete revolution of the link AB.

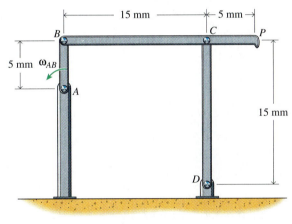

Fig. P14-116

14-6   MOTION RELATIVE TO ROTATING AXES

Thus far in the chapter, the position, velocity, and acceleration of the individual particles have all been described using a single fixed coordinate system. The relative position, relative velocity, and relative acceleration have also all been described using the same fixed coordinate system. For the types of problems considered thus far, this approach has been adequate, straightforward, and relatively simple to use.

There are several other types of problems, however, for which it is convenient to describe the position or motion of one of the particles relative to a rotating coordinate system. Problems of this type include the following:

1. The motion is observed from a coordinate system that is rotating. For example, the earth rotates, and coordinate systems fixed to the earth are rotating coordinate systems. The effect of the earth's rotation in describing the motion of swings, baseballs, bicycles, airplanes, and so on is so small that it is not even considered. However, the effect is not small in describing the motion of rockets and spacecraft as observed from a rotating earth.
2. The motions of two points are somehow related but are not equal and are not on the same rigid body. For example, some mechanisms are connected by pins that slide in grooves or slots. Relative motion is conveniently specified by giving the translational and rotational motion of the member containing the slot, the shape of the slot, and the rate of travel of the pin along the slot.
3. The solution of kinetic problems involving the rotation of irregularly shaped rigid bodies. Moments and products of inertia depend on the coordinate system used to describe them. If the axes are fixed but the body rotates, then the moments of inertia will change unless the body has certain symmetries. If the coordinate axes are allowed to rotate with the body, however, the moments and products of inertia will be constant.

Of course, the rotation of the coordinate system needs to be taken into account in the derivatives for velocity and acceleration.

14-6.1   Position

In order to see how the rotation of the coordinate system affects the description of the motion, let A and B be any two points undergoing plane motion. In terms of a fixed XY-coordinate system, the locations of A and B are given by the position vectors

$$\mathbf{r}_A = X_A\mathbf{i} + Y_A\mathbf{j} \qquad (a)$$
$$\mathbf{r}_B = X_B\mathbf{i} + Y_B\mathbf{j} \qquad (b)$$

where $\mathbf{i}$ and $\mathbf{j}$ are unit vectors along the X- and Y-axes, respectively. Then the triangle law of addition of two vectors gives $\mathbf{r}_B = \mathbf{r}_A + \mathbf{r}_{B/A}$. If $\mathbf{r}_{B/A}$ is measured in the fixed XY-coordinate system, this is exactly the result used in the first part of this chapter.

However, suppose that point A represents a particle on a rigid body that is rotating with angular velocity $\boldsymbol{\omega} = \dot{\theta}\mathbf{k}$ and angular acceleration $\boldsymbol{\alpha} = \ddot{\theta}\mathbf{k}$ (Fig. 14-40). Further suppose that the motion (position, velocity, and acceleration) of A is easy to describe in terms of the fixed coordinate system. Suppose that point B, on the other hand, represents

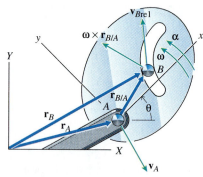

Figure 14-40   The pin B moves along a slot whose shape (a quarter circle) is easily described in terms of a coordinate system that is fixed in and rotates with the plate. If the relative position vector $\mathbf{r}_{B/A}$ is described in terms of this rotating coordinate system, then the derivatives of the unit vectors $\mathbf{e}_x$ and $\mathbf{e}_y$ will not be zero and must be included in the calculation of relative velocity and relative acceleration.

a particle that moves in a prescribed manner relative to the rotating rigid body—perhaps a pin that slides in a slot. Although it may be easy to describe the motion (position, velocity, and acceleration) of point B relative to the rotating body, it may not be very easy to describe its motion relative to the fixed XY-coordinate system.

Instead, let x–y be a rotating coordinate system that is attached to and rotates with the rigid body. Point A will be chosen for the origin of the rotating coordinate system. Then, in terms of this rotating coordinate system, the relative position vector is

$$\mathbf{r}_{B/A} = x\mathbf{e}_x + y\mathbf{e}_y \qquad (c)$$

where the unit vectors along the rotating x- and y-axes have been denoted $\mathbf{e}_x$ and $\mathbf{e}_y$ to distinguish them from the fixed unit vectors $\mathbf{i}$ and $\mathbf{j}$ and to emphasize that they are functions of time. Therefore, the position of B is given by

$$\mathbf{r}_B = \mathbf{r}_A + \mathbf{r}_{B/A} = \mathbf{r}_A + (x\mathbf{e}_x + y\mathbf{e}_y) \qquad (14\text{-}11)$$

in which x, y, $\mathbf{r}_A = X_A\mathbf{i} + Y_A\mathbf{j}$, $\mathbf{e}_x = \cos\theta\mathbf{i} + \sin\theta\mathbf{j}$, and $\mathbf{e}_y = -\sin\theta\mathbf{i} + \cos\theta\mathbf{j}$ are all presumed known as functions of time.

14-6.2 Velocity

The relationship between the absolute and relative velocities is obtained by differentiating the relative position equation (Eq. 14-11) with respect to time to get

$$\begin{aligned}
\mathbf{v}_B &= \mathbf{v}_A + \frac{d\mathbf{r}_{B/A}}{dt} = \mathbf{v}_A + \frac{d(x\mathbf{e}_x + y\mathbf{e}_y)}{dt} \\
&= \mathbf{v}_A + \frac{dx}{dt}\mathbf{e}_x + x\frac{d\mathbf{e}_x}{dt} + \frac{dy}{dt}\mathbf{e}_y + y\frac{d\mathbf{e}_y}{dt} \\
&= \mathbf{v}_A + \mathbf{v}_{B\text{rel}} + x\frac{d\mathbf{e}_x}{dt} + y\frac{d\mathbf{e}_y}{dt} \qquad (14\text{-}12a)
\end{aligned}$$

in which $\mathbf{v}_{B\text{rel}} = \dot{x}\mathbf{e}_x + \dot{y}\mathbf{e}_y$ is the velocity of B relative to (as measured in) the rotating xy-coordinate system. The last two terms arise because the directions of the unit vectors $\mathbf{e}_x$ and $\mathbf{e}_y$ vary with time due to the rotation of the xy-axes.

The derivatives of the unit vectors $\mathbf{e}_x$ and $\mathbf{e}_y$ are evaluated using the chain rule of differentiation, which gives

$$\frac{d\mathbf{e}_x}{dt} = \frac{d\mathbf{e}_x}{d\theta}\frac{d\theta}{dt} \qquad (d)$$

where

$$\frac{d\mathbf{e}_x}{d\theta} = \lim_{\Delta\theta\to 0}\frac{\mathbf{e}_x(\theta + \Delta\theta) - \mathbf{e}_x(\theta)}{\Delta\theta} \qquad (e)$$

But in the limit as $\Delta\theta\to 0$, the distance $|\mathbf{e}_x(\theta + \Delta\theta) - \mathbf{e}_x(\theta)|$ tends to the arc length along a unit circle, $\Delta s = 1\,\Delta\theta$, and the angle β tends to 90° (Fig. 14-41). Therefore, the vector $\mathbf{e}_x(\theta + \Delta\theta) - \mathbf{e}_x(\theta)$ has magnitude $\Delta\theta$ and points in the $\mathbf{e}_y$-direction and

$$\frac{d\mathbf{e}_x}{dt} = \dot{\theta}\lim_{\Delta\theta\to 0}\frac{\Delta\theta\mathbf{e}_y}{\Delta\theta} = \dot{\theta}\mathbf{e}_y = \boldsymbol{\omega}\times\mathbf{e}_x \qquad (f)$$

where $\boldsymbol{\omega} = \omega\mathbf{k}$ and $\omega = \dot{\theta} = d\theta/dt$.

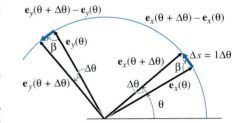

Figure 14-41 In the limit as $\Delta t \to 0$, the angle $\Delta\theta \to 0$, the displacement $\delta\mathbf{e}_x$ points in the direction of $\mathbf{e}_y$, the displacement $\delta\mathbf{e}_y$ points in the direction of $-\mathbf{e}_x$, and the lengths of the displacement vectors $|\delta\mathbf{e}_x| = |\delta\mathbf{e}_y| \to \Delta s = 1\,\Delta\theta$.

Similarly, the derivative of $\mathbf{e}_y$ with respect to time can be calculated using the chain rule

$$\frac{d\mathbf{e}_y}{dt} = \frac{d\mathbf{e}_y}{d\theta}\frac{d\theta}{dt} \tag{g}$$

where

$$\frac{d\mathbf{e}_y}{d\theta} = \lim_{\Delta\theta\to0}\frac{\mathbf{e}_y(\theta + \Delta\theta) - \mathbf{e}_y(\theta)}{\Delta\theta} \tag{h}$$

But in the limit as $\Delta\theta \to 0$, the distance $|\mathbf{e}_y(\theta + \Delta\theta) - \mathbf{e}_y(\theta)|$ again tends to the arc length along a unit circle, $\Delta s = 1\Delta\theta$, and the angle β again tends to 90° (Fig. 14-41). Therefore, the vector $\mathbf{e}_y(\theta + \Delta\theta) - \mathbf{e}_y(\theta)$ has magnitude $\Delta\theta$ and points in the negative $\mathbf{e}_x$-direction and

$$\frac{d\mathbf{e}_y}{dt} = \dot{\theta}\lim_{\Delta\theta\to0}\frac{-\Delta\theta\mathbf{e}_x}{\Delta\theta} = -\dot{\theta}\mathbf{e}_x = \boldsymbol{\omega}\times\mathbf{e}_y \tag{i}$$

Substituting these results back into the relative velocity equation (Eq. 14-12a) gives

$$\mathbf{v}_B = \mathbf{v}_A + \mathbf{v}_{Brel} + (x\boldsymbol{\omega}\times\mathbf{e}_x + y\boldsymbol{\omega}\times\mathbf{e}_y)$$
$$= \mathbf{v}_A + \boldsymbol{\omega}\times\mathbf{r}_{B/A} + \mathbf{v}_{Brel} \tag{14-12b}$$

where $\mathbf{v}_A$, $\mathbf{v}_B$, and $\boldsymbol{\omega}$ are all measured relative to the fixed XY-coordinate system, and $\mathbf{r}_{B/A}$ and $\mathbf{v}_{Brel}$ are measured relative to the rotating xy-coordinate system. Of course, all of the vectors in Eq. 14-12b must be expressed in a common coordinate system before the vector sums and product can be performed. Either $\mathbf{r}_{B/A}$ and $\mathbf{v}_{Brel}$ must be expressed in the fixed XY-coordinate system (using $\mathbf{e}_x = \cos\theta\,\mathbf{i} + \sin\theta\,\mathbf{j}$ and $\mathbf{e}_y = -\sin\theta\,\mathbf{i} + \cos\theta\,\mathbf{j}$) or $\mathbf{v}_A$ and $\mathbf{v}_B$ must be expressed in the rotating xy-coordinate system (using $\mathbf{i} = \cos\theta\,\mathbf{e}_x - \sin\theta\,\mathbf{e}_y$ and $\mathbf{j} = \sin\theta\,\mathbf{e}_x + \cos\theta\,\mathbf{e}_y$). The choice is based solely on the form in which the data are given and the form in which the results are desired.

If A and B are two points fixed on the same rigid body, then $\mathbf{v}_{Brel} = 0$, $\boldsymbol{\omega}$ is the angular velocity of the body, and Eq. 14-12b reduces to Eq. 14-9c. The added complexity of the rotating coordinate system is not needed, nor is it useful for this type of problem.

If A is a point fixed in a rotating rigid body and B is a pin sliding in a slot in the body (Fig. 14-40), then $\mathbf{v}_A + \boldsymbol{\omega}\times\mathbf{r}_{B/A}$ is the velocity point B would have if it were attached to the rigid body rather than moving relative to it. The last term $\mathbf{v}_{Brel}$ is the additional velocity point B has because of its motion along the slot. The direction of $\mathbf{v}_{Brel}$ is tangent to the slot as shown.

14-6.3   Acceleration

The relationship between the absolute and relative accelerations is obtained by differentiating the relative velocity equation (Eq. 14-12b) with respect to time to get

$$\mathbf{a}_B = \mathbf{a}_A + \frac{d\boldsymbol{\omega}}{dt}\times\mathbf{r}_{B/A} + \boldsymbol{\omega}\times\frac{d\mathbf{r}_{B/A}}{dt} + \frac{d\mathbf{v}_{Brel}}{dt} \tag{14-13a}$$

From the calculation for relative velocity,

$$\frac{d\mathbf{r}_{B/A}}{dt} = \mathbf{v}_{Brel} + \boldsymbol{\omega} \times \mathbf{r}_{B/A} \qquad (j)$$

A similar calculation for the derivative of $\mathbf{v}_{Brel}$ gives

$$\begin{aligned}
\frac{d\mathbf{v}_{Brel}}{dt} &= \frac{d(\dot{x}\mathbf{e}_x + \dot{y}\mathbf{e}_y)}{dt} \\
&= (\ddot{x}\mathbf{e}_x + \ddot{y}\mathbf{e}_y) + \left(\dot{x}\frac{d\mathbf{e}_x}{dt} + \dot{y}\frac{d\mathbf{e}_y}{dt}\right) \\
&= \mathbf{a}_{Brel} + (\dot{x}\boldsymbol{\omega} \times \mathbf{e}_x + \dot{y}\boldsymbol{\omega} \times \mathbf{e}_y) \\
&= \mathbf{a}_{Brel} + \boldsymbol{\omega} \times \mathbf{v}_{Brel} \qquad (k)
\end{aligned}$$

where $\mathbf{a}_{Brel} = \ddot{x}\mathbf{e}_x + \ddot{y}\mathbf{e}_y$ is the acceleration of B relative to (as measured in) the rotating xy-coordinate system. Substituting Eqs. j and k into Eq. 14-13a and rearranging terms yields

$$\boxed{\begin{aligned}
\mathbf{a}_B = \mathbf{a}_A + \boldsymbol{\alpha} \times \mathbf{r}_{B/A} + \boldsymbol{\omega} \times (\boldsymbol{\omega} \times \mathbf{r}_{B/A}) \\
+ \mathbf{a}_{Brel} + 2\boldsymbol{\omega} \times \mathbf{v}_{Brel}
\end{aligned}} \qquad (14\text{-}13b)$$

where $\mathbf{a}_A$, $\mathbf{a}_B$, $\boldsymbol{\omega}$, and $\boldsymbol{\alpha}$ are all measured relative to the fixed XY-coordinate system, and $\mathbf{r}_{B/A}$, $\mathbf{v}_{Brel}$, and $\mathbf{a}_{Brel}$ are measured relative to the rotating xy-coordinate system. Again, the vectors in Eq. 14-13b must be expressed in a common coordinate system before the vector sums and products can be performed. Either the fixed XY-coordinate system or the rotating xy-coordinate system can be used. The choice is based solely on the form in which the data are given and the form in which the results are desired.

If A and B are two points on the same rigid body, then $\mathbf{v}_{Brel} = \mathbf{a}_{Brel} = \mathbf{0}$, $\boldsymbol{\omega}$ and $\boldsymbol{\alpha}$ are the angular velocity and angular acceleration of the body, and Eq. 14-13b reduces to Eq. 14-10c. The added complexity of the rotating coordinate system is not needed, nor is it useful for this type of problem.

If A is a point fixed in a rotating rigid body and B is a pin sliding in a slot in the body (Fig. 14-40), then $\mathbf{a}_A + \boldsymbol{\alpha} \times \mathbf{r}_{B/A} + \boldsymbol{\omega} \times (\boldsymbol{\omega} \times \mathbf{r}_{B/A})$ is the acceleration point B would have if it were attached to the rigid body rather than moving relative to it. The term $\mathbf{a}_{Brel}$ is the additional acceleration point B has because of its motion along the slot. The remaining term $2\boldsymbol{\omega} \times \mathbf{v}_{Brel}$, called the **Coriolis acceleration,** has no simple interpretation. As indicated by the vector cross product, the Coriolis acceleration will always be perpendicular to both $\boldsymbol{\omega}$ (it will be in the plane of motion) and $\mathbf{v}_{Brel}$ (it will be perpendicular to the slot along which the pin travels).

The orientation, location of the origin, angular velocity, and angular acceleration of the rotating coordinate system should be selected to simplify the calculation of the various terms in the relative velocity and relative acceleration equations. For example, the origin A should be a point whose absolute velocity and absolute acceleration are easily obtained. The angular velocity and angular acceleration of the rotating frame should be chosen so that the velocity and acceleration of point B are easy to calculate relative to the rotating coordinate system. The orientation of the rotating coordinate system relative to the fixed coordinate system should be chosen such that the components of the various vectors are easy to describe.

Finally, Eqs. 14-12 and 14-13 are equally valid for describing the relative motion of individual particles and of points on rigid bodies. Although they were derived for plane motion ($\boldsymbol{\omega} = \omega\mathbf{k}$, $\mathbf{r}_{B/A} = x\mathbf{e}_x + y\mathbf{e}_y$, etc.), it will be seen in the next section that the vector forms of these equations are valid for general three-dimensional motion as well.

EXAMPLE PROBLEM 14-15

Car B is traveling along a straight road at a constant speed of 60 mi/h while car A is traveling around a circular curve of radius 500 ft at a constant speed of 45 mi/h (Fig. 14-42). Determine the velocity and acceleration that car B appears to have to an observer riding in and turning with car A at the instant shown.

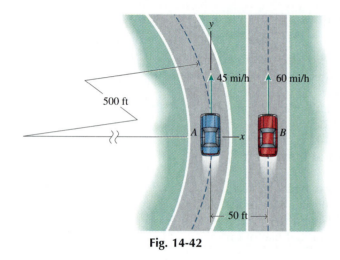

Fig. 14-42

SOLUTION

In terms of fixed r–θ coordinates with origin at the center of the turn,

$$\mathbf{v}_A = r\dot{\theta}\mathbf{e}_\theta \qquad 66\mathbf{e}_\theta = 500\dot{\theta}\mathbf{e}_\theta$$

Therefore, $\dot{\theta} = 0.1320$ rad/s. Also, since the speed of car A is constant, $\ddot{\theta} = 0$ and the acceleration of car A is

$$\mathbf{a}_A = (\ddot{r} - r\dot{\theta}^2)\mathbf{e}_r + (r\ddot{\theta} + 2\dot{r}\dot{\theta})\mathbf{e}_\theta = -(500)(0.1320)^2\mathbf{e}_r = -8.712\mathbf{e}_r \ \text{ft/s}^2$$

The moving xy-coordinate system is fixed in and moving with car A. Since these coordinates are always aligned with the r–θ coordinates above, the rotation rates of the xy-coordinate system are $\boldsymbol{\omega} = 0.1320\mathbf{k}$ rad/s and $\boldsymbol{\alpha} = \mathbf{0}$. Then the relative velocity equation (Eq. 14-12b) is

$$\mathbf{v}_B = \mathbf{v}_A + \boldsymbol{\omega} \times \mathbf{r}_{B/A} + \mathbf{v}_{Brel}$$

where

$$\mathbf{v}_A = 45\mathbf{e}_y \ \text{mi/h} = 66\mathbf{e}_y \ \text{ft/s} \qquad \mathbf{v}_B = 60\mathbf{e}_y \ \text{mi/h} = 88\mathbf{e}_y \ \text{ft/s}$$
$$\boldsymbol{\omega} \times \mathbf{r}_{B/A} = (0.1320\mathbf{k}) \times (50\mathbf{e}_x) = 6.600\mathbf{e}_y \ \text{ft/s}$$

Solving for $\mathbf{v}_{Brel}$ gives

$$\mathbf{v}_{Brel} = 15.40\mathbf{e}_y \ \text{ft/s} = 10.50\mathbf{e}_y \ \text{mi/h} \qquad\qquad \text{Ans.}$$

The relative acceleration equation (Eq. 14-13b) is

$$\mathbf{a}_B = \mathbf{a}_A + \boldsymbol{\alpha} \times \mathbf{r}_{B/A} + \boldsymbol{\omega} \times (\boldsymbol{\omega} \times \mathbf{r}_{B/A}) + \mathbf{a}_{Brel} + 2\boldsymbol{\omega} \times \mathbf{v}_{Brel}$$

where

$$\mathbf{a}_B = \mathbf{0} \qquad \mathbf{a}_A = -8.712\mathbf{e}_x \text{ ft/s}^2 \qquad \boldsymbol{\alpha} \times \mathbf{r}_{B/A} = \mathbf{0}$$
$$\boldsymbol{\omega} \times (\boldsymbol{\omega} \times \mathbf{r}_{B/A}) = (0.1320\mathbf{k}) \times (6.600\mathbf{e}_y) = -0.871\mathbf{e}_x \text{ ft/s}^2$$
$$2\boldsymbol{\omega} \times \mathbf{v}_{Brel} = 2(0.1320\mathbf{k}) \times (15.40\mathbf{e}_y) = -4.066\mathbf{e}_x \text{ ft/s}^2$$

Solving for $\mathbf{a}_{Brel}$ gives

$$\mathbf{a}_{Brel} = 13.65\mathbf{e}_x \text{ ft/s}^2 \qquad\qquad \text{Ans.}$$

As the 400-mm-long arm BC of the mechanism shown in Fig. 14-43 oscillates, the collar C slides up and down the arm AD. Given $\phi = 1.5 \sin \pi t$ rad where t is in seconds, determine the rotation rate ω of arm AD and the speed v of the slider along arm AD when $t = \frac{1}{3}$s.

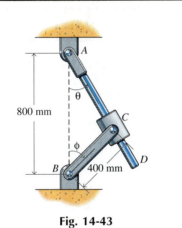

Fig. 14-43

SOLUTION

The xy-coordinate system will be chosen to rotate with arm AD and to have its origin at A as shown in Fig. 14-44. When $t = \frac{1}{3}$s

$$\phi = 1.299 \text{ rad} = 74.43° \qquad \dot{\phi} = 2.356 \text{ rad/s}$$
$$\ddot{\phi} = -12.821 \text{ rad/s}^2$$

and

$$\mathbf{v}_C = (400)(2.356)\mathbf{e}_\phi = 942.48 \text{ mm/s} \;\diagdown\; 74.43°$$

Also, by the law of cosines

$$L_{AC}^2 = 800^2 + 400^2 - 2(800)(400) \cos 74.43°$$

or

$$L_{AC}^2 = 792.60 \text{ mm}$$

and then by the law of sines

$$\frac{\sin \theta}{400} = \frac{\sin 74.43°}{792.60} \qquad \theta = 29.09°$$

Now the relative velocity equation (Eq. 14-12b) is

$$\mathbf{v}_C = \mathbf{v}_A + \boldsymbol{\omega} \times \mathbf{r}_{C/A} + \mathbf{v}_{Crel}$$

where in terms of the rotating coordinate system (see Fig. 14-44)

$$\mathbf{v}_A = \mathbf{0} \qquad \mathbf{v}_{Crel} = v\mathbf{e}_x$$
$$\boldsymbol{\omega} \times \mathbf{r}_{C/A} = \omega\mathbf{k} \times 792.60\mathbf{e}_x = 792.60\omega\mathbf{e}_y \text{ mm/s}$$
$$\mathbf{v}_C = 942.48(\cos 13.52°\mathbf{e}_x - \sin 13.52°\mathbf{e}_y) \text{ mm/s}$$

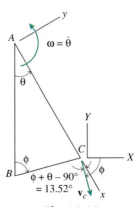

Fig. 14-44

Then the e_y component of the relative velocity equation gives

$$\omega = \dot{\theta} = -0.278 \text{ rad/s} \qquad \omega = 0.278 \text{ rad/s} \downarrow \qquad \text{Ans.}$$

and the e_x component gives

$$v = 916.4 \text{ mm/s (outward)} \qquad \text{Ans.}$$

Alternatively, the components of the relative velocity equation can be written in terms of the fixed XY-coordinate system (see Fig. 14-44)

$$\mathbf{v}_A = 0$$
$$\mathbf{v}_{Crel} = v \sin 29.09° \mathbf{i} - v \cos 29.09° \mathbf{j}$$
$$\omega \times \mathbf{r}_{C/A} = \omega \mathbf{k} \times 792.60(\sin 29.09 \mathbf{i} - \cos 29.09 \mathbf{j})$$

and

$$\mathbf{v}_C = 942.48(\cos 74.43 \mathbf{i} - \sin 74.43 \mathbf{j})$$

Then the $\mathbf{i}$ and $\mathbf{j}$ components of the relative velocity equation are

$$692.6 \, \omega + 0.486 \, v = 253.0$$
$$385.3 \, \omega - 0.874 \, v = -908.0$$

Solving these simultaneous equations gives the same results as before.

> The motion of the slider relative to A must be along the rod AD. In terms of the xy-coordinate system which rotates with the rod AD, $\mathbf{v}_{Crel} = v\mathbf{e}_x$. The absolute velocity of the slider can be written in either the fixed XY- or the rotating xy-coordinate systems, $\mathbf{v}_C = 942.48$ mm/s $\searrow 74.43° = 253.0 \mathbf{i} - 908.0 \mathbf{j}$ mm/s $= 916.4 \mathbf{e}_x - 220.3 \mathbf{e}_y$ mm/s.

EXAMPLE PROBLEM 14-17

In the mechanism of Fig. 14-45, arm AB rotates clockwise at a constant rate of 6 rev/min while the pin P moves outward along a radial slot in the rotating disk at a constant rate of 1.0 in./s. At the instant shown, $r = 3$ in., $\omega = 12$ rev/min, and $\alpha = 0.1$ rad/s², both clockwise. Determine the absolute velocity and acceleration of the pin P at this instant.

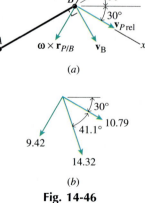

Fig. 14-45

SOLUTION

Choose the rotating xy-coordinate system with origin at B and x-axis aligned with the slot as shown in Fig. 14-46a. Then the angular velocity of the xy coordinate system is

$$\omega = (-12\mathbf{k} \text{ rev/min})\left(\frac{2\pi \text{ rad/rev}}{60 \text{ sec/min}}\right) = -1.2566\mathbf{k} \text{ rad/s}.$$

The relative velocity equation (Eq. 14-12b) is

$$\mathbf{v}_P = \mathbf{v}_B + \omega \times \mathbf{r}_{P/B} + \mathbf{v}_{Prel}$$

where in terms of the rotating coordinate system (Fig. 14-46a)

$$\mathbf{v}_B = (18)\frac{(6)(2\pi)}{60} \searrow 60°$$
$$= 9.795\mathbf{e}_x - 5.655\mathbf{e}_y \text{ in./s}$$
$$\mathbf{v}_{Prel} = 1.0\mathbf{e}_x \text{ in./s}$$
$$\omega \times \mathbf{r}_{P/B} = (-1.2566\mathbf{k}) \times (3\mathbf{e}_x) = -3.770\mathbf{e}_y \text{ in./s}$$

Therefore,

$$\mathbf{v}_P = 10.79\mathbf{e}_x - 9.42\mathbf{e}_y \text{ in./s} \qquad \text{Ans.}$$

or (see Fig. 14-46b)

$$\mathbf{v}_P = 14.32 \text{ in./s} \searrow 71.1° \qquad \text{Ans.}$$

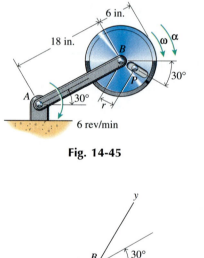

Fig. 14-46

The relative acceleration equation (Eq. 14-13b) is

$$\mathbf{a}_P = \mathbf{a}_B + \boldsymbol{\alpha} \times \mathbf{r}_{P/B} + \boldsymbol{\omega} \times (\boldsymbol{\omega} \times \mathbf{r}_{P/B})$$
$$+ \mathbf{a}_{Prel} + 2\boldsymbol{\omega} \times \mathbf{v}_{Prel}$$

where in terms of the rotating coordinate system (Fig. 14-46c)

$$\mathbf{a}_B = (18)\left[\frac{(6)(2\pi)}{60}\right]^2 \nearrow 30°$$
$$= -3.553\mathbf{e}_x - 6.154\mathbf{e}_y \text{ in./s}^2$$
$$\boldsymbol{\alpha} \times \mathbf{r}_{P/B} = (-0.1\mathbf{k}) \times (3\mathbf{e}_x) = -0.3\mathbf{e}_y \text{ in./s}^2$$
$$\boldsymbol{\omega} \times (\boldsymbol{\omega} \times \mathbf{r}_{P/B}) = (-1.2566\mathbf{k}) \times (-3.770\mathbf{e}_y)$$
$$= -4.737\mathbf{e}_x \text{ in./s}^2$$
$$2\boldsymbol{\omega} \times \mathbf{v}_{Prel} = 2(-1.2566\mathbf{k}) \times (1.0\mathbf{e}_x)$$
$$= -2.513\mathbf{e}_y \text{ in./s}^2$$
$$\mathbf{a}_{Prel} = \mathbf{0}$$

Therefore

$$\mathbf{a}_P = -8.29\mathbf{e}_x - 8.97\mathbf{e}_y \text{ in./s}^2 \qquad \text{Ans.}$$

or (see Fig. 14-46d)

$$\mathbf{a}_P = 12.21 \text{ in./s}^2 \nearrow 17.3° \qquad \text{Ans.}$$

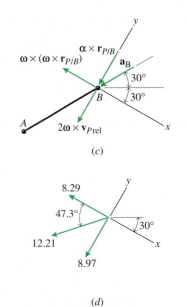

(c)

(d)

Fig. 14-46

The xy-coordinate system is fixed in and rotates with the disk such that the x-axis is always aligned with the slot. In the rotating coordinate system, the velocity and acceleration of the pin is simply $\mathbf{v}_{Prel} = 1.0 \ \mathbf{e}_x$ in./s and $\mathbf{a}_{Prel} = \mathbf{0}$.

PROBLEMS

Introductory Problems

14-117* The slotted plate of Fig. P14-117 has a constant clockwise angular velocity of 15 rad/s. The slider B has a constant speed relative to the slot of u = 10 in./s. If a = 0 in., determine the absolute velocity $\mathbf{v}_B$ and the absolute acceleration $\mathbf{a}_B$ of the slider. Repeat for a counterclockwise angular velocity of 15 rad/s.

14-118* The slotted plate of Fig. P14-117 has a constant clockwise angular velocity of 15 rad/s. The slider A has a constant speed relative to the slot of u = 100 mm/s. If a = 200 mm and b = 0 mm, determine the absolute velocity $\mathbf{v}_A$ and the absolute acceleration $\mathbf{a}_A$ of the slider. Repeat for a counterclockwise angular velocity of 15 rad/s.

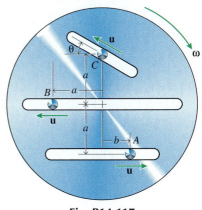

Fig. P14-117

14-119 The slotted plate of Fig. P14-117 has a constant clockwise angular velocity of 15 rad/s. The slider B has a constant speed relative to the slot of $u = 10$ in./s. If $a = 8$ in., determine the absolute velocity $\mathbf{v}_B$ and the absolute acceleration $\mathbf{a}_B$ of the slider. Repeat for a counterclockwise angular velocity of 15 rad/s.

14-120 The slotted plate of Fig. P14-117 has an angular velocity of 15 rad/s and an angular acceleration of 5 rad/s², both clockwise. The slider C has a constant speed relative to the slot of $u = 100$ mm/s. If $a = 200$ mm and $\theta = 20°$, determine the absolute velocity $\mathbf{v}_C$ and the absolute acceleration $\mathbf{a}_C$ of the slider. Repeat for a counterclockwise angular velocity and angular acceleration.

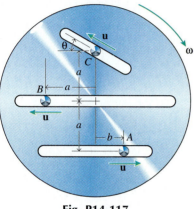

Fig. P14-117

14-121* The slotted plate of Fig. P14-121 has an angular velocity of 15 rad/s and an angular acceleration of 5 rad/s², both clockwise. The slider D has a constant speed relative to the slot of $u = 10$ in./s. If $a = 8$ in. and $\theta = 0°$, determine the absolute velocity $\mathbf{v}_D$ and the absolute acceleration $\mathbf{a}_D$ of the slider. Repeat for a counterclockwise angular velocity and angular acceleration.

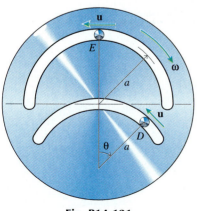

Fig. P14-121

14-122 The slotted plate of Fig. P14-121 has an angular velocity of 15 rad/s and an angular acceleration of 5 rad/s², both clockwise. The slider E has a constant speed relative to the slot of $u = 100$ mm/s. If $a = 200$ mm, determine the absolute velocity $\mathbf{v}_E$ and the absolute acceleration $\mathbf{a}_E$ of the slider. Repeat for a counterclockwise angular velocity and angular acceleration.

Intermediate Problems

14-123* As the arm of Fig. P14-123 rotates, a peg A in the underside of the slider follows the spiral groove in the fixed plate and pulls the slider outward. The spiral is given by $r = 0.035\ \theta^2$, where θ is in radians and r is in inches. The arm is rotating at a constant rate of $\dot{\theta} = 1.5$ rad/s. Determine the absolute velocity $\mathbf{v}_A$ and absolute acceleration $\mathbf{a}_A$ of the slider when $r = 6$ in.

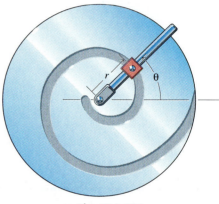

Fig. P14-123

14-124* Car A is traveling along a straight road at a constant speed of 90 km/h while car B is traveling around a circular curve of radius 150 m at a constant speed of 70 km/h (Fig. P14-124). Determine the velocity and acceleration that car A appears to have to an observer riding in and turning with car B at the instant shown.

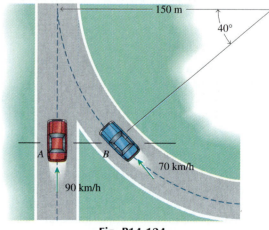

Fig. P14-124

14-125 Car A is traveling along a straight road at a constant speed of 65 mi/h while car B is traveling around a circular curve of radius 600 ft at a speed of 45 mi/h (Fig. P14-125). If the speed of car B is decreasing at the rate of 10 ft/s^2, determine the velocity and acceleration that car A appears to have to an observer riding in and turning with car B at the instant shown.

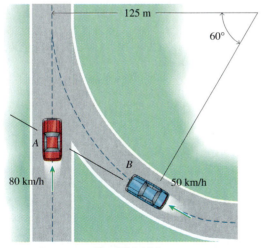

Fig. P14-125

14-126 Car A is traveling along a straight road at a constant speed of 80 km/h while car B is traveling around a circular curve of radius 125 m at a speed of 50 km/h (Fig. P14-126). If the speed of car B is increasing at the rate of 5 m/s^2, determine the velocity and acceleration that car A appears to have to an observer riding in and turning with car B at the instant shown.

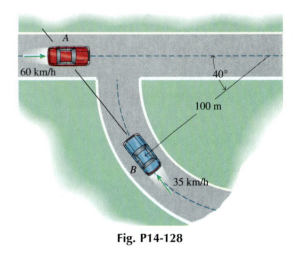

Fig. P14-126

14-127 Car A is traveling around a circular curve of radius 500 ft at a constant speed of 45 mi/h while car B is traveling around a circular curve of radius 750 ft at a constant speed of 60 mi/h (Fig. P14-127). Determine the velocity and acceleration that car A appears to have to an observer riding in and turning with car B at the instant shown.

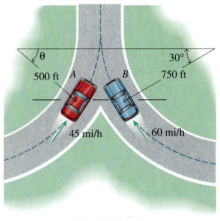

Fig. P14-127

14-128* Car A is traveling along a straight road at a constant speed of 60 km/h while car B is traveling around a circular curve of radius 100 m at a speed of 35 km/h (Fig. P14-128). If the speed of car B is decreasing at the rate of 1.5 m/s^2, determine the velocity and acceleration that car A appears to have to an observer riding in and turning with car B at the instant shown.

Fig. P14-128

14-129* In the mechanism of Fig. P14-129, arm AB is rotating counterclockwise with an angular velocity of 2 rad/s at the instant shown. The angular velocity is decreasing at a rate of 0.5 rad/s^2. For this instant, determine

a. The angular velocity ω_{CD} of the arm CD and the velocity v of the slider along the arm CD.

b. The angular acceleration of α_{CD} of the arm CD and the acceleration a of the slider along the arm CD.

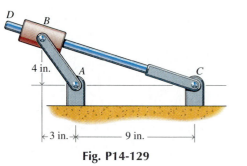

Fig. P14-129

14-130 In the mechanism of Fig. P14-130, arm AB is rotating counterclockwise with an angular velocity of 2 rad/s at the instant shown. The angular velocity is decreasing at a rate of 0.5 rad/s^2. For this instant, determine

a. The angular velocity ω_{CD} of the arm CD and the velocity v of the slider along the arm CD.

b. The angular acceleration of α_{CD} of the arm CD and the acceleration a of the slider along the arm CD.

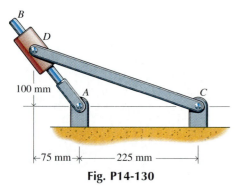

Fig. P14-130

14-131 As the 6-in.-long arm AB of the mechanism shown in Fig. P14-131 oscillates, the pin B slides up and down in the slot of arm CD. Given that $\theta = \cos \pi t$, where θ is in radians and t is in seconds, determine the angular velocity ω_{CD} and the angular acceleration α_{CD} of the arm CD when $t = \frac{1}{2}$ s.

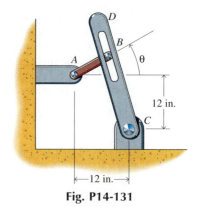

Fig. P14-131

14-132* As the 75-mm-long arm AB of the mechanism shown in Fig. P14-132 oscillates, the pin B slides up and down in the slot of arm CD. Given that $\theta = 3 \sin 2\pi t$ where θ is in radians and t is in seconds, determine the angular velocity ω_{CD} and the angular acceleration α_{CD} of the arm CD when $t = 0.1$ s.

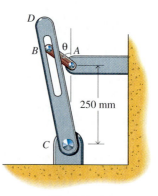

Fig. P14-132

14-133* The wheel of Fig. P14-133 is rotating clockwise at a constant rate of 120 rev/min. Pin D is fastened to the wheel at a point 5 in. from its center and slides in the slot of arm AB. Determine the angular velocity ω_{AB} and the angular acceleration α_{AB} of the arm AB at the instant shown.

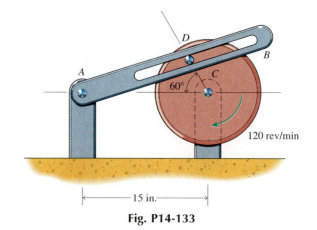

Fig. P14-133

14-134 The flywheel shown in Fig. P14-134 is rotating with a constant angular velocity of 30 rad/s. Pin C is fastened to the wheel at a point 250 mm from its center and slides in the slot of arm AB. For the instant shown,

a. Determine the angular velocity ω_{AB} and the angular acceleration α_{AB} of the arm AB.
b. Determine the acceleration $\mathbf{a}_G$ of the mass center G of the arm AB.

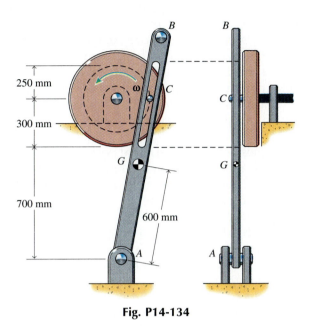

Fig. P14-134

14-135 The flywheel shown in Fig. P14-135 is rotating with a constant angular velocity of 50 rad/s. Pin C is fastened to the wheel at a point 5 in. from its center and slides in the slot of arm AB. For the instant shown where $\theta = 60°$,

a. Determine the angular velocity ω_{AB} and the angular acceleration α_{AB} of the arm AB.
b. Determine the acceleration $\mathbf{a}_G$ of the mass center G of the arm AB.

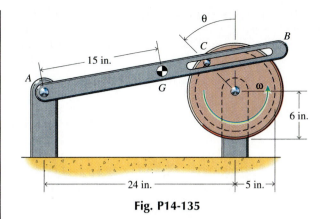

Fig. P14-135

Computer Problems

C14-136 As the 75-mm-long arm AB of the mechanism shown in Fig. P14-136 rotates, the pin B slides up and down in the slot of arm CD. If AB rotates with a constant angular velocity of $\dot{\theta} = 120$ rev/min, calculate and plot

a. The angular velocity ω_{CD} of the arm CD as a function of θ for one complete revolution of the control link ($0° < \theta < 360°$).
b. The angular acceleration α_{CD} of the arm CD as a function of θ for one complete revolution of the control link.

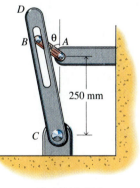

Fig. P14-136

14-7 THREE-DIMENSIONAL MOTION OF A RIGID BODY

The three-dimensional motion of rigid bodies studied in this section is considerably more complex than the two-dimensional motion treated in earlier sections of this chapter. Not only do particles of the body move in three-dimensional space, the directions of the angular velocity and the angular acceleration vectors vary with time. Rather than being simply useful for describing the motion, vector analysis will be absolutely required to describe the motion of bodies in three dimensions.

Before going on to the general three-dimensional motion of a rigid body or the special case of rotation about a fixed point, it is first necessary to consider some of the complicating aspects of rotations of rigid bodies in three dimensions.

14-7.1 Euler's Theorem

Euler's theorem states that *when a rigid body rotates about a fixed point, any position of the body is obtainable from any other position of the body by a single rotation about some axis through the fixed point.*

In order to prove Euler's theorem, consider the motion of a rigid body that rotates about a fixed point A. Point B represents the position of an arbitrary particle at some instant of time, and point B' represents its position at some later time (Fig. 14-47). Since the body is rigid, the particle B must move on a spherical surface of radius R centered at A. The spherical shell shown represents the possible positions of particle B during the motion.

The particle that occupies point B' in the initial position of the body will be called C and will be moved to C' in the final position by this same motion. Since the final position of particle B is the same as the initial position of particle C, both particles are the same distance from A and both particles move on the same spherical surface.

The configuration of a rigid body is fixed by any three points in it. Therefore, the proof of the theorem requires showing that the motion of the body that takes particle B to B' can be obtained by a single rotation about some axis though A and that the same rotation takes the particle C to C'.

Since points B, $B' = C$, and C' are all on the same sphere, the great-circle arcs $\overline{BC}$ (the distance between particles B and C in the initial position of the body) and $\overline{B'C'}$ (the distance between particles B and C in the final position of the body) must be equal by rigidity of the body. On the sphere with center A and radius R, construct the great circles that bisect the arcs $\overline{BB'}$ and $\overline{CC'}$ orthogonally. These two circles intersect at two points, one of which is labeled D on Fig. 14-47. Finally, draw the great-circle arcs $\overline{BD}$, $\overline{B'D} = \overline{CD}$, and $\overline{C'D}$. By their construction, these arcs will be equal. Therefore, the two spherical triangles $BB'D$ and $CC'D$ are congruent, and the angle ϕ between the tangents to $\overline{BD}$ and $\overline{B'D}$ at D is equal to the angle between the tangents to $\overline{CD}$ and $\overline{C'D}$ at D. Then a rotation of magnitude ϕ in the proper sense about AD will bring B to B' and C to C', determining the final position of the body from the initial position as stated in Euler's theorem.

14-7.2 Finite Rotations (Are Not Vectors)

It follows from Euler's theorem that the motion during a time interval Δt of a rigid body with a fixed point can be considered as a rotation of angle $\Delta\theta$ about a certain axis. This might be expressed much like a vector having its direction along the axis of rotation and magnitude equal to the amount of rotation. For example, the expression $\phi = \phi\mathbf{e}_{AD}$ might be used to designate the rotation of Fig. 14-47. However, even though these expressions have a magnitude and direction, they do not obey the rules of vector addition and are not vectors unless the rotations are infinitesimally small.

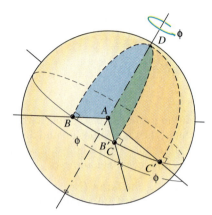

Figure 14-47 When a body rotates about a fixed point A, all other points in the body must move on spherical paths about the point A. No matter how the body spins and rotates about point A, if some particle in the body is initially at point B and ends up at point B', there exists an axis AD such that a single rotation about the axis AD will produce the same final position of the body.

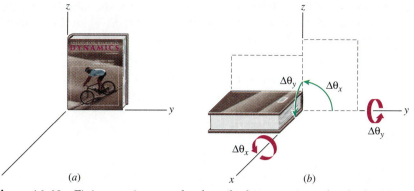

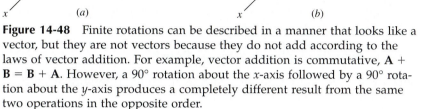

Figure 14-48 Finite rotations can be described in a manner that looks like a vector, but they are not vectors because they do not add according to the laws of vector addition. For example, vector addition is commutative, **A** + **B** = **B** + **A**. However, a 90° rotation about the x-axis followed by a 90° rotation about the y-axis produces a completely different result from the same two operations in the opposite order.

It is easily shown by means of a simple example that finite rotations do not obey the rules for addition of vectors. Take a book and set up a coordinate system as shown in Fig. 14-48a. Let $\Delta\boldsymbol{\theta}_x = 90°\,\mathbf{i}$ and $\Delta\boldsymbol{\theta}_y = 90°\,\mathbf{j}$ represent counterclockwise rotations of 90° about the x- and y-axes, respectively. Then the rotation $\Delta\boldsymbol{\theta}_x + \Delta\boldsymbol{\theta}_y$ (that is, the rotation $\Delta\boldsymbol{\theta}_x$ followed by the rotation $\Delta\boldsymbol{\theta}_y$) results in the final position shown in Fig. 14-48b. However, the rotation $\Delta\boldsymbol{\theta}_y + \Delta\boldsymbol{\theta}_x$ (that is, $\Delta\boldsymbol{\theta}_y$ followed by $\Delta\boldsymbol{\theta}_x$) results in the final position shown in Fig. 14-48c. Obviously, these final positions are not the same, and the "sum" of the rotations depends on the order in which they are written

$$\Delta\boldsymbol{\theta}_x + \Delta\boldsymbol{\theta}_y \neq \Delta\boldsymbol{\theta}_y + \Delta\boldsymbol{\theta}_x \qquad (a)$$

Therefore, finite rotations are not vectors.

Quantities such as finite rotations are known as "pseudo vectors." Although representable by directed line segments, they do not add properly as vectors and they are not vector quantities. For this and other reasons, finite rotations are difficult quantities to work with. Although they must be dealt with in advanced dynamics problems, the computation of finite rotations will not be required in this first course in dynamics.

14-7.3 Infinitesimal Rotations

Although finite rotations do not combine vectorially, rotations that are small enough in magnitude do combine properly and are vectors. As depicted in Fig. 14-49, the infinitesimal rotations $d\boldsymbol{\theta}_1$ and $d\boldsymbol{\theta}_2$ of a rigid body about the fixed point A would move point P first to Q_1 and then to S if rotation $d\boldsymbol{\theta}_1$ were applied first. If rotation $d\boldsymbol{\theta}_2$ were applied first, P would move first to Q_2 and then to S'. Although these motions take place on the surface of a sphere of radius R, for infinitesimal rotations

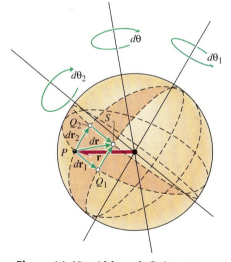

Figure 14-49 Although finite rotations do not add in a proper vector fashion, infinitesimal rotations do add properly. The point P moves to the same final position S regardless of the order in which the rotations are performed.

the curvature of the sphere is of negligible effect, the sides of the displacement figure are essentially parallel, and $S = S'$. Then the total displacement of point P is given by

$$
\begin{aligned}
d\mathbf{r} &= d\mathbf{r}_1 + d\mathbf{r}_2 = d\mathbf{r}_2 + d\mathbf{r}_1 \\
&= d\boldsymbol{\theta}_1 \times \mathbf{r} + d\boldsymbol{\theta}_2 \times \mathbf{r} = d\boldsymbol{\theta}_2 \times \mathbf{r} + d\boldsymbol{\theta}_1 \times \mathbf{r} \\
&= (d\boldsymbol{\theta}_1 + d\boldsymbol{\theta}_2) \times \mathbf{r} = (d\boldsymbol{\theta}_2 + d\boldsymbol{\theta}_1) \times \mathbf{r} \\
&= d\boldsymbol{\theta} \times \mathbf{r}
\end{aligned}
\tag{b}
$$

where

$$
d\boldsymbol{\theta} = d\boldsymbol{\theta}_1 + d\boldsymbol{\theta}_2
\tag{c}
$$

is a single resultant rotation about the axis shown and the order of the vector addition is not important.

14-7.4   Rotation about a Fixed Point

Since $d\boldsymbol{\theta}$ is now known to be a vector, its derivative with respect to time is also a vector. This vector

$$
\boldsymbol{\omega} = \dot{\boldsymbol{\theta}} = \frac{d\boldsymbol{\theta}}{dt}
\tag{d}
$$

is called the angular velocity vector. The direction of the angular velocity vector $\boldsymbol{\omega}$ represents the axis about which the body is rotating, and its magnitude is the rate of rotation about that axis. For rotation about a fixed point, however, the direction of this axis is not fixed. Therefore, both the magnitude and the direction of $\boldsymbol{\omega}$ will be functions of time.

Angular acceleration is the derivative of the angular velocity with respect to time:

$$
\boldsymbol{\alpha} = \dot{\boldsymbol{\omega}}
\tag{e}
$$

Since both the direction and magnitude of $\boldsymbol{\omega}$ are functions of time, the derivative must take into account variations in both quantities. In general, the direction of $\boldsymbol{\alpha}$ is not the same as the direction of $\boldsymbol{\omega}$.

The velocity of any point in the rigid body is given by the derivative of its position with respect to time. If the displacement of Eq. b takes place in time dt, then

$$
\frac{d\mathbf{r}}{dt} = \frac{d\boldsymbol{\theta}}{dt} \times \mathbf{r}
\tag{f}
$$

and the velocity of the particle at point P is given by

$$
\mathbf{v}_P = \boldsymbol{\omega} \times \mathbf{r}_P
\tag{14-14}
$$

The acceleration of point P is given by the derivative of the velocity with respect to time:

$$
\begin{aligned}
\mathbf{a}_P &= \frac{d\mathbf{v}_P}{dt} = \frac{d\boldsymbol{\omega}}{dt} \times \mathbf{r}_P + \boldsymbol{\omega} \times \frac{d\mathbf{r}_P}{dt} \\
&= \boldsymbol{\alpha} \times \mathbf{r}_P + \boldsymbol{\omega} \times \mathbf{v}_P = \boldsymbol{\alpha} \times \mathbf{r}_P + \boldsymbol{\omega} \times (\boldsymbol{\omega} \times \mathbf{r}_P)
\end{aligned}
\tag{14-15}
$$

Although these equations have the same form as the equations for plane motion, it is important to remember that both the magnitudes and directions of $\boldsymbol{\omega}$ and $\boldsymbol{\alpha}$ vary with time and that the direction of $\boldsymbol{\alpha}$ is not the same as the direction of $\boldsymbol{\omega}$.

14-7.5   General Motion of a Rigid Body

If A and B are any two particles, their positions are related by the tri-angle law of addition for vectors

$$\mathbf{r}_B = \mathbf{r}_A + \mathbf{r}_{B/A} \qquad (14\text{-}16)$$

where $\mathbf{r}_A$ and $\mathbf{r}_B$ are the absolute positions of the particles A and B, respectively, and $\mathbf{r}_{B/A}$ is the position of B relative to A. The time derivative of Eq. 14-16 gives the relative velocity equation

$$\mathbf{v}_B = \mathbf{v}_A + \mathbf{v}_{B/A} \qquad (14\text{-}17a)$$

where $\mathbf{v}_A$ and $\mathbf{v}_B$ are the absolute velocities of particles A and B, respectively, and $\mathbf{v}_{B/A}$ is the velocity of B relative to A. If particles A and B are two particles in a rigid body, however, then the distance between them is constant and particle B appears to travel on a spherical surface about A. Therefore, the relative velocity $\mathbf{v}_{B/A}$ is given by (Eq. 14-14)

$$\mathbf{v}_{B/A} = \boldsymbol{\omega} \times \mathbf{r}_{B/A} \qquad (14\text{-}17b)$$

Therefore,

$$\mathbf{v}_B = \mathbf{v}_A + \mathbf{v}_{B/A} = \mathbf{v}_A + \boldsymbol{\omega} \times \mathbf{r}_{B/A} \qquad (14\text{-}17c)$$

and the velocity of particle B consists of the sum of two parts: $\mathbf{v}_A$, which represents a translation of the entire body with particle A, and $\boldsymbol{\omega} \times \mathbf{r}_{B/A}$, which represents rotation about point A. Similarly, taking the time derivative of Eq. 14-17c gives

$$\mathbf{a}_B = \mathbf{a}_A + \mathbf{a}_{B/A} = \mathbf{a}_A + \boldsymbol{\alpha} \times \mathbf{r}_{B/A} + \boldsymbol{\omega} \times (\boldsymbol{\omega} \times \mathbf{r}_{B/A}) \qquad (14\text{-}18)$$

and the acceleration of particle B also consists of the sum of two parts: $\mathbf{a}_A$, which represents a translation of the entire body with particle A, and $\boldsymbol{\alpha} \times \mathbf{r}_{B/A} + \boldsymbol{\omega} \times (\boldsymbol{\omega} \times \mathbf{r}_{B/A})$, which represents rotation about point A.

Again it must be cautioned that the directions of $\boldsymbol{\omega}$ and $\boldsymbol{\alpha}$ are not fixed and the computation of the relative velocity and the relative acceleration must not be done in such a manner as to assume that they are fixed.

14-7.6   Three-Dimensional Motion Relative to Rotating Axes

The development of the relative velocity and relative acceleration equations for three-dimensional motion relative to rotating axes parallels that of planar motion in Section 14-6. The difference is that for three-dimensional motion, the directions of the angular velocity and angular acceleration vectors are not fixed as they were for two-dimensional motion. While the variation of these directions with time must be included in the differentiations, the vector form of Eqs. 14-12b and 14-13b properly accounts for them, and the equations are correct for both two- and three-dimensional motion.

In a manner analogous to the development of Section 14-6, consider the general three-dimensional motion of two points A and B whose locations are specified in the fixed XYZ-coordinate system by the position vectors

$$\mathbf{r}_A = X_A \mathbf{i} + Y_A \mathbf{j} + Z_A \mathbf{k} \qquad (g)$$
$$\mathbf{r}_B = X_B \mathbf{i} + Y_B \mathbf{j} + Z_B \mathbf{k} \qquad (h)$$

and **i**, **j**, and **k** are unit vectors along the X-, Y-, and Z-axes, respectively. The relative position, however, will be written relative to a coordinate system xyz, which has its origin attached to A and rotates with an angular velocity $\boldsymbol{\omega}$ and an angular acceleration $\boldsymbol{\alpha}$ relative to the fixed coordinate system XYZ. Therefore,

$$\mathbf{r}_{B/A} = x\mathbf{e}_x + y\mathbf{e}_y + z\mathbf{e}_z \tag{i}$$

where the unit vectors along the x-, y-, and z-axes have again been denoted $\mathbf{e}_x$, $\mathbf{e}_y$, and $\mathbf{e}_z$ to distinguish them from **i**, **j**, and **k** and to emphasize that they are functions of time. Then by vector addition the absolute position vectors $\mathbf{r}_A$ and $\mathbf{r}_B$ and the relative position vector $\mathbf{r}_{B/A}$ are related by

$$\mathbf{r}_B = \mathbf{r}_A + (x\mathbf{e}_x + y\mathbf{e}_y + z\mathbf{e}_z) \tag{14-19}$$

The relationship between the absolute and relative velocities is obtained by differentiating the relative position equation (Eq. 14-19) with respect to time to get

$$
\begin{aligned}
\mathbf{v}_B = \mathbf{v}_A + \frac{d\mathbf{r}_{B/A}}{dt} &= \mathbf{v}_A + \frac{d(x\mathbf{e}_x + y\mathbf{e}_y + z\mathbf{e}_z)}{dt} \\
&= \mathbf{v}_A + \frac{dx}{dt}\mathbf{e}_x + x\frac{d\mathbf{e}_x}{dt} + \frac{dy}{dt}\mathbf{e}_y + y\frac{d\mathbf{e}_y}{dt} + \frac{dz}{dt}\mathbf{e}_z + z\frac{d\mathbf{e}_z}{dt} \\
&= \mathbf{v}_A + \mathbf{v}_{Brel} + \left(x\frac{d\mathbf{e}_x}{dt} + y\frac{d\mathbf{e}_y}{dt} + z\frac{d\mathbf{e}_z}{dt} \right) \tag{14-20a}
\end{aligned}
$$

in which $\mathbf{v}_{Brel} = \dot{x}\mathbf{e}_x + \dot{y}\mathbf{e}_y + \dot{z}\mathbf{e}_z$ is the velocity of B relative to (as measured in) the rotating xyz-coordinate system.

The derivatives of the unit vectors in the parentheses of Eq. 14-20a could be evaluated in a manner analogous to that used in Section 14-6. Alternatively, the unit vector $\mathbf{e}_x$ can be thought of as the position of an imaginary particle that rotates about the point A with angular velocity $\boldsymbol{\omega}$. Then the derivative of $\mathbf{e}_x$ with respect to time would be the velocity of the imaginary particle and is given by

$$\frac{d\mathbf{e}_x}{dt} = \boldsymbol{\omega} \times \mathbf{e}_x$$

Similarly,

$$\frac{d\mathbf{e}_y}{dt} = \boldsymbol{\omega} \times \mathbf{e}_y \qquad \text{and} \qquad \frac{d\mathbf{e}_z}{dt} = \boldsymbol{\omega} \times \mathbf{e}_z$$

Substituting these results back into Eq. 14-20a gives

$$
\begin{aligned}
\mathbf{v}_B &= \mathbf{v}_A + \mathbf{v}_{Brel} + (x\boldsymbol{\omega} \times \mathbf{e}_x + y\boldsymbol{\omega} \times \mathbf{e}_y + z\boldsymbol{\omega} \times \mathbf{e}_z) \\
&= \mathbf{v}_A + \boldsymbol{\omega} \times \mathbf{r}_{B/A} + \mathbf{v}_{Brel} \tag{14-20b}
\end{aligned}
$$

where $\mathbf{v}_A$, $\mathbf{v}_B$, and $\boldsymbol{\omega}$ are all measured relative to the fixed XYZ-coordinate system; $\mathbf{r}_{B/A}$ and $\mathbf{v}_{Brel}$ are measured relative to the rotating xyz-coordinate system.

The relationship between the absolute and relative accelerations is obtained by differentiating the relative velocity equation (Eq. 14-20b) with respect to time to get

$$\mathbf{a}_B = \mathbf{a}_A + \frac{d\boldsymbol{\omega}}{dt} \times \mathbf{r}_{B/A} + \boldsymbol{\omega} \times \frac{d\mathbf{r}_{B/A}}{dt} + \frac{d\mathbf{v}_{Brel}}{dt} \tag{14-21a}$$

From the calculation for relative velocity,

$$\frac{d\mathbf{r}_{B/A}}{dt} = \mathbf{v}_{Brel} + \boldsymbol{\omega} \times \mathbf{r}_{B/A} \tag{j}$$

A similar calculation for the derivative of $\mathbf{v}_{Brel}$ with respect to time gives

$$\frac{d\mathbf{v}_{Brel}}{dt} = \frac{d(\dot{x}\mathbf{e}_x + \dot{y}\mathbf{e}_y + \dot{z}\mathbf{e}_z)}{dt}$$

$$= (\ddot{x}\mathbf{e}_x + \ddot{y}\mathbf{e}_y + \ddot{z}\mathbf{e}_z) + \left(\dot{x}\frac{d\mathbf{e}_x}{dt} + \dot{y}\frac{d\mathbf{e}_y}{dt} + \dot{z}\frac{d\mathbf{e}_z}{dt}\right)$$

$$= \mathbf{a}_{Brel} + (\dot{x}\boldsymbol{\omega} \times \mathbf{e}_x + \dot{y}\boldsymbol{\omega} \times \mathbf{e}_y + \dot{z}\boldsymbol{\omega} \times \mathbf{e}_z)$$

$$= \mathbf{a}_{Brel} + \boldsymbol{\omega} \times \mathbf{v}_{Brel} \tag{k}$$

where $\mathbf{a}_{Brel} = \ddot{x}\mathbf{e}_x + \ddot{y}\mathbf{e}_y + \ddot{z}\mathbf{e}_z$ is the acceleration of point B relative to (as measured in) the rotating xyz-coordinate system. Substituting Eqs. j and k into Eq. 14-21a and rearranging terms yields

$$\mathbf{a}_B = \mathbf{a}_A + \boldsymbol{\alpha} \times \mathbf{r}_{B/A} + \boldsymbol{\omega} \times (\boldsymbol{\omega} \times \mathbf{r}_{B/A})$$
$$+ 2\boldsymbol{\omega} \times \mathbf{v}_{Brel} + \mathbf{a}_{Brel} \tag{14-21b}$$

where $\mathbf{a}_A$, $\mathbf{a}_B$, $\boldsymbol{\omega}$, and $\boldsymbol{\alpha}$ are all measured relative to the fixed XYZ-coordinate system; $\mathbf{r}_{B/A}$, $\mathbf{v}_{Brel}$, and $\mathbf{a}_{Brel}$ are measured relative to the rotating xyz-coordinate system; and the term $2\boldsymbol{\omega} \times \mathbf{v}_{Brel}$ is called the Coriolis acceleration.

Equations 14-20b and 14-21b require the sum of several vectors. In order to add all these vectors together, their components need to be written in a common coordinate system. The components may be written in either the fixed XYZ-coordinate system or the rotating xyz-coordinate system. The choice is based solely on the form in which the data are given and the form in which the results are desired.

The orientation, location of the origin, angular velocity, and angular acceleration of the rotating coordinate system should be selected to simplify the calculation of the various terms in the relative velocity and relative acceleration equations. For example, the origin A should be a point whose absolute velocity and absolute acceleration are easily obtained. The angular velocity and angular acceleration of the rotating frame should be chosen so that the velocity and acceleration of point B are easy to calculate relative to the rotating coordinate system. The orientation of the rotating coordinate system relative to the fixed coordinate system should be chosen such that the components of the various vectors are easy to describe.

The 400-mm-diameter disk of Fig. 14-50 is rigidly attached to a 600-mm-long axle and rolls without slipping on a fixed surface in the xy-plane. The axle, which is perpendicular to the disk, is attached to a ball-and-socket joint at A and is free to pivot about A. As the disk and axle rotate about their own axis with angular velocity ω_1, the axle also rotates about a vertical axis with angular velocity ω_2. If $\omega_1 = 5$ rad/s and $\dot{\omega}_1 = 20$ rad/s^2 at the instant shown, determine

a. The total angular velocity $\boldsymbol{\omega}$ and total angular acceleration $\boldsymbol{\alpha}$ of the disk at this instant.

b. The velocity $\mathbf{v}_C$ and acceleration $\mathbf{a}_C$ of point C on the rim of the disk at this instant.

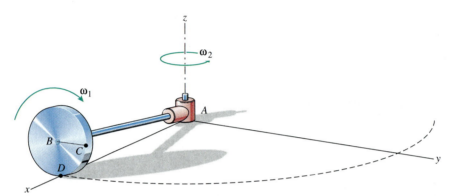

Fig. 14-50

SOLUTION

a. The rod AB makes an angle of 18.43° with the x-axis (Fig. 14-51a). In accordance with the right-hand rule, the angular velocity $\boldsymbol{\omega}_1$ points from B to A and has components

$$\boldsymbol{\omega}_1 = \omega_1 \mathbf{e}_{BA} = -5 \cos 18.43° \ \mathbf{i} - 5 \sin 18.43° \ \mathbf{k} \ \text{rad/s}$$
$$= -4.744\mathbf{i} - 1.581\mathbf{k} \ \text{rad/s} \qquad (a)$$

where $\mathbf{e}_{BA}$ is a unit vector pointing from B to A. Since the wheel rolls without slipping, the arc length $\overline{DD'}$ on the wheel is equal to the arc length $\overline{DD'}$ on the surface (Fig. 14-51b)

$$s = 200\theta_1 = 632.46\theta_2 \qquad (b)$$

Differentiating Eq. b with respect to time gives the relationship between the rotation rates ω_1 and ω_2

$$\omega_2 = 0.3162\omega_1 \qquad (c)$$

Then
$$\omega_2 = 1.581\mathbf{k} \ \text{rad/s}$$
$$\boldsymbol{\omega} = \boldsymbol{\omega}_1 + \boldsymbol{\omega}_2 = -4.74\mathbf{i} \ \text{rad/s} \qquad \text{Ans.}$$

The angular acceleration $\boldsymbol{\alpha}$ is the derivative of the angular velocity

$$\boldsymbol{\alpha} = \dot{\boldsymbol{\omega}} = \dot{\boldsymbol{\omega}}_1 + \dot{\boldsymbol{\omega}}_2$$

where the derivatives must account for changes in the directions as well as changes in the magnitudes of $\boldsymbol{\omega}_1$ and $\boldsymbol{\omega}_2$. Following the discussion of

Since the wheel rolls without slipping, all of the points on the wheel between D and D' had to have been in contact with points on the horizontal surface as the wheel rolled. Therefore, the arc length measured around the wheel from D to D' ($s = 200 \ \theta_1$) must be the same as the arc length measured along the surface from D to D' ($s = 632.46 \ \theta_2$).

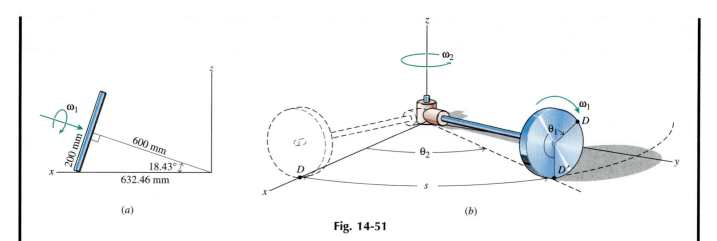

(a) *(b)*

Fig. 14-51

Section 14-7 for the derivative of the unit vector $\mathbf{e}_{BA}$, the derivative of $\boldsymbol{\omega}_1$ is

$$\dot{\boldsymbol{\omega}}_1 = \dot{\omega}_1 \mathbf{e}_{BA} + \omega_1 \dot{\mathbf{e}}_{BA} = \dot{\omega}_1 \mathbf{e}_{BA} + \omega_1 (\boldsymbol{\omega}_2 \times \mathbf{e}_{BA})$$
$$= (-20 \cos 18.43°\mathbf{i} - 20 \sin 18.43°\mathbf{k})$$
$$+ 5[(1.581\mathbf{k}) \times (-\cos 18.43°\mathbf{i} - \sin 18.43°\mathbf{k})]$$
$$= -18.97\mathbf{i} - 7.50\mathbf{j} - 6.323\mathbf{k} \text{ rad/s}^2$$

Differentiating Eq. *c* with respect to time gives

$$\dot{\omega}_2 = 0.3162\dot{\omega}_1$$

and, since the direction of ω_2 is constant,

$$\dot{\boldsymbol{\omega}}_2 = 6.324\mathbf{k} \text{ rad/s}^2$$

Therefore,

$$\boldsymbol{\alpha} = -18.97\mathbf{i} - 7.50\mathbf{j} \text{ rad/s}^2 \qquad \text{Ans.}$$

b. The position of point C relative to the point about which the disk is rotating is

$$\mathbf{r}_C = 600 \cos 18.43°\mathbf{i} + 200\mathbf{j} + 600 \sin 18.43°\mathbf{k}$$
$$= 569.2\mathbf{i} + 200\mathbf{j} + 189.69\mathbf{k} \text{ mm}$$

Then the velocity of point C is given by Eq. 14-14

$$\mathbf{v}_C = \boldsymbol{\omega} \times \mathbf{r}_C = -4.744\mathbf{i} \times (569.2\mathbf{i} + 200\mathbf{j} + 189.69\mathbf{k})$$
$$= 899.9\mathbf{j} - 948.8\mathbf{k} \text{ mm/s}$$
$$\cong 900\mathbf{j} - 949\mathbf{k} \text{ mm/s} \qquad \text{Ans.}$$

The acceleration of point C is given by Eq. 14-15

$$\mathbf{a}_C = \boldsymbol{\alpha} \times \mathbf{r}_C + \boldsymbol{\omega} \times \mathbf{v}_C$$
$$= (-18.97\mathbf{i} - 7.50\mathbf{j}) \times (569.2\mathbf{i} + 200\mathbf{j} + 189.69\mathbf{k})$$
$$-4.744\mathbf{i} \times (899.9\mathbf{j} - 948.8\mathbf{k})$$
$$= -1423\mathbf{i} - 903\mathbf{j} - 3790\mathbf{k} \text{ mm/s}^2 \qquad \text{Ans.}$$

The slender rod of Fig. 14-52 is connected to the sliders A and B by ball and socket joints. If slider A is moving in the negative x-direction at a constant speed of 6 in./s, determine

a. The velocity $\mathbf{v}_B$ and acceleration $\mathbf{a}_B$ of slider B at the instant shown.
b. The angular velocity $\boldsymbol{\omega}$ and angular acceleration $\boldsymbol{\alpha}$ of the rod at the instant shown. (Assume that the rod is not rotating about its own axis.)

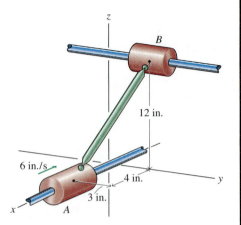

Fig. 14-52

SOLUTION

a. Letting x represent the position of slider A and y represent the position of slider B, the length of the rod AB at any instant of time is

$$\sqrt{(x^2 + y^2 + 12^2)} = 13 \qquad (a)$$

Squaring both sides of Eq. *a* and differentiating with respect to time gives

$$2x\dot{x} + 2y\dot{y} = 0 \qquad (b)$$

where $\dot{x} = v_A = -6$ in./s and $\dot{y} = v_B$. At the instant shown, $x = 4$ in. and $y = 3$ in. Therefore,

$$v_B = 8 \text{ in./s}$$

or

$$\mathbf{v}_B = 8\mathbf{j} \text{ in./s} \qquad \text{Ans.}$$

Differentiating Eq. *b* with respect to time gives

$$\dot{x}^2 + x\ddot{x} + \dot{y}^2 + y\ddot{y} = 0$$

where $\ddot{x} = a_A = 0$ and $\ddot{y} = a_B$. Therefore,

$$a_B = -100/3 \text{ in./s}^2$$

or

$$\mathbf{a}_B = -33.3\mathbf{j} \text{ in./s}^2 \qquad \text{Ans.}$$

b. The relative velocity equation is

$$\mathbf{v}_B = \mathbf{v}_A + \mathbf{v}_{B/A} = \mathbf{v}_A + \boldsymbol{\omega} \times \mathbf{r}_{B/A}$$

where $\mathbf{v}_A = -6\mathbf{i}$ in./s, $\mathbf{v}_B = 8\mathbf{j}$ in./s, and the position of B relative to A is

$$\mathbf{r}_{B/A} = -4\mathbf{i} + 3\mathbf{j} + 12\mathbf{k} \text{ in.}$$

Therefore,

$$8\mathbf{j} = -6\mathbf{i} + (\omega_x\mathbf{i} + \omega_y\mathbf{j} + \omega_z\mathbf{k}) \times (-4\mathbf{i} + 3\mathbf{j} + 12\mathbf{k})$$
$$= (12\omega_y - 3\omega_z - 6)\mathbf{i} - (4\omega_z + 12\omega_x)\mathbf{j} + (3\omega_x + 4\omega_y)\mathbf{k} \qquad (c)$$

Although Eq. *c* is a vector equation, its three components

$$\begin{aligned} x\text{:} \quad & 12\omega_y - 3\omega_z = 6 & (d) \\ y\text{:} \quad & 12\omega_x + 4\omega_z = -8 & (e) \\ z\text{:} \quad & 3\omega_x + 4\omega_y = 0 & (f) \end{aligned}$$

are not sufficient to find the three unknown components of the angular velocity. The relative velocity of the ends of bar AB are independent of the rotation of bar AB about its own axis, and Eqs. *d*, *e*, and *f* allow infinitely many solutions for the angular velocity $\boldsymbol{\omega}$ differing by the rate of rotation about AB. This ambiguity is removed by the assumption that the bar is not rotating about its own axis.

The assumption that bar AB is not rotating about its own axis is equivalent to saying that the component of $\boldsymbol{\omega}$ in the direction of the bar is zero:

$$\boldsymbol{\omega} \cdot \mathbf{r}_{B/A} = (\omega_x \mathbf{i} + \omega_y \mathbf{j} + \omega_z \mathbf{k}) \cdot (-4\mathbf{i} + 3\mathbf{j} + 12\mathbf{k})$$
$$= -4\omega_x + 3\omega_y + 12\omega_z = 0 \qquad (g)$$

Solving Eqs. d, e, and g simultaneously gives

$$\omega_x = -0.5680 \text{ rad/s} \qquad \omega_y = 0.4260 \text{ rad/s}$$
$$\omega_z = -0.2959 \text{ rad/s}$$

or

$$\boldsymbol{\omega} = -0.568\mathbf{i} + 0.426\mathbf{j} - 0.296\mathbf{k} \text{ rad/s} \qquad \text{Ans.}$$

Similarly, the relative acceleration equation

$$\mathbf{a}_B = \mathbf{a}_A + \mathbf{a}_{B/A} = \mathbf{a}_A + \boldsymbol{\alpha} \times \mathbf{r}_{B/A} + \boldsymbol{\omega} \times \mathbf{v}_{B/A} \qquad (h)$$

where $\mathbf{a}_A = \mathbf{0}$, $\mathbf{a}_B = -33.33\mathbf{j}$ in./s^2, and $\mathbf{v}_{B/A} = \mathbf{v}_B - \mathbf{v}_A = 6.000\mathbf{i} + 8.000\mathbf{j}$ in./s. Therefore, the relative acceleration equation, Eq. h,

$$-33.33\mathbf{j} = \mathbf{0} + (\alpha_x\mathbf{i} + \alpha_y\mathbf{j} + \alpha_z\mathbf{k}) \times (-4\mathbf{i} + 3\mathbf{j} + 12\mathbf{k})$$
$$+ (-0.5680\mathbf{i} + 0.4260\mathbf{j} - 0.2959\mathbf{k}) \times (6.00\mathbf{i} + 8.00\mathbf{j})$$
$$= [(12\alpha_y - 3\alpha_z)\mathbf{i} - (4\alpha_z + 12\alpha_x)\mathbf{j} + (3\alpha_x + 4\alpha_y)\mathbf{k}]$$
$$+ [2.367\mathbf{i} - 1.7754\mathbf{j} - 7.100\mathbf{k}] \qquad (i)$$

has the three components

$$x: \qquad 12\alpha_y - 3\alpha_z = -2.367 \qquad (j)$$
$$y: \qquad 12\alpha_x + 4\alpha_z = 31.555 \qquad (k)$$
$$z: \qquad 3\alpha_x + 4\alpha_y = 7.100 \qquad (l)$$

Again using the assumption that the bar is not rotating about its own axis,

$$\boldsymbol{\alpha} \cdot \mathbf{r}_{B/A} = (\alpha_x\mathbf{i} + \alpha_y\mathbf{j} + \alpha_z\mathbf{k}) \cdot (-4\mathbf{i} + 3\mathbf{j} + 12\mathbf{k})$$
$$= -4\alpha_x + 3\alpha_y + 12\alpha_z = 0 \qquad (m)$$

Finally, solving Eqs. k, l, and m simultaneously gives

$$\alpha_x = 2.367 \text{ rad/s}^2 \qquad \alpha_y = 0.000 \text{ rad/s}^2$$
$$\alpha_z = 0.789 \text{ rad/s}^2$$

or

$$\boldsymbol{\alpha} = 2.367\mathbf{i} + 0.789\mathbf{k} \text{ rad/s}^2 \qquad \text{Ans.}$$

Rotation of the slender rod AB about its own axis does not cause any motion of either slider. Similarly, the amount or rate of rotation of the rod about its own axis cannot be determined from the motion of the sliders. Since there does not appear to be anything causing the rod to rotate about its own axis, it will be assumed that it is not.

The fire truck ladder shown in Fig. 14-53 is being raised at a constant rate of $\dot{\theta}_2 = 0.5$ rad/s. Simultaneously, it is rotating about a vertical axis at a constant rate of $\dot{\theta}_1 = 0.8$ rad/s and is being extended at a constant rate of $\dot{s} = 1.5$ m/s. Determine the velocity $\mathbf{v}_B$ and acceleration $\mathbf{a}_B$ of the end of the ladder when $s = 10$ m and $\theta_2 = 30°$.

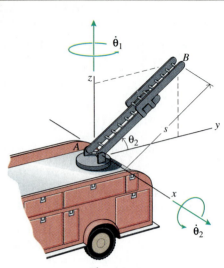

SOLUTION

The rotating xyz-coordinate system is chosen with its origin at A as shown on Fig. 14-53. The rotation rate of the xyz system is chosen so that the ladder is always in the yz-plane. Then the relative velocity equation (Eq. 14-20b) is

$$\mathbf{v}_B = \mathbf{v}_A + \boldsymbol{\omega} \times \mathbf{r}_{B/A} + \mathbf{v}_{Brel}$$

where

$$\mathbf{v}_A = \mathbf{0}$$
$$\boldsymbol{\omega} = \dot{\theta}_1 \mathbf{e}_z$$
$$\boldsymbol{\omega} \times \mathbf{r}_{B/A} = \dot{\theta}_1 \mathbf{e}_z \times (s \cos \theta_2 \mathbf{e}_y + s \sin \theta_2 \mathbf{e}_z)$$
$$= -(0.8)(10) \cos 30° \; \mathbf{e}_x = -6.928 \mathbf{e}_x \text{ m/s}$$

and (Fig. 14-54a)

$$\mathbf{v}_{Brel} = [\dot{s} \cos \theta_2 \; \mathbf{e}_y + \dot{s} \sin \theta_2 \; \mathbf{e}_z$$
$$\qquad -s\dot{\theta}_2 \sin \theta_2 \; \mathbf{e}_y + s\dot{\theta}_2 \cos \theta_2 \; \mathbf{e}_z]$$
$$= [1.5 \cos 30° - (10)(0.5) \sin 30°] \mathbf{e}_y$$
$$\qquad + [1.5 \sin 30° + (10)(0.5) \cos 30°] \mathbf{e}_z$$
$$= -1.201 \mathbf{e}_y + 5.080 \mathbf{e}_z \text{ m/s}$$

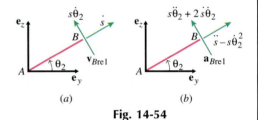

(a) *(b)*

Fig. 14-54

Therefore,

$$\mathbf{v}_B = -6.928 \mathbf{e}_x - 1.201 \mathbf{e}_y + 5.080 \mathbf{e}_z \text{ m/s} \qquad \text{Ans.}$$

The relative acceleration equation (Eq. 14-21b) is

$$\mathbf{a}_B = \mathbf{a}_A + \boldsymbol{\alpha} \times \mathbf{r}_{B/A} + \boldsymbol{\omega} \times (\boldsymbol{\omega} \times \mathbf{r}_{B/A}) + \mathbf{a}_{Brel} + 2\boldsymbol{\omega} \times \mathbf{v}_{Brel}$$

where

$$\mathbf{a}_A = \mathbf{0}$$
$$\boldsymbol{\alpha} \times \mathbf{r}_{B/A} = \mathbf{0}$$
$$\boldsymbol{\omega} \times (\boldsymbol{\omega} \times \mathbf{r}_{B/A}) = \dot{\theta}_1 \mathbf{e}_z \times (-6.928 \mathbf{e}_x) = -5.543 \mathbf{e}_y \text{ m/s}^2$$
$$2\boldsymbol{\omega} \times \mathbf{v}_{Brel} = 2\dot{\theta}_1 \mathbf{e}_z \times (-1.201 \mathbf{e}_y + 5.080 \mathbf{e}_z) = 1.922 \mathbf{e}_x \text{ m/s}^2$$

and (Fig. 14-54b)

$$\mathbf{a}_{Brel} = (\ddot{s} - s\dot{\theta}_2^2)(\cos \theta_2 \; \mathbf{e}_y + \sin \theta_2 \; \mathbf{e}_z)$$
$$\qquad + (s\ddot{\theta}_2 + 2\dot{s}\dot{\theta}_2)(-\sin \theta_2 \mathbf{e}_y + \cos \theta_2 \mathbf{e}_z)$$
$$= [0 - (10)(0.5)^2][\cos 30° \; \mathbf{e}_y + \sin 30° \; \mathbf{e}_z]$$
$$\qquad + [0 + 2(1.5)(0.5)][-\sin 30° \; \mathbf{e}_y + \cos 30° \; \mathbf{e}_z]$$
$$= -2.915 \mathbf{e}_y + 0.049 \mathbf{e}_z \text{ m/s}^2$$

Therefore

$$\mathbf{a}_B = 1.922 \mathbf{e}_x - 8.458 \mathbf{e}_y + 0.049 \mathbf{e}_z \text{ m/s}^2 \qquad \text{Ans.}$$

PROBLEMS

Introductory Problems

14-137 Sketch the final position of the book of Fig. P14-137 after successive rotations of

$$\Delta\theta_x = 90° \qquad \Delta\theta_y = 90° \qquad \Delta\theta_z = 90°$$

Also determine the single rotation (axis and angle) that is equivalent to this combination of rotations.

14-138 Sketch the final position of the book of Fig. P14-137 after successive rotations of

$$\Delta\theta_x = 180° \qquad \Delta\theta_y = 90° \qquad \Delta\theta_z = 180°$$

Also determine the single rotation (axis and angle) that is equivalent to this combination of rotations.

14-139 Sketch the final position of the book of Fig. P14-137 after successive rotations of

$$\Delta\theta_z = 90° \qquad \Delta\theta_y = 90° \qquad \Delta\theta_z = 90° \qquad \Delta\theta_x = 90°$$

Also determine the single rotation (axis and angle) that is equivalent to this combination of rotations.

14-140 Sketch the final position of the book of Fig. P14-137 after successive rotations of

$$\Delta\theta_x = 90° \qquad \Delta\theta_y = 90° \qquad \Delta\theta_x = -90° \qquad \Delta\theta_z = 90°$$

Also determine the single rotation (axis and angle) that is equivalent to this combination of rotations.

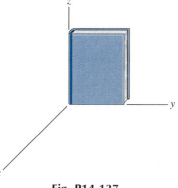

Fig. P14-137

14-141* Determine which of the following rotations of an object will result in the same final position of the object.

a. $\Delta\theta_x = 90°, \Delta\theta_y = 90°$
b. $\Delta\theta_x = 90°, \Delta\theta_z = 90°$
c. $\Delta\theta_x = 90°, \Delta\theta_z = -90°$
d. $\Delta\theta_y = -90°, \Delta\theta_x = 90°$
e. $\Delta\theta_x = 90°, \Delta\theta_y = 90°, \Delta\theta_z = 90°$
f. $\Delta\theta_y = 90°, \Delta\theta_z = 90°, \Delta\theta_x = -90°$
g. $\Delta\theta_z = -90°, \Delta\theta_x = 90°, \Delta\theta_y = 90°$
h. $\Delta\theta_y = -90°, \Delta\theta_x = 90°, \Delta\theta_y = 90°, \Delta\theta_x = 90°$
i. $\Delta\theta_x = 90°, \Delta\theta_y = 90°, \Delta\theta_x = -90°, \Delta\theta_y = -90°$

14-142* An electric motor spins the fan blades of Fig. P14-142 at a constant rate of $\omega_1 = 600$ rev/min. At the same time, the motor rotates about a vertical axis at a constant rate of $\omega_z = 5$ rev/min. For the instant shown, determine the total angular velocity $\boldsymbol{\omega}$ and total angular acceleration $\boldsymbol{\alpha}$ of the fan blades.

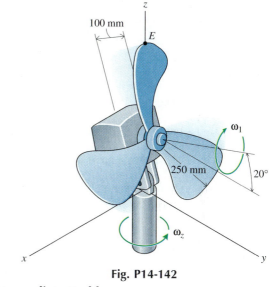

Fig. P14-142

Intermediate Problems

14-143 The uniform cylinder of Fig. P14-143 has a radius of 2 in. and is 8 in. long. The cylinder rotates about its own axis at a constant rate of $\omega_1 = 20$ rad/s. Simultaneously, the yoke holding the cylinder is rotating about a vertical axis at a constant rate of $\omega_2 = 8$ rad/s. For the instant shown, determine the total angular velocity $\boldsymbol{\omega}$ and total angular acceleration $\boldsymbol{\alpha}$ of the disk.

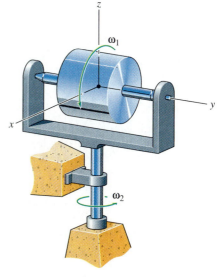

Fig. P14-143

14-144 A platform rotates at a constant angular speed of $\omega_1 = 10$ rad/s while a 100-mm-diameter cylinder mounted on the platform rotates relative to the platform at a constant angular speed of $\omega_2 = 25$ rad/s (Fig. P14-144). If $a = 50$ mm, determine the total angular velocity ω and total angular acceleration α of the disk at the instant shown.

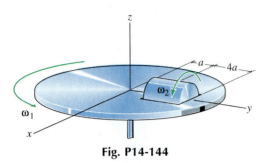

Fig. P14-144

14-145* The 15-in.-long pipe AB of Fig. P14-145 is rotating about a vertical axis at a rate of ω_z. At the same time, the 12-in.-long pipe BC is rotating about AB at a rate of ω_2 and the 10-in.-diameter disk rotates about pipe BC at a rate of ω_1. For the instant shown (when BC is in the horizontal plane, $\omega_1 = 5$ rad/s = constant, $\omega_2 = \dot{\omega}_2 = 0$, $\omega_z = 3$ rad/s, and $\dot{\omega}_z = -10$ rad/s²), determine

a. The total angular velocity ω and the total angular acceleration α of the disk.

b. The velocity $\mathbf{v}_D$ and acceleration $\mathbf{a}_D$ of point D on the rim of the disk.

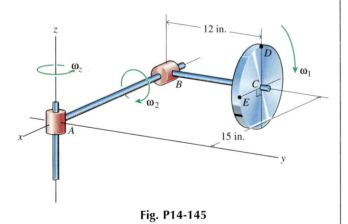

Fig. P14-145

14-146* The 400-mm-diameter disk of Fig. P14-146 is rigidly attached to a 750-mm-long axle and rolls without slipping on a fixed surface in the xy-plane. The axle, which is perpendicular to the disk, is attached to a ball-and-socket joint at A and is free to pivot about A. As the disk rotates about its axle with angular velocity ω_1, the axle also rotates about a vertical axis with angular velocity ω_2. If $\omega_1 = 2$ rad/s and $\dot{\omega}_1 = -5$ rad/s² at the instant shown, determine

a. The total angular velocity ω and the total angular acceleration α of the disk at this instant.

b. The velocity $\mathbf{v}_C$ and acceleration $\mathbf{a}_C$ of point C on the rim of the disk at this instant.

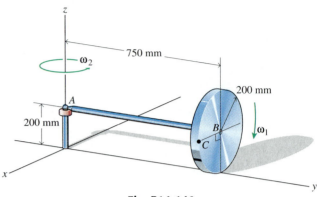

Fig. P14-146

14-147* The 48-in.-long slender rod of Fig. P14-147 is connected to sliders A and B by ball-and-socket joints. At the instant shown, $x = 20$ in. and slider A is moving in the negative x-direction at a constant speed of 18 in./s. Determine

a. The velocity $\mathbf{v}_B$ and acceleration $\mathbf{a}_B$ of slider B at this instant.

b. The angular velocity ω and the angular acceleration α of the rod at this instant. (Assume that the rod is not rotating about its own axis.)

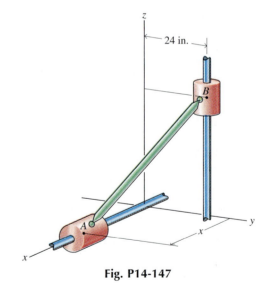

Fig. P14-147

14-148 The 1200-mm-long slender rod of Fig. P14-148 is connected to sliders A and B by ball-and-socket joints. At the instant shown, $y = 750$ mm and slider B is moving in the positive y-direction at a constant speed of 100 mm/s. Determine

a. The velocity $\mathbf{v}_A$ and acceleration $\mathbf{a}_A$ of slider A at this instant.

b. The angular velocity $\boldsymbol{\omega}$ and the angular acceleration $\boldsymbol{\alpha}$ of the rod at this instant. (Assume that the rod is not rotating about its own axis.)

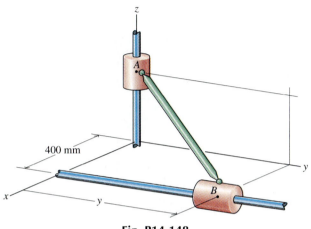

Fig. P14-148

14-149 The 50-in.-long slender rod of Fig. P14-149 is connected to the sliders A and B by ball-and-socket joints. When the slider B crosses the x-axis ($x_B = 24$ in., $y_B = 0$ in., $z_B = 0$ in.), the velocity and acceleration of slider A are $\dot{y} = 18$ in./s and $\ddot{y} = -6$ in./s^2, respectively. For this instant, determine

a. The velocity $\mathbf{v}_B$ and acceleration $\mathbf{a}_B$ of slider B.

b. The angular velocity $\boldsymbol{\omega}$ and the angular acceleration $\boldsymbol{\alpha}$ of the rod. (Assume that the rod is not rotating about its own axis.)

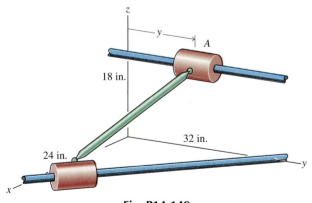

Fig. P14-149

14-150 The 600-mm-diameter wheel of Fig. P14-150 is rotating at a constant rate of $\omega_0 = 5$ rad/s. The 1000-mm-long slender rod AB is connected to the rim of the wheel at A and to the slider B by ball-and-socket joints. For the instant shown when $\theta = 90°$, determine

a. The velocity $\mathbf{v}_B$ and acceleration $\mathbf{a}_B$ of slider B.

b. The angular velocity $\boldsymbol{\omega}$ and the angular acceleration $\boldsymbol{\alpha}$ of the rod. (Assume that the rod is not rotating about its own axis.)

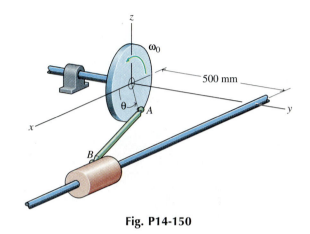

Fig. P14-150

14-151* Crank OA rotates about an axis parallel to the y-axis as shown in Fig. P14-151. The 36-in.-long slender rod is connected to the crank at A and to the slider B by ball-and-socket joints. At the instant when $\theta = 0°$, the angular velocity and angular acceleration of the crank are $\dot{\theta} = 3$ rad/s and $\ddot{\theta} = 10$ rad/s^2, respectively. For this instant, determine

a. The velocity $\mathbf{v}_B$ and acceleration $\mathbf{a}_B$ of slider B.

b. The angular velocity $\boldsymbol{\omega}$ and the angular acceleration $\boldsymbol{\alpha}$ of the rod. (Assume that the rod is not rotating about its own axis.)

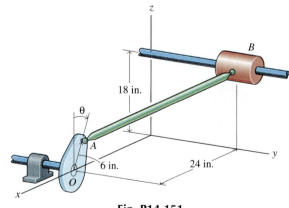

Fig. P14-151

Challenging Problems

14-152* The fire truck ladder of Fig. P14-152 is rotating about a vertical axis at a constant rate of $\omega_1 = 0.8$ rad/s with $\dot{s} = 0$ m/s, $\ddot{s} = -2.5$ m/s², $\dot{\theta}_2 = 0$ rad/s, and $\ddot{\theta}_2 = -1.5$ rad/s² when $s = 10$ m and $\theta_2 = 30°$. Determine the velocity $\mathbf{v}_B$ and acceleration $\mathbf{a}_B$ of the end of the ladder at this instant.

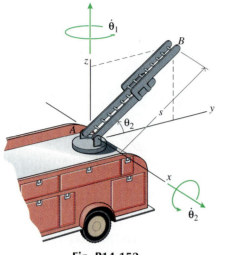

Fig. P14-152

14-153 A bead is sliding along a bent rod which is rotating about the x-axis (Fig. P14-153). At the instant shown, the rod is in the xz-plane and $\omega = 5$ rad/s, $\dot{\omega} = 18$ rad/s², $s = 8$ in., $\dot{s} = 1$ in./s, and $\ddot{s} = -2.5$ in./s². Determine the velocity $\mathbf{v}_B$ and acceleration $\mathbf{a}_B$ of the bead at this instant.

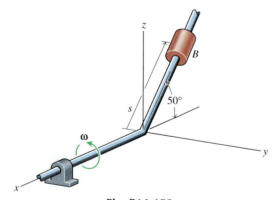

Fig. P14-153

14-154* A bead B slides along a slot in a 500-mm-diameter disk that is rotating about a vertical axis (Fig. P14-154). At the instant shown, $\omega = 3$ rad/s, $\dot{\omega} = 8$ rad/s², $s = 200$ mm, $\dot{s} = 250$ mm/s, and $\ddot{s} = -50$ mm/s². Determine the velocity $\mathbf{v}_B$ and acceleration $\mathbf{a}_B$ of the bead at this instant.

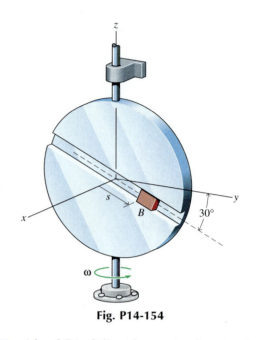

Fig. P14-154

14-155 A bead B is sliding along a circular wire ring that is rotating about the y-axis (Fig. P14-155). At the instant shown, the 20-in.-diameter ring is in the yz-plane and $\omega = 8$ rad/s, $\dot{\omega} = 12$ rad/s², $\theta = 30°$, and $\dot{\theta} = 10$ rad/s = constant. Determine the velocity $\mathbf{v}_B$ and acceleration $\mathbf{a}_B$ of the bead at this instant.

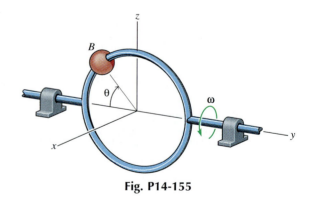

Fig. P14-155

Computer Problems

C14-156 A bead B slides along a slot in a 500-mm-diameter disk which is rotating about a vertical axis (Fig. P14-156). Let the xyz-coordinate system rotate with the disk at a constant angular speed of $\omega = 3$ rad/s so that the disk is always in the yz-plane. Then, if the bead slides along the slot at a constant speed of $\dot{s} = 200$ mm/s, compute and plot

a. The velocity components v_{Bx}, v_{By}, v_{Bz} as functions of s, the bead's position along the slot (-250 mm $< s <$ 250 mm).

b. The acceleration components a_{Bx}, a_{By}, a_{Bz} as functions of s, the bead's position along the slot (-250 mm $< s <$ 250 mm).

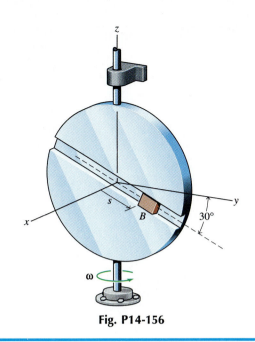

Fig. P14-156

SUMMARY

Kinematics is the study of how objects move. For solid bodies, a full description of the motion requires giving both the location and the orientation of the body. The kinematics of solid bodies involves both linear and angular quantities. A sound understanding of kinematics is a necessary background for the study of kinetics, which relates the motion to the forces that cause it.

All solid bodies will be considered to be rigid. In a rigid body, the distance between all pairs of particles is fixed and independent of time. Also, the angles between all triples of points is fixed.

In the translation of a rigid body, the orientation of every straight line in the body is fixed: horizontal lines remain horizontal, and vertical lines remain vertical. This means that the motion of every point in the rigid body is the same as every other point in the rigid body. The kinematics of the particles that make up a rigid body undergoing translation is identical to the kinematics of particle motion.

In a plane motion of a rigid body, each particle in the body remains in a single plane. Coplanar translation and rotation about a fixed axis are specific types of plane motion. A general plane motion is any plane motion for which lines in the body rotate but no point in the body is fixed. General plane motions of a rigid body consists of a translation of a body with some point plus a rotation of the body about that point.

There are two general approaches to solving general plane motion problems: absolute motion analysis and relative motion analysis. In the absolute motion approach, geometric relationships are written that describe the constraints that act on the body and its interaction with other bodies. These relationships are then used to describe the location and motion of other points on the body. The relative motion approach uses the rigidity of the body to relate the velocity and acceleration of two different points in the same rigid body. Since the distance between any two points in a rigid body is fixed, the expressions for the relative velocity and the relative acceleration take particularly simple forms, which depend only on the angular velocity and angular acceleration of the body.

Either approach may be used to solve a particular problem. Some problems are easily described geometrically and are easily handled by the absolute motion approach. Problems that are not easily described geometrically are usually solved using the relative motion approach. For many problems the choice is a matter of personal preference.

In the general plane motion of a rigid body, no point is fixed for all time. However, at any instant of time, it is always possible to find a point in the body (or the body extended) that has zero velocity. Once the instantaneous center is located, the velocity of any other point in the body is found using the relative velocity equation. Use of the instantaneous center is not required to solve any problem. It is just another way of expressing the relative velocity equation.

The instantaneous center of zero velocity for a rigid body in general plane motion is not fixed. Therefore, different points of the rigid body will be instantaneous centers at different instants of time, and the location of the instantaneous center of zero velocity will move with respect to time. The instantaneous center of zero velocity must not be used to compute accelerations.

There are several types of problems for which it is convenient to describe the position or motion of one particle relative to a rotating coordinate system. In particular, some mechanisms are connected by pins that slide in grooves or slots. Relative motion is conveniently specified by giving the translational and rotational motion of the member containing the slot, the shape of the slot, and the rate of travel of the pin along the slot. Differentiation of the relative position equation to get the relative velocity and relative acceleration equations must take into account the rotation of the coordinate system. This gives three new terms: $\mathbf{v}_{Brel}$, $\mathbf{a}_{Brel}$, and $2\boldsymbol{\omega} \times \mathbf{v}_{Brel}$ in the relative velocity and relative acceleration equations.

REVIEW PROBLEMS

14-157*   The wheel of Fig. P14-157 has a velocity of 2 ft/s up the inclined plane and a clockwise angular velocity of 5 rad/s. Determine the rate of slip at the contact point A (the relative velocity between the point on the wheel and the stationary surface).

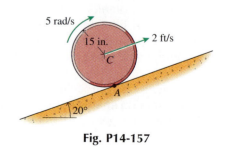

Fig. P14-157

14-158 The 2-m-long rod *AB* of Fig. P14-158 slides on a 1-m-high step. If end *B* of the rod is made to move to the right at a constant speed of 0.25 m/s,

a. Determine the angular velocity ω_{AB} and angular acceleration α_{AB} of the rod at the instant when $\theta = 50°$.

b. Determine the velocity $\mathbf{v}_C$ and acceleration $\mathbf{a}_C$ of point *C*, the point on the rod in contact with the corner of the step, at the instant when $\theta = 50°$.

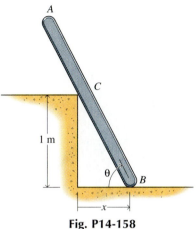

Fig. P14-158

14-159 The axles of two wheels are rigidly attached to a vehicle as shown in Fig. P14-159. The velocity and acceleration of the vehicle are 30 ft/s to the right and 3 ft/s² to the left, respectively. The wheels are further connected by a horizontal bar *AB*. If the smaller wheel rolls without slipping, determine the angular acceleration of both wheels and the relative acceleration between the larger wheel and the surface at point *C*.

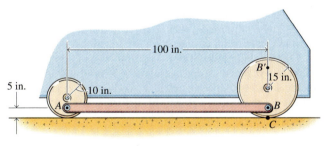

Fig. P14-159

14-160* The case of the roller bearing shown in Fig. P14-160 is fixed, whereas the inner shaft rotates at a constant rate of 5000 rev/min. If the roller bearings roll without slipping on the 50-mm-diameter shaft and on the 60-mm-diameter casing, determine the velocity **v** and acceleration **a**

a. Of point *A* on the surface of the shaft.
b. Of point *C* at the middle of a roller bearing.
c. Of point *B* on the surface of a roller bearing.
d. Of point *D* on the surface of a roller bearing.

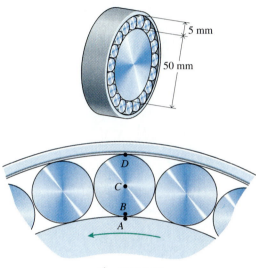

Fig. P14-160

14-161 A small block *B* rotates with the horizontal turntable *A* of Fig. P14-161. The distance between the block and the axis of rotation is $r = 5$ in., the angular acceleration of the turntable is $\alpha = -2$ rad/s² = constant, and the initial angular velocity is $\omega = 15$ rad/s. Determine the angle ϕ between the acceleration of the block $\mathbf{a}_B$ and the radial line *OB* at $t = 6$ s.

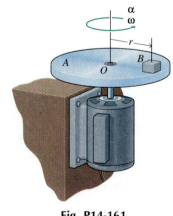

Fig. P14-161

14-162 An electric motor spins the fan blades of Fig. P14-162 at a constant rate of 600 rev/min. At the same time, the motor rotates about a vertical axis with an angular velocity ω_z and an angular acceleration α_z. Determine the total angular velocity $\boldsymbol{\omega}$ and angular acceleration $\boldsymbol{\alpha}$ of the fan blades for the instant when $\omega_z = 3$ rad/s and $\alpha_z = 12$ rad/s².

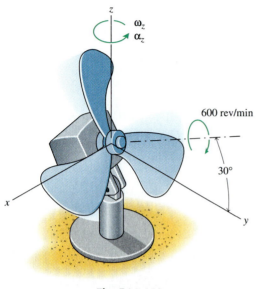

Fig. P14-162

14-163* The truck shown in Fig. P14-163 is initially at rest at a stop light. When the light turns green, the truck accelerates at a constant rate of 0.8 ft/s² and the 3-ft-diameter tank begins to roll backward with a constant angular acceleration of $\alpha = 0.025$ rad/s². Determine

a. How far the truck travels before the tank rolls off the back of the truck.

b. The velocity $\mathbf{v}_C$ of the center of the tank and the angular velocity ω of the tank when it falls off the back of the truck.

c. The slip velocity (relative velocity between the tank and the road) when the tank hits the road.

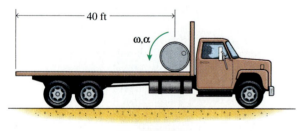

Fig. P14-163

14-164 The blades of the helicopter shown in Fig. P14-164 are rotating at a constant rate of 360 rev/min. In addition, the helicopter is traveling forward with a speed of $v = 250$ km/h. For the instant in which the blade is perpendicular to the velocity $\mathbf{v}$ ($\theta = 90°$).

a. Determine the velocity $\mathbf{v}$ and the acceleration $\mathbf{a}$ of both tips of the blades.

b. Find the instantaneous center of zero velocity of the blade.

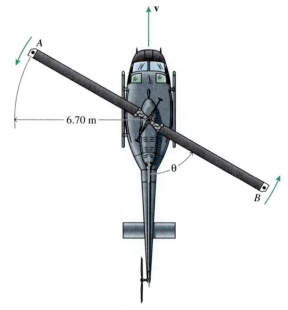

Fig. P14-164

14-165* Initially, disk B of Fig. P14-165 is rotating clockwise at 200 rev/min and disk A is rotating counterclockwise at 500 rev/min. When the two disks are brought together, they slip for a period of 5 s, during which the angular acceleration of each disk is constant. At the end of the 5 s, the disks roll without slipping on one another and disk B has reached a final angular velocity of 250 rev/min clockwise. Determine the angular acceleration of each disk and the final angular velocity of disk A.

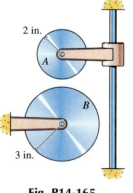

Fig. P14-165

14-166* The piston of Fig. P14-166 is connected to the crankshaft by a 650-mm-long connecting rod. At the instant shown, the crankshaft has an angular velocity of 360 rev/min and an angular acceleration of 5 rad/s², both clockwise. Determine a_C, the acceleration of the piston, and α_{BC}, the angular acceleration of the connecting rod BC, at this instant.

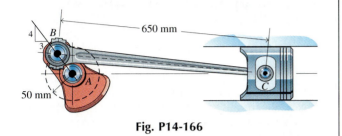

Fig. P14-166

14-167 The radar antenna of Fig. P14-167 is tracking an airplane. At the instant shown, the radar dish is rotating about a vertical axis at a constant rate of 0.4 rad/s, $\phi = 30°$, $\dot{\phi} = -0.5$ rad/s, and $\ddot{\phi} = 0.02$ rad/s². For this instant determine

a. The angular velocity $\boldsymbol{\omega}$ and the angular acceleration $\boldsymbol{\alpha}$ of the antenna.
b. The velocity $\mathbf{v}_H$ and the acceleration $\mathbf{a}_H$ of the signal horn H.

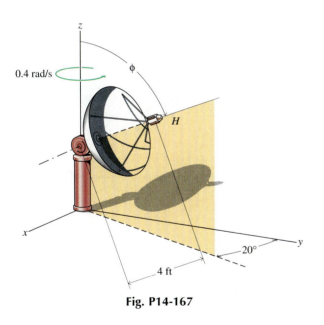

Fig. P14-167

14-168 When the 800-mm long arrow of Fig. P14-168 lodges in the vertical bar AB, the arrow and bar rotate as a single rigid body about the frictionless pivot at A. Express $\mathbf{a}_{Ga}$, the acceleration of the mass center of the arrow, and $\mathbf{a}_{GAB}$, the acceleration of the mass center of the bar AB, in terms of ω_{AB} the angular velocity and α_{AB} the angular acceleration of the bar.

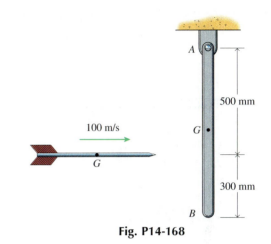

Fig. P14-168

14-169* The tank shown in Fig. P14-169 is moving forward at a constant speed of 30 mi/h. At the instant shown, the turret is facing forward and rotating at a constant rate of $\omega_1 = 2$ rad/s while the gun barrel is being raised at a rate of $\omega_2 = 0.5$ rad/s and $\dot{\omega}_2 = 0.03$ rad/s². If the gun barrel is 10 ft long, determine the velocity $\mathbf{v}_A$ and acceleration $\mathbf{a}_A$ of the end of the gun barrel.

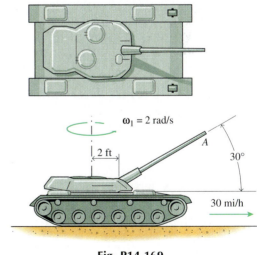

Fig. P14-169

14-170* The shuttle C of Fig. P14-170 is made to oscillate back and forth by the rotation of the 0.50-m-diameter wheel D. If the wheel is rotating at a constant angular rate of 30 rev/min, determine the velocity $\mathbf{v}_C$ and acceleration $\mathbf{a}_C$ of the shuttle for the instant shown in which arm AB is horizontal.

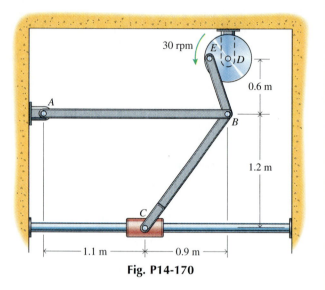

Fig. P14-170

14-171 At the instant shown, the truck of Fig. P14-171 has a velocity of 30 mi/h and is accelerating at a rate of 5 ft/s². If end A of the 10-ft-long bar AB is sliding backward at a constant rate of 2 ft/s relative to the truck,

a. Determine the velocity $\mathbf{v}_G$ and the acceleration $\mathbf{a}_G$ of the center of the bar AB.

b. Locate C, the instantaneous center of zero velocity of the bar AB.

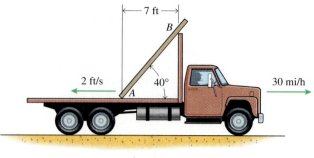

Fig. P14-171

14-172 The two gears shown in Fig. P14-172 rotate about their respective shafts. Bars AB and BC are each 125 mm long. At the instant shown, the larger gear is rotating counterclockwise with a constant angular velocity of $\omega_A = 12$ rad/s. Determine the angular acceleration of the two bars α_{AB} and α_{BC} and the acceleration of pin B at this instant.

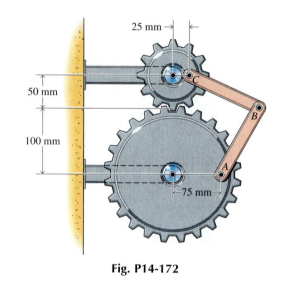

Fig. P14-172

*O*n thrill rides such as a roller coaster, riders experience speed, excitement, and a sense of impending danger. For example, if the cars go too slowly through the loop, they will tend to fall away from the track. Similarly, if the cars go too fast over a hill, they will tend to rise up from the track. In both cases, the track must exert pulling forces on the cars to hold them to the track.

Chapter 15

KINETICS OF PARTICLES: NEWTON'S LAWS

15-1 INTRODUCTION

Previously, in Chapters 13 (kinematics of particles) and 14 (kinematics of rigid bodies), the motions of particles and rigid bodies were studied without considering the forces required to produce the motion. Relationships were developed that describe how the velocity and the acceleration of a body change either with time or with a change in position. In an earlier course in statics, methods were developed for determining the resultant force **R** and the resultant couple **C** for any force system that can act on a body. When the resultant of the force system is zero, the body is in equilibrium (at rest or moving with a constant velocity). When the resultant of the force system is not zero, the body experiences accelerated motion. Unbalanced forces and the motions they produce (kinetics) will be the topic for the remaining chapters of this book.

The motion a body experiences when it is subjected to an unbalanced system of forces can be established by using three different methods: (1) the force, mass, and acceleration method, (2) the work and energy method, and (3) the impulse and momentum method. The most useful method for a particular problem depends on the nature of the force system (constant or variable) and the information sought (reactions, velocities, accelerations, etc.).

15-2 EQUATIONS OF MOTION

Prior to the time of Galileo and Newton, it was generally assumed that a body at rest was in its natural state; therefore, some kind of force was required to keep it moving. The great contribution of Newton to the science of mechanics was his realization that a force is not needed to keep a body moving once it has been set in motion and that the effect of a force is to change the velocity, not to maintain the velocity.

15-2.1   Newton's Second Law

Newton's three laws of motion, as they are commonly expressed today, were listed in Section 12-2. The first law, pertaining to a particle at rest or moving with a constant velocity, and the third law, governing action and reaction between interacting bodies, were used in developing the concepts of statics. Newton's second law of motion, which relates the accelerated motion of a particle to the forces producing the motion, forms the basis for studies in dynamics. It has previously been noted that Newton's first law, which covers the case of particle equilibrium, is a special case of his second law. When the resultant force on a particle is zero ($\mathbf{R} = \mathbf{0}$), the acceleration of the particle is zero ($\mathbf{a} = \mathbf{0}$); therefore, the particle is either at rest or moving with a constant velocity (in equilibrium). A modern statement of Newton's second law, as presented in Section 12-2, is

Law 2.   When a particle is acted on by an unbalanced force, the particle will be accelerated in the direction of the force; the magnitude of the acceleration will be directly proportional to the force and inversely proportional to the mass of the particle.

Mathematically, Newton's second law is expressed as

$$\mathbf{a} = k\frac{\mathbf{F}}{m} \tag{15-1}$$

where

   $\mathbf{a}$ is the acceleration of the particle
   $\mathbf{F}$ is the force acting on the particle
   m is the mass of the particle
   k is a proportionality constant, which depends on the units selected for the acceleration, force, and mass. A system in which $k = 1$ has consistent kinetic units.

With $k = 1$, Eq. 15-1 can be written in its familiar form

$$\mathbf{F} = m\mathbf{a} \tag{15-2}$$

At the present time, two systems with consistent kinetic units are used by engineers in the United States: the International System of Units (SI units), and the U.S. customary system of units. In the SI system, the base quantities are length (m), mass (kg), and time (s). The unit of force, called a newton (N), is defined as the force required to give a mass of 1 kg an acceleration of 1 m/s^2. The SI system is an absolute system, since the three base units are the same in any environment (on the earth, on the moon, anywhere in space, etc.).

A system in which the base quantities are length (ft), force (lb), and time (s) continues to be widely used in engineering in the United States. The unit of time (the second) is the same as the corresponding SI unit. The unit of length (the foot) is defined as 0.3048 m. The unit of force (the pound) is defined as the weight at sea level and at a latitude of 45° of a platinum standard having a mass of 0.453 592 43 kg. Since the unit of force depends on the gravitational attraction of the earth, U.S. customary units do not form an absolute system. The unit of mass in the U.S. customary system is the slug. By definition, a slug of mass receives an acceleration of 1 ft/s^2 when a force of 1 lb is applied to it.

Equation 15-2 expresses the fact that the magnitudes of **F** and **a** are proportional and that the vectors **F** and **a** have the same direction. Equation 15-2 is valid both for constant forces and for forces that vary with time (in either magnitude or direction).

When Eq. 15-2 is used to solve kinetics problems, the reference axes for the acceleration measurements must be fixed in space (have a constant orientation with respect to the stars). Such a system of axes is called a newtonian frame of reference or a primary inertial system. When a system of reference axes is attached to the earth, the measured acceleration is not the absolute acceleration required for application of Eq. 15-2, owing to rotation of the earth on its axis and the acceleration of the earth with respect to the sun as the earth moves in its orbit around the sun. For most engineering problems on the surface of the earth, the corrections required to compensate for the acceleration of the earth with respect to the primary system are negligible and the accelerations measured with respect to axes attached to the surface of the earth may be treated as absolute. Equation 15-2 is not valid, however, if **a** represents a relative acceleration measured with respect to a system of moving axes on earth. Also, the acceleration components of the earth's motion must be considered when dealing with such problems as flight paths for spacecraft and ballistic missile trajectories.

15-2.2 Newton's Law of Gravitation and the Weight of a Body

The law that describes the mutual attraction between two isolated bodies was formulated by Newton and is known as the *law of gravitation*. This law is expressed mathematically by

$$F = \frac{Gm_1m_2}{r^2} \tag{15-3}$$

where

F is the magnitude of the mutual force of attraction between the two bodies
G is the universal gravitational constant
m_1 is the mass of one of the bodies
m_2 is the mass of the second body
r is the distance between the centers of mass of the two bodies

The mutual forces of attraction between the two bodies represent the action of one body on the other; therefore, they obey Newton's third law, which requires that they be equal in magnitude, opposite in direction, and collinear (lie along the line joining the centers of mass of the two bodies).

The mass m of a body is an absolute quantity that is independent of the position of the body and independent of the surroundings in which the body is placed. The weight W of a body is the gravitational attraction exerted on the body by the planet earth (or by any other massive body such as the moon). Therefore, the weight of a body depends on the position of the body relative to some other body. For a body of mass m at the surface of the earth, the weight W of the body is

$$W = \frac{Gm_e m}{r_e^2} = mg \qquad (15\text{-}4)$$

where

m_e is the mass of the earth
r_e is the mean radius of the earth
$g = Gm_e/r_e^2$ is the gravitational constant for the earth

In the SI system, the weight $\mathbf{W}$ of a body (the force of gravity), like any other force, is expressed in newtons. In the U.S. Customary system, the weight of a body is expressed in pounds. The internationally accepted values of g relative to the earth at sea level and at a latitude of 45° are 9.80665 m/s^2 and 32.1740 ft/s^2. These values of g are commonly rounded off for routine engineering work to 9.81 m/s^2 and 32.2 ft/s^2.

15-2.3   Equations of Motion for a Single Particle

When a system of forces $\mathbf{F}_1$, $\mathbf{F}_2$, $\mathbf{F}_3$, . . ., $\mathbf{F}_n$ acts on a particle, the resultant is a force $\mathbf{R}$ with a line of action through the mass center of the particle, since any system of forces acting on a particle must be a concurrent force system. The motion of the particle resulting from the action of the resultant $\mathbf{R}$ is given by Newton's second law of motion as

$$\mathbf{R} = \Sigma\mathbf{F} = m\mathbf{a} \qquad (15\text{-}5)$$

Equation 15-5 is a vector equation, and it must be satisfied regardless of the coordinate system used to express its components. For example, the rectangular cartesian components of Eq. 15-5 are

$$R_x = \Sigma F_x = ma_x \qquad (15\text{-}6a)$$
$$R_y = \Sigma F_y = ma_y \qquad (15\text{-}6b)$$
$$R_z = \Sigma F_z = ma_z \qquad (15\text{-}6c)$$

Similarly, the plane, polar-coordinate components of Eq. 15-5 are

$$R_r = \Sigma F_r = ma_r \qquad (15\text{-}7a)$$
$$R_\theta = \Sigma F_\theta = ma_\theta \qquad (15\text{-}7b)$$

and the plane, normal, and tangential-coordinate components of Eq. 15-5 are

$$R_n = \Sigma F_n = ma_n \qquad (15\text{-}8a)$$
$$R_t = \Sigma F_t = ma_t \qquad (15\text{-}8b)$$

For many problems in particle kinetics, it is convenient to have the acceleration of the particle expressed as a function of the position (x, y, z) of the particle. For these cases, combining Eqs. (15-6) and (13-8) yields

$$\Sigma F_x = ma_x = m\frac{d^2x}{dt^2} = m\ddot{x} \qquad (15\text{-}9a)$$

$$\Sigma F_y = ma_y = m\frac{d^2y}{dt^2} = m\ddot{y} \qquad (15\text{-}9b)$$

$$\Sigma F_z = ma_z = m\frac{d^2z}{dt^2} = m\ddot{z} \qquad (15\text{-}9c)$$

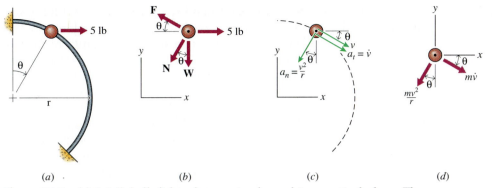

F

5 lb

θ

N W

y

x

5 lb

θ

θ

v

$a_t = \dot{v}$

θ

$a_n = \frac{v^2}{r}$

y

x

y

θ

x

θ

$m\dot{v}$

$\frac{mv^2}{r}$

θ

r

(a)   (b)   (c)   (d)

Figure 15-1   (a) A 1-lb ball slides along a circular rod in a vertical plane. The ball is pulled by a constant 5-lb force that acts in the horizontal direction.
(b) The free-body diagram of the ball includes the weight force $W = 1$ lb, the 5-lb applied force, and the normal reaction force N.
(c) Since the ball travels in a circular path, the velocity and acceleration components are easily represented in terms of polar coordinates. The components can be converted to xy-coordinates if necessary.
(d) The kinetic diagram shows the components of the $m\mathbf{a}$ vector—the right-hand side of Newton's second law.

It is important to draw a free-body diagram (such as Fig. 15-1b) so that all forces that act on the particle are identified. As in statics, the free-body diagram should show only the actual forces which act on the particle. Forces whose direction is known (such as the weight of the body, tensions in ropes, normal and frictional forces of contact with a surface, etc.) should be drawn in their known direction. Forces of known magnitude should be labeled with the known value. Totally unknown forces are often expressed in terms of their components. The components of these unknown forces may be drawn in either the positive or negative coordinate directions but must be labeled appropriately. If the solution of the problem gives the component a positive value, then the direction drawn on the free-body diagram is correct; if the value is negative, then the direction drawn on the free-body diagram is opposite the correct direction of the force.

Velocities and accelerations may be shown on a separate (kinematic) diagram (Fig. 15-1c) if necessary or useful. The kinematic diagram shows the relationship between the position of the particle and the components of the velocity and acceleration vectors. Students may also find it useful to show the $m\mathbf{a}$ vectors on a kinetic diagram (Fig. 15-1d) next to the free-body diagram used for the forces. Like the forces on the free-body diagram, if the magnitude and/or direction of the acceleration is known, it should be drawn accordingly on the kinetic diagram. If neither the magnitude nor the direction of the acceleration is known, its components may be drawn in either the positive or the negative coordinate direction. If the solution of the problem gives the component a positive value, then the direction drawn on the kinetic diagram is correct; if the value is negative, then the direction drawn on the kinetic diagram is opposite the correct direction of the acceleration.

Regardless of the coordinate system used to express Newton's second law, it is important to use the same components on both sides of the equation. That is, for the x-component of Newton's second law (with the positive direction to the right), the 5-lb force shown on Fig.

15-1b would be entered on the left-hand side of Eq. 15-6a with a positive sign, while the normal and friction forces would be entered on the left-hand side as $-N \sin \theta$ and $-F \cos \theta$, respectively. Similarly, for the y-component of Newton's second law (with the positive direction upward), the friction force would be entered on the left-hand side of Eq. 15-6b as $+F \sin \theta$, while the normal and weight forces would be entered on the left-hand side as $-N \cos \theta$ and $-W$, respectively. Keeping the positive directions for Eq. 15-6a to the right and for Eq. 15-6b upward, the acceleration vectors shown on Fig. 15-1d would be entered on the right-hand side of Eq. 15-6a as $+m\dot{v} \cos \theta$ and $-(mv^2/r) \sin \theta$ and on the right-hand side of Eq. 15-6b as $-m\dot{v} \sin \theta$ and $-(mv^2/r) \cos \theta$, respectively. The resulting equations of motion would then be

$$+\rightarrow \Sigma F_x = ma_x: \quad 5 - N \sin \theta - F \cos \theta = m\dot{v} \cos \theta - \frac{mv^2}{r} \sin \theta$$

$$+\uparrow \Sigma F_y = ma_y: \quad F \sin \theta - N \cos \theta - W = -m\dot{v} \sin \theta - \frac{mv^2}{r} \cos \theta$$

15-2.4 Equations of Motion for a System of Particles

The equations of motion for a system of particles can be obtained by applying Newton's second law to each particle within the system. As an example, consider the collection of n particles shown in Fig. 15-2a. The ith particle has a mass m_i, and its location is specified with respect to an appropriate system of reference axes by using the position vector $\mathbf{r}_i$ from the origin of coordinates. Each particle in the system (see Fig. 15-2b) may be subjected to a system of external forces with a resultant $\mathbf{R}_i$ and a system of internal forces $\mathbf{f}_{i1}, \mathbf{f}_{i2}, \mathbf{f}_{i3}, \mathbf{f}_{ij}, \ldots, \mathbf{f}_{in}$. The internal forces result from elastic interactions between particles and from electric or magnetic effects. The internal force exerted by particle p_j on particle p_i is denoted as $\mathbf{f}_{ij}$. Applying Newton's second law to the ith particle yields

$$\mathbf{R}_i + \sum_{j=1}^{n} \mathbf{f}_{ij} = m_i \mathbf{a}_i \qquad (a)$$

In the summation of internal forces, $\mathbf{f}_{ii}$ is zero since particle p_i does not exert a force on itself.

If a force $\mathbf{f}_{ij}$ is exerted on particle p_i by particle p_j, Newton's third law requires that a force $\mathbf{f}_{ji}$, which is equal in magnitude, opposite in direction, and collinear with $\mathbf{f}_{ij}$, is exerted on particle p_j by particle p_i. Thus,

$$\mathbf{f}_{ij} + \mathbf{f}_{ji} = 0 \qquad (b)$$

An equation of motion for the system is obtained by summing the equations of motion for the n individual particles. Thus,

$$\sum_{i=1}^{n} \mathbf{R}_i + \sum_{i=1}^{n} \left(\sum_{j=1}^{n} \mathbf{f}_{ij} \right) = \sum_{i=1}^{n} m_i \mathbf{a}_i \qquad (c)$$

Since all the internal forces of the system exist as equal, opposite, and collinear pairs, they sum to zero and Eq. c reduces to

$$\mathbf{R} = \sum_{i=1}^{n} \mathbf{R}_i = \sum_{i=1}^{n} m_i \mathbf{a}_i \qquad (15\text{-}10)$$

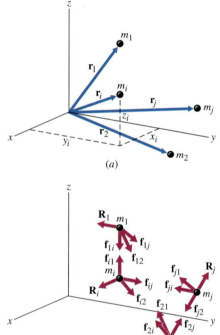

(a)

(b)

Figure 15-2 Both external forces **R** and internal forces **f** act on each particle in the interacting system of particles shown. When the equations of Newton's second law for each particle are added together for all of the particles in the system, the internal forces will cancel in pairs.

Equation 15-10 indicates that the resultant **R** of the external system of applied forces acting on a system of particles is equal to the resultant of the $m\mathbf{a}$ inertia vectors of the particles of the system. The quantity $m\mathbf{a}$ is sometimes referred to as an inertia force; however, since it is neither a contact force nor a gravitational force (weight), most people avoid using the word *force* with the inertia vector $m\mathbf{a}$.

Equation 15-10 can be written in an alternate form if the mass center of the system of particles is considered. The mass center of the system is the point G defined by the position vector $\mathbf{r}_G$, which satisfies the relation

$$m\mathbf{r}_G = \sum_{i=1}^{n} m_i \mathbf{r}_i \qquad (d)$$

where

$$m = \sum_{i=1}^{n} m_i \qquad (e)$$

is the total mass of the system of particles. Differentiating Eq. d with respect to time yields

$$m\dot{\mathbf{r}}_G = \sum_{i=1}^{n} m_i \dot{\mathbf{r}}_i \qquad (f)$$

$$m\ddot{\mathbf{r}}_G = \sum_{i=1}^{n} m_i \ddot{\mathbf{r}}_i \qquad (g)$$

which can be written

$$m\mathbf{a}_G = \sum_{i=1}^{n} m_i \mathbf{a}_i \qquad (h)$$

Combining Eqs. 15-10 and h yields

$$\mathbf{R} = m\mathbf{a}_G \qquad (15\text{-}11)$$

Equation 15-11 is a mathematical expression of the "principle of motion of the mass center" of a system of particles. Equation 15-11 for a system of particles is identical in form to Eq. 15-5 for a single particle. This correspondence indicates that a system of particles can be treated as a single particle with the mass of the system concentrated at the mass center G if the resultant **R** of the external forces applied to the particles of the system has a line of action that passes through the mass center G of the system. In fact, any body can be considered a particle when applying Eq. 15-11. In general, however, the line of action of the resultant force **R** will not pass through the mass center of the system, and the resultant will consist of a resultant force **R**, which passes through the mass center G, and a resultant couple **C**. Rotational motion resulting from the couple **C** is discussed in the next chapter on rigid-body kinetics.

15-3 RECTILINEAR MOTION

The kinematics of a particle subjected to rectilinear motion was described in Section 13-3. Motion in this case is along a straight line and, if the coordinate system is oriented such that the x-axis coincides with the line of motion, the position, velocity, and acceleration of the particle are completely described by their x-components,

$$\mathbf{r} = x\mathbf{i} \tag{15-12a}$$
$$\mathbf{v} = \dot{\mathbf{r}} = \dot{x}\mathbf{i} \tag{15-12b}$$
$$\mathbf{a} = \ddot{\mathbf{r}} = \ddot{x}\mathbf{i} \tag{15-12c}$$

In this case Eqs. 15-6 for the particle reduce to

$$R_x = \Sigma F_x = ma_x \tag{15-13a}$$
$$R_y = \Sigma F_y = ma_y = 0 \tag{15-13b}$$
$$R_z = \Sigma F_z = ma_z = 0 \tag{15-13c}$$

Equations 15-13 indicate that for rectilinear motion in the x-direction forces in the y- and z-directions must be in equilibrium. If the motion of the particle is known, Eq. 15-13a gives the forces necessary to produce the given motion. If the forces acting in the x-direction are known, then Eq. 15-13a gives the acceleration in the x-direction

$$a_x = \ddot{x} = \frac{R_x}{m} \tag{i}$$

If the resultant force R_x is constant, then the acceleration a_x will also be constant; if R_x is a function of time, then a_x will be a function of time; if R_x is a function of position, then a_x will be a function of position; and so on. Once the acceleration a_x is known as a function of time or of position or of velocity, it can be integrated as in Section 13-3 to get the velocity and/or position of the particle at any other time, as illustrated in the following examples.

CONCEPTUAL EXAMPLE 15-1: WEIGHT AND GRAVITATIONAL FORCE

What is the weight of a body and how is it different from the gravitational force acting on the body?

SOLUTION

According to Newton's law of gravitation, the earth exerts an attractive force

$$F_g = \frac{GM_e m}{r^2} \tag{a}$$

on a body of mass m, where G is the universal gravitational constant, M_e is the mass of the earth (constant), and r is the distance from the center of the earth to the body. If the body remains near the surface of the earth, then $r \cong r_e =$ constant and Eq. a can be written

$$F_g = mg \tag{b}$$

where

$$g = \frac{GM_e}{r_e^2} = \text{constant} \tag{c}$$

is called the gravitational constant. If the body is to be in equilibrium, then the body must be acted on by an equal upward force. For example, when a man steps on a bathroom scale, the scale must exert an equal upward force $F_S = mg$

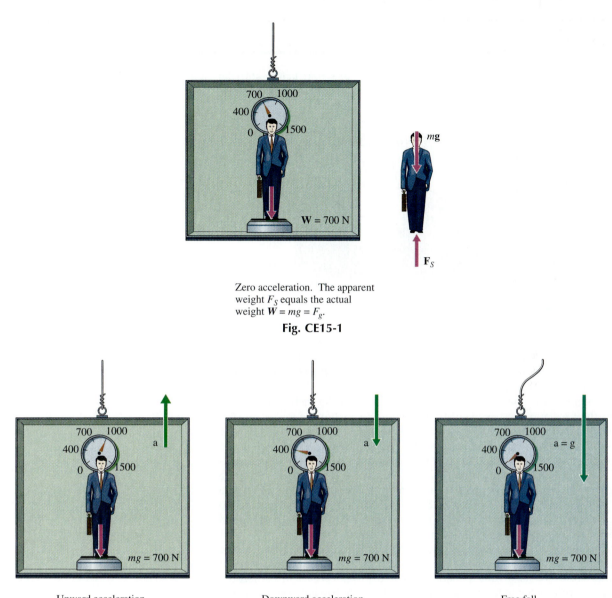

Zero acceleration. The apparent
weight F_S equals the actual
weight $W = mg = F_g$.

Fig. CE15-1

Upward acceleration.
$F_S > mg$.

Downward acceleration.
$F_S < mg$.

Free fall.
$F_S = 0$

Fig. CE15-2

on the man. It is this force on our feet that we perceive as our weight, and we usually define the weight of a body as $W = mg$.

However, if the man stands on the scale in an elevator that is accelerating upward, then the scale must exert a force upward on the man that is greater than the gravitational force mg, since

$$\uparrow\Sigma F = F_S - mg = ma \qquad (d)$$

and

$$F_S = mg + ma = W + ma \qquad (e)$$

That is, we perceive that we are *heavier* (that we *weigh more*) when we ride up in an elevator (*a*, positive) and that we are *lighter* (that we *weigh less*) when we ride down in an elevator (*a*, negative).

CONCEPTUAL EXAMPLE 15-2: ACCELERATION OF A DRAG RACER

Why do drag racers spin their tires and create a lot of smoke before starting a race? Is this strictly show for the crowd? Or does it serve some other useful purpose?

SOLUTION

According to Newton's second law, the acceleration of a race car is equal to the ratio of its propulsive force and its mass ($a = P/m$). Therefore, increasing the acceleration requires that either the propulsive force P must be increased or the mass m must be decreased.

The propulsive force on the race car is provided by the frictional force between the tires and the pavement ($P = F$), and the frictional force is directly proportional to the normal force on the tires ($F = \mu N$). Therefore, cutting the mass of the car in half does not increase the acceleration of the car, since it also cuts both the normal force and the frictional force (which is equal to the propulsive force) in half. Similarly, doubling the propulsive force by doubling the weight of the car (which doubles the normal force and the friction force) also does not increase the acceleration of the car ($a = 2P/2m = P/m$). In order to increase the acceleration of the car, the propulsive force (the friction force) must be increased without increasing the mass of the car.

The constant of proportionality between the friction and the normal force is the coefficient of friction, which for rubber and concrete varies between about 0.5 and 0.8. Cold, hard rubber tends to have a lower coefficient of friction, warm, soft rubber tends to have a higher coefficient of friction. Therefore, the drag racer can increase the propulsive force (the friction force) by spinning the tires, which warms and softens the rubber and thus increases the coefficient of friction.

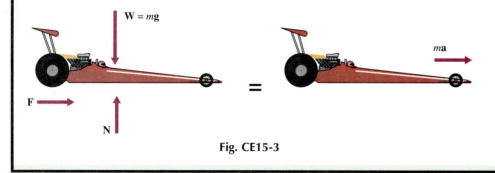

Fig. CE15-3

EXAMPLE PROBLEM 15-1

A 90-lb block is initially at rest on a horizontal surface when a 50-lb force is suddenly applied to the block as shown in Fig. 15-3. Determine the displacement, velocity, and acceleration of the block 3 s after the 50-lb force is applied if the horizontal surface is smooth.

Fig. 15-3

SOLUTION

A free-body diagram of the block resting on the smooth surface is shown in Fig. 15-4a. There is no motion in the direction perpendicular to the surface; $v_y = a_y = 0$. Therefore, the acceleration of the block must be in the x-direction as shown in Fig. 15-4b. The equations of motion for the block (Eq. 15-6) give

$$+ \rightarrow \Sigma F_x = ma_x: \qquad 50 \cos 30° = \frac{90}{32.2}a$$

$$+ \uparrow \Sigma F_y = ma_y: \qquad N - 90 - 50 \sin 30° = 0$$

Therefore, $N = 115.0$ lb and

$$a = 15.49 \text{ ft/s}^2\rightarrow \qquad\qquad\qquad \text{Ans.}$$

Integrating with respect to time to get the velocity gives

$$v = 15.49t + C_1 = 15.49t$$

where the constant of integration $C_1 = 0$, since $v = 0$ when $t = 0$. Then, at $t = 3$ s,

$$v = 15.49(3) = 46.47 \cong 46.5 \text{ ft/s}\rightarrow \qquad\qquad \text{Ans.}$$

Finally, integrating the velocity $v = dx/dt = 15.49t$ with respect to time gives the position

$$x = 7.746t^2 + C_2$$

where the constant of integration $C_2 = 0$, since $x = 0$ when $t = 0$. Then, at $t = 3$ s,

$$x = (7.746)(3)^2 = 69.7 \text{ ft} \qquad\qquad\qquad \text{Ans.}$$

> Since all of the forces used to calculate a are constant, the acceleration $a = dv/dt$ is also constant.

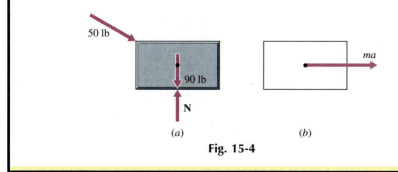

(a)   (b)

Fig. 15-4

Two bodies A and B with masses $m_A = 50$ kg and $m_B = 60$ kg are connected by a rope that passes over a pulley as shown in Fig. 15-5. Assume that the rope and pulley have negligible mass and that the length of the rope remains constant. The kinetic coefficient of friction μ_k between block A and the inclined surface is 0.25. Determine the tension in the cord and the acceleration of block A after the blocks are released from rest.

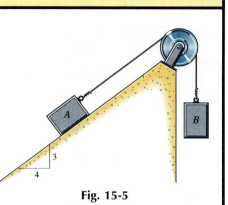

Fig. 15-5

SOLUTION

A quick check shows that if block A were on a frictionless surface, a force of 294 N acting down the incline would be necessary to hold the system in equilibrium. Since no such force is present, block A must be pulled up the incline and friction must act down the incline (as shown on Fig. 15-6a) to oppose this motion. Assuming that block B is sufficiently heavy to cause motion, the magnitude of the friction force will be $\mu_k N$. Block A has no motion in the direction perpendicular to the surface; $v_{Ay} = a_{Ay} = 0$. The acceleration of block A must be in the x-direction as shown in Fig. 15-6b. The equations of motion for block A (Eq. 15-6) give

$$+\nearrow \Sigma F_x = ma_x: \quad T - \mu_k N_A - (3/5)(50)(9.81) = 50a_A$$
$$+\nwarrow \Sigma F_y = ma_y: \quad N_A - (4/5)(50)(9.81) = 0$$

Therefore, $N_A = 392.4$ N and

$$T - 392.4 = 50a_A \qquad (a)$$

The free-body diagram and kinetic diagram for block B are shown in Figs. 15-6c and d, respectively. Since the pulley has negligible mass and it rotates freely, the tension in the rope is the same on both sides of the pulley. Also, since the length of the rope is constant, $a_B = -a_A$ and the equations of motion for block B (Eq. 15-6) give

$$+\rightarrow \Sigma F_x = ma_x: \qquad 0 = 0$$
$$+\uparrow \Sigma F_y = ma_y: \quad T - (60)(9.81) = -60a_A$$

Therefore

$$T - 588.6 = -60a_A \qquad (b)$$

Solving Eqs. a and b simultaneously gives

$$T = 482 \text{ N} \qquad \text{Ans.}$$
$$a_A = 1.784 \text{ m/s}^2 \nearrow \qquad \text{Ans.}$$

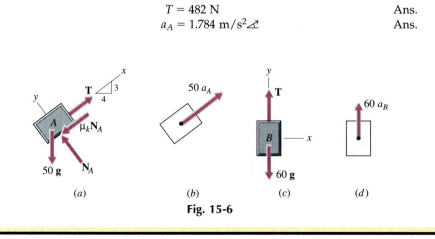

(a)   (b)   (c)   (d)

Fig. 15-6

Since the acceleration of block A came out positive, the assumptions that block B is sufficiently heavy to cause motion of block A up the incline and that the magnitude of the friction force is $\mu_k N_A$ are correct. Had the acceleration of block A come out negative, the assumptions would have been incorrect. (A negative acceleration of block A would mean that the friction force $\mu_k N_A$ would be pushing block A down the incline.) In this case, the blocks would be in equilibrium with the friction force F_A smaller than $\mu_k N_A$. Solution of the equilibrium equations would then give the unknown forces N_A, F_A, and T.

The two blocks shown in Fig. 15-7 are at rest on a horizontal surface when a force **F** is applied to block B. Blocks A and B weigh 45 lb and 75 lb, respectively. The static coefficient of friction μ_s between the two blocks is 0.25, and the kinetic coefficient of friction μ_k between the horizontal surface and block B is 0.20. Determine the maximum force **F** that can be applied to block B before the blocks no longer move together.

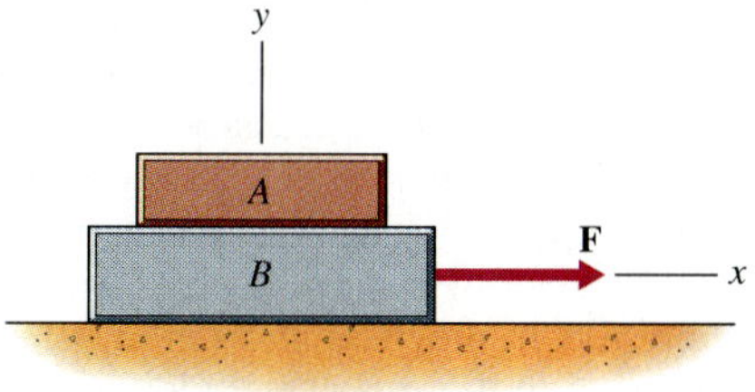

Fig. 15-7

SOLUTION

Free-body diagrams for blocks A and B are shown in Figs. 15-8a and c. Since block B is known to move to the right (in the direction of the force **F**), the friction force on the bottom of block B must act to the left and will have a magnitude of $\mu_k N_B = 0.20 N_B$. The maximum force **F** will correspond to impending slip between the two blocks, so the magnitude of the friction force between the two blocks will be $\mu_s N_A = 0.25 N_A$. Neither block has any motion in the direction perpendicular to the surface; $v_y = a_y = 0$. Therefore, the accelerations of the blocks must be in the x-direction as shown in Figs. 15-8b and d. The equations of motion for block A (Eq. 15-6) give

$$+\rightarrow \Sigma F_x = ma_x: \qquad 0.25 N_A = \frac{45}{32.2} a_A$$

$$+\uparrow \Sigma F_y = ma_y: \qquad N_A - 45 = 0$$

Therefore $N_A = 115.0$ lb and $a_A = 8.05$ ft/s^2.

Since the two blocks move together, $a_B = a_A = 8.05$ ft/s^2, and applying the equations of motion (Eqs. 15-6) to block B yields

$$+\rightarrow \Sigma F_x = ma_x: \qquad F - 0.25(115.0) - 0.2 N_B = \frac{45}{32.2}(8.05)$$

$$+\uparrow \Sigma F_y = ma_y: \qquad N_B - 115.0 - 45 = 0$$

Therefore, $N_B = 120.0$ lb and

$$F = 54.0 \text{ lb} \qquad\qquad\qquad \text{Ans.}$$

The friction forces on block B act to the left (opposing its motion). The friction force on block A acts to the right (causing block A to move to the right with block B and preventing it from slipping off of block B).

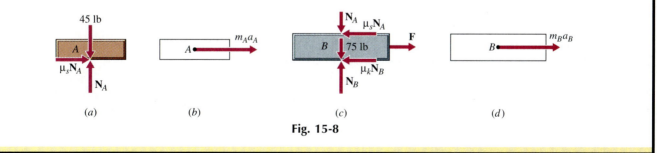

(a)   *(b)*   *(c)*   *(d)*

Fig. 15-8

The sled shown in Fig. 15-9 is used to test small solid-propellant rockets. The combined mass of the sled and rocket is 1000 kg. From the burn characteristics of the propellant, it is known that the thrust provided by the rocket during motion of the sled can be expressed as

$$F = A + Bt - Ct^2$$

where F is in newtons and t is in seconds. If the sled is released at rest when the thrust of the rocket is 10 kN and the sled travels 700 m and attains a velocity of 150 m/s during a 10-s test run, determine

a. Values for the constants A, B, and C,
b. The maximum and minimum accelerations experienced by the sled during the test.

Neglect friction between the sled and the rails and the reduction in mass of the propellant during the test.

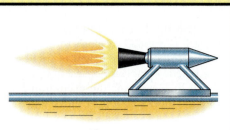

Fig. 15-9

SOLUTION

A free-body diagram of the rocket sled is shown in Fig. 15-10a. There is no motion in the direction perpendicular to the surface; $v_y = a_y = 0$. Therefore, the acceleration of the sled must be in the x-direction as shown in Fig. 15-10b. The equations of motion for the sled (Eq. 15-6) give

$$+\rightarrow \Sigma F_x = ma_x: \qquad A + Bt - Ct^2 = 1000a$$
$$+\uparrow \Sigma F_y = ma_y: \qquad N - 1000(9.81) = 0$$

where $a = \ddot{x}$. Therefore, $N_A = 9810$ N and

$$1000\ddot{x} = A + Bt - Ct^2 \qquad (a)$$

a. Since the acceleration (Eq. a) is expressed as a function of time, the velocity and displacement of the sled can be obtained simply by integrating Eq. a with respect to time, giving

$$1000\dot{x} = At + \frac{1}{2}Bt^2 - \frac{1}{3}Ct^3 + C_1 \qquad (b)$$

$$1000x = \frac{1}{2}At^2 + \frac{1}{6}Bt^3 - \frac{1}{12}Ct^4 + C_1 t + C_2 \qquad (c)$$

But since the sled starts from rest ($\dot{x} = 0$ at $t = 0$), Eq. b gives the first constant of integration $C_1 = 0$. Also, since the sled starts at $x = 0$ when $t = 0$,

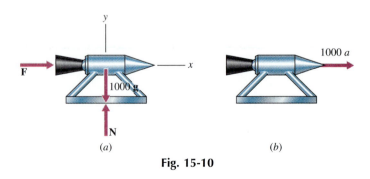

(a) (b)

Fig. 15-10

Eq. *c* gives the second constant of integration $C_2 = 0$. Finally, since the initial thrust of the rocket is 10 kN,

$$10(10^3) = A + B(0) - C(0)^2$$

the constant *A* is

$$A = 10,000 \text{ N} \qquad\qquad \text{Ans.}$$

From the end of test condition, $\dot{x} = 150$ m/s at $t = 10$ s, Eq. *b* gives

$$1000(150) = 10,000(10) + \frac{1}{2}B(10)^2 - \frac{1}{3}C(10)^3$$

which simplifies to

$$15B - 100C = 15,000 \qquad\qquad (d)$$

From the other end-of-test condition, $x = 70$ m at $t = 10$ s, Eq. *c* gives

$$1000(700) = \frac{1}{2}(10,000)(10)^2 + \frac{1}{6}B(10)^3 - \frac{1}{12}C(10)^4$$

which simplifies to

$$B - 5C = 1200 \qquad\qquad (e)$$

Simultaneous solution of Eqs. *d* and *e* yields

$$B = 1800 \text{ N/s} \qquad\qquad \text{Ans.}$$
$$C = 120 \text{ N/s}^2 \qquad\qquad \text{Ans.}$$

b. The acceleration of the sled (Eq. *a*) can now be written

$$a = \ddot{x} = 10 + 1.8t - 0.12t^2 \text{ m/s}^2 \qquad\qquad (f)$$

Equation *f* is a quadratic function of time starting at 10 m/s^2 when $t = 0$ s and ending at 16.0 m/s^2 when $t = 10$ s. The maximum or minimum of the quadratic function must occur where

$$\frac{da}{dt} = 1.8 - 0.24t = 0$$

or at $t = 7.50$ s. But, at $t = 7.50$ s, the acceleration is 16.75 m/s^2. Therefore, for $0 \le t \le 10$ s,

$$a_{\text{max}} = 16.75 \text{ m/s}^2 \quad \text{at} \quad t = 7.50 \text{ s} \qquad \text{Ans.}$$
$$a_{\text{min}} = 10.00 \text{ m/s}^2 \quad \text{at} \quad t = 0 \text{ s} \qquad \text{Ans.}$$

Recall that setting the slope of a function equal to zero ($da/dt = 0$) finds a relative extrema (maximum or minimum) of the function. On a finite interval such as $0 \le t \le 10$ s, the value of the function at an end point of the interval may be more positive or more negative than an internal extrema even though the slope is not zero at either end of the interval.

Motion of block A ($W = 805$ lb) on the inclined surface of Fig. 15-11 is resisted by friction ($\mu = 0.10$) and a linear spring ($k = 25$ lb/ft). If the block is released at rest with the spring unstretched, determine, during the first phase of motion down the inclined surface,

a. The maximum displacement of the block from its rest position.
b. The velocity of the block when it is 15 ft from its rest position.
c. The time required for the block to move 15 ft from its rest position.
d. The acceleration of the block as it begins to move up the inclined surface.

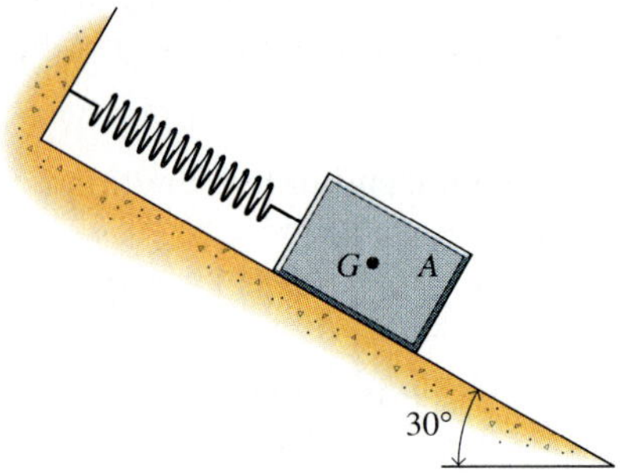

Fig. 15-11

SOLUTION

A free-body diagram of the block for the first phase of the motion (down the inclined surface) is shown in Fig. 15-12a. When the block moves down the inclined surface, friction acts up the surface to oppose the motion; the magnitude of the friction is $\mu N = 0.10N$. There is no motion in the direction perpendicular to the surface; $v_y = a_y = 0$. Therefore, the acceleration of the block must be in the x-direction as shown in Fig. 15-12b. The equations of motion for the block (Eq. 15-6) give

$$+\nwarrow \Sigma F_x = ma_x: \qquad 805 \sin 30° - 0.1N - 25x = \frac{805}{32.2}a$$

$$+\nearrow \Sigma F_y = ma_y: \qquad N - 805 \cos 30° = 0$$

where the spring force $F = kx = 25x$ and $x = 0$ when the spring is unstretched. Therefore $N = 697.2$ lb and

$$a = 13.31 - x \qquad\qquad (a)$$

Since the acceleration is a function of position rather than of time, the definition of acceleration must be rewritten using the chain rule of differentiation

$$a = \frac{dv}{dt} = \frac{dv}{dx}\frac{dx}{dt} = v\frac{dv}{dx} = 13.31 - x \qquad (b)$$

Then, integrating

$$\int v \, dv = \int (13.31 - x) \, dx \qquad\qquad (c)$$

gives

$$\frac{1}{2}v^2 = 13.31x - \frac{1}{2}x^2 + C_1 \qquad\qquad (d)$$

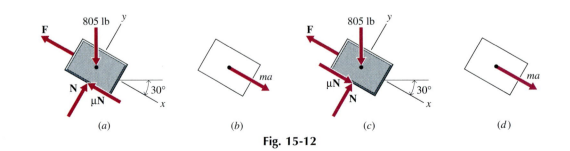

(a) $\qquad$ (b) $\qquad$ (c) $\qquad$ (d)

Fig. 15-12

where the constant of integration $C_1 = 0$ since the block starts from rest ($v = 0$) when the spring is unstretched ($x = 0$). Rearranging Eq. d gives

$$v = \sqrt{26.62x - x^2} \qquad (e)$$

a. The maximum displacement occurs when $dx/dt = v = 0$. Therefore,

$$26.62x - x^2 = 0$$
$$x = x_{max} = 26.62 \text{ ft} \searrow \qquad \text{Ans.}$$

b. When the displacement of the block is 15 ft,

$$v = \sqrt{26.62(15) - (15)^2} = 13.20 \text{ ft/s} \searrow \qquad \text{Ans.}$$

c. To get a relationship between the position of the block and time requires integrating the velocity $v = dx/dt$. Rearranging Eq. e

$$dt = \frac{dx}{\sqrt{26.62x - x^2}}$$

and integrating gives

$$t = \sin^{-1}\left(\frac{x - 13.31}{13.31}\right) + C_2$$

where the constant of integration $C_2 = \pi/2$, since the spring is initially unstretched. Then, the time at which $x = 15$ ft is

$$t = \sin^{-1}\left(\frac{15 - 13.31}{13.31}\right) + \frac{\pi}{2} = 1.698 \text{ s} \qquad \text{Ans.}$$

d. A free-body diagram of the block for the second phase of the motion (up the inclined surface) is shown in Fig. 15-12c. When the block moves up the inclined surface, friction acts down the surface to oppose the motion; the magnitude of the friction is still $\mu N = 0.10N$. There is still no motion in the direction perpendicular to the surface; $v_y = a_y = 0$. Therefore, the acceleration of the block must be in the x-direction as shown in Fig. 15-12d. The equations of motion for the block (Eq. 15-6) now give

$$+\searrow \Sigma F_x = ma_x: \qquad 805 \sin 30° + 0.1N - 25(26.62) = \frac{805}{32.2}a$$
$$+\nearrow \Sigma F_y = ma_y: \qquad N - 805 \cos 30° = 0$$

where $F = kx_{max} = 25(26.62)$. Therefore, now $N = 697.2$ lb and

$$a = -7.73 \text{ ft/s}^2 = 7.73 \text{ ft/s}^2 \searrow \qquad \text{Ans.}$$

A particle falls under the force of gravity in a medium that exerts a resisting force proportional to the velocity of the particle. Develop equations for the velocity and displacement of the particle. The velocity and displacement of the particle are zero at time $t = 0$.

SOLUTION

A free-body diagram of the particle is shown in Fig. 15-13a. Although there is no restraint to motion in the horizontal direction, there is also no force to cause any motion in the horizontal direction. Therefore, the acceleration of the particle will be in the vertical (x-) direction as shown in Fig. 15-13b. The equations of motion for the particles (Eq. 15-6) give

$$+\downarrow\Sigma F_x = ma_x: \qquad mg - kv = ma_x$$
$$+\rightarrow\Sigma F_y = ma_y: \qquad 0 = 0$$

Therefore, the y-component of the equations of motion is satisfied identically and

$$ma_x = m\frac{dv}{dt} = mg - kv \qquad (a)$$

Rearranging Eq. a

$$\frac{m\,dv}{mg - kv} = dt$$

and integrating gives

$$-\frac{m}{k}\ln(mg - kv) = t + C_1 \qquad (b)$$

where the constant of integration $C_1 = -(m/k)\ln(mg)$, since the particle starts from rest ($v = 0$ when $t = 0$). Rearranging Eq. b gives

$$\ln(mg - kv) - \ln(mg) = -\frac{kt}{m}$$

$$\frac{mg - kv}{mg} = e^{-kt/m}$$

or

$$v = \frac{dx}{dt} = \frac{mg}{k}(1 - e^{-kt/m}) \qquad \text{Ans.}$$

Integrating again to get the position as a function of time yields

$$x = \frac{mg}{k}t + \frac{m^2g}{k^2}e^{-kt/m} + C_2$$

where the constant of integration $C_2 = -m^2g/k^2$, since the position of the particle is zero at $t = 0$. Therefore,

$$x = \frac{mg}{k}t + \frac{m^2g}{k^2}(e^{-kt/m} - 1) \qquad \text{Ans.}$$

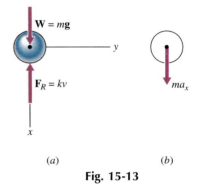

(a)   (b)

Fig. 15-13

Since the acceleration a_x is not constant, Eq. a must be rearranged before it can be integrated. All of the terms involving the variable v must be on one side of the equation, and all of the terms involving the variable t must be on the other side of the equation. Constant terms may appear on either side of the equation.

As the time gets large ($t \to \infty$), the exponential gets small ($e^{-kt/m} \to 0$) and the velocity becomes constant ($v \to mg/k$). This limiting velocity is called the *terminal velocity* of the body.

PROBLEMS

Introductory Problems

15-1* A box weighing 322 lb rests on the floor of an elevator (Fig. P15-1). Determine the force exerted on the box by the floor of the elevator

a. If the elevator starts upward with an acceleration of 10 ft/s^2.
b. If the elevator starts downward with an acceleration of 8 ft/s^2.

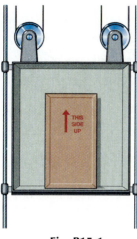

Fig. P15-1

15-2* Determine the constant force **F** required to accelerate a 1000-kg automobile on a level road from rest to 20 m/s in 10 s (Fig. P15-2).

Fig. P15-2

15-3 A 200-lb block rests on a horizontal surface as shown in Fig. P15-3. Determine

a. The magnitude of the force **F** required to produce an acceleration of 5 ft/s^2 if the surface is smooth.
b. The acceleration that a 100-lb force **F** would produce if the kinetic coefficient of friction between the block and the surface is $\mu_k = 0.25$.

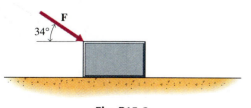

Fig. P15-3

15-4* A 15-kg block of ice slides on a horizontal surface for 20 m before it stops (Fig. P15-4). If the initial speed of the block was 15 m/s, determine

a. The force of friction between the block of ice and the surface.
b. The kinetic coefficient of friction μ_k between the block of ice and the surface.

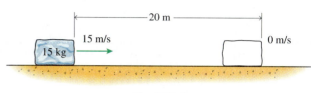

Fig. P15-4

15-5 A television set is placed on a spring-operated platform scale on the floor of an elevator (Fig. P15-5). If the scale indicates a weight of 50 lb when the elevator is at rest, determine

a. The acceleration of the elevator when the scale indicates a weight of 40 lb.
b. The weight indicated by the scale when the elevator accelerates upward at 10 ft/s^2.

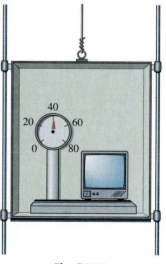

Fig. P15-5

15-6 A 1500-kg automobile is moving along a level road at a constant speed of 60 km/h (Fig. P15-6). If the automobile accelerates at a constant rate and reaches a speed of 80 km/h in 5 s, determine

a. The force required to produce this acceleration.
b. The distance traveled by the automobile during the 5-s interval that it is accelerating.

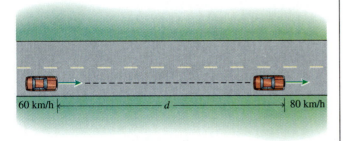

60 km/h |← d →| 80 km/h

Fig. P15-6

15-7* A force of 20 lb is applied to a 25-lb block as shown in Fig. P15-7. Let $x = 0$ and $v = 0$ when $t = 0$, and determine the velocity and displacement of the block at $t = 5$ s if

a. The inclined plane supporting the block is smooth.
b. The kinetic coefficient of friction between the inclined plane and the block is $\mu_k = 0.25$.

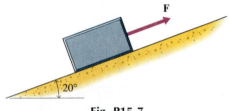

F

20°

Fig. P15-7

15-8* A 20-kg block is pushed up an inclined plane by a horizontal force **F** of 200 N as shown in Fig. P15-8. The kinetic coefficient of friction between the inclined plane and the block is $\mu_k = 0.10$. If $v = 0$ and $x = 0$ when $t = 0$, determine

a. The acceleration of the block.
b. The time required for the block to travel 15 m up the incline.
c. The velocity of the block after it has traveled 10 m up the incline.

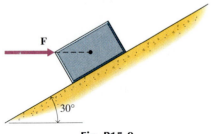

F

30°

Fig. P15-8

15-9 A 25,000-lb navy fighter plane launched from a carrier increases in speed from 20 mi/h (the carrier speed) to 150 mi/h in 2.2 s (Fig P15-9). Determine

a. The constant force applied by the catapult.
b. The distance traveled by the plane during the launch.

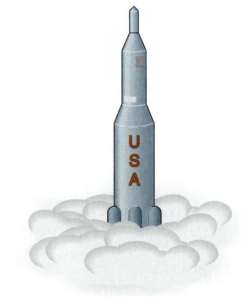

150 mi/h

Fig. P15-9

15-10 A Saturn V rocket has a mass of $2.75(10^6)$ kg and a thrust of $33(10^6)$ N (Fig. P15-10). Determine

a. The initial vertical acceleration of the rocket.
b. The rocket's velocity 10 s after liftoff.
c. The time required to reach an altitude of 10,000 m.

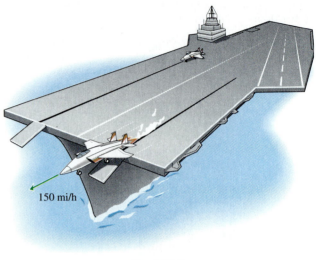

USA

Fig. P15-10

15-11 A child is sledding on a hill as shown in Fig. P15-11. The coefficient of kinetic friction between the sled and the snow is $\mu_k = 0.15$ and the combined weight of the child and the sled is 64.4 lb. If the child has a speed of 5 ft/s at the top of the hill, determine

a. The time required to reach the bottom of the hill.
b. The speed of the sled at the bottom of the hill.

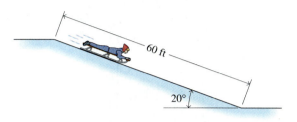

Fig. P15-11

15-12* An 80-kg skydiver is falling at 85 m/s when she opens her parachute (Fig. P15-12). If her speed is reduced to 5 m/s during the next 60 m of fall, determine the average force exerted on her body by the parachute during this interval.

Fig. P15-12

Intermediate Problems

15-13* The 2000-lb elevator cage of Fig. P15-13 is brought to rest from an initial speed of 25 ft/s in a distance of 50 ft. Determine the uniform deceleration and the tension in the elevator cable while the cage is coming to rest.

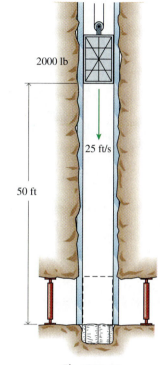

Fig. P15-13

15-14* A 5-kg particle released from an elevated platform reaches a speed of 10 m/s before a constant resisting force is applied. During the next 10 m of fall, the speed is reduced to 5 m/s. Determine the force exerted on the particle.

15-15 The chute shown in Fig. P15-15 is 20 ft long and is used to transfer boxes from the street into the basement of a store. The kinetic coefficients of friction are $\mu_k = 0.25$ between the box and the chute and $\mu_k = 0.40$ between the box and the basement floor. If a 30-lb box is given an initial velocity of 10 ft/s when it is placed on the chute, determine

a. The velocity of the box as it leaves the end of the chute.
b. The distance d that the box will slide on the basement floor after it leaves the end of the chute.

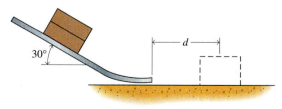

Fig. P15-15

15-16 The 200-kg cart shown in Fig. P15-16 is traveling to the right with an initial speed of 5 m/s when it comes to an inclined plane. Determine

a. The acceleration of the cart during its travel up the inclined plane.
b. The distance d that the cart will travel up the inclined plane before coming to rest.

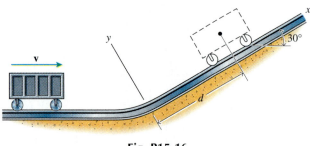

Fig. P15-16

15-17* Blocks A and B, which weigh 30 lb and 60 lb, respectively, are connected by a rope as shown in Fig. P15-17. The coefficients of friction are $\mu_k = 0.20$ for block A and $\mu_k = 0.15$ for block B. If the blocks are initially at rest and the force **F** applied to the cable is 40 lb, determine

a. The acceleration of block B.
b. The velocity of block A after 5 s.

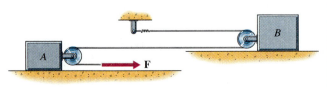

Fig. P15-17

15-18* A freight elevator contains three crates as shown in Fig. P15-18. The mass of the elevator cage is 750 kg, and the masses of crates A, B, and C are 300 kg, 200 kg, and 100 kg, respectively. During a short interval of the lift, the elevator experiences an upward acceleration of 8 m/s². During this interval, determine

a. The tension T in the elevator cable.
b. The force exerted on crate A by the floor of the elevator.
c. The force exerted by crate B on crate C.

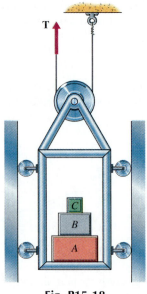

Fig. P15-18

15-19 Three boxes connected with cables rest on a horizontal surface as shown in Fig. P15-19. The weights of boxes A, B, and C are 200 lb, 150 lb, and 300 lb, respectively. The kinetic coefficient of friction between the surface and the boxes is $\mu_k = 0.20$. If the force **F** applied to the boxes is 175 lb, determine the acceleration of the boxes and the tensions in the cables between the boxes.

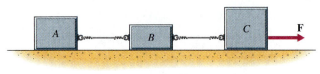

Fig. P15-19

15-20 The 5400-kg truck shown in Fig. P15-20 is carrying a 2200-kg crate. If the truck accelerates uniformly from 40 km/h to 80 km/h in a distance of 60 m, determine

a. The uniform friction force exerted on the truck's wheels by the road.
b. The minimum coefficient of friction between the crate and the truck bed for which the crate will not slide off the truck.

Fig. P15-20

15-21* Blocks A and B, which weigh 30 lb and 50 lb, respectively, are connected by a rope as shown in Fig. P15-21. The kinetic coefficients of friction μ_k are 0.35 for block A and 0.15 for block B. During motion of the blocks down the inclined plane, determine

a. The acceleration of block B.
b. The tension in the rope.

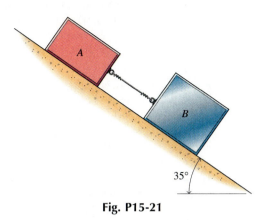

Fig. P15-21

15-22 Two bodies A and B, with masses of 25 kg and 30 kg, respectively, are shown in Fig. P15-22a. During motion of the bodies,

a. Determine the acceleration of body A and the tension in the cable connecting the bodies.
b. Determine the acceleration of body B if body A is replaced with a constant force of 245 N as shown in Fig. P15-22b.

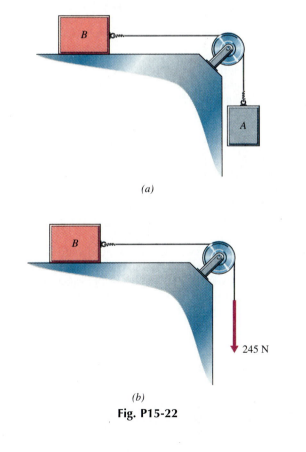

(a)

(b)

Fig. P15-22

15-23 Blocks A and B, which weigh 200 lb and 120 lb, respectively, are connected by a rope as shown in Fig. P15-23a. During motion of the bodies,

a. Determine the acceleration of block A and the tension in the rope connecting the bodies.
b. Determine the acceleration of block A if block B is replaced with a constant force of 120 lb as shown in Fig. P15-23b.

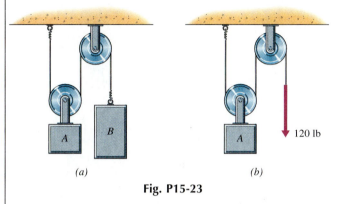

(a)

(b)

120 lb

Fig. P15-23

15-24* Blocks A ($m_A = 25$ kg) and B ($m_B = 40$ kg) of Fig. P15-24 are connected by flexible cables to sheaves that have diameters of 300 mm and 150 mm, respectively. The two sheaves are fastened together and are both weightless and frictionless. Find the tensions in the cables after the bodies are released from rest.

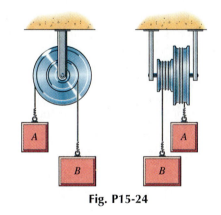

Fig. P15-24

15-25* Two blocks A and B are connected by a flexible cable as shown in Fig. P15-25. The kinetic coefficient of friction between block A and the inclined surface is $\mu_k = 0.15$, and the system is released from rest. If the weight of body A is 50 lb and block B strikes the horizontal surface 3 s after being released, determine

a. The acceleration of body B.
b. The weight of body B.
c. The tension in the cable while the blocks are in motion.

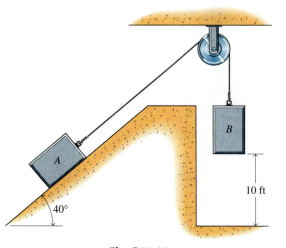

Fig. P15-25

15-26 Two bodies A ($m_A = 40$ kg) and B ($m_B = 30$ kg) are connected by a flexible cable as shown in Fig. P15-26. The kinetic coefficient of friction between body A and the inclined surface is $\mu_k = 0.25$, and the system is released from rest. During motion of the bodies, determine

a. The acceleration of body A.
b. The tension in the cable connecting the bodies.
c. The velocity of body B after 5 s of motion.

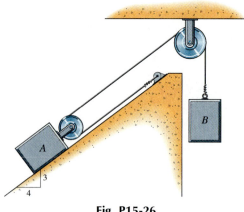

Fig. P15-26

15-27 Two bodies A ($W_A = 30$ lb) and B ($W_B = 20$ lb) are connected by a flexible cable as shown in Fig. P15-27. The kinetic coefficient of friction between body A and the inclined surface is $\mu_k = 0.30$ and the horizontal surface supporting body B is smooth. When the bodies are in the position shown, body B is moving to the right with a speed of 5 ft/s. Determine

a. The tension in the cable connecting the bodies.
b. The time required for body B to come to rest.
c. The distance traveled by body B before it comes to rest.

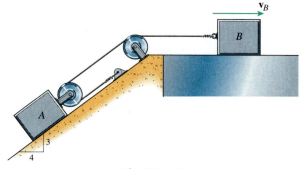

Fig. P15-27

15-28* Two bodies A ($m_A = 25$ kg) and B ($m_B = 30$ kg) are connected by a flexible cable as shown in Fig. P15-28. The kinetic coefficient of friction between both bodies and the horizontal surface is $\mu_k = 0.20$ and the system is initially at rest. If a 100-N force $\mathbf{F}$ is suddenly applied to body B, determine

a. The acceleration of body A.
b. The tension in the cable connecting the bodies.
c. The distance traveled by body A during the first 5 s that the force $\mathbf{F}$ is applied.

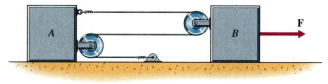

Fig. P15-28

15-29 Cardboard packages are placed onto a conveyor belt from above as shown in Fig. P15-29. If the coefficient of friction between the belt and the packages is 0.30 and the packages slip for their first 4 in. of travel, determine the speed of the conveyor belt.

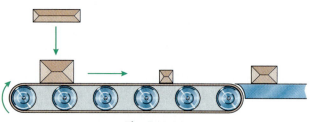

Fig. P15-29

Challenging Problems

15-30 The 20-kg collar shown in Fig. P15-30a is subjected to a horizontal force that varies with time t as shown in Fig. P15-30b. The coefficients of static and kinetic friction between the collar and the horizontal rod are 0.35 and 0.30, respectively. Determine

a. The time t_1 at which the collar starts to move.
b. The velocity v_5 of the collar when $t = 5$ s.
c. The distance d that the collar moves along the rod in the first 5 s.

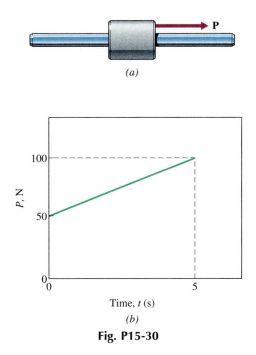

(a)

(b)

Fig. P15-30

15-31* The 644-lb block shown in Fig. P15-31 is sliding to the left on a smooth surface with a speed of 20 ft/s when the force **F** is applied. The magnitude of the force varies with time according to the expression $F = 40 + 12t$ where F is in pounds and t is in seconds. Determine

a. The time required to bring the block to rest.
b. The distance moved while coming to rest.
c. The position of the block 10 s after the force **F** is applied.

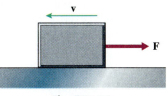

Fig. P15-31

15-32* The speed of a 12,500-kg navy fighter plane landing on a carrier is reduced from 216 km/h to rest by the plane's brakes and the cable arrester system (Fig. P15-32). The arresting force provided by the plane's brakes is a constant 90 kN. If the arresting force provided by the cable system can be expressed in equation form as $F = 850,000t - 425,000t^2$ where F is in Newtons and t is in seconds, determine

a. The maximum deceleration experienced by the pilot.
b. The time required for the arresting operation.
c. The distance traveled by the plane during the arrest.

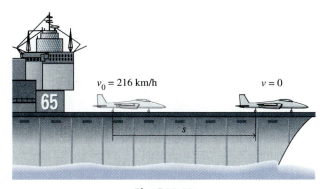

Fig. P15-32

15-33 The 0.33-lb ball shown in Fig. P15-33 is expelled from the tube by a spring which has a constant $k = 5$ lb/in. The unstretched length of the spring is 20 in. and the ball is released from rest in the position shown. If friction between the ball and the tube can be neglected, determine

a. The velocity of the ball as it exits the tube.
b. The time required for the ball to exit the tube.

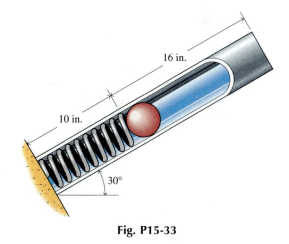

Fig. P15-33

15-34 A 55-kg skydiver jumps from a stationary balloon at a height of 3000 m (Fig. P15-34). The drag force exerted on her body by the air when she is in a spread-eagle position can be expressed as $F_D = 0.180v^2$, where F_D is in Newtons and v is in meters per second. Determine her terminal velocity and the time required to reach 95 percent of her terminal velocity.

$F_D = 0.180v^2$

Fig. P15-34

15-35 The block shown in Fig. P15-35 weighs 50 lb. The coefficient of kinetic friction between the block and the inclined surface is $\mu_k = 0.20$. At the instant shown, the velocity of the block is 20 ft/s down the inclined surface. If a resisting force $F_R = 0.50v$, where F_R is in pounds and v is in feet per second, is applied at this instant, determine

a. The velocity of the block after 5 s.
b. The distance that the block moves during the first 5 s that the resisting force is applied.
c. The terminal velocity of the block.

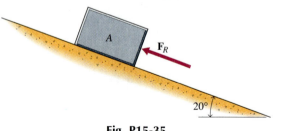

A F_R

$20°$

Fig. P15-35

15-36* A 5-kg projectile is fired vertically upward from the earth with a speed of 300 m/s. Determine

a. The maximum height attained by the projectile if air resistance is neglected.
b. The maximum height attained by the projectile if the drag force exerted by the air is $F_D = 0.006v^2$, where F_D is in Newtons and v is in meters per second.
c. The velocity of the projectile when it returns to earth if the drag force of the air is acting.

15-37 The parachute and 400-lb crate shown in Fig. P15-37a are falling with a constant speed v. When the crate hits the ground, the parachute cords become slack (they exert no more force on the crate) and the ground exerts a force on the crate that varies according to time as shown in Fig. P15-37b. Determine the maximum constant descent speed for which the maximum force exerted on the crate by the ground P_{max} does not exceed 2000 lb.

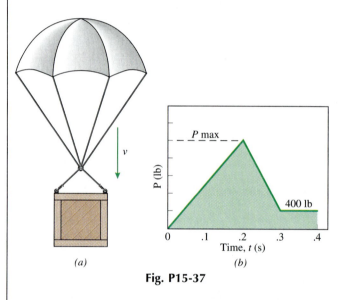

(a)   *(b)*

Fig. P15-37

15-38 A 57-kg woman bungee jumper jumps from a high bridge as shown in Fig. P15-38. The bungee cord has an elastic constant of 171 N/m and a free length of 40 m. Determine

a. The duration of the free-fall portion of the jump (the elapsed time until the bungee cord becomes taut).
b. The speed of the woman when the bungee cord becomes taut and starts to exert a force on her body.
c. The stretch d in the bungee cord when the woman stops falling.

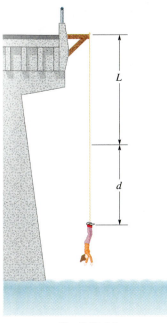

Fig. P15-38

15-39 The 1100-lb dragster shown in Fig. P15-39 is traveling at a speed of 100 mi/h when the driver pops the parachute. If the parachute exerts a drag force on the dragster that is proportional to the square of the speed of the dragster, $F_D = 0.025v^2$, where F is in pounds and v is in feet per second, determine

a. The deceleration of the dragster at the instant the parachute opens.

b. The distance traveled by the dragster as it slows to a speed of 40 mi/h.

Fig. P15-39

15-40* A 2-kg circular disk is supported by two identical springs ($k = 400$ N/m) as shown in Fig. P15-40. The free length of each spring is 300 mm. If the disk is released at rest with the springs horizontal, determine the velocity of the disk when it has fallen 100 mm below its initial position.

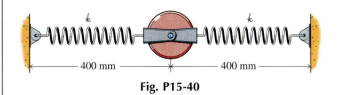

Fig. P15-40

15-41* Blocks A ($W_A = 25$ lb) and B ($W_B = 50$ lb) are at rest, and the spring ($k = 25$ lb/ft) is unstretched when the blocks are in the position shown in Fig. P15-41. Determine the velocity and acceleration of block B when it is 1 ft below its initial position.

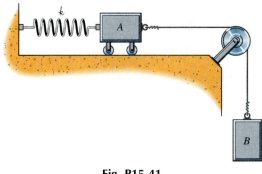

Fig. P15-41

15-42 Block A ($m_A = 10$ kg) is at rest, and the spring ($k = 25$ N/m) is unstretched when the block is in the position shown in Fig. P15-42. One second after the block is released to move, determine

a. The velocity and acceleration of the block.
b. The tension in the cable.

Fig. P15-42

211

15-43 The flexible chain shown in Fig. P15-43 weighs 0.50 lb/ft. The kinetic coefficient of friction between the chain and the horizontal surface is 0.20. If the chain is released at rest in the position shown, determine the velocity of the chain at the instant it becomes completely vertical and the time required for end A to leave the surface.

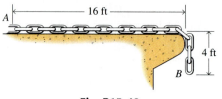

Fig. P15-43

15-44 Blocks A (m_A = 30 kg) and B (m_B = 20 kg) are connected by a flexible cable as shown in Fig. P15-44. The pulleys have negligible mass and are very small. Determine the acceleration of both blocks and the tension in the cable for the position shown

a. If $v_A = 0$ m/s.
b. If $v_A = 5$ m/s downward.

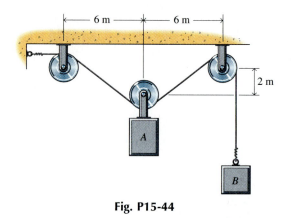

Fig. P15-44

15-45* Carts A (W_A = 200 lb) and B (W_B = 300 lb) are connected by a flexible cable as shown in Fig. P15-45. The pulley has negligible mass and is very small. Determine the acceleration of both carts and the tension in the cable if F = 50 lb and

a. $v_B = 0$ ft/s at the instant shown.
b. $v_B = 10$ ft/s at the instant shown.

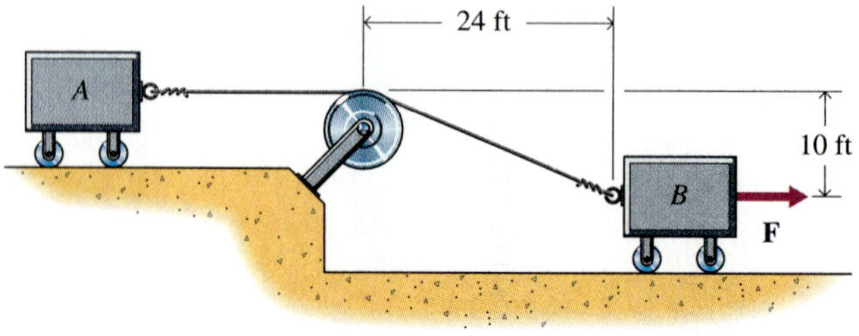

Fig. P15-45

Computer Problems

C15-46 The 10-kg cart shown in Fig. P15-46 is being pulled to the left by a constant force P = 10 N applied to the end of the cord. If the cart starts from rest when x = 8 m, calculate and plot

a. The speed of the cart v as a function of its position x ($-3 < x < 8$ m).
b. The position of the cart x as a function of time t ($0 < t < 5$ s). (You may need to solve the differential equation numerically.)

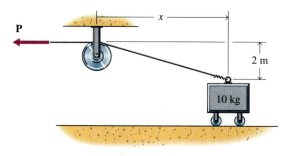

Fig. P15-46

C15-47 The pair of blocks shown in Fig. P15-47 are suspended by a cord which passes around three small pulleys. Block A weighs 60 lb, block B weighs 40 lb, and the system starts from rest when y = 6 ft. Use the Euler method for solving differential equations to calculate the position and velocity of both blocks as a function of time. Then plot for $0 < t < 15$ s

a. The position y of block A as a function of time t.
b. The velocities of both blocks (v_A and v_B) as functions of time t.
c. The tension T in the connecting cord as a function of time t.

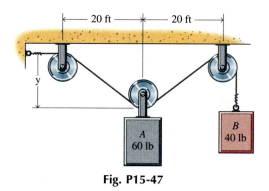

Fig. P15-47

C15-48 A 30-kg mass is suspended by a cord which passes around three small pulleys as shown in Fig. P15-48. A constant force $P = 300$ N is applied to the other end of the cord. If the system starts from rest when $y = 2$ m, use the Euler method for solving differential equations to calculate the position and velocity of both blocks as a function of time. Then plot for $0 < t < 10$ s

a. The position y of the 30-kg mass as a function of time t.

b. The velocity v of the 30-kg mass as a function of time t.

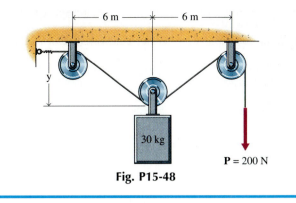

Fig. P15-48

15-4 CURVILINEAR MOTION

The kinematics of a particle subjected to curvilinear motion was described in Sections 13-5 and 13-7. Motion in this case is along a curved path. As was the case with rectilinear motion if the motion of a particle is known or a desired motion is specified, then Newton's second law (Eq. 15-5)

$$\mathbf{R} = \Sigma \mathbf{F} = m\mathbf{a}$$

can be used to find the resultant force necessary to produce the motion. Alternatively, if the forces acting on a particle are known at some instant, Newton's second law gives the acceleration of the particle at that instant. If the velocity or position of the particle at some later time is desired, then Newton's second law must be written for a general position or instant of the motion and then integrated.

The expression of Newton's second law in terms of components is often called the equations of motion. For example, in terms of rectangular cartesian components, the equations of motion are (Eq. 15-6)

$$R_x = \Sigma F_x = ma_x$$
$$R_y = \Sigma F_y = ma_y$$
$$R_z = \Sigma F_z = ma_z$$

For planar curvilinear motion, Newton's second law may be expressed in rectangular coordinates, polar coordinates, or normal/tangential coordinates. For space curvilinear motion, Newton's second law may be expressed in rectangular, cylindrical, or spherical coordinates. The choice of which coordinate system to use is based on the nature in which the data is given or on the type of results desired.

Integration of the equations of motion is usually much more difficult for curvilinear motion than it is for rectilinear motion. Rectilinear motion involves the derivative of a single independent variable ($v = dx/dt$, $a_x = dv/dt = d^2x/dt^2$), and the integration is usually straightforward. In curvilinear motion, however, the equations of motion may involve the derivatives of two or even three different independent variables. If the equations are interdependent, the solution can be very difficult. Sometimes, rewriting the equations in a different coordinate system will make the solution easier. Other times, a numerical solution of the equations may be necessary.

The following examples illustrate the procedure for solving problems involving curvilinear motion of a particle in a plane and in space.

A 30-lb projectile is fired horizontally with an initial velocity of 750 ft/s from the top of a hill, which is 500 ft above the surrounding area (Fig. 15-14). Determine the range R of the projectile (horizontal distance traveled) and the elapsed time before it strikes the ground. Neglect air resistance.

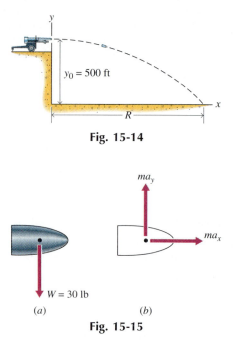

Fig. 15-14

SOLUTION

A free-body diagram for the projectile is shown in Fig. 15-15a. The projectile moves in a vertical plane, and the only force acting on the projectile during its flight is the Earth's gravity. Since the magnitude ($W = 30$ lb) and direction ($-\mathbf{j}$) of the gravitational force are both constant, rectangular coordinates are appropriate and the rectangular components of the inertia vector are shown on Fig. 15-15b. Then, the equations of motion for the projectile (Eqs. 15-6) are

$$+\rightarrow\Sigma F_x = ma_x: \qquad 0 = \frac{30}{32.2}a_x$$

$$+\uparrow\Sigma F_y = ma_y: \qquad -30 = \frac{30}{32.2}a_y$$

Fig. 15-15

Therefore, $a_x = dv_x/dt = 0$ ft/s^2, $a_y = dv_y/dt = -32.2$ ft/s$^2 = -g$, and the motion is independent of the weight of the projectile. Integrating the acceleration components to get the velocity and position of the projectile as a function of time gives

$$v_x = \frac{dx}{dt} = C_1 = 750 \text{ ft/s} \qquad (a)$$

$$x = 750t + C_2 = 750t \text{ ft} \qquad (b)$$

$$v_y = \frac{dy}{dt} = -32.2t + C_3 = -32.2t \text{ ft/s} \qquad (c)$$

$$y = -16.1t^2 + C_4 = 500 - 16.1t^2 \text{ ft} \qquad (d)$$

where the constants of integration are $C_1 = 750$ ft/s (since $v_x = 750$ ft/s when $t = 0$), $C_2 = 0$ ft (since $x = 0$ when $t = 0$), $C_3 = 0$ ft/s (since $v_y = 0$ when $t = 0$), and $C_4 = 500$ ft (since $y = 500$ ft when $t = 0$).

The projectile will strike the ground when $y = 0$. Therefore, from Eq. d

$$t = \sqrt{500/16.1} = 5.573 \text{ s} \cong 5.57 \text{ s} \qquad \text{Ans.}$$

At this time, the x-position of the projectile is (Eq. b)

$$R = x = 750(5.573) = 4180 \text{ ft} \qquad \text{Ans.}$$

When air resistance may be neglected, objects moving freely near the surface of the earth experience an acceleration equal to the gravitational constant and directed toward the center of the earth. The solution of these *projectile motion* problems involves only kinematic principles.

A sphere of mass m is attached to the top end of a slender vertical rod of negligible mass as shown in Fig. 15-16. When the sphere is given a small displacement, rotation of the system about the pin at point O is initiated. Determine the linear velocity $\mathbf{v}$ of the sphere and the force $\mathbf{P}$ in the rod when the rod is in a horizontal position if $m = 5$ kg and $R = 2$ m.

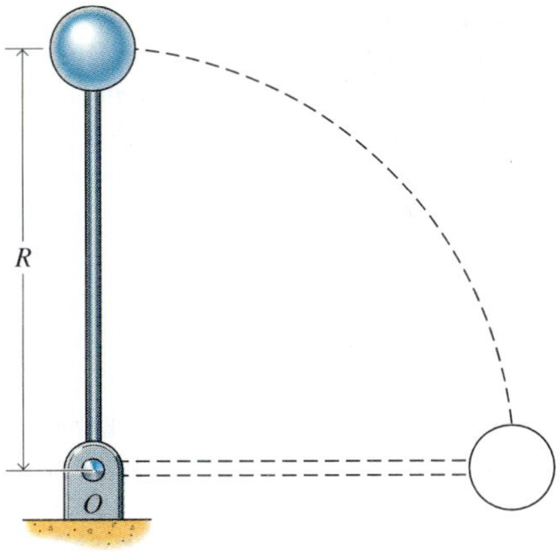

SOLUTION

A free-body diagram of the sphere when the rod is horizontal (Fig. 15-17a) shows that two forces act on the sphere; the weight $\mathbf{W}$ and the reaction of the rod $\mathbf{P}$. Since the sphere moves in a circular path in a vertical plane, polar coordinates are appropriate and the polar components of the inertia vector are shown in Fig. 15-17b. Then, the equations of motion for the sphere (Eqs. 15-7) are

$$+\rightarrow \Sigma F_r = ma_r: \qquad P = 5(-2\dot{\theta}^2) \qquad (a)$$
$$+\downarrow \Sigma F_\theta = ma_\theta: \qquad 5(9.81) = 5(2\ddot{\theta}) \qquad (b)$$

where $r = 2$ m = constant, $\dot{r} = \ddot{r} = 0$, the acceleration components are $a_r = \ddot{r} - r\dot{\theta}^2 = -2\dot{\theta}^2$ and $a_\theta = r\ddot{\theta} + 2\dot{r}\dot{\theta} = 2\ddot{\theta}$, and the angle θ is increasing in the clockwise direction. Equation b gives the angular acceleration $\ddot{\theta}$, but the solution of Eq. a for the force P requires knowing the angular velocity $\omega = \dot{\theta}$. Therefore, the equations of motion must be rewritten for an arbitrary angle θ and integrated for $0 \le \theta \le 90°$ to determine the angular velocity of the rod when $\theta = 90°$.

A free-body diagram of the sphere at an arbitrary angle θ is shown in Fig. 15-17c, and the polar components of the inertia vector are shown in Fig. 15-17d. Now, the equations of motion for the sphere (Eqs. 15-7) are

$$+\nearrow \Sigma F_r = ma_r: \qquad P - 5(9.81)\cos\theta = 5(-2\dot{\theta}^2) \qquad (c)$$
$$+\searrow \Sigma F_\theta = ma_\theta: \qquad 5(9.81)\sin\theta = 5(2\ddot{\theta}) \qquad (d)$$

Using $\omega = \dfrac{d\theta}{dt}$ and the chain rule of differentiation $\ddot{\theta} = \dfrac{d\omega}{dt} = \dfrac{d\omega}{d\theta}\dfrac{d\theta}{dt} = \omega\dfrac{d\omega}{d\theta}$,

Eq. d can be written

$$2\omega \, d\omega = 9.81 \sin\theta \, d\theta \qquad (e)$$

and integrating gives

$$\omega^2 = -9.81 \cos\theta + C_1 = 9.81(1 - \cos\theta) \qquad (f)$$

where the constant of integration $C_1 = 9.81$, since the sphere starts from rest ($\omega = \dot{\theta} = 0$) when $\theta = 0$. Therefore, when $\theta = 90°$ the angular velocity of the sphere is

$$\omega = \dot{\theta} = \sqrt{9.81(1 - \cos 90°)} = 3.132 \text{ rad/s}$$

Then, the velocity of the sphere is

$$v = r\dot{\theta} = 2(3.132) = 6.26 \text{ m/s}\downarrow \qquad \text{Ans.}$$

and the force P is (Eq. a or c)

$$P = 5[-2(3.132^2)] = -98.1 \text{ N} = 98.1 \text{ N}\leftarrow \qquad \text{Ans.}$$

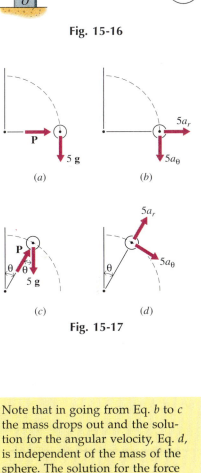

Fig. 15-16

(a) (b)

(c) (d)

Fig. 15-17

Note that in going from Eq. b to c the mass drops out and the solution for the angular velocity, Eq. d, is independent of the mass of the sphere. The solution for the force P, however, does depend on the mass of the sphere.

A sphere with a mass of 3 kg slides along a rod (see Fig. 15-18), which is bent in a vertical plane into a shape that can be described by the equation $y = 8 - \frac{1}{2}x^2$, where x and y are measured in meters. When $x = 2$ m, the collar is moving along the rod at a speed of 5 m/s and the speed is increasing at a rate of 3 m/s^2. Determine the normal $\mathbf{F}_n$ and tangential $\mathbf{F}_t$ components of the force being exerted on the sphere by the rod at this time.

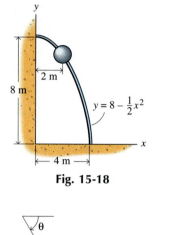

Fig. 15-18

SOLUTION

A free-body diagram for the sphere is shown in Fig. 15-19a. The sphere moves on the curved path in the vertical plane under the influence of the forces F_n, F_t, and the weight $W = 3g$. The motion of the sphere is given in terms of path variables. Therefore, it will be easier to get the acceleration components in normal and tangential coordinates than in polar or rectangular cartesian coordinates. Also, writing the equations of motion in terms of normal and tangential coordinates will separate the unknown variables F_n and F_t, for an easier solution. The normal and tangential components of the inertia vector are shown on Fig. 15-19b. Then, the equations of motion for the sphere (Eqs. 15-8) are

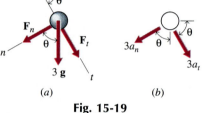

$+\nwarrow \Sigma F_t = ma_t$: $F_t + 3(9.81) \sin \theta = 3a_t$ *(a)*
$+\nearrow \Sigma F_n = ma_n$: $F_n + 3(9.81) \cos \theta = 3a_n$ *(b)*

Fig. 15-19

where $a_t = \dot{v} = 3$ m/s^2, and $a_n = \dfrac{v^2}{\rho}$. For the curved path, $y = 8 - \frac{1}{2}x^2$, $\dfrac{dy}{dx} = -x$, $\dfrac{d^2y}{dx^2} = -1$, and the radius of curvature is (see any calculus book)

$$\rho = \left| \frac{[1 + (dy/dx)^2]^{3/2}}{d^2y/dx^2} \right| = (1 + x^2)^{3/2}$$

(the absolute value is necessary to ensure that ρ is positive). Then, when $x = 2$, $dy/dx = -2$, $\rho = 11.180$ m, and $a_n = 5^2/11.180 = 2.236$ m/s^2. Also, since dy/dx is the slope of the curve

$$\theta = \tan^{-1}\left(-\frac{dy}{dx}\right) = \tan^{-1}2 = 63.43°$$

(the negative sign is necessary because θ is measured below the x-axis rather than above the x-axis). Then, Eq. *a* gives

$$F_t = 3(3) - 3(9.81) \sin 63.43° = -17.32 \text{ N} = 17.32 \text{ N}\searrow \qquad \text{Ans.}$$

and Eq. *b* gives

$$F_n = 3(2.236) - 3(9.81) \cos 63.43° = -6.46 \text{ N} = 6.46 \text{ N}\swarrow \qquad \text{Ans.}$$

In a carnival ride, chairs swing about a vertical axis with a constant angular velocity so that the swing rope is inclined at 65° to the vertical as shown in Fig. 15-20. The rope is 40 ft long, the combined weight of the rider and chair is 180 lb, and the weight of the rope may be neglected. Determine the tension T in the rope and the linear velocity v of the chair and rider.

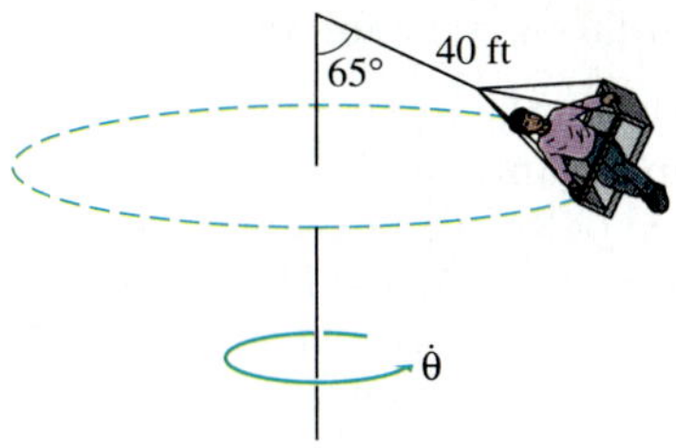

SOLUTION

A free-body diagram for the chair is shown in Fig. 15-21a. The chair moves on a circular path of radius $r = 40 \sin 65° = 36.25$ ft = constant in a horizontal plane under the influence of the two forces $\mathbf{T}$ and $\mathbf{W}$. Since the forces have components perpendicular to the plane of motion, cylindrical coordinates are appropriate, and the cylindrical components of the inertia vector are shown on Fig. 15-21b. Then, the equations of motion for the chair are

$$\Sigma F_r = ma_r: \qquad -T \sin 65° = \frac{180}{32.2}(-36.250\dot{\theta}^2) \qquad (a)$$

$$\Sigma F_\theta = ma_\theta: \qquad\qquad 0 = 0 \qquad (b)$$

$$\Sigma F_z = ma_z: \qquad T \cos 65° - 180 = 0 \qquad (c)$$

where $r = 36.25$ ft = constant, $\dot{r} = \ddot{r} = 0$, $\dot{\theta}$ = constant, $\ddot{\theta} = 0$, and the acceleration components are $a_r = \ddot{r} - r\dot{\theta}^2 = -36.25\dot{\theta}^2$, $a_\theta = r\ddot{\theta} + 2\dot{r}\dot{\theta} = 0$, and $a_z = \ddot{z} = 0$ (since the chair has no motion in the z-direction). Equation b is satisfied automatically, and Eq. c gives the tension T

$$T = 180/\cos 65° = 425.92 \text{ lb} \cong 426 \text{ lb} \qquad \text{Ans.}$$

Then Eq. a gives the angular velocity

$$\omega = \dot{\theta} = 1.3801 \text{ rad/s}$$

Finally, the velocity of the chair is

$$v = r\omega = 36.25(1.3801) = 50.0 \text{ ft/s} \qquad \text{Ans.}$$

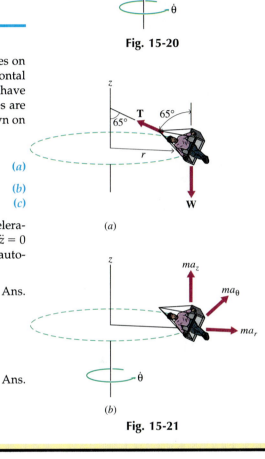

Fig. 15-20

(a)

(b)

Fig. 15-21

PROBLEMS

Introductory Problems

15-49* An airplane in level flight drops a 1000-lb bomb from an altitude of 30,000 ft, as shown in Fig. P15-49. If the speed of the plane is 450 mi/h when the bomb is released, determine the horizontal distance from the point of release to the point of impact and the time of flight for the bomb. Neglect air resistance.

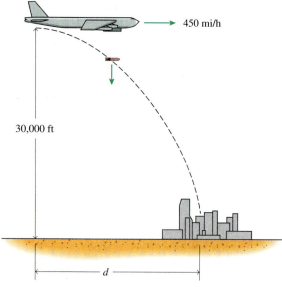

450 mi/h

30,000 ft

d

Fig. P15-49

15-50* An airplane is descending at an angle of 20° with respect to the horizontal when it drops a bomb (Fig. P15-50). If the altitude at the time of release is 5000 m and the speed of the plane is 750 km/h, determine the range (horizontal distance traveled) of the bomb and the elapsed time before it strikes the ground. Neglect air resistance.

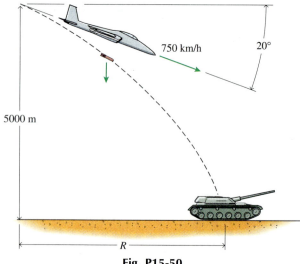

750 km/h

20°

5000 m

R

Fig. P15-50

15-51 A water ski ramp is set at an angle of 25° as shown in Fig. P15-51. The speed of a 180-lb skier as he leaves the end of the ramp after releasing his tow rope is 20 mi/h. If air resistance is negligible, determine

a. The maximum height *h* attained by the skier.
b. The distance *R* from the end of the ramp to his point of landing.

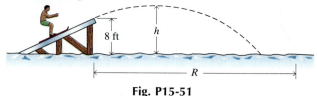

8 ft

h

R

Fig. P15-51

15-52 The toboggan shown in Fig. P15-52 is given an initial velocity of 5 m/s down the snow packed hill. The coefficient of kinetic friction between the toboggan and snow is 0.10, and the mass of the toboggan and its occupants is 80 kg. Treat the toboggan as a particle and determine the distance *d* to where the toboggan lands after leaving the hill.

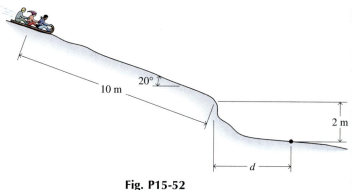

10 m

20°

2 m

d

Fig. P15-52

15-53 A 2-oz tennis ball is thrown vertically upward with an initial speed of 35 ft/s. If a strong cross wind exerts a constant horizontal force of 0.01 lb on the ball, determine the horizontal distance traveled by the tennis ball before it returns to the ground.

15-54* A 100-g apple falls from a tree branch 3 m above the ground. As it falls, a strong cross wind exerts a constant horizontal force of 0.05 N on the apple. Determine the horizontal distance traveled by the apple as it falls to the ground.

15-55* A 50-lb particle slides on a frictionless horizontal surface. At some point in its path, the velocity of the particle is $\mathbf{v}_1 = 25\mathbf{i} - 40\mathbf{j}$ ft/s. Thirty seconds later, the velocity of the particle has changed to $\mathbf{v}_2 = -75\mathbf{i} + 82\mathbf{j}$ ft/s. Determine the magnitude and direction of the constant force required to produce this change in motion.

15-56 A 5-kg particle slides on a frictionless horizontal surface. At some point in its path, the velocity of the particle is $\mathbf{v}_1 = 96\mathbf{i} + 72\mathbf{j}$ m/s. Twenty-five seconds later, the velocity of the particle has changed to $\mathbf{v}_2 = 36\mathbf{i} - 12\mathbf{j}$ m/s. Determine the magnitude and direction of the constant force required to produce this change in motion.

Intermediate Problems

15-57 The circular disk shown in Fig. P15-57 rotates in a horizontal plane. A 3-lb block rests on the disk 8 in. from the axis of rotation. The static coefficient of friction between the block and the disk is 0.50. If the disk starts from rest with a constant angular acceleration of 0.5 rad/s², determine the length of time required for the block to begin to slip.

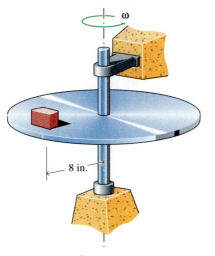

Fig. P15-57

15-58* A 5-kg block rests on a smooth frame that can be rotated about a vertical axis as shown in Fig. P15-58. When the frame is not rotating, the tension in the spring is 80 N. Determine the force exerted on the block by the stop when the frame is rotating at a constant angular velocity of 30 rev/min.

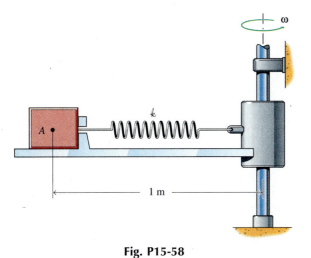

Fig. P15-58

15-59* Two bodies, A ($W_A = 50$ lb) and B ($W_B = 75$ lb), and the frame on which they rest rotate about a vertical axis at a constant angular velocity of 50 rev/min as shown in Fig. P15-59. If friction between the bodies and the frame is negligible, determine

a. The tension T in the cable connecting the bodies.
b. The force exerted on body B by the stop.

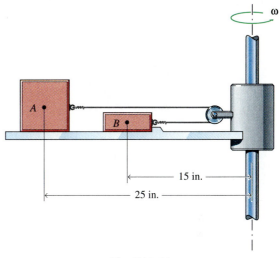

Fig. P15-59

15-60 The car shown in Fig. P15-60 maintains a constant speed of 100 km/h. At both the bottom and top of the hill, determine the force that the car seat exerts on an 80-kg driver.

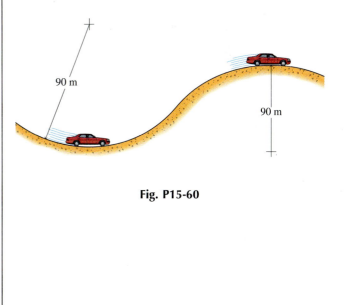

Fig. P15-60

219

15-61 A highway is designed for traffic moving at 65 mi/h. Along a certain portion of the highway, the radius of a curve is 900 ft. If the curve is banked as shown in Fig. P15-61 so that no friction is required to keep cars on the road, determine

a. The required angle of banking (angle θ) of the road.
b. The minimum coefficient of friction between the tires and the road that would keep traffic from skidding at this speed if the curve were not banked.

Fig. P15-61

15-62 A curve of radius 200 m on a level road is banked at the correct angle for a speed of 65 km/h (Fig. P15-61). If an automobile rounds this curve at 100 km/h, determine the minimum coefficient of friction required between the tires and the road so that the automobile will not skid.

15-63 A 1-lb bead is sliding along a rod that rotates with constant angular velocity $\dot{\theta}$ = 60 rev/min in a horizontal plane as shown in Fig. P15-63. If the bead moves along the rod with a constant velocity relative to the rod of v = 6 in./s, determine the transverse component of the force exerted by the rod on the bead when r = 18 in.

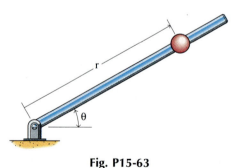

Fig. P15-63

15-64* A small sphere (m = 0.50 kg) is mounted on a circular hoop as shown in Fig. P15-64. Friction between the sphere and the hoop is negligible, and the sphere is free to slide when the hoop is rotated. Determine the angle θ and the force **P** exerted by the hoop on the sphere when the hoop is rotating about a vertical diameter at a constant angular velocity of ω = 120 rev/min.

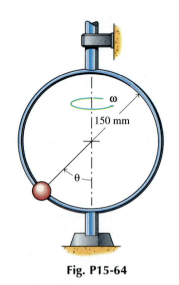

Fig. P15-64

15-65* A 10-lb sphere is attached to a vertical rod with two cords as shown in Fig. P15-65. When the system rotates about the axis of the rod, the cords extend as shown. Determine

a. The tensions in the two cords when the angular velocity of the system is ω = 5 rad/s.
b. The angular velocity ω for which cord B is taut but carries no load.

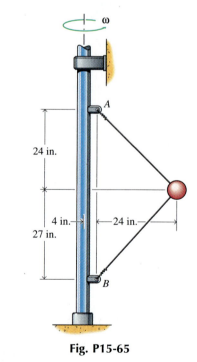

Fig. P15-65

15-66 The container shown in Fig. P15-66 is rotating about a vertical axis with a constant angular velocity. Determine the angular velocity $\dot{\theta}$ for which the 2-kg ball will maintain a fixed position relative to the frictionless side of the container.

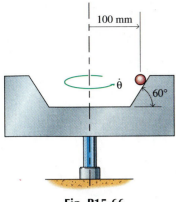

Fig. P15-66

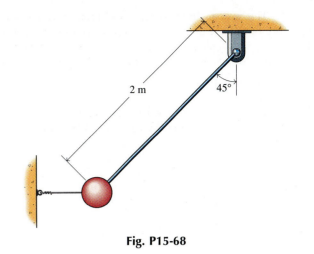

Fig. P15-68

Challenging Problems

15-67* A 5-lb ball attached to a 6-ft cord swings through a full circle in a vertical plane as shown in Fig. P15-67. If the velocity of the ball is 15 ft/s at the top of the circle, determine the tension in the cord and the linear velocity of the ball

a. When the angle θ is 45°.
b. When the angle θ is 270°.

15-69 A small ($W = 2$ lb) chunk of ice is initially at rest at the very top of the hemispherical observatory (Fig. P15-69). If a gust of wind gives the ice an initial speed of $v_0 = 1$ ft/s, determine the angle θ where the ice will lose contact with the roof. Assume that friction between the ice and the roof of the observatory is negligible.

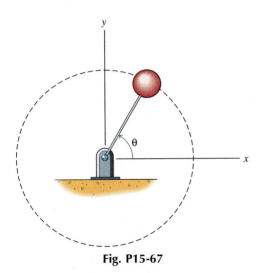

Fig. P15-67

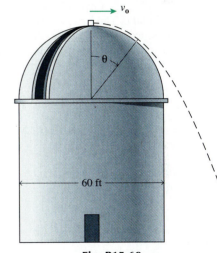

Fig. P15-69

15-68* A 3-kg sphere is supported by a rod of negligible mass and a cord as shown in Fig. P15-68. Determine the tension T in the rod

a. When the sphere is in the position shown in the figure.
b. Immediately after the cord is cut.
c. When the sphere is at the bottom of its swing.

15-70 A child is playing with a 150-g bar of soap in a hemispherical sink of radius 200 mm (Fig. P15-70). If she releases the soap from rest when $\theta = 30°$ and friction may be neglected, determine the soap bar's speed at $\theta = 90°$.

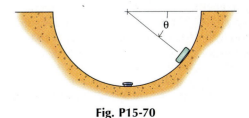

Fig. P15-70

15-71 The hopper shown in Fig. P15-71 releases balls onto the smooth circular track with essentially zero velocity. The balls leave the track at C in a horizontal direction and are collected in the basket B. Determine

a. The speed of a ball at an arbitrary position θ along the track.
b. The speed v_C of a ball as it leaves the track at $\theta = 90°$.
c. The distance x_B so that the balls land in the basket.

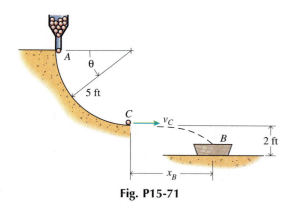

Fig. P15-71

15-72* The 1.5-kg particle P shown in Fig. P15-72 slides on a rod that has been bent to form a circular arc of radius $R = 2$ m in a vertical plane. The circular portion of the rod is smooth but the kinetic coefficient of friction between the particle and the rod is 0.10 on the straight portion of the rod. If the particle is released from rest in the position shown, determine

a. The force exerted by the rod on the particle at a point 1 m below the point of release.
b. The distance d traveled by the particle along the straight portion of the rod before coming to rest.

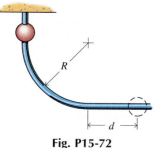

Fig. P15-72

15-73* A 5-lb sphere slides along a rod that is bent in a vertical plane into a shape that can be described by the equation $x^2 = 8y$, where x and y are both measured in feet. When the sphere is at the point $x = -8$ ft and $y = 8$ ft as shown in Fig. P15-73, it is moving along the rod at a speed of 15 ft/s and is slowing down at a rate of 3 ft/s². Determine the normal and tangential components of the force being exerted on the sphere by the rod at this time.

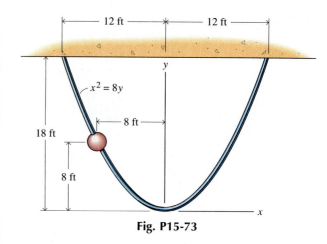

Fig. P15-73

15-74 A 10-kg block A is held in place on a 20-kg cart B, which rests against an obstruction C as shown in Fig. P15-74. If all surfaces are frictionless, determine

a. The acceleration a_A of the 10-kg block when it is released from rest.
b. The acceleration a_A of the block and the acceleration a_B of the cart if the obstruction C is removed at the same time that the block A is released from rest.

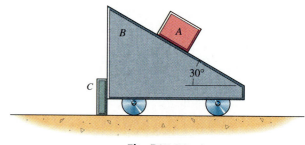

Fig. P15-74

Computer Problems

C15-75 When air resistance is neglected, it is easily shown that the trajectory of objects moving near the surface of the earth are parabolas. However, all objects moving through a fluid (such as air) experience a drag force that is proportional to the square of their speed and that acts in a direction opposite to their velocity.

Suppose that a small ball is thrown upward with an initial speed v_0 and an initial angle of θ_0 to the horizontal. At some point in its trajectory the ball will be acted on by the forces of gravity W and wind drag

$$D = C_D \frac{1}{2} \rho v^2 A$$

where C_D is the drag coefficient (may be taken as approximately one-half for spheres at moderate speeds), ρ is the mass density of the air through which the ball is moving, and $A = \pi r^2$ is the cross-sectional area of the ball (Fig. P15-75).

a. If a tennis ball ($W = 2.0$ oz, $r = 1.25$ in.) is thrown with an initial speed $v_0 = 60$ ft/s through air ($\rho = 0.002377$ slug/ft^3), use the Euler method for solving differential equations to compute the position of the ball as a function of time until it returns to the ground for initial angles $\theta_0 = 15°, 30°, 45°$, and $60°$.

b. Plot the trajectory (y versus x) for each of the initial angles above on a single graph. Also plot on the same graph the trajectory of the ball neglecting wind resistance for each initial angle.

c. Compute and plot the range (the horizontal distance between the initial and final points of the trajectory) for various initial angles θ_0, ($30° < \theta_0 < 60°$). When wind resistance is neglected, the maximum range is attained when $\theta_0 = 45°$ (independent of the initial speed). What angle gives the maximum range when wind resistance is included?

d. Repeat the calculations for an initial velocity $v_0 = 30$ ft/s. Does the angle that gives the maximum range depend on the initial speed when wind resistance is included?

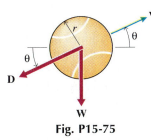

Fig. P15-75

C15-76 A small block slides down the inside of a circular bowl as shown in Fig. P15-76. If the mass of the block $m = 2$ kg, the radius of the bowl is $r = 1.5$ m, the coefficients of friction are $\mu_s = \mu_k = 0.3$, and the block is released from rest with $\theta = \theta_0$,

a. Use the Euler method for solving differential equations to compute the position of the block as a function of time until it comes to rest for initial angles $\theta_0 = 80°, 60°, 45°$, and $30°$.

b. For each of the initial angles, plot the angular position of the block θ, the speed of the block v, and the friction force acting on the block F as functions of the time t ($0 < t < 5$ s). (Be careful to ensure that the friction force always opposes the motion.)

c. For what initial angle θ_0 will the block stop just as it gets to the bottom of the bowl? For what initial angle θ_0 will the block stop at the highest position on the opposite side of the bowl? For what initial angle θ_0 will the block stop at the highest position on the same side of the bowl (possibly after sliding past the bottom and coming back)? Do the results depend on the weight of the block?

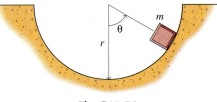

Fig. P15-76

C15-77 A small paperweight ($W = 2.5$ lb) slides on the outside of a 2-ft-radius cylinder as shown in Fig. P15-77. If friction may be neglected and the weight starts from rest when $\theta = 0°$, compute and plot

a. The speed of the weight v as a function of the angular position θ ($0 < \theta < \beta$) where β is the angle at which the weight loses contact with the cylinder (the normal force becomes zero).

b. The normal force N between the weight and the cylinder as a function of the angular position θ ($0 < \theta < \beta$).

c. Plot the angular position θ as a function of time t ($0 < t < t_\beta$). (You may need to use the Euler method for solving differential equations.)

d. Repeat the problem for $\mu_s = \mu_k = 0.3$ and ($\phi_s < \theta < \beta$) where $\phi_s = \tan^{-1}\mu_s$ is the angle of static friction. (Assume that the weight starts from rest at $\theta = \phi_s$.)

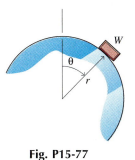

Fig. P15-77

C15-78 A 2-kg particle P is driven along a circular slot in a vertical plane by a slotted bar that is rotating about a fixed point A as shown in Fig. P15-78. If all surfaces are smooth and the bar is rotating counterclockwise at a constant rate of 25 rad/s, compute and plot

a. The normal force N exerted on the particle by the circular slot as a function of the angular position θ for one complete revolution of the arm AB ($0 < \theta < 360°$).

b. The normal force B exerted on the particle by the arm AB as a function of the angular position θ for one complete revolution of the arm AB.

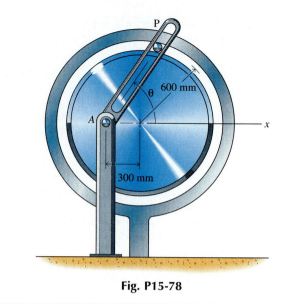

Fig. P15-78

15-5   CENTRAL FORCE MOTION

The motion of a particle moving under the influence of a force directed toward a fixed point is called central-force motion. Common examples of central-force motion include the motion of planets about the sun and the motion of the moon and artificial satellites about the Earth. From observations of the motions of planets about the sun, J. Kepler (1571–1630) deduced the following three laws, which govern central-force motion.[1]

Kepler's Laws of Planetary Motion

Law 1: The planets move about the sun in elliptical orbits with the sun at one focus.

Law 2: The radius vector joining each planet with the sun describes equal areas in equal times.

Law 3: The cubes of the mean distances of the planets from the sun are proportional to the squares of their times of revolution.

Newton's law of universal gravitation gives the magnitude of the force **F** between two masses separated by a distance r as (Eq. 15-3)

$$F = \frac{Gm_1m_2}{r^2}$$

[1]Dr. Ernst Mach, "The Science of Mechanics," 9th ed., The Open Court Publishing Company, LaSalle, Illinois, 1942. Originally published in German in 1893 and translated from German to English by Thomas J. McCormack in 1902.

Consider the case where m_1 is a very large mass that can be considered fixed in space and m_2 is a small mass that moves in the xy-plane under the action of the force $\mathbf{F}$ exerted by mass m_1 on mass m_2. By using polar coordinates with the origin fixed at mass m_1, the motion of mass m_2 is given by Eqs. 15-7 as

$$\Sigma F_r = m_2 a_r = m_2(\ddot{r} - r\dot{\theta}^2) = -\frac{Gm_1m_2}{r^2} \qquad (a)$$

$$\Sigma F_\theta = m_2 a_\theta = m_2(r\ddot{\theta} + 2\dot{r}\dot{\theta}) = 0 \qquad (b)$$

Rewriting Eq. b

$$\frac{m_2}{r}\frac{d}{dt}(r^2\dot{\theta}) = 0 \qquad (c)$$

and integrating yields

$$r^2\dot{\theta} = h \qquad \textbf{(15-14)}$$

where h is a constant. The physical significance of Eq. 15-14 can be visualized (see Fig. 15-22) by considering the area generated by the radius vector r as it turns through an angle $d\theta$ in time dt. The shaded area dA, shown in Fig. 15-22, is a triangle; therefore,

$$dA = \frac{1}{2}(r)(r\,d\theta) = \frac{1}{2}r^2 d\theta \qquad (d)$$

The area dA and Eq. 15-14 are related by the expression

$$2\frac{dA}{dt} = r^2\frac{d\theta}{dt} = r^2\dot{\theta} = h \qquad \textbf{(15-15)}$$

The quantity $dA/dt = h/2$ is called the areal speed and is a constant for any central-force system. Equation 15-15 is a mathematical statement of Kepler's second law of planetary motion.

The equation for the path of a particle subjected to a central force is obtained from Eqs. a and 15-14. The derivatives are simplified by using the substitution $u = 1/r$. From Eq. 15-15,

$$\dot{\theta} = \frac{d\theta}{dt} = \frac{h}{r^2} = hu^2 \qquad (e)$$

$$\dot{r} = \frac{dr}{dt} = \frac{dr}{d\theta}\frac{d\theta}{dt} = \frac{h}{r^2}\frac{dr}{d\theta} = -h\frac{du}{d\theta} \qquad (f)$$

$$\ddot{r} = \frac{d\dot{r}}{dt} = \frac{d\dot{r}}{d\theta}\frac{d\theta}{dt} = \frac{h}{r^2}\frac{d\dot{r}}{d\theta} = -h^2u^2\frac{d^2u}{d\theta^2} \qquad (g)$$

Substituting Eqs. e and g into Eq. a yields

$$\frac{d^2u}{d\theta^2} + u = \frac{Gm_1}{h^2} \qquad (h)$$

The solution of this differential equation (which can be verified by direct substitution) is

$$u = \frac{1}{r} = C\cos(\theta + \beta) + \frac{Gm_1}{h^2} \qquad (i)$$

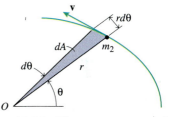

Figure 15-22 The area swept out by the radius vector in a central force motion is a triangle whose area is approximately $dA \cong r^2\dot{\theta}/2$.

where C and β are constants of integration to be determined from the initial conditions of the problem. Choosing the x-axis so that $\theta = 0$ when r is a minimum (u is a maximum, assuming C is positive) makes $\beta = 0$. Thus,

$$\frac{1}{r} = C \cos \theta + \frac{Gm_1}{\hbar^2} \qquad (j)$$

Solving Eq. j for r yields

$$r = \frac{\hbar^2/(Gm_1)}{1 + [C\hbar^2/(Gm_1)] \cos \theta} \qquad (15\text{-}16)$$

Equation 15-16 is the equation, in polar form, of a conic section (ellipse, parabola, or hyperbola). The origin of the coordinate system (the force center O) is a focus of the conic section, and the polar axis ($\theta = 0$) is an axis of symmetry.

The eccentricity e of a conic section is defined as

$$e = \frac{C}{Gm_1/\hbar^2} = \frac{C\hbar^2}{Gm_1} \qquad (15\text{-}17)$$

Equation 15-16 for r can be written in terms of the eccentricity e as

$$r = \frac{\hbar^2}{Gm_1} \frac{1}{1 + e \cos\theta} \qquad (15\text{-}18)$$

Equation 15-18 predicts three different types of paths for the particle, depending on the eccentricity e,

1. When $e > 1$, the radius $r \to \infty$ as $\cos \theta \to -1/e$. The path is a hyperbola. Many comets follow hyperbolic trajectories through the solar system.
2. When $e = 1$, the radius $r \to \infty$ as $\cos \theta \to -1$ ($\theta = \pm 180°$). The path is a parabola. Spacecraft leaving the earth for other points in the solar system may follow a parabolic path.
3. When $e < 1$, the radius r remains finite for all values of θ. The path is an ellipse. For the particular case when $e = 0$, the radius r is constant and the path is a circle. Spacecraft and other satellites in earth orbit follow elliptical or circular paths.

The different types of paths are shown in Fig. 15-23. A second branch of the hyperbola (not shown on Fig. 15-23) corresponds to a repulsive central force field rather than an attractive central force field. Equation 15-18 is a mathematical statement of Kepler's first law.

The components of velocity in polar coordinates are $v_r = \dot{r}$ and $v_\theta = r\dot{\theta}$ where from Eq. 15-18

$$\dot{r} = \frac{\hbar^2}{Gm_1} \frac{e \sin \theta}{(1 + e \cos \theta)^2} \dot{\theta} \qquad (k)$$

and from Eqs. 15-14 and 15-18

$$\dot{\theta} = \frac{\hbar}{r^2} = \frac{G^2 m_1^2}{\hbar^3}(1 + e \cos \theta)^2 \qquad (l)$$

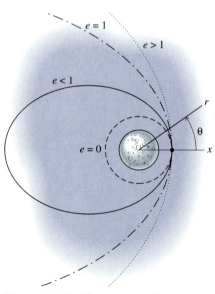

Figure 15-23   The motion of a particle in central force motion is always a conic section; either a circle (if the eccentricity $e = 0$), an ellipse (if $e < 1$), a parabola (if $e = 1$), or a hyperbola (if $e > 1$).

Therefore

$$v_r = \dot{r} = \frac{Gm_1}{h} e \sin \theta \qquad (m)$$

$$v_\theta = r\dot{\theta} = \frac{Gm_1}{h}(1 + e \cos \theta) \qquad (n)$$

and the velocity of the particle at any point on its path is

$$v = \sqrt{v_r^2 + v_\theta^2} = \frac{Gm_1}{h}\sqrt{e^2 + 2e \cos \theta + 1} \qquad (15\text{-}19)$$

For planetary motion about the sun and for artificial earth satellites, e is less than 1 and the orbits are ellipses (see Fig. 15-24). The minimum distance from the focus to the particle is called the perigee r_p of the orbit and occurs when $\theta = 0°$. The maximum distance is called the apogee r_a and occurs when $\theta = 180°$. Thus, from Eq. 15-18

$$r_p = r_{\min} = \frac{h^2}{Gm_1(1 + e)} \qquad (15\text{-}20a)$$

$$r_a = r_{\max} = \frac{h^2}{Gm_1(1 - e)} \qquad (15\text{-}20b)$$

The semimajor axis a, the semiminor axis b, and the area A of the ellipse (see Figs. 15-24 and 15-25) are

$$a = \frac{1}{2}(r_a + r_p)$$

$$= \frac{h^2}{2Gm_1(1 - e)} + \frac{h^2}{2Gm_1(1 + e)} = \frac{h^2}{Gm_1(1 - e^2)} \qquad (15\text{-}21)$$

$$b = \sqrt{a^2 - c^2}$$

$$= \sqrt{a^2 - (a - r_p)^2} = a\sqrt{1 - e^2} \qquad (15\text{-}22)$$

$$A = \pi ab = \pi a^2 \sqrt{1 - e^2} \qquad (15\text{-}23)$$

The period T (time for one revolution) can be obtained by using Eq. 15-15:

$$dA = \frac{h}{2} dt \qquad (p)$$

$$\int_0^A dA = \frac{h}{2}\int_0^T dt \qquad (q)$$

$$T = \frac{2A}{h} = \frac{2\pi a^2}{h}(1 - e^2)^{1/2} = \frac{2\pi a^2}{h}\left(\frac{h^2}{Gm_1 a}\right)^{1/2} \qquad (15\text{-}24)$$

From which

$$\frac{T^2}{a^3} = \frac{4\pi^2}{Gm_1} \qquad \text{or} \qquad T = \left[\frac{4\pi^2 a^3}{Gm_1}\right]^{1/2} \qquad (15\text{-}25)$$

Equation 15-25 is a mathematical statement of Kepler's third law.

For space probes and satellites launched from the earth, the very large mass m_1 in the previous equations is the mass of the earth m_e. The astronomical data required for solution of the following example problems is listed in Appendix B (Table B-8).

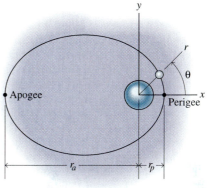

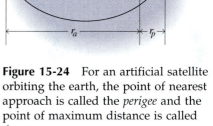

Figure 15-24 For an artificial satellite orbiting the earth, the point of nearest approach is called the *perigee* and the point of maximum distance is called the *apogee*.

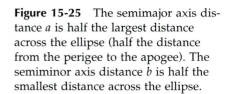

Figure 15-25 The semimajor axis distance a is half the largest distance across the ellipse (half the distance from the perigee to the apogee). The semiminor axis distance b is half the smallest distance across the ellipse.

Suppose that two satellites are in the same circular orbit with satellite A leading satellite B by some angle θ. Explain why satellite B must slow down rather than speed up if it wants to catch satellite A.

SOLUTION

According to Kepler's second law of planetary motion, the radius vector joining each planet to the sun describes equal areas in equal times. That is, a planet whose orbit encloses a larger area will take longer to orbit the sun than will a planet whose orbit encloses a smaller area. Kepler's laws also describe the motion of artificial satellites about the Earth.

If satellite B speeds up, it will go into an elliptical orbit that has its perigee at the initial height and an apogee that is higher than the initial circular orbit. This elliptical orbit clearly encloses more area than does the circle that satellite A stays in. Therefore, satellite B will take longer to complete one trip around the elliptical orbit, and when it gets back to its perigee (the height of the circular orbit), it will be even further behind satellite A than when it started.

On the other hand, if satellite B slows down, it will go into an elliptical orbit that has its apogee at the initial height and a perigee that is lower than the initial circular orbit. This elliptical orbit will enclose less area than the circular orbit that satellite A stays in. Therefore, satellite B will take less time to complete one trip around the elliptical orbit than it would have had it stayed in the circular orbit. When satellite B gets back to its apogee (the height of the circular orbit), it will be closer to satellite A than it was when it started. If the trip around the ellipse is fast enough, satellite B will come back to the circular orbit just in time to meet satellite A.

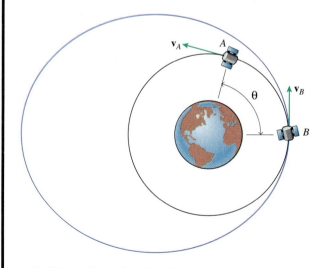

Satellite A will complete 1.2 orbits around the circular orbit in the same time that satellite B completes 1 orbit around the elliptical trajectory.

Fig. CE15-4

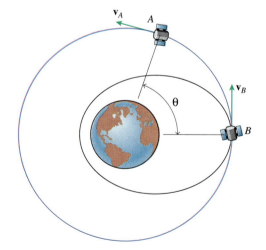

Satellite A will complete only 0.8 orbits around the circular orbit in the same time that satellite B completes 1 orbit around the elliptical trajectory.

Fig. CE15-5

EXAMPLE PROBLEM 15-11

A rocket transports a satellite to a point 250 mi above the earth's surface, as shown in Fig. 15-26. Determine the velocity (parallel to the earth's surface) required to place the satellite in a circular orbit.

SOLUTION

After the powered flight of the rocket ends and the satellite is given an initial velocity $\mathbf{v}_0$ parallel to the earth's surface, the satellite is in free flight and subjected only to the gravitational attraction of the earth. The motion of the satellite is described by Eq. 15-16, and the velocity at any point in its flight path is given by Eq. 15-19. For a circular orbit, $e = 0$ and Eq. 15-19 becomes

$$v = \frac{Gm_1}{h}\sqrt{e^2 + 2e \cos \theta + 1} = \frac{Gm_e}{h} = v_0 \qquad (a)$$

Equation a indicates that the velocity is constant and equal to the initial velocity v_0. The constant h can be determined by using Eq. 15-15 and the initial conditions for the launch; namely, when $r = r_0$, $v = v_0$. Thus,

$$h = r^2\dot{\theta} = r_0^2 \frac{v_0}{r_0} = r_0 v_0 \qquad (b)$$

From Eqs. a and b

$$v_0 = \frac{Gm_e}{h} = \frac{Gm_e}{r_0 v_0} \qquad \text{or} \qquad v_0^2 = \frac{Gm_e}{r_0}$$

At an altitude $h = 250$ miles,

$$r_0 = r_e + h = 3960 + 250 = 4210 \text{ mi}$$

Therefore

$$v_0 = \left[\frac{Gm_e}{r_0}\right]^{1/2} = \left[\frac{3.439(10^{-8})(4.095)(10^{23})}{4210(5280)}\right]^{1/2} = 2.517(10^4) \text{ ft/s}$$
$$= 17{,}160 \text{ mi/h} \qquad \text{Ans.}$$

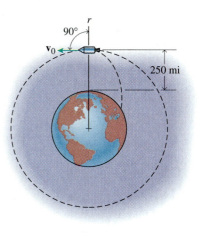

Fig. 15-26

Alternatively, the velocity of a satellite traveling in a circular orbit is easily obtained directly from Newton's second law. Since gravitational attraction is the only force acting on the satellite and the acceleration is $a_n = v^2/r$ directed toward the center of the earth, Newton's second law gives

$$F_g = \frac{Gm_e m}{r^2} = m\frac{v^2}{r}$$

or

$$v^2 = \frac{Gm_e}{r}$$

A rocket transports a satellite to a point 800 km above the earth's surface (Fig. 15-27). Determine the velocity (parallel to the earth's surface) required to place the satellite

a. In an elliptical orbit with a maximum altitude of 8000 km
b. On a parabolic flight path out of the earth's gravitational field.

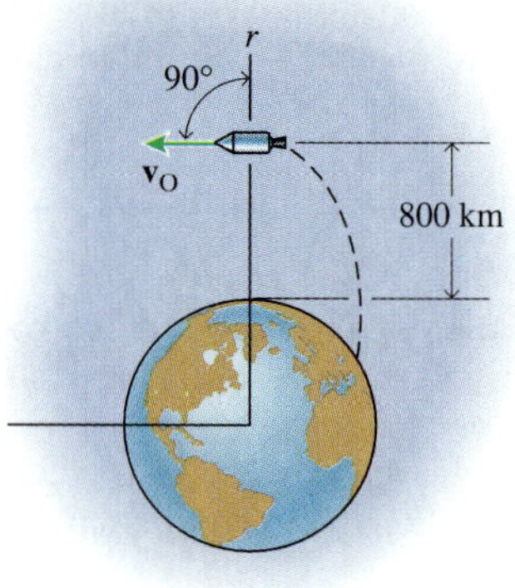

Fig. 15-27

SOLUTION

Once the satellite is in free flight and subjected only to the earth's gravitational field, the motion is described by Eq. 15-16, and the velocity at any point in its flight path is given by Eq. 15-19.

a. For an elliptical orbit with

$$r_p = r_e + h_p = 6.371(10^6) \text{ m} + 0.800(10^6) \text{ m} = 7.171(10^6) \text{ m}$$
$$r_a = r_e + h_a = 6.371(10^6) \text{ m} + 8.000(10^6) \text{ m} = 14.371(10^6) \text{ m}$$

Eqs. 15-20 yield

$$r_a = \frac{\hbar^2}{Gm_1(1-e)} \qquad r_p = \frac{\hbar^2}{Gm_1(1+e)}$$

$$e = \frac{r_a - r_p}{r_a + r_p} = \frac{14.371(10^6) - 7.171(10^6)}{14.371(10^6) + 7.171(10^6)} = 0.3342$$

At $\theta = 0°$, Eq. 15-19 becomes

$$v = \frac{Gm_1}{\hbar}\sqrt{e^2 + 2e \cos \theta + 1} = \frac{Gm_1}{\hbar}(1 + e) = v_0 \qquad (a)$$

The constant $\hbar$ can be determined by using Eq. 15-15 and the initial conditions for the launch; namely, when $r = r_0$, $v = v_0$. Thus,

$$\hbar = r^2 \dot{\theta} = r_0^2 \frac{v_0}{r_0} = r_0 v_0 \qquad (b)$$

From Eqs. a and b

$$v_0 = \frac{Gm_e}{\hbar}(1 + e) = \frac{Gm_e}{r_0 v_0}(1 + e) \qquad \text{or} \qquad v_0^2 = \frac{Gm_e}{r_0}(1 + e)$$

With $r_0 = r_p = 7.171(10^6)$ m,

$$v_0 = \left[\frac{Gm_e}{r_0}(1 + e)\right]^{1/2} = \left[\frac{6.673(10^{-11})(5.976)(10^{24})}{7.171(10^6)}(1 + 0.3342)\right]^{1/2}$$
$$= 8.614(10^3) \text{ m/s} = 8.61 \text{ km/s} \qquad \text{Ans.}$$

b. For a parabolic flight trajectory, $e = 1$; therefore,

$$v_0 = \left[\frac{Gm_e}{r_0}(1 + e)\right]^{1/2} = \left[\frac{2Gm_e}{r_0}\right]^{1/2}$$
$$= \left[\frac{2(6.673)(10^{-11})(5.976)(10^{24})}{7.171(10^6)}\right]^{1/2}$$
$$= 10.546(10^3) \text{ m/s} = 10.55 \text{ km/s} \qquad \text{Ans.}$$

The velocity v_0 associated with a parabolic flight trajectory is the minimum velocity required for escape from the earth's gravitational field and is commonly referred to as the escape velocity v_{esc}.

The maximum speed of a satellite in an elliptical orbit ($e = 0.25$) is 16,000 mi/h. Determine

a. The maximum and minimum distances (in miles) from the surface of the earth to the satellite's trajectory.

b. The period of the elliptical orbit.

SOLUTION

a. For an elliptical orbit, the maximum velocity occurs at r_p, where it is parallel to the surface of the earth. Thus, from Eq. 15-15

$$h = r^2\dot{\theta} = r_p^2 \frac{v_p}{r_p}$$
$$= r_p v_p = r_p v_{max}$$

Then from Eqs. 15-20, with $v_{max} = 16{,}000$ mi/h $= 23{,}467$ ft/s,

$$r_p = \frac{Gm_e(1 + e)}{v_{max}^2}$$
$$= \frac{3.439(10^{-8})(4.095)(10^{23})(1 + 0.25)}{(23{,}467)^2}$$
$$= 3.197(10^7) \text{ ft} = 6055 \text{ mi}$$
$$r_a = \frac{1 + e}{1 - e} r_p$$
$$= \frac{1 + 0.25}{1 - 0.25}(6055) = 10{,}090 \text{ mi}$$
$$h_p = r_p - r_e$$
$$= 6055 - 3960 = 2095 \text{ mi} \qquad \text{Ans.}$$
$$h_a = r_a - r_e$$
$$= 10{,}090 - 3960 = 6130 \text{ mi} \qquad \text{Ans.}$$

b. For the ellipse

$$a = \frac{1}{2}(r_a + r_p)$$
$$= \frac{1}{2}[5.328(10^7) + 3.197(10^7)]$$
$$= 4.263(10^7) \text{ ft}$$

From Eq. 15-25

$$T = \left[\frac{4\pi^2 a^3}{Gm_e}\right]^{1/2}$$
$$= \left\{\frac{4\pi^2[4.263(10^7)]^3}{3.439(10^{-8})(4.095)(10^{23})}\right\}^{1/2}$$
$$= 1.4737(10^4) \text{ s} = 4.09 \text{ h} \qquad \text{Ans.}$$

EXAMPLE PROBLEM 15-14

A space probe launched from the surface of the earth (see Fig. 15-28) is traveling parallel to the surface of the earth, at an altitude of 1200 km, when the powered portion of the flight ends with a velocity of 12.00 km/s. Determine

a. The eccentricity e of the trajectory.
b. The velocity of the space probe when the distance from the center of the earth to the space probe is 100,000 km.
c. The maximum angle θ for this flight trajectory.

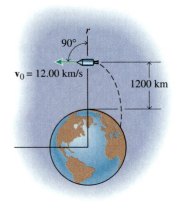

Fig. 15-28

SOLUTION

a. Since the space probe is traveling parallel to the surface of the earth when the powered portion of the flight ends, the initial conditions for the flight (when $\theta = 0°$) are

$$r_0 = r_e + h_0 = 6.371(10^6) + 1.200(10^6) = 7.571(10^6) \text{ m}$$
$$v_0 = 12.00 \text{ km/s} = 12.00(10^3) \text{ m/s}$$

From Eq. 15-15

$$h = r_0^2 \dot{\theta} = r_0^2 \frac{v_0}{r_0} = r_0 v_0 = 7.571(10^6)12.00(10^3)$$
$$= 9.085(10^{10}) \text{ m}^2/\text{s}$$

From Eq. 15-18 with $r = r_0$ when $\theta = 0°$

$$e = \frac{h^2}{Gm_e r_0} - 1$$
$$= \frac{[9.085(10^{10})]^2}{6.673(10^{-11})(5.976)(10^{24})(7.571)(10^6)} - 1 = 1.734 \qquad \text{Ans.}$$

b. The velocity is determined by using Eqs. 15-18 and 15-19. From Eq. 15-18

$$\cos\theta = \frac{h^2 - Gm_e r}{Gm_e r e}$$
$$= \frac{[9.085(10^{10})]^2 - 6.673(10^{-11})(5.976)(10^{24})(100)(10^6)}{6.673(10^{-11})(5.976)(10^{24})(100)(10^6)(1.734)}$$
$$= -0.4573 \qquad \theta = 117.2°$$

From Eq. 15-19

$$v = \frac{Gm_e}{h} \sqrt{e^2 + 2e\cos\theta + 1}$$
$$= \frac{6.673(10^{-11})(5.976)(20^{24})}{9.085(10^{10})} \sqrt{1.734^2 + 2(1.734)\cos 117.2° + 1}$$
$$= 6.830(10^3) \text{ m/s} = 6.83 \text{ km/s} \qquad \text{Ans.}$$

c. The maximum angle θ_{max} occurs as $r \to \infty$. From Eq. 15-18, $r \to \infty$ as

$$1 + e\cos\theta \to 0$$

Therefore

$$\cos\theta_{max} = -\frac{1}{e} = -\frac{1}{1.734}$$
$$\theta_{max} = 125.2° \qquad \text{Ans.}$$

Since the eccentricity is greater than 1 ($e = 1.734$), the resulting trajectory is an hyperbola. The angle θ_{max} is the direction of the asymptote of the hyperbola.

PROBLEMS

Introductory Problems

15-79* Determine the orbital speed and period of a satellite in a circular orbit 1000 mi above the earth's surface.

15-80* A rocket transports a satellite to a point 1500 km above the earth's surface (Fig. P15-80). Determine the velocity (parallel to the earth's surface) required to place the satellite

a. In a circular orbit.
b. On a parabolic flight path.

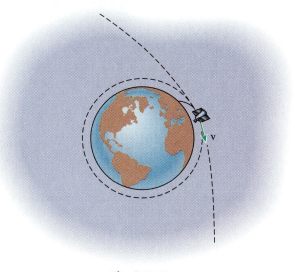

Fig. P15-80

15-81 A satellite is inserted into orbit at an altitude of 750 mi above the surface of the earth and with a speed of 18,000 mi/h parallel to the surface of the earth (Fig. P15-81). Determine

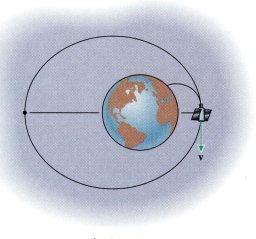

Fig. P15-81

a. The eccentricity e of the orbit.
b. The maximum and minimum altitudes for the satellite's trajectory.

15-82* The altitude of a satellite in an elliptical orbit around the earth is 1600 km at apogee and 600 km at perigee (Fig. P15-82). Determine

a. The eccentricity e of the orbit.
b. The orbital speeds at apogee and perigee.
c. The period of the orbit.

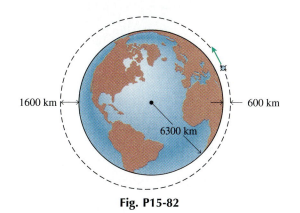

Fig. P15-82

15-83 A satellite is inserted into orbit at an altitude of 500 mi above the surface of the earth and with a speed of 20,000 mi/h parallel to the surface of the earth (Fig. P15-83). Determine

a. The eccentricity e of the orbit.
b. The orbital speeds at apogee and perigee.
c. The period of the orbit.

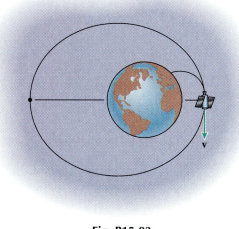

Fig. P15-83

15-84 The Apollo lunar lander is in an orbit about the moon with a perigee altitude of 50 km and an apogee altitude of 500 km above the lunar surface (Fig. P15-84). Determine

a. The eccentricity e of the orbit.
b. The orbital speeds at apogee and perigee.
c. The period of the orbit.

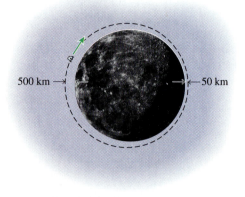

500 km → ← 50 km

Fig. P15-84

Intermediate Problems

15-85 A satellite is to be placed into an equatorial circular orbit so that it always remains over the same point on the earth's surface. Determine the radius of the orbit in miles and the orbital speed of the satellite in miles per hour.

15-86* A satellite is in a circular polar orbit 750 km above the earth's surface. If the satellite's orbit is fixed in space and the earth rotates under the satellite at a constant rate, determine the separation (in degrees of longitude and in kilometers) between successive southerly passes across the earth's equator (Fig. P15-86).

Fig. P15-86

15-87* It is desired to place a satellite in a circular polar orbit such that successive ground tracks at the equator are spaced $d = 2000$ mi. apart (Fig. P15-86). Determine the required altitude of the circular orbit.

15-88 A rocket is in a circular orbit at an altitude of 500 km. During a very short interval of an orbit, the engines increase the velocity by 1000 m/s (Fig. P15-88). Determine

a. The eccentricity e of the new orbit.
b. The altitude and orbital speed of the rocket at the highest point of its new orbit.

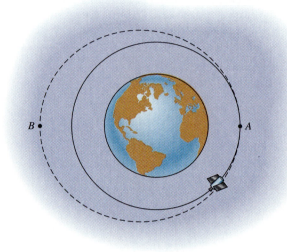

B •   • A

Fig. P15-88

15-89 A satellite is traveling in a circular orbit 500 miles above the surface of the earth (Fig. P15-88). Determine

a. The change in velocity required to place the satellite in an elliptical orbit with an eccentricity of 0.30.
b. The altitude and orbital speed of the satellite at the highest point of its new elliptical orbit.

15-90* A satellite is traveling in an elliptical orbit ($e = 0.25$) above the surface of the earth. If the maximum distance from the center of the earth to the satellite is 15,000 km, determine

a. The altitude of the satellite at perigee.
b. The velocity of the satellite at apogee and perigee.
c. The period of the elliptical orbit.

234

15-91 The altitude of a satellite in an elliptical orbit around the earth is 21,000 mi at apogee and 2500 mi at perigee as shown in Fig. P15-91. Determine

a. The eccentricity e of the orbit.
b. The orbital speeds at apogee and perigee.
c. The radial distance r from the center of the earth and the velocity v when $\theta = 150°$.

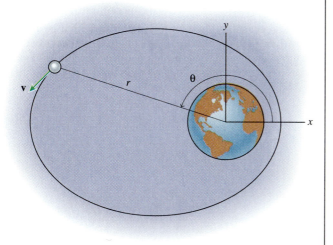

Fig. P15-91

15-92 A satellite is in an elliptic orbit around the earth with perigee and apogee altitudes of 1000 km and 2500 km, respectively. At the apogee of the elliptic orbit an on-board engine changes the satellite's speed by 500 m/s. Determine the new perigee and apogee altitudes if the change in velocity is

a. Tangentially in the direction of the satellite motion.
b. Tangentially opposite to the direction of the satellite motion.

15-93* A space vehicle is traveling in a circular orbit 300 mi above the surface of the moon (Fig. P15-93). If air resistance can be neglected, determine

a. The change in speed required to bring the vehicle to the surface of the moon at a point 180° from the point of retrorocket firing.
b. The time required for the descent.

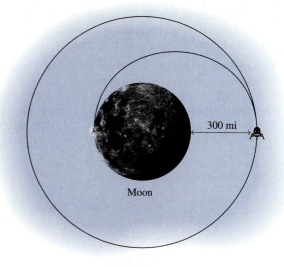

Fig. P15-93

15-94* A satellite is traveling in a circular orbit 5000 km above the surface of the Earth as shown in Fig. P15-94. At point A, the velocity is increased to put the satellite in an elliptical orbit with a maximum altitude of 10,000 km at point B. Determine

a. The eccentricity e of the elliptical orbit.
b. The change in velocity required at point A to place the satellite in the elliptical orbit.
c. The change in velocity required at point B to change the elliptical orbit to the higher circular orbit.
d. The period of the higher circular orbit.

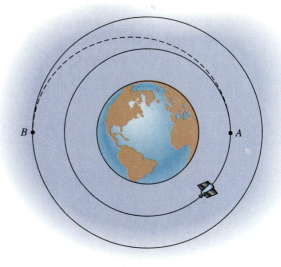

Fig. P15-94

15-95 Satellites *A* and *B* are in circular orbits 100 mi and 800 mi, respectively, above the surface of the earth (Fig. P15-95). If satellite *A* is to rendezvous with satellite *B* at point *C* using the elliptic orbit shown, determine

a. The amount by which the speed of satellite *A* must be increased to put it on the elliptic orbit.
b. The amount by which the speed of satellite *A* must be increased at point *C* to complete the maneuver.
c. The relative angle ϕ which should exist between the two satellite positions at the start of the maneuver.

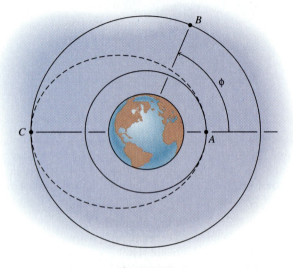

Fig. P15-95

15-96 At some instant the earth is aligned with Mars as shown in Fig. P15-96. Assuming that both orbits are in exactly the same plane and are perfect circles, determine the time it will take for the planets to become aligned again.

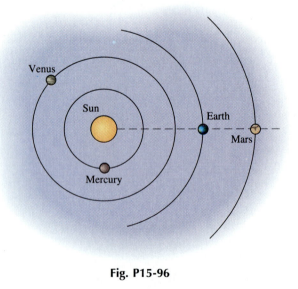

Fig. P15-96

Challenging Problems

15-97* A satellite is in a circular orbit 250 mi above the surface of the earth. Determine the new perigee and apogee altitudes of the satellite if an on-board maneuvering engine

a. Increases the orbital speed of the satellite by 800 ft/s.
b. Gives the satellite a radial (outward) component of velocity of 800 ft/s.
c. Gives the satellite a radial (inward) component of velocity of 800 ft/s.

15-98* A satellite is in an elliptic orbit with a perigee altitude of 1000 km and an apogee altitude of 9000 km. Determine the *r*- and θ-components of the satellite's velocity as it crosses the minor axis of the ellipse (point *B* of Fig. P15-98).

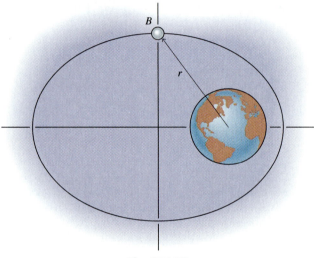

Fig. P15-98

15-99 A satellite is being placed into a low earth orbit by a launch vehicle. When the rocket motor of the launch vehicle shuts off, the satellite is at an altitude of 80 mi above the Earth and has a velocity of 26,000 ft/s. As a result of guidance error, the satellite is injected into orbit at an angle of 85° with respect to the radius vector (Fig. P15-99) rather than parallel to the surface of the earth as planned. Determine

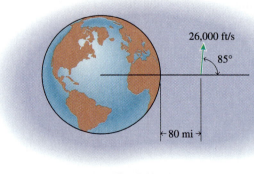

Fig. P15-99

236

a. The equation of the planned (parallel injection) orbit.
b. The equation of the actual orbit.
c. The velocity and altitude at apogee of the planned orbit.
d. The velocity and altitude at apogee of the actual orbit.
e. The altitude at perigee of the actual orbit.

15-100 Two satellites are in the same circular orbit 1000 km above the surface of the earth with satellite *A* leading satellite *B* by 2500 km (Fig. P15-100). Satellite *B* proposes to "catch up" to satellite *A* by "slowing down" into the elliptic orbit shown. Determine the amount by which satellite *B* must slow down to catch satellite *A* after

a. One period in the elliptic orbit.
b. Two periods in the elliptic orbit.

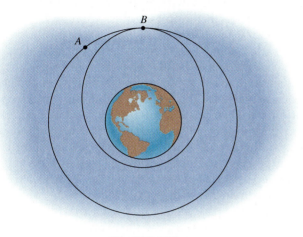

Fig. P15-100

15-101 Two satellites are in the same circular orbit 150 mi above the surface of the earth with satellite *A* leading satellite *B* by 1500 mi (Fig. P15-100). Satellite *B* proposes to "catch up" to satellite *A* by "slowing down" into the ellip-

tic orbit shown. If satellite *B* cannot come closer than 50 mi to the surface of the Earth, determine

a. The minimum number of orbits satellite *B* must spend in the elliptic orbit in order to catch satellite *A*.
b. The amount by which satellite *B* must slow down to get into the elliptic orbit.

15-102* The lunar lander, which is sitting on the surface of the moon, wants to return to the command module, which is in a circular orbit 80 km above the moon's surface (Fig. P15-102). Determine

a. The velocity (magnitude and direction) with which the lunar lander must leave the moon's surface to rendezvous with the command module as shown.
b. The amount by which the speed of the lunar lander must be increased at its apogee to complete its rendezvous with the command module.

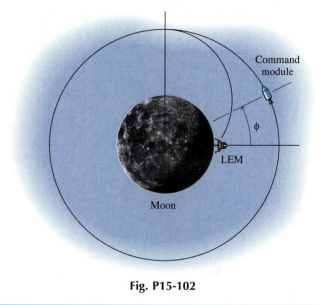

Fig. P15-102

SUMMARY

The basic law governing the motion of a particle is Newton's second law, which relates the accelerated motion of a particle to the forces producing the motion. Mathematically, Newton's second law is expressed as

$$\mathbf{F} = m\mathbf{a} \tag{15-2}$$

Equation 15-2 expresses the fact that the magnitude of **F** and **a** are proportional and that the vectors **F** and **a** have the same direction. Equation 15-2 is valid both for constant forces and for forces that vary with time.

The system of axes used for the acceleration measurements must be a primary inertial system. Any *nonrotating system of axes*, which translates with a constant velocity with respect to the primary system, is equally satisfactory. For most engineering problems *on the surface of the earth,* the corrections required to compensate for the acceleration of the earth with respect to the primary system are negligible, and the accelerations measured with respect to axes attached to the surface of the earth may be treated as absolute.

The equation of motion for a system of particles can be obtained by using Newton's second law for each individual particle of the system and summing the results to obtain an equation for the motion of the mass center G of the system. The result is

$$\mathbf{R} = m\mathbf{a}_G \qquad (15\text{-}11)$$

Equation 15-11 is valid for any type of motion and shows that the equation of motion for a system of particles is the same as the equation of motion for a single particle located at the mass center of the system and having a mass equal to the total mass of the system. Any body can be considered to be a particle when applying this equation.

Newton's second law (Eqs. 15-2 and 15-11) may be expressed in any convenient coordinate system. Rectilinear motion (motion of a particle along a straight line) is conveniently described in terms of rectangular cartesian components. If the coordinate system is oriented such that the x-axis coincides with the line of motion, the position, velocity, and acceleration of the particle are completely described by their x-components. Since there is no motion perpendicular to the line, the components of forces in the y- and z-directions must be in equilibrium. Curvilinear motion may be described in terms of either rectangular cartesian components, polar coordinates, normal and tangential coordinates, cylindrical coordinates, or spherical coordinates. The best choice will depend on how the particle moves or how the data is given.

Regardless of the coordinate system used to express Newton's second law, it is important to use the same components and consistent coordinate directions on both sides of the equations. A free-body diagram should always be drawn to identify all forces which act on the particle and to aid in the expression of the force components in Eqs. 15-2 and 15-11. Students may also find it useful to show the inertia vectors ($m\mathbf{a}$) on a separate diagram next to the free-body diagram used for the forces. However, the inertia vectors are not forces and must never be included on the free-body diagram.

The motion of a particle moving under the influence of a force directed toward a fixed point is called central-force motion. Common examples include the motion of planets about the sun and the motion of the moon and artificial satellites about the Earth. The force involved in these examples is given by Newton's law of universal gravitation, which states that the force $\mathbf{F}$ between two masses m_1 and m_2 separated by a distance r has the magnitude

$$F = \frac{Gm_1m_2}{r^2} \qquad (15\text{-}3)$$

where G is the universal gravitational constant. All particles moving in central-force motion travel in either an elliptical path (if the eccentricity of the motion is $e < 1$), a parabolic path (if $e = 1$), or a hyperbolic path (if $e > 1$). Also, for all particles moving in central-force motion, the quantity $\hbar = r^2\dot{\theta}$ is a constant.

REVIEW PROBLEMS

15-103* Two carts are connected by a cable that passes over a small pulley as shown in Fig. P15-103. The weights of carts A and B are 500 lb and 400 lb, respectively. After the carts are released from rest, determine

a. The acceleration of the carts.
b. The tension in the cable.
c. The distance moved during the first 10 s of motion.

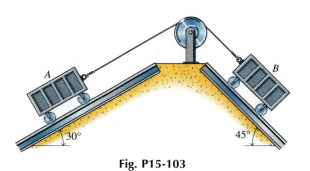

Fig. P15-103

15-104* A 5-kg block rests on a smooth conical surface that revolves about a vertical axis with a constant angular velocity ω. The block is attached to the rotating shaft with a cable as shown in Fig. P15-104. Determine

a. The tension in the cable when the system is rotating at 20 rev/min.
b. The angular velocity in revolutions per minute when the force between the conical surface and the block is zero.

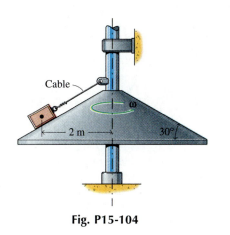

Fig. P15-104

15-105 The airplane shown in Fig. P15-105 is at the lowest point of a curved path where the radius of curvature is 2500 ft. Determine the "apparent" weight of the pilot if her true weight is 140 lb and the speed of the airplane is 160 mi/h.

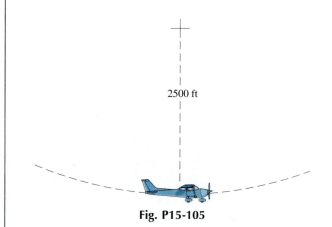

Fig. P15-105

15-106* A 0.50-kg block moves along a smooth circular slot in a vertical plane as shown in Fig. P15-106. When the block is in the position shown, its speed is 20 m/s up and to the left. If the unstretched length of the spring ($k =$ 25 N/m) is 300 mm, determine the acceleration of the block and the force exerted on the block by the surface of the slot.

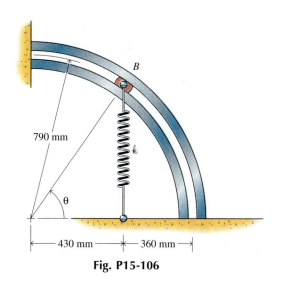

Fig. P15-106

239

15-107 The lunar lander is in a circular orbit 50 mi above the surface of the moon (Fig. P15-107). Determine

a. The amount by which the velocity must be decreased to land after one-quarter orbit (at *B*).
b. The speed with which the lander would strike the moon if it did not use its retrorockets to slow its descent.
c. The angle that the velocity of part *b* makes with the radial direction.

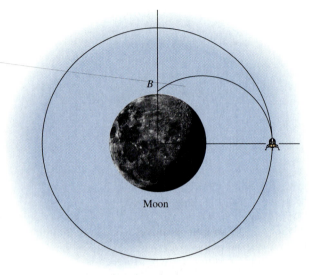

Fig. P15-107

15-108 Two 1-kg beads are connected by a massless, inextensible cord, and are constrained to move along a frictionless wire in a vertical plane as shown in Fig. P15-108. If the beads are released from rest in the position shown, determine the initial tension in the cord.

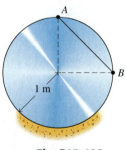

Fig. P15-108

15-109* The weight of block *A* of Fig. P15-109 is 50 lb. The block is at rest, and the spring ($k = 10$ lb/ft) is unstretched when the block is released to move. Determine

a. The velocity of the block after it has moved 3 ft.
b. The maximum distance that the block moves from its initial position.

Fig. P15-109

15-110 When a mortar launches a 5-kg projectile, the angle of inclination θ (with respect to the horizontal) can be set, but the muzzle velocity v_0 is fixed (Fig. P15-110). If $v_0 = 100$ m/s and air resistance may be neglected, determine

a. The angle of inclination required to hit a target that is located 770 m horizontally and 150 m vertically (above) from the mortar position.
b. The elapsed time from when the mortar is fired until the shell hits the target.

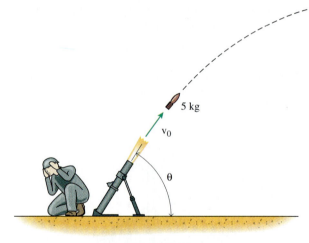

Fig. P15-110

15-111* A car on the track of a roller coaster ride at an amusement park is shown in Fig. P15-111. The car and its four occupants weigh 900 lb. The speed of the car is 35 mi/h when it passes point *A* on the track. If the car and its occupants are treated as a particle, determine

a. The speed of the car at the bottom of the loop.
b. The force exerted on the car by the track at point *B*.
c. The force exerted on the car by the track at point *C*.
d. The minimum speed required at the top of the loop to keep the car in contact with the track.

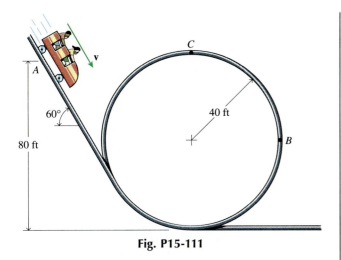

Fig. P15-111

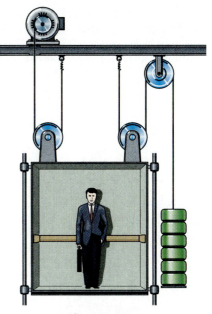

Fig. P15-113

15-112 Force P causes the 1.5-kg collar to move to the right as shown in Fig. P15-112. The coefficient of friction between the collar and the stationary rod is 0.40. The mass $m = 1.5$ kg is attached to the collar by a light, inextensible cord. Determine the value of P for which the cord will maintain a constant angle of $\theta = 5°$.

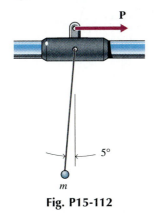

Fig. P15-112

15-113 An elevator contains a passenger who weighs 190 lb (Fig. P15-113). Determine the "apparent" weight of the passenger if

a. The elevator starts upward with an acceleration of 9 ft/s².
b. The elevator starts downward with an acceleration of 7 ft/s².

15-114* Satellites A and B are in circular orbits 200 km and 800 km, respectively, above the surface of the earth (Fig. P15-114). If satellite A is to rendezvous with satellite B at point C by using the elliptic orbit shown, determine

a. The amount by which the speed of satellite A must be increased to put it on the elliptic orbit.
b. The amount by which the speed of satellite A must be increased at point C to complete the maneuver.
c. The relative angle ϕ that should exist between the two satellite positions at the start of the maneuver.

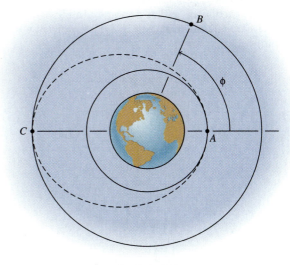

Fig. P15-114

15-115 The velocity of a baseball ($W = 0.33$ lb) when it is hit by a bat changes from 80 mi/h in one direction to 90 mi/h in the opposite direction (Fig. P15-115). If the ball and the bat are in contact for 0.005 s, what constant force must be exerted on the ball by the bat?

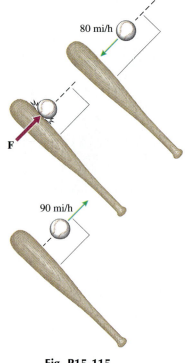

Fig. P15-115

15-116 Two blocks of mass $m_A = 10$ kg and $m_B = 15$ kg are connected by a light inextensible rope that passes over a massless, frictionless pulley as shown in Fig. P15-116. If the blocks are released from rest in the position shown, determine

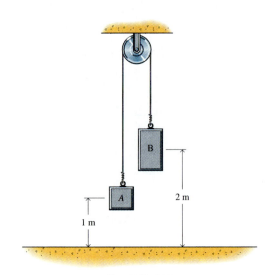

Fig. P15-116

a. The velocity of block B just as it hits the ground.
b. The time it takes for the block to hit the ground.
c. The tension in the rope while the blocks are moving.

15-117* A 1-lb particle P is driven along a circular slot in a vertical plane by a slotted bar, which is rotating about a fixed point A as shown in Fig. P15-117. The angular velocity of the bar is 25 rad/s clockwise, and the angular acceleration is 20 rad/s² counterclockwise. If all surfaces are smooth, determine the forces exerted on the particle

a. When $\theta = 60°$.
b. When $\theta = 120°$.

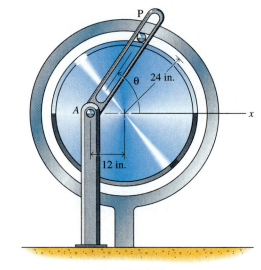

Fig. P15-117

15-118* A rocket launched from the surface of the earth has a speed of 8.85 km/s when powered flight ends at an altitude of 550 km. The flight path of the rocket at this time is inclined at an angle of 84° with respect to a radial line through the center of the earth (Fig. P15-118). Determine

a. The eccentricity e of the trajectory.
b. The altitude of the rocket at perigee.
c. The velocity of the rocket at apogee and perigee.
d. The period of the orbit.

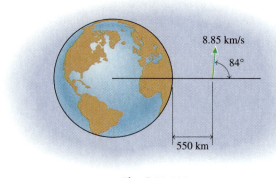

Fig. P15-118

15-119 A 50-lb projectile is fired as shown in Fig. P15-119 with an initial velocity of 1500 ft/s. If air resistance can be neglected, determine the radius of curvature of the projectile's path when it is at its peak.

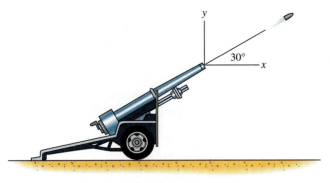

Fig. P15-119

15-120 A 1-kg collar that slides along a smooth horizontal rod has an attached pin that is constrained to move in the slot of arm AB as shown in Fig. P15-120. The arm oscillates with angular position given by $\theta(t) = (\pi/2) - (\pi/4)$ cos nt, where $\theta(t)$ is in radians, t is the time in seconds, and $n = 1$ rad/s. Determine the horizontal component of the force of the arm on the pin when $t = 10$ s.

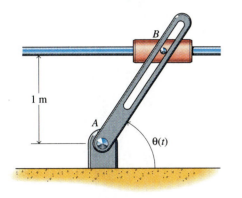

Fig. P15-120

15-121* A meteoroid is first observed approaching the earth along a hyperbolic flight path when it is 250,000 mi from the center of the earth (Fig. P15-121). If the speed of the meteoroid at this time is 5000 mi/h at $\theta = 150°$, determine

a. The eccentricity e of the trajectory.
b. The closest approach of the meteoroid to the earth's surface.
c. The velocity of the meteoroid when it is closest to the earth's surface.

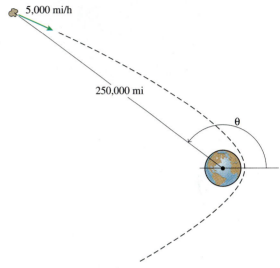

Fig. P15-121

15-122 A 5-kg sphere S is attached to a 1-kg block B, which is free to slide in a smooth horizontal slot as shown in Fig. P15-122. The mass of the rod connecting the sphere to the block is negligible. If the system is released from rest in the position shown, determine

a. The tension in the rod as motion begins.
b. The acceleration of the block as motion begins.

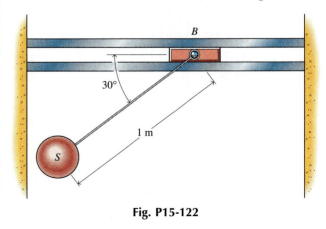

Fig. P15-122

15-123 A 30-lb crate of eggs rests on the seat of a truck as shown in Fig. P15-123. The coefficient of friction between the crate and the seat is 0.70. If the driver of the truck is forced to suddenly apply the brakes, what maximum deceleration will cause the crate to slide on the seat? (The seat may be represented as a straight line at a 5° slope.)

Fig. P15-123

15-124* The rocket engine of a 15,000-kg missile produces a thrust of 200 kN. If the missile is launched in a vertical direction, determine its velocity and vertical height after one minute

a. If air resistance is negligible.
b. If air resistance produces a drag force $F_D = 0.25v^2$, where F_D is in newtons and v is in meters per second.

15-125 A worker at B walks backward at a constant rate of 2 ft/s and raises a 40-lb basket of tools to a co-worker on the third floor of a building (Fig. P15-125). The 59-ft long rope passes over a small frictionless pulley at C. If the person at B walks with the end of the rope at a constant height above the ground, determine the tension in the rope as the basket passes the second-floor window, which is 16 ft above the ground.

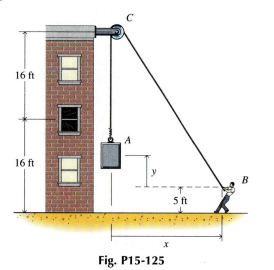

Fig. P15-125

15-126 The system shown in Fig. P15-126 consists of a 20-kg block and a massless spring. The force in the spring is $F = k\delta$, where $k = 100$ N/m is the modulus of the spring and δ is the stretch of the spring measured from its unloaded position. If the block is moved 200 mm to the right of the undeformed position of the spring and released from rest, determine the speed of the block as it passes the position of zero spring force. Friction may be neglected.

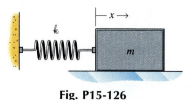

Fig. P15-126

15-127* Blocks A and B of Fig. P15-127 weigh 60 lb and 40 lb, respectively. If the blocks are released from rest in the position shown, determine

a. The velocity of block B after it has moved 10 ft.
b. The tension in the cable supporting block A.

Fig. P15-127

15-128 The circular disk shown in Fig. P15-128 is rotating in a vertical plane. The motion of body A ($m = 1$ kg) in the smooth radial slot is resisted by a spring attached to the hub of the disk. When the disk is in the position shown, its angular velocity is 100 rad/s clockwise and its angular acceleration is 25 rad/s² counterclockwise. At this instant, determine

a. The force exerted on body A by the disk.
b. The spring constant k if the rest position of body A is 350 mm from the axis of the disk.

Fig. P15-128

A motorcycle is propelled by friction between its rear wheel and the ground.

Chapter 16

KINETICS OF RIGID BODIES: NEWTON'S LAWS

16-1   INTRODUCTION

Rigid bodies can be viewed as a collection of particles; therefore, the relationships developed in Chapter 15 for the motion of a system of particles can be utilized for rigid bodies. In this chapter, extensive use is made of Eq. 15-11, which relates the resultant $\mathbf{R}$ of the applied external forces to the acceleration $\mathbf{a}_G$ of the mass center G of the system. For the general case, where the resultant of the applied external forces consists of a resultant force $\mathbf{R}$ that passes through the mass center G plus a couple $\mathbf{C}$, the body experiences both rotation and translation, and additional equations are needed to relate the moments of the external forces to the angular motion of the body.

16-2   EQUATIONS FOR PLANE MOTION

Newton's laws apply only to the motion (translation) of a single particle; therefore, they are inadequate to describe the complete motion of a rigid body, which may include both translation and rotation. In this section, Newton's laws of motion are extended to cover plane motion of a rigid body. Later, in Section 16-5, Newton's laws will be further extended to cover the general case of three-dimensional motion of a rigid body. These laws (either planar or three-dimensional) provide differential equations that relate the linear and angular accelerated motion of the body to the forces and moments producing the motion. These equations can be used to determine

1.   The instantaneous accelerations due to known forces and moments, or

2.   The forces and moments required to produce a prescribed motion.

   In Chapter 15 the "principle of motion of the mass center" of a system of particles was developed. Since a rigid body can be viewed as a collection of particles that remain at fixed distances with respect to each other, the motion of the mass center G of a rigid body is given by Eq. 15-11

$$\mathbf{R} = m\mathbf{a}_G$$

where
   $\mathbf{R}$ is the resultant of the forces acting on the rigid body at a given instant of time.
   m is the mass of the rigid body.
   $\mathbf{a}_G$ is the instantaneous linear acceleration of the mass center of the rigid body in the direction of the resultant force $\mathbf{R}$.

   Since Eq. 15-11 was obtained by the simple process of summing forces, no information is provided regarding the location of the line of action of the resultant force $\mathbf{R}$. The mass center G of a rigid body moves (translates) as though the rigid body were a single particle of mass m subjected to the resultant force $\mathbf{R}$. The actual motion of most rigid bodies consists of the superposition of a translation produced by the resultant force $\mathbf{R}$ and a rotation produced by the moment of the resultant force $\mathbf{R}$ when its line of action does not pass through the mass center G of the body.

   Consider the rigid body of arbitrary shape shown in Fig. 16-1a. The XYZ-coordinate system is fixed in space. The xyz-coordinate system is attached to the body at point A. The displacement of an element of mass dm with respect to the moving point A is given by the vector $\mathbf{r}$ and with respect to the origin O of the XYZ-coordinate system by the vector $\mathbf{R}$. The displacement of point A with respect to the origin O of the XYZ system is given by the vector $\mathbf{r}_A$. The resultant external and internal forces acting on the element of mass dm are $\mathbf{F}$ and $\mathbf{f}$, respectively. The moment produced about point A by the forces $\mathbf{F}$ and $\mathbf{f}$ is

$$d\mathbf{M}_A = \mathbf{r} \times (\mathbf{F} + \mathbf{f}) \qquad (a)$$

But from Newton's second law

$$\mathbf{F} + \mathbf{f} = dm\, \mathbf{a}_{dm} = dm\, \ddot{\mathbf{R}} \qquad (b)$$

Thus, from Eqs. a and b,

$$d\mathbf{M}_A = \mathbf{r} \times (\mathbf{F} + \mathbf{f}) = (\mathbf{r} \times \mathbf{a}_{dm})\, dm \qquad (c)$$

The acceleration $\mathbf{a}_{dm}$ for a rigid body in either plane motion or general three-dimensional motion can be written (Eq. 14-10 or 14-18)

$$\mathbf{a}_{dm} = \mathbf{a}_A + (\boldsymbol{\alpha} \times \mathbf{r}) + [\boldsymbol{\omega} \times (\boldsymbol{\omega} \times \mathbf{r})] \qquad (d)$$

Substituting Eq. d into Eq. c and integrating yields

$$\mathbf{M}_A = \int_m (\mathbf{r} \times \mathbf{a}_A)\, dm + \int_m [\mathbf{r} \times (\boldsymbol{\alpha} \times \mathbf{r})]\, dm$$
$$+ \int_m \{\mathbf{r} \times [\boldsymbol{\omega} \times (\boldsymbol{\omega} \times \mathbf{r})]\}\, dm \qquad (16\text{-}1)$$

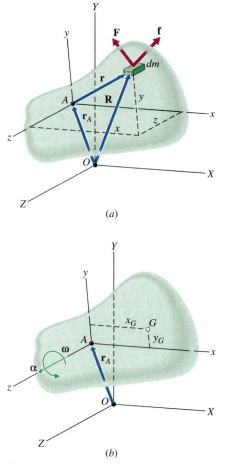

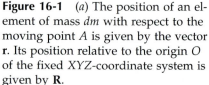

Figure 16-1   (a) The position of an element of mass dm with respect to the moving point A is given by the vector $\mathbf{r}$. Its position relative to the origin O of the fixed XYZ-coordinate system is given by $\mathbf{R}$.
(b) For planar motion all points of the body travel in parallel planes. The parallel plane that passes through the mass center G is called the *plane of motion*. The angular velocity and angular acceleration vectors are both perpendicular to the plane of motion. It is assumed that the point A is in the plane of motion and that the xy-plane coincides with the plane of motion.

Plane motion of a rigid body is defined as motion in which all elements of the body move in parallel planes. The parallel plane that passes through the mass center G of the body is known as the *plane of motion*, and point A is assumed to be in the plane of motion. Thus, as shown in Fig. 16-1b, the angular velocity and angular acceleration vectors $\boldsymbol{\omega}$ and $\boldsymbol{\alpha}$, respectively, are parallel to each other and are perpendicular to the plane of motion. If the xyz-coordinate system is chosen so that the motion is parallel to the xy-plane, then $a_{Az} = \omega_x = \omega_y = \alpha_x = \alpha_y = 0$ and the angular velocity and angular acceleration of the body are

$$\boldsymbol{\omega} = \omega_z\mathbf{k} = \omega\mathbf{k}$$
$$\boldsymbol{\alpha} = \alpha_z\mathbf{k} = \alpha\mathbf{k}$$

Then, denoting the rectangular components of the vector $\mathbf{r}$ by

$$\mathbf{r} = x\mathbf{i} + y\mathbf{j} + z\mathbf{k}$$

the rectangular components of Eq. 16-1 become

$$M_{Ax} = -a_{Ay}\int_m z\,dm - \alpha\int_m zx\,dm + \omega^2\int_m yz\,dm$$
$$M_{Ay} = a_{Ax}\int_m z\,dm - \alpha\int_m yz\,dm - \omega^2\int_m zx\,dm$$
$$M_{Az} = a_{Ay}\int_m x\,dm - a_{Ax}\int_m y\,dm + \alpha\int_m (x^2+y^2)\,dm \qquad (16\text{-}2)$$

Integrals of the form $\int_m x\,dm$ are first moment expressions that are normally studied in detail in most statics courses. Integrals of the form $\int_m x^2\,dm$ and $\int_m xy\,dm$ are similar to the expressions previously encountered in statics for second moments of area and mixed second moments of area. The integrals in Eqs. 16-2 represent the inertia properties of the rigid body and are known as moments of inertia and products of inertia, respectively. A complete discussion of moments and products of inertia is presented in Appendix A, together with solved examples. A brief review of moments and products of inertia is provided in the following section for those students who have covered the topic in some detail in a previous statics course. Section 16-3 can be omitted by those students who have a firm understanding of the subject.

The first moments, moment of inertia, and products of inertia appearing in Eqs. 16-2 are

$$\int_m x\,dm = x_G m \qquad\qquad \int_m zx\,dm = I_{Azx}$$
$$\int_m y\,dm = y_G m \qquad\qquad \int_m yz\,dm = I_{Ayz}$$
$$\int_m z\,dm = z_G m = 0 \qquad \int_m (x^2+y^2)\,dm = I_{Az} \qquad (e)$$

Equations 16-2 written in terms of the first moments and the moments and products of inertia given in Eqs. e become

$$M_{Ax} = -\alpha I_{Azx} + \omega^2 I_{Ayz} \qquad (16\text{-}3a)$$
$$M_{Ay} = -\alpha I_{Ayz} - \omega^2 I_{Azx} \qquad (16\text{-}3b)$$
$$M_{Az} = a_{Ay}x_G m - a_{Ax}y_G m + \alpha I_{Az} \qquad (16\text{-}3c)$$

This set of equations relates the moments of the external forces acting on the rigid body to the angular velocities and inertia properties of the body. The moments of the forces and the moments and products of inertia are with respect to x-, y-, and z-axes through point A that are fixed in the body. If the x-, y-, and z-axes are not fixed in the body, the moments and products of inertia will be functions of time. Equations 16-3 clearly show the dependence of the moment about a given axis on the angular velocity ω about the z-axis. Alternatively, the equations show that moments M_{Ax} and M_{Ay} may be required to maintain planar motion about the z-axis.

For most dynamics problems involving planar motion, a considerable simplification of Eqs. 16-3 is possible. If the body is symmetric about the plane of motion, the product of inertia terms vanish ($I_{Ayz} = I_{Azx} = 0$) and Eqs. 16-3 become

$$M_{Ax} = 0$$
$$M_{Ay} = 0$$
$$M_{Az} = a_{Ay}x_{G}m - a_{Ax}y_{G}m + \alpha I_{Az} \qquad (f)$$

If the body is symmetric about the plane of motion and the acceleration of point A is zero (for example, if A is on a fixed axis of rotation of a body), then $a_{Ax} = a_{Ay} = 0$ and Eqs. 16-3 reduce to

$$M_{Ax} = 0 \qquad (16\text{-}4a)$$
$$M_{Ay} = 0 \qquad (16\text{-}4b)$$
$$M_{Az} = I_{Az}\alpha \qquad (16\text{-}4c)$$

Similarly, if the body is symmetric about the plane of motion and the point A coincides with the center of mass G of the body, then $x_G = y_G = 0$ and Eqs. 16-3 become

$$M_{Gx} = 0 \qquad (16\text{-}5a)$$
$$M_{Gy} = 0 \qquad (16\text{-}5b)$$
$$M_{Gz} = I_{Gz}\alpha \qquad (16\text{-}5c)$$

Equations 16-3 through 16-5 together with Eqs. 15-11 provide the relationships required to solve a wide variety of plane motion problems. An extensive discussion of plane motion is presented in Section 16-4, following the discussion of moments and products of inertia in the next section.

16-3   MOMENTS AND PRODUCTS OF INERTIA

In the previous analysis of the motion of a rigid body, integrals were encountered that involve the product of the mass of a small element of the body and the square of its distance from a line of interest. This product is called the second moment of the mass of the element or, more frequently, the moment of inertia of the element.

16-3.1   Moment of Inertia

The moment of inertia dI of the element of mass dm shown in Fig. 16-2 about the axis OO is defined as

$$dI = r^2\, dm \qquad (g)$$

The moment of inertia of the entire body about axis OO is defined as

$$I = \int_m r^2\, dm \qquad (16\text{-}6)$$

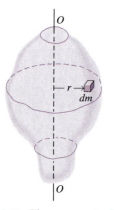

Figure 16-2   The moment of inertia dI of the element of mass dm about the axis OO is defined as $dI = r^2\, dm$.

Since both the mass of the element and the distance squared from the axis to the element are always positive, the moment of inertia of a mass is always a positive quantity.

Moments of inertia have the dimensions of mass multiplied by length squared, ML^2. Common units for the measurement of moment of inertia in the SI system are $kg \cdot m^2$. In the U.S. customary system, force, length, and time are selected as the fundamental quantities, and mass has the dimensions FT^2L^{-1}. Therefore, moment of inertia has the units $lb \cdot s^2 \cdot ft$. If the mass of the body W/g is expressed in slugs ($lb \cdot s^2/ft$), the units of measurement of moment of inertia in the U.S. customary system are $slug \cdot ft^2$.

The moments of inertia of a body with respect to an xyz-coordinate system can be determined by considering an element of mass, as shown in Fig. 16-3. From the definition of moment of inertia,

$$dI_x = r_x^2 \, dm = (y^2 + z^2) \, dm$$

Similar expressions can be written for the y- and z-axes. Thus,

$$I_x = \int_m r_x^2 \, dm = \int_m (y^2 + z^2) \, dm \qquad (16\text{-}7a)$$

$$I_y = \int_m r_y^2 \, dm = \int_m (z^2 + x^2) \, dm \qquad (16\text{-}7b)$$

$$I_z = \int_m r_z^2 \, dm = \int_m (x^2 + y^2) \, dm \qquad (16\text{-}7c)$$

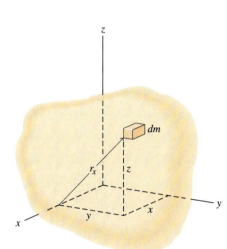

Figure 16-3 The perpendicular distance from the x-axis to the element of mass dm is $r_x = \sqrt{x^2 + y^2}$. Therefore, the moment of inertia of the element of mass with respect to the x-axis is $dI_x = r_x^2 dm = (y^2 + z^2) \, dm$.

When integration methods are used to determine the moment of inertia of a body with respect to an axis, the mass of the body can be divided into elements in various ways. Depending on how the element is chosen, single, double, or triple integration may be required. The geometry of the body usually determines whether Cartesian or polar coordinates are used.

In some instances, a body can be regarded as a system of particles. The moment of inertia of a system of particles with respect to a line of interest is the sum of the moments of inertia of the particles with respect to the given line. Thus, if the masses of the particles of a system are denoted by $m_1, m_2, m_3, \dots, m_n$ and the distances of the particles from a given line are denoted by $r_1, r_2, r_3, \dots, r_n$, the moment of inertia of the system can be expressed as

$$I = \Sigma mr^2 = m_1 r_1^2 + m_2 r_2^2 + m_3 r_3^2 + \cdots + m_n r_n^2$$

Moments of inertia for thin plates are relatively easy to determine. For example, consider the thin plate shown in Fig. 16-4. The plate has a uniform density ρ, a uniform thickness t, and a cross-sectional area A. The z-coordinate of every piece of mass dm in the plate is $z \leq t/2 \cong 0$, so the moments of inertia about x-, y-, and z-axes are

$$I_{xm} = \int_m (y^2 + z^2) \, dm \cong \int_m y^2 \, dm = \int_V y^2 \rho \, dV$$

$$= \int_A y^2 \rho t \, dA = \rho t \int_A y^2 \, dA = \rho t I_{xA} \qquad (16\text{-}8a)$$

$$I_{ym} = \int_m (x^2 + z^2) \, dm \cong \int_m x^2 \, dm = \int_V x^2 \rho \, dV$$

$$= \int_A x^2 \rho t \, dA = \rho t \int_A x^2 \, dA = \rho t I_{yA} \qquad (16\text{-}8b)$$

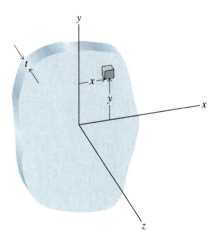

Figure 16-4 If the body consists of a thin slab of uniform thickness, then the moments of inertia simplify considerably. Not only can the uniform thickness be factored out of the integrals ($dm = \rho \, dV = \rho t \, dA$), but the z-coordinate of every element of mass will be very small and can be neglected when compared to the other terms in the integrals ($x^2 + z^2 \cong x^2$ and $y^2 + z^2 \cong y^2$).

$$I_{zm} = \int_m (x^2 + y^2)\, dm \cong \rho t I_{yA} + \rho t I_{xA}$$
$$= \rho t(I_{yA} + I_{xA}) \tag{16-8c}$$

where the subscripts m and A denote moments of inertia and second moments of area, respectively. Since the equations for the moments of nertia of thin plates contain the expressions for the second moments of area, the results listed in Appendix B (Table B-3) for second moments of areas can be used for moments of inertia by simply multiplying the results listed in the table by ρt.

Moments of inertia with respect to the x-, y-, and z-axes for the general three-dimensional body can be determined by using Eqs. 16-7. If the density of the body is uniform, the element of mass dm can be expressed in terms of the element of volume dV of the body as $dm = \rho\, dV$. Equations 16-7 then become

$$I_x = \rho \int_V (y^2 + z^2)\, dV \tag{16-9a}$$

$$I_y = \rho \int_V (z^2 + x^2)\, dV \tag{16-9b}$$

$$I_z = \rho \int_V (x^2 + y^2)\, dV \tag{16-9c}$$

If the density of the body is not uniform, it must be expressed as a function of position and retained within the integral sign.

Frequently in engineering practice, a body of interest can be broken up into a number of simple shapes, such as cylinders, spheres, plates, and rods, for which the moments of inertia have been evaluated and tabulated. The moment of inertia of the composite body, with respect to any axis, is equal to the sum of the moments of inertia of the separate parts of the body with respect to the specified axis. For example,

$$I_x = \int_m (y^2 + z^2)\, dm$$
$$= \int_{m_1} (y^2 + z^2)\, dm_1 + \int_{m_2} (y^2 + z^2)\, dm_2 + \cdots + \int_{m_n} (y^2 + z^2)\, dm_n$$
$$= I_{x1} + I_{x2} + I_{x3} + \cdots + I_{xn}$$

When one of the component parts is a hole, its moment of inertia must be subtracted from the moment of inertia of the larger part to obtain the moment of inertia for the composite body. Appendix B (Table B-5) contains a listing of the moments of inertia for some frequently encountered shapes such as rods, plates, cylinders, spheres, and cones.

16-3.2 Radius of Gyration

The definition of moment of inertia (Eq. g) indicates that the dimensions of moment of inertia are mass multiplied by a length squared. As a result, the moment of inertia of a body can be expressed as the product of the mass m of the body and a length k squared. This length k is defined as the radius of gyration of the body. Thus, the moment of inertia I of a body with respect to a given line can be expressed as

$$I = mk^2 \quad \text{or} \quad k = \sqrt{\frac{I}{m}} \tag{16-10}$$

The radius of gyration of the mass of a body with respect to any axis can be viewed as the distance from the axis to the point where the total mass must be concentrated to produce the same moment of inertia with respect to the axis as does the actual (or distributed) mass.

The radius of gyration for masses is very similar to the radius of gyration for areas discussed in Section 10-2.3. The radius of gyration the body such as the mass center. The radius of gyration of the mass of a body with respect to any axis is always greater than the distance from the axis to the mass center of the body. There is no useful physical interpretation for a radius of gyration; it is merely a convenient means of expressing the moment of inertia of the mass of a body in terms of its mass and a length.

16-3.3 Parallel-Axis Theorem for Moments of Inertia

The parallel-axis theorem for moments of inertia is very similar to the parallel-axis theorem for second moments of area discussed in Section 10-2.1. Consider the body shown in Fig. 16-5, which has an xyz-coordinate system with its origin at the mass center G of the body and a parallel $x'y'z'$-coordinate system with its origin at point O'. Observe in the figure that

$$x' = x_G + x$$
$$y' = y_G + y$$
$$z' = z_G + z$$

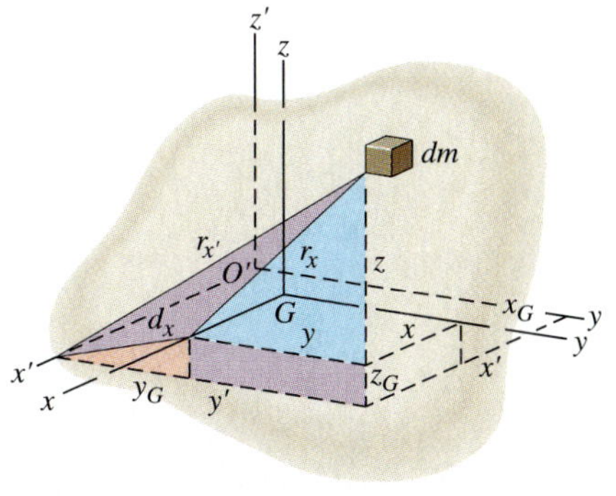

The distance d_x between the x'- and x-axes is

$$d_x = \sqrt{y_G^2 + z_G^2}$$

The moment of inertia of the body about an x'-axis that is parallel to the x-axis through the mass center is by definition

$$I_{x'} = \int_m r_{x'}^2 \, dm = \int_m [(y_G + y)^2 + (z_G + z)^2] \, dm$$
$$= \int_m (y^2 + z^2) \, dm + y_G^2 \int_m dm + 2y_G \int_m y \, dm$$
$$+ z_G^2 \int_m dm + 2z_G \int_m z \, dm$$

However,

$$\int_m (y^2 + z^2) \, dm = I_{xG}$$

and, since the x- and y-axes pass through the mass center G of the body,

$$\int_m y \, dm = 0 \qquad \int_m z \, dm = 0$$

Therefore,

$$I_{x'} = I_{xG} + (y_G^2 + z_G^2)m = I_{xG} + d_x^2 m \qquad (16\text{-}11a)$$
$$I_{y'} = I_{yG} + (z_G^2 + x_G^2)m = I_{yG} + d_y^2 m \qquad (16\text{-}11b)$$
$$I_{z'} = I_{zG} + (x_G^2 + y_G^2)m = I_{zG} + d_z^2 m \qquad (16\text{-}11c)$$

Equation 16-11 is the parallel-axis theorem for moments of inertia. The subscript G indicates that the x-axis passes through the mass center G of the body. Thus, if the moment of inertia of a body with respect to an axis passing through its mass center is known, the moment of inertia of the body with respect to any parallel axis can be found, without integrating, by use of Eqs. 16-11.

Figure 16-5 The $x'y'z'$-axes are for an arbitrary coordinate system with its origin at an arbitrary point O'. The xyz-axes are parallel to the $x'y'z'$-axes but have their origin at the mass center G of the body. The perpendicular distance between the x- and x'-axes is $d_x = \sqrt{y_G^2 + z_G^2}$.

A similar relationship exists between the radii of gyration for the two axes. Thus, if the radii of gyration for the two parallel axes are denoted by k_{xG} and $k_{x'}$, the equation may be written

$$k_{x'}^2 m = k_{xG}^2 m + d_x^2 m$$

Hence

$$k_{x'}^2 = k_{xG}^2 + d_x^2 \qquad (16\text{-}12a)$$
$$k_{y'}^2 = k_{yG}^2 + d_y^2 \qquad (16\text{-}12b)$$
$$k_{z'}^2 = k_{zG}^2 + d_z^2 \qquad (16\text{-}12c)$$

(*Note:* Equations 16-11 and 16-12 are valid only for transfers to or from *xyz*-axes passing through the mass center of the body. **They are not valid for transfers between two arbitrary axes.**)

16-3.4   Product of Inertia

In analyses of the motion of rigid bodies, expressions are sometimes encountered that involve the product of the mass of a small element and the coordinate distances from a pair of orthogonal coordinate planes. This product, which is similar to the mixed second moment of an area, is called the product of inertia of the element. For example, the product of inertia of the element shown in Fig. 16-6 with respect to the *xz*- and *yz*-planes is by definition

$$dI_{xy} = xy \, dm \qquad (16\text{-}13)$$

The sum of the products of inertia of all elements of mass of the body with respect to the same orthogonal planes is defined as the product of inertia of the body. The three products of inertia for the body shown in Fig. 16-6 are

$$I_{xy} = \int_m xy \, dm \qquad (16\text{-}14a)$$
$$I_{yz} = \int_m yz \, dm \qquad (16\text{-}14b)$$
$$I_{zx} = \int_m zx \, dm \qquad (16\text{-}14c)$$

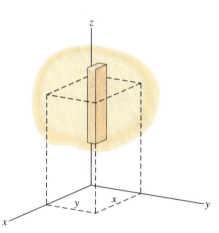

Figure 16-6   Every point of the rectangular element shown is at the same distance *y* from the *xz*-plane and is also at the same distance *x* from the *yz*-plane. The product of inertia of this element with respect to the *xz*- and *yz*-planes is $dI_{xy} = xy \, dm$.

Products of inertia, like moments of inertia, have the dimensions of mass multiplied by a length squared ML^2. Common units for the measurement of product of inertia in the SI system are kg · m². In the U.S. customary system, common units are slug · ft².

The product of inertia of a body can be positive, negative, or zero, since the two coordinate distances have independent signs. The product of inertia will be positive for coordinates with the same sign and negative for coordinates with opposite signs. The product of inertia will be zero if either of the planes is a plane of symmetry, since pairs of elements on opposite sides of the plane of symmetry will have positive and negative products of inertia that will add to zero in the summation process.

The integration methods used to determine moments of inertia apply equally well to products of inertia. Depending on how the element is chosen, single, double, or triple integration may be required. Moments of inertia for thin plates were related to second moments of area for the same plate. Likewise, products of inertia can be related to the mixed second moments for the plates. If the plate has a uniform density ρ, a uniform thickness t, and a cross-sectional area A, the products of inertia are by definition

$$I_{xym} = \int_m xy\, dm = \int_v xy\, \rho\, dV = \int_A xy\, \rho t\, dA$$

$$= \rho t \int_A xy\, dA = \rho t\, I_{xyA} \qquad (16\text{-}15a)$$

$$I_{yzm} = \int_m yz\, dm = 0 \qquad (16\text{-}15b)$$

$$I_{zxm} = \int_m zx\, dm = 0 \qquad (16\text{-}15c)$$

where the subscripts m and A denote products of inertia of mass and mixed second moments of area, respectively. The products of inertia I_{yzm} and I_{zxm} for a thin plate are zero, since the x- and y-axes are assumed to lie in the midplane of the plate (plane of symmetry).

A parallel-axis theorem for products of inertia can be developed that is very similar to the parallel-axis theorem for mixed second moments of area discussed in Section 10-2.5. Consider the body shown in Fig. 16-7, which has an xyz-coordinate system with its origin at the mass center G of the body and a parallel $x'y'z'$-coordinate system with its origin at point O' of the body. Observe in the figure that

$$x' = x_G + x$$
$$y' = y_G + y$$
$$z' = z_G + z$$

The product of inertia $I_{x'y'}$ of the body with respect to the $x'z'$- and $y'z'$-planes is by definition

$$I_{x'y'} = \int_m x'y'\, dm = \int_m (x_G + x)(y_G + y)\, dm$$

$$= \int_m x_G y_G\, dm + \int_m x_G y\, dm + \int_m y_G x\, dm + \int_m xy\, dm$$

Since x_G and y_G are the same for every element of mass dm,

$$I_{x'y'} = x_G y_G \int_m dm + x_G \int_m y\, dm + y_G \int_m x\, dm + \int_m xy\, dm$$

However,

$$\int_m xy\, dm = I_{xy}$$

and, since the x- and y-axes pass through the mass center G of the body,

$$\int_m y\, dm = 0 \qquad \int_m x\, dm = 0$$

Therefore,

$$I_{x'y'} = I_{xyG} + x_G y_G m \qquad (16\text{-}16a)$$
$$I_{y'z'} = I_{yzG} + y_G z_G m \qquad (16\text{-}16b)$$
$$I_{z'x'} = I_{zxG} + z_G x_G m \qquad (16\text{-}16c)$$

Equations 16-16 are the parallel-axis theorem for products of inertia. The subscript G indicates that the x- and y-axes pass through the mass center G of the body. Thus, if the product of inertia of a body with respect to a pair of orthogonal planes that pass through its mass center is known, the product of inertia of the body with respect to any other pair of parallel planes can be found, without integrating, by use of Eqs. 16-16.

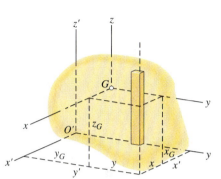

Figure 16-7 The $x'y'z'$-axes are for an arbitrary coordinate system with its origin at an arbitrary point O'. The xyz-axes are parallel to the $x'y'z'$-axes but have their origin at the mass center G of the body.

16-3.5 Principal Moments of Inertia

In some instances, in the dynamic analysis of bodies, principal axes and maximum and minimum moments of inertia, which are similar to maximum and minimum second moments of an area, must be determined. Again, the problem is one of transforming known or easily calculated moments and products of inertia with respect to one coordinate system (such as an xyz-coordinate system along the edges of a rectangular prism) to a second $x'y'z'$-coordinate system through the same origin O but inclined with respect to the xyz system. A complete development of the equations required for determining principal moments of inertia is presented in Appendix A. Also included in Appendix A is a selection of solved example problems involving moment of inertia, radius of gyration, product of inertia, and principal moment of inertia determinations.

16-4   TRANSLATION, ROTATION, AND GENERAL PLANE MOTION OF RIGID BODIES

Plane motion problems can be classified in three distinct categories, which depend on the nature of the motion: (1) translation, (2) fixed-axis rotation, and (3) general plane motion. The equations for general plane motion were developed in Section 16-2. Translation and fixed-axis rotation are special cases of general plane motion.

For a body of arbitrary shape, the equations for general plane motion developed in Section 16-2 are given by Eqs. 15-11 and 16-3. If the origin of the xyz-coordinate system is at the mass center G of the body, then $x_G = y_G = 0$ and the equations become

$$\mathbf{R} = \Sigma\mathbf{F} = m\mathbf{a}_G \tag{16-17a}$$
$$\Sigma M_{Gx} = -\alpha I_{Gzx} + \omega^2 I_{Gyz} \tag{16-17b}$$
$$\Sigma M_{Gy} = -\alpha I_{Gyz} - \omega^2 I_{Gzx} \tag{16-17c}$$
$$\Sigma M_{Gz} = I_{Gz}\alpha \tag{16-17d}$$

16-4.1   Translation

The motion of a rigid body is defined as translation when every straight line in the body remains parallel to its initial position during the motion. That is, during translation, there is no angular motion ($\boldsymbol{\omega} = \boldsymbol{\alpha} = \mathbf{0}$). However, the accelerations of any two points in a rigid body are related by (Eq. 14-14 or 14-23).

$$\mathbf{a}_B = \mathbf{a}_A + \boldsymbol{\alpha} \times \mathbf{r}_{B/A} + \boldsymbol{\omega} \times (\boldsymbol{\omega} \times \mathbf{r}_{B/A})$$

But if $\boldsymbol{\omega} = \boldsymbol{\alpha} = \mathbf{0}$, then $\mathbf{a}_B = \mathbf{a}_A$, and for a body in translation all parts of the body experience the same linear acceleration.

Translation can occur only when the resultant of the external forces acting on the body is a force $\mathbf{R}$ whose line of action passes through the mass center G of the body. That is, if $\boldsymbol{\omega} = \boldsymbol{\alpha} = \mathbf{0}$, then Eqs. 16-17 require that $\Sigma\mathbf{M}_G = \mathbf{0}$. However, if the resultant force $\mathbf{R}$ passes through the mass center G, then the moment of the resultant force about some other point A will not in general be zero. The general equations for plane motion (Eq. 16-17) simplify to

$$\mathbf{R} = \Sigma\mathbf{F} = m\mathbf{a}_G \qquad \Sigma\mathbf{M}_G = \mathbf{0}$$

only when moments are summed about the mass center G of a translating body. If moments are summed about any other point A, then the moment equation must include the acceleration of the point A and the form of the moment equation will be more complex (Eq. 16-3).

The problem of summing moments about a point other than the mass center can be simplified using the concept of equivalent force-couple systems. For example, the rigid plate of Fig. 16-8 is in translational motion. The free-body diagram of the plate is drawn in Fig. 16-9a. In Fig. 16-9b the forces on the free-body diagram have been reduced to an equivalent force-couple system. However, the equivalent force-couple system of Fig. 16-9b is equipollent to (has the same vector components and moment as) the inertia vectors shown in Fig. 16-9c. Clearly, summing vector components in the x- and y-directions on Figs. 16-9a and c and setting them equal satisfies the equations of motion for the plate (Eq. 16-17a). Also, summing moments about the mass center G on Figs. 16-9a and c clearly satisfy the equations of motion for the plate (Eq. 16-17b, c, and d). However, summing moments about point B on Figs. 16-9a and c also satisfies the general form of the equations of motion (Eq. 16-3) since

$$+\curvearrowleft \Sigma M_B = b(ma_{Gy}) + c(ma_{Gx})$$

and $b = x_G$, $c = -y_G$, $a_{Gx} = a_{Bx}$, $a_{Gy} = a_{By}$, and $\alpha = 0$. Although the sum of moments about B is not zero, it involves only one of the unknown forces and may lead to the solution more directly than summing moments about the mass center G.

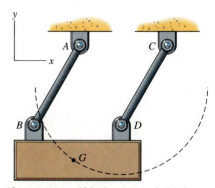

Figure 16-8 If links AB and CD have the same length and are parallel, then the top edge of the plate will always be parallel to the ceiling and the plate moves in curvilinear translation.

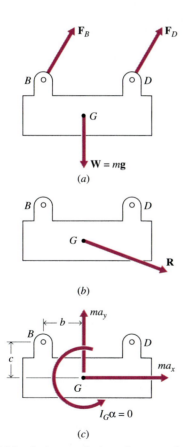

Figure 16-9 For a rigid body in translation, there can be no moment about the mass center G. Therefore, the forces on the free-body diagram (a) must be equivalent to the single resultant force **R** acting through the mass center (b) and equipollent to the inertia vectors shown on the kinetic diagram (c).

Why do drag racers have large rear tires and tiny front tires on the end of a long, flimsy, front-end structure?

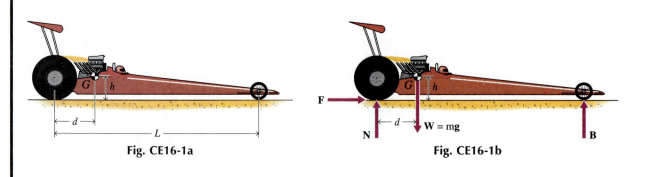

Fig. CE16-1a   **Fig. CE16-1b**

SOLUTION

According to Newton's second law, the acceleration of a race car is equal to the ratio of its propulsive force and its mass ($a = P/m$). Therefore, increasing the acceleration requires that either the propulsive force P must be increased or the mass m must be decreased. However, as noted in Conceptual Example 15-2, the propulsive force on the race car is provided by the frictional force between the tires and the pavement ($P = F$), and the frictional force is directly proportional to the normal force on the tires ($F = \mu N$). Therefore, attempting to increase the propulsive force by simply increasing the weight or decreasing the mass of the car has no effect on the acceleration of the car.

For any given weight, however, the normal force on the drive tires is a maximum when the center of gravity of the race car is directly over the drive tires ($d = 0$). Of course, the weight can't be directly over the drive tires, since then the car would flip over backwards as soon as the driver stepped on the gas. That is, the moment equation

$$\curvearrowleft\Sigma M_G = Fh - Wd + BL = I_G\alpha = 0 \qquad (a)$$

cannot be satisfied if $d = 0$ unless $F = 0$ (since B cannot be negative). But if $F = P = 0$, then $ma = 0$ and the car cannot accelerate.

Instead, the motor (which is effectively the center of mass) is placed just far enough in front of the drive wheels so that the car does not flip over backward at maximum acceleration. At the point of impending tip, $B = 0$ and

$$d = \frac{Fh}{W} = \frac{ma_{max}h}{mg} = \frac{a_{max}h}{g} \qquad (b)$$

The front tires are then placed well in front of the motor, using a minimum amount of structure so that they carry only a small fraction of the total weight (when the car is not accelerating) and the drive wheels carry the bulk of the weight. This results in the largest possible normal force on the drive tires and hence the largest propulsive force and the largest acceleration of the race car for any given weight.

A 16 × 20-ft hangar door, which weighs 800 lb, is supported by two rollers as shown in Fig. 16-10. A force **F** of 300 lb is applied to open the door. Determine the acceleration of the door and the support forces exerted on the door by the two rollers. Neglect frictional forces and the mass of the rollers.

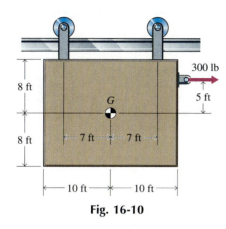

Fig. 16-10

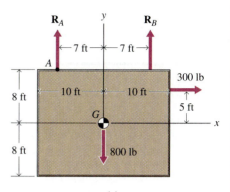

(a)

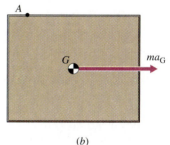

(b)

Fig. 16-11

SOLUTION

A free-body diagram for the door is shown in Fig. 16-11a. As long as the two rollers remain in contact with the rail, the motion of the mass center of the door will be along a horizontal straight line. The motion is translation ($\omega = \alpha = \mathbf{a}_{Gy} = 0$), and the only nonzero inertia vector is $m\mathbf{a}_{Gx}$. Therefore, the forces on the free-body diagram of Fig. 16-11a are equipollent to the inertia vectors shown on Fig. 16-11b, and the equations of motion for the door (Eqs. 16-17) are

$$+\rightarrow\Sigma F_x = ma_{Gx}: \qquad 300 = \frac{800}{32.2}a_G \qquad (a)$$

$$+\uparrow\Sigma F_y = ma_{Gy}: \qquad R_A + R_B - 800 = 0 \qquad (b)$$

$$+\curvearrowleft\Sigma M_{Gz} = I_{Gz}\alpha: \qquad R_B(7) - R_A(7) - 300(5) = 0 \qquad (c)$$

Equation a gives

$$a_G = 12.075 \text{ ft/s}^2 \cong 12.08 \text{ ft/s}^2\rightarrow \qquad \text{Ans.}$$

and solving Eqs. b and c simultaneously gives

$$R_A = 292.86 \text{ lb} \cong 293 \text{ lb}\uparrow \qquad \text{Ans.}$$
$$R_B = 507.14 \text{ lb} \cong 507 \text{ lb}\uparrow \qquad \text{Ans.}$$

Alternatively, summing moments about A on Figs. 16-11a and b gives the set of equations

$$+\rightarrow\Sigma F_x = ma_{Gx}: \qquad 300 = \frac{800}{32.2}a_G \qquad (a)$$

$$+\uparrow\Sigma F_y = ma_{Gy}: \qquad R_A + R_B - 800 = 0 \qquad (b)$$

$$+\curvearrowleft\Sigma M_{Az}: \qquad R_B(14) + 300(3) - 800(7) = 8\left(\frac{800}{32.2}a_G\right) \qquad (d)$$

Using this set of equations, Eq. a gives the linear acceleration

$$a_G = 12.075 \text{ ft/s}^2 \cong 12.08 \text{ ft/s}^2\rightarrow \qquad \text{Ans.}$$

Even though the sum of moments about the mass center is zero, the moment equation is necessary to solve the problem. It gives another relationship between the roller forces R_A and R_B.

Note that although the angular acceleration of the door is zero, the sum of moments about the moving point A is not zero. The sum of moments about point A equals $I_A\alpha$ only if the point A is the mass center of the body or the point A is on a fixed axis of rotation. For a body moving with pure translational motion, there can never be an axis of rotation.

as before. Then Eq. *d* gives R_B directly

$$R_B = 507.14 \text{ lb} \cong 507 \text{ lb}\uparrow \qquad\qquad \text{Ans.}$$

and Eq. *b* gives R_A directly

$$R_A = 292.86 \text{ lb} \cong 293 \text{ lb}\uparrow \qquad\qquad \text{Ans.}$$

Summing moments about *A* rather than about the mass center *G* has eliminated the need to solve simultaneous equations.

EXAMPLE PROBLEM 16-2

The 1400-kg automobile shown in Fig. 16-12 has a wheel base of 3 m. Its center of mass is located 1.30 m behind the front axle and 0.5 m above the surface of the pavement. If the automobile has a rear-wheel drive and the coefficient of friction between the tires and the pavement is 0.80, determine the maximum acceleration that the vehicle can develop while traveling up the 15° incline.

Fig. 16-12

SOLUTION

A free-body diagram of the automobile is shown in Fig. 16-13*a*. Since the vehicle has a rear-wheel drive, a frictional driving force is shown only for the rear wheels. For maximum acceleration, the frictional force will be the maximum it can be ($= \mu R_R = 0.80R_R$). As long as the tires remain in contact with the pavement, the motion of the mass center of the automobile will be along a straight line parallel to the pavement. The motion is translation ($\omega = \alpha = \mathbf{a}_{Gy} = 0$), and the only nonzero inertia vector is $m\mathbf{a}_{Gx}$. Therefore, the forces on the free-body diagram of Fig. 16-13*a* are equipollent to the inertia vectors shown on Fig. 16-13*b*, and the equations of motion for the automobile (Eq. 16-17) are

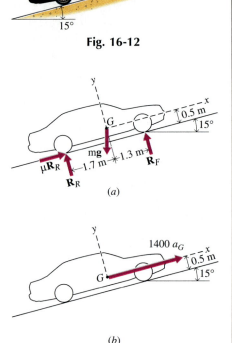

(a)

(b)

Fig. 16-13

$$+\nearrow \Sigma F_x = ma_{Gx}: \qquad 0.8R_R - 1400(9.81)\sin 15° = 1400a_G \qquad (a)$$
$$+\nwarrow \Sigma F_y = ma_{Gy}: \qquad R_F + R_R - 1400(9.81)\cos 15° = 0 \qquad (b)$$
$$+\curvearrowleft \Sigma M_{Gz} = I_{Gz}\alpha: \qquad 1.3R_F - 1.7R_R + 0.5(0.80R_R) = 0 \qquad (c)$$

Rewriting Eqs. *b* and *c*

$$R_F + R_R = 13{,}266 \text{ N}$$
$$R_F - R_R = 0$$

and solving them simultaneously gives

$$R_F = R_R = 6633 \text{ N}$$

Then, Eq. *a* gives

$$a_G = 1.251 \text{ m/s}^2 \angle 15° \qquad\qquad \text{Ans.}$$

Note that while summing moments about the rear wheel rather than about the mass center eliminates one of the unknowns from the left-hand side of the moment equation (Eq. *c*) it adds an unknown to the right-hand side of the moment equation. Therefore, the resulting set of equations

$$+\nearrow \Sigma F_x = ma_{Gx}: \qquad 0.8R_R - 1400(9.81)\sin 15° = 1400a_G$$
$$+\nwarrow \Sigma F_y = ma_{Gy}: \qquad R_F + R_R - 1400(9.81)\cos 15° = 0$$
$$+\curvearrowleft \Sigma M_{Gz} = I_{Gz}\alpha: \qquad 3R_F - 1.7[(1400)(9.81)\cos 15°]$$
$$+ 0.5[(1400)(9.81)\sin 15°] = -0.5(1400a_G)$$

is no easier to solve than the original set.

EXAMPLE PROBLEM 16-3

A 90-lb triangular plate is supported by two cables as shown in Fig. 16-14. When the plate is in the position shown, the angular velocity of the cables is 4 rad/s counterclockwise. At this instant, determine

a. The acceleration of the mass center of the plate.
b. The tension in each of the cables.

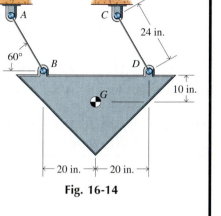

Fig. 16-14

SOLUTION

The figure $ABDC$ is a parallelogram. Therefore, during the motion of the plate, side BD will remain horizontal (parallel to AC) and the motion of the plate is translation ($\boldsymbol{\omega} = \boldsymbol{\alpha} = 0$). Using kinematics to determine the acceleration of the mass center of the plate (Fig. 16-15a)

$$
\begin{aligned}
\mathbf{a}_G = \mathbf{a}_B = \mathbf{a}_A + \mathbf{a}_{B/A} &= 0 + r\alpha_{AB}\mathbf{e}_t + r\omega_{AB}^2\mathbf{e}_n \\
&= 0 + 2\alpha_{AB}(\sin 60°\mathbf{i} + \cos 60°\mathbf{j}) + 2(4^2)(-\cos 60°\mathbf{i} + \sin 60°\mathbf{j}) \\
&= (1.7321\alpha_{AB} - 16)\mathbf{i} + (\alpha_{AB} + 27.713)\mathbf{j} \text{ ft/s}^2 \qquad (a)
\end{aligned}
$$

where α_{AB} and $\omega_{AB} = 4$ rad/s are the rotation rates of the wires (α and ω of the plate are both zero) and $\mathbf{a}_G = \mathbf{a}_B = \mathbf{a}_D$, since the plate moves in translation.

The free-body diagram of the plate is shown in Fig. 16-15b. The forces on the free-body diagram are equipollent to the inertia vectors shown on Fig. 16-15c, and the equations of motion for the plate (Eqs. 16-17) are

$$
+\rightarrow \Sigma F_x = ma_{Gx}: \qquad -(T_{AB} + T_{CD})\cos 60° = \frac{90}{32.2}(1.7321\alpha_{AB} - 16) \qquad (b)
$$

$$
+\uparrow \Sigma F_y = ma_{Gy}: \qquad (T_{AB} + T_{CD})\sin 60° - 90 = \frac{90}{32.2}(\alpha_{AB} + 27.713) \qquad (c)
$$

$$
+\curvearrowleft \Sigma M_{Gz} = I_{Gz}\alpha: \qquad \frac{10}{12}(T_{AB} + T_{CD})\cos 60° + \frac{20}{12}(T_{CD} - T_{AB})\sin 60° = 0 \qquad (d)
$$

Equations b and c can be rewritten

$$
T_{AB} + T_{CD} = 89.44 - 9.682\alpha_{AB}
$$
$$
T_{AB} + T_{CD} = 193.36 + 3.227\alpha_{AB}
$$

which gives

$$
\alpha_{AB} = -8.049 \text{ rad/s}^2
$$
$$
T_{AB} + T_{CD} = 167.37 \qquad (e)
$$

It is important to carefully draw and label both the free-body diagram and the kinetic diagram. Similar but different entities such as the angular acceleration α (which is zero) of the plate and the angular acceleration α_{AB} (which is not zero) of the wire AB must be clearly distinguished either by using subscripts or by using completely different symbols.

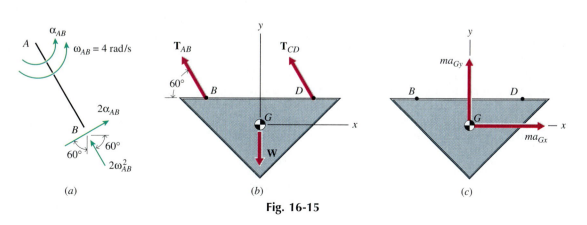

Fig. 16-15

a. Substituting α_{AB} into Eq. *a* gives the acceleration of the mass center

$$\mathbf{a}_G = -29.9\mathbf{i} + 19.66\mathbf{j} \text{ ft/s}^2 = 35.8 \text{ ft/s}^2 \angle 33.30° \qquad \text{Ans.}$$

b. Substituting Eq. *e* into Eq. *d* gives

$$T_{CD} - T_{AB} = -48.32 \qquad (f)$$

Then solving Eqs. *e* and *f* simultaneously gives the tensions

$$T_{AB} = 107.8 \text{ lb} \qquad \text{Ans.}$$
$$T_{CD} = 59.5 \text{ lb} \qquad \text{Ans.}$$

EXAMPLE PROBLEM 16-4

A 60-kg cabinet is being moved with a 10-kg skid as shown in Fig. 16-16. The coefficients of static and kinetic friction are 0.6 and 0.5, respectively, at all surfaces of contact. Determine the largest force P that can be applied to the skid without causing the cabinet to move relative to the skid.

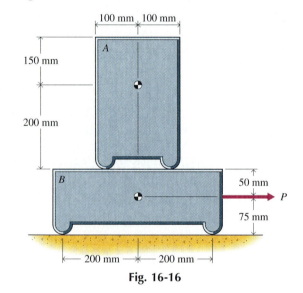

Fig. 16-16

SOLUTION

A free-body diagram of the cabinet is shown in Fig. 16-17a. The type of impending motion is not specified and could be either by slipping or by tipping. If the impending motion is by slipping, then the frictional forces will be $A_f = \mu_s A_n$ and $C_f = \mu_s C_n$ where A_n and C_n must both be positive. If the impending motion is by tipping, then $C_f = C_n = 0$ and $A_f \leq \mu_s A_n$. As long as the cabinet moves with the skid, the motion of the mass center of the cabinet will be along a straight line parallel to the floor. The motion is translation ($\boldsymbol{\omega} = \boldsymbol{\alpha} = \mathbf{a}_{Gy} = \mathbf{0}$), and the only nonzero inertia vector is $m\mathbf{a}_{Gx}$. Therefore, the forces on the free-body diagram of Fig. 16-17a are equipollent to the inertia vectors shown on Fig. 16-17b, and the equations of motion for the cabinet (Eqs. 16-17) are

$$+\rightarrow \Sigma F_x = ma_{Gx}: \qquad A_f + C_f = 60a_G \qquad (a)$$
$$+\uparrow \Sigma F_y = ma_{Gy}: \qquad A_n + C_n - 60(9.81) = 0 \qquad (b)$$
$$+\circlearrowleft \Sigma M_{Gz} = I_{Gz}\alpha: \qquad 0.2(A_f + C_f) + 0.1(C_n - A_n) = 0 \qquad (c)$$

Whenever a friction force is replaced with its maximum value μN, it is absolutely necessary that the friction force be drawn in the correct direction. The skid B is moving to the right, so the friction forces must act to the left on the skid to oppose the motion. This means that the friction forces must act to the right on the cabinet A. Although the friction forces on the bottom of the cabinet are the direct cause of its motion, they do so only to prevent the cabinet from sliding on the skid.

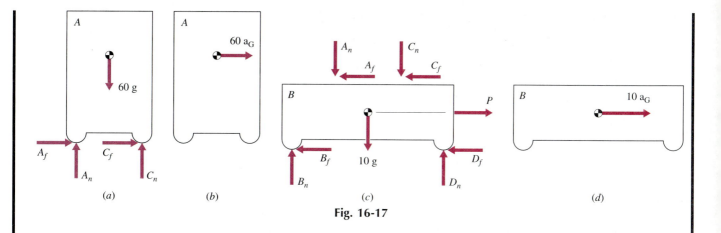

(a) (b) (c) (d)

Fig. 16-17

The three equations of motion for the cabinet contain five unknowns and cannot be solved without additional information. The three equations of motion for the skid will involve three new unknowns and will not help in the solution of these three equations. Therefore, an assumption must be made about the type of impending motion in order to solve this problem.

Guess that impending motion is by slipping:

Equation b gives

$$A_n + C_n = 588.6 \text{ N} \qquad\qquad (d)$$

and if impending motion is by slipping, then $A_f = \mu_s A_n$, $C_f = \mu_s C_n$, and

$$A_f + C_f = \mu_s(A_n + C_n) = 0.6(588.6) = 353.16 \text{ N}$$

Then Eq. c gives

$$C_n - A_n = -706.32 \text{ N} \qquad\qquad (e)$$

Solving Eqs. d and e simultaneously gives

$$A_n = 647.46 \text{ N} \qquad C_n = -58.86 \text{ N}$$

But C_n cannot be negative, since that would mean a force is required to pull down on the cabinet to keep it from tipping over and a pulling force is not available at that point. Therefore, the guess that the cabinet slips before tipping is incorrect.

Guess that impending motion is by tipping:

If impending motion is by tipping, then $C_f = C_n = 0$. Now Eqs. b, c, and a give

$$A_n = 588.6 \text{ N}$$
$$A_f = 0.5A_n = 294.3 \text{ N}$$
$$a_G = 4.905 \text{ m/s}^2$$

Note that the friction force $A_f < \mu_s A_n$, so the assumption that the cabinet does not slip before the impending tip is correct.

The free-body diagram of the skid is drawn in Fig. 16-17c. The skid is known to be sliding on the floor, so the friction forces there are simply $\mu_k B_n$ and $\mu_k D_n$. The motion of the mass center of the skid will also be along a straight line parallel to the floor. The motion is translation ($\boldsymbol{\omega} = \boldsymbol{\alpha} = \mathbf{a}_{Gy} = \mathbf{0}$), and the only nonzero inertia vector is $m\mathbf{a}_{Gx}$. Therefore, the forces on the free-body diagram of Fig. 16-17c are equipollent to the inertia vectors shown in Fig. 16-17d, and the equations of motion for the skid (Eqs. 16-17) are

$$+\rightarrow \Sigma F_x = ma_{Gx}: \qquad P - (294.3) - 0.5(B_n + D_n) = 10(4.905) \qquad (f)$$
$$+\uparrow \Sigma F_y = ma_{Gy}: \qquad (B_n + D_n) - (588.6) - 10(9.81) = 0 \qquad (g)$$
$$+\backslash \Sigma M_{Gz} = I_{Gz}\alpha: \qquad 0.05(294.3) - 0.075[0.5(B_n + D_n)]$$
$$+ 0.1(588.6) + 0.2(D_n - B_n) = 0 \qquad (h)$$

Equation g gives

$$B_n + D_n = 686.7 \text{ N} \qquad (i)$$

Solving Eqs. h and i simultaneously gives

$$B_n = 462.91 \text{ N} \qquad D_n = 223.79 \text{ N}$$

Finally, Eq. g gives

$$P = 687 \text{ N} \rightarrow \qquad\qquad \text{Ans.}$$

PROBLEMS

In the following problems, all ropes, cords, and cables are assumed to be flexible, inextensible, and of negligible mass. All pins and pulleys have negligible mass and are frictionless unless specified otherwise.

Introductory Problems

16-1* A 250-lb force **P** is applied to the 900-lb block shown in Fig. P16-1. If the kinetic coefficient of friction between the block and the horizontal surface is $\mu_k = 0.20$, determine the acceleration of the block and the reactions at the contact points A and B.

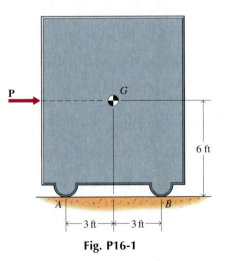

Fig. P16-1

16-2* A 75-kg cabinet is moved across a horizontal surface as shown in Fig. P16-2. Determine the maximum force **P** that can be applied without tipping the cabinet.

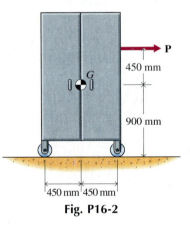

Fig. P16-2

16-3 The center of gravity of the 2300-lb dragster shown in Fig. P16-3 is located 24 in. in front of the rear wheels and 20 in. above the pavement. Determine the maximum possible acceleration of the dragster (without tipping over backwards) and the minimum coefficient of friction between the tires and the pavement required to achieve this acceleration.

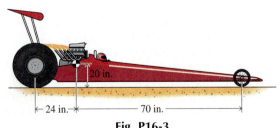

Fig. P16-3

16-4* A 750-N force **P** is applied to the 350-kg block shown in Fig. P16-4. If the kinetic coefficient of friction between the block and the horizontal surface is $\mu_k = 0.15$, determine the acceleration of the block and the reactions at the contact points A and B.

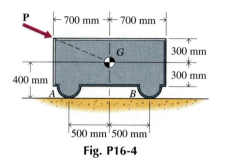

Fig. P16-4

16-5 A 450-lb crate 18 in. wide by 45 in. tall is sitting on the back of a truck as shown in Fig. P16-5. The truck is traveling at 65 mi/h when the driver suddenly applies the brakes. Determine the minimum distance required to stop the truck (without causing the crate to tip over) and the minimum static coefficient of friction required between the crate and the truck to prevent the crate from slipping.

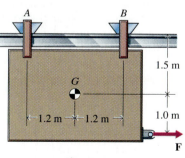

Fig. P16-5

16-6 A 2.50-kN force **F** is applied to a 1000-kg slab of material that is supported on a rail by two shoes as shown in Fig. P16-6. If the kinetic coefficient of friction between the rail and the shoes is $\mu_k = 0.25$, determine the vertical reactions of the rail on the shoes.

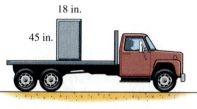

Fig. P16-6

16-7* A 600-lb crate slides down an inclined surface as shown in Fig. P16-7. If the kinetic coefficient of friction between the crate and the inclined surface is $\mu_k = 0.30$, determine the normal and frictional forces exerted on the crate at the contact surfaces A and B.

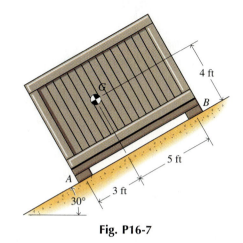

Fig. P16-7

16-8 A 250-kg slab of material is supported on an inclined rail by shoes A and B as shown in Fig. P16-8. If the kinetic coefficient of friction between the rail and the shoes is $\mu_k = 0.20$, determine the normal and the frictional forces exerted by the rail on the shoes while the slab is moving along the rail.

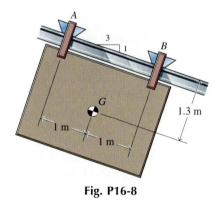

Fig. P16-8

Intermediate Problems

16-9* The front-wheel-drive automobile shown in Fig. P16-9 has a weight of 3100 lb. If the static coefficient of friction between the tires and the pavement is $\mu_s = 0.70$, determine the minimum time required for the automobile to accelerate uniformly from rest to a speed of 60 mi/h.

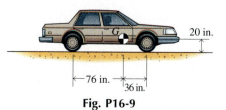

Fig. P16-9

16-10 The 40-kg plate shown in Fig. P16-10 is supported by two small rollers A and B and by the wire C. Determine the acceleration of the plate and the normal force on both wheels immediately after the cord C is cut.

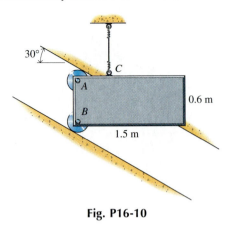

Fig. P16-10

16-11* A homogeneous 4-ft-diameter cylinder that weighs 2000 lb rests on the bed of a flatbed truck as shown in Fig. P16-11a. The blocks shown in Fig. P16-11b are used to prevent the cylinder from rolling as the truck accelerates. Determine the acceleration of the truck required to start the cylinder rolling over the block.

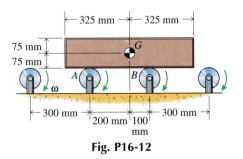

(a)   (b)

Fig. P16-11

16-12* The 100-mm-diameter rollers of the conveyor system shown in Fig. P16-12 are driven at a constant angular velocity of 25 rad/s. The block being transported by the system has a mass of 250 kg and is moving to the right with a velocity of 1.0 m/s at the instant shown. If the coefficient of friction between the block and the rollers is 0.25, determine the acceleration of the mass center of the block and the vertical components of the forces exerted on the block by the rollers at A and B.

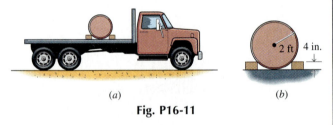

Fig. P16-12

16-13 The 60-lb boxes shown in Fig. P16-13 are placed on the moving conveyor belt from above. The boxes are 2 ft wide and the conveyor belt is moving at a constant speed of 5 ft/s. If the coefficient of friction between the boxes and the conveyor belt is $\mu_k = 0.4$, determine the maximum height h that the boxes can be and the distance d that the boxes move while they slip.

Fig. P16-13

16-14 A 2-m-long board with a mass of 20 kg rests on the back of a truck as shown in Fig. P16-14. If the bottom of the board is blocked so it cannot slide, determine the maximum acceleration that the truck can have without causing the board to tip over.

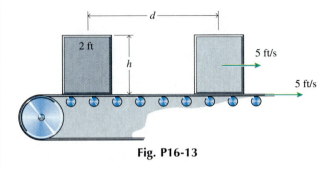

Fig. P16-14

16-15* A 100-lb homogeneous plate is suspended by two cables A and B of equal length as shown in Fig. P16-15. The plate swings in a vertical plane and is subjected to a 20-lb horizontal force F. In the position shown, the cables are rotating counterclockwise with an angular velocity of 5 rad/s. At this instant, determine the angular acceleration of the cables and the tensions in the cables.

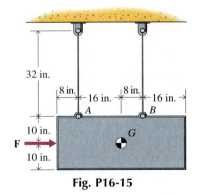

Fig. P16-15

16-16* The 10-kg thin plate shown in Fig. P16-16 is supported in a vertical plane by two links A and B and by a flexible cord C. Determine the acceleration of the mass center G of the plate and the force in each link immediately after the cord C is cut. Neglect the mass of the links.

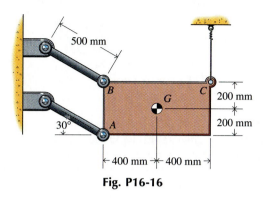

Fig. P16-16

16-17 A 14-ft-long log weighing 400 lb is suspended by three wires as shown in Fig. P16-17. Determine the acceleration of the mass center G of the log and the force in wires AB and CD immediately after wire E is cut.

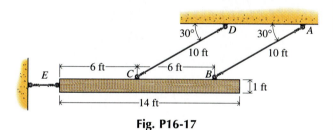

Fig. P16-17

16-18 The platform and lever system shown in Fig. P16-18 is used to transfer boxes between floors in a factory. In the position shown, lever AB is rotating clockwise with an angular velocity of 0.5 rad/s and the angular velocity is decreasing at a rate of 1.5 rad/s². If the mass of the box is 500 kg, determine the vertical components of the forces exerted on the box at supports C and D and the minimum coefficient of friction required to prevent slipping of the box at this instant.

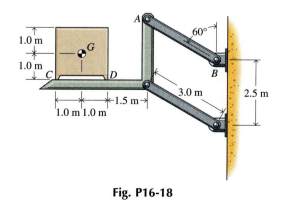

Fig. P16-18

16-19* When the two blocks A and B shown in Fig. P16-19 are released from rest, block A ($W_A = 500$ lb) slides up the inclined plane without tipping. If the kinetic coefficient of friction between block A and the inclined plane is $\mu_k = 0.2$, determine the maximum permissible weight for block B.

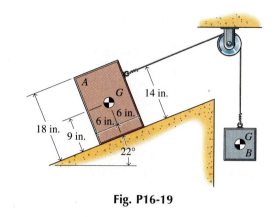

Fig. P16-19

Challenging Problems

16-20* The small rear-wheel-drive truck shown in Fig. P16-20 has a mass of 1750 kg and is carrying a 400-kg load. The center of mass of the truck is 0.75 m behind the front axle; the center of mass of the crate is 0.5 m in front of the rear axle. If the static coefficient of friction between the pavement and the tires is $\mu_s = 0.85$ and the crate is securely tied down, determine the minimum time required for the truck to accelerate uniformly from rest to 90 km/h.

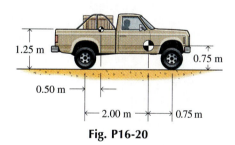

Fig. P16-20

267

16-21 A crate rests on a small cart as shown in Fig. P16-21. The weights of the crate and the cart are 850 lb and 100 lb, respectively. The static coefficient of friction between the crate and the cart is $\mu_s = 0.25$. If a 150-lb force **P** is applied to the cart, determine the support reactions at wheels A and B.

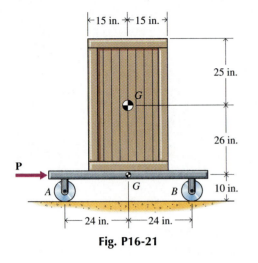

Fig. P16-21

16-22* A crate rests on a small cart as shown in Fig. P16-22. The masses of the crate and the cart are 150 kg and 25 kg, respectively. The static coefficient of friction between the crate and the cart is $\mu_s = 0.10$. If the crate does not slip or tip, determine the maximum permissible mass for block B.

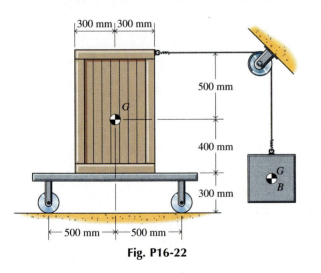

Fig. P16-22

16-23 A 50-lb block rests on a 120-lb platform that is supported by four weightless rods as shown in Fig. P16-23. The coefficient of friction between the platform and the block is 0.10, and the platform is released at rest in the position shown by cutting the cable at A. Determine the acceleration of the mass center of the block and the forces exerted on the platform by the rods at A and C at the instant the motion begins.

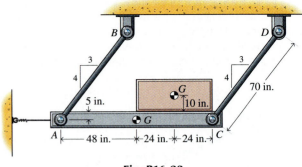

Fig. P16-23

16-24 Two crates A ($m_A = 50$ kg) and B ($m_B = 40$ kg) are connected by a cable as shown in Fig. P16-24. Both crates are symmetric with respect to the plane of motion, and the cable AB and the force **P** are in the plane of motion. If the surface supporting the crates is smooth, determine the maximum force **P** that can be applied to crate B before either crate is on the verge of tipping.

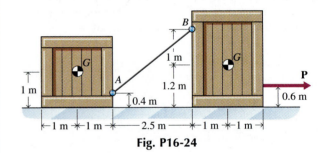

Fig. P16-24

Computer Problems

C16-25 A 16-lb rectangular plate swings at the end of two identical slender links as shown in Fig. P16-25. If the mass of the links can be neglected and the system starts from rest when $\theta = \theta_0 = 20°$, calculate and plot

a. The angular velocity of the links $\dot{\theta}$ as a function of their angular position θ ($20° < \theta < 160°$).

b. The tensions T_{AB} and T_{CD} in the two links as a function of their angular position θ ($20° < \theta < 160°$).

c. What is the smallest initial angle θ_0 for which both tensions T_{AB} and T_{CD} will always be positive?

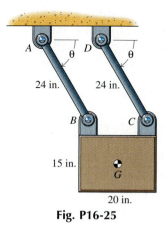

Fig. P16-25

a. The required torque T as a function of the angular position θ ($5° < \theta < 175°$).
b. The forces exerted on the platform by the lever arms at A and C as a function of θ ($5° < \theta < 175°$).
c. The normal and friction forces exerted on the bottom of the crate by the platform as a function of θ ($5° < \theta < 175°$).

C16-26 The platform and lever system shown in Fig. P16-26 is used to transfer boxes between floors in a factory. The mass of the crate is 120 kg, the mass of the platform is 30 kg, and their combined mass center is at G. If the platform is slowly lowered at a constant rate of $\dot\theta = 0.2$ rad/s by means of a torque T applied to the lever arm AB, calculate and plot

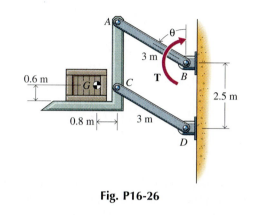

Fig. P16-26

16-4.2 Fixed-Axis Rotation

When a body rotates about a fixed axis as in Fig. 16-18, all elements of the body not on the axis of rotation travel in circular paths about the fixed axis while points directly on the axis of rotation do not move at all. In Fig. 16-18 point A is the intersection of the axis of rotation and the plane of motion of the body. Since the acceleration of point A is zero, the equations of motion for the body are conveniently expressed by Eqs. 15-11 and 16-3

$$\mathbf{R} = \Sigma\mathbf{F} = m\mathbf{a}_G \tag{a}$$
$$\Sigma M_{Ax} = -\alpha I_{Azx} + \omega^2 I_{Ayz} \tag{b}$$
$$\Sigma M_{Ay} = -\alpha I_{Ayz} - \omega^2 I_{Azx} \tag{c}$$
$$\Sigma M_{Az} = I_{Az}\alpha \tag{d}$$

If the body is symmetric with respect to the plane of motion, then the products of inertia will be zero ($I_{Azx} = I_{Ayz} = 0$) and the right-hand sides of Eqs. b and c will both be zero. If in addition all forces are applied in and act in the plane of motion, then the moments of the forces will have only a z-component and the left-hand sides of Eqs. b and c will also both be zero. Therefore, *for the fixed-axis rotation of a rigid body symmetric with respect to the plane of motion (and for which the forces are applied in the plane of motion), the equations of motion reduce to*

$$\mathbf{R} = \Sigma\mathbf{F} = m\mathbf{a}_G \tag{16-18a}$$
$$\Sigma M_{Az} = I_{Az}\alpha \tag{16-18b}$$

where A is the point of intersection of the axis of rotation and the plane of motion of the body. Equation 16-18b is not valid if the acceleration of point A is not zero.

Equations 16-18 represent three scalar equations and can be solved for no more than three unknowns. Generally, however, they will con-

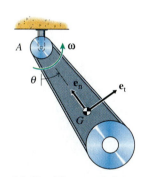

Figure 16-18 The pendulum rotates in a vertical plane about the pivot A. The acceleration of A is zero and the acceleration of the mass center is just $\mathbf{a}_G = \mathbf{a}_{G/A} = r_G\alpha\mathbf{e}_t + r_G\omega^2\mathbf{e}_n$.

tain more than three unknowns, so additional equations will be necessary. The additional equations are the kinematic relations between the acceleration of the mass center $\mathbf{a}_G$ and the angular acceleration α. Since the velocity and acceleration of point A are zero, the velocity and acceleration of the mass center G are given by the relative velocity and relative acceleration equations (Eqs. 14-9c and 14-10c)

$$\mathbf{v}_G = \mathbf{v}_A + \mathbf{v}_{G/A} = 0 + r_G\omega\mathbf{e}_t$$
$$\mathbf{a}_G = \mathbf{a}_A + \mathbf{a}_{G/A} = 0 + r_G\alpha\mathbf{e}_t + r_G\omega^2\mathbf{e}_n$$

The solution of fixed-axis rotation problems consists of a combination of Eqs. 16-18 and the kinematic relationships as illustrated in the following examples.

CONCEPTUAL EXAMPLE 16-2: EXAMPLES OF PLANAR MOTION

What is the difference between simple planar motion, such as a disk rotating on a shaft, and general planar motion, such as a rotating crankshaft?

SOLUTION

In planar motions, all points in the body move in parallel planes. The plane that contains the mass center of the body is called the plane of motion. For simple planar motion, all forces acting on the body must be applied in the plane of motion and all moments acting on the body must act perpendicular to the plane of motion.

In the case of the rotating disk, the plane of motion is parallel to the face of the disk. The mass center of the disk travels a circular path in this plane and will have acceleration components $a_n = r\omega^2$ directed toward the center of the shaft on which the disk is mounted and $a_t = r\alpha$ directed tangent to the circular path of motion. The equations of motion are satisfied in this case if the shaft exerts a force on the disk of $N = mr\omega^2$ in the plane of motion and a moment of $M = I\alpha$ about the axis of the shaft (assuming that the weight of the disk is negligible compared to the inertia term $mr\omega^2$). No forces out of the plane of motion are required to prevent the disk from twisting out of its plane of motion. This system represents simple planar motion.

The rotating system composed of two disks and the segment of shaft connecting them also represents simple planar motion if the disks are both mounted on the same side of the shaft. In this case, the plane of motion (the plane containing the mass center of the system) is midway between the two disks. The mass center of the system and the mass centers of the two disks rotate in identical circular paths. The equations of motion are satisfied in this case if a single bearing in the plane of motion exerts a force on the shaft of $N = mr\omega^2$ in the plane of motion and a moment $M = I\alpha$ acts about the axis of the shaft. Again, no forces out of the plane of motion are required to prevent the system from twisting out of its plane of motion.

The rotating system composed of two disks mounted on opposite sides of the shaft (which is similar to a rotating crankshaft) is still executing a planar motion. In this case, however, either forces acting outside of the plane of

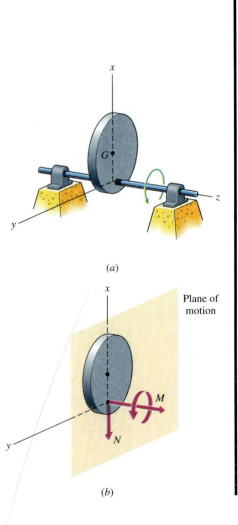

(a)

(b)

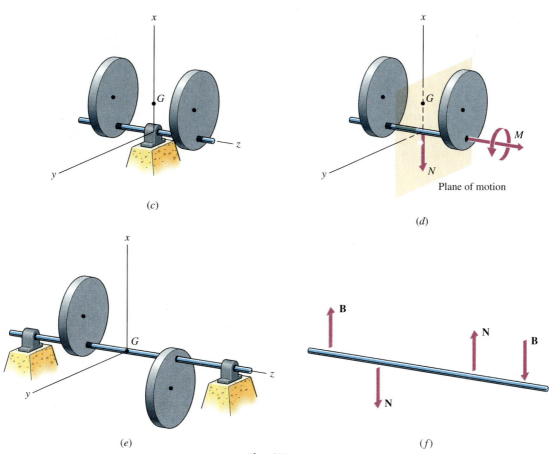

(c)

(d)

Plane of motion

(e)

(f)

Fig. CE16-2

motion or a moment acting parallel to the plane of motion is required to keep the system from twisting out of its plane of motion. The easiest way to see this is to consider the shaft alone. Individually, each disk moves in simple planar motion and will exert a simple force $N = mr\omega^2$ on the shaft in the plane of motion of the disk. However, these two forces form a couple perpendicular to the axis of the shaft. This couple would cause the system to twist out of its plane of motion if it were not opposed by bearing forces outside the plane of motion.

In terms of the equations of motion, the plane of motion is a plane of symmetry and the axis of rotation is a principal axis in the first two cases. Therefore, the products of inertia I_{yz} and I_{zx} are both zero in the first two cases. Relative to the axis of the shaft, the moment equation can be written simply $\Sigma M = I_z \alpha$ (Eq. 16-18b). For the rotating crankshaft, however, the plane of motion is not a plane of symmetry and the axis of rotation is not a principal axis. In this case, the products of inertia I_{yz} and I_{zx} are not both zero and the more general equations of planar motion (Eqs. 16-3) must be used to calculate the bearing forces required to maintain the planar motion.

The rod *AB* shown in Fig. 16-19 has a constant cross section and a mass of 10 kg. As a result of the rotation of crank *C*, rod *AB* oscillates in a vertical plane. In the position shown, its angular velocity ω is 10 rad/s clockwise and its angular acceleration α is 40 rad/s² counterclockwise. Determine the force exerted on the rod *AB* by the lightweight link *DG* and by the pin at support *A*.

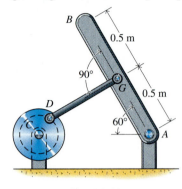

Fig. 16-19

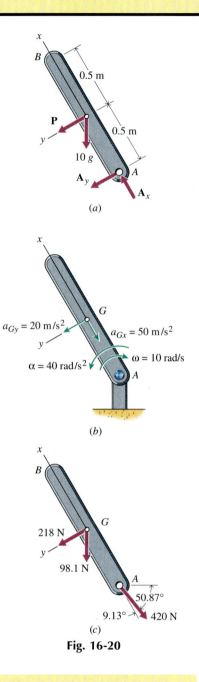

(a)

(b)

(c)

Fig. 16-20

SOLUTION

A free-body diagram of the rod is shown in Fig. 16-20a. The rod rotates about a fixed axis through the pin *A*. The equations of motion for the rod (Eqs. 16-18) are

$$+\nwarrow\Sigma F_x = ma_{Gx}: \qquad A_x - 10(9.81)\sin 60° = 10a_{Gx} \qquad (a)$$
$$+\nearrow\Sigma F_y = ma_{Gy}: \qquad A_y + 10(9.81)\cos 60° + P = 10a_{Gy} \qquad (b)$$
$$+\curvearrowright\Sigma M_{Az} = I_{Az}\alpha: \qquad 0.5[P + 10(9.81)\cos 60°] = I_{Az}(40) \qquad (c)$$

where the moment of inertia of the rod about the pin *A* is

$$I_{Az} = I_{Gz} + md^2 = \frac{1}{12}mL^2 + m(L/2)^2$$

$$= \frac{1}{12}(10)(1)^2 + 10(0.5)^2 = 3.333 \text{ kg} \cdot \text{m}^2$$

and the acceleration of the mass center of the rod is (Fig. 16-20b)

$$\mathbf{a}_G = \mathbf{a}_A + \mathbf{a}_{G/A} = 0 + r_G\alpha\mathbf{e}_t + r_G\omega^2\mathbf{e}_n$$
$$= (0.5)(40)\mathbf{j} + 0.5(10^2)(-\mathbf{i}) = -50\mathbf{i} + 20\mathbf{j} \text{ m/s}^2 \qquad (d)$$

Therefore, $a_{Gx} = -50$ m/s², $a_{Gy} = 20$ m/s², and

$$A_x = -415.0 \text{ N} \qquad A_y = -66.66 \text{ N}$$
$$\mathbf{A} = 420 \text{ N} \nwarrow 50.87° \qquad\qquad\qquad \text{Ans.}$$
$$P = 218 \text{ N} \nearrow 30° \qquad\qquad\qquad \text{Ans.}$$

Note that the force exerted on the rod by the pin at *A* (Fig. 16-20c) does not act along the rod ($A_y \neq 0$). From Eq. 16-17d

$$\Sigma M_{Gz} = I_{Gz}\alpha$$

for a general plane motion, since there is an angular acceleration, there must be a moment about the mass center to cause the angular acceleration. Furthermore, the angular acceleration is counterclockwise, so the moment about the mass center must also be counterclockwise, which is consistent with the result that A_y is negative.

It is important to make sure that the positive directions for a_{Gx}, a_{Gy}, and α used in the kinematic diagram (Fig. 16-20b) and the relative acceleration equation (Eq. *d*) agree with the positive directions used in the equations of motion (Eqs. *a*, *b*, and *c*).

The weight of the unbalanced wheel shown in Fig. 16-21 is 64.4 lb, and its radius of gyration with respect to the axis of rotation is 6 in. The wheel is driven at a constant speed of 60 rev/min by a torque C. Determine the horizontal and vertical components of the force exerted on the axle A by the wheel and the torque required to drive the wheel as it rotates one full turn.

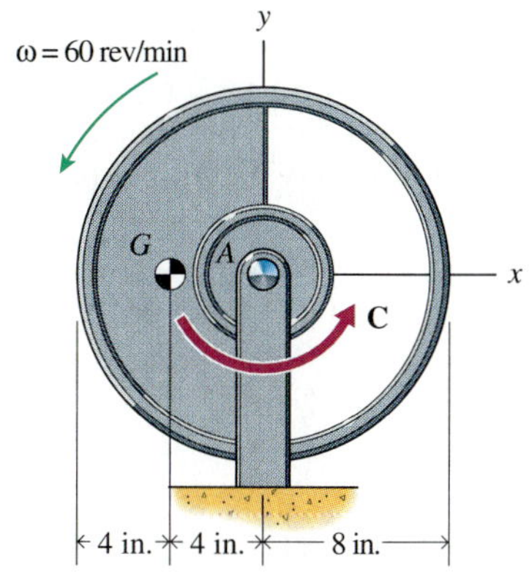

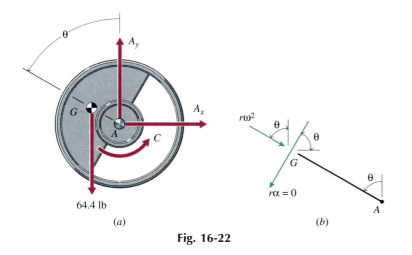

$\omega = 60$ rev/min

← 4 in. →← 4 in. →←—— 8 in. ——→

Fig. 16-21

SOLUTION

A free-body diagram of the wheel is shown in Fig. 16-22a. The wheel is acted on by the axle forces A_x and A_y, by the weight $W = 64.4$ lb, and by the driving torque C. The wheel rotates about a fixed axis through the axle A.

Fig. 16-22

(a)   (b)

The equations of motion for the wheel (Eqs. 16-18) are

$$+\rightarrow \Sigma F_x = ma_{Gx}: \qquad A_x = \frac{64.4}{32.2}a_{Gx}$$

$$+\uparrow \Sigma F_y = ma_{Gy}: \qquad A_y - 64.4 = \frac{64.4}{32.2}a_{Gy}$$

$$+\curvearrowleft \Sigma M_{Az} = I_{Az}\alpha: \qquad 64.4\left(\frac{4}{12}\sin\theta\right) + C = I_{Az}\alpha$$

where $\theta = 0$ when the mass center is directly above the axle, the moment of inertia of the wheel about the axle A is

$$I_{Az} = mk_A^2 = \frac{64.4}{32.2}\left(\frac{6}{12}\right)^2 = 0.5 \text{ slug} \cdot \text{ft}^2$$

$\omega = 60$ rev/min $= 2\pi$ rad/s $=$ constant, $\alpha = 0$ rad/s^2, and the acceleration of the mass center of the wheel is (Fig. 16-22b)

$$\mathbf{a}_G = \mathbf{a}_A + \mathbf{a}_{G/A} = \mathbf{0} + r_G\alpha\mathbf{e}_t + r_G\omega^2\mathbf{e}_n$$

$$= \frac{4}{12}(2\pi)^2 \sin\theta\mathbf{i} - \frac{4}{12}(2\pi)^2 \cos\theta\mathbf{j}$$

The center of mass G travels a circular path about the axle A. Since the wheel rotates at a constant speed, $\alpha = 0$ and the acceleration of the center of mass is just $a_{Gn} = r_G\omega^2$ directed toward the axle.

Therefore, $a_{Gx} = 13.1595 \sin \theta$ ft/s^2, $a_{Gy} = -13.1595 \cos \theta$ ft/s^2, and

$$A_x = 26.32 \sin \theta \text{ lb} \rightarrow \qquad \text{Ans.}$$
$$A_y = 64.4 - 26.32 \cos \theta \text{ lb} \uparrow \qquad \text{Ans.}$$
$$C = -21.47 \sin \theta \text{ lb} \cdot \text{ft} \downarrow \qquad \text{Ans.}$$

Actually, A_x and A_y are the forces exerted on the wheel by the axle. Equal and opposite forces are exerted on the axle by the wheel. The variation of the forces A_x and A_y and the driving torque C are shown in Fig. 16-22c.

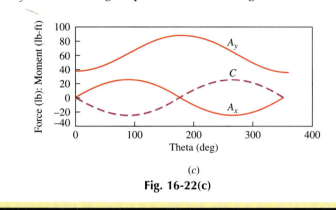

(c)

Fig. 16-22(c)

EXAMPLE PROBLEM 16-7

The uniform bar shown in Fig. 16-23 is 2 m long, has a mass of 10 kg, and is initially at rest in a vertical position. If the bar is disturbed and starts to fall to the right, determine the angular velocity of the bar and the horizontal and vertical components of the force exerted on the bar by the frictionless pivot at A when the bar is horizontal.

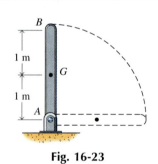

Fig. 16-23

SOLUTION

A free-body diagram of the bar in the horizontal position is shown in Fig. 16-24a. The bar rotates about a fixed axis through the pin A. The equations of motion for the rod (Eqs. 16-18) are

$$+\rightarrow \Sigma F_x = ma_{Gx}: \qquad A_x = m(-\omega^2) \qquad (a)$$
$$+\uparrow \Sigma F_y = ma_{Gy}: \qquad A_y - mg = m(-\alpha) \qquad (b)$$
$$+\downarrow \Sigma M_{Az} = I_{Az}\alpha: \qquad 1mg = I_{Az}\alpha \qquad (c)$$

where the moment of inertia of the bar about the pin A is

$$I_{Az} = I_{Gz} + md^2 = \frac{1}{12}mL^2 + m(L/2)^2 = 4m/3$$

and the acceleration of the mass center of the bar is (Fig. 16-24b)

$$\mathbf{a}_G = \mathbf{a}_A + \mathbf{a}_{G/A} = 0 + r_G\alpha\mathbf{e}_t + r_G\omega^2\mathbf{e}_n$$
$$= 1\alpha(-\mathbf{j}) + 1\omega^2(-\mathbf{i}) = -\omega^2\mathbf{i} - \alpha\mathbf{j}$$

Equations a, b, and c involve four unknowns (A_x, A_y, α, and ω) and cannot be solved directly. The angular acceleration of the bar will have to be determined for an arbitrary angle θ and then integrated with respect to time to determine the angular velocity when $\theta = 90°$. Then the equations of motion can be solved for the remaining three unknowns.

A free-body diagram of the bar is shown in Fig. 16-24c for an arbitrary angle θ. The equations of motion for the bar (Eqs. 16-18) are now

$$+\rightarrow \Sigma F_x = ma_{Gx}: \qquad A_x = ma_{Gx} \qquad (d)$$
$$+\uparrow \Sigma F_y = ma_{Gy}: \qquad A_y - mg = ma_{Gy} \qquad (e)$$
$$+\downmdash \Sigma M_{Az} = I_{Az}\alpha: \qquad mg(1\ \sin\ \theta) = (4m/3)\alpha \qquad (f)$$

where the acceleration of the mass center of the rod is (Fig. 16-24d)

$$\mathbf{a}_G = \mathbf{a}_A + \mathbf{a}_{G/A} = \mathbf{0} + r_G\alpha\mathbf{e}_t + r_G\omega^2\mathbf{e}_n$$
$$= 1\alpha(\sin\ \theta\mathbf{i} - \cos\ \theta\mathbf{j}) + 1\omega^2(-\cos\ \theta\mathbf{i} - \sin\ \theta\mathbf{j})$$

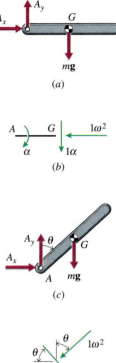

(a)

Therefore, $a_{Gx} = \alpha \sin\ \theta - \omega^2 \cos\ \theta$, $a_{Gy} = -\alpha \cos\ \theta - \omega^2 \sin\ \theta$, and

$$A_x = m(\alpha \sin\ \theta - \omega^2 \cos\ \theta) \qquad (g)$$
$$A_y = mg - m(\alpha \cos\ \theta + \omega^2 \sin\ \theta) \qquad (h)$$
$$\alpha = 0.75g \sin\ \theta \qquad (i)$$

Rewriting the angular acceleration using the chain-rule of differentiation

(b)

$$\alpha = \frac{d\omega}{dt} = \frac{d\omega}{d\theta}\frac{d\theta}{dt} = \omega\frac{d\omega}{d\theta} = 0.75g \sin\ \theta \qquad (j)$$

and integrating gives

$$\omega^2 = -1.5g \cos\ \theta + C_1 = 1.5g(1 - \cos\ \theta) \qquad (k)$$

where the constant of integration $C_1 = 1.5g$, since the bar starts from rest ($\omega = 0$ when $\theta = 0$). Finally, when the bar is horizontal ($\theta = 90°$)

$$\alpha = 0.75g = 7.538 \text{ rad/s}^2$$
$$\omega = \sqrt{1.5g} = 3.836 \text{ rad/s} \qquad \text{Ans.}$$
$$A_x = ma = 10(7.358) = 73.6 \text{ N} \qquad \text{Ans.}$$
$$A_y = mg - m\omega^2 = 10(9.81) - 10(3.836)^2$$
$$= -49.1 \text{ N} = 49.1 \text{ N}\leftarrow \qquad \text{Ans.}$$

(c)

(Note that the angular velocity and angular acceleration are independent of the mass of the bar.)

The variation of the forces A_x and A_y is shown in Fig. 16-24e.

(d)

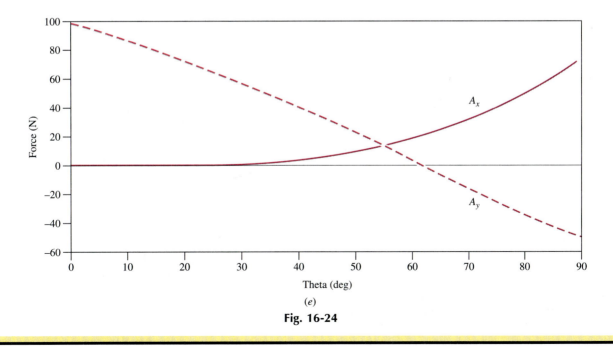

(e)

Fig. 16-24

PROBLEMS

In the following problems, all ropes, cords, and cables are assumed to be flexible, inextensible, and of negligible mass. All pins and pulleys have negligible mass and are frictionless unless specified otherwise.

Introductory Problems

16-27* The 20-lb slender bar *AB* shown in Fig. P16-27 is supported by a frictionless pin at *A* and a cord at *B*. If the cord at *B* breaks, determine the acceleration of the mass center of the bar and the reaction at support *A* at the instant that motion begins.

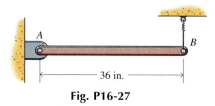

Fig. P16-27

16-28* The 15-kg slender bar shown in Fig. P16-28 is rotating counterclockwise in a vertical plane about a smooth pin at *A*. When the bar is in the position shown, its angular velocity is 10 rad/s. At this instant, determine the angular acceleration of the bar and the magnitude and direction of the force exerted on the bar by the pin at the support *A*.

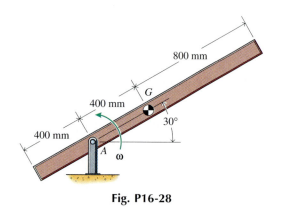

Fig. P16-28

16-29 A thin circular ring of uniform thickness rotates in a vertical plane about a frictionless pin at point *A* as shown in Fig. P16-29. The ring has a diameter of 24 in. and weighs 5 lb. In the position shown (the diameter through pin *A* is horizontal), the angular velocity of the ring is 10 rad/s, counterclockwise. At this instant, determine the angular acceleration of the ring and the horizontal and vertical components of the force exerted on the ring by the pin at support *A*.

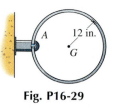

Fig. P16-29

16-30 A 25-kg circular disk of uniform thickness rotates in a vertical plane about a frictionless pin at point *A* as shown in Fig. P16-30. In the position shown (the diameter through pin *A* is vertical), the angular velocity of the disk is 20 rad/s counterclockwise and the magnitude of the couple **C** is 50 N · m. At this instant, determine the angular acceleration of the disk and the horizontal and vertical components of the force exerted on the disk by the pin at support *A*.

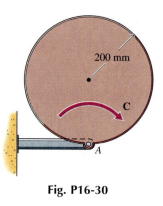

Fig. P16-30

16-31* A 100-lb rectangular plate of uniform thickness is supported by a smooth pin and a cable as shown in Fig. P16-31. If the cable at *B* breaks, determine the acceleration of the mass center of the plate and the magnitude of the reaction at support *A* at the instant that motion begins.

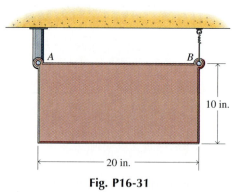

Fig. P16-31

16-32* An 80-kg semicircular plate of uniform thickness is supported by a smooth pin and a cable as shown in Fig. P16-32. If the cable at B breaks, determine the acceleration of the mass center of the plate and the magnitude of the reaction at support A at the instant that motion begins.

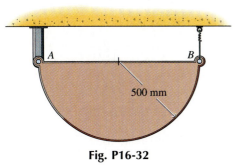

500 mm

Fig. P16-32

16-33 A 10-lb uniform wheel is at rest when it is placed in contact with a moving belt as shown in Fig. P16-33. The kinetic coefficient of friction between the belt and the 16-in.-diameter wheel is $\mu_k = 0.1$, and the belt moves with a constant speed of 30 ft/s. Determine the number of revolutions that the wheel turns before it rolls without slipping on the moving belt.

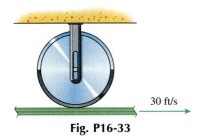

30 ft/s

Fig. P16-33

16-34 A 10-kg uniform flywheel is connected to a constant torque motor with a flexible belt as shown in Fig. P16-34. The diameter of the flywheel is 400 mm, and the mass of the motor armature may be neglected. If the flywheel starts from rest, determine the torque necessary to rotate the flywheel at 4200 rev/min after 5 revolutions.

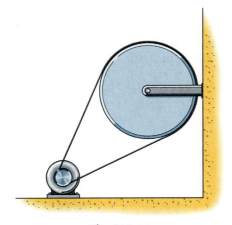

Fig. P16-34

Intermediate Problems

16-35 A 300-lb force is applied to the end of a rope that is wrapped around the outside of a 6-ft-diameter cable drum as shown in Fig. P16-35a.

a. If the 600-lb cable drum can be treated as a uniform cylinder, determine the angular acceleration of the cable drum and the time required to rotate the cable drum 3 revolutions.

b. Repeat part a if the 300-lb force is replaced with a 300-lb block as in Fig. P16-35b.

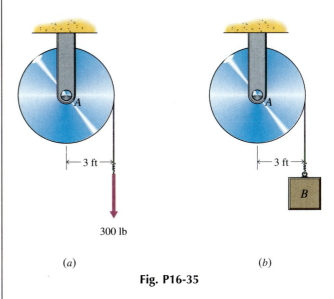

3 ft

300 lb

(a)

3 ft

B

(b)

Fig. P16-35

16-36* A horizontal force **F** of 250 N is applied to a cable that is wrapped around the inner drum of the compound pulley being used to lift block B in Fig. P16-36. The 20-kg pulley has a radius of gyration with respect to the axis of rotation of 160 mm. If block B has a mass of 10 kg, determine the angular acceleration of the pulley and the tension in the cable connected to block B.

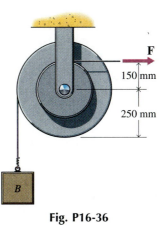

F

150 mm

250 mm

B

Fig. P16-36

16-37* Two blocks A ($W_A = 50$ lb) and B ($W_B = 90$ lb) are supported by cables wrapped around a compound cable drum as shown in Fig. P16-37. The 60-lb drum has a radius of gyration of 7.50 in. with respect to its axis of rotation. During motion of the system, determine the tensions in the two cables and the angular acceleration of the cable drum.

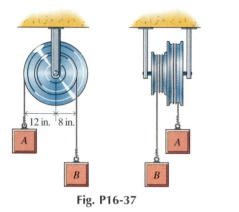

Fig. P16-37

16-38 A torque **M** of 300 N · m is applied to pulley A of the belt drive shown in Fig. P16-38. Pulley A is a solid circular disk with a mass of 15 kg. Pulley B and the cable drum have a combined mass of 75 kg and a radius of gyration with respect to the axis of rotation of 150 mm. The mass of block C is 150 kg. If the system starts from rest, determine

a. The tension T_C in the cable supporting block C.
b. The acceleration a_C of block C.
c. The time t_{10} required to raise block C a distance of 10 m.

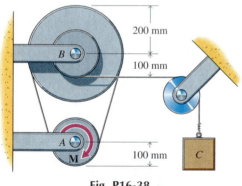

Fig. P16-38

16-39 The 50-lb solid circular disk A of Fig. P16-39 rotates about a smooth pin at O. Block B weighs 20 lb. During motion of the system, determine

a. The angular acceleration α_A of disk A.
b. The tension T in the cable.
c. The horizontal and vertical components of the force exerted on disk A by the pin at O.

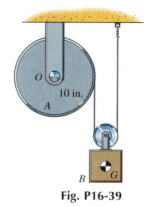

Fig. P16-39

16-40 Bar AB of Fig. P16-40 rotates in a horizontal plane with a constant angular velocity of 15 rad/s. At the end of bar AB is a 2-kg slender bar C, which has a uniform cross section and supports a 4-kg sphere D. Bar C is held in a vertical position by a cable. Determine the tension T in the cable and the magnitude of the force exerted by pin B on bar C.

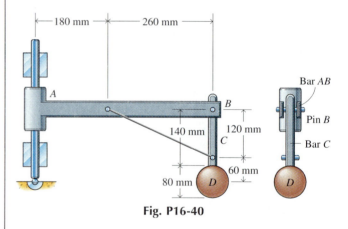

Fig. P16-40

16-41* The 24-in.-diameter unbalanced disk shown in Fig. P16-41 weighs 125 lb, and its radius of gyration with respect to the fixed axis of rotation is 8.50 in. A cable wrapped around the circumference of the disk passes over a small pulley of negligible mass and is attached to a 50-lb block B. If the system is released from rest in the position shown, determine the tension in the cable and the acceleration of block B as motion begins.

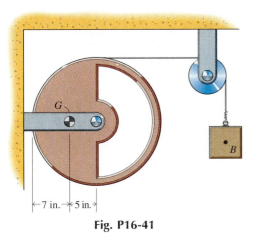

Fig. P16-41

←7 in.→*←5 in.→*

16-42* The speed of a rotating system is controlled with a brake as shown in Fig. P16-42. The rotating parts of the system have a mass of 300 kg and a radius of gyration with respect to the axis of rotation of 200 mm. The kinetic coefficient of friction between the brake pad and the brake drum is $\mu_k = 0.50$. When a force **P** of 500 N is being applied to the brake lever, determine the horizontal and vertical components of the reaction at support B of the brake lever and the time t required to reduce the speed of the system from 1000 rev/min to rest.

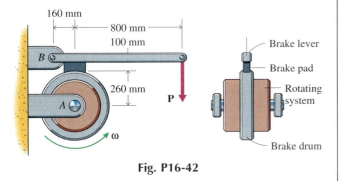

Fig. P16-42

Challenging Problems

16-43* A solid circular disk of uniform thickness rotates in a vertical plane about a frictionless pin at point A as shown in Fig. P16-43. The disk has a diameter of 16 in. and weighs 50 lb. If the disk starts from rest when $\theta = 0°$, determine the angular velocity of the disk and the magnitude of the force exerted on the disk by the pin at support A

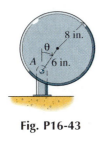

Fig. P16-43

a. When $\theta = 90°$.
b. When $\theta = 180°$.

16-44 A 50-kg rectangular plate of uniform thickness is supported by a smooth pin and a cable as shown in Fig. P16-44. If the cable at B breaks, determine the angular velocity of the plate and the magnitude of the force exerted on the plate by the pin at support A

a. When the mass center of the plate is directly below A.
b. When side AB of the plate is vertical (corner B is directly below A).

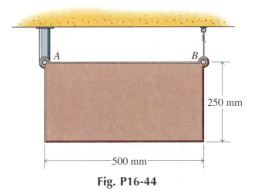

Fig. P16-44

16-45 The 24-in.-diameter disk shown in Fig. P16-45 weighs 125 lb. The mass center of the disk is located 5 in. from the center of the disk, and its radius of gyration with respect to the fixed axis of rotation is 8.50 in. A cable wrapped around the circumference of the disk passes over a small pulley of negligible mass and is attached to a 60-lb block B. If the system is released from rest when $\theta = 0°$, determine the angular velocity of the disk and the magnitude of the force exerted on the disk by the pin at support O

a. When $\theta = 90°$.
b. When $\theta = 180°$.

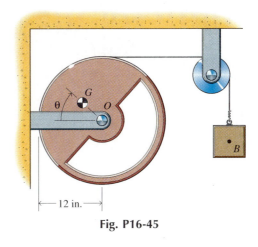

Fig. P16-45

16-46* The 1-m-diameter disk shown in Fig. P16-46 has a mass of 20 kg. The mass center of the disk is located 250 mm from the center of the disk, and its radius of gyration with respect to the fixed axis of rotation is 375 mm. A cable wrapped around the circumference of the disk passes over a small pulley of negligible mass and is attached to a 25-kg block B. If the disk has a counterclockwise angular velocity of 10 rad/s when $\theta = 0°$, determine

a. The angle θ_{max} when the disk stops and begins to rotate clockwise.

b. The angular velocity of the disk and the magnitude of the force exerted on the disk by the pin at support O when $\theta = 90°$.

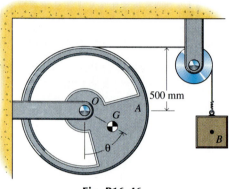

Fig. P16-46

16-47* The flywheel shown in Fig. P16-47 is rotating at a constant angular velocity of 50 rad/s in a counterclockwise direction. Bar AB weighs 20 lb and has a radius of gyration with respect to its mass center of 9.15 in., and the slot in bar AB is smooth. Determine the horizontal and vertical components of the force exerted by the pin at support A on bar AB when the angle $\theta = 60°$.

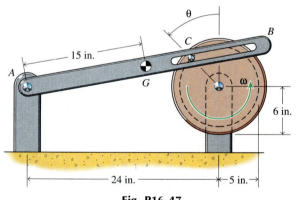

Fig. P16-47

16-48 The flywheel shown in Fig. P16-48 is rotating with a constant angular velocity of 30 rad/s. Bar AB has a mass of 15 kg and a radius of gyration with respect to its mass center of 325 mm. The coefficient of friction between the pin at C and the slot in the bar is 0.10. When the bar is in the position shown, determine the horizontal and vertical components of the force exerted on the bar by the pin at support A.

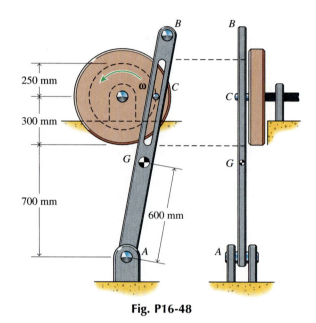

Fig. P16-48

Computer Problems

C16-49 The 16-in.-diameter unbalanced wheel shown in Fig. P16-49 weighs 20 lb, and its radius of gyration with respect to the fixed axis of rotation is 5 in. A cable wrapped around the circumference of the disk passes over a small pulley of negligible mass and is attached to a 10-lb block B. If the system is released from rest when $\theta = 0°$, calculate and plot

a. The force exerted on the wheel by the axle at A as a function of its angular position θ ($0° < \theta < 600°$).

b. The tension in the cable as a function of θ ($0° < \theta < 600°$).

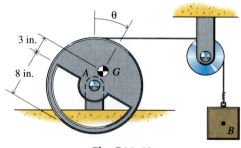

Fig. P16-49

C16-50 The flywheel shown in Fig. P16-50 is rotating at a constant angular velocity of 5 rad/s in a counterclockwise direction. The 500-mm-long bar AB has a mass of 2.5 kg and a radius of gyration of 270 mm relative to point A. If the slot in AB is smooth, calculate and plot

a. The force exerted on the pin P by the bar AB as a function of the angular position of the flywheel θ for one complete revolution of the flywheel $(0° < \theta < 360°)$.
b. The magnitude of the force exerted on the bar AB by the support at A as a function of the angular position of the flywheel θ for one complete revolution of the flywheel $(0° < \theta < 360°)$.

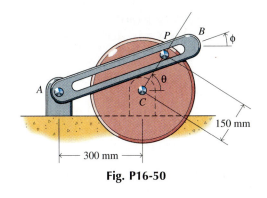

Fig. P16-50

16-4.3 General Plane Motion of Symmetric Bodies

Translation and rotation about a fixed axis are two special forms of planar motion. In general, however, bodies translate and rotate at the same time. For a body of arbitrary shape, this general planar motion is best described by Eqs. 16-17

$$\mathbf{R} = \Sigma\mathbf{F} = m\mathbf{a}_G \tag{a}$$
$$\Sigma M_{Gx} = -\alpha I_{Gzx} + \omega^2 I_{Gyz} \tag{b}$$
$$\Sigma M_{Gy} = -\alpha I_{Gyz} - \omega^2 I_{Gzx} \tag{c}$$
$$\Sigma M_{Gz} = I_{Gz}\alpha \tag{d}$$

where G is the center of mass of the body.

If the body is symmetric with respect to the plane of motion, then the products of inertia will be zero ($I_{Azx} = I_{Ayz} = 0$) and the right-hand sides of Eqs. b and c will both be zero. If in addition all forces are applied in and act in the plane of motion, then the moments of the forces will have only a z-component and the left-hand sides of Eqs. b and c will also both be zero. Therefore, *for the general plane motion of a rigid body symmetric with respect to the plane of motion (and for which the forces are applied in the plane of motion), the equations of motion reduce to*

$$\mathbf{R} = \Sigma\mathbf{F} = m\mathbf{a}_G \tag{16-19a}$$
$$\Sigma M_{Gz} = I_{Gz}\alpha \tag{16-19b}$$

where G is the center of mass of the body.

Equations 16-19 are valid for all planar motions of symmetric bodies including translation and fixed-axis rotation. In fact, Eqs. 16-19 (with $\alpha = 0$) are the equations that were used to describe translation in Section 16-4.1. However, although Eqs. 16-18b and 16-19b have a similar form, it is not obvious that they are equivalent for fixed-axis rotation. To see that Eqs. 16-18b and 16-19b are equivalent (for fixed-axis rotation only), consider the pendulum of Fig. 16-25 which rotates about a frictionless pin at A and has its center of mass at G. The free-body diagram of the pendulum is drawn in Fig. 16-26a. Since the mass center rotates in a circular path about the fixed pin A, the acceleration of the mass center is (Fig. 16-26b)

$$\mathbf{a}_G = \mathbf{a}_A + \mathbf{a}_{G/A} = 0 + r_{G/A}\alpha\mathbf{e}_t + r_{G/A}\omega^2\mathbf{e}_n$$

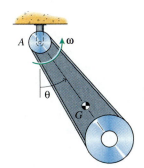

Figure 16-25 The pendulum rotates in a vertical plane about the pivot A.

281

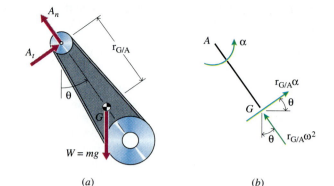

(a)   (b)

Figure 16-26   (a) *Free-body diagram of the pendulum. The axle force is shown in terms of its normal and tangential components but could just as easily have been shown in terms of its horizontal and vertical components.*
(b) *Kinematic diagram of the pendulum. The acceleration of A is zero and the acceleration of the mass center is just $\mathbf{a}_G = \mathbf{a}_{G/A} = r_G\alpha\mathbf{e}_t + r_G\omega^2\mathbf{e}_n$.*

Then the equations of motion (Eq. 16-19) for the pendulum can be written

$$+\nearrow \Sigma F_t = ma_{Gt}: \qquad A_t - W \sin\theta = mr_{G/A}\alpha \qquad (e)$$
$$+\searrow \Sigma F_n = ma_{Gn}: \qquad A_n - W \cos\theta = mr_{G/A}\omega^2 \qquad (f)$$
$$+\curvearrowleft \Sigma M_G = I_{Gz}\alpha: \qquad -r_{G/A}A_t = mk_G^2\alpha \qquad (g)$$

where the moment of inertia about the mass center $I_{Gz} = mk_G^2$ and k_G is the radius of gyration relative to the mass center. But, substituting Eq. *e* into Eq. *g* and rearranging gives

$$-r_{G/A}W \sin\theta = (mk_G^2 + mr_{G/A}^2)\alpha = I_{Az}\alpha$$

which is exactly what Eq. 16-18b gives for the pendulum.

Therefore, either Eqs. 16-18 or Eqs. 16-19 can be used for the fixed-axis rotation of a symmetric rigid body. Many instructors prefer to use Eqs. 16-19, since they are more general and can be used for translation, fixed-axis rotation, and general plane motion. However, many others prefer to use Eqs. 16-18 for fixed-axis rotation, since the moment equation does not involve the unknown force components at the axis of rotation and the equations are usually a little simpler to solve.

Although either Eqs. 16-18 or Eqs. 16-19 can be used for the fixed-axis rotation of a symmetric rigid body, *Eqs. 16-18 are not valid for the general motion of a symmetric rigid body* and Eqs. 16-19 must be used. For those problems in which it is more convenient to sum moments about a point other than the mass center, the concept of equivalent force-couple systems can be used. For example, the bar shown in Fig. 16-27 moves in general plane motion. The free-body diagram of the bar is drawn in Fig. 16-28a. In Fig. 16-28b the forces on the free-body diagram have been reduced to an equivalent force-couple system. However, the equivalent force-couple system of Fig. 16-28b is equipollent to (has the same vector components and the same moment as) the inertia vectors shown in Fig. 16-28c. Clearly, summing vector components in the x- and y-directions on Figs. 16-28a and c and setting them equal satisfies the equations of motion for the rod (Eq. 16-19a). Also, summing moments about the mass center G in Figs. 16-28a and c clearly satisfies the equations of motion for the rod (Eq. 16-19b). However, summing mo-

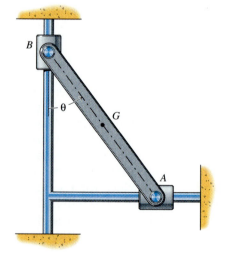

Figure 16-27   The ends of the rod *AB* slide without friction on the horizontal and vertical rods.

ments about point B in Figs. 16-28a and c also satisfies the general form of the equations of motion (Eq. 16-3), since

$$+\curvearrowleft\Sigma M_B = 2bA - bW = b(ma_{Gy}) + c(ma_{Gx}) + I_{Gz}\alpha \qquad (h)$$

and the acceleration of the mass center is

$$\begin{aligned}
\mathbf{a}_G &= \mathbf{a}_B + \mathbf{a}_{G/B} = \mathbf{a}_B + r_{G/B}\alpha\mathbf{e}_t + r_{G/B}\omega^2\mathbf{e}_n \\
&= \mathbf{a}_B + r_{G/B}\alpha\,(c/r_{GB}\mathbf{i} + b/r_{G/B}\mathbf{j}) + r_{G/B}\omega^2(-b/r_{G/B}\mathbf{i} + c/r_{G/B}\mathbf{j})
\end{aligned}$$

Therefore,

$$\begin{aligned}
a_{Gx} &= a_{Bx} + c\alpha - b\omega^2 \\
a_{Gy} &= a_{By} + b\alpha + c\omega^2
\end{aligned}$$

and Eq. h becomes

$$\begin{aligned}
\Sigma M_B = 2bA - bW &= bm(a_{By} + b\alpha + c\omega^2) + cm(a_{Bx} + c\alpha - b\omega^2) + I_{Gz}\alpha \\
&= bma_{By} + cma_{Bx} + [I_{Gz} + m(b^2 + c^2)]\alpha \\
&= b(ma_{By}) + c(ma_{Bx}) + I_{Bz}\alpha
\end{aligned}$$

which is identical to Eq. 16-3c, since $b = x_G$ and $c = -y_G$.

Equations 16-19 represent three scalar equations and can be solved for no more than three unknowns. Generally, however, they will contain more than three unknowns, so additional equations will be necessary. The additional equations are the kinematic relations between the acceleration of the mass center $\mathbf{a}_G$ and the angular acceleration α given by the relative velocity and relative acceleration equations (Eq. 14-9c and 14-10c)

$$\begin{aligned}
\mathbf{v}_G &= \mathbf{v}_A + \mathbf{v}_{G/A} \\
\mathbf{a}_G &= \mathbf{a}_A + \mathbf{a}_{G/A}
\end{aligned}$$

The solution of general plane motion problems consists of a combination of Eqs. 16-19 and the kinematic relationships as illustrated in the following examples.

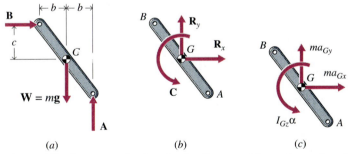

(a)   (b)   (c)

Figure 16-28   The forces on the free-body diagram (a) can be replaced with an equivalent force-couple at the mass center (b) and are equipollent to the inertia vectors shown on the kinetic diagram (c).

EXAMPLE PROBLEM 16-8

The rod AB shown in Fig. 16-29 has a constant cross section and a mass of 10 kg. As a result of the rotation of crank C, rod AB oscillates in a vertical plane. In the position shown, its angular velocity ω is 10 rad/s clockwise and its angular acceleration α is 40 rad/s² counterclockwise. Determine the forces exerted on the rod AB by the lightweight link DG and by the pin at support A.

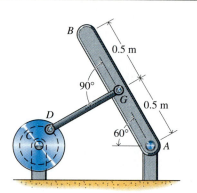

Fig. 16-29

SOLUTION

The motion of bar AB is a fixed-axis rotation about pin A. Although this problem could be solved using the special equations of motion for fixed-axis rotation (see Example Problem 5), it can also be solved using the more general equations of motion, Eqs. 16-19. The free-body diagram of the rod is shown in Fig. 16-30a.

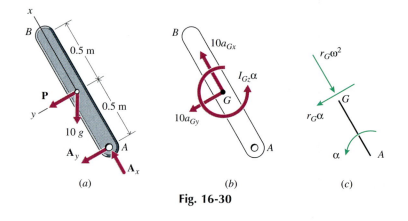

Fig. 16-30

The forces on the free-body diagram are equipollent to the inertia vectors shown on the kinetic diagram of Fig. 16-30b, and the equations of motion for the rod (Eqs. 16-19) are

$$+\searrow\Sigma F_x = ma_{Gx}: \qquad A_x - 10(9.81)\sin 60° = 10a_{Gx} \qquad (a)$$
$$+\nearrow\Sigma F_y = ma_{Gy}: \qquad A_y + 10(9.81)\cos 60° + P = 10a_{Gy} \qquad (b)$$
$$+\curvearrowleft\Sigma M_{Gz} = I_{Gz}\alpha: \qquad -0.5A_y = I_{Gz}(40) \qquad (c)$$

where the moment of inertia of the rod about the mass center G is

$$I_{Gz} = \frac{1}{12}mL^2 = \frac{1}{12}(10)(1)^2 = 0.83333 \text{ kg} \cdot \text{m}^2$$

and the acceleration of the mass center of the rod is (Fig. 16-30c)

$$\mathbf{a}_G = \mathbf{a}_A + \mathbf{a}_{G/A} = 0 + r_G\alpha\mathbf{e}_t + r_G\omega^2\mathbf{e}_n$$
$$= (0.5)(40)\mathbf{j} + 0.5(10^2)(-\mathbf{i}) = -50\mathbf{i} + 20\mathbf{j} \text{ m/s}^2 \qquad (d)$$

Therefore, $a_{Gx} = -50$ m/s², $a_{Gy} = 20$ m/s², and

$$A_x = -415.0 \text{ N} \qquad A_y = -66.66 \text{ N}$$
$$\mathbf{A} = 420 \text{ N} \searrow 50.87° \qquad\qquad \text{Ans.}$$
$$P = 218 \text{ N} \nearrow 30° \qquad\qquad \text{Ans.}$$

The equations of motion could also have been written by summing moments about pin A in the free-body diagram of Fig. 16-30a and in the kinetic

> It is important to make sure that the positive directions for a_{Gx}, a_{Gy}, and α are the same in the equations of motion (Eqs. a, b, and c), in the kinetic diagram (Fig. 16-30b), in the kinematic diagram (Fig. 16-30c), and in the relative acceleration equation (Eq. d).

diagram of Fig. 16-30b. This would give the equations of motion

$+\searrow\Sigma F_x = ma_{Gx}$: $A_x - 10(9.81)\sin 60° = 10a_{Gx}$ (a)
$+\nearrow\Sigma F_y = ma_{Gy}$: $A_y + 10(9.81)\cos 60° + P = 10a_{Gy}$ (b)
$+\curvearrowleft\Sigma M_{Az} = I_{Gz}\alpha + 0.5(10a_{Gy})$:
    $0.5[P + 10(9.81)\cos 60°] = 0.8333(40) + 0.5(10)(20)$
    $= 133.333$ (e)

which are identical to the equations used to solve Example Problem 16-5.

EXAMPLE PROBLEM 16-9

The slender bar AB shown in Fig. 16-31 has a uniform cross section and weighs 50 lb. It is fastened to collars at ends A and B that slide on smooth horizontal and vertical rods. When the bar is in the position shown, the collar at A has a velocity of 5 ft/s to the right and is accelerating at a rate of 4 ft/s². Determine the force **F**, the angular velocity $\boldsymbol{\omega}$ and the angular acceleration $\boldsymbol{\alpha}$ of the bar, and the forces exerted on the bar by the pins at A and B.

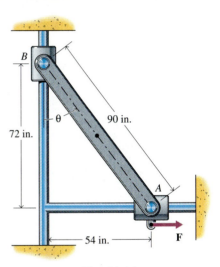

SOLUTION

As collar A moves to the right, the bar AB rotates counterclockwise. Since there is no fixed axis of rotation for the bar, Eqs. 16-19 must be used. A free-body diagram of the bar is shown in Fig. 16-32a. Since the horizontal and vertical rods are smooth, the forces exerted on the bar by the rods must be normal to the rods. The forces on the free-body diagram are equipollent to the inertia vectors shown on the kinetic diagram of Fig. 16-32b, and the equations of motion for the bar (Eqs. 16-19) are

$+\rightarrow\Sigma F_x = ma_{Gx}$: $B_x + F = \dfrac{50}{32.2}a_{Gx}$ (a)

$+\uparrow\Sigma F_y = ma_{Gy}$: $A_y - 50 = \dfrac{50}{32.2}a_{Gy}$ (b)

$+\curvearrowleft\Sigma M_{Gz} = I_{Gz}\alpha$: $(F - B_x)(7.5\cos\theta) + A_y(7.5\sin\theta) = I_{Gz}\alpha$ (c)

Fig. 16-31

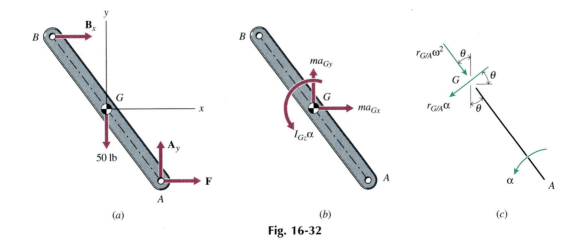

(a)   (b)   (c)

Fig. 16-32

285

where $L = 90$ in. $= 7.5$ ft and

$$I_{Gz} = \frac{1}{12}mL^2 = \frac{1}{12}\left(\frac{50}{32.2}\right)(7.5)^2 = 7.279 \text{ slug} \cdot \text{ft}^2$$

The three equations of motion contain five unknowns (a_{Gx}, a_{Gy}, α, F, and N), so two more equations are needed to complete the solution of this problem. These two equations come from the kinematics describing the motion of the sliders and the motion of the mass center relative to the sliders. Relative to coordinates centered at the junction of the horizontal and vertical rods, the positions of the sliders are

$$x_A = 7.5 \sin \theta \qquad y_B = 7.5 \cos \theta$$

Taking two time derivatives of x_A gives

$$v_A = \dot{x}_A = 7.5\dot{\theta} \cos \theta = 7.5\omega \cos \theta$$
$$a_A = \dot{v}_A = 7.5\dot{\omega} \cos \theta + 7.5\omega\,(-\dot{\theta} \sin \theta) = 7.5\alpha \cos \theta - 7.5\omega^2 \sin \theta$$

Therefore, at the given position, $\sin \theta = 0.6$, $\cos \theta = 0.8$, $x_A = 54$ in. $= 4.5$ ft, $v_A = 5$ ft/s, $a_A = 4$ ft/s^2, and

$$\omega = 0.8333 \text{ rad/s}\nwarrow \qquad \text{Ans.}$$
$$\alpha = 1.1875 \text{ rad/s}^2\nwarrow \qquad \text{Ans.}$$

Then using the relative acceleration equation and Fig. 16-32c

$$\mathbf{a}_G = \mathbf{a}_A + \mathbf{a}_{G/A} = \mathbf{a}_A + r_{G/A}\alpha\mathbf{e}_t + r_{G/A}\omega^2\mathbf{e}_n$$
$$= 4\mathbf{i} + 3.75\alpha(-\cos\theta\mathbf{i} - \sin\theta\mathbf{j}) + 3.75\omega^2(\sin\theta\mathbf{i} - \cos\theta\mathbf{j})$$
$$= 2.000\mathbf{i} - 4.755\mathbf{j} \text{ ft/s}^2 \qquad (d)$$

Combining Eqs. b and d gives

$$A_y = 42.616 \text{ lb} \cong 42.6 \text{ lb}\uparrow \qquad \text{Ans.}$$

while Eqs. a, c, and d give

$$B_x + F = 3.106 \qquad (a')$$
$$F - B_x = -29.08 \qquad (c')$$

Solving these two equations simultaneously gives

$$B_x = 16.093 \text{ lb} \cong 16.09 \text{ lb}\rightarrow \qquad \text{Ans.}$$
$$F = -12.99 \text{ lb} = 12.99 \text{ lb}\leftarrow \qquad \text{Ans.}$$

Relationships between the velocity and acceleration of collar A and the angular velocity and angular acceleration of the bar could also have been obtained using the relative velocity and relative acceleration equations between points A and B. For example, the relative velocity equation

$$v_B\mathbf{j} = v_A\mathbf{i} + 7.5\omega(-\cos\theta\mathbf{i}\ \sin\theta\mathbf{j})$$

would give

$$v_A = 7.5\omega \cos \theta$$

which is identical to the result obtained using the absolute motion approach.

EXAMPLE PROBLEM 16-10

A 10-kg uniform cylinder 150 mm in diameter rolls down an incline as shown in Fig. 16-33. If the cylinder is released from rest and rolls without slipping, determine

a. The initial angular acceleration of the cylinder.
b. The minimum coefficient of friction necessary to prevent slipping.
c. The time required for the cylinder to roll 5 m down the incline.

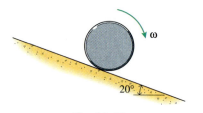

Fig. 16-33

SOLUTION

A free-body diagram of the cylinder is shown in Fig. 16-34a. The forces on the free-body diagram are equipollent to the inertia vectors shown on the kinetic diagram of Fig. 16-34b, and the equations of motion for the cylinder (Eqs. 16-19) are

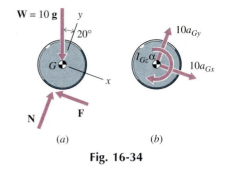

(a)   (b)

Fig. 16-34

$$+\searrow\Sigma F_x = ma_{Gx}: \qquad 10(9.81)\sin 20° - F = 10a_{Gx} \qquad (a)$$
$$+\nearrow\Sigma F_y = ma_{Gy}: \qquad N - 10(9.81)\cos 20° = 10a_{Gy} \qquad (b)$$
$$+\downarrow\Sigma M_{Gz} = I_{Gz}\alpha: \qquad 0.075F = I_{Gz}\alpha \qquad (c)$$

where

$$I_{Gz} = \frac{1}{2}mr^2 = \frac{1}{2}(10)(0.075)^2 = 0.02813 \text{ kg} \cdot \text{m}^2$$

The three equations of motion contain five unknowns (a_{Gx}, a_{Gy}, α, F, and N), so two more equations are needed to complete the solution of this problem. These come from the kinematics describing the motion of the mass center of the cylinder and the point of contact between the cylinder and the surface. The center of the cylinder moves along a straight line parallel to the surface. Therefore, $a_{Gy} = 0$ and the relative acceleration equation $\mathbf{a}_G = \mathbf{a}_A + \mathbf{a}_{G/A}$ gives

$$a_{Gx}\mathbf{i} = a_{Ay}\mathbf{j} + r\alpha\mathbf{i} + r\omega^2(-\mathbf{j})$$

and

$$a_{Gx} = r\alpha = 0.075\alpha \qquad (d)$$

a. Equation b can be solved immediately to get

$$N = 92.184 \text{ N}$$

Then, adding 0.075 times Eq. a to Eq. c gives

$$2.5164 = 0.08438\alpha$$
$$\alpha = 29.822 \text{ rad/s}^2 \cong 29.8 \text{ rad/s}^2\downarrow \qquad \text{Ans.}$$

b. Next, Eqs. d and a give the acceleration of the mass center and the friction required to prevent slipping

$$a_{Gx} = 2.237 \text{ m/s}^2$$
$$F = 11.185 \text{ N}$$

so that the minimum coefficient of friction required is

$$\mu_{\text{req}} = F/N = 0.1213 \qquad \text{Ans.}$$

c. Equations a, b, c, and d are equally valid for when the wheel is rolling as when it is just starting to roll. Nothing in the equations depends on the angular velocity of the wheel being zero. Therefore, the acceleration of the wheel is constant and it can be integrated with respect to time to get

$$v_{Gx} = 2.237t + C_1 = 2.237t \text{ m/s}$$
$$x_G = 1.118t^2 + C_2 = 1.118t^2 \text{ m}$$

where the constants of integration $C_1 = C_2 = 0$, since the cylinder starts from rest. Finally, when $x_G = 5$ m,

$$t = 2.11 \text{ s} \qquad \text{Ans.}$$

The equations of motion could also have been written summing moments about the contact point A using the free-body diagram of Fig.

16-34a and the kinetic diagram of Fig. 16-34b. This would give the equations

$$+\nwarrow\Sigma F_x = ma_{Gx}: \qquad 10(9.81)\sin 20° - F = 10a_{Gx} \qquad (a)$$
$$+\nearrow\Sigma F_y = ma_{Gy}: \qquad N - 10(9.81)\cos 20° = 10a_{Gy} \qquad (b)$$
$$+\!\!\!\downarrow\!\Sigma M_{Az} = I_{Gz}\alpha + 0.075(10a_{Gx}):$$
$$0.075[(10)(9.81)\sin 20°] = I_{Gz}\alpha + 0.075(10a_{Gx}) \qquad (e)$$

where $I_{Gz} = 0.02813$ kg · m² and $a_{Gx} = 0.075\alpha$. The advantage of this approach is that Eq. e has only one unknown and can be solved immediately. The rest of the solution is exactly as before.

EXAMPLE PROBLEM 16-11

A block and a spool are supported by cables wound around the spool as shown in Fig. 16-35. The block weighs 95 lb; the spool weighs 50 lb and has a radius of gyration of 4 in. with respect to its mass center. If the system is released from rest in the position shown, determine the acceleration of the mass center G of the spool and the tensions in the three cables.

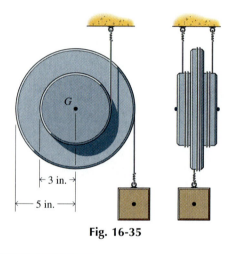

3 in.

5 in.

Fig. 16-35

SOLUTION

Free-body diagrams of the block and spool are shown in Fig. 16-36a. The forces on the free-body diagrams are equipollent to the inertia vectors shown on the kinetic diagrams of Fig. 16-36b. First write the equations of motion for the block (Eqs. 16-19)

$$+\rightarrow\Sigma F_x = ma_{Gx}: \qquad 0 = m_b a_{Gxb}$$
$$+\uparrow\Sigma F_y = ma_{Gy}: \qquad T_2 - 95 = \frac{95}{32.2}a_{Gyb}$$
$$+\!\!\!\uparrow\!\Sigma M_{Gz} = I_{Gz}\alpha: \qquad 0 = I_{Gzb}\alpha_b$$

Therefore, the motion of the block is a translation ($\boldsymbol{\omega} = \boldsymbol{\alpha} = \mathbf{0}$) in the y-direction ($a_{Gxb} = 0$), and

$$T_2 - 2.950a_{Gyb} = 95 \qquad (a)$$

As the block moves up 2 in., the spool rotates and its mass center moves down 3 in. Since the motion of the block and the spool are different, separate free-body diagrams must be drawn of the block and the spool.

288

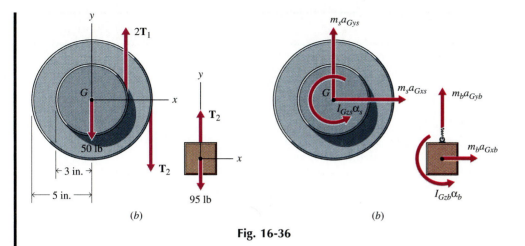

(b)   (b)

Fig. 16-36

Similarly, the equations of motion for the spool (Eqs. 16-19) are

$+\rightarrow\Sigma F_x = ma_{Gx}$: $0 = m_s a_{Gxs}$

$+\uparrow\Sigma F_y = ma_{Gy}$: $2T_1 - T_2 - 50 = \dfrac{50}{32.2}a_{Gys}$

$+\curvearrowleft\Sigma M_{Gz} = I_{Gz}\alpha$: $2T_1(3/12) - T_2(5/12) = I_{Gzs}\alpha_s$

where

$$I_{Gzs} = mk^2 = \left(\dfrac{50}{32.2}\right)\left(\dfrac{4}{12}\right)^2 = 0.17253 \text{ slug} \cdot \text{ft}^2$$

Therefore, $a_{Gxs} = 0$ and

$$2T_1 - T_2 - 1.5528a_{Gys} = 50 \qquad (b)$$
$$6T_1 - 5T_2 = 2.070\alpha_s \qquad (c)$$

The three equations a, b, and c still contain five unknowns (T_1, T_2, a_{Gyb}, a_{Gys}, and α_s), so two more equations are needed to complete the solution of this problem. These come from the kinematics describing the motion of the mass centers of the spool and the block. If the spool rotates counterclockwise (as assumed in the moment equation for the spool), then the mass center of the spool moves down, the block moves up, and

$$a_{Gyb} = (2/12)\alpha_s \qquad (d)$$
$$a_{Gys} = -(3/12)\alpha_s \qquad (e)$$

Substituting Eqs. d and e back into Eqs. a, b, and c gives

$$T_2 - 0.49167\alpha_s = 95 \qquad (a')$$
$$2T_1 - T_2 + 0.38820\alpha_s = 50 \qquad (b')$$
$$6T_1 - 5T_2 - 2.070\alpha_s = 0 \qquad (c')$$

Solving these three equations simultaneously gives

$$T_1 = 72.009 \text{ lb} \cong 72.0 \text{ lb} \qquad\qquad \text{Ans.}$$
$$T_2 = 90.337 \text{ lb} \cong 90.3 \text{ lb} \qquad\qquad \text{Ans.}$$
$$\alpha_s = -9.4833 \text{ rad/s}^2 \cong 9.48 \text{ rad/s}^2 \downcurvearrow$$
$$a_{Gyb} = -1.581 \text{ ft/s}^2 = 1.581 \text{ ft/s}^2 \downarrow$$
$$a_{Gys} = 2.37 \text{ ft/s}^2 \uparrow \qquad\qquad\qquad \text{Ans.}$$

Note that the tension T_2 is not equal to the weight of the block since the acceleration of the block a_{Gb} is not zero. It is the difference between the tension and the weight that provides the force which causes the block to accelerate.

A solid homogeneous cylinder with a mass of 100 kg rests on an inclined surface as shown in Fig. 16-37. The coefficient of friction between the cylinder and the inclined surface is 0.40. A cable wrapped around a shallow groove in the cylinder connects it to a block, which has a mass of 75 kg. The pulley that the cable passes over has a mass of 10 kg. If the system is released from rest in the position shown, determine the acceleration of the mass center G of the block, the acceleration of the mass center G of the cylinder, and the tensions in the two parts of the cable.

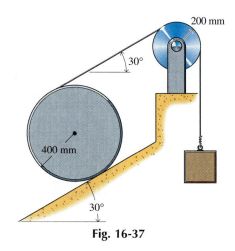

Fig. 16-37

SOLUTION

Free-body diagrams of the cylinder, block, and pulley are shown in Fig. 16-38a. The forces on the free-body diagrams are equipollent to the inertia vectors shown on the kinetic diagrams in Fig. 16-38b. First write the equations of motion for the block (Eqs. 16-19)

$$+\rightarrow\Sigma F_x = ma_{Gx}: \qquad\qquad 0 = m_b a_{Gxb}$$
$$+\uparrow\Sigma F_y = ma_{Gy}: \qquad T_2 - 735.75 = 75a_{Gyb}$$
$$+\circlearrowleft\Sigma M_{Gz} = I_{Gz}\alpha: \qquad\qquad 0 = I_{Gzb}\alpha_b$$

Therefore, the motion of the block is a translation ($\boldsymbol{\omega} = \boldsymbol{\alpha} = \mathbf{0}$) in the y-direction ($a_{Gxb} = 0$), and

$$T_2 - 75a_{Gyb} = 735.75 \qquad\qquad (a)$$

Similarly, the equations of motion for the cylinder (Eqs. 16-19) are

$$+ \nearrow\Sigma F_x = ma_{Gx}: \quad T_1 + F - 981\sin 30° = 100a_{Gxc} \qquad (b)$$
$$+ \nwarrow\Sigma F_y = ma_{Gy}: \quad N - 981\cos 30° = 0 \qquad\qquad (c)$$
$$+ \circlearrowleft\Sigma M_{Gz} = I_{Gz}\alpha: \qquad 0.4(F - T_1) = I_{Gzc}\alpha_c \qquad (d)$$

where the center of the cylinder moves in a straight line parallel to the inclined surface ($a_{Gyc} = 0$) and

$$I_{Gzc} = \frac{1}{2}mr^2 = \frac{1}{2}(100)(0.4)^2 = 8.000 \text{ kg} \cdot \text{m}^2$$

Note that the tension T_2 is not equal to the weight of the block since the acceleration of the block a_{Gyb} is not zero. Also, the tensions T_1 and T_2 are not equal since the mass of the pulley is not zero. It is the difference between the tensions that provides the moment which causes the mass of the pulley to rotate with an angular acceleration.

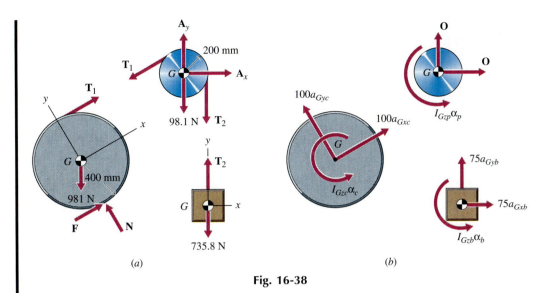

Fig. 16-38

The pulley rotates about a fixed axis through its mass center. Of the equations of motion (Eqs. 16-19), only the moment equation provides useful information in this problem

$$+\circlearrowleft\Sigma M_{Gz} = I_{Gz}\alpha:\qquad 0.2(T_1 - T_2) = I_{Gzp}\alpha_p \qquad (e)$$

where

$$I_{Gzp} = \frac{1}{2}mr^2 = \frac{1}{2}(10)(0.2)^2 = 0.2000 \text{ kg} \cdot \text{m}^2$$

The five equations a, b, c, d, and e still contain eight unknowns (F, N, T_1, T_2, a_{Gyb}, a_{Gxc}, α_c, and α_p), so three more equations are needed to complete the solution of this problem. These come from the kinematics describing the motion of the mass centers of the cylinder and the block. If the cylinder moves down the incline, then the cylinder and the pulley both rotate counterclockwise and the block moves up. If the cable does not slip on the pulley, then the acceleration of the cable (which is the same as the acceleration of the block) is the same as the tangential acceleration of a point on the rim of the pulley

$$a_{\text{cable}} = a_{GyB} = r_p\alpha_p = 0.2\alpha_p \qquad (f)$$

If the cylinder does not slip on the inclined surface, then the acceleration of the cable and the mass center of the cylinder can be written

$$a_{\text{cable}} = a_{GyB} = 0.8\alpha_c \qquad (g)$$
$$a_{Gxc} = -0.4\alpha_c \qquad (h)$$

Finally, Eqs. f, g, and h must be substituted into Eqs. a through e and the resulting equations solved. Equation c may be solved directly for the normal force

$$N = 849.57 \text{ N}$$

and the other four equations can be written (using $\alpha_p = 5a_{Gyb}$, $\alpha_c = 1.25a_{Gyb}$, and $a_{Gxc} = -0.4\alpha_c = -0.5a_{Gyb}$)

$$T_2 - 75a_{Gyb} = 735.75 \qquad (a')$$
$$T_1 + F + 50a_{Gyb} = 490.5 \qquad (b')$$
$$0.4(F - T_1) - 10a_{Gyb} = 0 \qquad (d')$$
$$0.2(T_1 - T_2) - 1a_{Gyb} = 0 \qquad (e')$$

Solving these four equations simultaneously gives

$$T_1 = 401.79 \text{ N} \cong 402 \text{ N} \qquad \text{Ans.}$$
$$T_2 = 422.66 \text{ N} \cong 423 \text{ N} \qquad \text{Ans.}$$
$$F = 297.43 \text{ N} \cong 297 \text{ N}$$
$$a_{Gyb} = -4.1745 \text{ m/s}^2 \cong 4.17 \text{ m/s}^2 \downarrow \qquad \text{Ans.}$$
$$a_{Gxc} = +2.0872 \text{ m/s}^2 \cong 2.09 \text{ m/s}^2 \nearrow \qquad \text{Ans.}$$
$$\alpha_c = -5.2181 \text{ rad/s}^2 \cong 5.22 \text{ rad/s}^2 \downarrow$$
$$\alpha_p = -20.872 \text{ rad/s}^2 \cong 20.9 \text{ rad/s}^2 \downarrow$$

A check is required to make sure that the coefficient of friction is sufficient to develop the frictional force required to prevent slipping between the cylinder and the inclined surface. The maximum available friction

$$F_{\text{avail}} = \mu N = 0.40(849.57) = 339.8 \text{ N}$$

is greater than $F = 297.43 \text{ N} = F_{\text{req}}$, so slipping will not occur. The actual friction force provided by the surface is $F = 297.43 \text{ N} = F_{\text{req}}$, not the maximum amount of friction available.

If F_{avail} had been less than F_{req}, then slipping would occur and the problem would have to be re-solved with $F = F_{\text{avail}} = \mu N$ in place of Eq g and using $0.4\alpha_c - a_{Gxc} = a_{Gyb}$ in place of Eq. h.

EXAMPLE PROBLEM 16-13

A system consisting of a flywheel, a connecting rod, and a piston is shown in Fig. 16-39. The connecting rod AB has a uniform cross section and a mass of 10 kg. The mass of the piston is 15 kg. A couple $\mathbf{T}$ rotates the flywheel counterclockwise at a constant angular velocity of 500 rev/min. Determine the vertical and horizontal components of the forces exerted on rod AB by the pins at A and B when $\theta = 60°$. Neglect friction between the cylinder wall and the piston.

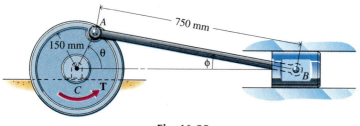

Fig. 16-39

SOLUTION

Free-body diagrams of the rod AB and the piston are shown in Fig. 16-40a. The forces on the free-body diagrams are equipollent to the inertia vectors shown on the kinetic diagram of Fig. 16-40b. First write the equations of motion for the rod (Eqs. 16-19)

$$+ \rightarrow \Sigma F_x = ma_{Gx}: \qquad\qquad B_x - A_x = 10a_{Gxr} \quad (a)$$
$$+ \uparrow \Sigma F_y = ma_{Gy}: \qquad\qquad A_y - B_y - 10(9.81) = 10a_{Gyr} \quad (b)$$
$$+ \circlearrowleft \Sigma M_{Gz} = I_{Gz}\alpha: \qquad 0.06495(A_x + B_x) - 0.3693(A_y + B_y) = I_{Gzr}\alpha_r \quad (c)$$

where

$$I_{Gzr} = \frac{1}{12}mL^2 = \frac{1}{12}(10)(0.750)^2 = 0.46875 \text{ kg} \cdot \text{m}^2$$

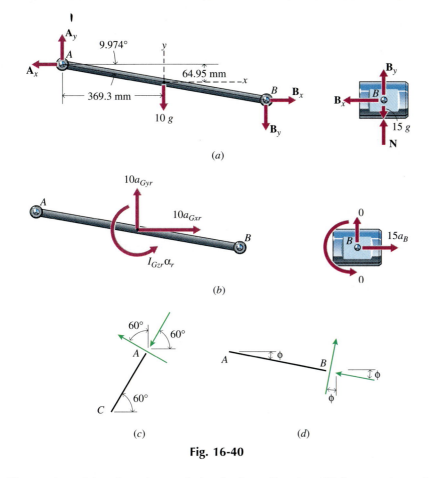

(a)

(b)

(c)   (d)

Fig. 16-40

The motion of the piston is translation in the x-direction. Of the equations of motion (Eqs. 16-19), only the x-component of the force equation provides useful information in this problem

$$+\rightarrow\Sigma F_x = ma_{Gx}: \qquad -B_x = 15a_B \qquad (d)$$

The four equations a, b, c, and d contain eight unknowns (A_x, A_y, B_x, B_y, a_{Gxr}, a_{Gyr}, a_B, and α_r), so four more equations are needed to complete the solution of this problem. These come from the kinematics describing the motion of the flywheel and the mass centers of the rod and the piston. Since the flywheel is rotating with a constant angular velocity ($\omega_f = 500$ rev/min = 52.36 rad/s, $\alpha_f = 0$ rad/s^2), the velocity and acceleration of the pin A are obtained using the relative velocity and relative acceleration equations. For the flywheel (and referring to Fig. 16-40c)

$$\mathbf{v}_A = \mathbf{v}_C + \mathbf{v}_{A/C} = \mathbf{0} + r_{A/C}\omega_f\mathbf{e}_t$$
$$= 0.150(52.36)(-\sin 60°\mathbf{i} + \cos 60°\mathbf{j})$$
$$= -6.802\mathbf{i} + 3.927\mathbf{j} \text{ m/s}$$

$$\mathbf{a}_A = \mathbf{a}_C + \mathbf{a}_{A/C} = \mathbf{0} + r_{A/C}\alpha_f\mathbf{e}_t + r_{A/C}\omega_f^2\mathbf{e}_n$$
$$= 0.150(52.36)^2(-\cos 60°\mathbf{i} - \sin 60°\mathbf{j})$$
$$= -205.62\mathbf{i} - 356.14\mathbf{j} \text{ m/s}^2$$

When the flywheel is at $\theta = 60°$, the rod makes an angle $\phi = \sin^{-1} 129.9/750 = 9.974°$ with the horizontal. Then for the connecting rod (and referring to Fig. 16-40d) the relative velocity equation is

$$\mathbf{v}_B = \mathbf{v}_A + \mathbf{v}_{B/A} = \mathbf{v}_A + r_{B/A}\omega_r\mathbf{e}_t$$
$$v_B\mathbf{i} = (-6.802\mathbf{i} + 3.927\mathbf{j}) + 0.750\omega_r(\sin \phi\mathbf{i} + \cos \phi\mathbf{j})$$
$$= (-6.802 + 0.12990\omega_r)\mathbf{i} + (3.927 + 0.73867\omega_r)\mathbf{j}$$

The x- and y-components of this equation give

$$\omega_r = -3.927/0.73867 = -5.316 \text{ rad/s} = 5.136 \text{ rad/s} \downarrow$$
$$v_B = [-6.802 + 0.12990(-5.316)] = -7.493 \text{ m/s} = 7.493 \text{ m/s} \leftarrow$$

Similarly, the relative acceleration equation is

$$\mathbf{a}_B = \mathbf{a}_A + \mathbf{a}_{B/A} = \mathbf{a}_A + r_{B/A}\alpha_r\mathbf{e}_t + r_{B/A}\omega_r^2\mathbf{e}_n$$
$$a_B\mathbf{i} = (-205.62\mathbf{i} - 356.14\mathbf{j}) + 0.750\alpha_r(\sin \phi\mathbf{i} + \cos \phi\mathbf{j})$$
$$+ 0.750(5.316)^2(-\cos \phi\mathbf{i} + \sin \phi\mathbf{j})$$
$$= (-226.49 + 0.12990\alpha_r)\mathbf{i} + (-352.47 + 0.73867\alpha_r)\mathbf{j}$$

The x- and y-components of this equation give

$$\alpha_r = 352.47/0.73867 = 477.17 \text{ rad/s}^2 \curvearrowleft \qquad (e)$$
$$a_B = [-226.49 + 0.12990(477.17)]$$
$$= -164.41 \text{ m/s}^2 = 164.41 \text{ m/s}^2 \leftarrow \qquad (f)$$

Finally, the acceleration of the mass center of the rod is

$$\mathbf{a}_{Gr} = \mathbf{a}_A + \mathbf{a}_{G/A} = \mathbf{a}_A + r_{G/A}\alpha_r\mathbf{e}_t + r_{G/A}\omega_r^2\mathbf{e}_n$$
$$= (-205.62\mathbf{i} - 356.14\mathbf{j}) + 0.375(477.17)(\sin \phi\mathbf{i} + \cos \phi\mathbf{j})$$
$$+ 0.375(5.316)^2(-\cos \phi\mathbf{i} + \sin \phi\mathbf{j})$$
$$= -185.07\mathbf{i} - 178.07\mathbf{j} \text{ m/s}^2 \qquad (g)$$

Substituting the accelerations (Eqs. e, f, and g) into the equations of motion (Eqs. a, b, c, and d) gives

$$B_x - A_x = -1850.7 \qquad (a')$$
$$A_y - B_y = -1682.60 \qquad (b')$$
$$(A_x + B_x) - 5.686(A_y + B_y) = 3443.8 \qquad (c')$$
$$B_x = 2466.2 \qquad (d')$$

Solving these four equations simultaneously gives

$$A_x = 4316.9 \text{ N} \cong 4320 \text{ N} \leftarrow \qquad \text{Ans.}$$
$$B_x = 2466.2 \text{ N} \cong 2470 \text{ N} \rightarrow \qquad \text{Ans.}$$
$$A_y = -547.7 \text{ N} \cong 548 \text{ N} \downarrow \qquad \text{Ans.}$$
$$B_y = 1134.9 \text{ N} \cong 1135 \text{ N} \downarrow \qquad \text{Ans.}$$

> Note that neither the force **A** nor the force **B** act along the connecting rod AB. The force **A** makes an angle of 7.23° with the horizontal while the force **B** makes an angle of 24.68° with the horizontal.

PROBLEMS

In the following problems, all ropes, cords, and cables are assumed to be flexible, inextensible, and of negligible mass. All pins and pulleys have negligible mass and are frictionless unless specified otherwise.

Introductory Problems

16-51* A 10-lb solid homogeneous sphere rolls without slipping down a plane that is inclined at an angle of 28° with the horizontal (Fig. P16-51). Determine the acceleration $\mathbf{a}_G$ of the mass center of the sphere and the minimum coefficient of friction μ_s required to prevent slipping.

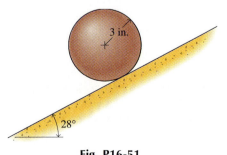

Fig. P16-51

16-52* A 15-kg solid circular disk rolls without slipping on the inclined surface shown in Fig. P16-52. In the position shown, the angular velocity of the disk is 10 rad/s clockwise. Determine the angular acceleration α of the disk and the minimum coefficient of friction μ_s required to prevent slipping.

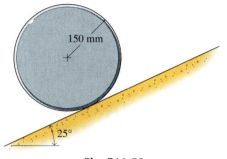

Fig. P16-52

16-53 Two 16-in.-diameter disks and an 8-in.-diameter cylinder are fastened together to form a spool as shown in Fig. P16-53. The 50-lb spool has a radius of gyration of 5 in. with respect to the axis of the spool and rolls without slipping on the horizontal surface. If a force **P** of 50 lb is applied to the spool through the cord that is wrapped around the cylinder, determine the acceleration $\mathbf{a}_G$ of the mass center of the spool and the minimum coefficient of friction μ_s required to prevent slipping.

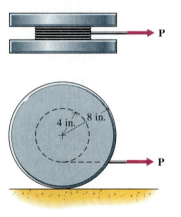

Fig. P16-53

16-54* Two 400-mm-diameter disks and a 240-mm-diameter disk are joined to form a spool as shown in Fig. P16-54. The 125-kg spool has a radius of gyration of 125 mm with respect to the axis of the spool. If a force **P** of 500 N is applied to the spool through a cord that is wrapped around the 240-mm disk, determine the acceleration $\mathbf{a}_G$ of the mass center of the spool and the angular acceleration α of the spool if

a. The horizontal surface is smooth ($\mu = 0$).
b. The horizontal surface is rough ($\mu = 0.25$).

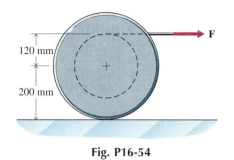

Fig. P16-54

16-55* A spool is supported by a cord wrapped around the inner core of the spool as shown in Fig. P16-55. The 10-lb spool has a radius of gyration of 4 in. with respect to its mass center G. When the spool is released to move in a vertical plane, determine the angular acceleration α of the spool and the tension T in the cord.

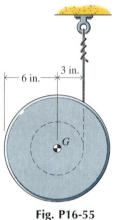

Fig. P16-55

16-56 The spoked wheel shown in Fig. P16-56 rolls without slipping on a horizontal surface. Because it has lost a couple of its spokes, the center of mass of the 12-kg wheel is 50 mm away from its middle and the radius of gyration relative to the mass center is 0.6 m. When $\theta = 30°$, the center of the wheel has a speed of 3 m/s to the right. For this instant, determine the angular acceleration of the wheel and the force acting on the wheel at its point of contact with the horizontal surface.

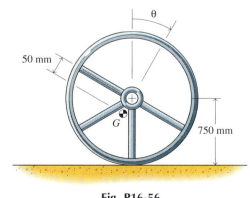

Fig. P16-56

16-57* The unbalanced wheel shown in Fig. P16-57 rolls without slipping down the inclined surface. The 32-lb wheel has a radius of gyration of 4.25 in. with respect to its axis of rotation. When the wheel is in the position shown, it has a counterclockwise angular velocity of 5 rad/s. At this instant, determine the angular acceleration of the wheel and the normal and frictional forces exerted on the wheel by the inclined surface.

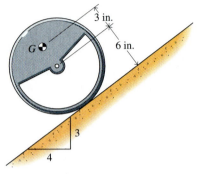

Fig. P16-57

16-58 The 8-kg homogeneous half-cylinder of Fig. P16-58 has a diameter of 500 mm. The cylinder is released from rest when $\theta = 30°$ and rolls without slipping on the horizontal surface. At this instant, determine the angular acceleration of the cylinder and the force acting on the cylinder at its point of contact with the horizontal surface.

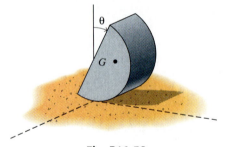

Fig. P16-58

16-59 The 3-lb, 16-in.-diameter uniform circular disk of Fig. P16-59 rolls without slipping on a horizontal surface. A 4-lb slender rod penetrates the disk 7 in. from its center. When $\theta = 60°$, the disk has a clockwise angular velocity of 5 rad/s. At this instant, determine the angular acceleration of the disk and the force acting on the disk at its point of contact with the horizontal surface.

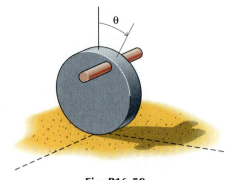

Fig. P16-59

16-60 The compound wheel of Fig. P16-60 consists of a 5-kg half-circular disk sandwiched between two circular 2-kg disks, each 600 mm in diameter. The wheel rolls without slipping and has an angular velocity of 10 rad/s clockwise (as viewed from the right end) when $\theta = 25°$. At this instant, determine the angular acceleration of the wheel and the force acting on the wheel at its point of contact with the horizontal surface.

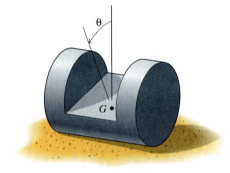

Fig. P16-60

Intermediate Problems

16-61* The 100-lb spool A shown in Fig. P16-61 has a radius of gyration of 4.75 in. with respect to its mass center G. A cord connects the spool to a 25-lb block B, which rests on an inclined surface. The kinetic coefficient of friction between the block and the inclined surface is $\mu_k = 0.10$. If the spool rolls without slipping on the horizontal surface, determine

a. The acceleration $\mathbf{a}_B$ of the block.
b. The tension T in the cable.
c. The minimum static coefficient of friction μ_s required to ensure rolling without slipping of the spool.

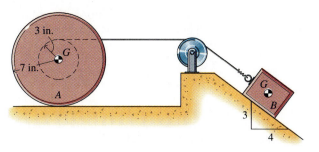

Fig. P16-61

16-62* The solid circular cylinder *A* shown in Fig. P16-62 has a radius *R* of 200 mm and a mass of 75 kg. A cord connects the cylinder to a 50-kg block *B*. If the cylinder rolls without slipping on the inclined surface, determine

a. The acceleration $\mathbf{a}_B$ of the block.
b. The tension *T* in the cable.

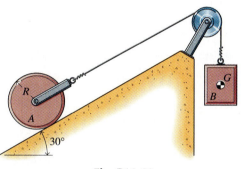

Fig. P16-62

16-63 A 500-lb block *A* is supported on an inclined surface by a 100-lb platform *B* and a pair of wheels *C* (one on the front side of the platform and one on the back side of the platform) as shown in Fig. P16-63. Each of the 100-lb wheels has a radius of gyration about the axis of rotation of 8.50 in. The static and kinetic coefficients of friction between the inclined surface and the contacting bodies are 0.20 and 0.15, respectively. During motion of the system, determine the acceleration of the mass center of the wheels and the normal and frictional forces exerted on the platform and wheels by the inclined surface.

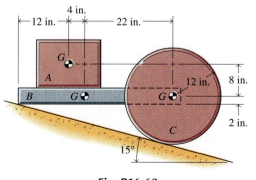

Fig. P16-63

16-64 A 200-mm-diameter cylinder *A* is mounted on a 50-mm-diameter axle as shown in Fig. P16-64. The mass of the cylinder and axle is 50 kg and the radius of gyration with respect to the axis of the axle is 70 mm. Flexible cords wrapped around the axle on both sides of the cylinder are connected to a 100-kg block that slides on a horizontal surface. The kinetic coefficient of friction between the horizontal surface and the block is 0.25. If the cylinder rolls without slipping, determine the acceleration of the block and the tensions in the two cords.

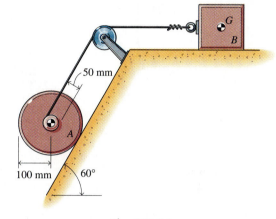

Fig. P16-64

16-65* A 12-in.-diameter disk mounted on a 4-in.-diameter shaft rolls without slipping on a pair of inclined rails. A cord wrapped around a shallow groove on the outer surface of the disk supports a 20-lb body *B* as shown in Fig. P16-65. The weight of the disk and shaft is 75 lb and the radius of gyration with respect to an axis through the mass center is 4.25 in. If the system is released from rest, determine

a. The initial tension *T* in the cable.
b. The acceleration $\mathbf{a}_B$ of body *B*.
c. The angular acceleration α of the disk.

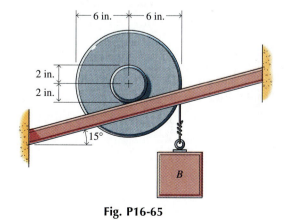

Fig. P16-65

16-66* A 100-kg cylinder A and a 50-kg body B are attached with cables to the compound cable sheave shown in Fig. P16-66. The sheave has a mass of 20 kg and a radius of gyration with respect to its axis of rotation of 110 mm. If the cylinder rolls without slipping on the inclined surface, determine the acceleration of body B and the tensions in the two cables.

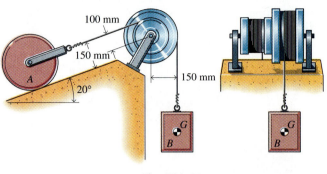

Fig. P16-66

16-67 A 15-lb stepped cylinder rolls on a pair of horizontal rails as shown in Fig. P16-67. The radius of gyration of the stepped cylinder is 8 in., and the static coefficient of friction between the axle and the rails is $\mu_s = 0.8$. If the system is released from rest with a stretch of 18 in. in spring 1 ($k_1 = 10$ lb/ft) and a stretch of 4 in. in spring 2 ($k_2 = 10$ lb/ft), verify that the cylinder does not slip when motion starts and determine

a. The angular acceleration of the cylinder when motion begins.
b. The force acting on the axle at its point of contact with the horizontal rails when motion begins.

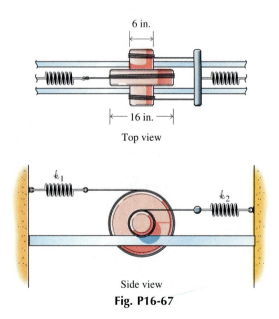

Top view

Side view
Fig. P16-67

16-68 A 12-kg stepped cylinder C has a radius of gyration of 150 mm and rolls without slipping on a horizontal surface as shown in Fig. P16-68. A spring having $k =$

2 kN/m is attached to a pair of cords wrapped around the axle of the stepped cylinder. A second cord wrapped around a shallow groove in the center part of the cylinder is attached to a 15-kg crate A. If the crate has a speed of 1.5 m/s downward when the spring is stretched 100 mm, determine

a. The acceleration $\mathbf{a}_A$ of the crate at this instant.
b. The force acting on the cylinder at its point of contact with the horizontal surface at this instant.

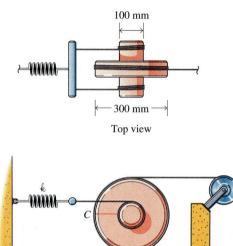

Top view

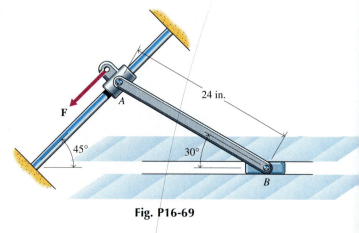

Side view
Fig. P16-68

16-69* The 25-lb bar AB shown in Fig. P16-69 has a uniform cross section. The collar at end A and the slider at end B have negligible mass and move on smooth surfaces. When in the position shown, bar AB has an angular velocity of 1 rad/s clockwise and an angular acceleration of 2 rad/s² counterclockwise. Determine the force $\mathbf{F}$ acting on the collar A and the force exerted by the slot on the slider at end B of the bar at this instant.

Fig. P16-69

16-70 Bars AB (m_{AB} = 20 kg) and BC (m_{BC} = 15 kg) of Fig. P16-70 have uniform cross sections. When the bars are in the position shown, collar C (which has negligible mass) is moving to the left with a velocity of 2 m/s and its speed is decreasing at a rate of 4 m/s^2. Determine the magnitude of the force $\mathbf{F}$ and the horizontal and vertical components of the force exerted on bar AB by the pins at A and B.

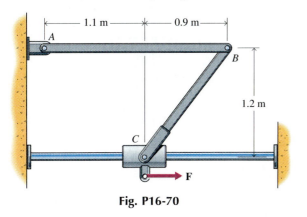

Fig. P16-70

16-71 Bars AB (W_{AB} = 7 lb) and BC (W_{BC} = 10 lb) of Fig. P16-71 have uniform cross sections. When the bars are in the position shown, the angular velocity of bar AB is 5 rad/s and its angular acceleration is 2 rad/s^2, both counterclockwise. If the surface at C is smooth, determine the forces exerted on bar BC by the pin at B and the surface at C.

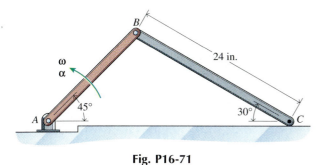

Fig. P16-71

Challenging Problems

16-72* The slender bar AB shown in Fig. P16-72 rests on a smooth surface at B and is attached to a collar at A that slides freely on the smooth vertical rod. The bar has a uniform cross section and a mass of 20 kg; the collar has negligible mass. The bar is initially at rest with $\theta = 0°$ when it is disturbed and rotates in a vertical plane under the action of gravity. Determine the angular acceleration of the bar and the reactions at A and B when $\theta = 60°$.

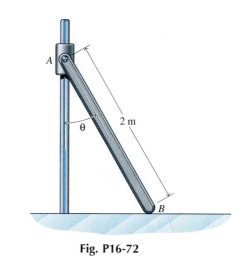

Fig. P16-72

16-73 A 16-lb ladder 12 ft long slides in a smooth corner as shown in Fig. P16-73. The ladder is initially at rest with $\theta = 0°$ when the lower end is disturbed slightly. Determine the angle θ, the angular velocity ω, and the velocity $\mathbf{v}_B$ of end B when end A loses contact with the vertical wall.

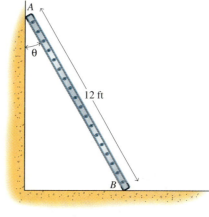

Fig. P16-73

16-74* The 220-mm-diameter bowling ball shown in Fig. P16-74 has a mass of 7.25 kg. The instant that the ball comes in contact with the alley, it has a forward velocity $\mathbf{v}$ of 7 m/s and a backspin ω of 6 rad/s. If the kinetic coefficient of friction between the ball and the alley is 0.15, determine the elapsed time and the distance traveled before the ball begins to roll without slipping.

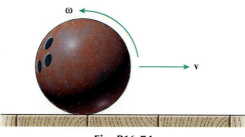

Fig. P16-74

16-75* A 4-ft-diameter homogeneous cylinder weighing 300 lb rests on the bed of a flatbed truck as shown in Fig. P16-75. The truck accelerates from rest at a rate of 3 ft/s² for 20 s and then moves with a constant velocity. If the cylinder rolls without slipping on the bed of the truck, determine the distance traveled by the truck before the cylinder rolls off of the truck.

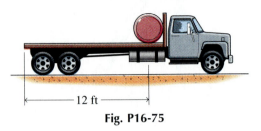

Fig. P16-75

16-76 The 10-kg spool C of Fig. P16-76 has a centroidal radius of gyration of 75 mm. A cord connects the spool to a 25-kg crate. If the system is released from rest and the spool rolls without slipping, determine the speed v_C and the angular velocity ω_C of the spool and the speed v_A of the crate after the crate has dropped 2 m.

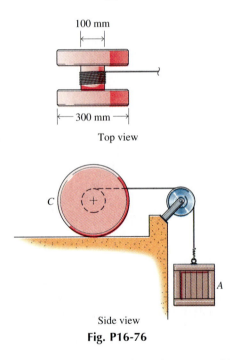

Side view
Fig. P16-76

16-77 A 12-lb spool C has a radius of gyration of 8 in. and rolls without slipping on a horizontal surface as shown in Fig. P16-77. A spring having $k = 24$ lb/ft is attached to a pair of cords wrapped around the outer disks of the spool. A second cord wrapped around the center of the spool is attached to a 30-lb crate A. If the crate has a speed of 4 ft/s downward when the spring is stretched 4 in., determine

a. The maximum distance that the crate will drop.

b. The speed v_C and angular velocity ω_C of the spool and the speed v_A of the crate when the stretch in the spring is zero.

c. The maximum distance that the crate will rise above its initial position.

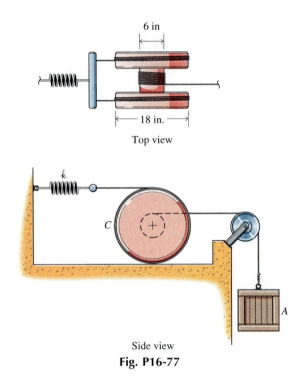

Top view

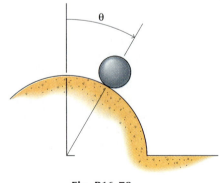

Side view
Fig. P16-77

16-78 A solid, homogeneous sphere (2 kg, 100 mm in diameter) rolls without slipping on the outside of a fixed cylinder 400 mm in diameter (Fig. P16-78). The static coefficient of friction between the sphere and the cylinder is $\mu_s = 0.7$. If the sphere is released from rest when $\theta = 0°$, determine the angle θ at which the sphere will begin to slip. (Does the sphere lose contact with the cylinder before or after this angle?)

Fig. P16-78

16-79 The 40-lb slender bar *AB* shown in Fig. P16-79 has a uniform cross section. The bar is supported by two flexible cables and is held in position by the horizontal cord at *B*. Determine the acceleration $\mathbf{a}_G$ of the mass center of the bar, the angular acceleration α_{AB} of the bar, and the tensions T_A and T_B in the two cables immediately after the horizontal cord at *B* is cut.

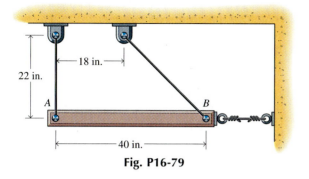

Fig. P16-79

16-80* The 10-kg slender bar *AB* shown in Fig. P16-80 has a uniform cross section. The bar is initially held fixed by two cords. Determine the tension T_B in the cord at *B* and the angular acceleration α_{AB} of the bar immediately after the cord at *A* is cut. Assume that the horizontal surface in contact with end *A* of the bar is smooth.

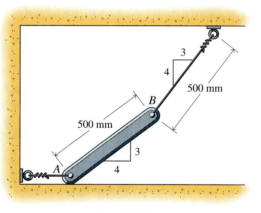

Fig. P16-80

Computer Problems

C16-81 The 10-lb uniform bar *AB* shown in Fig. P16-81 rotates in a vertical plane. The coefficient of friction between the bar and the surface at *A* is 0.6. If the bar is released from rest when $\theta = 0°$, calculate and plot

a. The normal and friction forces exerted on the bar at *A* as functions of its angular position θ ($0° < \theta < 90°$).
b. The location of the mass center of the bar; that is, x_G versus y_G as the bar falls ($0° < \theta < 90°$).
c. The motion of end *A* of the bar; that is, x_A and y_A as functions of time *t* as the bar falls ($0° < \theta < 90°$).

d. At what angle θ_s does the bar begin to slip on the surface? When it begins to slip, does *A* slip to the left or to the right? Does *A* ever lift away from the surface? (Note that as the bar falls, the normal force decreases and the friction eventually is not sufficient to keep the bar from sliding. Therefore, it will probably be necessary to solve the differential equations of motion numerically.)

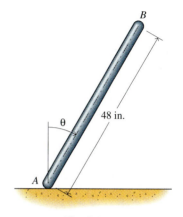

Fig. P16-81

C16-82 The spoked wheel shown in Fig. P16-82 rolls without slipping on a horizontal surface. Because it has lost a couple of its spokes, the center of mass of the 12-kg wheel is 50 mm away from its middle and the radius of gyration relative to the mass center is 0.6 m. If the center of the wheel has a speed of 3 m/s when $\theta = 0°$, calculate and plot

a. The normal and friction forces exerted on the wheel as functions of θ ($0° < \theta < 360°$).
b. The angular velocity $\dot{\theta}$ of the wheel as a function of θ ($0° < \theta < 360°$).

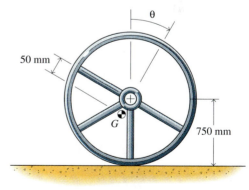

Fig. P16-82

C16-83 The mechanism shown in Fig. P16-83 is a simplification of a printing press. The print drum is a solid circular cylinder weighing 16 lb, bar AB rotates counterclockwise with a constant angular velocity of $\dot{\theta} = 15$ rev/min, and the weight of bar BC may be neglected. If the lengths are $L_{AB} = 2.5$ ft and $L_{BC} = 4$ ft and the radius of the drum is 1 ft, calculate and plot

a. The normal and friction forces exerted on the print drum as functions of θ ($0° < \theta < 360°$).

b. The force exerted on the print drum by arm BC as a function of θ ($0° < \theta < 360°$).

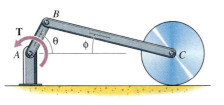

Fig. P16-83

C16-84 The slider crank mechanism of Fig. P16-84 is an idealization of an automobile crank shaft AB, connecting rod BC, and piston C. Treat the connecting rod as a uniform rod $L_{BC} = 175$ mm long with a mass of 0.12 kg. The crank throw is $L_{AB} = 75$ mm long. The mass of the piston is 0.17 kg. If the crank shaft is rotating at a constant rate $\dot{\theta} = 4800$ rev/min, calculate and plot

a. The magnitude of the force exerted on the connecting rod at B by the crank shaft as a function of θ ($0° < \theta < 360°$).

b. The magnitude of the force exerted on the piston by the connecting rod as a function of θ ($0° < \theta < 360°$).

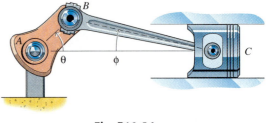

Fig. P16-84

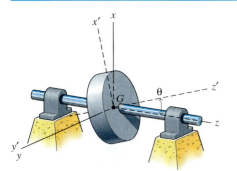

Figure 16-41 Although all points of the disk rotate in planes parallel to the xy-plane, the x-axis is not a principal axis. Therefore, $I_{Gzx} \neq 0$ and the bearings must provide moments about the x- and y-axes to keep the shaft aligned with the z-axis.

16-4.4 General Plane Motion of Nonsymmetric Bodies

For those cases where the body is not symmetric with respect to the plane of motion, Eqs. 16-3 must be carefully considered and reduced as appropriate for the selection of the xyz-coordinate system attached to the body. For example, consider a solid circular disk (see Fig. 16-41) mounted on a shaft with its axis inclined at an angle θ with respect to the axis of the shaft. For an xyz-coordinate system with its origin at the mass center G of the disk, $x_G = y_G = 0$, $I_{Gyz} = 0$, and $\mathbf{a}_G = \mathbf{0}$. Thus, Eqs. 16-3 reduce to

$$
\begin{array}{ll}
\Sigma F_x = ma_{Gx} = 0 & \Sigma M_{Gx} = -\alpha I_{Gzx} \\
\Sigma F_y = ma_{Gy} = 0 & \Sigma M_{Gy} = -\omega^2 I_{Gzx} \\
\Sigma F_z = 0 & \Sigma M_{Gz} = \alpha I_{Gz}
\end{array} \quad (a)
$$

A second example of a body that is not symmetric with respect to the plane of motion is illustrated in Fig. 16-42. In this case, a triangular plate of uniform thickness is attached to and rotates with a circular shaft. For an xyz-coordinate system with its origin A on the axis of the shaft, $y_G = 0$, $I_{Ayz} = 0$, and $\mathbf{a}_A = \mathbf{0}$. Thus, Eqs. 16-3 reduce to

$$
\begin{array}{ll}
\Sigma F_x = ma_{Gx} = -mx_G\omega^2 & \Sigma M_{Ax} = -\alpha I_{Azx} \\
\Sigma F_y = ma_{Gy} = mx_G\alpha & \Sigma M_{Ay} = -\omega^2 I_{Azx} \\
\Sigma F_z = 0 & \Sigma M_{Az} = \alpha I_{Az}
\end{array} \quad (b)
$$

These six equations provide sufficient information for the determination of six unknowns, which may include the bearing components B_x, B_y, C_x, and C_y needed at any instant to produce the moments M_x and M_y required to maintain the body in a state of plane motion.

The procedure for solving motion problems involving general plane motion is illustrated in the following examples.

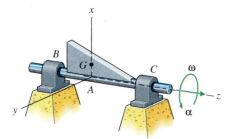

Figure 16-42 Although all points of the plate rotate in planes parallel to the xy-plane, neither the x-axis nor the y-axis is a principal axis. Therefore, $I_{Gzx} \neq 0$ and the bearings must provide moments about the x- and y-axes to keep the shaft aligned with the z-axis.

Two 120-mm-diameter spheres are attached to a shaft and rotated as shown in Fig. 16-43. Each sphere has a mass of 7.50 kg. The bars connecting the spheres to the shaft have 30-mm diameters, are 220 mm long, and have masses of 1.20 kg. The 40-mm-diameter shaft has a mass of 8.50 kg. Determine the components of the bearing reactions at the supports and the applied torque **T** when the shaft is rotating counterclockwise at 600 rev/min and increasing in speed at the rate of 60 rev/min per second. Assume that the bearing at A resists any motion of the shaft in the z-direction.

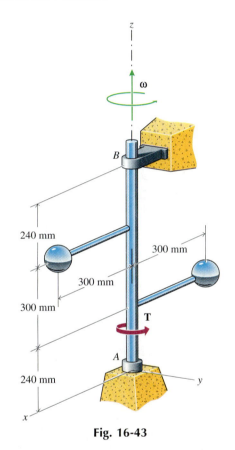

240 mm

300 mm

300 mm

300 mm

T

240 mm

Fig. 16-43

SOLUTION

The free-body diagram of the system is shown in Fig. 16-44a. As the vertical shaft rotates, the two spheres move in horizontal circular paths. Although the motion is planar (all points of the system move in parallel horizontal planes), the system is not symmetric with respect to the yz-plane (at the instant shown). Therefore, Eqs. 16-17 must be used to solve this problem

$$\Sigma F_x = ma_{Gx}: \qquad A_x + B_x = 0 \qquad (a)$$
$$\Sigma F_y = ma_{Gy}: \qquad A_y + B_y = 0 \qquad (b)$$
$$\Sigma F_z = ma_{Gz}: \qquad A_z - 2m_1g - 2m_2g - m_3g = 0 \qquad (c)$$
$$\Sigma M_{Gx} = -\alpha I_{Gzx} + \omega^2 I_{Gyz}: \qquad 0.390(A_y - B_y) = -\alpha I_{Gzx} \qquad (d)$$
$$\Sigma M_{Gy} = -\alpha I_{Gyz} - \omega^2 I_{Gzx}: \qquad 0.390(B_x - A_x) = -\omega^2 I_{Gzx} \qquad (e)$$
$$\Sigma M_{Gz} = I_{Gz}\alpha: \qquad T = I_{Gz}\alpha \qquad (f)$$

where $\omega = 600$ rev/min $= 62.83$ rad/s, $\alpha = 60$ rev/min/s $= 6.283$ rad/s^2, $\mathbf{a}_G = 0$, and at the instant shown $I_{Gyz} = 0$ (since the xy-plane is a plane of symme-

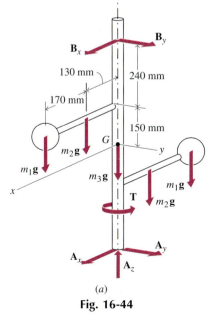

B$_x$

130 mm

170 mm

240 mm

150 mm

$m_1\mathbf{g}$

$m_2\mathbf{g}$

G

$m_3\mathbf{g}$

B$_y$

$m_1\mathbf{g}$

$m_2\mathbf{g}$

T

A$_x$

A$_z$

A$_y$

(a)

Fig. 16-44

try). Also at the instant shown, the moments and products of inertia of each of the two spheres are

$$I_{Gz1} = \frac{2}{5}m_1R_1^2 + m_1d_1^2$$

$$= \frac{2}{5}(7.50)(0.060)^2 + 7.50(0.300)^2 = 0.68580 \text{ kg} \cdot \text{m}^2$$

$$I_{Gzx1} = 0 + m_1 z_{G1} x_{G1} = 0 + 7.50(0.150)(0.300) = 0.33750 \text{ kg} \cdot \text{m}^2$$

For each of the two horizontal bars the moments and products of inertia are

$$I_{Gz2} = \left(\frac{1}{4}m_2R_2^2 + \frac{1}{12}m_2L_2^2\right) + m_2d_2^2$$

$$= \frac{1}{4}(1.20)(0.015)^2 + \frac{1}{12}(1.20)(0.220)^2\right) + 1.20(0.130)^2 = 0.02519 \text{ kg} \cdot \text{m}^2$$

$$I_{Gzx2} = 0 + m_2 z_{G2} x_{G2} = 0 + 1.20(0.150)(0.130) = 0.02340 \text{ kg} \cdot \text{m}^2$$

For the vertical shaft the moments and products of inertia are

$$I_{Gz3} = \frac{1}{2}m_3R_3^2 = \frac{1}{2}(8.50)(0.020)^2 = 0.00170 \text{ kg} \cdot \text{m}^2$$

$$I_{Gzx3} = 0 \text{ kg} \cdot \text{m}^2$$

Therefore, for the entire system

$$I_{Gz} = 2I_{Gz1} + 2I_{Gz2} + I_{Gz3} = 1.4237 \text{ kg} \cdot \text{m}^2$$
$$I_{Gzx} = 2I_{Gzx1} + 2I_{Gzx2} + I_{Gzx3} = 0.7218 \text{ kg} \cdot \text{m}^2$$

When the angular velocity, angular acceleration, and moments and products of inertia are substituted into the equations of motion, they become

$$A_x + B_x = 0 \qquad (a')$$
$$A_y + B_y = 0 \qquad (b')$$
$$A_z = 254.08 \text{ N} \qquad (c')$$
$$(A_y - B_y) = -11.629 \qquad (d')$$
$$(B_x - A_x) = -116.287 \qquad (e')$$
$$T = 8.945 \text{ N} \cdot \text{m} \qquad (f')$$

Solving these six equations simultaneously gives

$A_x = +3653 \text{ N} \cong +3650 \text{ N}$	Ans.
$A_y = -5.814 \text{ N} \cong -5.81 \text{ N}$	Ans.
$A_z = +254.08 \text{ N} \cong +254 \text{ N}$	Ans.
$B_x = -3653 \text{ N} \cong -3650 \text{ N}$	Ans.
$B_y = +5.814 \text{ N} \cong +5.81 \text{ N}$	Ans.
$T = +8.945 \text{ N} \cdot \text{m} \cong +8.95 \text{ N} \cdot \text{m}$	Ans.

The results are shown in Fig. 16-44b.

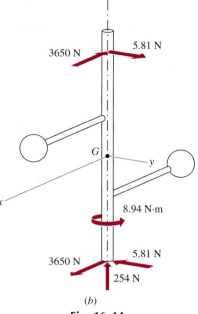

(b)

Fig. 16-44

A torque of 8.95 N · m about the z-axis is required to rotate the system as specified. In addition, however, the bearing forces must exert moments about the x- and y-axes to keep the motion in the xy-plane.

A solid circular steel ($\gamma = 0.284$ lb/in.^{3}) disk is mounted on a shaft as shown in Fig. 16-45. Determine the components of the bearing reactions and the applied torque $\mathbf{T}$ when the disk is in the position shown (the x-axis is vertical) and the shaft is rotating counterclockwise at 500 rev/min and increasing in speed at the rate of 50 rev/min per second. Assume that the bearing at B resists any motion of the shaft in the z-direction and that the mass of the shaft is negligible.

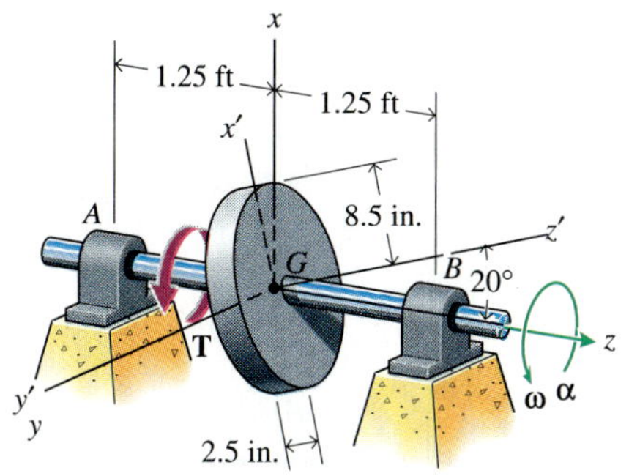

Fig. 16-45

SOLUTION

The free-body diagram of the disk and shaft is shown in Fig. 16-46. Although the motion is planar (all points of the disk move in parallel vertical planes), the system is not symmetric with respect to the yz-plane (at the instant shown). Therefore, Eqs. 16-17 must be used to solve this problem

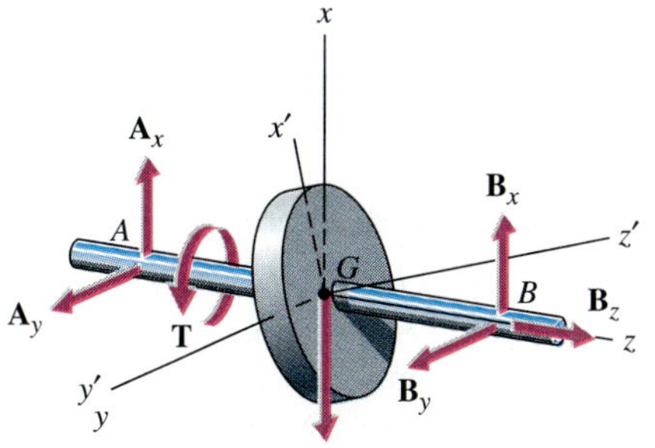

$\Sigma F_x = m a_{Gx}$:	$A_x + B_x - 161.16 = 0$	(a)
$\Sigma F_y = m a_{Gy}$:	$A_y + B_y = 0$	(b)
$\Sigma F_z = m a_{Gz}$:	$B_z = 0$	(c)
$\Sigma M_{Gx} = -\alpha I_{Gzx} + \omega^2 I_{Gyz}$:	$1.25(A_y - B_y) = -\alpha I_{Gzx}$	(d)
$\Sigma M_{Gy} = -\alpha I_{Gyz} - \omega^2 I_{Gzx}$:	$1.25(B_x - A_x) = -\omega^2 I_{Gzx}$	(e)
$\Sigma M_{Gz} = I_{Gz}\alpha$:	$T = I_{Gz}\alpha$	(f)

where $W = \gamma V = \gamma(\pi R^2 L) = 0.284\pi(8.5)^2(2.5) = 161.16$ lb, $\omega = 500$ rev/min = 52.36 rad/s, $\alpha = 50$ rev/min/s = 5.236 rad/s^2, $\mathbf{a}_G = \mathbf{0}$, and at the instant shown $I_{Gyz} = 0$ (since the xz-plane is a plane of symmetry). The moment of inertia I_{Gz} and the product of inertia I_{Gzx} can be determined by using the principal moments of inertia $I_{x'}$, $I_{y'}$, and $I_{z'}$ from Table B-5

$$I_{x'} = I_{y'} = \frac{1}{4}mR^2 + \frac{1}{12}mL^2$$

$$= \frac{1}{4}\left(\frac{161.16}{32.2}\right)\left(\frac{8.5}{12}\right)^2 + \frac{1}{12}\left(\frac{161.16}{32.2}\right)\left(\frac{2.5}{12}\right)^2 = 0.6459 \text{ slug} \cdot \text{ft}^2$$

$$I_{z'} = \frac{1}{2}mR^2 = \frac{1}{2}\left(\frac{161.16}{32.2}\right)\left(\frac{8.5}{12}\right)^2 = 1.2556 \text{ slug} \cdot \text{ft}^2$$

and the transformation equations A-13a and A-13b

$$I_z = I_{x'}\cos^2\theta_{x'z} + I_{y'}\cos^2\theta_{y'z} + I_{z'}\cos^2\theta_{z'z}$$
$$= 0.6459\cos^2 110° + 0.6459\cos^2 90° + 1.2556\cos^2 20° = 1.1843 \text{ slug} \cdot \text{ft}^2$$
$$I_{zx} = -I_{x'}\cos\theta_{x'z}\cos\theta_{x'x} - I_{y'}\cos\theta_{y'z}\cos\theta_{y'x} - I_{z'}\cos\theta_{z'z}\cos\theta_{z'x}$$
$$= -0.6459\cos 110°\cos 20° - 0.6459\cos 90°\cos 90°$$
$$\qquad - 1.2556\cos 20°\cos 70° = -0.19595 \text{ slug} \cdot \text{ft}^2$$

When the angular velocity, angular acceleration, and moments and products of inertia are substituted into the equations of motion, they become

$$A_x + B_x = 161.16 \qquad (a')$$
$$A_y + B_y = 0 \qquad (b')$$
$$B_z = 0 \qquad (c')$$
$$(A_y - B_y) = 0.82079 \qquad (d')$$
$$(B_x - A_x) = 429.77 \qquad (e')$$
$$T = 6.201 \qquad (f')$$

W = 161.2 lb

Fig. 16-46

Equations a', c', e', and f' give

$$A_x = -134.31 \text{ lb} \cong -134.3 \text{ lb} \qquad \text{Ans.}$$
$$B_x = +295.47 \text{ lb} \cong +295 \text{ lb} \qquad \text{Ans.}$$
$$B_z = 0 \text{ lb} \qquad \text{Ans.}$$
$$T = +6.201 \text{ lb} \cdot \text{ft} = +6.20 \text{ lb} \cdot \text{ft} \qquad \text{Ans.}$$

Then, Eqs. b' and d' give

$$A_y = +0.41040 \text{ lb} \cong +0.410 \text{ lb} \qquad \text{Ans.}$$
$$B_y = -0.41040 \text{ lb} \cong -0.410 \text{ lb} \qquad \text{Ans.}$$

Although this is a planar motion, problems of this type can also be solved as three-dimensional problems by resolving the angular velocity ω and the angular acceleration α into x'- and y'-components. As shown in Section 16-5, this approach simplifies the solution by eliminating the requirement for determining the nonprincipal moments of inertia and the products of inertia.

PROBLEMS

Intermediate Problems

16-85* Two thin rectangular plates (each weighing 15 lb) are mounted on a slender horizontal shaft as shown in Fig. P16-85. Determine the bearing reactions when the plates lie in a vertical plane (as shown) and the shaft is rotating at a constant angular velocity of $\omega = 75$ rad/s. Assume that the bearing at B resists any motion of the shaft in the axial direction and that the mass of the shaft may be neglected.

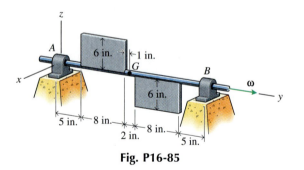

Fig. P16-85

16-86* A thin triangular plate ($m = 10$ kg) is mounted on a slender horizontal shaft as shown in Fig. P16-86. Determine the bearing reactions when the plate lies in a vertical plane (as shown) and the shaft is rotating at a constant angular velocity of $\omega = 75$ rad/s. Assume that the bearing at B resists any motion of the shaft in the axial direction and that the mass of the shaft may be neglected.

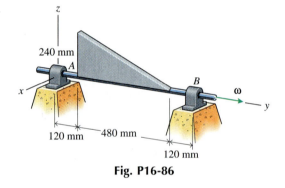

Fig. P16-86

16-87 Two thin quarter-circular plates (each weighing 20 lb) are mounted on a slender horizontal rod as shown in Fig. P16-87. Determine the bearing reactions when the plates lie in a vertical plane (as shown) and the angular velocity and angular acceleration of the shaft are $\omega = 100$ rad/s and $\alpha = 25$ rad/s^2, respectively. Assume that the bearing at B resists any motion of the shaft in the axial direction and that the mass of the shaft may be neglected.

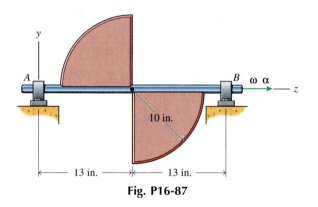

Fig. P16-87

16-88 The slender branched rod shown in Fig. P16-88 rotates at a constant rate of $\omega = 120$ rev/min. If the rod has a mass density of $\rho = 0.25$ kg/m, determine the bearing reactions when the rod lies in a horizontal plane (as shown). Assume that the bearing at B resists any motion of the shaft in the axial direction and that the portions of the shaft that extend beyond the bearings may be neglected.

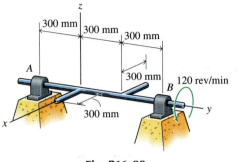

Fig. P16-88

16-89 The three identical spheres of Fig. P16-89 each weigh 2 lb and have diameters of 4 in. The centers of the spheres are 10 in. from the center of the shaft, and they are symmetrically located around the shaft. If the shaft is rotating at a constant rate of 60 rev/min, determine the bearing reactions when the shaft is in the position shown. Assume that the bearing at B resists any motion of the shaft in the axial direction and that the masses of the shafts may be neglected.

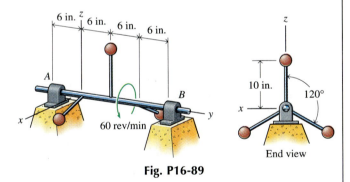

Fig. P16-89

16-90 The 1.5-kg cylinder AB of Fig. P16-90 has a diameter of 50 mm and a length of 200 mm. It is mounted on a thin circular disk having a radius of 400 mm and a mass of 0.5 kg. The distance between the axis of the cylinder and the axis of the shaft is 300 mm. If the shaft is rotating at a constant rate of 180 rev/min, determine the bearing reaction when the cylinder is aligned with the y-axis as shown. Assume that the mass of the vertical shaft may be neglected.

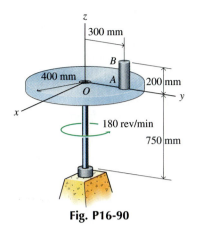

Fig. P16-90

16-91* Two 4-in.-diamater spheres (each weighs 10 lb) are attached to a vertical shaft and rotated as shown in Fig. P16-91. The rods connecting the spheres to the shaft have 1 in. diameters, are 7 in. long, and weigh 1.50 lb each. The 2-in.-diameter vertical shaft weighs 20 lb. Determine the components of the bearing reactions at the supports and the applied torque T when the angular velocity ω of the shaft is 100 rad/s and is increasing at a rate of 20 rad/s². Assume that the bearing at A resists any motion of the shaft in the z-direction.

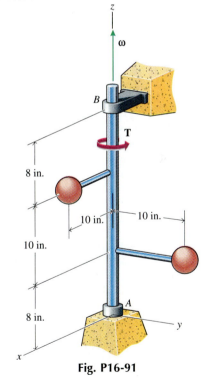

Fig. P16-91

16-92* Two rectangular bars (each of mass 5 kg) are attached to a vertical shaft and rotated as shown in Fig. P16-92. The mass of the 40-mm-diameter vertical shaft is 6.5 kg. Determine the components of the bearing reactions at the supports and the applied torque T when the angular velocity ω of the shaft is 150 rad/s and is decreasing at a rate of 25 rad/s². Assume that the bearing at A resists any motion of the shaft in the z-direction.

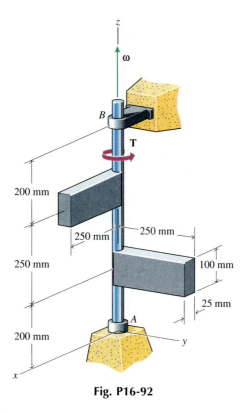

Fig. P16-92

16-93 The thin disk of Fig. P16-93 has a radius of 10 in., weighs 2 lb, and is mounted on a slender horizontal shaft at a point 7 in. from its center. Determine the bearing reactions when the disk is in the position shown and the shaft is rotating at a constant angular velocity of $\omega = 150$ rev/min. Assume that the bearing at B resists any motion of the shaft in the axial direction and that the mass of the shaft is negligible.

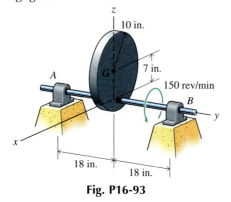

Fig. P16-93

Challenging Problems

16-94* A 130-mm-diameter solid circular cylinder ($m = 50$ kg) is mounted on a slender horizontal shaft as shown in Fig. P16-94. Determine the bearing reactions when the axis of the cylinder lies in a vertical plane (as shown) and the shaft is rotating at a constant angular velocity of $\omega = 60$ rad/s. Assume that the bearing at B resists any motion of the shaft in the axial direction and that the mass of the shaft is negligible.

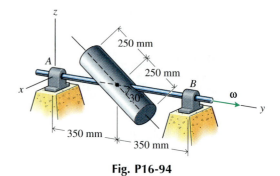

Fig. P16-94

16-95* A thin rectangular plate ($W = 60$ lb) is mounted on a slender horizontal shaft as shown in Fig. P16-95. Determine the bearing reactions when the plate lies in a vertical plane (as shown) and the shaft is rotating at a constant angular velocity of $\omega = 90$ rad/s. Assume that the bearing at B resists any motion of the shaft in the axial direction and that the mass of the shaft is negligible.

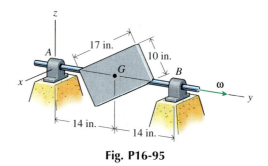

Fig. P16-95

16-96 A 75-mm-diameter solid circular cylinder with hemispherical ends is mounted on a slender horizontal shaft as shown in Fig. P16-96. The mass of the cylinder is 6 kg and the mass of each hemisphere is 1 kg. Determine the bearing reactions when the axis of the cylinder lies in a vertical plane (as shown) and the angular velocity and angular acceleration of the shaft are $\omega = 50$ rad/s and $\alpha = 15$ rad/s², respectively. Assume that the bearing at B resists any motion of the shaft in the axial direction and that the mass of the shaft is negligible.

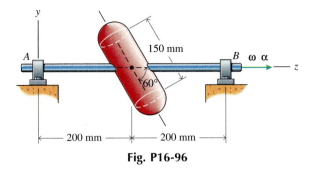

Fig. P16-96

16-97 The bent rod shown in Fig. P16-97 rotates at a constant rate of 90 rev/min about a vertical axis. If the rod has a specific weight of $\gamma = 1.5$ lb/ft, determine the bearing reaction when the rod lies in the vertical yz-plane (as shown). Neglect the small segment of the rod that is aligned with the vertical axis.

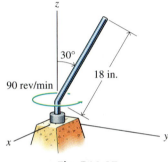

Fig. P16-97

16-98 The thin circular plate of Fig. P16-98 has a radius of 225 mm, a mass of 7.5 kg, and rotates at a constant rate of 150 rev/min on a slender horizontal shaft. Determine the bearing reactions when the shaft is in the position shown. Assume that the bearing at B resists any motion of the shaft in the axial direction and that the mass of the shaft is negligible.

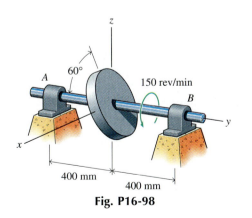

Fig. P16-98

Computer Problems

C16-99 The three identical spheres of Fig. P16-99 each weigh 2 lb and have diameters of 4 in. The centers of the spheres are 10 in. from the center of the shaft, and they are symmetrically located around the shaft. If the shaft is rotating at a constant rate of 60 rev/min, calculate and plot A and B, the magnitudes of the bearing reactions, as functions of θ, the rotation angle, for one complete revolution of the shaft. Assume that the bearing at B resists any motion of the shaft in the axial direction, that the masses of the shafts may be neglected, and that $\theta = 0°$ at the instant shown.

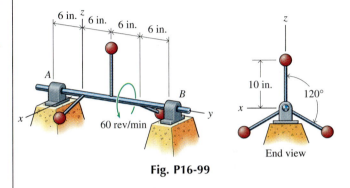

Fig. P16-99

C16-100 A 130-mm-diameter solid circular cylinder ($m = 50$ kg) is mounted on a slender horizontal shaft as shown in Fig. P16-100. If the shaft is rotating at a constant rate of 60 rad/s, calculate and plot A and B, the magnitudes of the bearing reactions, as functions of θ, the rotation angle, for one complete revolution of the shaft. Assume that the bearing at B resists any motion of the shaft in the axial direction, that the mass of the shaft may be neglected, and that $\theta = 0°$ at the instant shown.

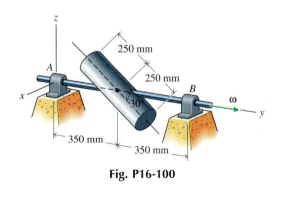

Fig. P16-100

16-5 THREE-DIMENSIONAL MOTION OF A RIGID BODY

The moment $\mathbf{M}_A$ about an arbitrary point A in a body subjected to a system of external forces was developed in Section 16-2 and is given by Eq. 16-1

$$\mathbf{M}_A = \int_m (\mathbf{r} \times \mathbf{a}_A)\, dm + \int_m [\mathbf{r} \times (\boldsymbol{\alpha} \times \mathbf{r})]\, dm$$
$$+ \int_m \{\mathbf{r} \times [\boldsymbol{\omega} \times (\boldsymbol{\omega} \times \mathbf{r})]\}\, dm \qquad (a)$$

Then, denoting the rectangular components of the vector $\mathbf{r}$ by

$$\mathbf{r} = x\, \mathbf{i} + y\, \mathbf{j} + z\, \mathbf{k}$$

the rectangular components of Eq. a become

$$
\begin{aligned}
M_{Ax} = {} & a_{Az}\int_m y\, dm - a_{Ay}\int_m z\, dm + \alpha_x\int_m y^2\, dm \\
& - \alpha_y\int_m xy\, dm - \alpha_z\int_m zx\, dm + \alpha_x\int_m z^2\, dm \\
& + \omega_z\omega_x\int_m xy\, dm - \omega_y^2\int_m yz\, dm + \omega_y\omega_z\int_m y^2\, dm \\
& - \omega_y\omega_z\int_m z^2\, dm + \omega_z^2\int_m yz\, dm - \omega_x\omega_y\int_m zx\, dm
\end{aligned} \qquad (b)
$$

$$
\begin{aligned}
M_{Ay} = {} & a_{Ax}\int_m z\, dm - a_{Az}\int_m x\, dm + \alpha_y\int_m z^2\, dm \\
& - \alpha_z\int_m yz\, dm - \alpha_x\int_m xy\, dm + \alpha_y\int_m x^2\, dm \\
& + \omega_x\omega_y\int_m yz\, dm - \omega_z^2\int_m zx\, dm + \omega_z\omega_x\int_m z^2\, dm \\
& - \omega_z\omega_x\int_m x^2\, dm + \omega_x^2\int_m zx\, dm - \omega_y\omega_z\int_m xy\, dm
\end{aligned} \qquad (c)
$$

$$
\begin{aligned}
M_{Az} = {} & a_{Ay}\int_m x\, dm - a_{Ax}\int_m y\, dm + \alpha_z\int_m x^2\, dm \\
& - \alpha_x\int_m zx\, dm - \alpha_y\int_m yz\, dm + \alpha_z\int_m y^2\, dm \\
& + \omega_y\omega_z\int_m zx\, dm - \omega_x^2\int_m xy\, dm + \omega_x\omega_y\int_m x^2\, dm \\
& - \omega_x\omega_y\int_m y^2\, dm + \omega_y^2\int_m xy\, dm - \omega_z\omega_x\int_m yz\, dm
\end{aligned} \qquad (d)
$$

When Eqs. b, c, and d are written in terms of first moments, moments of inertia, and products of inertia they become

$$
\begin{aligned}
M_{Ax} = {} & a_{Az}y_Gm - a_{Ay}z_Gm + I_{Ax}\alpha_x \\
& - (I_{Ay} - I_{Az})\omega_y\omega_z + I_{Axy}(\omega_z\omega_x - \alpha_y) \\
& - I_{Ayz}(\omega_y^2 - \omega_z^2) - I_{Azx}(\omega_x\omega_y + \alpha_z)
\end{aligned} \qquad (16\text{-}20a)
$$

$$
\begin{aligned}
M_{Ay} = {} & a_{Ax}z_Gm - a_{Az}x_Gm + I_{Ay}\alpha_y \\
& - (I_{Az} - I_{Ax})\omega_z\omega_x + I_{Ayz}(\omega_x\omega_y - \alpha_z) \\
& - I_{Azx}(\omega_z^2 - \omega_x^2) - I_{Axy}(\omega_y\omega_z + \alpha_x)
\end{aligned} \qquad (16\text{-}20b)
$$

$$
\begin{aligned}
M_{Az} = {} & a_{Ay}x_Gm - a_{Ax}y_Gm + I_{Az}\alpha_z \\
& - (I_{Ax} - I_{Ay})\omega_x\omega_y + I_{Azx}(\omega_y\omega_z - \alpha_x) \\
& - I_{Axy}(\omega_x^2 - \omega_y^2) - I_{Ayz}(\omega_z\omega_x + \alpha_y)
\end{aligned} \qquad (16\text{-}20c)
$$

Equations 16-20 relate the rotational behavior of the rigid body and the moments acting on the body. If the angular velocity $\boldsymbol{\omega}$ and angular acceleration $\boldsymbol{\alpha}$ (and the acceleration $\mathbf{a}_A$ of point A) are specified, then Eqs. 16-20 give the moment $\mathbf{M}_A$ required to produce the motion. If the moment $\mathbf{M}_A$ is specified (and the angular velocity $\boldsymbol{\omega}$ and acceleration $\mathbf{a}_A$ are known), then Eqs. 16-20 give the angular acceleration of the body.

The orientation of the *xyz*-coordinate system used to express Eqs. 16-20 may be chosen in whatever manner is convenient for the evaluation of the moments and products of inertia. For example, the form of the equations can be simplified considerably by choosing the *xyz*-coordinate system to coincide with the principal axes through the mass center *G* of the body. With the origin at the center of mass

$$x_G = y_G = z_G = 0$$

and for principal axes,

$$I_{xy} = I_{yz} = I_{zx} = 0$$

Therefore, if the *xyz*-coordinate axes are principal axes centered at the mass center *G* of the body, Eqs. 16-20 reduce to

$$M_{Gx} = I_{Gx}\alpha_x - (I_{Gy} - I_{Gz})\omega_y\omega_z \qquad (16\text{-}21a)$$
$$M_{Gy} = I_{Gy}\alpha_y - (I_{Gz} - I_{Gx})\omega_z\omega_x \qquad (16\text{-}21b)$$
$$M_{Gz} = I_{Gz}\alpha_z - (I_{Gx} - I_{Gy})\omega_x\omega_y \qquad (16\text{-}21c)$$

Equations 16-21 are known as Euler's[1] equations.

Of course, since the body is rotating and the orientation of the *xyz*-coordinate system must be fixed, the evaluation of the moments and products of inertia will not, in general, be constant. Only for highly symmetric bodies will these moments and products of inertia be independent of the orientation of the body. Therefore, if it is desired to integrate the angular acceleration to obtain the angular velocity of the body, then the moments and products of inertia must include any dependence on time or orientation of the body.

Equations 15-11 ($\mathbf{R} = \Sigma\mathbf{F} = m\mathbf{a}_G$) together with Eqs. 16-21 provide the relationships required to solve a variety of three-dimensional motion problems as illustrated in the following examples.

[1]Leonhard Euler (1707–1783), a Swiss mathematician.

The combined mass of the shaft and cylinder shown in Fig. 16-47 is 20 kg. The moments of inertia of the combined shaft and cylinder about the x-, y-, and z-axes through the center of mass G are $I_{Gx} = I_{Gz} = 0.1595$ kg · m² and $I_{Gy} = 0.0625$ kg · m². If the cylinder is rotating at a constant angular velocity of 75 rad/s and the frame is rotating at a constant angular velocity of 25 rad/s, determine the reactions at supports A and B of the shaft. Assume that the bearing at B resists any force along the axis of the shaft.

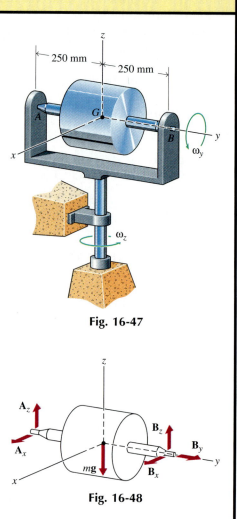

SOLUTION

A free-body diagram for the shaft and cylinder is shown in Fig. 16-48. The xyz-coordinate system will be chosen with its origin at the mass center of the cylinder (which is stationary, $\mathbf{a}_G = \mathbf{0}$) and with the y-axis along the shaft at the instant shown in Fig. 16-47. These axes are principal axes for the system, so the equations of motion (Eqs. 15-11 and 16-21) are

$\Sigma F_x = ma_{Gx}:$ $\qquad A_x + B_x = 0$ $\qquad$ (a)
$\Sigma F_y = ma_{Gy}:$ $\qquad B_y = 0$ $\qquad$ (b)
$\Sigma F_z = ma_{Gz}:$ $\qquad A_z + B_z - mg = 0$ $\qquad$ (c)
$\Sigma M_{Gx} = I_{Gx}\alpha_x - (I_{Gy} - I_{Gz})\omega_y\omega_z:$ $\quad 0.25(B_z - A_z) = I_{Gx}\alpha_x - (I_{Gy} - I_{Gz})\omega_y\omega_z$ $\quad$ (d)
$\Sigma M_{Gy} = I_{Gy}\alpha_y - (I_{Gz} - I_{Gx})\omega_z\omega_x:$ $\qquad 0 = I_{Gy}\alpha_y$ $\qquad$ (e)
$\Sigma M_{Gz} = I_{Gz}\alpha_z - (I_{Gx} - I_{Gy})\omega_x\omega_y:$ $\quad 0.250(A_x - B_x) = I_{Gz}\alpha_z$ $\quad$ (f)

where $a_{Gx} = a_{Gy} = a_{Gz} = \omega_x = 0$.

Although the magnitude of the angular velocity vector is constant ($\dot{\omega}_x = \dot{\omega}_y = \dot{\omega}_z = 0$), its direction is changing. Therefore, the angular acceleration $\boldsymbol{\alpha} \neq \mathbf{0}$. One component of $\boldsymbol{\omega}$ is always in the z -direction, one component is along the shaft (along the line AB), which is instantaneously aligned with the y-axis, and the third component is zero. Therefore, the angular acceleration is (see Example Problem 14-18)

$$\boldsymbol{\alpha} = \frac{d\boldsymbol{\omega}}{dt} = \frac{d}{dt}(75\mathbf{e}_{AB} + 25\mathbf{e}_z) = 75\frac{d\mathbf{e}_{AB}}{dt}$$
$$= 75(25\mathbf{e}_z \times \mathbf{e}_{AB}) = 1875(-\mathbf{e}_x) = -1875\mathbf{i} \text{ rad/s}^2$$

and $\alpha_x = -1875$ rad/s², $\alpha_y = \alpha_z = 0$ rad/s².

Then Eq. e is satisfied identically, and Eqs. a, b, and f give

$$A_x = B_x = B_y = 0 \qquad \text{Ans.}$$

Equations c and d can be rewritten

$$A_z + B_z = 20(9.81) = 196.2 \text{ N} \qquad (c')$$
$$0.25(B_z - A_z) = \frac{1}{0.250}[0.1595(-1875) - (0.0625 - 0.1595)(75)(25)]$$
$$= -468.75 \qquad (d')$$

Solving these two equations simultaneously gives

$$A_z = 332.48 \text{ N} \cong 332 \text{ N}\uparrow \qquad \text{Ans.}$$
$$B_z = -136.28 \text{ N} \cong 136.3\text{N}\downarrow \qquad \text{Ans.}$$

That is, the shaft wants to rise up at B and a downward force is required at B to keep the shaft horizontal.

Fig. 16-47

Fig. 16-48

The x- and y-axes do not rotate with the cylinder but are only aligned with the cylinder at the instant shown.

This cylinder exhibits gyroscopic action. The cylinder rotating about its own axis is the gyroscope and axis of the cylinder is the gyroscopic axis. When the gyroscope is turned about an axis perpendicular to the gyroscopic axis, the gyroscope reacts by trying to rotate about a third axis which is perpendicular to the first two.

The solid circular steel ($\gamma = 0.284$ lb/in.^{3}) disk of Example Problem 16-15 is repeated here (Fig. 16-49). Determine the components of the bearing reactions and the applied torque **T** when the cylinder is in the position shown (the x-axis is vertical) and the shaft is rotating counterclockwise at 500 rev/min and increasing in speed at the rate of 50 rev/min per second. Assume that the bearing at B resists any motion of the shaft in the z-direction and that the mass of the shaft is negligible.

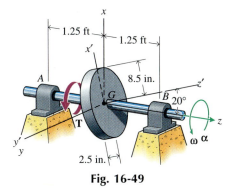

Fig. 16-49

SOLUTION

A free-body diagram of the disk and shaft is shown in Fig. 16-50. The xyz-coordinate system is chosen with its origin at the mass center of the disk (which is stationary, $\mathbf{a}_G = 0$) and with the z-axis along the shaft ($\boldsymbol{\omega} = \omega\mathbf{k}$, $\boldsymbol{\alpha} = \alpha\mathbf{k}$) at the instant shown. The $x'y'z'$-coordinate system is chosen with its origin also at the mass center of the cylinder, with the y'-axis along the y-axis and with the z'-axis perpendicular to the disk at the instant shown. The $x'y'z'$-axes are principal axes for the system.

From Example Problem 16-15

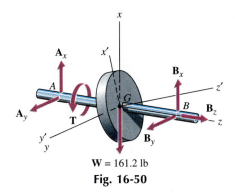

$$W = 161.2 \text{ lb} \qquad I_{Gx'} = 0.6459 \text{ slug} \cdot \text{ft}^2$$
$$\omega_z = 52.36 \text{ rad/s} \qquad I_{Gy'} = 0.6459 \text{ slug} \cdot \text{ft}^2$$
$$\alpha_z = 5.236 \text{ rad/s}^2 \qquad I_{Gz'} = 1.2556 \text{ slug} \cdot \text{ft}^2$$

W = 161.2 lb

Fig. 16-50

Relative to the principal axes the angular velocity and angular acceleration can be written

$$\boldsymbol{\omega} = 52.36\mathbf{k} = 52.36(-\sin 20°\ \mathbf{i}' + \cos 20°\ \mathbf{k}')$$
$$= -17.91\mathbf{i}' + 49.20\mathbf{k}' \text{ rad/s}$$
$$\boldsymbol{\alpha} = 5.236\mathbf{k} = 5.236(-\sin 20°\ \mathbf{i}' + \cos 20°\ \mathbf{k}')$$
$$= -1.791\mathbf{i}' + 4.920\mathbf{k}' \text{ rad/s}^2$$

and the components of the moments about the principal axes are

$$\boldsymbol{\Sigma}\mathbf{M}_G = \Sigma(\mathbf{r} \times \mathbf{F}) = T(-\sin 20°\ \mathbf{i}' + \cos 20°\ \mathbf{k}')$$
$$+ 1.25(-\sin 20°\ \mathbf{i}' + \cos 20°\ \mathbf{k}') \times B_x (\cos 20°\ \mathbf{i}' + \sin 20°\ \mathbf{k}')$$
$$+ 1.25(\sin 20°\ \mathbf{i}' - \cos 20°\ \mathbf{k}') \times A_x (\cos 20°\ \mathbf{i}' + \sin 20°\ \mathbf{k}')$$
$$+ 1.25(-\sin 20°\ \mathbf{i}' + \cos 20°\ \mathbf{k}') \times B_y\mathbf{j}'$$
$$+ 1.25(\sin 20°\ \mathbf{i}' - \cos 20°\ \mathbf{k}') \times A_y\mathbf{j}'$$
$$= [1.17462(A_y - B_y) - 0.34202T]\mathbf{i}' + [1.25(B_x - A_x)]\mathbf{j}'$$
$$+ [0.42753(A_y - B_y) + 0.93969T]\mathbf{k}'$$

Therefore the equations of motion (Eqs. 15-11 and 16-21) are

$$\Sigma F_x = ma_{Gx}: \qquad A_x + B_x - 161.2 = 0 \qquad\qquad (a)$$
$$\Sigma F_y = ma_{Gy}: \qquad\qquad A_y + B_y = 0 \qquad\qquad (b)$$
$$\Sigma F_z = ma_{Gz}: \qquad\qquad B_z = 0 \qquad\qquad (c)$$
$$\Sigma M_{Gx'} = I_{Gx'}\alpha_{x'} - (I_{Gy'} - I_{Gz'})\omega_{y'}\omega_{z'}:$$
$$1.17462(A_y - B_y) - 0.34202T = (0.6459)(-1.791) \qquad (d)$$
$$\Sigma M_{Gy'} = I_{Gy'}\alpha_{y'} - (I_{Gz'} - I_{Gx'})\omega_{z'}\omega_{x'}:$$
$$1.25(B_x - A_x) = -(1.2556 - 0.6459)(49.20)(-17.91) \qquad (e)$$
$$\Sigma M_{Gz'} = I_{Gz'}\alpha_{z'} - (I_{Gx'} - I_{Gy'})\omega_{x'}\omega_{y'}:$$
$$0.42753(A_y - B_y) + 0.93969T = 1.2556(4.920) \qquad (f)$$

Equations a, c, and e give

$$A_x = -134.31 \text{ lb} \cong -134.3 \text{ lb} \qquad\qquad \text{Ans.}$$
$$B_x = +295.47 \text{ lb} \cong +295 \text{ lb} \qquad\qquad \text{Ans.}$$
$$B_z = 0 \text{ lb} \qquad\qquad\qquad\qquad \text{Ans.}$$

Then, Eqs. d and f give

$$T = +6.201 \text{ lb} \cdot \text{ft} \cong +6.20 \text{ lb} \cdot \text{ft} \qquad\qquad \text{Ans.}$$

Finally, Eqs. b and d give

$$A_y = +0.41040 \text{ lb} \cong +0.410 \text{ lb} \qquad\qquad \text{Ans.}$$
$$B_y = -0.41040 \text{ lb} \cong -0.410 \text{ lb} \qquad\qquad \text{Ans.}$$

These answers are identical to those obtained in Example Problem 16-15.

PROBLEMS

Intermediate Problems

16-101* A thin rectangular plate ($W = 60$ lb) is mounted on a slender horizontal shaft as shown in Fig. P16-101. Determine the bearing reactions when the plate lies in a vertical plane (as shown) and the shaft is rotating at a constant angular velocity of $\omega = 90$ rad/s. Assume that the bearing at B resists any motion of the shaft in the axial direction and that the mass of the shaft is negligible.

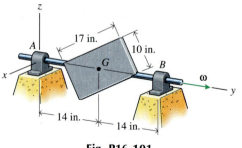

Fig. P16-101

16-102* A 130-mm-diameter solid circular cylinder ($m = 50$ kg) is mounted on a slender horizontal shaft as shown in Fig. P16-102. Determine the bearing reactions when the axis of the cylinder lies in a vertical plane (as shown) and the shaft is rotating at a constant angular velocity of $\omega = 60$ rad/s. Assume that the bearing at B resists any motion of the shaft in the axial direction and that the mass of the shaft is negligible.

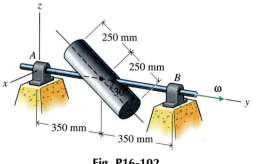

Fig. P16-102

16-103 The 12-in.-long circular cylinder of Fig. P16-103 has a diameter of 6 in., weighs 4 lb, and is mounted on a slender horizontal shaft along its "diagonal." Determine the bearing reactions when the axis of the cylinder lies in a vertical plane (as shown) and the shaft is rotating at a constant angular velocity of $\omega = 60$ rev/min. Assume that the bearing at B resists any motion of the shaft in the axial direction and that the mass of the shaft is negligible.

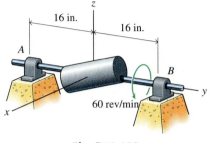

Fig. P16-103

16-104* A 75-mm-diameter solid circular cylinder with hemispherical ends is mounted on a slender horizontal shaft as shown in Fig. P16-104. The mass of the cylinder is 6 kg, and the mass of each hemisphere is 1 kg. Determine the bearing reactions when the axis of the cylinder lies in a vertical plane (as shown) and the angular velocity and angular acceleration of the shaft are $\omega = 50$ rad/s and $\alpha = 15$ rad/s^2, respectively. Assume that the bearing at B resists any motion of the shaft in the axial direction and that the mass of the shaft is negligible.

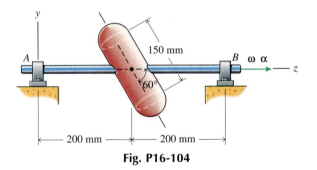

Fig. P16-104

16-105 The thin circular plate of Fig. P16-105 has a radius of 9 in., weighs 15 lb, and rotates at a constant rate of 150 rev/min about a slender horizontal axis. Determine the bearing reactions when the shaft is in the position shown. Assume that the bearing at B resists any motion of the shaft in the axial direction and that the mass of the shaft is negligible.

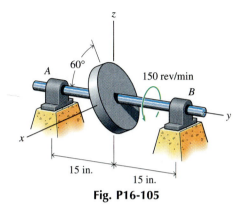

Fig. P16-105

16-106 A 5-kg slender rod AB and a 6-kg uniform sphere are supported on a vertical shaft as shown in Fig. P16-106. Determine the reaction at A and the tension in cable CD when the shaft is rotating counterclockwise with a constant angular velocity of 30 rad/s.

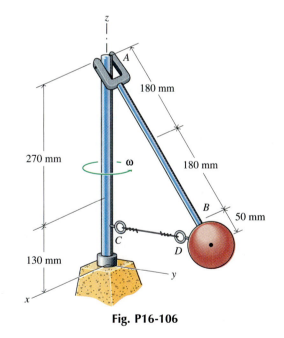

Fig. P16-106

Challenging Problems

16-107* The solid circular disk shown in Fig. P16-107 weighs 25 lb. The disk is rotating on the shaft AB with a constant angular velocity of 500 rev/min. At the same time, the shaft AB is rotating in a vertical plane about a pin at support A. Determine the reaction at support A when the system is in the position shown and the angular velocity and angular acceleration of the shaft are $\omega_z = 20$ rad/s and $\alpha_z = 5$ rad/s^2, respectively. Assume that the mass of the shaft may be neglected.

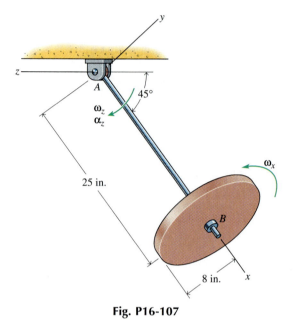

Fig. P16-107

16-108* The solid circular disk shown in Fig. P16-108 rolls without slipping along a circular path on the horizontal surface as the shaft on which it is mounted rotates about the vertical post. The mass of the disk is 50 kg. Assume that the mass of the shaft supporting the disk may be neglected and that the bearings at A and B both slide freely on the vertical post. Determine the bearing reactions and the force between the disk and the horizontal surface when the shaft is rotating with a constant angular velocity $\omega_y = 25$ rad/s.

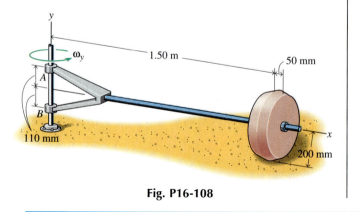

Fig. P16-108

16-109 The crank shown in Fig. P16-109 is rotating counterclockwise with a constant angular velocity of 20 rad/s. Bar AB weighs 20 lb and is connected to the crank at A and to the slider at B with ball and socket joints. Determine the reactions at ends A and B of the bar when the crank is in the position shown in the figure.

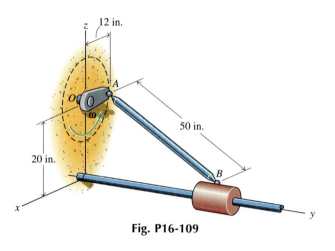

Fig. P16-109

16-110 The horizontal turntable shown in Fig. P16-110 is rotating about a vertical axis with a constant angular velocity of $\omega = 20$ rad/s. Bar AB has a mass of 2 kg and is connected to the slider at A and to the turntable at B with ball-and-socket joints. Determine the reactions at ends A and B of the bar when the turntable is in the position shown in the figure.

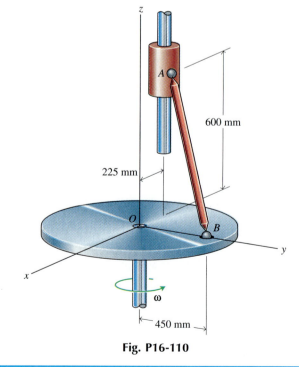

Fig. P16-110

16-6 D'ALEMBERT'S PRINCIPLE—REVERSED EFFECTIVE FORCES

Newton's second law of motion as applied to a particle or to the mass center of a rigid body is given by Eq. 15-16 as

$$\mathbf{R} = \Sigma\mathbf{F} = m\mathbf{a}_G$$

Jean Le Rond d'Alembert (1717–1783) was the first to suggest that a system of inertia forces ($m\mathbf{a}_G$) can be added to a system of actual forces in a dynamics problem to obtain a force system that is in equilibrium.[2] The process, known as d'Alembert's principle, can be expressed mathematically as

$$\mathbf{R} + (-m\mathbf{a}_G) = \mathbf{R} + \mathbf{F}_{re} = 0 \qquad (a)$$

The term $\mathbf{F}_{re} = (-m\mathbf{a}_G)$ in Eq. a is known as a reversed effective force. Reversed effective forces are not true forces, since they do not represent the action of a different body on the body of interest.

For problems involving translation of a rigid body, solution by use of d'Alembert's principle is accomplished by placing the reversed effective force $\mathbf{F}_{re} = (-m\mathbf{a}_G)$ at the mass center of the body when the free-body diagram is drawn. The equilibrium equations $\Sigma\mathbf{F} = 0$ and $\Sigma\mathbf{M} = 0$ are then applied using all forces on the free-body diagram (including the reversed effective force). Moment equations used for solution of the problem can be written with respect to points on or off the body. By selecting a moment center to eliminate a number of unknowns from the moment equation, the need to solve simultaneous equations is frequently avoided.

Application of d'Alembert's principle becomes involved when the body has angular motion. For a rigid body experiencing plane motion, reversed effective couples in addition to reversed effective forces must be applied to the free-body diagram. With the xy-plane as the plane of motion and the mass center G as the origin of the xyz-coordinate system, the reversed effective forces and couples that must be placed on the free-body diagram are

$$
\begin{aligned}
\mathbf{F}_{rex} &= -ma_{Gx}\mathbf{i} & \mathbf{C}_{rex} &= -(-\alpha I_{Gzx} + \omega^2 I_{Gyz})\mathbf{i} \\
\mathbf{F}_{rey} &= -ma_{Gy}\mathbf{j} & \mathbf{C}_{rey} &= -(-\alpha I_{Gyz} - \omega^2 I_{Gzx})\mathbf{j} \\
\mathbf{F}_{rez} &= 0 & \mathbf{C}_{rez} &= -(\alpha I_{Gz})\mathbf{k} \qquad (b)
\end{aligned}
$$

The reversed effective forces must be placed at the mass center of the body. The reversed effective couples can be placed anywhere on the body. Note that the moments and products of inertia in Eqs. b are with respect to axes through the mass center of the body. Again, the equilibrium equations $\Sigma\mathbf{F} = 0$ and $\Sigma\mathbf{M} = 0$ can be used to solve the motion problem by using the reversed effective forces, the reversed effective couples, and the applied forces and couples shown on the free-body diagram. Proper selection of moment centers for moment equations can simplify the solution.

[2]Dr. Ernst Mach, *The Science of Mechanics,* 9th ed., The Open Court Publishing Company, LaSalle, Ill., 1942. Originally published in German in 1893 and translated from German to English by Thomas H. McCormack in 1902.

For general three-dimensional motion problems with principal axes and the origin of the *xyz*-coordinate system at the mass center *G* of the body, the reversed effective forces and couples required to solve a dynamics problem by using d'Alembert's principle are

$$\mathbf{F}_{rex} = -ma_{Gx}\mathbf{i} \qquad \mathbf{C}_{rex} = -[I_{Gx}\alpha_x - (I_{Gy} - I_{Gz})\omega_y\omega_z]\mathbf{i}$$
$$\mathbf{F}_{rey} = -ma_{Gy}\mathbf{j} \qquad \mathbf{C}_{rey} = -[I_{Gy}\alpha_y - (I_{Gz} - I_{Gx})\omega_z\omega_x]\mathbf{j}$$
$$\mathbf{F}_{rez} = -ma_{Gz}\mathbf{k} \qquad \mathbf{C}_{rez} = -[I_{Gz}\alpha_z - (I_{Gx} - I_{Gy})\omega_x\omega_y]\mathbf{k} \qquad (c)$$

D'Alembert's principle provides an alternative method for solving dynamics problems, which some instructors find appealing. Since the method does not provide additional information, it will not be pursued further in this text. The following two example problems are provided to illustrate the procedure for those with an interest in the method.

EXAMPLE PROBLEM 16-18

An automobile with a wheelbase of 114 in. weighs 3500 lb. The center of mass of the car is 48 in. behind the front axle and 22 in. above the surface of the road (Fig. 16-51). Determine the normal forces exerted by the road on the front and rear wheels as the speed of the car is reduced uniformly from 60 mi/h to 30 mi/h in a distance of 150 ft on a level stretch of road.

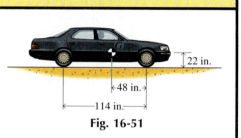

Fig. 16-51

SOLUTION

As long as the tires remain in contact with the pavement, the motion of the mass center of the automobile will be along a straight line parallel to the pavement. The motion is translation ($\boldsymbol{\omega} = \boldsymbol{\alpha} = \mathbf{a}_{Gy} = 0$) with $\mathbf{a}_G = a_{Gx}\mathbf{i} = $ constant. Rewriting the acceleration using the chain rule of differentiation

$$a_{Gx} = \frac{dv}{dt} = \frac{dv}{dx}\frac{dx}{dt} = v\frac{dv}{dx}$$

and integrating

$$\int v\,dv = a_{Gx}\int dx$$

gives

$$\frac{v^2}{2} = a_{Gx}x + C_1 = a_{Gx}x + 3872$$

where the constant of integration $C_1 = (88.00)^2/2 = 3872 \text{ ft}^2/\text{s}^2$, since $v = 60 \text{ mi/h} = 88.00 \text{ ft/s}$ when $x = 0$. Then, if the car slows to $v = 30 \text{ mi/h} = 44.00 \text{ ft/s}$ when $x = 150 \text{ ft}$, the acceleration of the car is

$$\frac{44^2}{2} = a_{Gx}(150) + 3872$$
$$a_{Gx} = -19.360 \text{ ft/s}^2$$

and the reversed effective force is

$$\mathbf{F}_{re} = -m\mathbf{a}_G = -\frac{3500}{32.2}(-19.360\mathbf{i})$$
$$= 2104\mathbf{i} \text{ lb} = 2104 \text{ lb}\rightarrow$$

A free-body diagram of the automobile is shown in Fig. 16-52. Assuming four-wheel braking, there will be frictional forces at both the front and back wheels. However, it is not given that the wheels are skidding or even that slip is impending. Therefore, the frictional forces are simply labeled F_A and F_B. The reversed effective force is directed toward the front of the car ($\mathbf{F}_{re} = 2104\mathbf{i}$ lb), since the acceleration of the car is toward the rear ($\mathbf{a}_G = -19.360\mathbf{i}$).

The front-wheel reaction N_B is determined using a moment equation with the moment center at A to eliminate the frictional forces F_A and F_B

$$+\lfloor\Sigma M_A = N_B(114) - 3500(66) - 2104(22) = 0$$
$$N_B = 2432 \text{ lb} \cong 2430 \text{ lb}\uparrow \qquad\qquad \text{Ans.}$$

Similarly, the rear-wheel reaction N_A is calculated by using a moment equation with the moment center at B

$$+\lceil\Sigma M_B = N_A(114) - 3500(48) + 2104(22) = 0$$
$$N_A = 1068 \text{ lb}\uparrow \qquad\qquad \text{Ans.}$$

As a check

$$+\uparrow\Sigma F = N_A + N_B - 3500 = 2432 + 1068 - 3500 = 0$$

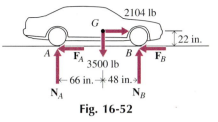

Fig. 16-52

Since the automobile is slowing down, the acceleration $a_G = 19.360$ ft/s² and the inertia vector $ma_G = 2104$ slug · ft/s² both point toward the rear of the car. Therefore, the reversed effective force $F_{re} = 2104$ lb points toward the front of the car.

EXAMPLE PROBLEM 16-19

A thin 600-mm-diameter circular disk with a mass of 60 kg is supported on an inclined surface by a block and a cable wrapped around a shallow groove in the curved surface of the disk as shown in Fig. 16-53. Determine the tension T in the cable and the acceleration a_G of the mass center of the disk once the block is removed and the disk is free to slide on the inclined surface. The kinetic coefficient of friction between the disk and the inclined surface is 0.20.

SOLUTION

A free-body diagram for the disk is shown in Fig. 16-54. Motion of the disk is possible only if slipping occurs; therefore, $F = \mu N$. Since the center of the disk is always a distance R away from the flat surface, the acceleration of the center of the disk is $\mathbf{a}_G = -a_G\mathbf{i}$ (down the incline) and the reversed effective force (which is applied at the mass center G)

$$\mathbf{F}_{re} = -m\mathbf{a}_G = -60(-a_G\mathbf{i}) = 60a_G\mathbf{i}$$

acts up the incline on the free-body diagram. Since the disk does not slip on the cable, the acceleration of point A can have no x-component ($a_{Ax} = 0$), the disk must rotate clockwise about point A ($\boldsymbol{\alpha} = -\alpha\mathbf{k}$), and the x-component of the relative acceleration equation $\mathbf{a}_G = \mathbf{a}_A + \mathbf{a}_{G/A}$ gives

$$-a_G = 0 + (-0.300\alpha)$$

where a_G is the acceleration of the disk down the incline and α is the clockwise angular acceleration of the disk. Then the reversed effective couple

$$\mathbf{C}_{re} = -I_G\boldsymbol{\alpha} = -I_G(-\alpha\mathbf{k}) = I_G(3.333a_G\mathbf{k}) = 3.333I_Ga_G\mathbf{k}$$

is drawn as a counterclockwise couple on the free-body diagram.

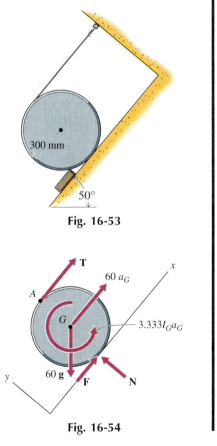

Fig. 16-53

Fig. 16-54

The equilibrium equations can now be written for the disk

$+\angle\Sigma F_x = 0:$ $\qquad F + T + 60a_G - 60(9.81)\sin 50° = 0$

$+\Sigma F_y = 0:$ $\qquad N - 60(9.81)\cos 50° = 0$

$+\langle\Sigma M_A = 0:$ $\quad 3.333 I_G a_G + 0.300(60a_G) + 0.600(0.2N)$
$\qquad\qquad\qquad - 0.300[60(9.81)\sin 50°] = 0$

where the moment of inertia of the disk is

$$I_G = \frac{1}{2}mR^2 = \frac{1}{2}(60)(0.300)^2 = 2.700 \text{ kg} \cdot \text{m}^2$$

Solving these equations gives

$N = 378.3 \text{ N}\searrow$

$a_G = 3.329 \text{ m/s}^2 \cong 3.33 \text{ m/s}^2\nearrow$ $\qquad\qquad\qquad$ Ans.

$T = 175.55 \text{ N} \cong 175.6 \text{ N}$ $\qquad\qquad\qquad$ Ans.

SUMMARY

Any system of forces acting on a rigid body can be replaced by an equivalent system that consists of a resultant force **R**, whose line of action passes through the mass center G of the body, and a resultant couple **M**$_G$. The motion of most rigid bodies consists of the superposition of a translation produced by the resultant force **R** and a rotation produced by the couple **M**$_G$.

The translational motion of the mass center G of a rigid body is identical to the motion of a single particle located at the mass center and whose mass is equal to the total mass of the rigid body

$$\mathbf{R} = m\mathbf{a}_G \qquad\qquad (15\text{-}16)$$

The general, three-dimensional, rotational motion of the rigid body is most conveniently expressed by Euler's equations

$$M_{Gx} = I_{Gx}\alpha_x - (I_{Gy} - I_{Gz})\omega_y\omega_z \qquad (16\text{-}21a)$$
$$M_{Gy} = I_{Gy}\alpha_y - (I_{Gz} - I_{Gx})\omega_z\omega_x \qquad (16\text{-}21b)$$
$$M_{Gz} = I_{Gz}\alpha_z - (I_{Gx} - I_{Gy})\omega_x\omega_y \qquad (16\text{-}21c)$$

where the *xyz*-coordinate system has its origin at the center of mass G of the body and the coordinate directions are selected to coincide with the principal axes through the mass center. Since the body is rotating and the orientation of the *xyz*-coordinate system must be fixed, the evaluation of the moments and products of inertia will not, in general, be constant. Only for highly symmetric bodies will these moments and products of inertia be independent of the orientation of the body. Therefore, if it is desired to integrate the angular acceleration to obtain the angular velocity of the body, then the moments and products of inertia must include any dependence on time or orientation of the body.

A very large number of important dynamics problems involve plane motion—a motion in which all elements of the body move in parallel planes. The parallel plane that passes through the mass center G of the body is known as the *plane of motion*. When a rigid body is experiencing plane motion, the angular velocity and angular acceleration vectors $\boldsymbol{\omega}$ and $\boldsymbol{\alpha}$, respectively, are parallel to each other and are both perpendicular to the plane of motion. If the xyz-coordinate system is chosen so that the motion is parallel to the xy-plane and the body is symmetric with respect to the plane of motion, Euler's equations reduce to

$$M_{Gz} = I_{Gz}\alpha$$

For these symmetric bodies, the x- and y-components of Euler's equations are satisfied identically and provide no useful information.

The equations of motion (force and moment) generally contain more unknowns than there are equations, so additional equations are usually necessary. The additional equations are the kinematic relations between the acceleration of the mass center $\mathbf{a}_G$ and the angular acceleration α, and are generally given by the relative velocity and relative acceleration equations.

REVIEW PROBLEMS

16-111* A brake for regulating the descent of a body is shown in Fig. P16-111. The rotating parts of the brake (cable sheave and brake drum) weigh 250 lb and have a radius of gyration with respect to their axis of rotation of 4.25 in. The kinetic coefficient of friction between the brake pad and the drum is $\mu_k = 0.50$. The weight of body C is 1000 lb. When a force $\mathbf{P}$ of 150 lb is being applied to the brake lever, determine

a. The tension in the cable T.
b. The acceleration a_C of body C.
c. The horizontal and vertical components of the reaction at support B of the brake lever.

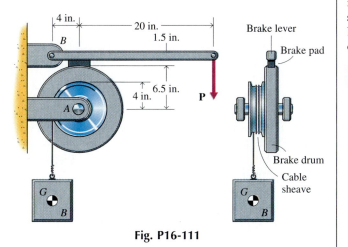

Fig. P16-111

16-112* A 50-kg solid circular cylinder A rests on a 35-kg block B as shown in Fig. P16-112. When a force $\mathbf{F}$ is applied to the block, the block accelerates at a rate of 5 m/s². If the coefficients of friction between the block and the cylinder and between the block and the horizontal surface are both $\mu_s = \mu_k = 0.25$, determine the magnitude of $\mathbf{F}$ and the frictional force exerted by the block on the cylinder.

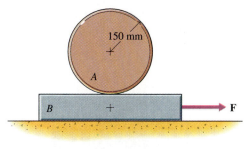

Fig. P16-112

16-113 Bar AB of Fig. P16-113 weighs 60 lb and has a uniform cross section. It is attached to the 50-lb cart with a smooth pin at A and rests against a smooth surface at B. Determine the magnitude of force $\mathbf{P}$ and the forces exerted on the bar at supports A and B

a. If the acceleration of the cart is 15 ft/s² to the right.
b. If the reaction at support B is zero.

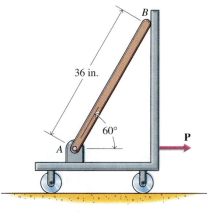

Fig. P16-113

16-114 Bars AB ($m_{AB} = 10$ kg) and BC ($m_{BC} = 15$ kg) of Fig. P16-114 have uniform cross sections. When the bars are in the position shown, collar C is moving to the left with a velocity of 1 m/s and its speed is increasing at a rate of 2 m/s^2. Determine the forces exerted on bar AB by the pins at A and B and the force **F** necessary to produce this motion. Neglect the mass of the collar C.

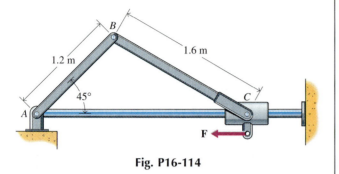

Fig. P16-114

16-115* The speed of the 3100-lb automobile shown in Fig. P16-115 is reduced uniformly from 60 mi/h to 12 mi/h in a distance of 150 ft. Determine the normal forces exerted by the pavement on the front and rear wheels during the braking action and the minimum coefficient of friction required between the tires and the pavement. The car has four-wheel brakes.

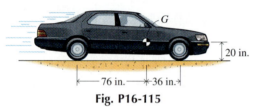

Fig. P16-115

16-116* The 50-kg unbalanced wheel shown in Fig. P16-116 rolls without slipping on the horizontal surface. The radius of gyration of the wheel with respect to an axis through its mass center is 160 mm. In the position shown, the angular velocity of the wheel is $\omega = 6$ rad/s. For this instant, determine the angular acceleration of the wheel and the force acting on the wheel at its point of contact with the horizontal surface.

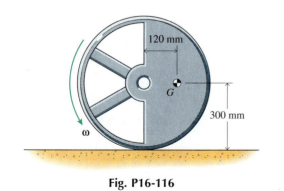

Fig. P16-116

16-117 A thin, circular ring of uniform thickness rotates in a vertical plane about a frictionless pin at point A as shown in Fig. P16-117. The ring has a diameter of 24 in. and weighs 5 lb. When the center of the ring is directly above the support ($\theta = 0°$), the angular velocity of the ring is 10 rad/s, clockwise. Determine the angular acceleration of the ring and the horizontal and vertical components of the force exerted on the ring by the pin at support A when $\theta = 90°$.

Fig. P16-117

16-118 The solid circular disk A shown in Fig. P16-118 has a mass of 50 kg and a radius of 200 mm. A cable wrapped around a shallow groove in the disk is attached to bar BC, which has a uniform cross section and a mass of 25 kg. If the system is released from rest in the position shown (with bar BC horizontal), determine for this instant

a. The angular acceleration of the disk α_A.
b. The tension T in the cable.
c. The horizontal and vertical components of the reaction at support C.

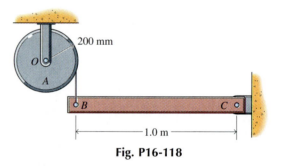

Fig. P16-118

16-119 The 30-lb slender bar shown in Fig. P16-119 has a uniform cross section. Determine the acceleration of the mass center of the bar and the reaction at support A immediately after the cord at support B is cut

a. If the horizontal surface at A is smooth ($\mu = 0$).
b. If the horizontal surface at A is rough ($\mu = 0.25$).

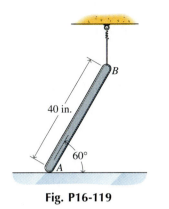

Fig. P16-119

16-120 A 1000-kg uniform crate rests on the back of a 2500-kg rear-wheel-drive flatbed truck as shown in Fig. P16-120. The center of mass of the truck is located 2 m behind the front axle and 0.85 m above the pavement. The static coefficient of friction between the crate and the truck bed is $\mu_s = 0.25$. If the crate does not slip or tip, determine the maximum permissible acceleration and the minimum static coefficient of friction between the tires and the pavement required to achieve this acceleration.

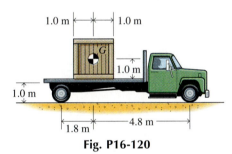

Fig. P16-120

16-121* A 64-lb block is supported by a 15-lb bar with a uniform cross section and a 25-lb solid circular disk of uniform thickness as shown in Fig. P16-121. The vertical surfaces in contact with the block are smooth, and the disk rolls without slipping on the horizontal surface. In the position shown, the velocity of pin A is 2.5 ft/s to the right. If the magnitude of force **F** is 75 lb, determine the angular velocity and angular acceleration of bar AB and the force exerted by the pin at B on the block.

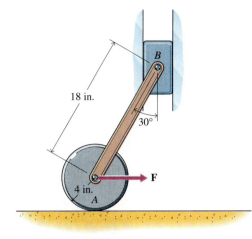

Fig. P16-121

16-122* A 40-kg solid circular disk A is connected with a link to a 50-kg block B. Both rest on a rough ($\mu = 0.25$) inclined plane as shown in Fig. P16-122. During motion of the system, determine

a. The force in the link.
b. The normal components of the reactions at supports C and D of the block.
c. The angular acceleration α_A of the disk.

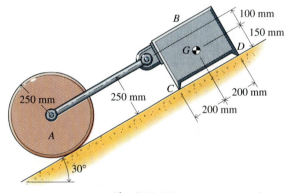

Fig. P16-122

16-123 A uniform cylinder is balanced on the edge of a ledge as shown in Fig. P16-123. The corner is rough so that the cylinder will not slip. If the cylinder is disturbed and begins to roll off the edge, determine the maximum angle θ_m through which the cylinder will rotate without losing contact with the corner.

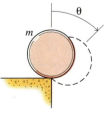

Fig. P16-123

16-124* A 25-kg frame is supported in a vertical plane by two pairs of links and a cable as shown in Fig. P16-124. A 10-kg block rests on the frame. If the cable breaks, determine the force carried by each pair of links and the force exerted on the block by the frame after the links have rotated 30° from their initial horizontal position. Neglect the masses of the links.

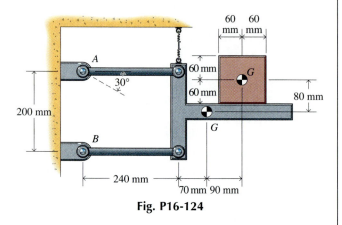

Fig. P16-124

16-125 The 25-lb horizontal turntable shown in Fig. P16-125 has a radius of gyration with respect to the axis of rotation of 14 in. A 3-lb block B rests on the turntable in the position shown. The coefficient of friction between the block and the table is $\mu_s = 0.55$. If a constant 2 lb · ft couple C is applied when the system is at rest, determine the number of rotations before the block begins to slip and the angular velocity of the table when slipping begins.

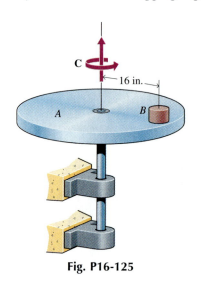

Fig. P16-125

16-126* A 25-kg half disk is supported by a cable at A as shown in Fig. P16-126. When the cable at A is cut, the disk rolls without slipping on the horizontal surface. At the instant motion begins, determine

a. The acceleration a_G of the mass center of the half disk.
b. The angular acceleration α of the half disk.
c. The normal and frictional forces exerted on the half disk by the horizontal surface.

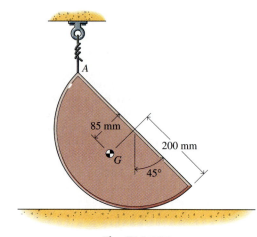

Fig. P16-126

16-127 A continuous cable supports an 80-lb disk A and a 50 lb block B as shown in Fig. P16-127. There is no slipping between the disk and the cable. During motion of the system, determine

a. The acceleration a_B of the block B.
b. The angular acceleration α_A of the disk A.
c. The tension T_C in the cable at the support C.

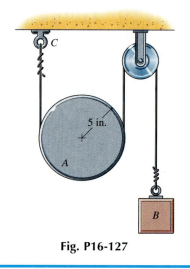

Fig. P16-127

As the skiers ski down the mountain, they trade potential energy (height) for kinetic energy (speed).

Chapter 17

KINETICS OF PARTICLES: WORK AND ENERGY METHODS

17-1   INTRODUCTION

 The preceding two chapters solved kinetics problems using Newton's second law. This chapter and the next will present an alternate method—the method of work and kinetic energy—that is useful for solving certain types of kinetics problems.

In the Newton's second law approach, the instantaneous equations of motion were used to relate the forces acting on a body to the acceleration of the body. If the equations were applied to a specific position of the body, only instantaneous relationships were obtained. If the equations were applied to an arbitrary position of the body, then the resulting acceleration could be integrated using the principles of kinematics discussed in Chapters 13 and 14.

The work–energy method combines the principles of kinematics with Newton's second law to directly relate the position and speed of a body. In this approach, Newton's second law is integrated in a general sense with respect to position. Clearly, for this method to be useful, the forces acting on a body must be a function of position only. For some types of forces, however, the resulting integrals can be evaluated explicitly. The result is a simple algebraic equation that relates the speed of the body at two different positions of its motion.

Since the work–energy method combines the principles of kinematics with Newton's second law, it is not a new or independent principle. There are no problems that can be solved with the work–energy

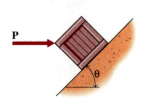

Figure 17-1 A crate of mass m is pushed up a frictionless incline by a constant horizontal force **P**.

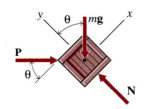

Figure 17-2 *Free-body diagram of the crate.* Since the crate translates in the x-direction, $ma_y = 0$.

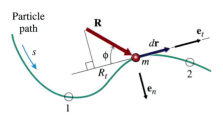

Figure 17-3 A particle of mass m moves along a general curved path. The force **R** represents the sum of all external forces acting on the particle. At every point along the curve, a normal and tangential coordinate system is set up with $\mathbf{e}_t$ in the direction of motion ($d\mathbf{r}$).

method cannot be solved using Newton's second law. However, when the work–energy method does apply, it is usually the easiest method of solving a problem.

17-2 FORCE, DISTANCE, AND VELOCITY

Many problems in dynamics consist of finding the speed of an object after it has been pushed or pulled a certain distance by some force. Solving such problems with Newton's second law always consists of the same set of steps. For example, consider the crate of Fig. 17-1, which is being pushed up an inclined plane by a constant horizontal force P. If friction can be neglected and the crate has an initial speed of v_i, what will be the speed of the crate after it has moved a distance d up the incline?

The free-body diagram of the crate is shown in Fig. 17-2. The component of Newton's second law along the inclined plane (in the direction of the motion) is

$$+\nearrow\Sigma F_x = ma_x: \qquad P\cos\theta - mg\sin\theta = ma \qquad (a)$$

Rewriting the acceleration using the chain rule of differentiation

$$a = \frac{dv}{dt} = \frac{dv}{dx}\frac{dx}{dt} = v\frac{dv}{dx}$$

Equation a can be rearranged and integrated

$$\int (P\cos\theta - mg\sin\theta)dx = m\int v\,dv$$

to get

$$(P\cos\theta - mg\sin\theta)d = m\frac{v_f^2 - v_i^2}{2} \qquad (b)$$

where the constant of integration has been chosen so that $v = v_i$ when $x = 0$.

The left-hand side of Eq. b depends on many details about the particular problem being solved, including exactly what forces act on the crate, in what direction the forces act, and how far the crate moves. The form of the right-hand side, however, is independent of the particular problem being solved. It always involves the same manipulation with the chain rule of differentiation. It always involves the same integration, $\int v\,dv$. It always results in the same right-hand side, $mv^2/2$. In fact, the result doesn't even depend on the particle moving along a straight path.

17-3 KINETIC ENERGY

Suppose the crate of Fig. 17-1 is replaced with a particle moving on a general space curve as shown in Fig. 17-3. The force $\mathbf{R}$ on the free-body diagram of Fig. 17-3 represents the resultant of all external forces acting on the particle. The tangential component of Newton's second law for the particle is

$$\Sigma F_t = ma_t: \qquad R\cos\phi = m\frac{dv}{dt} \qquad (c)$$

where $\mathbf{v} = v\mathbf{e}_t = \dfrac{ds}{dt}\mathbf{e}_t$ and $d\mathbf{r} = ds\,\mathbf{e}_t$. Then, using the chain rule of differentiation, Eq. c can be written

$$R\cos\phi = m\frac{dv}{ds}\frac{ds}{dt} = mv\frac{dv}{ds} \qquad (d)$$

Finally, rearranging Eq. d and integrating along the particle's path from point 1 to point 2 gives

$$\int_{s_1}^{s_2} R\cos\phi\,ds = m\int_{v_1}^{v_2} v\,dv = m\frac{v_2^2 - v_1^2}{2} \qquad (e)$$

and the right-hand side of Eq. e for a particle moving along a general space curve is the same as the right-hand side of Eq. b for the crate moving up the inclined plane.

The combination of terms on the right-hand side of Eqs. b and e occurs so often it has been given a name and a symbol. The *kinetic energy* (T) of a particle of mass m moving with a speed v is

$$T = \frac{1}{2}mv^2 \qquad \textbf{(17-1)}$$

In terms of the kinetic energy, Eq. e can be written

$$\int_{s_1}^{s_2} R\cos\phi\,ds = T_2 - T_1 \qquad \textbf{(17-2)}$$

Since the mass m and the square of the speed v^2 are both positive, the kinetic energy of a particle will always be positive. Kinetic energy has dimensions of $ML^2/T^2 = FL$, or force times length. In the SI system of units, this combination of dimensions is called a *joule* (1 J = 1 N · m). In the U.S. Customary system of units, there is no special unit for kinetic energy. It is expressed simply as ft · lb.

17-4   WORK OF A FORCE

Unlike the kinetic energy terms on the right-hand side of Eqs. b and e, the exact form of the terms on the left-hand side of Eqs. b and e depends on what forces act on the particle and on the directions of the forces relative to the direction of the motion of the particle. The definition of work that follows is designed so that the left-hand side of Eqs. b and e will be identified as the sum of the work done by all forces acting on the particle.

In mechanics, a force does work only when the particle to which the force is applied moves. For example, when a constant force $\mathbf{P}$ is applied to a particle, which moves a distance d in a straight line as shown in Fig. 17-4, the work done on the particle by the force $\mathbf{P}$ is defined by the scalar product

$$U = \mathbf{P}\cdot\mathbf{d} = Pd\cos\phi$$
$$= P_x\,d_x + P_y\,d_y + P_z\,d_z \qquad \textbf{(17-3)}$$

where ϕ is the angle between the vectors $\mathbf{P}$ and $\mathbf{d}$. Equation 17-3 is usually interpreted as: *The work done by the force is the product of the magnitude of the force $\mathbf{P}$ and the rectangular component of the displacement in the direction of the force $d\cos\phi$* (Fig. 17-4). However, the $\cos\phi$ can be asso-

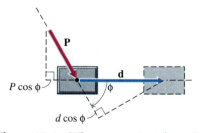

Figure 17-4   When a constant force $\mathbf{P}$ acts on a particle that moves a distance d in a straight line, the work done on the particle by the force is the product of the magnitude of the force (P) and the rectangular component of the displacement in the direction of the force ($d\cos\phi$).

ciated with the force **P** instead of with the displacement **d**. Then Eq. 17-3 would be interpreted as: *The work done by the force is the product of the magnitude of the displacement d and the rectangular component of the force in the direction of the displacement P cos ϕ* (Fig. 17-4).

When $0 \le \phi < 90°$, the force and displacement are in the same direction and the work done by the force is positive. When $90° < \phi \le 180°$, the force and displacement are in opposite directions and the work done by the force is negative. When $\phi = 90°$, the force and displacement are perpendicular and the work done by the force is zero. Of course, the work done by the force is also zero if the displacement is zero ($d = 0$).

Work has dimensions of force times length, the same as kinetic energy. It may be noted that work, kinetic energy, and the moment of a force all have the same dimensions: They are all force times length. However, work or energy and moment of a force are two totally different concepts, and the special unit *joule* should only be used to describe work and energy. The moment of a force must always be expressed as N · m in the SI system of units.

If the force is not constant or if the displacement is not a straight line, Eq. 17-3 gives the work done by the force only during an infinitesimal part of the displacement, $d\mathbf{r}$:

$$dU = \mathbf{P} \cdot d\mathbf{r} = P \, ds \cos \phi = P_t \, ds$$
$$= P_x \, dx + P_y \, dy + P_z \, dz \tag{17-4}$$

where $d\mathbf{r} = ds \, \mathbf{e}_t = dx \, \mathbf{i} + dy \, \mathbf{j} + dz \, \mathbf{k}$. Integrating Eq. 17-4 along the particle path from position 1 to position 2 gives the work done by the force $U_{1\to2}$

$$U_{1\to2} = \int_1^2 dU = \int_1^2 \mathbf{P} \cdot d\mathbf{r} = \int_{s_1}^{s_2} P_t \, ds$$
$$= \int_{x_1}^{x_2} P_x \, dx + \int_{y_1}^{y_2} P_y \, dy + \int_{z_1}^{z_2} P_z \, dz \tag{17-5}$$

17-4.1   Work Done by a Constant Force

When a constant force $\mathbf{P} = P_x\mathbf{i} + P_y\mathbf{j} + P_z\mathbf{k}$ is applied to a particle, Eq. 17-5 gives the work done by the force on the particle as

$$U_{1\to2} = P_x\int_{x_1}^{x_2}dx + P_y\int_{y_1}^{y_2}dy + P_z\int_{z_1}^{z_2}dz$$
$$= P_x(x_2 - x_1) + P_y(y_2 - y_1) + P_z(z_2 - z_1) \tag{f}$$

Note that evaluation of the work done by a constant force depends on the coordinates at the end points of the particle's path but not on the actual path traveled by the particle. For the constant force **P** shown in Fig. 17-5, it doesn't matter if the particle moves along path *a* from position 1 to position 2 or along path *b* or along some other path. The work done by the force **P** is always the same. Forces for which the work done is independent of the path are called conservative forces. They will be studied in more detail in Section 17-7.

The weight of a particle *W* is a particular example of a constant force. When bodies move near the surface of the earth, the force of the earth's gravity is essentially constant ($P_x = 0$, $P_y = 0$, and $P_z = -W$). Therefore, the work done on a particle by its weight is $-W(z_2 - z_1)$.

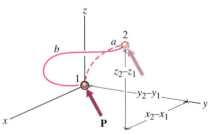

Figure 17-5   For a constant force **P**, the work done on a particle as the particle moves from point 1 to point 2 along path *a* is the same as the work done on the particle as it moves from point 1 to point 2 along path *b*. That is, the work done on a particle by a constant force is independent of the path followed and depends only on where the particle starts and where it ends.

When $z_2 > z_1$, the particle moves upward (opposite the gravitational force) and the work done by gravity is negative. When $z_2 < z_1$, the particle moves downward (in the direction of the gravitational force) and the work done by gravity is positive.

17-4.2   Work Done by a Massless Linear Spring Force

The force required to stretch a massless, linear spring is directly proportional to the stretch in the spring

$$P = k(\ell - \ell_0) = k\delta \qquad (g)$$

where k is a constant known as the *modulus* of the spring and ℓ, ℓ_0, and δ are the present length, unstretched length, and deformation of the spring from its unloaded position, respectively. When the present length is greater than the unstretched length ($\delta > 0$), the spring is stretched and the force P is positive, as shown in Fig. 17-6. When the present length is less than the unstretched length ($\delta < 0$), the spring is compressed and the force P is negative (the force actually pushes on the spring).

When the spring of Fig. 17-6 is connected to a particle, the force exerted on the particle is equal in magnitude and opposite in direction to that on the spring (Fig. 17-7). If the particle moves from position 1 (where the deformation of the spring is δ_1), to position 2 (where the deformation of the spring is δ_2), the work done on the particle by the spring will be

$$U_{1 \to 2} = \int_{\delta_1}^{\delta_2} - k\delta \, d\delta = -\frac{1}{2}k(\delta_2^2 - \delta_1^2) \qquad (h)$$

where the minus sign arises because the force $k\delta$ acts to the left on the particle and the displacement $d\delta$ is to the right. If $0 < \delta_1 < \delta_2$, then the net motion of the particle is to the right (opposite the spring force) and the work done on the particle is negative. If $0 < \delta_2 < \delta_1$, then the spring force and the motion are both to the left and the work done on the particle is positive.

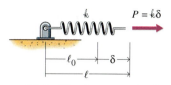

Figure 17-6   The force required to stretch an elastic spring is directly proportional to the stretch $P = k(\ell - \ell_0) = k\delta$. The constant of proportionality k is called the spring modulus.

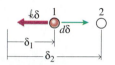

Figure 17-7   When a particle that is attached to an elastic spring moves to increase the stretch in the spring, the spring will do negative work on the particle. Since the force exerted on the particle by the spring is not constant, the work done on the particle must be computed using integration.

When the athlete leans against a wall to stretch his muscles, the man exerts a force on the wall and the wall exerts a force of equal magnitude back on the man. Why is it that these forces do no work?

Fig. CE17-1

SOLUTION

When the athlete leans against the wall, he exerts a force **P** on the wall. If he maintains that position long enough, his arms will eventually feel tired. Popular experience is that if doing something makes us feel tired, then it must be work. The mechanics definition of work, however, requires a motion in the direction of a force. Although the athlete pushes on the wall with some force **P**, the wall does not move. Since there is no motion of the wall in the direction of the force **P**, the force does no work on the wall.

Although the mechanics definition of work seems contrary to popular experience, it is a practical one. In mechanics, the name *work* is given to the integral of the force that arises when Newton's second law is integrated with respect to distance along the path of the motion

$$U_{1 \to 2} = \int_1^2 \mathbf{F} \cdot (ds\ \mathbf{e}_t) = \int_1^2 m\mathbf{a} \cdot (ds\ \mathbf{e}_t) = \frac{v_2^2 - v_1^2}{2} = T_2 - T_1$$

The scalar product $\mathbf{F} \cdot (ds\ \mathbf{e}_t)$ will be zero if the force is zero ($\mathbf{F} = \mathbf{0}$), if the displacement is zero ($ds = 0$), or if the force $\mathbf{F}$ is perpendicular to the displacement $ds\ \mathbf{e}_t$. All three cases are included in the statement that a force does no work if there is no motion in the direction of the force.

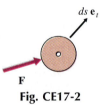

$ds\ \mathbf{e}_t$

F

Fig. CE17-2

A 16-lb crate slides down a ramp as shown in Fig. 17-8. If the crate is released from rest 10 ft above the bottom of the ramp and the coefficient of friction between the crate and the ramp is $\mu_k = 0.20$, determine

a. The work done on the crate by gravity as the crate slides to the bottom of the ramp.
b. The work done on the crate by the normal force N as the crate slides to the bottom of the ramp.
c. The work done on the crate by the frictional force F as the crate slides to the bottom of the ramp.

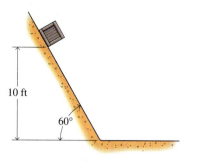

Fig. 17-8

SOLUTION

The free-body diagram of the crate is shown in Fig. 17-9 for a general position along the ramp. As the crate slides down the ramp, it moves 10 ft in the vertical direction, $10/\tan 60° = 5.774$ ft in the horizontal direction, and $10/\sin 60° = 11.547$ ft along the ramp.

a. The gravitational force is constant in both magnitude (16 lb) and in direction (vertical downward). Therefore, the work done on the crate by the gravitational force is just

$$(U_{1 \to 2})_g = Wd = 16(10) = 160.0 \text{ lb} \cdot \text{ft} \qquad \text{Ans.}$$

b. The normal force N does no work because it is perpendicular to the motion

$$(U_{1 \to 2})_N = 0 \text{ lb} \cdot \text{ft} \qquad \text{Ans.}$$

However, even though it does not affect the work–energy equation, the normal force must still be found in order to determine the work done by friction. Since the crate does not move in the y-direction, the y-component of Newton's second law

$$+\nearrow \Sigma F_y = ma_y: \qquad N - 16 \cos 60° = 0$$

is easily solved, giving

$$N = 16 \cos 60° = 8.00 \text{ lb} = \text{constant}$$

Therefore, the friction force is

$$F = \mu_k N = (0.20)(8.00) = 1.600 \text{ lb} = \text{constant}$$

c. The friction force acts along the ramp (constant direction), and it also has a constant magnitude. Therefore, the work done on the crate by friction as the crate slides 11.547 ft down the ramp is

$$(U_{1 \to 2})_F = Fd = -1.6(11.547) = -18.48 \text{ lb} \cdot \text{ft} \qquad \text{Ans.}$$

where the negative sign arises because the friction force acts up the ramp while the motion is down the ramp.

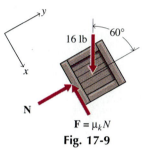

$\mathbf{F} = \mu_k N$
Fig. 17-9

It is important to always draw a complete and proper free-body diagram. A free-body diagram identifies the body to which the work–energy equation will be applied and the forces whose work must be computed. A complete and proper free-body diagram also includes forces that do no work. Forces that do no work are important because they may be needed for other calculations.

The work done by a constant force is the product of the magnitude of the force and the distance the particle moves in the direction of the force $U = Fd$. When the particle moves in the same direction as the force, the distance d and the work U are both positive. When the particle moves in the $-\mathbf{F}$ direction, the distance d and the work U are both negative.

A 24,000-kg jet is launched from the deck of an aircraft carrier by a steam cat-apult as shown in Fig. 17-10a. The engines of the jet exert a constant thrust of 140 kN on the jet, and the force of the catapult on the jet is shown in Fig. 17-10b. If the catapult pushes the jet for a distance of 90 m along the carrier deck, determine

a. The work done on the jet by its engines during launch of the jet.
b. The work done on the jet by the catapult during launch of the jet.

(a)

(b)

Fig. 17-10

SOLUTION

A free-body diagram of the jet is shown in Fig. 17-11. Both the engine thrust T and the catapult force P act in the direction of motion of the jet. The weight of the jet and the normal force on the wheels both act perpendicular to the mo-tion, and neither of these two forces do any work during the launch of the jet.

a. During launch of the jet, the engine thrust is constant in both magnitude and direction. Therefore, the work done on the jet by the engine thrust is just

$$(U_{1\rightarrow2})_T = Td = 140,000(90) = 12.60(10^6)\ \text{J} \qquad \text{Ans.}$$

b. The catapult thrust is constant in direction but does not have a constant magnitude. Therefore, the work done on the jet by the catapult thrust is

$$(U_{1\rightarrow2})_P = \int \mathbf{F} \cdot d\mathbf{r} = \int_0^{90} P\,ds$$

But the value of the integral $\int P\,ds$ is just the area under the graph of Fig. 17-10b and

$$(U_{1\rightarrow2})_P = \frac{1,100,000 + 65,000}{2}(90) = 52.4(10^6)\ \text{J} \qquad \text{Ans.}$$

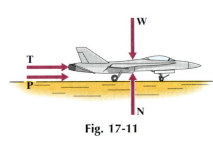

Fig. 17-11

It is often the case that a force will have a constant direction and the variation of its magnitude with dis-tance will be given graphically. Computation of the work done by such a force using integration, $U = \int \mathbf{F} \cdot d\mathbf{r} = \int P\,ds$ requires determin-ing the equation of the curve P versus s. However, the value of the integral $\int P\,ds$ is just the area under the graph of P versus s. If the area is composed of simple shapes such as triangles or rectangles whose ar-eas are easily determined, then the work done by the force is simply the sum of these areas.

PROBLEMS

In the following problems, all ropes, cords, and cables are assumed to be flexible, inextensible, and of negligible mass. All pins and pulleys are small, have negligible mass, and are frictionless.

Introductory Problems

17-1* In a shipping warehouse, packages slide down a chute and fall on the floor as shown in Fig. P17-1. If the coefficient of friction between a 25-lb package and the ramp is $\mu_k = 0.40$ and $\theta = 20°$, determine the work done on the package

a. By the friction force on the bottom of the package as the package slides 10 ft down the chute.

b. By the gravitational force W as the package slides 10 ft down the chute.

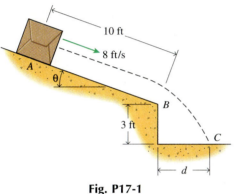

Fig. P17-1

17-2* A tow rope is used to pull skiers up a ski slope as shown in Fig. P17-2. If the coefficient of friction between the skis and the snow is $\mu_k = 0.1$ and the weight of the middle skier is 650 N, determine the work done on the skier

a. By the friction force on the bottom of her skis as the rope pulls her 100 m up the slope.

b. By the gravitational force W as the rope pulls her 100 m up the slope.

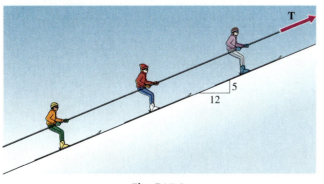

Fig. P17-2

17-3 The 4-lb pendulum bob shown in Fig. P17-3 is released from rest when $\theta = 75°$. Determine the work done on the pendulum bob

a. By the cord tension T as the bob swings from $\theta = 75°$ to $\theta = 0°$.

b. By the gravitational force W as the bob swings from $\theta = 75°$ to $\theta = 0°$.

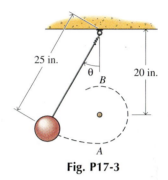

Fig. P17-3

17-4 A train is moving at 30 km/h when a coupling breaks and the last car separates from the train. As soon as the car separates, the brakes are automatically applied, locking all wheels of the runaway car. If the coefficient of friction between the wheels and the rails is $\mu_k = 0.2$ and the tracks slope downward at 5°, determine the work done on the 180,000-kg car

a. By the frictional force on the wheels as the car slides 25 m along the tracks.

b. By the gravitational force W as the car slides 25 m along the tracks.

17-5* A fully loaded Boeing 747 has a take-off weight of 660,000 lb, and its engines develop a combined thrust of 200,000 lb (Fig. P17-5). If the runway is flat and level, determine the work done on the aircraft

a. By the engines as the aircraft moves 1000 ft down the runway.

b. By the gravitational force W as the aircraft moves 1000 ft down the runway.

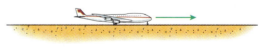

Fig. P17-5

17-6 A 1000-kg car is pushed by a constant force **F** as shown in Fig. P17-6. If the road is flat and level, determine the work done on the car

a. By the $F = 2$ kN force as the car moves 20 m along the road.
b. By the gravitational force W as the car moves 20 m along the road.

Fig. P17-6

17-7 A 200-lb block is pushed along a horizontal surface by a constant force **F** as shown in Fig. P17-7. If $F = 100$ lb and the kinetic coefficient of friction between the block and the surface is $\mu_k = 0.25$, determine the work done on the block

a. By the frictional force on the bottom of the block as the block moves 20 ft along the surface.
b. By the gravitational force W as the block moves 20 ft along the surface.
c. By the 100-lb force F as the block moves 20 ft along the surface.

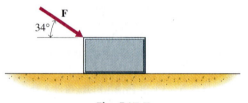

Fig. P17-7

17-8* A 20-kg block is pushed up an inclined plane by a horizontal force **F** as shown in Fig. P17-8. If $F = 200$ N and the kinetic coefficient of friction between the block and the inclined plane is $\mu_k = 0.10$, determine the work done on the block

a. By the frictional force on the bottom of the block as the block moves 15 m up the incline.
b. By the gravitational force W as the block moves 15 m up the incline.
c. By the horizontal force $F = 200$ N as the block moves 15 m up the incline.

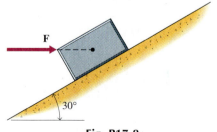

Fig. P17-8

Intermediate Problems

17-9* The pair of blocks shown in Fig. P17-9 are connected by a light inextensible cord and are released from rest when the spring is unstretched. The static and kinetic coefficients of friction are $\mu_s = 0.2$ and $\mu_k = 0.1$, respectively. If the 10-lb block moves 9 in. to the right, determine the work done

a. On the 5-lb block by the gravitational force.
b. On the 10-lb block by the frictional force acting on the bottom of the block.
c. On the 10-lb block by the spring force $F_s = k(\ell - \ell_0)$, where the spring constant is $k = 20$ lb/ft and the unstretched length of the spring is $\ell_0 = 12$ in.

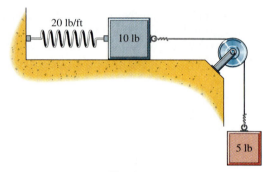

Fig. P17-9

17-10* The pair of blocks shown in Fig. P17-10 are connected by a light inextensible cord and are released from rest when the spring is stretched 600 mm. The static and kinetic coefficients of friction are $\mu_s = 0.3$ and $\mu_k = 0.2$, respectively. If the 5-kg block moves 300 mm to the left, determine the work done

a. On the 5-kg block by the frictional force acting on the bottom of the block.
b. On the 5-kg block by the spring force $F_s = k(\ell - \ell_0)$, where the spring constant is $k = 1$ kN/m and the unstretched length of the spring is $\ell_0 = 250$ mm.
c. On the 10-kg block by the gravitational force.
d. On the 10-kg block by the frictional force acting on the bottom of the block.

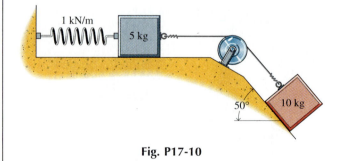

Fig. P17-10

17-11 The pair of blocks shown in Fig. P17-11 are released from the position shown with $d = 18$ in. If block B drops to the floor, determine the work done

a. On block B by the gravitational force $W_B = 10$ lb.
b. On block A by the gravitational force $W_A = 5$ lb.
c. On block A by the spring force $F_s = k(\ell - \ell_0)$, where the spring constant is $k = 20$ lb/ft and the unstretched length of the spring is $\ell_0 = 12$ in.

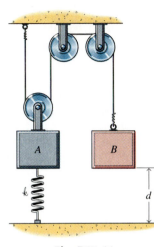

Fig. P17-11

17-12 The pair of blocks shown in Fig. P17-12 are connected by a light inextensible cord. The blocks are released from rest from the position shown with the spring stretched 150 mm. If the 2-kg block rises 100 mm, determine the work done

a. On the 2-kg block by the gravitational force.
b. On the 10-kg block by the spring force $F_s = k(\ell - \ell_0)$, where the spring constant is $k = 1.2$ kN/m.

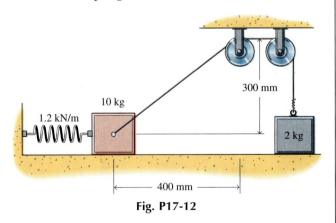

Fig. P17-12

17-13* The 0.33-lb ball shown in Fig. P17-13 is released from rest in the position shown. If the unstretched length of the spring is 20 in. and the spring constant is $k = 5$ lb/in., determine the work done on the ball as it travels the 16 in. along the tube

a. By the spring force.
b. By the gravitational force.

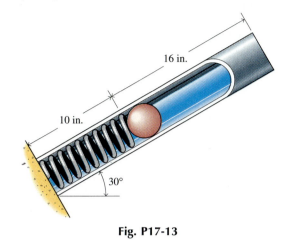

Fig. P17-13

17-14* A 2-kg package slides on a frictionless floor and strikes the bumpers shown in Fig. P17-14. The two linear springs are identical, with spring constants of $k = 1.5$ kN/m. Determine the work done on the package by the springs as spring 1 is compressed by 120 mm (and spring 2 is compressed by 20 mm).

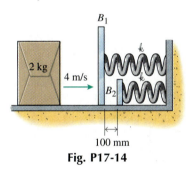

Fig. P17-14

17-15 The pressure in the cylindrical, gas-filled chamber of Fig. P17-15 varies inversely with volume ($p = $ constant/volume). The pressure inside the chamber is the same as the pressure outside the chamber ($p_{atm} = 14.7$ lb/in.^{2}) when the piston is at $x = 10$ in. Determine the net work done on the piston by the air pressure (the work done on the piston by the air pressure outside as well as inside the chamber)

a. As the piston moves from $x = 10$ in. to $x = 15$ in.
b. As the piston moves from $x = 10$ in. to $x = 6$ in.

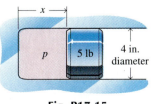

Fig. P17-15

17-16 When a bullet is fired from a rifle, the gas pressure in the barrel varies as shown in Fig. P17-16. The rifle barrel is 600 mm long and has a 7-mm diameter. Estimate the work done on the bullet by the gas pressure as the bullet travels the length of the barrel.

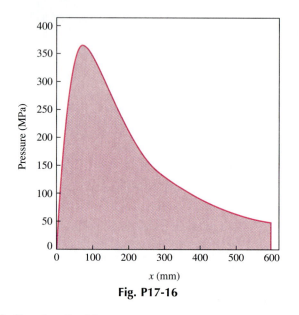

Fig. P17-16

Challenging Problems

17-17* A 3-lb collar slides on a frictionless vertical guide as shown in Fig. P17-17. A constant horizontal force of 12 lb is applied to the end of the light inextensible cord that is attached to the collar. If the collar is released from rest when $d = 32$ in, determine the work done on the collar

a. By the gravitational force as the collar rises 12 in.
b. By the tension in the cord as the collar rises 12 in.

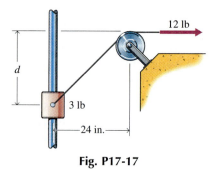

Fig. P17-17

17-18* A 5-kg block slides on the inside of a cylindrical shell as shown in Fig. P17-18. The radius of the cylinder is 3 m. Determine the work done on the block by the gravitational force as the block slides from $\theta = 30°$ to $\theta = 90°$.

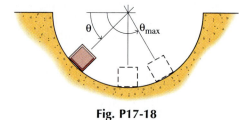

Fig. P17-18

17-19 The pair of blocks shown in Fig. P17-19 are connected by a light inextensible cord. The spring has a modulus of $k = 72$ lb/ft and an unstretched length of $\ell_0 = 12$ in. Determine the work done by the spring force $F_s = k(\ell - \ell_0)$ on the 4-lb block as it moves 4 in. to the right (from $x = 0$ to $x = 4$ in.).

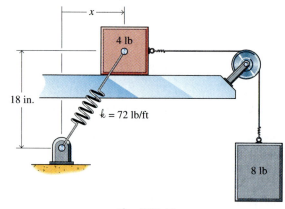

Fig. P17-19

17-20 A 5-kg mass slides on a frictionless vertical rod as shown in Fig. P17-20. The undeformed length of the spring is $\ell_0 = 200$ mm, the modulus of the spring is $k = 40$ N/m, and the distance $d = 300$ mm. Determine the work done on the slider by the spring force $F_s = k(\ell - \ell_0)$ as the slider falls 400 mm (from $b = 0$ to $b = 400$ mm).

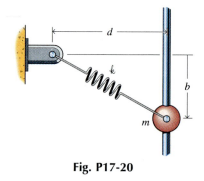

Fig. P17-20

17-5 PRINCIPLE OF WORK AND KINETIC ENERGY

The principle of work and energy is expressed by Eq. 17-2, which was obtained by integrating Newton's second law with respect to position. Replacing the integral on the left-hand side of Eq. 17-2 with $U_{1\rightarrow2}$, the work done by the resultant force **R** gives the *principle of work and energy*

$$T_2 - T_1 = U_{1\rightarrow2} \qquad \textbf{(17-6a)}$$

which states that *the net increase in the kinetic energy of a particle in a displacement from position 1 to position 2 is equal to the work done on the particle by external forces during the displacement.* Alternatively, the principle of work and energy is often written in the form

$$T_i + U_{i\rightarrow f} = T_f \qquad \textbf{(17-6b)}$$

which states that *the final energy of a particle is equal to the sum of its initial kinetic energy and the work done on the particle by external forces.*

Since the mass m and the square of the speed v^2 are both positive quantities, the kinetic energy of a particle will always be positive. If the work done on the particle is positive, the final kinetic energy will be larger than the initial kinetic energy ($0 < T_1 < T_2$). If the work done is negative, the final kinetic energy will be smaller than the initial ($0 < T_2 < T_1$).

The convenience of the method of work and energy is that it directly relates the speed of a particle at two different positions of its motion to the forces that do work on the particle during the motion. If Newton's second law were applied directly, the acceleration would have to be obtained for an arbitrary position of the particle. Then the acceleration would have to be integrated using the principles of kinematics. The work–energy method combines these two steps into one.

The limitations of the method of work–energy are that Eq. 17-6 is a scalar equation and can be solved for no more than one unknown; the acceleration cannot be calculated directly; and only forces that do work are involved. These are usually not serious limitations, however. The normal component of acceleration is a function of velocity, $a_n = mv^2/\rho$, and the velocity is easily found using the work–energy method. Then the normal component of Newton's second law can be used to determine forces that act normal to the path of motion and that do no work.

Finally, a complete free-body diagram must be drawn, to ensure that all forces are identified and considered. Although forces that do no work are not needed in the work–energy equation, they are needed for everything else. Always drawing complete free-body diagrams is just a good habit to get into.

A 16-lb crate slides down a ramp as shown in Fig. 17-12. If the crate is released from rest 10 ft above the bottom of the ramp and the coefficient of friction between the crate and the ramp is $\mu_k = 0.20$, determine the speed of the crate when it reaches the bottom of the ramp.

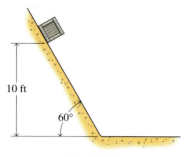

Fig. 17-12

SOLUTION

The free-body diagram of the crate is shown in Fig. 17-13 for a general position along the ramp. As the crate slides down the ramp, it moves 10 ft in the vertical direction, $10/\tan 60° = 5.774$ ft in the horizontal direction, and $10/\sin 60° = 11.547$ ft along the ramp. The normal force N does no work, since it is perpendicular to the motion. However, even though it does not affect the work–energy equation, the normal force must still be found in order to determine the work done by friction. Since the crate does not move in the y-direction, the y-component of Newton's second law

$$+\nwarrow\Sigma F_y = ma_y: \qquad N - 16 \cos 60° = 0$$

is easily solved, giving

$$N = 16 \cos 60° = 8.00 \text{ lb} = \text{constant}$$

Therefore, the friction force is

$$F = \mu_k N = (0.20)(8.00) = 1.600 \text{ lb} = \text{constant}$$

Since the crate starts from rest, its initial kinetic energy is zero $T_i = 0$. The final kinetic energy of the crate is

$$T_f = \frac{1}{2}mv^2 = \frac{1}{2}\frac{16}{32.2}v_f^2 = 0.2484v_f^2$$

and the work done on the crate by the normal, gravitational, and frictional forces (which are all constant) as the crate slides down the ramp are

$$(U_{i\to f})_N = 0 \text{ ft} \cdot \text{lb}$$
$$(U_{i\to f})_g = Wd = 16(10) = 160.00 \text{ ft} \cdot \text{lb}$$
$$(U_{i\to f})_F = Fd = -1.600(11.547) = -18.48 \text{ ft} \cdot \text{lb}$$

respectively. Substituting these values into the work–energy equation (Eq. 17-6)

$$T_i + U_{i\to f} = T_f$$

gives

$$0 + (160.00 - 18.48) = 0.2484v_f^2$$

or

$$v_f = 23.9 \text{ ft/s} \qquad\qquad\qquad \text{Ans.}$$

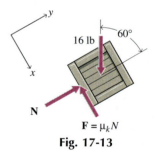

Fig. 17-13

EXAMPLE PROBLEM 17-4

The 5-kg block shown in Fig. 17-14 slides along a horizontal floor and strikes the bumper B. The coefficient of friction between the block and the floor is $\mu_k = 0.25$, and the mass of the bumper may be neglected. If the speed of the block is 10 m/s when it is 15 m from the bumper, determine

a. The speed v_C of the block at the instant it strikes the bumper.
b. The maximum deflection δ_{max} of the spring due to the motion of the block.

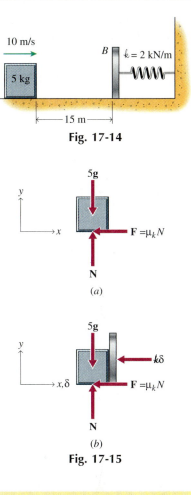

Fig. 17-14

SOLUTION

a. The free-body diagram of the block before it contacts the bumper is shown in Fig. 17-15a. Using the y-component of Newton's second law $\Sigma F_y = ma_y = 0$ gives $N = (5)(9.81) = 49.05$ N. Then the friction force is $F = (0.25)(49.05) = 12.263$ N.

The initial kinetic energy of the block is

$$T_i = \frac{1}{2}mv_i^2 = \frac{1}{2}(5)(10)^2 = 250.0 \text{ J}$$

and the kinetic energy when it is about to strike the bumper is

$$T_f = \frac{1}{2}mv_f^2 = \frac{1}{2}(5)v_C^2 = 2.500v_C^2$$

The work done on the block by friction as the block slides along the floor is

$$(U_{i \to f})_F = Fd = -12.263(15) = -183.95 \text{ J}$$

and since the normal force is perpendicular to the motion, it does no work. Therefore, the work–energy principle gives

$$250.0 - 183.95 = 2.500v_C^2 \quad \text{or} \quad v_C = 5.14 \text{ m/s} \quad \text{Ans.}$$

b. After the block contacts the bumper, the block and the bumper will move together. The free-body diagram of the system is shown in Fig. 17-15b. The normal and friction forces are still $N = 49.05$ N and $F = \mu_s N = 12.263$ N, respectively. The initial velocity of the block for this phase of the motion is $v_C = 5.14$ m/s and the initial kinetic energy is

$$T_i = \frac{1}{2}mv_i^2 = \frac{1}{2}(5)(5.14)^2 = 66.05 \text{ J}$$

At the point of maximum spring deflection, the velocity of the block is zero, so the final kinetic energy is zero ($T_f = 0$). The work done on the block by friction and by the spring force as the block slides along the floor an additional distance δ_{max} is

$$(U_{i \to f})_F = Fd = -12.263 \, \delta_{max}$$

$$(U_{i \to f})_s = \int_0^{\delta_{max}} - 2000 \, \delta \, d\delta = -1000 \, \delta_{max}^2$$

The work–energy equation then gives

$$66.05 - 12.263 \, \delta_{max} - 1000 \, \delta_{max}^2 = 0 \quad \text{or} \quad \delta_{max} = 0.251 \text{ m} \quad \text{Ans.}$$

(a)

(b)

Fig. 17-15

Maximum deflection of the spring occurs when the block stops moving to the right. At this instant the velocity (and therefore the kinetic energy) of the block is zero.

When the block stops moving to the right, it is still acted on by a spring force of $F_s = k\delta_{max} = 2000(0.251) = 502$ N which acts to push the block to the left. The friction force now must act to the right to oppose the impending motion to the left. Since the friction force available to prevent motion $F = \mu_s N \cong \mu_k N = 12.263$ N is much less than the force required to prevent motion, the block will slide back to the left.

A 10-lb package slides down a ramp and falls on the floor as shown in Fig. 17-16. If the initial speed of the package is 5 ft/s and friction can be neglected, determine

a. The distance d from the end of the ramp to where the package hits the floor.
b. The speed of the package v_f when it hits the floor.

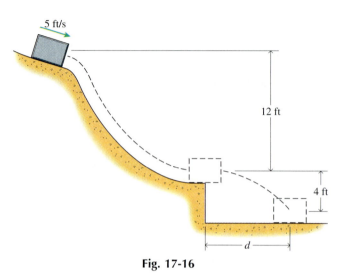

5 ft/s

12 ft

4 ft

d

Fig. 17-16

SOLUTION

The problem must be divided into two parts—what happens to the package while it is on the ramp and what happens to the package after it leaves the ramp. First, the work–energy principle will be used to determine the speed of the package when it leaves the end of the ramp. Then, Newton's second law will be used to determine the motion of the package after it leaves the end of the ramp.

a. The free-body diagram of the package for an arbitrary position along the ramp is shown in Fig. 17-17a. The normal force N always acts perpendicular to the motion and does no work; the work done by the gravitational force W as the package falls 12 ft is

$$(U_{i \to f})_g = Wd = 10(12) = 120.0 \text{ ft} \cdot \text{lb}$$

The initial kinetic energy of the package is

$$T_i = \frac{1}{2}mv_i^2 = \frac{1}{2}\frac{10}{32.2}(5)^2 = 3.882 \text{ ft} \cdot \text{lb}$$

and the kinetic energy of the package when it leaves the ramp is

$$T_f = \frac{1}{2}mv_f^2 = \frac{1}{2}\frac{10}{32.2}v_r^2 = 0.15528v_r^2$$

Therefore, the work–energy principle (Eq. 17-6)

$$T_i + U_{i \to f} = T_f$$

Since no horizontal forces act on the package after it leaves the end of the ramp, the distance d does not appear in any term of the work–energy equation and the work–energy principle cannot be used to determine the distance d to where the package hits the floor. However, the work–energy principle easily gives the speed of the package when it leaves the end of the ramp which is needed when Newton's second law is integrated to get the distance d.

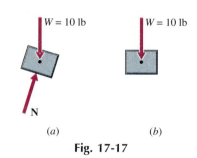

$W = 10$ lb $W = 10$ lb

N

(a)   (b)

Fig. 17-17

gives

$$3.882 + 120.0 = 0.15528v_r^2$$

or

$$v_r = 28.245 \text{ ft/s} \cong 28.2 \text{ ft/s} \rightarrow$$

The free-body diagram of the package for an arbitrary position after the package leaves the ramp is shown in Fig. 17-17b. The only force acting on the package is gravity and Newton's second law is

$$+\rightarrow\Sigma F_x = ma_x: \qquad 0 = 0$$
$$+\uparrow\Sigma F_y = ma_y: \qquad -mg = ma_y$$

Therefore,

$$\mathbf{a} = -g\mathbf{j} = -32.2\mathbf{j} \text{ ft/s}^2$$

Integrating twice with respect to time to get the velocity and position of the package

$$\mathbf{v} = 28.245\mathbf{i} - 32.2t\mathbf{j} \text{ ft/s}$$
$$\mathbf{r} = 28.245t\mathbf{i} + (4 - 16.1t^2)\mathbf{j} \text{ ft}$$

where the constants of integration have been chosen so that $\mathbf{v} = 28.245\mathbf{i}$ ft/s and $\mathbf{r} = 4\mathbf{j}$ ft when $t = 0$. The package hits the floor when $y = 4 - 16.1t^2 = 0$ or when $t = 0.49844$ s. At this time, the x-position of the package is

$$x = d = 28.245(0.49844) = 14.08 \text{ ft} \qquad \text{Ans.}$$

b. When the package hits the floor ($t = 0.49844$ s) the velocity of the package is

$$\mathbf{v} = 28.245\mathbf{i} - 32.2(0.49844)\mathbf{j}$$
$$= 28.245\mathbf{i} - 16.050\mathbf{j} \text{ ft/s} \cong 32.5 \text{ ft/s} \searrow 29.61° \qquad \text{Ans.}$$

The speed of the package could also have been obtained using the work–energy principle. The kinetic energy of the package when it leaves the end of the ramp is

$$T_i = \frac{1}{2}mv_i^2 = \frac{1}{2}\frac{10}{32.2}(28.245)^2 = 123.882 \text{ ft} \cdot \text{lb}$$

and the kinetic energy of the package when it lands on the floor is

$$T_f = \frac{1}{2}mv_f^2 = \frac{1}{2}\frac{10}{32.2}v_f^2 = 0.15528v_f^2$$

The work done by the gravitational force W as the package falls an additional 4 ft is

$$(U_{i\rightarrow f})_g = Wd = 10(4) = 40.0 \text{ ft} \cdot \text{lb}$$

Therefore, the work–energy principle (Eq. 17-6) gives

$$123.882 + 40.0 = 0.15528v_f^2$$

or

$$v_f = 32.49 \text{ ft/s} \cong 32.5 \text{ ft/s} \searrow \qquad \text{Ans.}$$

Although the work–energy principle also gives the speed of the particle when it hits the floor, it cannot resolve the speed into horizontal and vertical components. The work–energy principle cannot give the angle that the velocity makes with the floor when the package lands.

A 0.5-kg bead slides along a circular wire as shown in Fig. 17-18. The diameter of the wire ring is 800 mm, and friction between the bead and the wire may be neglected. If the bead is released from rest when $\theta = 30°$, determine the force exerted on the bead by the wire when $\theta = 180°$ (the bead is at the bottom of the ring).

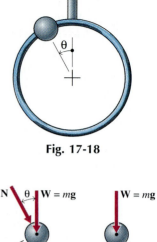

Fig. 17-18

SOLUTION

A free-body diagram of the bead for an arbitrary angle θ is shown in Fig. 17-19a. Since the normal force N is always perpendicular to the motion, it does not appear in any term of the work–energy equation, and the work–energy principle cannot be used to determine the normal force exerted on the bead by the wire. However, the work–energy principle easily gives the speed of the bead at the bottom of the ring, which is needed when Newton's second law is written for $\theta = 180°$.

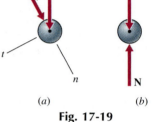

Since the bead starts from rest, the initial kinetic energy is $T_i = 0$, and the final kinetic energy of the bead is

$$T_f = \frac{1}{2}mv_f^2 = \frac{1}{2}(0.5)v_f^2 = 0.25v_f^2$$

The normal force does no work; the gravitational force is constant in both direction and magnitude. Therefore, the work done on the particle is

$$U_{i \to f} = W\Delta z = [0.5(9.81)](0.4 + 0.4 \cos 30°) = 3.661 \text{ J}$$

and the work–energy equation (Eq. 17-6)

$$T_i + U_{i \to f} = T_f$$

gives

$$0 + 3.661 = 0.25v_f^2$$

or

$$v_f = 3.827 \text{ m/s} \rightarrow$$

A free-body diagram of the bead for $\theta = 180°$ is shown in Fig. 17-19b. The vertical component of Newton's second law (with $a_y = a_n = v^2/\rho$)

$$+\uparrow\Sigma F_y = ma_y: \qquad N - 0.5(9.81) = 0.5\frac{v_f^2}{0.4}$$

gives

$$N = 23.2 \text{ N}\uparrow \qquad\qquad \text{Ans.}$$

(Fig. 17-19 caption located below images: labels (a), (b) and **Fig. 17-19**)

PROBLEMS

In the following problems, all ropes, cords, and cables are assumed to be flexible, inextensible, and of negligible mass. All pins and pulleys are small, have negligible mass, and are frictionless.

Introductory Problems

17-21* A fully loaded Boeing 747 has a take-off weight of 660,000 lb, and its engines develop a combined thrust of 200,000 lb. If air resistance and friction between the tires and runway are neglected, determine the required length of the runway for a take-off speed of 140 mi/h (Fig. P17-21).

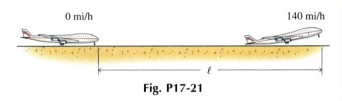

0 mi/h   140 mi/h

ℓ

Fig. P17-21

17-22* The 1200-kg car shown in Fig. P17-22 is traveling along a straight level road at 80 km/h when the driver jams on the brakes, forcing all four tires to skid. If the kinetic coefficient of friction is $\mu_k = 0.5$, determine the length of the skid marks when the car comes to rest.

Fig. P17-22

17-23 A 7500-lb truck is traveling on the freeway at 65 mi/h when the driver suddenly notices a moose standing on the road 200 ft straight ahead (Fig. P17-23). If it takes the driver 0.4 s to apply the brakes and the kinetic coefficient of friction between the tires and the road is $\mu_k = 0.5$,

a. Can the driver avoid hitting the moose without steering to one side?
b. Where will the truck come to rest relative to the moose?
c. If the driver must steer to one side, determine the speed of the truck as it passes the moose.

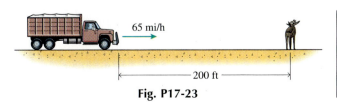

65 mi/h

200 ft

Fig. P17-23

17-24 A 1200-kg car is traveling along a mountain road at 90 km/h when a boulder rolls onto the highway 60 m in front of it (Fig. P17-24). The road is level, and the kinetic coefficient of friction between the tires and the road is $\mu_k = 0.5$. If the driver of the car takes 0.4 s to apply the brakes,

a. Can the driver of the car avoid hitting the boulder without steering to one side?
b. If the driver of the car must steer to one side, determine the speed of the car as it passes the boulder.

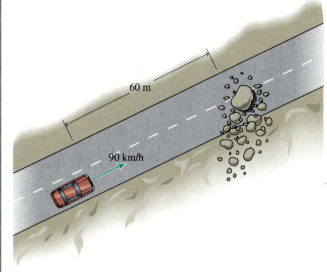

60 m

90 km/h

Fig. P17-24

17-25* A 25,000-lb jet is catapulted from the deck of an aircraft carrier by a hydraulic ram (Fig. P17-25). Determine the average force exerted on the jet if it accelerates from rest to 160 mi/h in 300 ft.

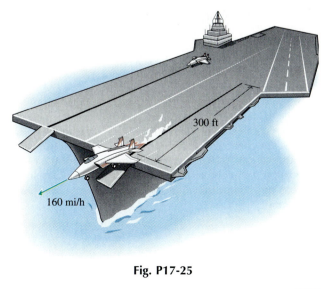

300 ft

160 mi/h

Fig. P17-25

17-26* A train is moving at 30 km/h when a coupling breaks and the last car separates from the train. As soon as the car separates, the brakes are automatically applied, locking all wheels of the runaway car. If the kinetic coefficient of friction between the wheels and the rails is $\mu_k = 0.2$, determine the distance that the 180,000-kg car will travel before coming to a stop

a. If the tracks are level.
b. If the tracks slope downward at 5°.

17-27 A 2000-lb elevator cage is descending a mine shaft at a speed of 25 ft/s when a constant braking force is applied to the cable drum (causing a constant tension in the hoisting cable). If the weight of the cable can be ignored and the elevator comes to rest in a distance of 50 ft, determine the tension in the hoisting cable (Fig. P17-27).

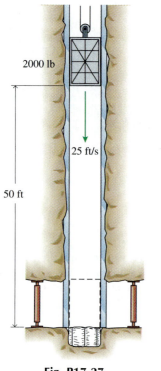

Fig. P17-27

17-28* In a shipping warehouse, packages are moved between levels by sliding them down a chute as shown in Fig. P17-28. The coefficient of friction between the package and the ramp is $\mu_k = 0.20$, the corner at the bottom of the ramp is abrupt but smooth, and $\theta = 30°$. If a 10-kg package starts at $\ell = 3$ m with an initial speed of 5 m/s down the ramp, determine

a. The speed of the package when it reaches the bottom of the ramp.
b. The distance d that the package will slide along the horizontal surface before coming to rest.

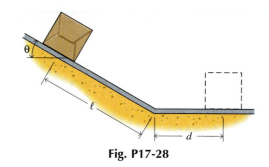

Fig. P17-28

Intermediate Problems

17-29* If the packages of Problem 17-28 come off the chute with too great a velocity, a bumper as shown in Fig. P17-29 may be required to "catch" them. The spring modulus is $k = 6.0$ lb/ft, and the mass of the bumper B may be neglected. If the static and kinetic coefficients of friction between a 15-lb package and the floor are $\mu_s = 0.6$ and $\mu_k = 0.4$, respectively, determine the maximum initial speed v_0 of the package when $\ell = 5$ ft such that the package will not rebound off the bumper.

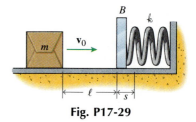

Fig. P17-29

17-30 A 2-kg package slides on a frictionless floor and strikes the bumpers shown in Fig. P17-30. The two linear springs are identical, with spring constants of $k = 1.5$ kN/m, and the mass of the bumpers may be neglected. If the initial speed of the package is 4 m/s, determine the maximum deformation of the springs.

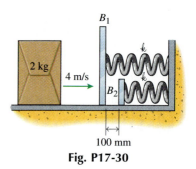

Fig. P17-30

17-31* The pressure of the cylindrical gas-filled chamber of Fig. P17-31 varies inversely with the volume of the gas (p = constant/volume). Initially, the piston is at rest, $x = 6$ in., and $p = 2p_{atm}$ where $p_{atm} = 14.7$ lb/in.^{2} If air pressure on the outside surface of the piston is constant ($= p_{atm}$), determine for the subsequent motion

a. The maximum speed v_{max} of the piston.
b. The maximum displacement x_{max} of the piston.
c. The minimum constant force F that must be applied to the piston to limit its motion so that $x_{max} < 18$ in.

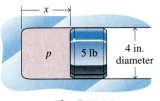

Fig. P17-31

17-32   The 5-kg block shown in Fig. P17-32 is released from rest when the spring is compressed 200 mm. The block is not attached to the spring. The block travels 2 m along a smooth surface and a distance d along the rough surface ($\mu_k = 0.30$). Determine the distance d at which the block comes to rest.

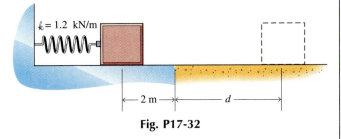

Fig. P17-32

17-33   The 3-lb collar shown in Fig. P17-33a is moving along a horizontal smooth rod with a velocity of 3 ft/s when it strikes a spring. The nonlinear spring has the force–displacement relationship shown in Fig. P17-33b (where x is measured from the undeformed position of the spring). Determine the value of the constant c if the collar is to come to rest when $x = 6$ in.

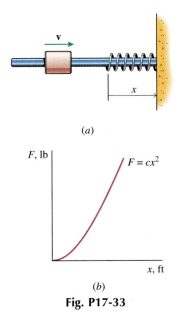

(a)

(b)

Fig. P17-33

17-34*   In a carnival game of skill, players slide a 60-g hockey puck across a horizontal wood floor. The object is to have the puck stop as close to the wall as possible without hitting it. The coefficient of friction between the puck and the floor is $\mu_k = 0.4$, and the "foul line" (the point at which the player must release the puck) is 2 m from the wall. If a winning play must stop within 100 mm of the wall, determine the range of initial velocities that will win a prize.

17-35   The ball shown in Fig. P17-35 weighs 0.33 lb, and the free length of the spring ($k = 5$ lb/in.) is 20 in. If the ball is released at rest in the position shown and friction between the ball and the tube is negligible, determine the speed of the ball as it exits the tube.

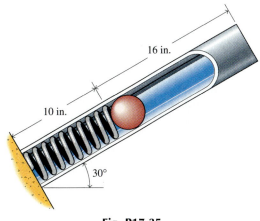

Fig. P17-35

17-36   In a carnival game of skill, players shoot nickels up an inclined playing surface and into a 50-mm-wide slot using a spring operated plunger as shown in Fig. P17-36. The coefficient of friction between the 5-g nickels and the wood floor is $\mu_k = 0.2$, and the spring modulus is $k = 75$ N/m. If the plunger is pulled back a distance δ and released from rest, determine the range of δ's that will win a prize in this game. (Neglect the small size of the nickel when deciding if the nickel will fall through the slot.)

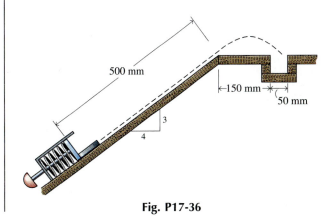

Fig. P17-36

Challenging Problems

17-37* A 3-lb collar slides on a frictionless vertical guide as shown in Fig. P17-37. A constant horizontal force of 12 lb is applied to the end of the light inextensible cord that is attached to the collar. If the collar is released from rest when $d = 32$ in., determine the velocity of the collar when $d = 18$ in.

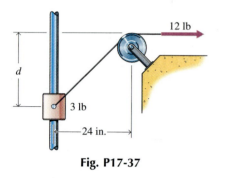

Fig. P17-37

17-38* A 10-kg block slides on a frictionless horizontal surface as shown in Fig. P17-38. A constant horizontal force of 50 N is applied to the end of the light inextensible cord that is attached to the block. If the block is released from rest from the position shown in which the spring is unstretched, determine for the ensuing motion

a. The maximum velocity of the block and the stretch in the spring at that position.
b. The maximum stretch in the spring.

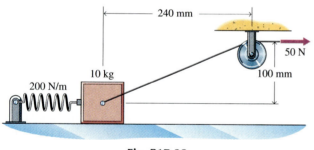

Fig. P17-38

17-39 A singer swings a 12-oz microphone in a vertical plane at the end of an 18-in.-long cord (Fig. P17-39). If the speed of the microphone is 12 ft/s at position A, determine

a. The angle θ at which the cord becomes slack (tension equals zero).
b. The maximum tension in the cord and the angle at which it occurs.

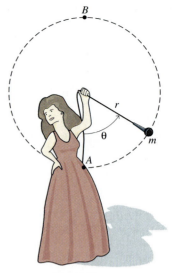

Fig. P17-39

17-40 A small sack of marbles is tied to the end of a 500-mm-long cord as shown in Fig. P17-40. If the maximum tension that the cord can withstand is $P_{max} = 50$ N, determine the maximum weight of marbles that would not break the cord when the boy slowly pulls the sack off of the shelf.

Fig. P17-40

17-41* A small box is sliding along a frictionless horizontal surface when it comes upon a circular ramp as shown in Fig. P17-41. If the initial speed of the box is $v_0 = 5$ ft/s and $r = 15$ in., determine the angle θ at which the box will lose contact with the circular ramp.

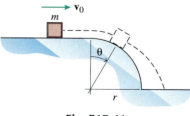

Fig. P17-41

17-42* A small toy car rolls down a ramp and through a vertical loop as shown in Fig. P17-42. The mass of the car is $m = 50$ g, and the diameter of the vertical loop is $d = 300$ mm. If the car is released from rest from a height of h, determine

a. The minimum release height h such that the car will travel all the way around the loop.
b. The force exerted on the track by the car when it is at point B (one-quarter of the way around the loop).

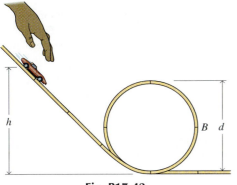

Fig. P17-42

17-43 A 1.5-lb weight slides on a frictionless vertical rod as shown in Fig. P17-43. The undeformed length of the spring is $\ell_0 = 8$ in., the spring modulus is $k = 80$ lb/ft, and the distance $d = 12$ in. If the slider is released from rest when $b = 9$ in., determine the speed of the slider when $b = 0$.

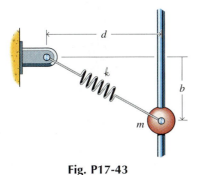

Fig. P17-43

17-44 A 2-kg particle slides along a rod in a vertical plane as shown in Fig. P17-44. The circular portion of the rod is smooth, but the straight portion ($d = 2$ m) is rough, with a kinetic coefficient of friction of $\mu_k = 0.15$. If the particle is released from rest in the position shown, determine the distance a where the particle hits the ground.

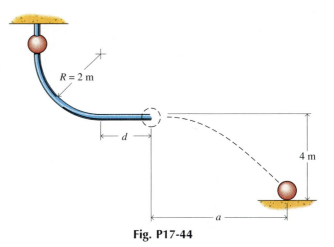

Fig. P17-44

17-45 A 3-lb sphere is supported in a vertical plane by a light rod as shown in Fig. P17-45. If the sphere is released from rest at position A, determine

a. The speed of the sphere when it is at position B.
b. The tension in the rod when the sphere is at position B.
c. The maximum height h to which the sphere will rise.

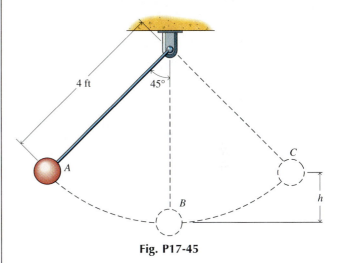

Fig. P17-45

17-46 A 32-kg child steps off a rock and swings at the end of a 6-m-long rope over a river as shown in Fig. P17-46. Determine the speed of the child when he lets go of the rope and the speed of the child when he enters the water if he lets go of the rope when

a. $\theta = 90°$.
b. $\theta = 120°$.

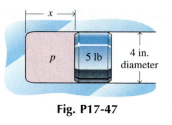

Fig. P17-47

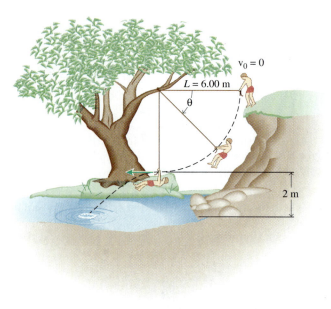

Fig. P17-46

Computer Problems

C17-47 The pressure in the cylindrical gas-filled chamber of Fig. P17-47 varies inversely with the volume of the gas (p = constant/volume). The pressure inside the chamber is the same as atmospheric ($p_{atm} = 14.7$ lb/in.^{2}) when the 5-lb piston is at $x = 10$ in. If $\dot{x} = -4$ ft/s when $x = 10$ in.,

a. Determine the range of travel (x_{min} and x_{max}) of the piston.
b. Calculate and plot the speed of the piston as a function of its position, $x_{min} < x < x_{max}$.

17-48 A singer swings a 0.35-kg microphone in a vertical plane at the end of a cord 750 mm long (Fig. P17-48). If the speed of the microphone is 7 m/s at position A, calculate and plot

a. The speed of the microphone as a function of the angle θ for one full revolution of the microphone ($0 < \theta < 360°$).
b. The tension in the microphone cord as a function of the angle θ for one full revolution of the microphone ($0 < \theta < 360°$).

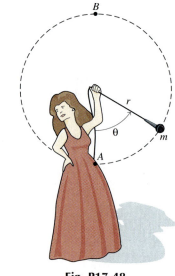

Fig. P17-48

17-6 SYSTEMS OF PARTICLES

Applying the principle of work and energy to a particle yields a single scalar equation. Therefore, when a group of particles interact with one another, the principle of work and energy may be applied to each individual particle and the equations added together. The result will be a single scalar equation that can be solved for at most one unknown. Unless the motions of most of the particles are known, this equation will not be very useful. One exception for which the equation is useful is when the particles are rigidly connected so that kinematics can be used to relate the motions of the particles.

17-6.1   Two Particles Connected by a Massless Rigid Link

Consider the two particles of Fig. 17-20a, which are connected by a massless, rigid link. The free-body diagrams of the particles and the link are shown in Fig. 17-20b, in which $\mathbf{R}_1$ and $\mathbf{R}_2$ represent the resultants of all external forces acting on the particles, and $\mathbf{f}_1$ and $\mathbf{f}_2$ represent the internal forces between the link and the particles. Since the mass of the link is assumed to be negligible, application of Newton's second law to the free-body diagram of the link gives $\Sigma \mathbf{F} = m\mathbf{a} = \mathbf{0}$ and $\Sigma \mathbf{M}_G = I_G \boldsymbol{\alpha} = \mathbf{0}$. Therefore, assuming that no external forces act on the rigid link, forces $\mathbf{f}_1$ and $\mathbf{f}_2$ must be directed along the link, equal in magnitude and opposite in direction ($\mathbf{f}_2 = -\mathbf{f}_1$), as shown in Fig. 17-20b.

During some infinitesimal motion of the pair of particles, particle 1 will move a distance dx along the link and dy_1 perpendicular to the link. Since the link is rigid, particle 2 will move the same distance dx along the link but a different distance dy_2 perpendicular to the link. The work done on particle 2 by the link $[dU_2 = \mathbf{f}_2 \cdot (dx\,\mathbf{i}) = -\mathbf{f}_1 \cdot (dx\,\mathbf{i})]$ will be the negative of the work done on particle 1 by the link $[dU_1 = \mathbf{f}_1 \cdot (dx\,\mathbf{i})]$ during the infinitesimal motion. When the equations expressing the work–energy principle for the two particles are added together, the work done by the internal forces $\mathbf{f}_1$ and $\mathbf{f}_2$ will cancel at every instant of the motion, leaving

$$\left(\frac{1}{2}m_1 v_{1i}^2 + \int_i^f \mathbf{R}_1 \cdot d\mathbf{r}_1 \right) + \left(\frac{1}{2}m_2 v_{2i}^2 + \int_i^f \mathbf{R}_2 \cdot d\mathbf{r}_2 \right) = \frac{1}{2}m_1 v_{1f}^2 + \frac{1}{2}m_2 v_{2f}^2$$

That is, Eq. 17-6 also applies to the pair of connected particles if T_i and T_f are interpreted as the sum of the initial and final kinetic energies of both particles and $U_{i \to f}$ is the work done on both particles by all external forces acting on the pair of particles. *So long as the particles are rigidly connected and are treated together, the work done by the internal forces between the link and the particles will cancel and may be neglected.*

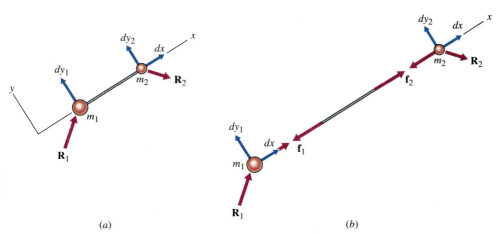

(a)   (b)

Figure 17-20   When two particles are connected by a rigid link, the displacement of the particles in the direction of the link must be the same for both particles. If the mass of the link can be neglected, then the forces exerted on the ends of the link (and therefore, on the two particles) will be equal in magnitude and opposite in direction ($\mathbf{f}_1 = -\mathbf{f}_2$). Therefore, the sum of the works done on the two particles by the rigid link will be zero.

Similarly, when two bodies are connected by a flexible inextensible cable, the resultant work done on the bodies by the force in the cable is zero. Assuming that the mass of the cable is negligible and that any pulleys are small, massless, and/or frictionless, the two forces at the ends of the cable will have the same magnitude. Then, since the cable is inextensible, the components of the displacements of the two ends in the direction of the forces must have the same magnitude, and the resultant work done by the cable is zero.

17-6.2   Two Particles Connected by a Massless Spring

Suppose that the rigid link of Fig. 17-20 is replaced by a massless spring as shown in Fig. 17-21. Then the internal forces at the ends of the spring (Fig. 17-21b) will again be equal in magnitude and opposite in direction. However, suppose that at the instant depicted, particle 1 is moving in a direction perpendicular to the spring ($dx_1 = 0$, $dy_1 \neq 0$) and particle 2 is moving in a direction parallel to the spring ($dx_2 \neq 0$, $dy_2 = 0$). Then, the work done on particle 2 by the spring will be $dU_2 = \mathbf{f}_2 \cdot (dx_2\ \mathbf{i}) = f_2\ dx_2$, and the work done on particle 1 by the spring will be zero. Therefore, when the two equations representing the work–energy principle for the two particles are added together, the work done by the internal spring forces will not cancel out. *Whenever particles are not rigidly connected together, the work done on the particles by any internal forces must be included along with the work done on the particles by any external forces.*

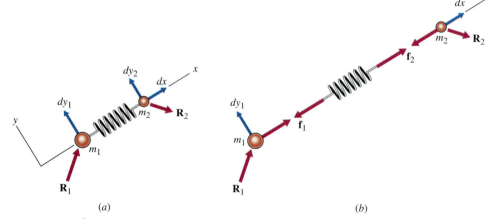

(a)   (b)

Figure 17-21   When two particles are connected by a non-rigid link such as a spring, the work done on the pair of particles by the link will not necessarily be zero. If the mass of the spring can be neglected, then the forces exerted on the ends of the spring (and therefore, on the two particles) will still be equal in magnitude and opposite in direction ($\mathbf{f}_1 = -\mathbf{f}_2$). However, since a spring is not rigid, the displacement of the particles in the direction of the spring may be different for each of the two particles. For example, mass m_1 may move only in the y-direction and mass m_2 may move only in the x-direction, and the works done on the two particles by $\mathbf{f}_1$ and $\mathbf{f}_2$ will not cancel.

17-6.3   General System of Interacting Particles

For a general system of interacting particles, the equations of work and kinetic energy for each of the particles can also be added together. The result will again be the same as Eq. 17-6 if T_i and T_f are interpreted as the sum of the initial and final kinetic energies of all the particles making up the system and $U_{i \to f}$ is the work done on the particles by *all forces*, both *internal and external*. Even though the internal forces always occur in equal-magnitude but opposite-direction pairs, the displacements will usually be different unless the particles are rigidly connected. Therefore, the work done by all the internal forces will usually not be zero.

EXAMPLE PROBLEM 17-7

Two blocks are connected by a light rope, which passes over a small frictionless pulley of negligible mass (Fig. 17-22). The coefficient of friction between the block A and the horizontal surface is $\mu_k = 0.4$. If the system is released from rest, determine the speed of the two blocks when block A has moved 6 ft to the right.

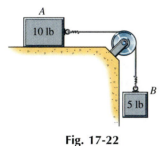

Fig. 17-22

SOLUTION

The free-body diagrams of the blocks are shown in Fig. 17-23. When block A moves a distance x to the right, block B moves down the same distance. Also, when block A moves to the right with a speed v, block B moves downward with the same speed v. Applying the y-component of Newton's second law to block A

$$+\uparrow \Sigma F_y = ma_y: \qquad N - 10 = 0$$

gives $N = 10$ lb = constant. Then the friction force acting on block A during the motion is $F = (0.4)(10) = 4$ lb = constant.

Block A: The initial kinetic energy of block A is zero ($T_i = 0$), and its final kinetic energy is

$$T_f = \frac{1}{2}mv^2 = \frac{1}{2}\left(\frac{10}{32.2}\right)v^2 = 0.15528v^2$$

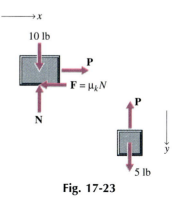

Fig. 17-23

Neither the weight nor the normal force do work on block A, since they are both perpendicular to the motion. The work done on block A by the constant forces F and P during the motion is

$$U_{i \to f} = Fd = (P - 4)(6) = 6P - 24$$

where the minus sign occurs because the friction force is to the left and the motion is to the right. Then, the work–energy equation (Eq. 17-6)

$$T_i + U_{i \to f} = T_f$$

for block A is

$$0 + (6P - 24) = 0.15528v^2 \qquad\qquad (a)$$

Block B: The initial kinetic energy of block B is also zero $T_i = 0$ and its final kinetic energy is

$$T_f = \frac{1}{2}mv^2 = \frac{1}{2}\left(\frac{5}{32.2}\right)v^2 = 0.07764v^2$$

The work done on block B by the constant weight and rope tension forces during the motion is

$$U_{i \to f} = Fd = (5 - P)(6) = 30 - 6P$$

where the minus sign occurs since the rope tension is upward and the motion is downward. Then, the work–energy equation (Eq. 17-6) for block B is

$$0 + (30 - 6P) = 0.07764v^2 \qquad\qquad (b)$$

Finally, adding Eqs. a and b together gives

$$6 = 0.23292v^2 \qquad\qquad (c)$$

or

$$v = 5.08 \text{ ft/s} \qquad\qquad \text{Ans.}$$

Note that when Eqs. a and b were added together to get Eq. c, the internal rope tension force P canceled out. This step could have been avoided by considering the two blocks as a system. Then, the work done by the internal force P can be ignored, because the cord is inextensible and the work done on the system is just the sum of the work done by the external forces

$$U_{i \to f} = 5(6) - 4(6) = 6 \text{ ft} \cdot \text{lb}$$

The initial kinetic energy is the sum of the kinetic energy of the system $T_i = 0$; the final kinetic energy is the sum of the kinetic energy of the system

$$T_f = 0.15528v^2 + 0.07764v^2 = 0.23292v^2$$

and the work–energy equation (Eq. 17-6) is

$$0 + 6 = 0.23292v^2$$

as before (Eq. c).

Because the work–energy principle can give no information about forces which do no work, it must often be supplemented with one or more components of Newton's second law. Since the same free-body diagram is used for both the work–energy principle and for Newton's second law, it is important to draw a complete and proper free-body diagram; that is, all forces must be included whether or not they do work.

Two blocks are connected by a light rope which passes around a pair of small frictionless pulleys of negligible mass (Fig. 17-24). The coefficient of friction between the block A and the horizontal surface is $\mu_k = 0.4$. If the system is released from rest, determine the speed of the two blocks when block A has moved 3 m to the left.

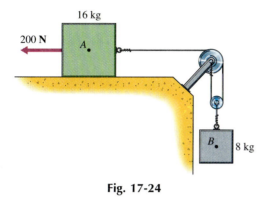

Fig. 17-24

SOLUTION

The free-body diagrams of the blocks are shown in Fig. 17-25. When block A moves a distance x to the left, block B moves up half as far ($y = 0.5x$). Also, when block A moves to the left with a speed v_A, block B moves up at half the speed ($v_B = 0.5v_A$). Applying the y-component of Newton's second law to block A

$$+ \uparrow \Sigma F_y = ma_y: \qquad N - 16(9.81) = 0$$

gives $N = 156.96$ N = constant. Then the friction force acting on block A during the motion is $F = (0.4)(156.96) = 62.784$ N = constant.

Block A: The initial kinetic energy of block A is zero ($T_i = 0$), and is final kinetic energy is

$$T_f = \frac{1}{2} mv^2 = \frac{1}{2}(16)v_A^2 = 8v_A^2$$

Neither the weight nor the normal force do work on block A because they are both perpendicular to the motion. The work done on block A by the constant forces F, P, and 200 N during the motion is

$$U_{i \to f} = Fd = (200 - P - 62.784)(3) = 411.648 - 3P$$

where the minus signs occur since the friction and rope tension forces are to the right and the motion is to the left. Then, the work–energy equation (Eq. 17-6)

$$T_i + U_{i \to f} = T_f$$

for block A is

$$0 + (411.648 - 3P) = 8v_A^2 \qquad\qquad (a)$$

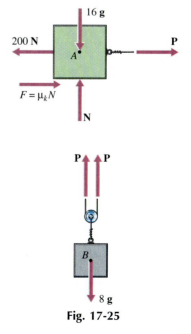

Fig. 17-25

Block B: The initial kinetic energy of block B is also zero ($T_i = 0$), and its final kinetic energy is

$$T_f = \frac{1}{2} mv^2 = \frac{1}{2}(8)v_B^2 = 4\left(\frac{v_A}{2}\right)^2 = v_A^2$$

The work done on block B by the constant weight and rope tension forces during the motion is

$$U_{i \to f} = Fd = (2P - 78.48)(1.5) = 3P - 73.575$$

where the minus sign occurs because the weight is downward and the motion is upward. Then, the work–energy equation (Eq. 17-6) for block B is

$$0 + (3P - 117.72) = v_A^2 \qquad\qquad (b)$$

Finally, adding Eqs. *a* and *b* together gives

$$293.928 = 9v_A^2 \qquad\qquad (c)$$

or

$$v_A = 5.71 \text{ m/s} \qquad\qquad \text{Ans.}$$

Note that when Eqs. *a* and *b* were added together to get Eq. *c*, the internal rope tension force P canceled out. This step could have been avoided by considering the two blocks as a system. Then, the work done by the internal force P can be ignored because the cord is inextensible and the work done on the system is just the sum of the work done by the external forces

$$U_{i \to f} = 200(3) - 62.784(3) - 78.48(1.5) = 293.928 \text{ J}$$

The initial kinetic energy is the sum of the kinetic energy of the system $T_i = 0$; the final kinetic energy is the sum of the kinetic energy of the system

$$T_f = 8v_A^2 + v_A^2 + 9v_A^2$$

and the work–energy equation (Eq. 17-6) is

$$293.928 = 9v_A^2$$

as before (Eq. *c*).

PROBLEMS

In the following problems, all ropes, cords, and cables are assumed to be flexible, inextensible, and of negligible mass. All pins and pulleys are small, have negligible mass, and are frictionless.

Introductory Problems

17-49* The two blocks A and B shown in Fig. P17-49 weigh $W_A = 5$ lb and $W_B = 10$ lb, respectively. If the blocks are released from rest and the horizontal plane is smooth, determine the speed of both blocks when B has moved downward 3 ft.

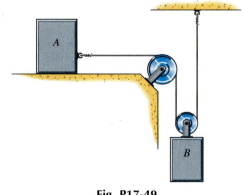

Fig. P17-49

17-50* The two blocks A and B shown in Fig. P17-50 have masses $m_A = 10$ kg and $m_B = 5$ kg, respectively. If the blocks are released from rest in the position where B barely touches the ground, determine

a. The speed of A just as it hits the horizontal surface.
b. The height that B rises, measured from the horizontal surface.

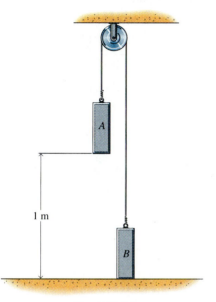

Fig. P17-50

17-51 The two blocks shown in Fig. P17-51 weigh $W_A = 60$ lb and $W_B = 40$ lb. If the blocks are released from rest from the position shown, determine the velocity of block B after it has moved 10 ft.

Fig. P17-51

17-52* The pair of blocks shown in Fig. P17-52 are connected by a light inextensible bar. Both the horizontal and vertical guide slots are frictionless. If the blocks are released from rest from the position shown, determine the velocity of the 3-kg block when

a. It is at the same level as the 2-kg block.
b. It is 150 mm below the 2-kg block.

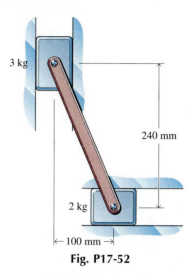

Fig. P17-52

17-53 The two blocks A and B shown in Fig. P17-53 weigh $W_A = 50$ lb and $W_B = 60$ lb, respectively. The kinetic coefficient of friction between block A and the inclined surface is $\mu_k = 0.15$. If the blocks are released from rest in the position shown, determine the speed of block B when it strikes the horizontal surface.

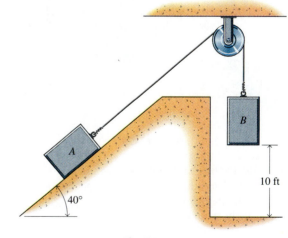

Fig. P17-53

17-54 The two blocks A and B shown in Fig. P17-54 have masses $m_A = 40$ kg and $m_B = 30$ kg, respectively. The kinetic coefficient of friction between block A and the inclined surface is $\mu_k = 0.25$. If the blocks are released from rest in the position shown, determine the speed of block A after it has moved 3 m.

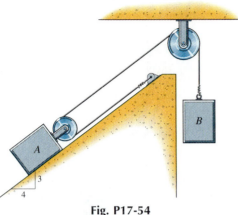

Fig. P17-54

17-55* The two blocks A and B shown in Fig. P17-55 weigh $W_A = 30$ lb and $W_B = 20$ lb, respectively. The kinetic coefficient of friction between block A and the inclined surface is $\mu_k = 0.15$, and the horizontal surface supporting block B is smooth. When the blocks are in the position shown, block B is moving to the right with a velocity of 5 ft/s. Determine the distance traveled by block B before it comes to rest.

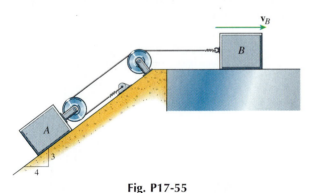

Fig. P17-55

17-56 The two blocks A and B shown in Fig. P17-56 have masses $m_A = 25$ kg and $m_B = 30$ kg, respectively. The static and kinetic coefficients of friction are $\mu_s = 0.25$ and $\mu_k = 0.20$, and the horizontal force $\mathbf{F}$ has a magnitude of 100 N. If the blocks are released from rest in the position shown, determine the speed of block B after it has moved 3 m.

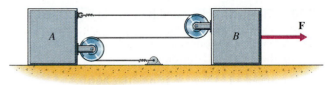

Fig. P17-56

Intermediate Problems

17-57 The two blocks A and B shown in Fig. P17-57 weigh $W_A = 25$ lb and $W_B = 50$ lb, respectively. The blocks are at rest, and the spring ($k = 25$ lb/ft) is unstretched when the blocks are in the position shown. Determine

a. The speed of block B when it is 1 ft below its initial position.
b. The maximum distance that block B will drop.

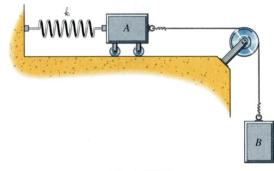

Fig. P17-57

17-58* The pair of blocks shown in Fig. P17-58 are released from rest with $d = 500$ mm. The masses of the blocks are $m_A = 6$ kg and $m_B = 4$ kg, and the spring is initially unstretched. Determine the minimum spring modulus such that block B does not hit the floor in the ensuing motion.

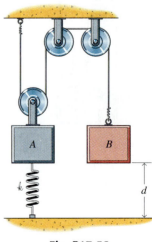

Fig. P17-58

17-59* The pair of blocks shown in Fig. P17-59 are released from rest when the spring is unstretched. The static and kinetic coefficients of friction are $\mu_s = 0.2$ and $\mu_k = 0.1$, respectively. For the ensuing motion, determine

a. The maximum velocity of the blocks and the stretch in the spring at that position.
b. The maximum distance that the 5-lb block will drop.
c. If the blocks will rebound from the position of part b.

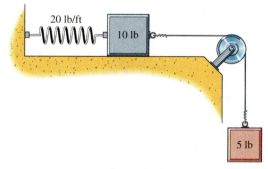

Fig. P17-59

17-60 The pair of blocks shown in Fig. P17-60 are released from rest when the spring is unstretched. The static and kinetic coefficients of friction are $\mu_s = 0.3$ and $\mu_k = 0.2$, respectively. For the ensuing motion, determine

a. The maximum velocity of the blocks and the stretch in the spring at that position.
b. The maximum distance that the 10-kg block will slide down the inclined surface.
c. If the blocks will rebound from the position of part b.

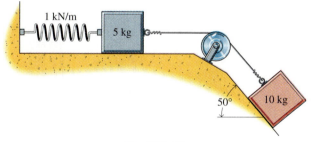

Fig. P17-60

17-61 The pair of blocks shown in Fig. P17-61 are released from rest when the spring is stretched 15 in. The kinetic coefficient of friction between the 10-lb block and the floor is $\mu_k = 0.6$. Determine for the ensuing motion the maximum velocity of the 10-lb block and the stretch in the spring at that position.

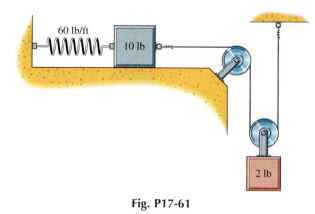

Fig. P17-61

Challenging Problems

17-62* The pair of blocks shown in Fig. P17-62 are released from rest from the position shown with the spring stretched 150 mm. Friction between the 10-kg block and the floor may be neglected. Determine for the ensuing motion the maximum distance that the 2-kg block will rise above the floor.

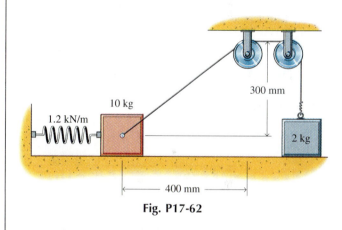

Fig. P17-62

17-63* When the pair of blocks shown in Fig. P17-63 are in the position shown, the 2-lb block has a speed of 5 ft/s to the right. If the horizontal surface and the vertical pole are both frictionless, determine the maximum distance that the 5-lb block will rise above its initial position.

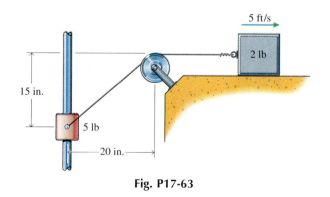

Fig. P17-63

17-64 The pair of blocks shown in Fig. P17-64 have masses $m_A = 30$ kg and $m_B = 20$ kg. If the blocks are released from rest from the position shown, determine the speed of block B after it has risen 2 m.

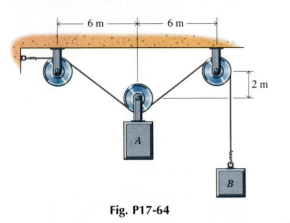

Fig. P17-64

17-65 The pair of carts shown in Fig. P17-65 are being pulled to the left by a constant force $P = 2$ lb. Cart A weighs 10 lb, cart B weighs 20 lb, and the system starts from rest when $x = 24$ ft. Determine the speed of cart B after it has moved 12 ft to the left.

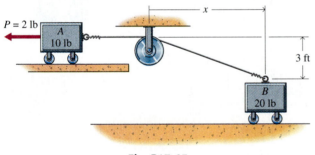

Fig. P17-65

17-66* The pair of blocks shown in Fig. P17-66 are released from rest when $x = -800$ mm. The spring has a modulus of $k = 500$ N/m and an unstretched length of $\ell_0 = 400$ mm. Friction may be neglected. For the ensuing motion, determine

a. The speed of the 2-kg block when $x = 0$ mm.
b. The maximum displacement x.

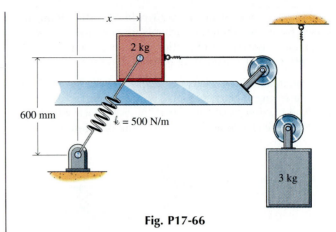

Fig. P17-66

Computer Problems

C17-67 The pair of blocks shown in Fig. P17-67 are released from rest when the spring is unstretched. The static and kinetic coefficients of friction are $\mu_s = 0.2$ and $\mu_k = 0.1$, respectively. For the ensuing motion,

a. Determine x_{max}, the maximum distance that the 10-lb block will move to the right.
b. Calculate and plot the speed of the 10-lb block as a function of its position x from its initial position $x = 0$ to its maximum position $x = x_{max}$.

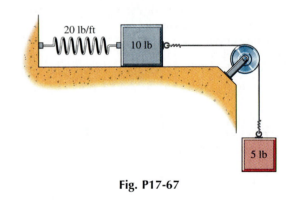

Fig. P17-67

C17-68 The pair of blocks shown in Fig. P17-68 are released from rest when the spring is unstretched. The static and kinetic coefficients of friction are $\mu_s = 0.3$ and $\mu_k = 0.2$, respectively. For the ensuing motion,

a. Determine x_{max}, the maximum distance that the 5-kg block will move to the right.

b. Calculate and plot the speed of the 5-kg block as a function of its position x from its initial position $x = 0$ to its maximum position $x = x_{max}$.

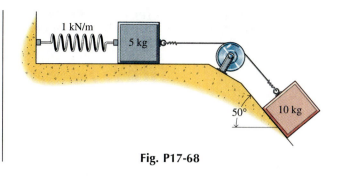

Fig. P17-68

17-7 CONSERVATIVE FORCES AND POTENTIAL ENERGY

Although the kinetic energy of a particle is always calculated as $T = \frac{1}{2}mv^2$, the work done on a particle must be calculated for each force acting on the particle. For a special category of forces called conservative forces, however, the work can also be calculated in a general sense, thus simplifying considerably the computations associated with the work–energy principle. For these conservative forces, the work done on a particle is calculated from a potential energy function, which depends only on the position of the particle.

17-7.1 Potential Energy of a Constant Force

A constant force is a trivial example of a force for which the work done can be replaced with a potential energy function. Consider a constant force $\mathbf{P}$ applied to the particle of Fig. 17-26. The coordinate system will be chosen with the positive x-axis in the direction of the force. Then, the work done on the particle by the force $\mathbf{P}$ as the particle moves from point 1 to point 2 is

$$U_{1\rightarrow2} = \int_1^2 \mathbf{P} \cdot d\mathbf{r} = \int_{x_1}^{x_2} P \, dx = P\int_{x_1}^{x_2} dx$$
$$= Px_2 - Px_1 \qquad (a)$$

The integral always comes out the same independent of the path that the particle follows. Therefore, the work done by the constant force $\mathbf{P}$ is evaluated by subtracting the value of Px at the initial position (position 1) from the value of Px at the final position (position 2).

For the constant force $\mathbf{P}$ the scalar function[1]

$$V_P = -Px \qquad (17\text{-}7)$$

is called the *potential energy* of the force. The value of the potential energy depends on the location of the origin from which x is measured. For a given particle position, the potential energy may be positive, negative, or zero depending on the location of the origin. However, the

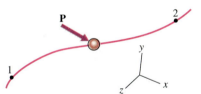

Figure 17-26 The work done on a particle by a constant force $\mathbf{P}$ as the particle moves from point 1 to point 2 is independent of the path followed by the particle and depends only on where the particle starts and where it ends. The work done can be computed simply as the difference in value of the potential function ($V_P = -Px$) at the beginning and end of the path.

---

[1]The choice $V_P = -Px$ rather than $V_P = Px$ is arbitrary. The minus sign is included so that T_1 and $(V_P)_1$ will have the same sign in Eq. 17-9.

work done on a particle by a constant force **P** is given by the difference in the potential energy

$$U_{1 \to 2} = (V_P)_1 - (V_P)_2 \qquad \text{(17-8)}$$

and the difference is the same regardless of the location from which x is measured. Note that if the final position of the particle is the same as its initial position, then the initial and final potentials are the same and the work done on the particle by the constant force P is zero. This is a characteristic of all conservative forces. The location for which the potential energy is zero is called the datum location. The datum is often chosen to make either the initial or the final potential energy zero.

The basic unit for potential energy is the same as for work and kinetic energy: joules (J) in the SI system of units or ft · lb in the U.S. Customary system of units.

In terms of the potential energy, then, the work–energy principle can be written

$$T_1 + (V_P)_1 - (V_P)_2 + \tilde{U}_{1 \to 2} = T_2$$

or

$$T_1 + (V_P)_1 + \tilde{U}_{1 \to 2} = T_2 + (V_P)_2 \qquad \text{(17-9)}$$

where $\tilde{U}_{1 \to 2}$ is the work done on the particle by all forces other than the constant force **P**.

17-7.2   Gravitational Potential Energy (Constant *g*)

The force of gravity acting on bodies near the surface of the earth may be considered constant. Therefore, the force of gravity acting on a particle is just a particular example of a constant force, and the work done on a particle by the force of gravity may be replaced by a potential energy function

$$U_{1 \to 2} = (V_g)_1 - (V_g)_2$$

where

$$V_g = mgh = Wh \qquad \text{(17-10)}$$

and h is the height of the particle.[2] Therefore, the work done by gravitational forces can be included in the work–energy principle simply as

$$T_1 + (V_g)_1 + \tilde{U}_{1 \to 2} = T_2 + (V_g)_2 \qquad \text{(17-11)}$$

where $\tilde{U}_{1 \to 2}$ is the work done on the particle by all forces other than the constant gravitational force.

The choice of the datum level (the height at which the gravitational potential energy is zero) is arbitrary. Although it is often taken to be the surface of the earth, it is just as often taken to be either the initial or the final height of the particle.

---

[2]The gravitational potential energy is *positive mgh* because the height of the particle increases in the direction opposite that of the direction of the gravitational force. In the previous section, the distance x increased in the same direction as the force **P**, so the potential energy was *negative Px*.

17-7.3 Gravitational Potential Energy (Inverse Square Law)

When particles move such that their height varies a great deal, the gravitational force $\mathbf{W} = -(GMm/r^2)\mathbf{e}_r$ can no longer be approximated as a constant. The value of GM is often replaced by noting that the weight of a body is $W = mg$ at the earth's surface. Comparing these two expressions for weight gives $GM = gR^2$, where R is the radius of the earth. Then if the displacement is expressed in cylindrical coordinates

$$d\mathbf{r} = dr\,\mathbf{e}_r + rd\theta\,\mathbf{e}_\theta + dz\,\mathbf{e}_z \qquad (b)$$

the work done by gravity becomes

$$U_{1\rightarrow2} = \int_{r_1}^{r_2} \mathbf{W}\cdot d\mathbf{r} = -\int_{r_1}^{r_2}\frac{mgR^2}{r^2}\,dr = \left[\frac{mgR^2}{r}\right]_{r_1}^{r_2}$$

$$= \frac{mgR^2}{r_2} - \frac{mgR^2}{r_1} \qquad (c)$$

The work done by the gravitational force is again independent of the path followed by the particle and depends only on the position of the particle at the beginning and end of the motion. The gravitational potential energy is then defined as

$$V_g = -\frac{mgR^2}{r} = -\frac{GMm}{r} \qquad (17\text{-}12)$$

and the work done by the gravitational force is

$$U_{1\rightarrow2} = (V_g)_1 - (V_g)_2$$

Except for the definition of the potential energy function, the work–energy equation for this case is the same as Eq. 17-11.

Note that the datum for the gravitational potential energy defined by Eq. 17-12 is at $r = \infty$ and that V_g is negative for $r < \infty$. Of course, a constant can always be added to the potential energy,[3] thus giving a different datum level, if desired.

17-7.4 Potential Energy of a Linear Elastic Spring Force

Next, consider a particle attached to a linear spring as shown in Fig. 17-27a. The force exerted on the particle by the spring is $\mathbf{F}_s = -k\delta\mathbf{e}_r$, as shown in Fig. 17-27b. Therefore, when the spring is stretched, δ is positive and the spring pulls on the particle in the negative $\mathbf{e}_r$ direction as drawn. On the other hand, when the spring is compressed, δ will be negative and the spring will actually push in the positive $\mathbf{e}_r$ direction.

Again expressing the displacement in cylindrical coordinates, the work done on the particle by the spring force $\mathbf{F}_s$ is

$$U_{1\rightarrow2} = \int_{r_1}^{r_2} \mathbf{F}_s\cdot d\mathbf{r} = -\int_{r_1}^{r_2} k\delta\,dr \qquad (d)$$

---

[3]It is important to distinguish between adding a constant to the potential energy function and adding a constant to the radius r. The correct way to set the datum (zero level) of potential energy) at the earth's surface is $V_g = mgR - mgR^2/r$ (not $-mgR^2/h$) where $h = r - R$ is the height above the surface of the earth.

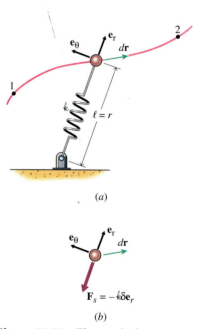

(a)

(b)

Figure 17-27 The work done on a particle by a spring force as the particle moves from point 1 to point 2 is independent of the path followed by the particle and depends only on where the particle starts and where it ends. The work done can be computed simply as the difference in value of the potential function ($V_s = \frac{1}{2}k\delta^2$) at the beginning and end of the path.

But the deformation of the spring is its present length minus its undeformed length ($\delta = \ell - \ell_0 = r - \ell_0$), so $d\delta = dr$. Then the work becomes

$$U_{1 \to 2} = -\int_{\delta_1}^{\delta_2} k\delta \, d\delta = -\left[\frac{1}{2}k\delta^2\right]_{\delta_1}^{\delta_2}$$

$$= -\frac{1}{2}k\delta_2^2 + \frac{1}{2}k\delta_1^2 \qquad (e)$$

which again is independent of the path of the particle and depends only on the stretch of the spring at the initial and final positions. The potential energy for the spring force is then defined as[4]

$$V_s = \frac{1}{2}k\delta^2 \qquad (17\text{-}13)$$

and the work done by the spring force can be included in the work–energy principle simply as

$$T_1 + (V_s)_1 + \tilde{U}_{1 \to 2} = T_2 + (V_s)_2 \qquad (17\text{-}14)$$

where $\tilde{U}_{1 \to 2}$ is the work done on the particle by all forces other than the spring force.

Note that the potential energy of the spring force defined by Eq. 17-13 is zero when the spring is undeformed. The datum can be set to any level by setting the potential energy $V_s = \frac{1}{2}k\delta^2 - \frac{1}{2}k\delta_d^2$, where δ_d is the deformation of the spring at the new datum.[5]

17-7.5   Friction

As an example of a nonconservative force, the work done by friction forces depends on the path. Frictional forces generally oppose motion, so the work done by friction forces is usually negative. Then, the longer the path traveled by the particle, the greater the work done by friction. Since the work done by frictional forces is not independent of the path, frictional forces are not conservative forces, and the work done by frictional forces cannot be represented by a potential function.

17-7.6   Conservative Forces

The concept of potential energy may be used whenever the work of the force considered is independent of the path followed by its point of application as this point moves from an initial position to a final position. Such forces are said to be *conservative forces*.

Consider a general force **F** for which the work done as the point of application moves from point 1 to point 2 is

$$U_{1 \to 2} = \int_1^2 \mathbf{F} \cdot d\mathbf{r} = \int_1^2 (F_x \, dx + F_y \, dy + F_z \, dz) \qquad (17\text{-}15)$$

If the integral is to be independent of the path and depends only on

---

[4]Note that it is the *deformation* of the spring and *not* the *length* of the spring that is squared. That is $V_s = \frac{1}{2}k\delta^2 = \frac{1}{2}k(\ell - \ell_0)^2 \ne \frac{1}{2}k(\ell^2 - \ell_0^2)$.

[5]Note that $V_s = \frac{1}{2}k\delta^2 - \frac{1}{2}k\delta_d^2$ is *not* the same as $\frac{1}{2}k(\delta - \delta_d)^2$, nor do these expressions differ by a constant.

the end points of the path, then the integrand of Eq. 17-15 must be an exact differential

$$U_{1 \to 2} = \int_1^2 (-dV) = V_1 - V_2 \tag{17-16}$$

That is,

$$-dV = F_x \, dx + F_y \, dy + F_z \, dz \tag{17-17}$$

where again the minus sign is included so that T_1 and V_1 will have the same sign in Eq. 17-9. Since the potential energy is a function of all three position variables $V = V(x, y, z)$, its differential is given by

$$dV = \frac{\partial V}{\partial x} dx + \frac{\partial V}{\partial y} dy + \frac{\partial V}{\partial z} dz \tag{17-18}$$

where $\partial V/\partial x$, $\partial V/\partial y$, and $\partial V/\partial z$ are the partial derivatives of the potential energy function $V(x, y, z)$. Therefore, comparing Eqs. 17-17 and 17-18 gives

$$\frac{\partial V}{\partial x} dx + \frac{\partial V}{\partial y} dy + \frac{\partial V}{\partial z} dz = -(F_x \, dx + F_y \, dy + F_z \, dz) \tag{f}$$

or

$$\left(\frac{\partial V}{\partial x} + F_x \right) dx + \left(\frac{\partial V}{\partial y} + F_y \right) dy + \left(\frac{\partial V}{\partial z} + F_z \right) dz = 0 \tag{g}$$

But if the work is in fact independent of the path, then Eq. g must be satisfied for arbitrary choices of dx, dy, and dz, from which it follows that

$$F_x = -\frac{\partial V}{\partial x} \qquad F_y = -\frac{\partial V}{\partial y} \qquad F_z = -\frac{\partial V}{\partial z} \tag{17-19}$$

That is, the components of the conservative force **F** are all derivable from the potential energy function $V(x, y, z)$.

Equation 17-19 may be expressed in vector notation

$$\mathbf{F} = -\left(\frac{\partial V}{\partial x} \mathbf{i} + \frac{\partial V}{\partial y} \mathbf{j} + \frac{\partial V}{\partial z} \mathbf{k} \right) = -\nabla V \tag{17-20}$$

in which ∇ is the vector operator

$$\nabla = \mathbf{i} \frac{\partial}{\partial x} + \mathbf{j} \frac{\partial}{\partial y} + \mathbf{k} \frac{\partial}{\partial z} \tag{h}$$

The vector ∇V is called the *gradient of V*.

Since the components of **F** are all derivable from a single scalar function, they are related to each other. For example, taking the partial derivatives of F_x and F_y with respect to y and x, respectively, gives

$$\frac{\partial F_x}{\partial y} = -\frac{\partial^2 V}{\partial y \, \partial x} \quad \text{and} \quad \frac{\partial F_y}{\partial x} = -\frac{\partial^2 V}{\partial x \, \partial y} \tag{i, j}$$

The order of the partial derivatives is immaterial, however, so

$$\frac{\partial F_x}{\partial y} = \frac{\partial F_y}{\partial x} \tag{17-21a}$$

Similar relations hold between F_y and F_z and between F_x and F_z:

$$\frac{\partial F_y}{\partial z} = \frac{\partial F_z}{\partial y} \quad \text{and} \quad \frac{\partial F_z}{\partial x} = \frac{\partial F_x}{\partial z} \quad \text{(17-21b,c)}$$

Therefore, it is easy to determine whether a force is conservative or not. If its components satisfy Eq. 17-21, the force is conservative and a potential energy function can be found. If its components do not satisfy Eq. 17-21, the force is not conservative and a potential energy function does not exist.

17-8 GENERAL PRINCIPLE OF WORK AND ENERGY

When the work–energy principle (Eq. 17-6) is used to solve a problem, the work done on a particle must be computed for each force that acts on the particle. Computation of the work term $U_{1 \to 2}$ is simplified considerably by using the concept of potential energy. The work term in the work–energy principle can be divided into two parts

$$U_{1 \to 2} = U_{1 \to 2}^{(c)} + U_{1 \to 2}^{(o)} \quad \text{(k)}$$

The term $U_{1 \to 2}^{(c)}$ represents the work done by all the conservative forces acting on the particle; that is, by all the forces whose potential energy function is known. Similarly, the term $U_{1 \to 2}^{(o)}$ represents the work done by all the other forces acting on the particle; that is, by all of the forces that either have no potential or whose potential energy function is not known. Then the work–energy principle becomes

$$T_1 + U_{1 \to 2}^{(c)} + U_{1 \to 2}^{(o)} = T_2 \quad \text{(17-22)}$$

But the conservative part of the work term can be replaced using the potential energy $U_{1 \to 2}^{(c)} = V_1 - V_2$, where $V = V_g + V_e + \cdots$ is the sum of the potential energies of the conservative forces. This gives

$$T_1 + V_1 + U_{1 \to 2}^{(o)} = T_2 + V_2 \quad \text{(17-23)}$$

The combination of terms $T + V$ is called the *total mechanical energy*[6] E. Equation 17-23 expresses that the net increase in the total mechanical energy between the initial and the final position of the particle is equal to the work done on the particle by nonconservative forces.

Equation 17-23 is commonly used to determine the speed or position of a particle before or after a particular displacement has occurred. Since the reference (or datum) level from which to determine the potential energy is arbitrary, the calculations are simplified if the datum location is chosen so as to make one or more of the terms in Eq. 17-23 zero. In fact, different datum locations can be chosen for each of the potential energy terms.

17-9 CONSERVATION OF ENERGY

In many problems of interest, friction forces are often negligible and the only forces acting on a particle are elastic springs and gravity. In these cases, $U_{1 \to 2}^{(o)} = 0$ and Eq. 17-23 becomes simply

$$T_1 + V_1 = T_2 + V_2 \quad \text{(17-24)}$$

---

[6]The total mechanical energy is a relative term. Since the datum or zero level of potential energy is arbitrary, a moving particle may have a zero or even negative value of total mechanical energy.

That is, when the only forces acting on a particle are conservative forces, the total mechanical energy remains constant

17-10   POWER AND EFFICIENCY

$$E_1 = E_2 = \text{constant} \qquad (17\text{-}25)$$

Equation 17-24 is often referred to as the *Principle of Conservation of Energy*. The total mechanical energy of the particle is conserved in the sense that whatever value $E = T + V$ has at position 1, it has the same value at position 2. Any decrease in the potential energy is accompanied by an equivalent increase in the kinetic energy and vice versa. The constant in Eq. 17-25 is determined from the known position and velocity of the particle at some instant. Since the location of the datum for the potential energy functions is arbitrary, the constant can be positive, zero, or negative.

Conservation of energy is not really a new principle. It is simply a special case of the general principle of work and energy, Eq. 17-23.

17-10   POWER AND EFFICIENCY

Power and efficiency are concepts that are closely related to the concepts of mechanical work and mechanical energy. Any measure of the mechanical output of a machine must take into account the rate at which the work is done (power) as well as the total amount of work done by the machine. Also, the rating of the machine should account for the amount of energy the machine requires to do the work.

17-10.1   Power

Two machines (two forces) that do the same amount of work are often not equivalent. A machine that requires 20 hours to do some amount of work is not an acceptable substitute for a machine that requires only 2 minutes to do the same amount of work. That is, the amount of time required to do an amount of work or the rate at which work is done is often an important consideration when choosing or designing a machine.

The output power of a force $\mathbf{F}$ is given by the rate at which it does work

$$\text{Power} = \frac{dU}{dt} = \frac{\mathbf{F} \cdot d\mathbf{r}}{dt} = \mathbf{F} \cdot \mathbf{v} \qquad (17\text{-}26)$$

Since $d\mathbf{r}$ is the displacement of the particle (the point to which the force is applied), $d\mathbf{r}/dt = \mathbf{v}$ is the velocity of the particle.

Power has dimensions of *work* (*force* times *length*) divided by *time*. In the SI system of units, this combination of units is called a *watt* ($1 \text{ W} = 1 \text{ J/s} = 1 \text{ N} \cdot \text{m/s}$). In the U.S. Customary system of units it is called *horsepower* ($1 \text{ hp} = 550 \text{ ft} \cdot \text{lb/s} = 33{,}000 \text{ ft} \cdot \text{lb/min}$).

17-10.2   Mechanical Efficiency

All real mechanical systems lose energy to friction, so the amount of work done by a machine (output work) is always less than the amount of work done on a machine (input work). The *mechanical efficiency* η_{mech} of the machine is defined by the ratio of these two quantities

$$\eta_{mech} = \frac{\text{output work}}{\text{input work}} \qquad (17\text{-}27a)$$

Dividing the numerator and denominator of Eq. 17-27a by dt gives

$$\eta_{mech} = \frac{\text{power output}}{\text{power input}} \qquad (17\text{-}27b)$$

Since all real systems lose energy to friction, the output work is always less than the input work and η_{mech} is always less than one.

Of course, the energy is not really lost in a thermodynamic sense. In a typical machine, the mechanical energy lost due to the negative work of friction forces between moving parts is converted to heat energy, which in turn is dissipated to the surroundings of the machine. When all forms of energy (mechanical, electrical, thermal, chemical, etc.) are accounted for, however, energy is conserved.

EXAMPLE PROBLEM 17-9

A 5-kg crate slides down a ramp as shown in Fig. 17-28 and strikes the bumper B. The coefficient of friction between the crate and the floor is $\mu_k = 0.25$ and the mass of the bumper may be neglected. If the crate is released from rest when it is 15 m from the bumper, determine:

a.   The speed v_C of the crate at the instant it strikes the bumper.
b.   The maximum deflection δ_{max} of the spring due to the motion of the crate.

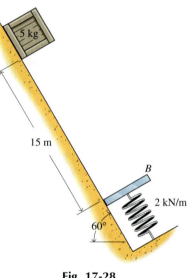

Fig. 17-28

SOLUTION

a.   The free-body diagram of the crate is shown in Fig. 17-29a for a general position along the ramp before the crate contacts the bumper. Since the crate does not move in the y-direction (the direction perpendicular to the surface), the y-component of Newton's second law

$$+ \nearrow \Sigma F_y = ma_y: \qquad N - 5(9.81)\cos 60° = 0$$

gives $N = 24.53$ N, and it is constant. The friction force during the motion is $F = \mu_k N = 0.25(24.53) = 6.131$ N, and it is also constant.

The gravitational force $W = mg$ is conservative, and its work can be determined using its potential energy. Using the initial height of the crate as the datum, the initial gravitational potential energy is zero $V_1 = 0$, and the gravitational potential energy when the crate strikes the bumper is

$$V_2 = mgh = (5)(9.81)(-15 \sin 60°) = -637.2 \text{ J}$$

The friction force is nonconservative, and its work must be computed directly. Since the friction force is constant, its work is simply

$$(U_{1\rightarrow 2})_F = Fd = -6.13(15) = -91.95 \text{ J}$$

The remaining force N acts normal to the motion and so does no work on the crate: $(U_{i\rightarrow f})_N = 0$.

Since the crate starts from rest, its initial kinetic energy is zero ($T_1 = 0$) and the kinetic energy of the crate when it is about to strike the bumper is

$$T_2 = \frac{1}{2}mv^2 = \frac{1}{2}(5)v_C^2 = 2.50 \ v_C^2$$

The work done by the gravitational force $W = mg$ is the difference between its initial and final potential energies $(U_{1\rightarrow 2})_g = (V_g)_1 - (V_g)_2 = (mgh)_1 - (mgh)_2 = mg(h_1 - h_2)$. Since only the difference in heights is important, it doesn't matter where the two heights are measured from as long as they are both measured from the same location (called the datum).

Substituting these values into the work–energy equation (Eq. 17-23)

$$T_1 + V_1 + U^{(o)}_{1 \to 2} = T_2 + v_2$$

gives

$$0 + 0 - 91.95 = 2.50v_C^2 - 637.2$$

or

$$v_C = 14.77 \text{ m/s} \qquad \text{Ans.}$$

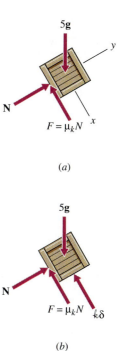

(a)

b. After the crate contacts the bumper, the crate and the bumper will move together. The free-body diagram of the system is shown in Fig. 17-29b. The normal and friction forces are still $N = 24.53$ N and $F = 6.13$ N, respectively. The initial velocity of the crate for this phase of the motion is $v_C = 14.77$ m/s, and the initial kinetic energy is

$$T_2 = \frac{1}{2}mv^2 = \frac{1}{2}(5)(14.77)^2 = 545.4 \text{ J}$$

(b)

Fig. 17-29

At the point of maximum spring deflection, the velocity of the crate is zero, so the final kinetic energy is zero ($T_3 = 0$).

The work done on the crate by friction as the crate slides along the floor an additional distance δ_{max} is

$$(U_{i \to f})_F = Fd = -6.13 \, \delta_{max}$$

Both the spring force and the gravitational force are conservative, and their work can be determined using their potential energies. Using the position when the crate first contacts the bumper as the datum, the initial gravitational potential energy is zero, $(V_g)_2 = 0$, and the final gravitational potential energy is

$$(V_g)_3 = mgh = 5g(-\delta_{max} \sin 60°) = -42.48 \, \delta_{max}$$

Since the spring is undeformed when the crate first contacts the crate, the same position will be used as the datum for the potential energy of the spring force. Then the initial potential energy of the spring force is zero, $(V_s)_2 = 0$, and the final potential energy of the spring force is

$$(V_s)_3 = \frac{1}{2} k\delta^2 = \frac{1}{2}(2000)\delta^2_{max} = 1000 \, \delta^2_{max}$$

The work–energy equation (Eq. 17-23) then becomes

$$545.4 + 0 + 0 - 6.13 \, \delta_{max} = 0 - 42.48 \, \delta_{max} + 1000 \, \delta^2_{max}$$

which gives

$$\delta_{max} = 0.757 \text{ m} \qquad \text{Ans.}$$

A 0.5-lb slider moves along a semicircular wire in a horizontal plane as shown in Fig. 17-30. The undeformed length of the spring is 8 in., and friction may be neglected. If the slider is released from rest at position A, determine

a. The velocity of the slider at position B.
b. The force exerted on the slider by the wire at position B.

Fig. 17-30

SOLUTION

a. The free-body diagram of the slider is shown in Fig. 17-31a for a general position along the wire. The weight force acts perpendicular to the motion (perpendicular to the figure) and does no work. The normal force N also acts perpendicular to the motion and does no work. The spring force is a conservative force, and its work can be calculated using its potential energy. The length of the spring at position A is $\ell_A = \sqrt{24^2 + 12^2} = 26.83$ in.; the deformation of the spring at position A is $\delta_A = (26.83 - 8)/12 = 1.5692$ ft. Therefore, the potential energy of the spring force at position A is

$$(V_s)_A = \frac{1}{2}k\delta^2 = \frac{1}{2}(0.75)(1.5692)^2 = 0.9234 \text{ ft} \cdot \text{lb}$$

At position B the deformation of the spring is $\delta_B = (12 - 8)/12 = 0.3333$ ft, and the potential energy of the spring force at B is

$$(V_s)_B = \frac{1}{2}k\delta^2 = \frac{1}{2}(0.75)(0.3333)^2 = 0.04166 \text{ ft} \cdot \text{lb}$$

There are no nonconservative forces, so $U_{1\to2}^{(o)} = 0$. Since the slider starts from rest, its initial kinetic energy is zero ($T_A = 0$); at position B the kinetic energy of the slider is

$$T_B = \frac{1}{2}mv^2 = \frac{1}{2}\left(\frac{0.5}{32.2}\right)v_B^2 = 0.007764v_B^2$$

Substituting these values in the work–energy equation (Eq. 17-23)

$$T_A + V_A + U_{A\to B}^{(o)} = T_B + V_B$$

gives

$$0 + 0.9234 + 0 = 0.007764v_B^2 + 0.04166$$

or

$$v_B = 10.66 \text{ ft/s} \qquad\qquad \text{Ans.}$$

b. The free-body diagram of the slider is shown in Fig. 17-31b for position B. The component of the slider's acceleration that is normal to the wire is $a_n = v^2/r = 10.66^2/1 = 113.64$ ft/s^2. Then the component of Newton's second law in the direction perpendicular to the wire ($\Sigma F_n = ma_n$) gives

$$N_B - (0.75)(0.3333) = \frac{0.5}{32.2}(113.64)$$

or

$$N_B = 2.015 \text{ lb} \qquad\qquad \text{Ans.}$$

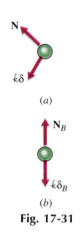

(a)

(b)

Fig. 17-31

Since the normal force N is always perpendicular to the motion, it does not appear in any term of the work–energy equation, and the work–energy principle cannot be used to determine the normal force exerted on the slider by the wire. However, the work–energy principle easily gives the speed of the slider at any position along the wire as needed in the normal component of Newton's second law which is used to calculate N.

Two blocks are connected by a light inextensible cord, which passes around small massless pulleys as shown in Fig. 17-32. If block B is pulled down 500 mm from the equilibrium position and released from rest, determine its speed when it returns to the equilibrium position.

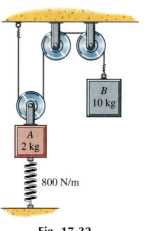

Fig. 17-32

SOLUTION

The two blocks A and B are a system of interacting particles. The work–energy equations for each particle individually can be added together to get a similar equation for the system

$$T_i + V_i + U^{(o)}_{i \to f} = T_f + V_f \qquad (a)$$

In Eq. a, T represents the sum of the kinetic energies of both particles, V represents the sum of the potential energies of all conservative forces acting on both particles, and $U^{(o)}_{i \to f}$ represents the sum of the work done by all other forces acting on both particles.

The free-body diagram of the two blocks are shown in Figs. 17-33a and 17-33b. In the equilibrium position the sum of forces is zero for both blocks:

$$\uparrow \Sigma F = 2T_{st} - (2)(9.81) - 800\, \delta_{st} = 2a_A = 0$$
$$\uparrow \Sigma F = T_{st} - (10)(9.81) = 10a_B = 0$$

Therefore, the static tension in the rope is $T_{st} = 98.1$ N and the static deformation of the spring is $\delta_{st} = 0.2207$ m.

Since the length of the cord is a constant, block A must rise whenever block B falls and vice versa. Referring to Fig. 17-33c, the length of the cord in the equilibrium position ($y_A = y_B = 0$) is given by

$$\ell = 2d + b + c \qquad (b)$$

When block A has moved upward a distance y_A and block B has moved downward a distance y_B, the length of the cord is given by

$$\ell = 2(d - y_A) + b + (c + y_B) \qquad (c)$$

Subtracting Eq. b from Eq. c gives the position relationship $y_B = 2y_A$; differentiating this equation gives the velocity relationship $v_B = \dot{y}_B = 2\dot{y}_A = 2v_A$.

Since the system is released from rest, the initial kinetic energies of both blocks are zero: $(T_A)_i = (T_B)_i = 0$. When the blocks return to the equilibrium position, the sum of their kinetic energies will be

$$(T_A)_f + (T_B)_f = \tfrac{1}{2}(2)v_A^2 + \tfrac{1}{2}(10)v_B^2 = 5.25v_B^2$$

Since the cord is inextensible, the work done by the tension force cancels out when the work is added up for both particles. Both the gravitational and the spring forces are conservative, so the work done by nonconservative forces is zero $U^{(o)}_{i \to f} = 0$. Measuring the potential energies of the forces acting on each block from its equilibrium position gives

$$(V_A)_i = (V_{Ag})_i + (V_{As})_i = (2)(9.81)\left(\frac{0.5}{2}\right) + \frac{1}{2}(800)\left(0.2207 + \frac{0.5}{2}\right)^2 = 93.53$$

$$(V_A)_f = (V_{Ag})_f + (V_{As})_f = \frac{1}{2}(800)(0.2207)^2 = 19.48 \text{ J}$$

$$(V_B)_i = (V_{Bg})_i = (10)(9.81)(-0.5) = -49.05 \text{ J}$$

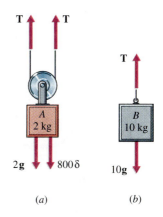

T ↑   T ↑

T ↑

2g ↓   800δ ↓

10g ↓

(a)   (b)

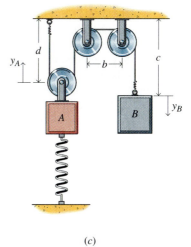

(c)

Fig. 17-33

and

$$(V_B)_f = (V_{Bg})_f = 0 \text{ J}$$

Substituting these values into Eq. *a* gives

$$0 + 93.53 - 49.05 + 0 = 5.25v_B^2 + 19.48$$

or

$$v_B = 2.18 \text{ m/s} \qquad \text{Ans.}$$

EXAMPLE PROBLEM 17-12

The drag force due to air resistance on a cyclist (Fig. 17-34) moving with speed *v* is given by

$$F_D = 0.006v^2$$

where F_D is in lbs and *v* is in ft/s. If the combined weight of the cyclist and bicycle is 180 lb, determine

a. The power required to maintain a steady speed of 20 mi/h on level ground.
b. The maximum speed the cyclist could maintain up a 5° incline for the same amount of power.

Fig. 17-34

SOLUTION

a. A free-body diagram of the cyclist is shown in Fig. 17-35a. The propulsive force *R* is generated by friction, which acts to prevent the wheel from slipping on the pavement. Ultimately, of course, the propulsive force (and the propulsive power) is provided by the cyclist when he pushes on the pedals, which causes the chain to pull on the rear sprocket, which causes the rear tire to turn.

To travel at a constant speed ($a = 0$ ft/s^2) on level ground requires a propulsive force of

$$+ \rightarrow \Sigma F_x = ma_x: \qquad R - F_D = 0$$

or

$$R = F_D = 0.006v^2$$

where $v = 20 \text{ mi/h} = 29.33 \text{ ft/s}$. Then the power is

$$\mathscr{P} = Rv = 0.006(29.33)^3$$
$$= 151.4 \text{ ft} \cdot \text{lb/s} = 0.275 \text{ hp} \qquad \text{Ans.}$$

Actually, $\mathscr{P} = 0.275$ hp is the amount of power that must be delivered to the wheel of the bicycle. The cyclist will have to provide more than this to overcome friction in the various parts of the drive train of the bicycle.

b. If the cyclist travels up a 5° incline, then (Fig. 17-35b)

$$+ \nearrow \Sigma F_t = ma_t: \qquad R - F_D - 180 \sin 5° = 0$$

which gives

$$R = 0.006v^2 + 180 \sin 5°$$

and for the same 151.4 ft · lb/s of power,

$$151.4 = 0.006v^3 + 180v \sin 5°$$
$$v = 9.339 \text{ ft/s} = 6.37 \text{ mi/h} \qquad \textbf{Ans.}$$

For a fixed amount of power, the speed of the cyclist going up a 5° incline is less than a third of his speed on level ground. In fact, to maintain the same 20 mi/h speed up the 5° incline, the cyclist would have to expend over four times as much power (about 612 ft · lb/s) as he would on level ground!

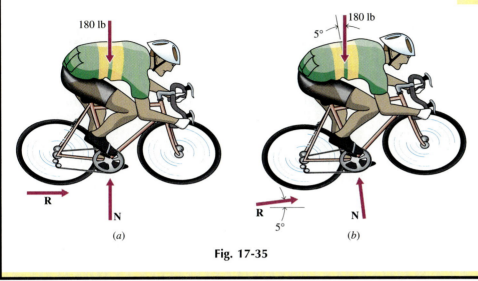

(a)   (b)

Fig. 17-35

PROBLEMS

In the following problems, all ropes, cords, and cables are assumed to be flexible, inextensible, and of negligible mass. All pins and pulleys are small, have negligible mass, and are frictionless.

Introductory Problems

17-69* In a shipping warehouse, packages are moved between levels by sliding them down a chute as shown in Fig. P17-69. If a 30-lb package starts at $\ell = 25$ ft with an initial speed of 15 ft/s down a $\theta = 10°$ chute, determine the kinetic coefficient of friction μ_k for which the package will reach the corner at the bottom of the ramp with zero speed.

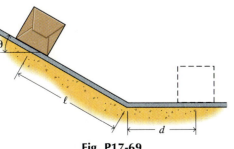

Fig. P17-69

17-70* If the packages of Problem 17-69 come off the ramp with too great a velocity, a bumper as shown in Fig. P17-70 may be required to "catch" them. The kinetic coefficient of friction between the package and the floor is $\mu_k = 0.25$, the spring modulus is $k = 1750$ N/m, and the mass of the bumper B may be neglected. If the speed of a 2.5 kg package is $v_0 = 8$ m/s when it is $\ell = 3$ m from the bumper, determine

a. The maximum deflection δ of the spring.
b. The final resting position of the package.

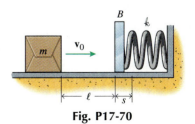

Fig. P17-70

17-71 The pair of blocks shown in Fig. P17-71 are released from rest with $d = 18$ in. The weights of the blocks are $W_A = 5$ lb and $W_B = 10$ lb, and the spring ($k = 20$ lb/ft) is unstretched in the initial position. Determine the speed of block B as it hits the floor.

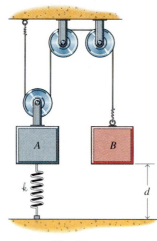

Fig. P17-71

17-72 The pair of blocks shown in Fig. P17-72 are released from rest when the spring is unstretched. The static and kinetic coefficients of friction are $\mu_s = 0.3$ and $\mu_k = 0.2$, respectively. For the ensuing motion, determine

a. The maximum velocity of the blocks and the stretch in the spring at that position.

b. The maximum distance that the 10-kg block will slide down the inclined surface.

c. If the blocks will rebound from the position of part b.

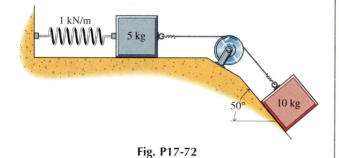

Fig. P17-72

17-73* The pair of blocks in Fig. P17-73 are released from rest when the spring is stretched 15 in. The kinetic coefficient of friction between the 10-lb block and the floor is $\mu_k = 0.6$. Determine for the ensuing motion the maximum velocity of the 10-lb block and the stretch in the spring at that position.

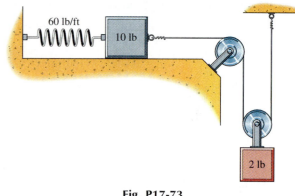

Fig. P17-73

17-74* A 0.5-kg mass slides on a frictionless rod in a vertical plane as shown in Fig. P17-74. The undeformed length of the spring is $\ell_0 = 250$ mm, the spring modulus is $k = 600$ N/m, and the distance $d = 800$ mm. If the slider is released from rest when $b = 300$ mm, determine the speed of the slider at positions A and B.

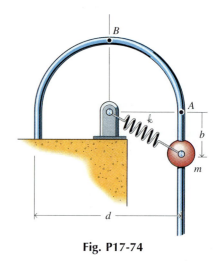

Fig. P17-74

17-75 A singer swings a 0.75-lb microphone in a vertical plane at the end of a 24-in.-long cord as shown in Fig. P17-75. If the tension in the cord when the microphone is at A is twice the tension in the cord when the microphone is at B, determine the velocity of the microphone and the tension in the cord when the microphone is at A.

Fig. P17-75

17-76 A small sack containing 1.5 kg of marbles is tied to the end of an 800-mm-long cord as shown in Fig. P17-76. The maximum tension that the cord can withstand is $P_{max} = 30$ N. If the boy slowly pulls the sack off of the shelf, determine the angle θ through which the sack will swing before the cord breaks.

Fig. P17-76

Intermediate Problems

17-77* A 3-lb collar slides on a frictionless vertical guide as shown in Fig. P17-77. A constant horizontal force of 12 lb is applied to the end of the light inextensible cord that is attached to the collar. If the collar is released from rest when $d = 32$ in., determine the velocity of the collar when $d = 18$ in.

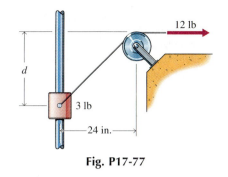

Fig. P17-77

17-78* A 10-kg block slides on a frictionless horizontal surface as shown in Fig. P17-78. A constant horizontal force of 50 N is applied to the end of the light inextensible cord that is attached to the block. If the block is released from rest from the position shown in which the spring is unstretched, determine for the ensuing motion

a.   The maximum velocity of the block and the stretch in the spring at that position.
b.   The maximum stretch in the spring.

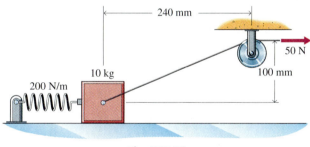

Fig. P17-78

17-79 When the pair of blocks shown in Fig. P17-79 are in the position shown, the 2-lb block has a speed of 5 ft/s to the right. If the horizontal surface and the vertical pole are both frictionless, determine the maximum distance that the 5-lb block will rise above its initial position.

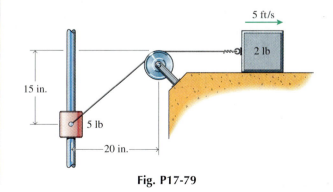

Fig. P17-79

17-80 The pair of blocks shown in Fig. P17-80 are released from rest from the position shown with the spring stretched 150 mm. Friction between the 10-kg block and the floor may be neglected. Determine for the ensuing motion the maximum distance that the 2-kg block will rise above the floor.

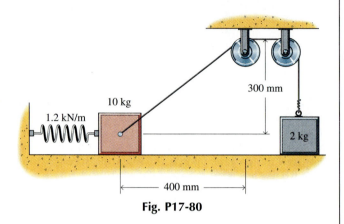

Fig. P17-80

17-81* A 1.5-lb collar slides on a frictionless vertical rod as shown in Fig. P17-81. The undeformed length of the spring is $\ell_0 = 8$ in., the spring modulus is $k = 80$ lb/ft, and the distance $d = 12$ in. If the collar is released from rest when $b = 9$ in., determine the speed of the collar when $b = 0$.

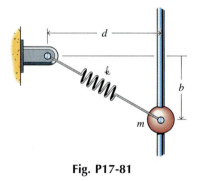

Fig. P17-81

17-82* A small box is sliding along a frictionless horizontal surface when it comes upon a circular ramp as shown in Fig. P17-82. If the radius of the ramp is $r = 750$ mm and the box loses contact with the ramp when $\theta = 25°$, determine the initial speed v_0 of the box.

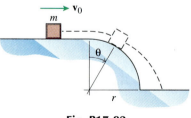

Fig. P17-82

17-83 A 1.2-lb roller B moves along a smooth circular slot in a vertical plane as shown in Fig. P17-83. The modulus of the spring is $k = 3$ lb/ft, and the unstretched length of the spring is $\ell_0 = 12$ in. If the roller is moved to the position where $\theta = 90°$ and released from rest, determine

a. The speed of the roller when $\theta = 0°$.
b. The force exerted on the roller by the surface of the slot when $\theta = 0°$.

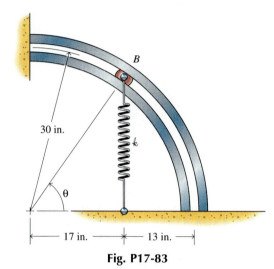

Fig. P17-83

17-84 An elevator E is attached by means of an inextensible cable to a 900-kg counterweight C (Fig. P17-84). The combined mass of the man and the elevator is 1000 kg. The elevator is raised and lowered using the motor M which is attached to the elevator by means of a second cable. Determine the power that the motor must supply if the elevator is

a. Raised at a constant rate of 0.5 m/s.
b. Lowered at a constant rate of 0.5 m/s.

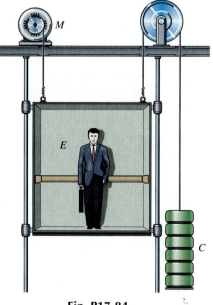

Fig. P17-84

17-85* A 400-lb crate C is attached to a power winch W as shown in Fig. P17-85. If the kinetic coefficient of friction between the crate and the 25° incline is $\mu_k = 0.2$ and the maximum power of the winch is 0.5 hp, determine the maximum constant speed at which the winch can raise the crate.

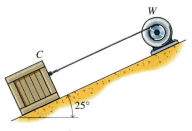

Fig. P17-85

17-86* A cyclist can maintain a speed of 30 km/h on level ground with a power output of 275 watts. The combined weight of the cyclist and bicycle is 800 N. Assuming that the retarding forces remain constant, determine

a. The maximum steady speed that the cyclist could maintain up a 5° incline for the same power.
b. The power that the cyclist would have to deliver to climb the 5° incline at 30 km/h.

17-87 A 2500-lb car requires 20 hp delivered to its wheels to maintain a steady 50 mi/h on a level road. Assuming that the retarding forces remain constant, determine

a. The maximum speed that the car could maintain up a 5° incline with the same 20 hp delivered to its wheels.
b. The horsepower that must be delivered to the wheels for the car to climb the 5° incline at a steady 50 mi/h.

17-88 The sum of all drag forces acting on a 1200-kg car moving with a speed v is given by

$$F_D = 200 + 0.8v^2$$

where F_D is in newtons and v is in meters per second (Fig. P17-88). Determine the power that must be delivered to the wheels to travel

a. At 40 km/h on a horizontal road.
b. At 80 km/h on a horizontal road.
c. At 40 km/h up a 5° incline.

Fig. P17-88

Challenging Problems

17-89* The pair of blocks shown in Fig. P17-89 are released from rest when $x = 0$. The spring has a modulus of $k = 72$ lb/ft and an unstretched length of $\ell_0 = 12$ in. Friction between the 4-lb block and the horizontal surface may be neglected. For the ensuing motion, determine

a. The speed of the blocks when $x = 4$ in.
b. The maximum displacement x_{max}.

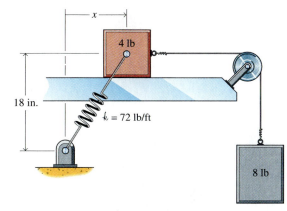

Fig. P17-89

17-90* The pair of blocks shown in Fig. P17-90 are released from rest when $x = -800$ mm. The spring has a modulus of $k = 500$ N/m and an unstretched length of $\ell_0 = 400$ mm. Friction between the 2 kg block and the horizontal surface may be neglected. For the ensuing motion, determine

a. The speed of the 2-kg block when $x = 0$ mm.
b. The maximum displacement x_{max}.

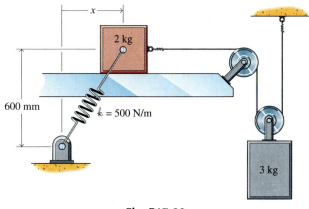

Fig. P17-90

17-91 The pair of blocks shown in Fig. P17-91 are suspended by a cord that passes around three small pulleys. If block A weighs 60 lb, block B weighs 40 lb, and the system starts from rest when $y = 6$ ft, determine the speed of block B after it has moved 10 ft.

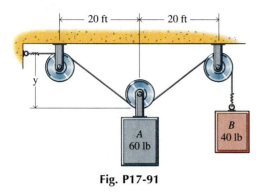

Fig. P17-91

Computer Problems

C17-92 An ice cube slides down the inside of a frictionless cylinder as shown in Fig. P17-92. The mass of the ice cube is 20 g and the radius of the cylinder is $r = 300$ mm. If the ice cube starts from rest when $\theta = 75°$, calculate and plot the kinetic energy T, the potential energy V, the total energy $E = T + V$, and the normal force N exerted on the ice cube by the cylinder, all as functions of the angle θ ($-60° < \theta < 75°$). (Let the zero of potential energy be at $\theta = 90°$).

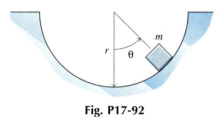

Fig. P17-92

C17-93 The pendulum shown in Fig. P17-93 consists of a 4-lb bob on an 18-in.-long string. The pendulum is released from rest when $\theta = \theta_0 = 60°$. Let the zero of potential energy be at $\theta = 90°$.

a. Calculate and plot the kinetic energy T, the potential energy V, the total energy $E = T + V$, and the tension P in the string, all as functions of the angle θ ($-45° < \theta < 60°$).

b. If the string breaks when the tension equals 7 lb, determine the angle θ_b when the string breaks. Calculate and plot the motion (y versus x) of the pendulum bob from the time when it is released until it strikes the floor.

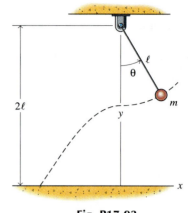

Fig. P17-93

C17-94 An ice cube is balanced on a beach ball when it is disturbed and starts to slide (Fig. P17-94). The mass of the ice cube is 20 g and the radius of the ball is 300 mm. Let the zero of potential energy be at the center of the ball $\theta = 90°$.

a. Calculate and plot the kinetic energy T, the potential energy V, the total energy $E = T + V$, and the normal force N between the ball and the ice cube, all as functions of the angle θ ($0° < \theta < \theta_{max}$). For what angle θ_{max} does the ice cube slip off of the ball?

b. Calculate and plot the motion (y versus x) of the ice cube from the time when it is released until it lands in the sand.

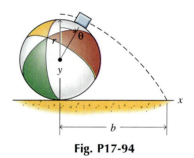

Fig. P17-94

C17-95 A 4-lb block A and an 8-lb block B are connected by a cord that passes over a small frictionless pulley as shown in Fig. P17-95. Block A is also attached to a spring which has a modulus of $k = 48$ lb/ft and an unstretched length of 12 in. When $x = 0$, block A is moving to the right with a speed of 10 ft/s. If the surface at A is smooth and $h = 18$ in.,

a. Determine the range of travel x_{min} and x_{max} of block A. Is $|x_{min}| = |x_{max}|$?
b. Calculate and plot the kinetic energy T, the potential energy V, and the total energy $E = T + V$ of the system, all as functions of x ($x_{min} < x < x_{max}$).

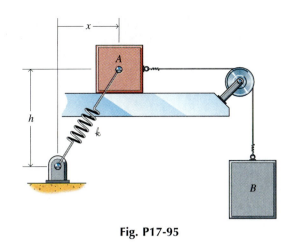

Fig. P17-95

SUMMARY

The work–energy method combines the principles of kinematics with Newton's second law to directly relate the position and speed of a body. For this method to be useful, the forces acting on a body must be a function of position only. For some types of these forces, however, the resulting integrals can be evaluated explicitly. The result is a simple algebraic equation that relates the speed of the body at two different positions of its motion.

The work done by a force on a particle is the product of the particle's displacement and the component of force in the direction of the displacement. If there is no displacement or no component of force in the direction of the displacement, then the force does no work on the particle.

For conservative forces, the amount of work done depends only on the position of the particle at the beginning and end of the motion. Examples of conservative forces are constant forces, gravitational forces, and linearly elastic spring forces.

Conservative forces can always be written as the gradient of a potential energy function. The work done by the force during some motion is the difference in the value of the potential energy function at the beginning and end of the motion.

The kinetic energy of a particle depends only on its speed ($T = \frac{1}{2}mv^2$). Since the mass m and the square of the speed v^2 are both positive quantities, the kinetic energy of a particle will always be positive.

The principle of work and energy

$$T_i + U_{i \to f} = T_f \qquad (17\text{-}6b)$$

states that the final kinetic energy of a particle is equal to the sum of its initial kinetic energy and the work done on the particle by external forces. If the work term is divided up into a part due to conservative forces $U_{i \to f}^{(c)} = V_i - V_f$ and a part due to all other forces $U_{i \to f}^{(o)}$, the work–energy principle can be written

$$T_i + V_i + U_{i \to f}^{(o)} = T_f + V_f \qquad (17\text{-}23)$$

The convenience of the method of work and energy is that it directly relates the speed of a particle at two different positions of its motion to the forces that do work on the particle during the motion. If Newton's second law were applied directly, the acceleration would have to be obtained for an arbitrary position of the particle. Then the acceleration would have to be integrated using the principle of kinematics. The work–energy method combines these two steps into one.

The limitations of the method of work and energy are that Eq. 17-6 is a scalar equation and can be solved for no more than one unknown, the acceleration cannot be calculated directly, and only forces that do work are involved. However, the normal component of acceleration is a function of velocity and the velocity is easily found using the work–energy method. Then the normal component of Newton's second law can be used to determine forces that act normal to the path of motion and those that do no work.

Since the work–energy method combines the principles of kinematics with Newton's second law, it is not a new or independent principle. There are no problems that can be solved with the work–energy method that cannot be solved using Newton's second law. However, when the work–energy method does apply, it is usually the easiest method of solving a problem.

REVIEW PROBLEMS

In the following problems, all ropes, cords, and cables are assumed to be flexible, inextensible, and of negligible mass. All pins and pulleys are small, have negligible mass, and are frictionless.

17-96* In a carnival game of skill, players slide nickels along a wooden playing surface as shown in Fig. P17-96. For a player to win a prize, her nickel must stop between lines C and D on the lower surface. The kinetic coefficient of friction between the 5-g nickels and the floor is $\mu_k = 0.2$, the corners are abrupt but smooth, and the "foul line" (the point at which the player must release the nickels) is 1 m from the corner B. Determine the range of initial velocities that will win a prize in this game.

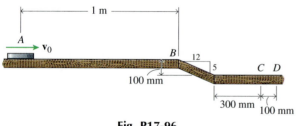

Fig. P17-96

17-97* A 4-lb pendulum bob has its motion interrupted by a small peg located directly under the support (Fig. P17-97). If the pendulum has an angular speed of 3 rad/s when $\theta = 75°$, determine the tension in the cord

a.   At position A.
b.   At position B.

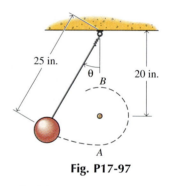

Fig. P17-97

17-98 The 2-kg block shown in Fig. P17-98 is lowered so that it just makes contact with the springs ($k = 100$ N/m each).

a.   Determine the static deflection of the springs (the distance that the block will drop if it is slowly lowered to its equilibrium position.)
b.   Find the maximum deflection of the springs (the distance that the block will drop if it is suddenly released).

Fig. P17-98

17-99 A 10-lb block is attached to a spring having $k = 48$ lb/ft and an unstretched length $\ell_0 = 18$ in., as shown in Fig. P17-99. The static and kinetic coefficients of friction between the block and the horizontal surface are $\mu_s = 0.5$ and $\mu_k = 0.4$, respectively. If $x = 0$ and the block has an initial speed of 7 ft/s when the spring is unstretched, determine the position x_s where the block will come to rest and the force F_s in the spring when the block stops for

a. Initial motion to the left.
b. Initial motion to the right.

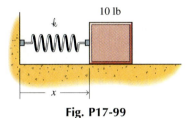

10 lb

Fig. P17-99

17-100 A 5-kg crate slides up a 25° inclined ramp with an initial speed of 10 m/s as shown in Fig. P17-100. If the kinetic coefficient of friction between the ramp and the crate is $\mu_k = 0.3$ and $\ell = 3$ m, determine

a. The speed of the crate when it reaches the top of the ramp.
b. The maximum height h attained by the crate.
c. The distance d at which the crate will strike the horizontal surface.

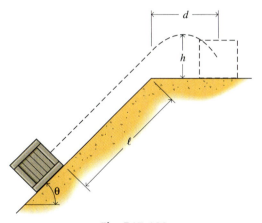

Fig. P17-100

17-101* The sum of all drag forces acting on a 3200-lb car moving with speed v is given by

$$F_D = a + bv^2$$

where a represents the rolling resistance of the tires and bv^2 represents air resistance (Fig. P17-101). If 8 hp must be delivered to the wheels of the car to maintain a constant speed of 30 mi/h and 14 hp must be delivered to the wheels to maintain a constant speed of 40 mi/h (both on a horizontal road), determine the horsepower required to travel

a. At 55 mi/h on a horizontal road.
b. At 40 mi/h up a 5° incline.

Fig. P17-101

17-102 The two frictionless sliders A and B shown in Fig. P17-102 each have a mass of 4 kg. The 600-mm rod connecting the sliders may be considered massless. The system is released from rest when $\theta = 70°$, and slider B strikes the spring when $x = 500$ mm. Determine the value of the spring constant if the maximum deflection of the spring is 50 mm.

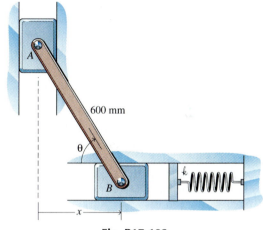

Fig. P17-102

17-103* A 0.5-lb weight slides on a frictionless rod in a vertical plane as shown in Fig. P17-103. The undeformed length of the spring is $\ell_0 = 6$ in., the spring modulus is $k = 5$ lb/ft, and the distance $d = 18$ in. If the weight is pulled down a distance b and released from rest, determine

a. The minimum distance b for which the weight will travel all of the way around the rod to C.
b. The speed of the weight when it reaches C.

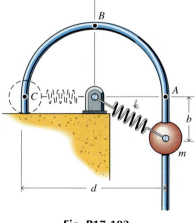

Fig. P17-103

17-104* A rope tow is used to pull skiers up a ski slope as shown in Fig. P17-104. If the kinetic coefficient of friction between the skis and the snow is $\mu_k = 0.1$ and the average weight of the skiers is 650 N, determine

a. The power required to operate the rope tow at 2 m/s if 50 skiers are holding onto the rope.
b. The speed of the rope tow if the power remains constant but an additional 25 skiers are holding onto the rope.

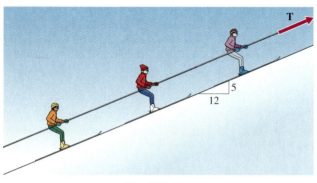

Fig. P17-104

17-105 The 4-lb block shown in Fig. P17-105 is released from rest when the spring is compressed 6 in. The spring and block are not connected. If the block moves a total of 5 ft before leaving the smooth surface, determine the distance a where the block hits the lower surface.

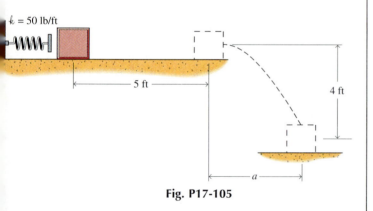

Fig. P17-105

17-106* A 1.5-kg package is dropped from a height of 600 mm onto the lightweight platform of a scale as shown in Fig. P17-106. The platform is supported by two identical springs ($k = 150$ N/m) that are compressed by 50 mm when no object is on the scale (the position shown). If no energy is lost when the package hits the scale, determine the maximum compression in the springs as a result of the package landing on the scale. Compare this value with the static compression of the scale (the package is slowly lowered onto the scale).

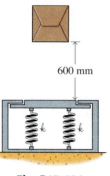

Fig. P17-106

17-107 The pendulum shown in Fig. P17-107 weighs 2 lb and has a length of 3 ft. The spring ($k = 20$ lb/ft) may be considered weightless. If the pendulum is released from rest when $\theta = 90°$, determine

a. The tension in the cord an instant before the pendulum hits the spring.
b. The maximum deflection of the spring.

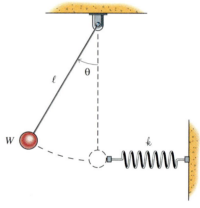

Fig. P17-107

17-108 The retarding force due to air resistance on a 70-kg cyclist is given by

$$F_D = 0.8v^2$$

where F_D is in newtons and v is in meters per second. If the cyclist can maintain a power output of 200 watts, determine the speed that the cyclist can maintain

a. On level ground.
b. Up a 5° incline.
c. Down a 5° incline.

17-109* The drag force on a vehicle moving with speed v is given by

$$F_D = 0.017W + 0.0012Cv^2$$

where F_D is the drag force in pounds, W is the weight of the vehicle in pounds, C is a constant which depends on the size and shape of the vehicle, and v is in feet per second. If a 3600-lb pickup carrying a 1500-lb load ($C = 18$) to market maintains a constant speed of 45 mi/h, determine

a. The power that must be delivered to the wheels.
b. The speed that the truck can maintain when returning from the market empty ($C = 14$) for the same amount of power.

17-110 A 57-kg woman bungee jumper jumps from a high bridge as shown in Fig. P17-110. The bungee cord has an elastic constant of 171 N/m and a free length of $L = 40$ m. Determine

a. The speed of the woman when the bungee cord becomes taut and starts to exert a force on her body.
b. The stretch d in the bungee cord when the woman stops falling.

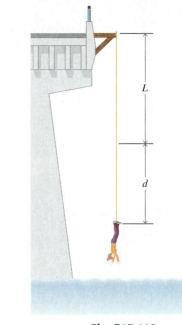

Fig. P17-110

A rotating wheel is the principle component of many exercise machines. Friction forces on the rim of the wheel remove energy from the wheel and tend to reduce its rotational speed. The person exercising must do work on the wheel to overcome the frictional work and keep the wheel spinning at a constant speed.

Chapter 18

KINETICS OF RIGID BODIES: WORK AND ENERGY METHODS

18-1   INTRODUCTION

The method of work and energy combines the principles of kinematics with Newton's second law to directly relate the position and speed of a body. Therefore, the method of work and energy is not a new principle but merely a special solution of the differential equations that arise when using Newton's second law. Nevertheless, the work–energy method greatly facilitates the solution of a certain class of problems.

In the work and energy approach, Newton's second law is integrated in a general sense with respect to position. For this method to be useful, the forces acting on a body must be a function of position only. For conservative forces, the resulting integrals can be evaluated explicitly in a general sense. The result is a simple algebraic equation that relates the speed of the body at two different positions of its motion.

For a general system of particles, the work–energy method was not found to be particularly useful. The principal reasons for this lack of usefulness were that (1) the motions of the particles are unrelated and must be specified independently; and (2) the work done by internal as well as external forces needed to be considered. However, it was shown in Section 17-6 that when two particles are connected by a rigid link, the works done by the internal forces cancel each other out. Therefore, the work–energy method is quite useful when the system of particles forms a rigid body as in this chapter.

The primary advantages of the work–energy method (the acceleration of a body need not be determined and integrated to determine the change in velocity of the body; forces that do no work have no effect on the work–energy equations and need not be included) are also the primary limitations of the work–energy method (the method of work and energy cannot determine accelerations or forces that do no work on a body).

Since Newton's second law is often used in conjunction with the principle of work and energy, a complete free-body diagram should be drawn. That is, all forces should be shown and not just those forces that do work during a particular motion of the body. Separate free-body diagrams showing the body in its initial and final positions may also be useful.

18-2   WORK OF FORCES AND COUPLES ACTING ON RIGID BODIES

Rigid bodies may be acted on by both forces and couples or pure moments. In addition, the body can rotate as well as translate. The work done by a force depends only on the motion of the point of application of the force. It does not depend on whether the motion is caused by the translation or the rotation of a rigid body. It will be shown, however, that a moment does no work due to a translation of the body on which it acts. Moments do work on a body only when the body rotates.

18-2.1   Work of Forces

The work done by a force **P** during a motion from point 1 to point 2 was defined in Chapter 17 as

$$U_{1 \to 2} = \int_1^2 \mathbf{P} \cdot d\mathbf{r} \qquad (18\text{-}1)$$

The calculation of work using Eq. 18-1 is independent of whether the force is applied to a particle, a translating rigid body, a rotating rigid body, or a translating and rotating rigid body. The work done by various types of forces was treated in detail in Section 17-4 and will not be repeated here.

For conservative forces, the potential energy V is also defined and determined in the same manner as for particles. The work done by conservative forces may be computed by direct integration using Eq. 18-1 or by using potential energy functions as discussed in Section 17-7.

18-2.2   Work of Internal Forces

Work done by internal forces in a rigid body does not have to be considered. Forces of interaction between two particles in a rigid body always occur in equal-magnitude but opposite-direction collinear pairs. Because the body is rigid, however, the two particles always undergo the same displacement in the direction of the forces. Therefore, the work done on one particle by one force always cancels the work done on the second particle by the second force, and the resultant work done by

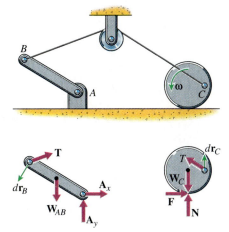

Figure 18-1 When two or more bodies are connected by frictionless pins or inextensible cables, the bodies can be treated as a system and the work done on the system by the internal pin or cable forces can be neglected because the work they do on one part of the system will cancel the work they do on some other part of the system.

the pair of internal forces is zero.[1] That is, the work done on a *rigid* body by a system of external forces is the algebraic sum of works done by the individual forces.

Work done on a pair of rigid bodies by smooth connecting pins or by flexible, inextensible cables also need not be considered. Again, the forces occur in equal-magnitude but opposite-direction pairs, and the points to which the forces are applied undergo equal displacements in the direction of the forces. Therefore, the resultant work done on the bodies by the connecting members of the system is zero.

For example, if the mass of the cable shown in Fig. 18-1 is negligible, then the tensions at the two ends of the cable will be the same. Since the cable is inextensible, however, the displacement in the direction of the cable at B and the displacement in the direction of the cable at C must also be equal. One of the forces will be in the direction of the displacement and will do positive work; the other force will be opposite the direction of the displacement and will do negative work. Therefore, the resultant work done on the pair of bodies by the cable must be zero.

18-2.3 Work of Couples and Moments

The work done by a couple is obtained by calculating the work done by each force of the couple separately and adding their works together. For example, consider a couple C acting on a rigid body as shown in Fig. 18-2*a* on page 388. During some small time dt the body translates and rotates. If the displacement of point A is $d\mathbf{r} = ds_t\,\mathbf{e}_t$, choose a second point B such that the line AB is perpendicular to $d\mathbf{r}$. Then the mo-

---

[1]If the body were not rigid, the internal forces would still occur in equal-magnitude and opposite-direction collinear pairs. However, the components of the displacements in the direction of the forces would, in general, not be the same. Therefore, the work done by the internal forces would not cancel and there would be a net work done by the internal forces.

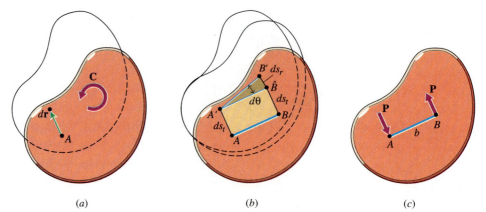

(a)   (b)   (c)

Figure 18-2   A couple **C** acting on a body does work on the body only when the body rotates. When a body that is acted on by a pure couple undergoes a translation, the works done by the two forces that make up the couple cancel and the couple does no net work on the body. When a body which is acted on by a pure couple rotates, however, the force at the pivot point does no work while the work done by the other force does work equal to $dU = P(b\,d\theta) = C\,d\theta$.

tion that takes point A to A' will take point B to B'. This motion may be considered in two parts: first a translation that takes the line AB to $A'\hat{B}$, followed by a rotation $d\theta$ about A' that takes $\hat{B}$ to B' (Fig. 18-2b).

Now represent the couple by the pair of forces of magnitude $P = C/b$ in the direction perpendicular to the line AB (Fig. 18-2c). During the translational part of the motion, one force will do positive work $P\,ds_t$ and the other will do negative work $-P\,ds_t$; therefore, the sum of the work done on the body by the pair of forces during the translational part of the motion is zero.

During the rotational part of the motion, A' is a fixed point and the force applied at A' does no work. The work done by the force at B is $dU = P\,dsr \cong Pb\,d\theta$, where $d\theta$ is in radians and $C = Pb$ is the magnitude of the moment of the applied couple. The work is positive if the couple is in the same sense as $d\theta$ and negative if the couple is in the opposite sense. Therefore, the total work done by the couple during the differential rotational motion is

$$dU = C\,d\theta = \mathbf{C}\cdot d\boldsymbol{\theta} \tag{18-2}$$

where $\mathbf{C} = C\,\mathbf{k}$ and $d\boldsymbol{\theta} = d\theta\,\mathbf{k}$.

The work done on the body by the couple as the body rotates through an angle $\Delta\theta = \theta_2 - \theta_1$ is obtained by integrating Eq. 18-2:

$$U_{1\to2} = \int_1^2 dU = \int_{\theta_1}^{\theta_2} C\,d\theta \tag{18-3}$$

If the couple is constant, then C can be taken outside the integral and Eq. 18-3 becomes

$$U_{1\to2} = C\int_1^2 d\theta = C(\theta_2 - \theta_1) = C\,\Delta\theta \tag{18-4}$$

18-2.4   Forces That Do No Work

One of the principal advantages of the method of work and energy is that forces that do no work do not enter the equation. Some of the more obvious forces that do no work are

1. Forces applied to points that do not move. For example, when a wheel rotates on a fixed, frictionless axle, the forces exerted on the wheel by the axle do no work.
2. Forces that act perpendicular to the motion. For example, the normal force acting on a body that slides or rolls along a surface does no work.

Not so obvious is the fact that the frictional force acting on a body that rolls without slipping does no work. The reason that the friction force does no work is because the contact point is an instantaneous center of zero velocity; it is instantaneously at rest. Since the point of application of the force is not moving (at the instant considered)

$$dU = \mathbf{F} \cdot d\mathbf{r}_{IC} = \mathbf{F} \cdot (\mathbf{v}_{IC}\, dt) = 0 \qquad (a)$$

and therefore the friction force does no work.

18-3 KINETIC ENERGY OF RIGID BODIES IN PLANE MOTION

The kinetic energy of a body is obtained by adding together the kinetic energies of the particles that make up the body. For a general body that is not rigid, there is no simple equation relating the motion of the various particles: There is no general expression for the kinetic energy of the body. When the body is rigid, however, the velocities of the various particles are related by the relative velocity equation. This relationship allows for a particularly simple formula expressing the kinetic energy of a rigid body in plane motion.

For example, consider the body shown in Fig. 18-3. Point A is any point in the body, and $\mathbf{r} = \mathbf{r}_{P/A} = x\mathbf{i} + y\mathbf{j} + z\mathbf{k}$ is the position vector from A to an arbitrary particle P of mass dm in the body. The velocity of dm is related to the velocity of point A by the relative velocity equation

$$\mathbf{v} = \mathbf{v}_A + \mathbf{v}_{P/A} = \mathbf{v}_A + \boldsymbol{\omega} \times \mathbf{r} \qquad (b)$$

and the kinetic energy of the particle can be written

$$
\begin{aligned}
dT &= \frac{1}{2}\, dm\, v^2 = \frac{1}{2}\, dm\, \mathbf{v} \cdot \mathbf{v} \\
&= \frac{1}{2}\, dm\, (\mathbf{v}_A + \boldsymbol{\omega} \times \mathbf{r}) \cdot (\mathbf{v}_A + \boldsymbol{\omega} \times \mathbf{r}) \qquad (c) \\
&= \frac{1}{2}\, dm\, v_A^2 + dm\, \mathbf{v}_A \cdot (\boldsymbol{\omega} \times \mathbf{r}) + \frac{1}{2}\, dm\, (\boldsymbol{\omega} \times \mathbf{r}) \cdot (\boldsymbol{\omega} \times \mathbf{r})
\end{aligned}
$$

Since the body has plane motion, the angular velocity of the body has only a z-component, $\boldsymbol{\omega} = \omega \mathbf{k}$, so

$$\boldsymbol{\omega} \times \mathbf{r} = \omega(x\mathbf{j} - y\mathbf{i}) \qquad (d)$$

and

$$(\boldsymbol{\omega} \times \mathbf{r}) \cdot (\boldsymbol{\omega} \times \mathbf{r}) = \omega^2(x^2 + y^2) \qquad (e)$$

Therefore, the kinetic energy of the piece of mass dm is

$$dT = \frac{1}{2}\, dm\, v_A^2 + dm\, \mathbf{v}_A \cdot (\boldsymbol{\omega} \times \mathbf{r}) + \frac{1}{2}\, dm\, \omega^2(x^2 + y^2) \qquad (f)$$

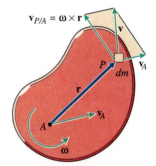

Figure 18-3 The velocity of every element of mass dm can be related to the motion of point A in the body using the relative velocity equation. When the kinetic energies of all of the elements of mass are added together, the result will be expressed in terms of the kinetic energy of point A and the energy of rotation about point A.

Integrating Eq. f over all the mass in the body and recognizing that $\mathbf{v}_A$ and $\boldsymbol{\omega}$ are independent of dm gives

$$T = \frac{1}{2} v_A^2 \int dm + \mathbf{v}_A \cdot \left(\boldsymbol{\omega} \times \int \mathbf{r}\, dm \right) + \frac{1}{2}\, \omega^2 \int (x^2 + y^2)\, dm \qquad (18\text{-}5)$$

But the integral in the first term of Eq. 18-5 is just the total mass of the body ($\int dm = m$), the integral in the second term is the location of the mass center measured from point A ($\int \mathbf{r}\, dm = m\mathbf{r}_{G/A}$), and the integral in the last term is the mass moment of inertia about an axis through point A parallel to the z-axis ($\int (x^2 + y^2)\, dm = I_{Az}$), so[2]

$$T = \frac{1}{2}\, mv_A^2 + m\mathbf{v}_A \cdot (\boldsymbol{\omega} \times \mathbf{r}_{G/A}) + \frac{1}{2}\, I_{Az}\omega^2 \qquad (18\text{-}6)$$

Although Eq. 18-6 is the general expression for the kinetic energy of a rigid body in plane motion, it is unnecessarily complex and is seldom used. Equation 18-6 simplifies significantly for the common types of plane motion to be discussed below. Even for general plane motion, Eq. 18-6 can be simplified considerably by an appropriate choice of the arbitrary point A.

18-3.1   Translation of a Rigid Body

When a rigid body moves without rotating, its angular velocity is zero ($\boldsymbol{\omega} = \mathbf{0}$) and the velocity of every point in the body is the same. In this case, Eq. 18-6 reduces to

$$T = \frac{1}{2}\, mv^2 \qquad (18\text{-}7)$$

where v is the speed of any point in the body. Clearly, the idealization of particle motion is just the pure translation of a rigid body.

18-3.2   Rotation of a Rigid Body About a Fixed Axis

If point A is a point on the axis of rotation for a rigid body that is rotating about a fixed axis, then $\mathbf{v}_A = \mathbf{0}$ and Eq. 18-6 simplifies to

$$T = \frac{1}{2} I_{Az}\omega^2 \qquad (18\text{-}8)$$

Note, however, that Eq. 18-8 is also true if point A is an instantaneous center of zero velocity. The reduction of Eq. 18-6 to 18-8 requires merely that point A have zero velocity at the instant of the calculation.

18-3.3   General Plane Motion of a Rigid Body

The general motion of a rigid body consists of a combination of translation and rotation. Even for the case of general plane motion of a rigid body, however, Eq. 18-6 can be simplified substantially by an appro-

---

[2]Although it is not obvious from the derivation at this point, the results obtained are not limited to the motion of plane slabs or to the motion of bodies that are symmetrical with respect to the reference plane. Equation 18-6 and the other equations derived from it (Eqs. 18-7, 18-8, and 18-9) may be applied to the study of the plane motion of any rigid body, regardless of its shape. (See Section 18-6.)

priate choice of point A. For example, when point A is chosen to be the mass center of the body G, then the position vector $\mathbf{r}_{G/A} = \mathbf{0}$, and

$$T = \frac{1}{2} mv_G^2 + \frac{1}{2} I_{Gz}\omega^2 \tag{18-9}$$

where v_G is the speed of the mass center of the body and I_{Gz} is the moment of inertia about an axis through the mass center G parallel to the z-axis (perpendicular to the plane of motion). The first term of Eq. 18-9 is just the kinetic energy associated with the translation of the body's mass center, and the second term is the kinetic energy associated with the rotation of the body about an axis through its center of mass.

The special case of rotation about a fixed axis through an arbitrary point A, Eq. 18-8, is also contained in the general expression for kinetic energy, Eq. 18-9. When the body rotates about a fixed axis through point A, the velocity of the mass center is given by

$$\mathbf{v}_G = \mathbf{v}_A + \mathbf{v}_{G/A} = \mathbf{0} + \boldsymbol{\omega} \times \mathbf{r}_{G/A} \tag{g}$$

so that

$$\begin{aligned}
v_G^2 = \mathbf{v}_G \cdot \mathbf{v}_G &= (\omega\mathbf{k} \times \mathbf{r}_{G/A}) \cdot (\omega\mathbf{k} \times \mathbf{r}_{G/A}) \\
&= (\omega x_G\mathbf{j} - \omega y_G\mathbf{i}) \cdot (\omega x_G\mathbf{j} - \omega y_G\mathbf{i}) \\
&= \omega^2(x_G^2 + y_G^2) = \omega^2 d^2
\end{aligned} \tag{h}$$

where $d^2 = x_G^2 + y_G^2$ is the square of the distance between the axis of rotation and the mass center G. Then the kinetic energy is

$$\begin{aligned}
T &= \frac{1}{2} mv_G^2 + \frac{1}{2}I_{Gz}\omega^2 = \frac{1}{2}(md^2 + I_{Gz})\omega^2 \\
&= \frac{1}{2} I_{Az}\omega^2
\end{aligned} \tag{18-8}$$

where $I_{Az} = I_{Gz} + md^2$ by the parallel-axis theorem for mass moments of inertia.

18-4 PRINCIPLE OF WORK AND ENERGY FOR THE PLANE MOTION OF RIGID BODIES

The principle of work and energy for a rigid body is obtained by adding together the equations of work and energy (Eq. 17-6) for each of the particles that make up the rigid body. This gives

$$T_1 + U_{1 \to 2} = T_2 \tag{18-10}$$

in which T_1 and T_2 are the total kinetic energies of all the particles that make up the body (given by Eq. 18-9), $U_{1 \to 2}$ is the total work done by all external forces and couples acting on all the particles, and the work done by the internal forces need not be considered. Equation 18-10, which gives the principle of work and energy for a rigid body, looks exactly the same as Eq. 17-6 for the work and energy of a particle. The difference between these equations is that the kinetic energy terms in Eq. 18-10 include the rotational kinetic energy of the rigid body as well as the translational kinetic energy and that the work term includes the work done by all external moments as well as the work done by all external forces that act on the rigid body.

Just as with a particle, the work term can be divided up into a part done by conservative forces (forces whose potential is known) $U_{1\to2}^{(c)}$ and a part done by all other forces (either nonconservative forces that have no potential or conservative forces whose potential is unknown) $U_{1\to2}^{(o)}$. The work done by the conservative forces can be expressed in terms of potential functions, so Eq. 18-10 can be written

$$T_1 + V_1 + U_{1\to2}^{(o)} = T_2 + V_2 \qquad \text{(18-11)}$$

When two or more rigid bodies are connected by inextensible cords or cables or by frictionless pins, Eq. 18-10 (or Eq. 18-11) can be written for each of the bodies. When the resulting equations are added together, the work done by the connection forces will cancel in pairs. Therefore, Eqs. 18-10 and 18-11 also express the work–energy principle for a system of connected rigid bodies. For such a system of rigid bodies, T is the kinetic energy of the entire system and $U_{1\to2}$ $(= V_1 + U_{1\to2}^{(o)} - V_2)$ includes the work done on the entire system by all external forces and moments.

For either a single rigid body or an interconnected system of rigid bodies, a complete free-body diagram must be drawn to ensure that all forces and moments are identified and considered. While including forces and moments that do no work on the free-body diagram may seem unnecessary, the work–energy principle is often used in conjunction with Newton's second law. Therefore, *all external forces and moments must be shown* on the free-body diagram and not just those forces and moments that do work on the body or bodies.

In addition to a complete free-body diagram of the body or bodies, it may also be helpful to draw diagrams that show the initial and final positions of the system for the given interval of motion.

18-5   POWER

Power, which is the time rate of doing work, was defined and discussed with respect to particle motion in Section 17-10. For a rigid body in plane motion, the work done must include the work done by couples as well as the work done by forces. If a rigid body is acted on simultaneously by a force $\mathbf{P}$ and a couple $\mathbf{C} = C\mathbf{k}$, the work done on the body is

$$dU = \mathbf{P} \cdot d\mathbf{r} + \mathbf{C} \cdot d\boldsymbol{\theta} \qquad (i)$$

in which $d\mathbf{r}$ is the displacement of the point of application of the force $\mathbf{P}$ and $d\boldsymbol{\theta} = d\theta\,\mathbf{k}$ is the rotation of the body. Dividing through by dt gives the total power supplied to the rigid body at some instant

$$\text{power} = \mathbf{P} \cdot \mathbf{v} + \mathbf{C} \cdot \boldsymbol{\omega} \qquad \text{(18-12)}$$

where $\mathbf{v} = d\mathbf{r}/dt$ is the velocity of the point of application of the force $\mathbf{P}$ and $\boldsymbol{\omega} = \omega\mathbf{k} = (d\theta/dt)\mathbf{k}$ is the angular velocity of the body.

Explain why the 16-lb block shown hits the floor with a higher speed if the 36-in.-diameter uniform drum is relatively light than it does if the drum has a large mass.

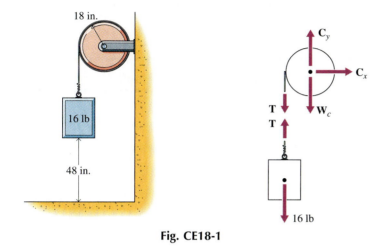

Fig. CE18-1

SOLUTION

Separate work–energy equations can be written for the 36-in. uniform drum and for the 16-lb block. For the block, both the cord tension T and the weight $W = mg$ do work on the block as it falls $d = 48$ in. $= 4$ ft. For the drum, the only force that does work is the cord tension T, and the work done on the drum by the cord tension ($U = Fd = 4T$) is equal in magnitude and opposite in sign to the work done on the block by the tension ($U = Fd = 24T$). Therefore, when the work–energy equations for the block and drum are added together, the work of the cord tension will drop out and the only work done on the system will be that of the weight force ($U = Fd = 16(4) = 64$ ft · lb).

Assuming that the system starts from rest, the work–energy equation $T_i + U_{i \to f} = T_f$ can be written

$$0 + 64 = \frac{1}{2} m_b v_b^2 + \frac{1}{2} \left(\frac{1}{2} m_d r^2 \right) \omega_d^2 \qquad (a)$$

or, since $v_b = r\omega_d$,

$$64 = \frac{1}{2} m_b v_b^2 + \frac{1}{4} m_d v_b^2 \qquad (b)$$

The first term on the right-hand side of Eqs. *a* and *b* is the kinetic energy of translation of the block. The second term on the right-hand side of Eqs. *a* and *b* is the kinetic energy of rotation of the drum. If the mass of the drum is small ($m_d \to 0$), then all of the work done by gravity will go into the translational energy of the block and the speed of the block when it hits the floor will be

$$v_b = \sqrt{\frac{128}{m_b}} \qquad (c)$$

If the mass of the drum is large, however, some of the work done by gravity will have to be used to make the drum rotate. Then the speed of the 16-lb block when it hits the floor is given by

$$v_b = \sqrt{\frac{256}{2m_b + m_d}} \qquad (d)$$

Equation d clearly shows that the more massive the drum, the slower the velocity of the block when it hits the floor.

CONCEPTUAL EXAMPLE 18-2: ROTATIONAL WORK-ENERGY-2

Suppose a 20-kg cart and a 20-kg disk are both released from rest at the top of a 10-m-long hill. That is, both the cart and the disk start with zero kinetic energy. If the cart rolls on small wheels and the disk rolls without slipping, the only force that does any work on either body is gravity and the work done on the two objects as they move down the hill is the same. So why is the 20-kg cart traveling faster than the 20-kg disk when they reach the bottom of the hill?

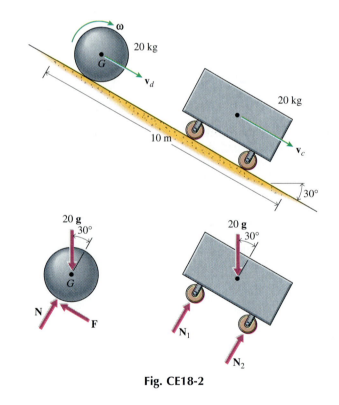

Fig. CE18-2

SOLUTION

The only forces acting on the cart are gravity ($W = mg$) and the two normal forces N_1 and N_2. The normal forces, however, always act perpendicular to the motion and do no work on the cart. Therefore, as the 20-kg cart rolls to the bottom of the 10-m-long hill, the work done on the cart is $U = Fd = 20(9.81)(10$

sin 30°) = 981 N · m. The speed of the cart v_c at the bottom of the hill is obtained from the work–energy equation, $T_i + U_{i \rightarrow f} = Tf$

$$0 + 981 = \frac{1}{2}(20)v_c^2 \qquad (a)$$

and the speed of the cart when it reaches the bottom of the hill is $v_c =$ 9.90 m/s.

When a body rolls without slipping on a rough surface, the contact point is an instantaneous center of zero velocity. Since the normal force and the friction force both act at the contact point, which is (instantaneously) at rest, neither force does any work on a body that is rolling without slipping. Therefore, the only force that does work on the disk is again the weight force $W = mg$. As the 20-kg disk rolls to the bottom of the 10-m-long hill, the work done on the disk is $U = Fd = 20(9.81)(10 \sin 30°) = 981$ N · m (the same as for the 20-kg cart). The speed of the disk $v_d = r\omega$ at the bottom of the hill is also obtained from the work–energy equation, $T_i + U_{i \rightarrow f} = T_f$,

$$0 + 981 = \frac{1}{2}(20)v_d^2 + \frac{1}{2}\left(\frac{1}{2}(20)r^2\right)\omega^2 \qquad (b)$$

and the speed of the 20-kg disk when it reaches the bottom of the hill is $v_d =$ 8.09 m/s, which is about 20 percent slower than the 20-kg cart rolling down the same 10-m-long hill.

The difference of course is that in the case of the cart, all of the work done on the cart was used to increase the kinetic energy of translation, $T = T_v = \frac{1}{2}mv^2$. In the case of the disk, only part of the work done on the disk was used to increase the kinetic energy of translation; the rest of the work done on the disk was needed to increase the kinetic energy of rotation, $T = T_v + T_\omega = \frac{1}{2}mv^2 + \frac{1}{2}I\omega^2$. Since less energy went into the $\frac{1}{2} mv^2$ term for the disk, its speed will be less at the bottom of the hill than that of the cart.

EXAMPLE PROBLEM 18-1

The turntable of a record player consists of a solid disk 12 in. in diameter and weighing 5 lb (Fig. 18-4). If a constant torque motor accelerates the turntable from rest to its operating speed of $33\frac{1}{3}$ rev/min in just one revolution, determine the torque **C** and the maximum power expended by the motor.

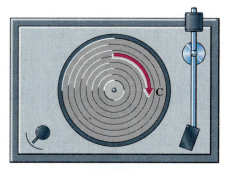

Fig. 18-4

SOLUTION

The free-body diagram of the turntable is shown in Fig. 18-5. Only the torque **C** does work $U_{1\rightarrow2} = C\,\Delta\theta$, where $\Delta\theta = 1$ rev $= 2\,\pi$ radians. Since the motion is fixed-axis rotation about an axis through the mass center of the turntable, the kinetic energy is given by $T = \dfrac{1}{2}I_G\omega^2$, where the moment of inertia is

$$I_G = \frac{1}{2}mr^2 = \frac{1}{2}\left(\frac{5}{32.2}\right)\left(\frac{6}{12}\right)^2 = 0.01941 \text{ slug} \cdot \text{ft}^2$$

the final angular velocity is $\omega_f = \dfrac{33\frac{1}{3}\text{ rev/min}}{60\text{ s/min}}(2\,\pi\,\text{rad/rev}) = 3.491$ rad/s, and the initial angular velocity is zero. Therefore, the work–energy equation (Eq. 18-10)

$$T_1 + U_{1\rightarrow2} = T_2$$

gives

$$0 + C(2\pi) = \frac{1}{2}(0.01941)(3.491)^2$$

or

$$\mathbf{C} = 0.01882 \text{ lb} \cdot \text{ft} \downarrow \qquad\qquad \text{Ans.}$$

The power of a couple is given by power $= C\omega$. Since the torque **C** is constant, the maximum power occurs where the angular velocity is maximum; that is

$$\text{power}_{max} = (0.01882)(3.491)$$
$$= 0.0657 \text{ lb} \cdot \text{ft/s} \qquad\qquad \text{Ans.}$$

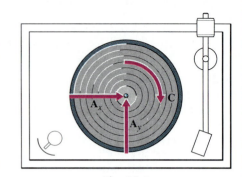

Fig. 18-5

EXAMPLE PROBLEM 18-2

A 4-kg concentrated mass is attached to the end of a 9-kg uniform slender rod that is rotating in a vertical plane as shown in Fig. 18-6. The rod AB is 2 m long and has an angular velocity of 3 rad/s clockwise when it is vertical. If the undeformed length of the spring is $\ell_0 = 0.25$ m, determine the spring modulus k such that the angular velocity of AB will be zero when the rod is horizontal.

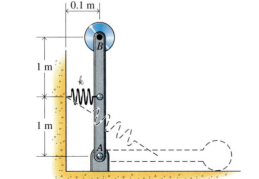

Fig. 18-6

SOLUTION

The free-body diagram of the system is shown in Fig. 18-7. The system is rotating about a fixed axis, so the kinetic energy is given by (Eq. 18-8)

$$T = \frac{1}{2}I_A\omega^2$$

where the moment of inertia is

$$I_A = (I_A)_{\text{rod}} + (I_A)_{\text{disk}} = \frac{1}{3}mL^2 + mL^2$$

$$= \frac{1}{3}9(2)^2 + 4(2)^2 = 28.00 \text{ kg} \cdot \text{m}^2$$

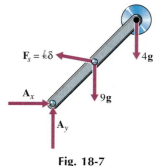

Fig. 18-7

Then the initial kinetic energy is $T_i = \frac{1}{2}(28.00)(3)^2 = 126.00$ J and the final kinetic energy is $T_f = 0$.

The two gravitational forces and the spring force each have a potential. Using the level of point A as the zero of gravitational potential energy gives

$$V_i = (V_i)_{4g} + (V_i)_{9g} + (V_i)_{k\delta} = mgh + mgh + \frac{1}{2}k\delta^2$$

$$= (9)(9.81)(1) + (4)(9.81)(2) + \frac{1}{2}k(0.1 - 0.25)^2$$

$$= 166.77 + 0.01125k \text{ J}$$

and

$$V_f = 0 + 0 + \frac{1}{2}k(1.4866 - 0.25)^2 = 0.7646k \text{ J}$$

The pin forces at A do no work, and there are no nonconservative forces acting on the system $U^{(o)}_{1\to2} = 0$. Substituting all these values in the work-energy equation (Eq. 18-11)

$$T_1 + V_1 + U^{(o)}_{1\to2} = T_2 + V_2$$

gives

$$126.00 + (166.77 + 0.01125k) + 0 = 0 + 0.7646k$$

or

$$k = 389 \text{ N/m} \qquad\qquad \text{Ans.}$$

The datum level, the level of zero potential energy, for gravitational potential energy can be chosen arbitrarily since only the difference between the potential energies is important $(U_{1\to2})_g = (V_g)_1 - (V_g)_2 = (mgh)_1 - (mgh)_2 = mg(h_1 - h_2)$. However, both heights must be measured from the same datum level and the location of the datum level should be clearly stated. The potential energy of spring forces $V_s = \frac{1}{2}k\delta^2$ is zero when the spring is unstretched ($\delta = 0$).

The wheel shown in Fig. 18-8 consists of a uniform half circle of wood weighing 20 lb encircled by an 18-in.-diameter circular steel band of negligible thickness and weight. If the wheel rolls without slipping on a horizontal floor and has an angular velocity of 15 rad/s clockwise when the mass center G is directly below the center of the wheel C, determine

a. The angular velocity of the wheel when G is directly to the left of C.
b. The normal and frictional components of the force exerted on the wheel by the floor when G is directly to the left of C.

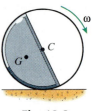

Fig. 18-8

SOLUTION

a. The free-body diagram of the wheel is shown in Fig. 18-9. The only force that does work on the wheel is the gravitational force. Using the level of the center of the wheel as the zero of potential energy, the initial and final gravitational potentials are

$$V_{gi} = mgh = (20)(-0.3183) = -6.366 \text{ lb} \cdot \text{ft} \qquad V_{gf} = 0$$

where $e = 4r/3\pi = 0.3183$ ft is the distance between the center of the wheel C and the center of gravity G.

Since the wheel is moving in general plane motion, the kinetic energy of the wheel is given by $T = \frac{1}{2}mv_G^2 + \frac{1}{2}I_G\omega^2$, where the velocity of the mass center G is $\mathbf{v}_G = \mathbf{v}_C + \mathbf{v}_{G/C} = r\omega\mathbf{i} + \mathbf{v}_{G/C}$, ω is the clockwise angular velocity, and the moment of inertia is

$$I_G = \left(\frac{1}{2} - \frac{16}{9\pi^2}\right)mr^2 = \left(\frac{1}{2} - \frac{16}{9\pi^2}\right)\left(\frac{20}{32.2}\right)\left(\frac{9}{12}\right)^2 = 0.11176 \text{ slug} \cdot \text{ft}^2$$

When G is directly below C, $\mathbf{v}_{G/C} = -e\omega_i\mathbf{i}$ and the initial kinetic energy of the wheel is

$$T_i = \frac{1}{2}\left(\frac{20}{32.2}\right)\left[\frac{9}{12}(15) - (0.3183)(15)\right]^2 + \frac{1}{2}(0.11176)(15)^2$$
$$= 25.60 \text{ lb} \cdot \text{ft}$$

When G is directly left of C, $\mathbf{v}_{G/C} = e\omega_f\mathbf{j}$ and the final kinetic energy of the wheel is

$$T_f = \frac{1}{2}\left(\frac{20}{32.2}\right)\left[\left(\frac{9}{12}\omega_f\right)^2 + \left(0.3183\ \omega_f\right)^2\right] + \frac{1}{2}0.1176\ \omega_f^2$$
$$= 0.2620\ \omega_f^2 \text{ lb} \cdot \text{ft}$$

Substituting all these quantities in the work–energy equation then gives

$$25.60 - 6.366 + 0 = 0.2620\ \omega_f^2 + 0 \qquad \omega_f = 8.57 \text{ rad/s} \qquad \text{Ans.}$$

b. When G is directly to the left of C, the acceleration of the mass center is $\mathbf{a}_G = \mathbf{a}_C + \mathbf{a}_{G/C} = (r\alpha\mathbf{i}) + (e\alpha\mathbf{j} + e\omega_f^2\mathbf{i})$. Then the equations of motion

$$\rightarrow \Sigma F_x = ma_x: \qquad -F = \frac{20}{32.2}\left(\frac{9}{12}\alpha + 23.37\right)$$

$$\uparrow \Sigma F_y = ma_y: \qquad N - 20 = \frac{20}{32.2}(0.3183\alpha)$$

$$\curvearrowright \Sigma M_G = I_G\alpha: \qquad \frac{9}{12}F - 0.3183N = 0.11176\alpha$$

give

$$N = 13.49 \text{ lb}\uparrow \qquad F = 0.820 \text{ lb}\leftarrow \qquad \text{Ans.}$$

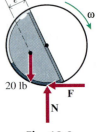

Fig. 18-9

Neither the normal force N nor the frictional force F do work since they both act at a point which is instantaneously at rest.

EXAMPLE PROBLEM 18-4

A 15-kg crate is attached to the end of an inextensible cord wrapped around a 40-kg, 600-mm-diameter, uniform drum as shown in Fig. 18-10. At the instant shown, the crate is falling at a rate of 9 m/s. Determine the constant braking couple **C** that must be applied to the drum to bring the crate to rest after it descends 3 m.

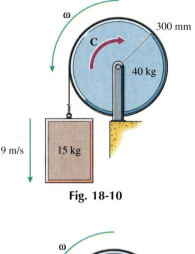

300 mm

40 kg

9 m/s   15 kg

Fig. 18-10

SOLUTION

Kinematics

Since the cord does not slip on the drum, when the crate drops $\Delta y = 3$ m the drum must rotate $\Delta\theta = \Delta y/r = 3/0.3 = 10$ rad. Also, if the crate is descending at a rate of $v = 9$ m/s, then the drum must be rotating at a rate of $\omega = v/r = 9/0.3 = 30$ rad/s.

Work–Energy

Since the crate is attached to the drum by an inextensible cord, the work done by the cord on the drum and on the crate will cancel and need not be considered. Therefore, the crate and the drum will be treated as a single system; the free-body diagram of the entire system is shown in Fig. 18-11. Neither the weight of the drum nor the forces exerted on the drum by the axle do work. The only nonconservative force acting on the system is the braking moment

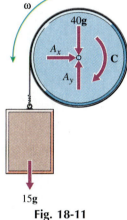

40g

A_x

C

A_y

15g

Fig. 18-11

$$U^{(o)}_{1\rightarrow2} = -C\,\Delta\theta = -10C$$

Using the initial height of the crate as the zero level of gravitational potential energy, the initial potential energy is zero ($V_i = 0$) and the final potential energy is

$$V_f = mgh = (15)(9.81)(-3) = -441.5 \text{ J}$$

The drum is moving in fixed-axis rotation about an axis through its mass center, so its initial kinetic energy is $T_d = \frac{1}{2}I_G\omega^2$, where $I_G = \frac{1}{2}(40)(0.3)^2 = 1.800$ kg · m². The crate is simply translating, so its initial kinetic energy is $T_C = \frac{1}{2}mv^2$. Therefore, the initial kinetic energy of the system is

$$T_i = \frac{1}{2}(15)(9)^2 + \frac{1}{2}(1.800)(30)^2 = 1417.5 \text{ J}$$

and the final kinetic energy is zero $T_f = 0$. Substituting all these values in the work–energy equation (Eq. 18-11)

$$T_1 + V_1 + U^{(o)}_{1\rightarrow2} = T_2 + V_2$$

gives

$$1417.5 + 0 - 10C = 0 - 441.5$$

or

$$\mathbf{C} = 185.9 \text{ N} \cdot \text{m}\downarrow \qquad\qquad \text{Ans.}$$

PROBLEMS

In the following problems, all ropes, cords, and cables are assumed to be flexible, inextensible, and of negligible mass. Unless stated otherwise, all pins and pulleys are small, have negligible mass, and are frictionless.

Introductory Problems

18-1* A 10-lb uniform wheel 16 in. in diameter is at rest when it is placed in contact with a moving belt as shown in Fig. P18-1. The kinetic coefficient of friction between the belt and the wheel is $\mu_k = 0.1$, and the belt moves with a constant speed of 30 ft/s. Determine the number of revolutions that the wheel turns before it begins to roll without slipping on the moving belt.

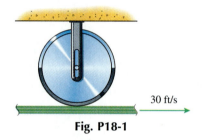

Fig. P18-1

18-2* A 10-kg uniform flywheel 400 mm in diameter is connected to a constant torque motor with a flexible belt as shown in Fig. P18-2. If the flywheel starts from rest, determine the torque necessary to rotate the flywheel at 4200 rev/min after 5 revolutions.

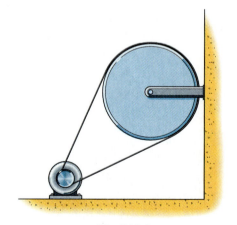

Fig. P18-2

18-3* A 16-lb crate hangs from the end of a cord wrapped around a 36-in.-diameter uniform drum as shown in Fig. P18-3. The system starts from rest when the crate is 48 in. above the floor. Determine the weight of the drum that will cause the crate to hit the floor at half the speed it would have if it were simply dropped from the same height.

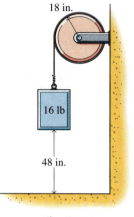

Fig. P18-3

18-4 A force of $P = 50$ N is applied to the end of a rope wrapped around the outside of a hollow drum as shown in Fig. P18-4a. The radius of gyration of the 20-kg drum is 175 mm, and axle friction may be neglected.

a. If the drum is released from rest, determine the downward velocity of point A on the rope after it has moved downward 3 m.
b. Repeat part a if the 50-N force is replaced with a 50-N weight as shown in Fig. P18-4b.

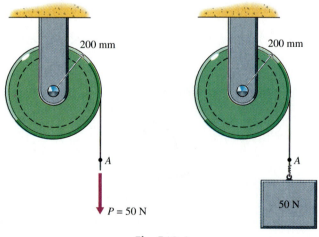

Fig. P18-4

18-5 A 100-lb rectangular plate of uniform thickness is supported by a pin and a cable as shown in Fig. P18-5. If the cable at B breaks, determine for the ensuing motion

a. The angular speed of the plate and the magnitude of the pin reaction at A when the mass center of the plate is directly below pin A.
b. The angular speed of the plate and the magnitude of the pin reaction at A when the plate has rotated 90° (when pin B is directly below pin A).
c. The maximum angle θ_{max} through which the plate will rotate.

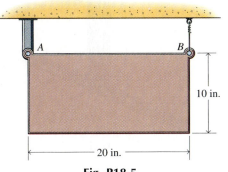

Fig. P18-5

18-6 An 80-kg semicircular plate of uniform thickness is supported by a pin and a cable as shown in Fig. P18-6. If the cable at B breaks, determine for the ensuing motion

a. The angular speed of the plate and the magnitude of the pin reaction at A when the mass center of the plate is directly below pin A.
b. The angular speed of the plate and the magnitude of the pin reaction at A when the plate has rotated 90° (when pin B is directly below pin A).
c. The maximum angle θ_{max} through which the plate will rotate.

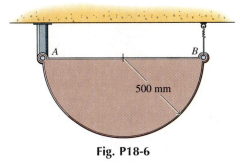

Fig. P18-6

18-7* A 5-lb uniform rod 36 in. long rotates in a vertical plane under the influence of a 2.5 lb · ft couple as shown in Fig. P18-7. If the rod is released from rest when it is horizontal, determine for the ensuing motion

a. The angular velocity of the rod when it is vertical.
b. The magnitude and direction of the force exerted on the rod by the pin at B when the rod makes an angle of 60° with the horizontal.

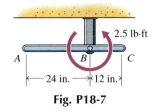

Fig. P18-7

18-8* The pendulum of Fig. P18-8 consists of a 30-kg concentrated mass on the end of a 45-kg uniform bar 2 m long. The pendulum swings in a vertical plane under the influence of a 500 N · m clockwise couple. If the pendulum has an angular velocity of 4 rad/s clockwise when $\theta = 90°$, determine for the ensuing motion

a. The angular velocity of the pendulum when $\theta = 330°$.
b. The magnitude and direction of the force exerted on the pendulum by the pin at A when $\theta = 330°$.

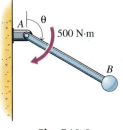

Fig. P18-8

18-9 The wheel shown in Fig. P18-9 is being pulled forward by a constant force **P** of magnitude 52 lb. The weight of the wheel is 75 lb, and its radius of gyration with respect to the axis of the wheel is 9.25 in. The wheel rolls without slipping on the horizontal surface and in the position shown has an angular velocity of 15 rad/s clockwise. Determine the velocity of the center of the wheel after it has rolled 20 ft along the horizontal surface.

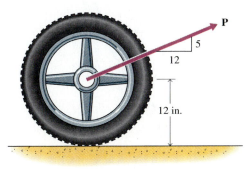

Fig. P18-9

18-10 The 5-kg flywheel of Fig. P18-10 has a diameter of 200 mm and a radius of gyration of 90 mm. A flexible rope is wrapped around the flywheel and attached to a spring that has a modulus $k = 120$ N/m. Initially, the flywheel is rotating clockwise at 20 rad/s and the spring is stretched by 800 mm. For the ensuing motion, determine

a. The maximum stretch in the spring.
b. The angular velocity of the flywheel when the rope becomes slack.

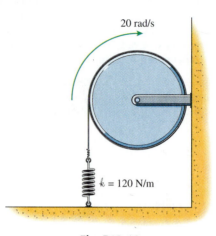

Fig. P18-10

Intermediate Problems

18-11* The slender bar shown in Fig. P18-11 has a uniform cross section and weighs 20 lb. The bar is released from rest when vertical and rotates in a vertical plane under the action of gravity. If the static coefficient of friction between the bar and the horizontal surface is $\mu_s = 0.50$, determine the angle θ_s when the bar begins to slip.

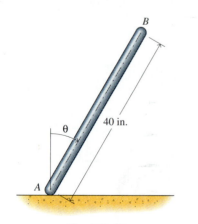

Fig. P18-11

18-12* A uniform rod is balanced on one end on a thin wire as shown in Fig. P18-12. A slight notch in the end of the rod prevents it from slipping off the wire. If the rod is disturbed and begins to tip over, determine the angle θ_m that the rod makes with the vertical when the rod loses contact with the wire.

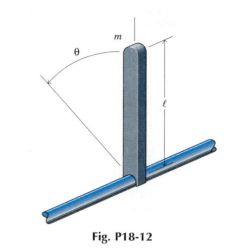

Fig. P18-12

18-13 A 16-lb bowling ball is placed on a 28° inclined surface and released from rest (Fig. P18-13). Assume that the ball is a uniform sphere 8.6 in. in diameter. If the static coefficient of friction between the ball and the surface is 0.25,

a. Verify that the ball will begin to roll without slipping.
b. Determine the speed v and the angular velocity ω of the ball after it has rolled 20 ft down the incline.
c. Compare the speed of part b with the speed of a 16-lb particle that slides (without friction) the same distance down the incline.

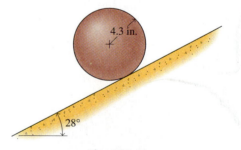

Fig. P18-13

18-14* A 12-kg uniform cylinder 300 mm in diameter is rolled up a 25° inclined surface (Fig. P18-14). If the cylinder has an initial speed of 10 m/s up the incline and rolls without slipping,

a. Determine the maximum distance that the cylinder will roll up the incline.
b. Compare the result of part a with the maximum distance that a 12-kg particle would slide (without friction) up the same incline.

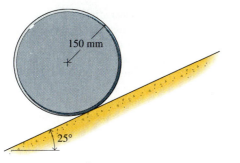

Fig. P18-14

18-15 If a bowling ball can be modeled as a 16-lb uniform sphere with a diameter of 8.6 in., determine the minimum speed v_0 that the ball must have to make it up the 30-in. ramp of the ball return shown in Fig. P18-15.

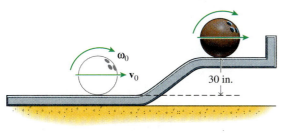

Fig. P18-15

18-16 Television cable is being removed from a 36-in. diameter spool as shown in Fig. P18-16. The spool (including the wire) weighs 450 lb, has a centroidal radius of gyration $k_G = 14$ in., and is initially at rest.

a. Determine the rotational speed of the spool after 20 feet of cable has been removed from it if a 12-lb force **P** pulls on the end of the cable and bearing friction exerts a resisting moment of 10 lb · ft on the spool,

b. If the force **P** is then removed, determine the additional amount of cable that will unwind before the spool stops rotating.

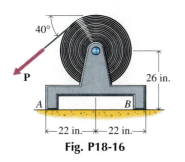

Fig. P18-16

18-17* A uniform cylinder is balanced on the edge of a ledge as shown in Fig. P18-17. The corner is rough, so the cylinder will not slip. If the cylinder is disturbed and begins to roll off the edge,

a. Determine the maximum angle θ_m through which the cylinder will rotate without losing contact with the corner.

b. Repeat for a thin hollow cylinder of the same mass and diameter.

c. Repeat for a uniform sphere of the same mass and diameter.

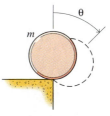

Fig. P18-17

18-18 The unbalanced disk A shown in Fig. P18-18 has a mass of 20 kg, its mass center is located 250 mm from the fixed axis of rotation O, and the radius of gyration with respect to the fixed axis of rotation is 350 mm. A cable wrapped around a shallow groove in the disk supports a 25-kg block. If the system is released from rest in the position shown, determine the angular velocity of the disk after it has rotated 270° (when the mass center G is directly to the right of O).

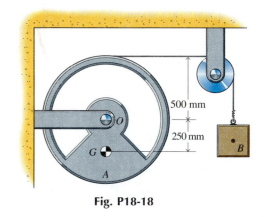

Fig. P18-18

403

18-19 A spool is supported by a cord wrapped around the inner core of the spool as shown in Fig. P18-19. The spool weighs 10 lb and has a radius of gyration of 4 in. with respect to its mass center G. At the instant shown, the spool has an angular velocity of 25 rad/s clockwise. Determine

a. The angular velocity of the spool when it has climbed 1 ft up the cord.
b. The maximum distance that the spool will climb up the cord.

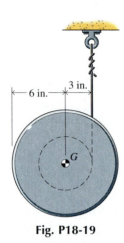

Fig. P18-19

18-20 Two 400-mm-diameter disks and a 240-mm-diameter cylinder are fastened together to form a spool that has a mass of 125 kg and a radius of gyration of 125 mm with respect to an axis through the mass center of the spool (Fig. P18-20). The spool is initially at rest when a constant force $\mathbf{F}$ of magnitude 500 N is applied to the spool through a cord wrapped around the 240-mm cylinder. If the spool rolls without slipping on the horizontal surface, determine the velocity v_G of the mass center of the spool and the angular velocity ω of the spool after it has rolled 3 m to the right.

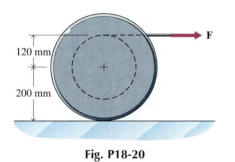

Fig. P18-20

18-21* Two 16-in.-diameter disks and an 8-in.-diameter cylinder are fastened together to form a spool weighing 50 lb that has a radius of gyration of 5 in. with respect to an axis through the mass center of the spool (Fig. P18-21). The spool is initially at rest when a constant force $\mathbf{P}$ of magnitude 50 lb is applied to the spool through a cord wrapped around the 8-in.-diameter cylinder. If the spool rolls without slipping on the horizontal surface, determine the velocity v_G of the mass center of the spool and the angular velocity ω of the spool after it has rolled 10 ft to the right.

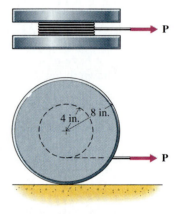

Fig. P18-21

18-22* The half disk shown in Fig. P18-22 has a mass of 25 kg. If the disk rolls without slipping when the cord at A is cut, determine the normal and frictional force exerted on the disk by the horizontal surface when the disk has rotated 45° (when the flat surface of the disk is horizontal).

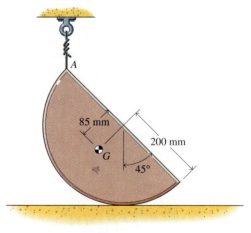

Fig. P18-22

18-23 A 25-lb uniform cylinder 16 in. in diameter rolls without slipping on an inclined surface as shown in Fig. P18-23. A spring having a modulus $k = 10$ lb/ft is attached to small frictionless pegs at the ends of the cylinder. If the cylinder has a speed of 4 ft/s down the incline when the spring is unstretched, determine

a. The speed v and the angular velocity ω of the cylinder when the spring is stretched 1 ft.
b. The maximum stretch in the spring.

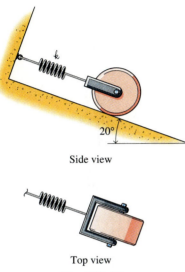

Side view

Top view

Fig. P18-23

18-24 A 15-kg uniform cylinder 800 mm in diameter rolls without slipping on an inclined surface as shown in Fig. P18-24. A cord wrapped around the cylinder is attached to a spring having $k = 150$ N/m. If the cylinder is released from rest when the spring is stretched by 1 m, determine

a. The speed v and angular velocity ω of the cylinder when the stretch in the spring is 0.5 m.
b. The stretch in the spring when the cylinder is again at rest.

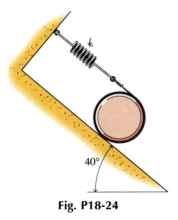

Fig. P18-24

18-25* A 30-lb, 12-in.-diameter, uniform cylinder C rolls without slipping on a horizontal surface as shown in Fig. P18-25. A cord wrapped around the cylinder passes over a small frictionless pulley and is attached to a 30-lb crate A. If the system is released from rest, determine the speed v_C and angular velocity ω_C of the cylinder and the speed v_A of the crate after the crate has dropped 5 ft.

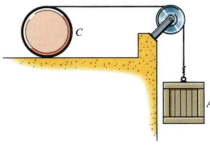

Fig. P18-25

18-26* The 10-kg spool C of Fig. P18-26 has a centroidal radius of gyration of 75 mm. A cord is attached to the center of the spool, passes over a small frictionless pulley, and is attached to a 25-kg crate A. If the system is released from rest and the spool rolls without slipping, determine the speed v_C and angular velocity ω_C of the spool and the speed v_A of the crate after the crate has dropped 2 m.

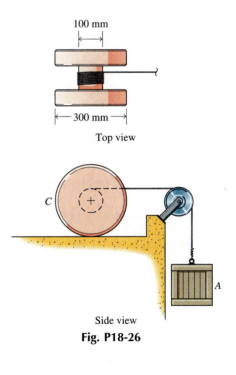

100 mm

Top view

300 mm

Side view

Fig. P18-26

18-27 The spool C of Fig. P18-27 consists of a uniform cylinder (8 lb, 6 in. in diameter) between two cylindrical disks (each 4 lb, 18 in. in diameter). A spring having $k = 24$ lb/ft is attached to cords wrapped around the disks. A second cord is wrapped around the center of the spool, passes over a small frictionless pulley, and is attached to a 30-lb crate A. If the crate has a speed of 4 ft/s downward when the spring is stretched by 4 in. and the spool rolls without slipping, determine

a. The maximum distance that the crate will drop.
b. The speed v_C and angular velocity ω_C of the spool and the speed v_A of the crate when the stretch in the spring is zero.
c. The maximum distance that the crate will rise above its initial position.

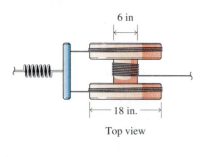

6 in

18 in.

Top view

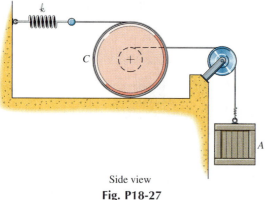

k

C

A

Side view
Fig. P18-27

18-28* The stepped cylinder C of Fig. P18-28 consists of a 5-kg cylindrical annulus (300-mm outer diameter and 100-mm inner diameter) and a 7-kg cylindrical axle (100-mm diameter). A spring having $k = 2$ kN/m is attached to cords wrapped around the axle. A second cord is wrapped around the center of the cylinder, passes over a small frictionless pulley, and is attached to a 15-kg crate A. If the crate has a speed of 1.5 m/s downward when the spring is stretched by 100 mm and the cylinder rolls without slipping, determine

a. The maximum distance that the crate will drop.
b. The speed v_C and angular velocity ω_C of the cylinder and the speed v_A of the crate when the stretch in the spring is zero.
c. The maximum distance that the crate will rise above its initial position.

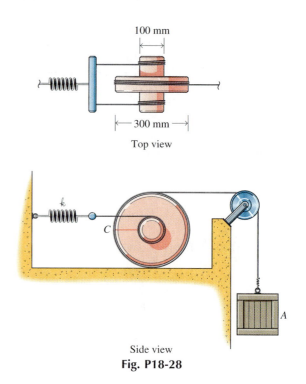

100 mm

300 mm

Top view

k

C

A

Side view
Fig. P18-28

18-29 The stepped cylinder of Fig. P18-29 consists of a 10-lb cylindrical annulus (16-in. outer diameter and 6-in. inner diameter) and a 5-lb cylindrical axle (6 in. in diameter). A spring having $k_1 = 10$ lb/ft is attached to a cord wrapped around the center of the cylinder. A second spring having $k_2 = 10$ lb/ft is attached to cords wrapped around the axle of the cylinder. The system is released from rest with a stretch of 18 in. in spring 1 and spring 2 unstretched. If the static coefficient of friction between the axle and the rails is $\mu_s = 0.8$,

a. Verify that the cylinder does not slip when motion starts.
b. Determine the speed v and the angular velocity ω of the cylinder when the stretch of spring 1 becomes zero.

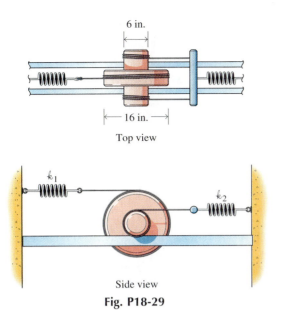

Top view

16 in.

Side view

Fig. P18-29

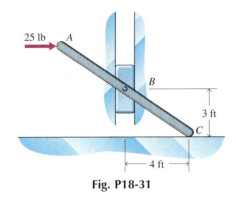

25 lb A

B

3 ft

C

4 ft

Fig. P18-31

18-32* A horizontal force of 150 N is applied to one end of a 3-kg uniform rod 4 m long (Fig. P18-32). A light slider attached to the middle of the rod moves in a frictionless vertical guide, and the surface at C is smooth. If the system is released from rest when $\theta = 20°$, determine the angular velocity ω of the rod and the velocity $\mathbf{v}_A$ of end A of the rod when $\theta = 80°$.

18-30 The unbalanced wheel shown in Fig. P18-30 has a mass of 50 kg and rolls without slipping on the horizontal surface. The radius of gyration of the wheel with respect to an axis through its mass center is 160 mm. In the position shown, the angular velocity of the wheel is 6 rad/s. Determine the normal and frictional forces exerted on the wheel by the horizontal surface when the wheel has rotated 210°.

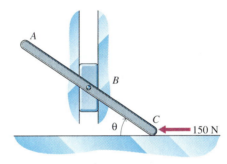

A

B

C

θ 150 N

Fig. P18-32

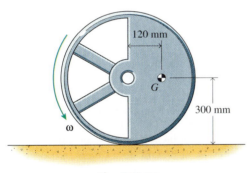

120 mm

G

300 mm

ω

Fig. P18-30

18-31* A horizontal force of 25 lb is applied to one end of a 5-lb uniform rod 10 ft long (Fig. P18-31). A light slider attached to the middle of the rod moves in a frictionless vertical guide, and the surface at C is smooth. If the system is released from rest in the position shown, determine the angular velocity ω of the rod and the velocity $\mathbf{v}_A$ of end A of the rod when the rod is vertical.

18-33 The 5-lb uniform rod of Fig. P18-33 is 3 ft long. The light sliders at the ends of the rod move in frictionless guides. A spring attached to the slider at A has a modulus of $k = 15$ lb/ft and is stretched 2 ft when the rod is vertical. If the system is released from rest when the rod is vertical, determine the angular velocity ω of the rod and the velocity $\mathbf{v}_B$ of end B when

a. The stretch in the spring is zero.
b. The bar is horizontal.

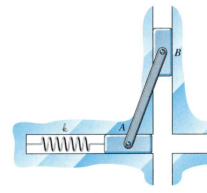

B

k

A

Fig. P18-33

18-34 The 5-kg uniform rod of Fig. P18-34 is 3 m long. The light sliders at the ends of the rod move in frictionless guides. A spring attached to the slider at B has a modulus of $k = 600$ N/m and is compressed by 500 mm when the rod is horizontal. If the system is released from rest when the rod is horizontal, determine the angular velocity ω of the rod and the velocity $\mathbf{v}_A$ of end A when the stretch in the spring is zero.

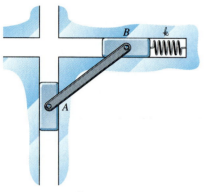

Fig. P18-34

18-35* A 15-lb weight A hangs from the frictionless axle of a pulley as shown in Fig. P18-35. The 25-lb pulley may be considered a uniform cylinder 16 in. in diameter. If a constant force $P = 10$ lb is applied to the end of the rope and the system starts from rest, determine the velocity $\mathbf{v}_A$ of the weight after it has fallen 10 ft.

Fig. P18-35

18-36* A 5-kg mass A hangs from the end of a rope that passes over a pulley as shown in Fig. P18-36. The 10-kg pulley may be considered a uniform cylinder 400 mm in diameter. If a constant force $P = 65$ N is applied to the axle of the pulley and the system starts from rest, determine the velocity $\mathbf{v}_A$ of the mass after it has fallen 2 m.

Fig. P18-36

18-37 A cart of mass m_1 rolls down a 30° incline as shown in Fig. P18-37. The two wheels are uniform cylinders with a mass of $m_2 = 2m_1$ each, are attached to the cart with frictionless axles, and roll without slipping. If the system is released from rest, determine the speed of the cart after it has traveled a distance d down the incline. Compare the answer to the speed of a single particle of mass $m_1 + 2m_2$ sliding the same distance down the incline in the absence of friction.

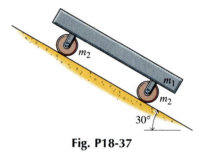

Fig. P18-37

18-38 A cart of mass m_1 rolls down a 30° incline as shown in Fig. P18-38. The two wheels, which are uniform cylinders with a mass of $m_2 = 2m_1$ each, are not attached to the cart. Friction is sufficient to prevent slipping between the wheels and the cart as well as between the wheels and the surface. If the system is released from rest, determine the speed of the cart after it has traveled a distance d down the incline. Compare the answer to the speed of a single particle of mass $m_1 + 2m_2$ sliding the same distance down the incline in the absence of friction.

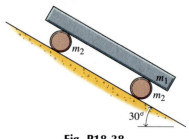

Fig. P18-38

Challenging Problems

18-39* A 16-lb ladder 12 ft long slides in a smooth corner as shown in Fig. P18-39. The ladder is at rest with $\theta = 0°$ when the lower end is disturbed slightly. Determine the angle θ_c, the angular velocity ω, and the velocity $\mathbf{v}_B$ of end B when end A loses contact with the vertical wall.

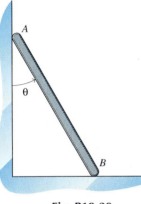

Fig. P18-39

18-40 The slender bar AB shown in Fig. P18-40 rests on a smooth surface at B and is attached to a collar at A that slides freely on the smooth vertical rod. The bar has a uniform cross section and a mass of 20 kg; the mass of the collar may be neglected. The bar is initially at rest with $\theta = 0°$ when the lower end is disturbed slightly. Determine the reactions at A and B when $\theta = 60°$.

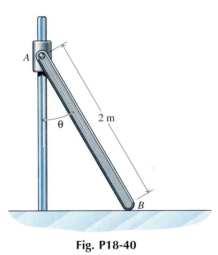

Fig. P18-40

18-41 A 20-lb uniform cylinder 24 in. in diameter rolls without slipping on a horizontal surface as shown in Fig. P18-41. The cylinder is joined to a 30-lb collar by a uniform slender rod (10 lb, 36 in. long). Friction between the collar and the vertical shaft may be neglected. If the system is released from rest when $h_A = 40$ in., determine the velocity $\mathbf{v}_A$ of the collar and the angular velocity ω_C of the cylinder when $h_A = 24$ in.

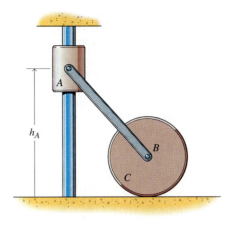

Fig. P18-41

18-42* A 5-kg uniform half cylinder 300 mm in diameter rotates about a frictionless pin as shown in Fig. P18-42. The half cylinder is joined to a 3-kg collar by a uniform slender rod (2 kg, 500 mm long). Friction between the collar and the horizontal shaft may be neglected. If the system is released from rest when $\theta = 0°$, determine the velocity of the collar $\mathbf{v}_A$ and the angular velocity of the half cylinder ω_C when $\theta = 120°$.

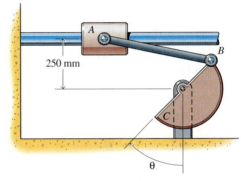

Fig. P18-42

18-43* A 5-lb uniform half cylinder 18 in. in diameter rolls without slipping on a horizontal surface as shown in Fig. P18-43. Friction between the uniform slender rod (8 lb, 30 in. long) and the horizontal surface at A may be neglected. If the system is released from rest when $\theta = 0°$, determine the velocity $\mathbf{v}_B$ and the angular velocity of the cylinder ω_C when $\theta = 90°$.

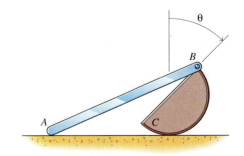

Fig. P18-43

409

18-44 The uniform half cylinder shown in Fig. P18-44 has a mass of 8 kg and a radius of 200 mm, and rolls without slipping on the horizontal surface. The uniform bar AB has a mass of 5 kg and is 600 mm long. The system starts from rest when $\theta = 0°$, and the angular velocity of the cylinder is 2 rad/s clockwise when $\theta = 90°$. Neglect friction between the bar and the surface at A, and determine the magnitude of the constant horizontal force $\mathbf{P}$.

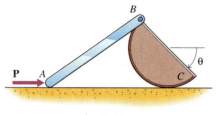

Fig. P18-44

18-45 A horizontal force of 25 lb is applied to one end of a 30-lb uniform rod 10 ft long (Fig. P18-45). A 20-lb slider attached to the middle of the rod moves in a frictionless vertical guide, and the surface at C is smooth. If end C of the rod has a velocity of 3 ft/s to the left in the position shown, determine

a. The maximum height attained by the slider B.
b. The angular velocity ω of the rod and the velocity $\mathbf{v}_C$ of end C of the rod when the slider height $h_B = 2$ ft.
c. The normal force $\mathbf{N}_C$ exerted on the rod by the surface at C when $h_B = 2$ ft.

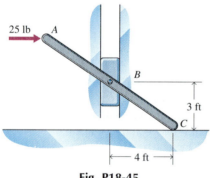

Fig. P18-45

18-46* A horizontal force of 150 N is applied to one end of a 20-kg uniform rod 4 m long (Fig. P18-46). A 15-kg slider attached to the middle of the rod moves in a frictionless vertical guide and the surface at C is smooth. If end C of the rod has a velocity of 1.35 m/s to the left when $\theta = 30°$, determine

a. The maximum height attained by the slider B.
b. The angular velocity ω of the rod and the velocity $\mathbf{v}_C$ of end C of the rod when $\theta = 20°$.
c. The normal force $\mathbf{N}_C$ exerted on the rod by the surface at C when $\theta = 20°$.

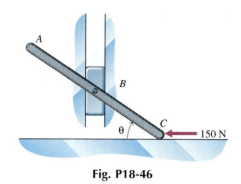

Fig. P18-46

18-47 The 5-lb uniform rod of Fig. P18-47 is 3 ft long. The sliders at the ends of the rod move in frictionless guides and weigh 2 lb each. A spring attached to the slider at A has a modulus of $k = 15$ lb/ft and is stretched 2 ft when the rod is vertical. If the system is released from rest when the rod is vertical, determine the angular velocity ω of the rod and the velocity $\mathbf{v}_B$ of end B

a. When the stretch in the spring is zero.
b. When the bar is horizontal.

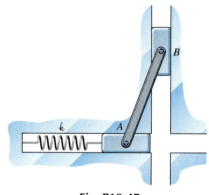

Fig. P18-47

18-48 A solid, homogeneous sphere (2 kg, 100 mm in diameter) rolls without slipping on the outside of a fixed cylinder 400 mm in diameter (Fig. P18-48). The coefficient of static friction between the sphere and the cylinder is $\mu_s = 0.7$. If the sphere is released from rest when $\theta = 0°$, determine the angle θ_s at which the sphere will begin to slip. Does the sphere lose contact with the cylinder before or after this angle?

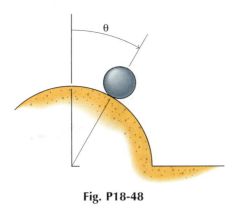

Fig. P18-48

18-49* A 12-lb uniform cylinder 16 in. in diameter rolls without slipping on a horizontal surface as shown in Fig. P18-49. The lightweight, slender bars AB and BC each have a length of 16 in. The system is at rest in the position shown when C is displaced slightly to the right. Determine the velocity $\mathbf{v}_C$ of the center of the wheel and the angular velocity ω_{AB} of the crank AB

a. When AB is horizontal.
b. When AB is vertical.

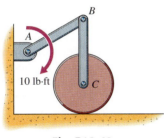

Fig. P18-49

18-50* The 2-kg slider shown in Fig. P18-50 moves in a frictionless guide. The crank arm AB has a mass of 1 kg and a length of 150 mm, BC has a mass of 3 kg and a length of 360 mm, and the spring has a modulus of $k = 1800$ N/m and an unstretched length of 150 mm. If the system is released from rest when $\theta = 0°$, determine the velocity $\mathbf{v}_C$ of the slider and the angular velocity ω_{AB} of the crank

a. When $\theta = 90°$.
b. When $\theta = 150°$.

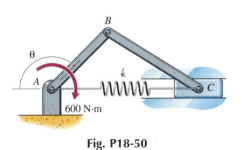

Fig. P18-50

18-51 A torsional spring attached to bar AB of Fig. P18-51 exerts a torque $M = k\theta$, where $k = 5$ ft · lb/rad. Bar AB weighs 10 lb and is 18 in. long, BC weighs 15 lb and is 30 in. long, and the surface at C is smooth. If the system is released from rest when $\theta = 0°$, determine the velocity $\mathbf{v}_C$ and the angular velocity ω_{AB}

a. When $\theta = 60°$.
b. When $\theta = 90°$.

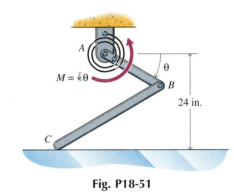

Fig. P18-51

18-52 A torsional spring attached to bar AB of Fig. P18-52 exerts a torque $M = k\theta$, where $k = 150$ N · m/rad. Bar AB has a mass of 25 kg and is 3 m long, BC has a mass of 50 kg and is 6 m long, and the surface at C is smooth. The system is initially at rest with $\theta = 60°$ and BC vertical when C is disturbed slightly to the right. Determine the velocity $\mathbf{v}_C$ and the angular velocity ω_{AB}

a. When $\theta = 120°$.
b. When $\theta = 180°$.

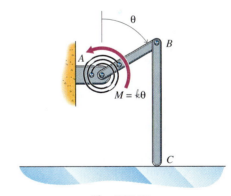

Fig. P18-52

Computer Problems

C18-53 A 16-lb bowling ball has a diameter of 8.6 in. The ball is released down the alley with an initial speed of 20 ft/s and no angular velocity. If the kinetic coefficient of friction between the ball and the alley is $\mu_k = 0.1$, calculate and plot the kinetic energy of the mass center $T_v = \frac{1}{2} mv_G^2$, the kinetic energy of rotation about the mass center $T_\omega = \frac{1}{2} I_G\omega^2$, and the total energy $E = T_v + T_\omega$, all as functions of the ball's position x_G from the moment the ball is released until it strikes the head pin 60 ft away.

C18-54 A slender uniform rod AB rotates in a vertical plane as shown in Fig. P18-54. The mass of the rod is 5 kg, and its length is 1.2 m. If the rod starts from rest when $\theta = 0°$, calculate and plot the kinetic energy of the mass center $T_v = \frac{1}{2} mv_G^2$, the kinetic energy of rotation about the mass center $T_\omega = \frac{1}{2} I_G\omega^2$, the gravitational potential energy V_g (use the level of A as the zero of potential energy), and the total energy $E = T_v + T_\omega + V_g$—all as functions of the angle θ $(0° < \theta < 90°)$.

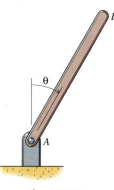

Fig. P18-54

C18-55 The slender uniform rod AB shown in Fig. P18-55 is attached to a lightweight collar at B and slides along a frictionless horizontal surface at A. The rod weighs 15 lb and is 5 ft long. If the rod starts from rest when $\theta = 0°$, calculate and plot

a. The angular velocity ω and the angular acceleration α of the rod as functions of θ $(0° < \theta < 90°)$.
b. The normal forces N_A and N_B exerted on the rod by the surface at A and by the collar at B as functions of the angle θ $(0° < \theta < 90°)$.
c. The kinetic energy of the mass center $T_v = \frac{1}{2} mv_G^2$, the kinetic energy of rotation about the mass center $T_\omega = \frac{1}{2} I_G\omega^2$, the gravitational potential energy V_g (use the level of the horizontal surface as the zero of potential energy), and the total energy $E = T_v + T_\omega + V_g$—all as functions of the angle θ $(0° < \theta < 90°)$.

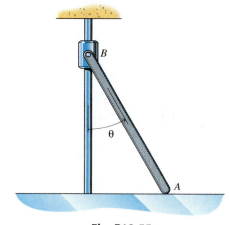

Fig. P18-55

C18-56 A 5-kg uniform half cylinder 800 mm in diameter rolls without slipping on a horizontal surface as shown in Fig. P18-56. If the cylinder is released from rest when $\theta = 0°$, calculate and plot

a. The angular velocity ω and the angular acceleration α of the half cylinder as functions of θ $(0° < \theta < 180°)$.
b. The normal force $\mathbf{N}$ and the friction force $\mathbf{F}$ exerted on the half cylinder by the surface as functions of the angle θ $(0° < \theta < 180°)$.
c. The kinetic energy of the mass center $T_v = \frac{1}{2} mv_G^2$, the kinetic energy of rotation about the mass center $T_\omega = \frac{1}{2} I_G\omega^2$, the gravitational potential energy V_g (use the level of the horizontal surface as the zero of potential energy), and the total energy $E = T_v + T_\omega + V_g$—all as functions of the angle θ $(0° < \theta < 180°)$.

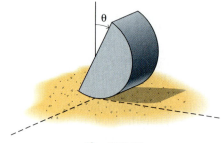

Fig. P18-56

18-6 KINETIC ENERGY OF A RIGID BODY IN THREE DIMENSIONS

In Section 18-3, the kinetic energy was calculated for a group of particles forming a rigid body and moving in a planar motion. In the present section, the restriction of planar motion will be removed.

As in Section 18-3, let point A be any point in the body and $\mathbf{r} = \mathbf{r}_{P/A} = x\mathbf{i} + y\mathbf{j} + z\mathbf{k}$ be the position vector from A to an arbitrary particle of mass dm in the body. Then the velocity of dm is related to the velocity of point A by the relative velocity equation

$$\mathbf{v} = \mathbf{v}_A + \mathbf{v}_{P/A} = \mathbf{v}_A + \boldsymbol{\omega} \times \mathbf{r} \qquad (a)$$

where $\omega = \omega_x\mathbf{i} + \omega_y\mathbf{j} + \omega_z\mathbf{k}$ is the angular velocity of the body. The kinetic energy of the particle is then

$$dT = \frac{1}{2} \, dm \, v^2 = \frac{1}{2} dm \, \mathbf{v} \cdot \mathbf{v}$$

$$= \frac{1}{2} \, dm(\mathbf{v}_A + \boldsymbol{\omega} \times \mathbf{r}) \cdot (\mathbf{v}_A + \boldsymbol{\omega} \times \mathbf{r})$$

$$= \frac{1}{2} \, dm \, v_A^2 + dm \, \mathbf{v}_A \cdot (\boldsymbol{\omega} \times \mathbf{r}) + \frac{1}{2} \, dm(\boldsymbol{\omega} \times \mathbf{r}) \cdot (\boldsymbol{\omega} \times \mathbf{r}) \qquad (b)$$

Expanding the vector and scalar products in Eq. b and rearranging slightly gives the kinetic energy of the piece of mass dm as

$$\begin{aligned} dT = & \frac{1}{2} \, dm \, v_A^2 + dm \, [(v_{Ay}\omega_z - v_{Az}\omega_y) \, r_x \\ & + (v_{Az}\omega_x - v_{Ax}\omega_z) \, r_y + (v_{Ax}\omega_y - v_{Ay}\omega_x) \, r_z] \\ & + \frac{1}{2} \, dm \, [\omega_x^2(r_y^2 + r_z^2) + \omega_y^2(r_x^2 + r_z^2) + \omega_z^2(r_x^2 + r_y^2) \\ & - 2\omega_x\omega_y r_x r_y - 2\omega_x\omega_z r_x r_z - 2\omega_y\omega_z r_y r_z] \end{aligned} \qquad (c)$$

Integrating Eq. c over all the mass in the body and recognizing that $\mathbf{v}_A$ and ω are independent of dm gives the total kinetic energy of the body

$$\begin{aligned} T = \int dT = & \frac{1}{2} \, v_A^2 \int dm + (v_{Ay}\omega_z - v_{Az}\omega_y) \int r_x \, dm \\ & + (v_{Az}\omega_x - v_{Ax}\omega_z) \int r_y \, dm + (v_{Ax}\omega_y - v_{Ay}\omega_x) \int r_z \, dm \\ & + \frac{1}{2} \, \omega_x^2 \int (r_y^2 + r_z^2) \, dm + \frac{1}{2} \, \omega_y^2 \int (r_x^2 + r_z^2) \, dm \\ & + \frac{1}{2} \, \omega_z^2 \int (r_x^2 + r_y^2) \, dm - \omega_x\omega_y \int r_x r_y \, dm \\ & - \omega_x\omega_z \int r_x r_z \, dm - \omega_y\omega_z \int r_y r_z \, dm \end{aligned} \qquad (d)$$

The first integral in Eq. d is just the mass of the body, the next three integrals give the location of the mass center of the body relative to point A, and the last six integrals are the moments and products of inertia relative to axes through point A. Therefore,

$$\begin{aligned} T = & \frac{1}{2} \, mv_A^2 + (v_{Ay}\omega_z - v_{Az}\omega_y) \, mr_{Gx} \\ & + (v_{Az}\omega_x - v_{Ax}\omega_z) \, mr_{Gy} + (v_{Ax}\omega_y - v_{Ay}\omega_x) \, mr_{Gz} \\ & + \frac{1}{2} \, \omega_x^2 \, I_{Ax} + \frac{1}{2} \, \omega_y^2 \, I_{Ay} + \frac{1}{2} \, \omega_z^2 \, I_{Az} \\ & - \omega_x\omega_y \, I_{Axy} - \omega_x\omega_z \, I_{Axz} - \omega_y\omega_z \, I_{Ayz} \end{aligned} \qquad \text{(18-13)}$$

Equation 18-13 can be simplified if point A coincides with the mass center G. Then $r_{Gx} = r_{Gy} = r_{Gz} = 0$ and the second, third, and fourth terms all vanish, leaving

$$T = \frac{1}{2} mv_G^2 + \frac{1}{2} \omega_x^2 I_{Gx} + \frac{1}{2} \omega_y^2 I_{Gy} + \frac{1}{2} \omega_z^2 I_{Gz}$$
$$- \omega_x\omega_y I_{Gxy} - \omega_x\omega_z I_{Gxz} - \omega_y\omega_z I_{Gyz} \qquad \text{(18-14)}$$

It can now be noted that Eq. 18-14 reduces to Eq. 18-9 for the special case of planar motion with $\omega_x = \omega_y = 0$. No assumptions about symmetry (that is about the moments or products of inertia) are needed. Equation 18-13 also simplifies for the special case of rotation about a fixed point O. When A coincides with a fixed point O, $\mathbf{v}_O = \mathbf{0}$ and

$$T = \frac{1}{2} \omega_x^2 I_{Ox} + \frac{1}{2} \omega_y^2 I_{Oy} + \frac{1}{2} \omega_z^2 I_{Oz}$$
$$- \omega_x\omega_y I_{Oxy} - \omega_x\omega_z I_{Oxz} - \omega_y\omega_z I_{Oyz} \qquad \text{(18-15)}$$

Finally, if the directions of the principal axes are chosen for the xyz-axes, then the products of inertia are all zero and Eqs. 18-14 and 18-15 become

$$T = \frac{1}{2} mv_G^2 + \frac{1}{2} \omega_x^2 I_{Gx} + \frac{1}{2} \omega_y^2 I_{Gy} + \frac{1}{2} \omega_z^2 I_{Gz} \qquad \text{(18-16)}$$

and

$$T = \frac{1}{2} \omega_x^2 I_{Ox} + \frac{1}{2} \omega_y^2 I_{Oy} + \frac{1}{2} \omega_z^2 I_{Oz} \qquad \text{(18-17)}$$

respectively.

Equation 18-14 can also be written in vector form

$$T = \frac{1}{2} m\mathbf{v}_G \cdot \mathbf{v}_G + \frac{1}{2} \boldsymbol{\omega} \cdot \mathbf{H}_G \qquad \text{(18-18)}$$

in which $\mathbf{H}_G$ is called the *angular momentum vector* and has the components

$$H_{Gx} = \omega_x I_{Gx} - \omega_y I_{Gxy} - \omega_z I_{Gxz} \qquad \text{(18-19}a\text{)}$$
$$H_{Gy} = -\omega_x I_{Gyx} + \omega_y I_{Gy} - \omega_z I_{Gyz} \qquad \text{(18-19}b\text{)}$$
$$H_{Gz} = -\omega_x I_{Gzx} - \omega_y I_{Gzy} + \omega_z I_{Gz} \qquad \text{(18-19}c\text{)}$$

The vector expression for the kinetic energy of a rigid body emphasizes that the kinetic energy is the scalar sum of the kinetic energy of translation $\frac{1}{2}m\mathbf{v}_G \cdot \mathbf{v}_G$ associated with the center of mass G and the kinetic energy of rotation about the mass center $\frac{1}{2}\boldsymbol{\omega} \cdot \mathbf{H}_G$. Equally important, it emphasizes that the reference system used for the computation of the inertia properties has its origin at the center of mass G.

The 16-lb, thin, homogeneous disk of Fig. 18-12 has a diameter of 20 in. and rotates freely on the 24-in.-long axle OG. As the disk rolls without slipping on the horizontal surface, the axle rotates freely about point O. Determine the kinetic energy of the disk when its angular speed ω_{GC} is 13 rad/s.

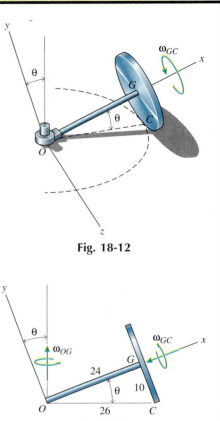

Fig. 18-12

SOLUTION

Coordinate axes are chosen with the x-axis along the axle OG (at the instant shown in Fig. 18-12) and the z-axis in the horizontal surface.[3] The y-axis then is inclined at an angle of $\theta = \tan^{-1}(10/24) = 22.62°$ to the vertical (Fig. 18-13). These are principal axes for the disk so that the moments of inertia relative to the mass center G are

$$I_{Gx} = \frac{1}{2}mr^2 = \frac{1}{2}\left(\frac{16}{32.2}\right)\left(\frac{10}{12}\right)^2 = 0.17253 \text{ slug} \cdot \text{ft}^2$$

$$I_{Gy} = \frac{1}{4}mr^2 = I_{Gz} = \frac{1}{4}\left(\frac{16}{32.2}\right)\left(\frac{10}{12}\right)^2 = 0.08627 \text{ slug} \cdot \text{ft}^2$$

$$I_{Gxy} = I_{Gxz} = I_{Gyz} = 0 \text{ slug} \cdot \text{ft}^2$$

As the disk rotates about the axle with an angular speed ω_{GC}, the axle rotates about a vertical axis with an angular speed of ω_{OG} (Fig. 18-13). Since the disk rolls without slipping, these angular speeds are related by

$$10\omega_{GC} = 26\omega_{OG}$$

Fig. 18-13

which gives $\omega_{OG} = 5$ rad/s. Then, in terms of the xyz-coordinate axes, the angular velocity of the disk is

$$\omega = -\omega_{GC}\mathbf{i} + \omega_{OG}(\sin\theta\,\mathbf{i} + \cos\theta\,\mathbf{j})$$
$$= -11.077\mathbf{i} + 4.615\mathbf{j} \text{ rad/s}$$

Finally, the kinetic energy of the disk is (Eq. 18-16)

$$T = \frac{1}{2}mv_G^2 + \frac{1}{2}\omega_x^2 I_{Gx} + \frac{1}{2}\omega_y^2 I_{Gy} + \frac{1}{2}\omega_z^2 I_{Gz}$$

where the speed of the mass center is $v_G = [(24/12)\cos\theta]\omega_{OG} = 9.231$ ft/s. Therefore,

$$T = \frac{1}{2}\left(\frac{16}{32.2}\right)(9.231)^2 + \frac{1}{2}(11.077)^2(0.17253)$$

$$+ \frac{1}{2}(4.615)^2(0.08627) + \frac{1}{2}(0)^2(0.08627)$$

$$= 32.7 \text{ lb} \cdot \text{ft} \qquad\qquad\qquad\qquad \text{Ans.}$$

> As the disk rotates through an angle θ_{GC}, the contact point C moves a distance $s = 10\theta_{GC}$ along the surface and the line OC rotates about the vertical axis by an amount $\theta_{OC} = s/26$. Since the axle OG is always directly above the line OC, the axle also rotates about the vertical axis through an angle $\theta_{OG} = \theta_{OC} = s/26$. Therefore, the rotations of the axle and the disk are related by $10\,\theta_{GC} = 26\,\theta_{OG}$. The time derivative of this relationship gives the relationship between the angular speeds, $10\,\omega_{GC} = 26\,\omega_{OG}$.

[3]It must be noted that the orientation of the coordinate system is fixed; the coordinate system does not rotate with the body. The coordinate system is chosen so that it coincides with the principal axes of the body at the instant shown simply for ease of computing the kinetic energy. However, since the kinetic energy is a scalar quantity, the number obtained is independent of the coordinate system used to compute it. Therefore, different fixed coordinate systems may be used to compute the kinetic energy at the initial and final instants when using the principle of work and energy.

The 5-kg, thin, homogeneous disk of Fig. 18-14 has a diameter of 200 mm and rotates freely on the 300-mm-long axle OG. As the disk rolls without slipping on the inclined surface, the axle rotates freely about point O. If the system is released from rest in the position shown (with the disk at its highest position on the inclined surface), determine the angular speed of the disk ω_{GC} when the disk is at its lowest position along the inclined surface.

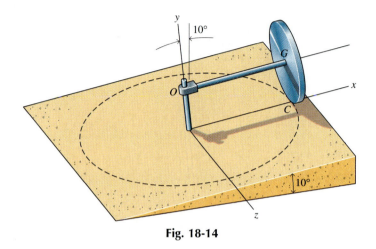

Fig. 18-14

SOLUTION

A free-body diagram of the disk and axle is shown in Fig. 18-15a. The normal force on the disk, the friction force on the disk, and the force at the fixed point O do no work because they all act at points that are instantaneously at rest. Also, the internal force between the disk and the axle does no work because friction there is negligible. Therefore, the only force that does work is gravity, and it is conservative $(U_{1\to2}^{(o)} = 0)$. The initial and final gravitational potential energies are

$$V_i = mgh = (5)(9.81)(0.3 \sin 10°) = 2.555 \text{ J}$$
$$V_f = mgh = (5)(9.81)(-0.3 \sin 10°) = -2.555 \text{ J}$$

respectively.

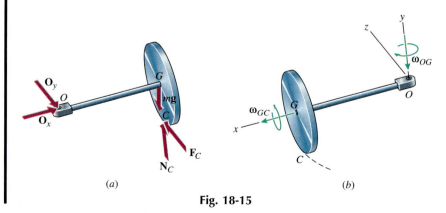

(a) (b)

Fig. 18-15

Since the system starts from rest, the initial kinetic energy is zero ($T_i = 0$). To compute the final kinetic energy, a system of axes will be chosen (Fig. 18-15b) in which the x-axis is oriented along the axle OG, the y-axis is perpendicular to the surface, and the z-axis is parallel to the surface. These are principal axes for the disk so that the moments of inertia relative to the fixed point O are

$$I_{Ox} = \frac{1}{2}mr^2 = \frac{1}{2}(5)(0.1)^2 = 0.02500 \text{ kg} \cdot \text{m}^2$$

$$I_{Oy} = I_{Gy} + md^2 = \frac{1}{4}mr^2 + md^2$$

$$= \frac{1}{4}(5)(0.1)^2 + (5)(0.3)^2 = 0.46250 \text{ kg} \cdot \text{m}^2 = I_{Oz}$$

$$I_{Oxy} = I_{Oxz} = I_{Oyz} = 0 \text{ kg} \cdot \text{m}^2$$

As the disk rotates about the axle with an angular speed ω_{GC}, the axle rotates about a vertical axis with an angular speed of ω_{OG} (Fig. 18-15b). Since the disk rolls without slipping, these angular speeds are related by

$$100\omega_{GC} = 300\omega_{OG}$$

Then, in terms of the xyz-coordinate axes, the angular velocity of the disk is

$$\boldsymbol{\omega} = \omega_{GC}\mathbf{i} - \omega_{OG}\mathbf{j} = \omega_{GC}\mathbf{i} - \frac{\omega_{GC}}{3}\mathbf{j}$$

and the final kinetic energy of the disk is (Eq. 18-17)

$$T_f = \frac{1}{2}\omega_x^2 \, I_{Ox} + \frac{1}{2}\omega_y^2 \, I_{Oy} + \frac{1}{2}\omega_z^2 \, I_{Oz}$$

$$= \frac{1}{2}\omega_{GC}^2(0.02500) + \frac{1}{2}\left(\frac{\omega_{GC}}{3}\right)^2(0.46250) + 0$$

$$= 0.03819\omega_{GC}^2$$

Finally, substituting all these values in the equation of work and energy (Eq. 18-11)

$$T_i + V_i + U_{i \to f}^{(o)} = T_f + V_f$$

gives

$$0 + 2.555 + 0 = 0.03819\omega_{GC}^2 - 2.555$$

or

$$\omega_{GC} = 11.57 \text{ rad/s} \qquad\qquad \text{Ans.}$$

Kinetic energy is a scalar quantity and the number obtained is independent of the orientation of the fixed coordinate system used to compute it. For ease of computing the final kinetic energy, the coordinate system is oriented so that it coincides with the principal axes of the disk at the instant of interest.

PROBLEMS

Introductory Problems

18-57* The 5-lb wheel of Fig. P18-57 is rotating with an angular speed of 20 rad/s on a smooth axle. Simultaneously, the arm holding the wheel is rotating about the horizontal shaft with an angular speed of 8 rad/s. If the wheel can be modeled as a uniform thin disk 12 in. in diameter, determine its kinetic energy.

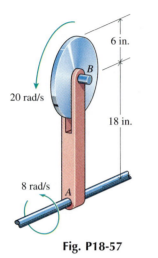

Fig. P18-57

18-58* The 5-kg wheel of Fig. P18-58 is rotating with an angular speed of 25 rad/s on a smooth axle. Simultaneously, the arm holding the wheel is rotating about the horizontal shaft with an angular speed of 10 rad/s. If the wheel can be modeled as a uniform thin disk 400 mm in diameter, determine its kinetic energy.

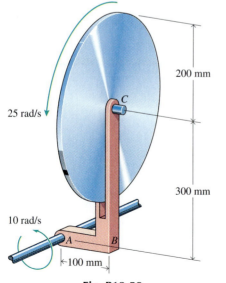

Fig. P18-58

18-59 The 4-lb disk of Fig. P18-59 is mounted at an angle of 40° to the horizontal shaft, which is rotating with an angular speed of 240 rev/min. If the disk can be modeled as a uniform thin disk 16 in. in diameter, determine its kinetic energy.

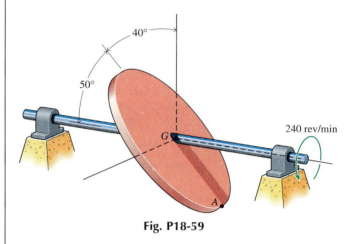

Fig. P18-59

18-60* The 4-kg rectangular plate of Fig. P18-60 is mounted at an angle of 40° to the horizontal shaft, which is rotating with an angular speed of 300 rev/min. If the plate can be modeled as a uniform thin rectangular plate, determine its kinetic energy.

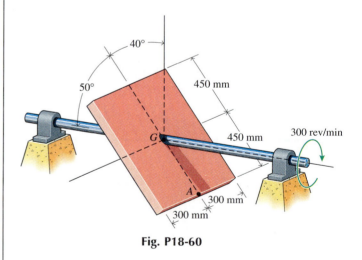

Fig. P18-60

18-61 The 4-lb disk of Problem 18-59 (a uniform thin disk 16 in. in diameter, mounted at an angle of 40° to the horizontal shaft) is initially at rest when a constant moment $\mathbf{M}_O$ is applied to the shaft, causing it to rotate. Determine the magnitude of the moment that will cause the disk to rotate at 240 rev/min after 4 revolutions.

18-62 The 4-kg rectangular plate of Problem 18-60 (a uniform thin rectangular plate, mounted at an angle of 40° to the horizontal shaft) is initially at rest when a constant moment M_O is applied to the shaft, causing it to rotate. Determine the magnitude of the moment that will cause the plate to rotate at 300 rev/min after 5 revolutions.

Intermediate Problems

18-63* The 8-lb disk of Fig. P18-63 rotates freely about the shaft OG as it rolls without slipping on a horizontal surface. If the disk can be modeled as a uniform thin disk 24 in. in diameter, determine the kinetic energy of the disk when ω = 240 rev/min.

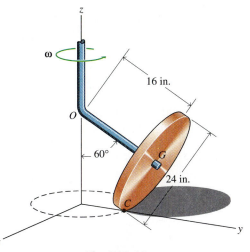

Fig. P18-63

18-64 A 2-kg, thin circular disk 200 mm in diameter (A) and an 8-kg, thin circular disk 400 mm in diameter (B) are rigidly joined by a light shaft AB, which is 120 mm long, as shown in Fig. P18-64. Determine the kinetic energy of the pair of disks when they are rotating with an angular speed of 5 rad/s.

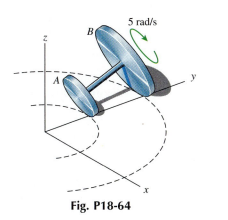

Fig. P18-64

18-65* The 16-lb, thin, homogeneous disk of Example Problem 18-5 (Fig. P18-65) has a diameter of 20 in. and is rigidly attached to the 4-lb axle OG, which is 24 in. long and 1 in. in diameter. If the disk rolls without slipping on a horizontal surface and the axle rotates freely about the fixed point O, determine the kinetic energy of the system when the angular speed of the disk ω_{GC} is 13 rad/s.

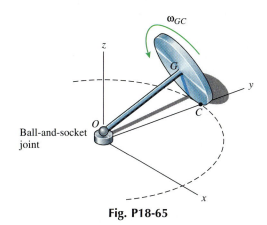

Fig. P18-65

18-66* The 5-kg, thin, homogeneous disk of Example Problem 18-6 (Fig. P18-66) has a diameter of 200 mm and is rigidly attached to the 2-kg axle OG, which is 300 mm long and 25 mm in diameter. If the disk rolls without slipping on a horizontal surface and the axle rotates freely about the vertical shaft at O, determine the kinetic energy of the system when the angular speed of the disk ω_{GC} is 15 rad/s.

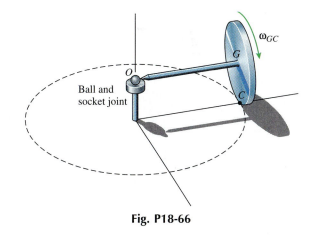

Fig. P18-66

18-67 The 8-lb disk of Problem 18-63 (a uniform thin disk 24 in. in diameter) is initially at rest when a constant moment $M_O = 10$ lb · ft is applied to the vertical shaft, causing it to rotate. Determine the angular speed of the vertical shaft after 3 revolutions.

18-68* The pair of disks of Problem 18-64 (a 2-kg, thin circular disk 200 mm in diameter and an 8-kg, thin circular disk 400 mm in diameter joined by a light shaft that is 120 mm long) is placed on an inclined surface and released from rest. If AB initially points directly up the 15° incline, determine the angular speed of the system when AB points directly down the incline.

18-69 The 16-lb, thin, homogeneous disk of Problem 18-65 is initially at rest when a constant moment $M_O = 15$ lb · ft (about a vertical axis) is applied to the hub O. Determine the angular speed ω_{GC} of the disk after it has rolled 1 revolution about the hub O.

18-70 The 5-kg, thin, homogeneous disk of Problem 18-66 is initially at rest when a constant moment $M_O = 12$ N · m (about a vertical axis) is applied to the hub O. Determine the angular speed ω_{GC} of the disk after it has rolled 2 revolutions about the vertical shaft.

Challenging Problems

18-71* A 3-lb uniform slender rod AB is attached by means of ball-and-socket joints to a rotating wheel and a slider as shown in Fig. P18-71. Determine the kinetic energy of the rod in the position shown if the angular speed of the wheel is $\omega = 20$ rad/s.

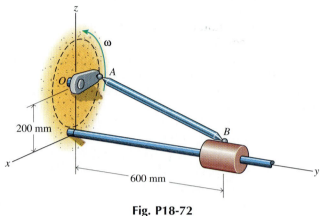

Fig. P18-71

18-72 A 2-kg uniform slender rod AB is attached by means of ball-and-socket joints to a rotating arm and a slider as shown in Fig. P18-72. Determine the kinetic energy of the rod in the position shown if the angular speed of the arm is $\omega = 15$ rad/s.

Fig. P18-72

18-73* A 5-lb uniform slender rod AB is attached by means of ball-and-socket joints to two sliders as shown in Fig. P18-73 ($a = 16$ in., $b = 32$ in., $c = 8$ in.). If slider A is moving upward with a speed of 15 ft/s at the instant shown, determine the kinetic energy of the rod.

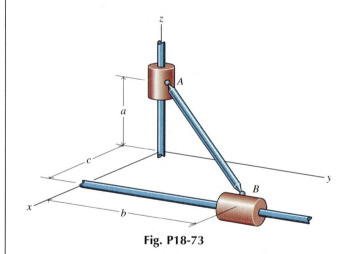

Fig. P18-73

18-74* A 4-kg uniform slender rod AB is attached by means of ball-and-socket joints to two sliders as shown in Fig. P18-73 ($a = 400$ mm, $b = 800$ mm, $c = 200$ mm). If slider B is moving to the right with a speed of 8 m/s at the instant shown, determine the kinetic energy of the rod.

18-75 The 5-lb slender rod of Problem 18-73 is released from rest when $a = 32$ in., $b = 16$ in., and $c = 8$ in. Determine the speed v_B of slider B when $b = 24$ in.

18-76 The 4-kg slender rod of Problem 18-74 is released from rest when $a = 800$ mm, $b = 400$ mm, and $c = 200$ mm. Determine the speed v_A of slider A when $a = 200$ mm.

SUMMARY

The work–energy method combines the principles of kinematics with Newton's second law to directly relate the position and speed of a body. For this method to be useful, the forces acting on a body must be a function of position only. For some types of these forces, however, the resulting integrals can be evaluated explicitly. The result is a simple algebraic equation that relates the linear and angular speed of the body at two different positions of its motion.

The work done on a rigid body must include the work done by forces and by couples. The work done by forces acting on a rigid body is calculated in exactly the same way as for forces acting on a particle. Individual forces do work only when the point at which they are applied translates and not when the body rotates about their point of application. Couples do work only when the body rotates.

For conservative forces, the potential energy V is also defined and determined in the same manner as for particles. The work done by conservative forces may be computed by direct integration using Eq. 18-1 or by using potential energy functions as discussed in Section 17-7.

The work done by internal forces in a rigid body occur in equal-magnitude but opposite-direction collinear pairs such that the work of one force cancels that of its counterpart; the resultant work done by the internal forces is zero. By the same token, when two or more rigid bodies are connected by smooth pins or by flexible, inextensible cables, the resultant work done on the bodies by the connecting members of the system is also zero.

The kinetic energy of a body is the sum of the kinetic energies of the particles that make up the body. For the case of a rigid body this gives

$$T = \frac{1}{2}mv_G^2 + \frac{1}{2}I_{Gz}\omega^2 \qquad \textbf{(18-9)}$$

where v_G is the speed of the mass center of the body and I_{Gz} is the moment of inertia about an axis through the mass center G parallel to the z-axis (perpendicular to the plane of motion). The first term of Eq. 18-9 is the kinetic energy associated with the translation of the mass center; the second term is the kinetic energy associated with the rotation of the body about an axis through the center of mass.

The principle of work and energy for a rigid body

$$T_1 + U_{1 \to 2} = T_2 \qquad \textbf{(18-10)}$$

looks exactly the same as Eq. 17-6b for the work and energy of a particle. The difference between these equations is that the kinetic energy terms in Eq. 18-10 include the rotational kinetic energy of the rigid body as well as the translational kinetic energy and that the work term includes the work done by all external moments as well as the work done by all external forces that act on the rigid body.

Just as with a particle, the work term can be divided up into a part done by conservative forces $U_{1 \to 2}^{(c)}$ and a part done by all other forces $U_{1 \to 2}^{(o)}$. The work done by the conservative forces can be expressed in terms of potential functions so that Eq. 18-10 can be written

$$T_1 + V_1 + U_{1 \to 2}^{(o)} = T_2 + V_2 \qquad \textbf{(18-11)}$$

The same advantages and limitations will be realized when the method of work and energy is applied to rigid-body motion as when it is applied to particle motion. The primary advantage is that the principle of work and energy directly relates the linear and angular speed of the body at two different positions of its motion to the forces and couples that do work on the body during its motion. The primary limitation is that Eq. 18-10 is a scalar equation and can be solved for no more than a single unknown.

Since the work–energy method combines the principles of kinematics with Newton's second law, it is not a new or independent principle. There are no problems that can be solved with the work–energy method that cannot be solved using Newton's second law. However, when the work–energy method does apply, it is usually the easiest method of solving a problem.

REVIEW PROBLEMS

18-77* The tractor shown in Fig. P18-77 is traveling at 20 mi/h up a 10 percent grade at the instant shown. The rear drive wheels are 6 ft in diameter and rotate as a unit having a combined weight of 2000 lb and a centroidal radius of gyration about the axle of $k_G = 2$ ft. The rest of the tractor weighs an additional 6000 lb. If the engine is suddenly disengaged,

a. Determine how much further up the incline the tractor will travel before coming to rest.

b. Determine how far up the incline the tractor would travel if the body weighed 8000 lb and the drive wheels were relatively light.

Fig. P18-77

18-78* A platform rotates at an angular speed of $\omega_1 = 10$ rad/s while a 5-kg cylinder of radius 50 mm mounted on the platform rotates relative to the platform at an angular speed of $\omega_2 = 25$ rad/s (Fig. P18-78). Determine the kinetic energy of the cylinder for $a = 50$ mm.

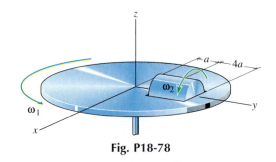

Fig. P18-78

18-79 The 3-lb, 16-in.-diameter uniform circular disk of Fig. P18-79 rolls without slipping on a horizontal surface. A 4-lb slender rod penetrates the disk 7 in. from its center. If this system has an angular velocity of 5 rad/s clockwise when $\theta = 0°$, determine the forces (magnitude and direction) exerted on the disk by the surface when $\theta = 60°$ and $120°$.

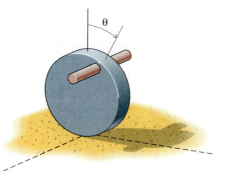

Fig. P18-79

18-80* A toy car is driven by a torsional spring as shown in Fig. P18-80. The drive wheel is a 750-g solid disk 150 mm in diameter, the body of the car has a mass of 150 g, and the mass of the front wheels may be neglected. The torsional spring exerts a moment on the drive wheel of $M = k\theta$ where $k = 0.01$ N · m/rad and θ is in radians. Determine the maximum speed of the car if it starts from rest with the spring wound up 8 revolutions. (The spring disengages when $\theta = 0°$.)

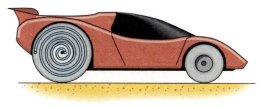

Fig. P18-80

18-81 An 8-lb uniform rectangular door opens upward and is counterbalanced with a spring as shown in Fig. P18-81. The modulus and undeformed length of the spring are $k = 24$ lb/ft and $\ell_0 = 23$ in., respectively. If the door has an angular velocity of 3 rad/s counterclockwise when it is vertical ($\theta = 0°$), determine the angular velocity of the door and the magnitude and direction of the hinge reaction on the door when it is horizontal ($\theta = 90°$).

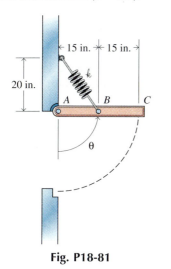

Fig. P18-81

18-82 The speed of a rotating system is controlled with a brake as shown in Fig. P18-82. The rotating parts of the system have a mass of 300 kg and a radius of gyration with respect to the axis of rotation of 200 mm. The kinetic coefficient of friction between the brake pad and the brake drum is $\mu_k = 0.50$. If the initial speed of the system is 1000 rev/min and the brake stops the rotating system after 25 revolutions, determine

a. The magnitude of the force **P** that is being applied to the brake lever.
b. The horizontal and vertical components of the reaction at support B of the brake lever.

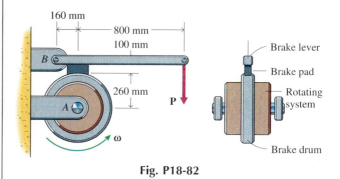

Fig. P18-82

18-83* A solid cylinder, a hollow cylinder, and a solid sphere each weigh 16 lb and have outside diameters of 14 in. The inside diameter of the hollow cylinder is 12 in. Determine the speeds that the three objects would attain in rolling 3 revolutions down a 30° incline.

18-84 The spoked wheel shown in Fig. P18-84 rolls without slipping on a horizontal surface. Because it has lost a couple of its spokes, the center of mass of the 12-kg wheel is 50 mm away from its middle and the radius of gyration relative to the mass center is 0.6 m. If the center of the wheel has a speed of 3 m/s when $\theta = 0°$, determine

a. The angular velocity ω of the wheel when $\theta = 90°$.
b. The normal and friction forces exerted on the wheel when $\theta = 90°$.

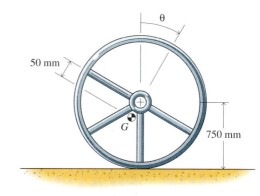

Fig. P18-84

18-85* A 15-lb uniform cylinder rests in a fixed trough as shown in Fig. P18-85. The diameter of the cylinder is 8 in., the coefficient of kinetic friction between the cylinder and the trough at A is $\mu_k = 0.15$, and the surface at B is smooth. If a constant couple of 3 lb · ft is suddenly applied to the cylinder, determine the angular velocity of the cylinder after it rotates 5 revolutions.

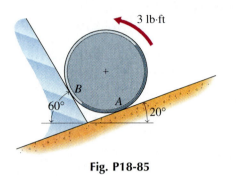

Fig. P18-85

18-86* A 45-kg post rotates in a vertical plane as shown in Fig. P18-86. The modulus and undeformed length of the spring are $k = 140$ N/m and $\ell_0 = 2$ m, respectively. If the angular velocity of the post is 3 rad/s clockwise when it is vertical, determine the angular velocity of the post and the magnitude and direction of the pin reaction on the post when the post is horizontal.

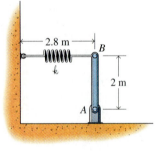

Fig. P18-86

18-87 A brake for regulating the descent of a body is shown in Fig. P18-87. The rotating parts of the brake (cable sheave and brake drum) weigh 250 lb and have a radius of gyration with respect to their axis of rotation of 4.25 in. The kinetic coefficient of friction between the brake pad and drum is $\mu_k = 0.50$. The weight of body C is 1000 lb. If body C is falling at 10 ft/s when a 150-lb force **P** is applied to the brake lever, determine the distance that the body C will fall before it is stopped by the brake.

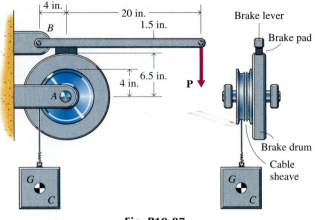

Fig. P18-87

18-88 At the instant shown in Fig. P18-88, the 70-g yo-yo has a counterclockwise angular velocity of 100 rad/s. If the yo-yo has a centroidal radius of gyration of $k_G = 14$ mm, determine

a. The height h that the yo-yo will climb up the string.
b. The number of revolutions that the yo-yo will rotate before coming to rest.

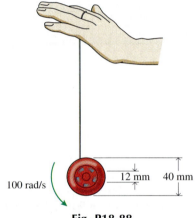

Fig. P18-88

18-89* The system shown in Fig. P18-89 is released from rest. Determine the speed of A after it has fallen 3 ft

a. If the mass of both pulleys is neglected
b. If pulley B is a uniform disk weighing 10 lb and pulley C is a uniform disk weighing 20 lb.

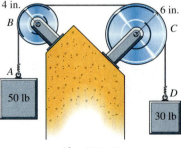

Fig. P18-89

18-90* The compound wheel of Fig. P18-90 consists of a 5-kg half circular disk sandwiched between two full circular disks (each 2 kg, 600 mm in diameter). If the wheel rolls without slipping and has an angular velocity of 10 rad/s clockwise when $\theta = 0°$, determine the force (magnitude and direction) exerted on the cylinder by the surface when $\theta = 60°$ and $120°$.

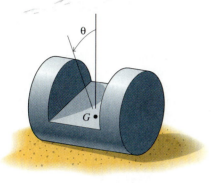

Fig. P18-90

18-91 The uniform cylinder of Fig. P18-91 has a radius of 2 in., a length of 8 in., and weight of 5 lb. The cylinder rotates about its own axis at a constant rate of $\omega_1 = 20$ rad/s. Simultaneously, the yoke holding the cylinder is rotating about a vertical axis at a rate of $\omega_2 = 8$ rad/s. Determine the kinetic energy of the cylinder at this instant.

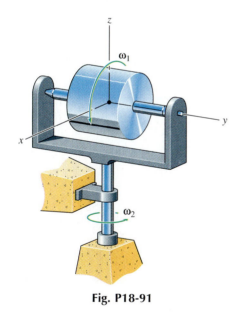

Fig. P18-91

18-92* A brake arm is used to control the motion of a drum and weight as shown in Fig. P18-92. The mass and centroidal radius of gyration of the drum are 40 kg and 120 mm, respectively. The coefficient of kinetic friction between the brake shoe and the drum is $\mu_k = 0.4$, and the system is initially at rest with the 50-kg mass 3 m above the floor.

a. If the brake handle is suddenly released, determine the speed of the 50-kg mass after it has fallen 2 m.
b. If the brake is suddenly reapplied after the 50-kg mass has fallen 2 m, determine the minimum force **P** that must be applied to the brake handle to prevent the 50-kg mass from hitting the floor.
c. If the maximum force that can be applied to the brake handle without breaking it is $P = 250$ N, determine the maximum distance that the 50-kg mass can be allowed to fall before the brake is reapplied.

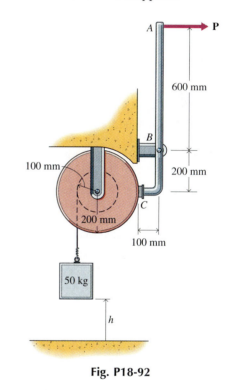

Fig. P18-92

18-93* A flywheel is rotated by a flexible rope as shown in Fig. P18-93. The flywheel is a 20-lb solid cylinder 18 in. in diameter, and the rope is initially wrapped five full turns around the wheel. If the flywheel starts from rest and the rope disengages when the end of the rope is at the edge of the flywheel A, determine the final angular velocity of the flywheel

a. When a constant force **P** having a magnitude of 25 lb is applied to the end of the flexible rope as shown in Fig. P18-93a.

b. When the force **P** is replaced by a load of 25 lb attached to the rope as shown in Fig. P18-93b.

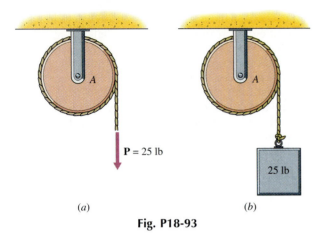

(a)   (b)

Fig. P18-93

18-94 Figure P18-94 shows a radial-arm wood saw that has an operating speed of 1500 rev/min. The blade and motor armature have a combined mass of 1.2 kg and a centroidal radius of gyration of $k_G = 25$ mm. When the saw is turned off, bearing friction and a magnetic brake exert a constant braking torque **T** on the blade and motor.

a. Determine the number of revolutions that the blade rotates before coming to rest if $T = 0.002$ N · m (bearing friction only).

b. Determine the torque **T** necessary to stop the blade in just 1 revolution.

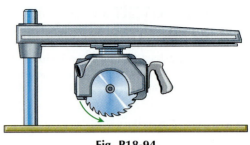

Fig. P18-94

18-95 A 7-lb uniform rod 6 ft long is attached to a light slider that moves in a frictionless vertical guide (Fig. P18-95). The other end of the rod slides on a frictionless surface. The system is at rest with $\theta = 0°$ when the lower end is disturbed slightly. Determine the angle θ, the angular velocity ω, and the velocity $\mathbf{v}_B$ of end B when the normal force **N** between the horizontal surface and end B becomes zero.

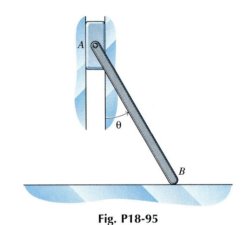

Fig. P18-95

P rior to the collision and after the collision, both the tennis ball and the tennis racket are generally treated as rigid bodies. During the collision, however, both the ball and the racket deform significantly. Fortunately, the precise details of this deformation can often be ignored.

Chapter 19

KINETICS OF PARTICLES: IMPULSE AND MOMENTUM

19-1   INTRODUCTION

The study of kinetics is based on Newton's second law of motion. In Chapters 15 and 16, Newton's second law was used directly to relate the forces acting on particles and rigid bodies with the resulting acceleration of the particles and rigid bodies. In fact, when information about the acceleration or the value of a force at an instant is desired, Newton's second law is usually the easiest method to use.

In Chapters 17 and 18, Newton's second law was integrated with respect to position to get the principle of work and energy. Since the principle of work and energy is just a combination of Newton's second law and the principles of kinematics, any problem that can be solved using the principle of work and energy can also be solved using Newton's second law. However, the principle of work and energy is particularly useful for solving problems in which the speeds of a body for two different positions of its motion are to be related and the forces involved can be expressed as functions of the position of the body.

The principles of impulse and momentum developed in this chapter and the next are obtained by integrating Newton's second law with respect to time. The resulting equations are useful for solving problems in which the velocities of a body for two different instants of time are to be related and the forces involved can be expressed as functions of time. Although the principles of impulse and momentum are not required to solve any particular problem, they will be found to be particularly useful for solving problems involving collisions between bodies and variable mass systems.

19-2   LINEAR IMPULSE AND MOMENTUM OF A PARTICLE

Let $\mathbf{R} = \Sigma\mathbf{F}$ be the resultant of all forces acting on a particle of mass m. Then Newton's second law for the particle can be written

$$\mathbf{R} = m\mathbf{a} = m\frac{d\mathbf{v}}{dt} \qquad (a)$$

Since the mass of the particle does not depend on time, it can be taken inside the derivative to get

$$\mathbf{R} = \frac{d}{dt}(m\mathbf{v}) \qquad (b)$$

When the forces are constants or functions of time only, Eq. b can be integrated to get

$$\int_{t_i}^{t_f} \mathbf{R}\, dt = \int_{mv_i}^{mv_f} d(m\mathbf{v}) = (m\mathbf{v})_f - (m\mathbf{v})_i \qquad (c)$$

or

$$(m\mathbf{v})_i + \int_{t_i}^{t_f} \mathbf{R}\, dt = (m\mathbf{v})_f \qquad (19\text{-}1)$$

where $\mathbf{v}_i$ is the velocity of the particle at some initial time t_i and $\mathbf{v}_f$ is the velocity of the particle at the final time t_f.

19-2.1   Linear Momentum

The vector $m\mathbf{v}$ in Eq. 19-1 is given the symbol $\mathbf{L}$ and is called the *linear momentum*, or simply the *momentum*, of the particle

$$\mathbf{L} = m\mathbf{v} \qquad (19\text{-}2)$$

Since m is a scalar, the linear momentum vector and the velocity of the particle are in the same direction. The magnitude of the linear momentum is equal to the product of the mass m and the speed v of the particle. In the SI system of measurement, the units of momentum are $kg \cdot m/s$ or equivalently, $N \cdot s$. In the U.S. Customary system of measurement, they are $slug \cdot ft/s$ or $lb \cdot s$.

19-2.2   Linear Impulse

The integral $\int_{t_i}^{t_f} \mathbf{R}\, dt$ is called the *linear impulse* of the force $\mathbf{R}$. The linear impulse is a vector that has units of force-time. In the SI system of measurement the magnitude of the linear impulse of a force is expressed in $N \cdot s = kg \cdot m/s$, which is the same as the unit obtained for the linear momentum of a particle. Therefore, Eq. 19-1 is dimensionally correct. If U.S. Customary units are used, the linear impulse is expressed in $lb \cdot s = slug \cdot ft/s$, which is also the same as the unit obtained for the linear momentum of a particle.

In general, the resultant force $\mathbf{R}(t) = R(t)\mathbf{e}_R$ is a vector that changes in both magnitude and direction during the time interval t_i to t_f. However, if the direction $\mathbf{e}_R$ of the force does not change during the time interval, it can be taken outside the integral. Then the value of the integral—which represents the magnitude of the impulse—is just the shaded area under the graph of R versus t (Fig. 19-1). If the magnitude

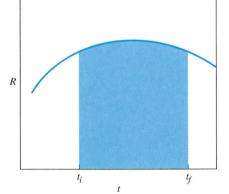

Figure 19-1   If the force $\mathbf{R}$ is constant in both magnitude and direction, then the magnitude of the impulse is just the area under the graph of R versus t.

of the force is also constant, then it can also be taken outside the integral, leaving

$$\int_{t_i}^{t_f} \mathbf{R}_c \, dt = \mathbf{R}_c \int_{t_i}^{t_f} dt = \mathbf{R}_c(t_f - t_i) \qquad (d)$$

Equation d is also used to define the time average impulse force $\mathbf{R}_{avg}$—the equivalent constant force that gives the same linear impulse as the original time-varying force $\mathbf{R}(t)$

$$\mathbf{R}_{avg} = \frac{1}{(t_f - t_i)} \int_{t_i}^{t_f} \mathbf{R} \, dt \qquad (e)$$

The average value of the force given by Eq. e (the time-average value) is usually different from the average value computed from the work done by a force (the distance-average value).

When both the magnitude and direction of the resultant force $\mathbf{R}(t)$ vary during the time interval, evaluation of the impulse integral must be carried out in component form. Rectangular Cartesian components are generally preferred because the unit vectors $\mathbf{i}$, $\mathbf{j}$, and $\mathbf{k}$ do not vary with time. Resolving $\mathbf{R}$ into its rectangular components gives

$$\int_{t_i}^{t_f} \mathbf{R} \, dt = \mathbf{i} \int_{t_i}^{t_f} R_x \, dt + \mathbf{j} \int_{t_i}^{t_f} R_y \, dt + \mathbf{k} \int_{t_i}^{t_f} R_z \, dt \qquad (f)$$

Although both the work done by a force (defined in Chapter 17) and the impulse of a force are integrals of a force, they are completely different concepts. Two important differences are as follows:

1. The work done by a force is a scalar quantity; the impulse of a force is a vector.
2. The work done by a force is zero if there is no component of the force in the direction of its displacement. The impulse of a force is never zero, even when it is applied at a stationary point.

19-2.3 Principle of Linear Impulse and Momentum

Equation 19-1 expresses the Principle of Linear Impulse and Momentum:

$$\mathbf{L}_i + \int_{t_i}^{t_f} \mathbf{R} \, dt = \mathbf{L}_f$$

The final linear momentum $\mathbf{L}_f$ of a particle is the vector sum of its initial linear momentum $\mathbf{L}_i$ and the impulse $\int \mathbf{R} \, dt$ of the resultant of all forces acting on the particle.

Unlike the work and energy equation, which is a scalar equation, Eq. 19-1 is a vector equation representing three scalar equations. Expressed in rectangular Cartesian coordinates, its three scalar components are

$$mv_{xi} + \int_{t_i}^{t_f} R_x \, dt = mv_{xf}$$

$$mv_{yi} + \int_{t_i}^{t_f} R_y \, dt = mv_{yf}$$

$$mv_{zi} + \int_{t_i}^{t_f} R_z \, dt = mv_{zf}$$

It should be noted at this point that the principle of linear impulse and momentum is not really a new principle. It is merely a combination of Newton's second law and the principles of kinematics for the special case in which force is a function of time. Nevertheless, it is useful for solving for the velocity of a particle when the force is known as a function of time and the acceleration is not of interest.

19-2.4   Conservation of Linear Momentum

It follows from Eq. b that the rate of change of the linear momentum $m\mathbf{v}$ is zero when $\mathbf{R} = \Sigma \mathbf{F} = 0$. When this happens, the linear momentum is *conserved*; that is, it is constant, in both magnitude and direction:

$$\mathbf{L}_i = \mathbf{L}_f \tag{g}$$

Linear momentum can be conserved in one direction (if the sum of forces in that direction is zero) independently of any other direction.

Although Eq. g is often called the *Principle of Conservation of Linear Momentum*, it is only a special case of the general principle of linear impulse and momentum. Conservation of linear momentum may be recognized as just an alternative statement of Newton's first law.

Conservation of linear momentum is not related to conservation of kinetic energy. For example, when an elastic sphere bounces off a hard surface, the speed of rebound is nearly the same as the speed at which it struck the surface. Therefore, the kinetic energy of the rebound is essentially the same as the kinetic energy before striking the surface, and kinetic energy is conserved. However, the direction of the velocity after the impact is opposite to what it was before the impact. Therefore, the linear momentum after the impact is the negative of what it was before the impact, and linear momentum is not conserved. Similarly, when two particles collide, it is possible to have linear momentum for the pair of particles conserved even though most of their kinetic energy is lost.

Finally, it should be noted that the mass m of the particle is assumed to be constant in Eqs. 19-1 and 19-2. Therefore, these equations should not be used to solve problems involving the motion of bodies, such as rockets, that gain or lose mass. Problems of that type will be considered in Section 19-7.

EXAMPLE PROBLEM 19-1

A Ping-Pong ball weighing 0.2 oz has an initial velocity $\mathbf{v}_i = 8\mathbf{j} + 6\mathbf{k}$ ft/s when a gust of wind exerts a force of $\mathbf{F} = 0.5t\,\mathbf{i}$ oz on the ball (t is in seconds). Determine the magnitude and direction of the velocity of the ball after 0.5 seconds. (The positive z-direction is up.)

SOLUTION

The free-body diagram of the ball shown in Fig. 19-2 includes the weight of the ball $\mathbf{W} = m\mathbf{g}$ and the force of the wind $\mathbf{F}$. The mass of the ball is

$$m = \frac{W}{g} = \frac{0.2}{(16)(32.2)} = 3.882(10^{-4}) \text{ slug}$$

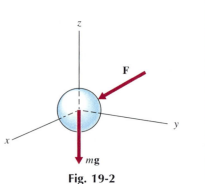

Fig. 19-2

and the linear impulse on the ball during the 0.5 seconds is

$$\int \mathbf{R}\, dt = \int_0^{0.5} \left[\frac{0.5t\mathbf{i} - 0.2\mathbf{k}}{16} \right] dt = 0.00391\mathbf{i} - 0.00625\mathbf{k} \text{ lb}$$

Substituting these values into the impulse–momentum equation (Eq. 19-1)

$$(m\mathbf{v})_i + \int_{t_i}^{t_f} \mathbf{R}\, dt = (m\mathbf{v})_f \qquad (a)$$

then gives

$$3.882(10^{-4})(8\mathbf{j} + 6\mathbf{k}) + 39.1(10^{-4})\mathbf{i} - 62.5(10^{-4})\mathbf{k} = 3.882(10^{-4})\mathbf{v}_f$$

which gives

$$\mathbf{v}_f = 10.07\mathbf{i} + 8\mathbf{j} - 10.10\mathbf{k} \text{ ft/s}$$
$$= 16.35(0.616\mathbf{i} + 0.489\mathbf{j} - 0.618\mathbf{k}) \text{ ft/s} \qquad \text{Ans.}$$

The steps followed here are exactly the same ones that would be followed when using Newton's second law approach. Rewriting Newton's second law ($\Sigma \mathbf{F} = \mathbf{R} = m\mathbf{a}$) as

$$m\, d\mathbf{v} = \mathbf{R}\, dt$$

and integrating gives Eq. *a*. The principle of impulse and momentum offers little or no advantage in solving problems of this type where the force $\mathbf{R}$ is a simple function of time and the integral $\int \mathbf{R}\, dt$ is integrated analytically. The real power and usefulness of the principle of impulse and momentum is realized when the force $\mathbf{R}$ is not a simple function of time and the integral $\int \mathbf{R}\, dt$ is evaluated as an area in a graph.

EXAMPLE PROBLEM 19-2

A 10-kg box is resting on a horizontal surface as shown in Fig. 19-3*a* when a horizontal force $\mathbf{P}$ is applied to it. The magnitude of $\mathbf{P}$ varies with time as shown in Fig. 19-3*b*. If the static and kinetic coefficients of friction are 0.4 and 0.3, respectively, determine

a. The velocity of the box at $t = 10$ s.
b. The velocity of the box at $t = 15$ s.
c. The time t_f at which the box stops sliding.

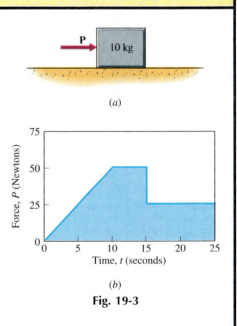

(a)

(b)

Fig. 19-3

SOLUTION

a. The free-body diagram of the box is shown in Fig. 19-4. Since there is no motion in the vertical direction, summing forces in the vertical direction gives

$$N = (10)(9.81) = 98.1 \text{ N}$$

   The friction force is then less than or equal to $0.4N$ until the box starts moving and is equal to $0.3N$ after the box starts moving. When the force $\mathbf{P}$ is initially applied to the box, the friction available is sufficiently great to keep the box from moving, and the friction force increases at the same rate as the force $\mathbf{P}$ (Fig. 19-5). When the force $\mathbf{P}$ reaches $0.4N = 39.24$ N (at $t = 7.848$ s), however, the friction available is no longer able to prevent the box from moving, the box begins to slide, and the friction force drops to $0.3N = 29.43$ N. The friction force then stays at 29.43 N until t_f, when the box again stops moving and the friction force drops to 25 N, which is the force required to hold the box in equilibrium.

Note that the *y*-component of the linear impulse-momentum equation could have been used to get the normal force N, although there is no advantage in using it.

The x-component of the linear impulse due to the force $\mathbf{P}$ acting on the box from $t = 0$ to $t = 10$ s is just the area under the curve

$$\int_0^{10} P \, dt = \frac{1}{2}(50)(10) = 250 \text{ N} \cdot \text{s}$$

and the x-component of the linear impulse due to the friction force acting on the box from $t = 0$ to $t = 10$ s is

$$\int_0^{10} F \, dt = \frac{1}{2}(-39.24)(7.848) + (-29.43)(10 - 7.848)$$
$$= -217.31 \text{ N} \cdot \text{s}$$

Then the x-component of the linear impulse–momentum equation (Eq. 19-1)

$$(m\mathbf{v})_i + \int_{t_i}^{t_f} \mathbf{R} \, dt = (m\mathbf{v})_f$$

gives

$$0 + (250 - 217.31) = 10v_{10}$$

or

$$v_{10} = 3.27 \text{ m/s} \qquad \text{Ans.}$$

b. From $t = 10$ s to $t = 15$ s the linear impulses are

$$\int_{10}^{15} P \, dt = (50)(15 - 10) = 250 \text{ N} \cdot \text{s}$$

$$\int_{10}^{15} F \, dt = (-29.43)(15 - 10) = -147.15 \text{ N} \cdot \text{s}$$

and the x-component of the linear impulse–momentum equation gives

$$(10)(3.27) + (250 - 147.15) = 10v_{15}$$

and

$$v_{15} = 13.55 \text{ m/s} \qquad \text{Ans.}$$

c. Between $t = 15$s and $t = t_f$ the linear impulses are

$$\int_{15}^{t_f} P \, dt = (25)(t_f - 15)$$

$$\int_{15}^{t_f} F \, dt = (-29.43)(t_f - 15)$$

and the x-component of the linear impulse–momentum equation gives

$$(10)(13.55) + (25 - 29.43)(t_f - 15) = 0$$

or

$$t_f = 45.6 \text{ s} \qquad \text{Ans.}$$

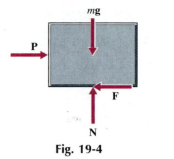

Fig. 19-4

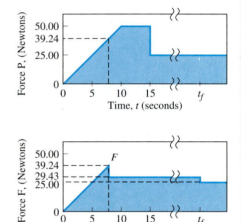

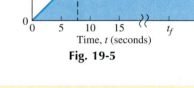

Fig. 19-5

Since the box starts from rest ($v_i = 0$), the linear impulse-momentum equation gives the velocity of the box at any time t as

$$0 + \int_0^t (P - F) \, dt = mv_t$$

For any t between 0 and 7.848 s, the pushing force P and the friction force F are equal so the box remains at rest ($v_t = 0$). For any t between 7.848 s and 15 s, the pushing force P is greater than the friction force F. Therefore, the integrand is positive, the impulse (that is, the integral) will get larger, and the velocity of the box will increase as t increases from 7.848 s to 15 s. For any t greater than 15 s, the integrand will be negative, the impulse will get smaller, and the velocity of the box will decrease as t increases past 15 s. Therefore, the maximum velocity of the box will occur when $t = 15$ s.

PROBLEMS

Introductory Problems

19-1* The speed of a 0.4-lb hockey puck sliding across the ice is observed to decrease from 60 ft/s to 40 ft/s in 3 s. Determine the average friction force acting on the puck and the corresponding kinetic coefficient of friction. (Assume that the ice surface is horizontal.)

19-2* A 1200-kg car is traveling at 75 km/h on an icy road when the driver suddenly applies the brakes. If the kinetic coefficient of friction is $\mu_k = 0.15$ and all four tires slide, determine the time it will take for the car to come to rest.

19-3 A 500-lb boat traveling at 20 mi/h comes to rest 10 s after the motor is shut off. Determine the average drag force of the water on the boat.

19-4* The speed of a toboggan sliding down a hill increases from 0 to 10 m/s in 6 s. The combined mass of the toboggan and riders is 110 kg, and the slope of the hill is 20°. Determine the average friction force between the toboggan and the snow and the corresponding coefficient of friction.

19-5 A 10-lb box slides down a ramp onto a horizontal floor as shown in Fig. P19-5. The speed of the box as it reaches the floor is 20 ft/s. If the kinetic coefficient of friction between the box and the floor is $\mu_k = 0.30$, determine the time required for the box to stop sliding.

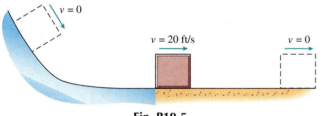

Fig. P19-5

19-6 The 10-kg block shown in Fig. P19-6 is at rest on a horizontal surface when the force $P = 3t^2$ is applied. If the static and kinetic coefficients of friction are $\mu_s = 0.30$ and $\mu_k = 0.25$, respectively, determine

a. The time at which the block begins to move.
b. The speed of the block when $t = 8$ s.

Fig. P19-6

19-7* The 100-lb cart shown in Fig. P19-7a is at rest when the force P is applied. The variation of P with time is shown in Fig. P19-7b. Determine the speed of the cart when $t = 6$ s

a. If the small weightless wheels are free to rotate.
b. If the wheels are locked and slide (the static and kinetic coefficients of friction are $\mu_s = 0.20$ and $\mu_k = 0.15$, respectively).

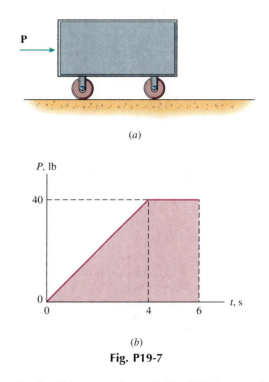

(a)

(b)

Fig. P19-7

19-8 As the 10-kg crate shown in Fig. P19-8 enters the incline, its speed is 10 m/s. If the kinetic coefficient of friction between the crate and the incline is $\mu_k = 0.35$, determine the time before the crate comes to rest.

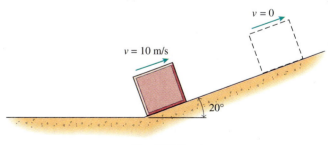

Fig. P19-8

19-9* The thrust of a 500-lb rocket sled varies with time as shown in Fig. P19-9. If the sled starts from rest and travels along a straight, horizontal track, determine its velocity when the rocket burns out. (Neglect friction.)

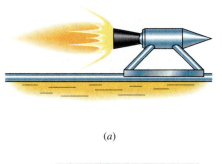

(a)

(b)

Fig. P19-9

Intermediate Problems

19-10* A 2.0-kg disk is sliding on a smooth, horizontal surface when it is acted on by a cross force **F** (Fig. P19-10a). The force makes a constant angle of $\theta = 50°$ with the initial direction of **v**, and its magnitude varies as shown in Fig. P19-10b. If the initial speed of the disk is $v = 10$ m/s, determine the magnitude and direction of the velocity of the disk when

a. $t = 5$ s.
b. $t = 10$ s.
c. $t = 15$ s.

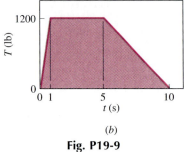

(a)   (b)

Fig. P19-10

19-11 The parachute and 400-lb crate shown in Fig. P19-11a are falling with a constant speed v. When the crate hits the ground, the parachute cords become slack (they exert no more force on the crate) and the ground exerts a force

on the crate that varies according to time as shown in Fig. P19-11b. Determine the maximum constant descent speed for which the maximum force exerted on the crate by the ground P_{max} does not exceed 2000 lb.

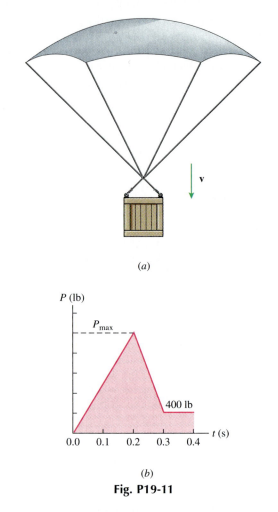

(a)

(b)

Fig. P19-11

19-12 A 60-g tennis ball has a horizontal speed of 10 m/s when it is struck by a tennis racket (Fig. P19-12). After the impact, the ball's velocity is 25 m/s (still horizontal) and makes an angle of 15° with the initial direction. If the time of contact is 0.05 s, determine the average force (magnitude and direction) of the tennis racket on the ball.

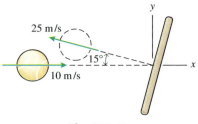

Fig. P19-12

19-13* A 5-oz baseball has a horizontal initial velocity of 90 ft/s just prior to being hit by a bat. After the impact, the ball's velocity is 110 ft/s at an angle of 30° above the horizontal (Fig. P19-13). If the time of the impact is 0.01 s, determine the average force (magnitude and direction) of the bat on the ball.

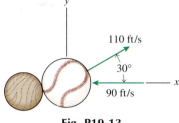

Fig. P19-13

19-14 A 0.4-kg disk is sliding on a smooth horizontal surface with an initial velocity of $v = 3i + 4j$ m/s when a force $F = 5i$ N is suddenly applied to the disk as shown in Fig. P19-14. Determine the velocity (magnitude and direction) of the disk at $t = 1.5$ s after the force is applied.

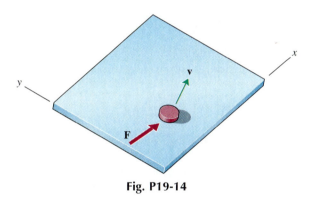

Fig. P19-14

Challenging Problems

19-15 A 0.5-lb ball has an initial velocity of $v_0 = 30i + 50k$ ft/s. If a constant force of $F = (-0.1i + 0.1j)$ lb acts on the ball, determine the velocity of the ball when the velocity is parallel to the yz-plane. (Gravity acts in the negative z-direction.)

19-16* A 10-kg box is resting on a horizontal surface when a horizontal force P is applied to it (Fig. P19-16a). The magnitude of P varies with time as shown in Fig. P19-16b. If the static and kinetic coefficients of friction are $\mu_s = 0.40$ and $\mu_k = 0.30$, respectively, determine

a. The time t_1 at which the box starts sliding.
b. The maximum velocity of the box v_{max} and the time t_m at which it occurs.
c. The time t_f at which the box stops sliding.

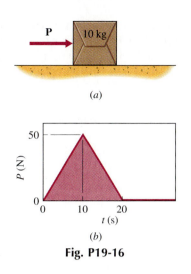

Fig. P19-16

19-17* A 5-lb box is resting on a horizontal surface when a force P is applied to it (Fig. P19-17a). The magnitude of P varies with time as shown in Fig. P19-17b. If the static and kinetic coefficients of friction are $\mu_s = 0.40$ and $\mu_k = 0.30$, respectively, determine

a. The time t_1 at which the box starts sliding.
b. The maximum velocity of the box v_{max} and the time t_m at which it occurs.
c. The time t_f at which the box stops sliding.

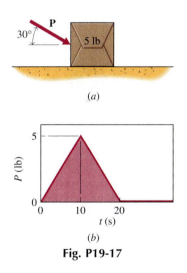

Fig. P19-17

19-18 A 10-kg box is resting on an inclined surface when a force **P** is applied to it (Fig. P19-18a). The magnitude of **P** varies with time as shown in Fig. P19-18b. If the static and kinetic coefficients of friction are $\mu_s = 0.60$ and $\mu_k = 0.40$, respectively, determine

a. The time t_1 at which the box starts sliding.
b. The maximum velocity of the box v_{max} and the time t_m at which it occurs.
c. The time t_f at which the box stops sliding.

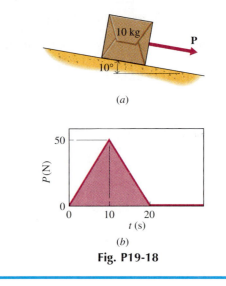

(a)

(b)

Fig. P19-18

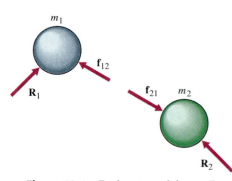

Figure 19-6 Both external forces **R** and internal forces **f** act on each particle of an interacting system. When the equations of impulse and momentum are added together for all of the particles in the system, the impulses of the internal forces will cancel in pairs.

19-3 INTERACTING SYSTEMS OF PARTICLES

When a problem involves two or more interacting particles as in Fig. 19-6, each particle may be considered separately and Eq. 19-1 may be written for each particle.

$$(\mathbf{L}_i)_1 + \int_{t_i}^{t_f} (\mathbf{R}_1 + \mathbf{f}_{12})\, dt = (\mathbf{L}_f)_1 \qquad (a)$$

$$(\mathbf{L}_i)_2 + \int_{t_i}^{t_f} (\mathbf{R}_2 + \mathbf{f}_{21})\, dt = (\mathbf{L}_f)_2 \qquad (b)$$

where $\mathbf{R}_1$ is the resultant of all external forces acting on particle 1, $\mathbf{f}_{12}$ is the force exerted on particle 1 by particle 2, and so on. Since the forces of action and reaction exerted by the particles on each other form pairs of equal and opposite forces ($\mathbf{f}_{21} = -\mathbf{f}_{12}$), and since the time interval from t_1 to t_2 is common to all the forces involved, the impulses of the forces of action and reaction always cancel out when these equations are added together. Therefore, the impulse–momentum equation for a pair of (or for N) interacting particles is

$$\sum_\ell (\mathbf{L}_i)_\ell + \sum_\ell \int_{t_i}^{t_f} \mathbf{R}_\ell\, dt = \sum_\ell (\mathbf{L}_f)_\ell \qquad (c)$$

where $(\sum \mathbf{L})_\ell = \sum (m\mathbf{v})_\ell$ is the vector sum of the momenta of both (or of all N) particles, $\sum (\int \mathbf{R}_\ell\, dt)$ is the vector sum of the impulses of all the external forces involved, and the internal forces need not be considered.

Therefore, for a system of N interacting particles: *The final momentum of the system of particles is the vector sum of their initial momenta and the sum of the impulses of the resultants of all external forces acting on the particles:*

$$\sum_\ell (\mathbf{L}_i)_\ell + \sum_\ell \int_{t_i}^{t_f} \mathbf{R}_\ell\, dt = \sum_\ell (\mathbf{L}_f)_\ell \qquad (19\text{-}3)$$

19-3.1 Motion of the Mass Center

The location of the mass center $\mathbf{r}_G$ of a system of N particles is calculated using the first moment of mass

$$m\mathbf{r}_G = \sum_{\ell=1}^{N} m_\ell \mathbf{r}_\ell \qquad (d)$$

where $m = \Sigma\, m_\ell$ is the total mass of all of the particles. Taking the time derivative of Eq. d and remembering that the mass of each particle is constant gives

$$m\mathbf{v}_G = \sum_{\ell=1}^{N} m_\ell \mathbf{v}_\ell \qquad (19\text{-}4)$$

That is, the total linear momentum of a system of particles is the same as if all the mass were concentrated in a single particle that was moving with the speed of the mass center of the system of particles. Using Eq. 19-4 allows Eq. 19-3 to be rewritten in the form

$$m(\mathbf{v}_G)_i + \sum_{\ell} \int_{t_i}^{t_f} \mathbf{R}_\ell\, dt = m(\mathbf{v}_G)_f \qquad (19\text{-}5)$$

19-3.2 Conservation of Linear Momentum for a System of Particles

If the sum of the impulses of all the external forces acting on the various particles is zero, the linear momentum of the system of particles is conserved

$$\sum_{\ell} (m\mathbf{v}_i)_\ell = \sum_{\ell} (m\mathbf{v}_f)_\ell \qquad (e)$$

or

$$(m\mathbf{v}_G)_i = (m\mathbf{v}_G)_f \qquad (f)$$

Dividing through by the total mass of the system of particles (which is constant) gives

$$\mathbf{v}_{Gi} = \mathbf{v}_{Gf} \qquad (g)$$

That is, *when the impulses of all the external forces acting on a system of particles is zero, the velocity* $\mathbf{v}_G$ *of the mass center of the particles is constant.* This situation occurs, for example, when two particles that are moving freely collide with one another. It should be noted, however, that although the total momentum of the colliding particles is conserved, their total energy is generally not conserved. Problems involving the collision or impact of two particles will be discussed in detail in Section 19-4.

19-3.3 Impulsive and Nonimpulsive Forces

When the linear impulse in a given direction is not zero but is known to be relatively small, it can frequently be neglected in order to obtain an approximate solution sufficiently accurate for many purposes. For example, if the linear momentum of the system composed of blocks A

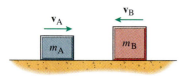

Figure 19-7 When two bodies collide, nonimpulsive forces can usually be neglected in the calculation of the change of momentum that occurs over the brief duration of the impact.

and B in Fig. 19-7 is large compared to the linear impulse of the frictional force, Eq. e is approximately true during the brief time Δt of impact even though friction exists between the plane and the blocks. Since the frictional forces cannot exceed μN, and the time of impact is small, the impulse of the friction on the blocks during the impact period would not change the large linear momentum of the blocks materially.

Forces that have very large magnitudes can produce a significant change in momentum even over very short time periods. Such forces are called *impulsive forces*. Motions that result from impulsive forces are called *impulsive motions*. The forces generated when one body strikes another is an example of an impulsive force.

Forces whose magnitudes are small compared to impulsive forces are called *nonimpulsive* forces. Examples are the weight of a body, friction, and spring forces. When the principle of impulse and momentum is applied over a short time interval, the impulse of nonimpulsive forces may be neglected compared to that of impulsive forces.

It is usually not known ahead of time whether unknown reaction forces are impulsive or not. Generally, *the reaction force of any support that acts to prevent motion in some direction is just as impulsive as the forces trying to cause motion in that direction.*

The final decision of whether or not the linear impulse of a force can be ignored must be based on the required accuracy of the result and on the estimated effect that the term has on the equation. If there is any doubt about whether the impulse of a force is important or not, it should be included in Eqs. 19-3 and 19-5.

19-3.4   Problems Involving Energy and Momentum

It must be remembered that the impulse–momentum principle is not an independent principle. Just like the work–energy principle, it is only a general, first integral of Newton's second law that is applicable for certain special situations. Any problem that can be solved by impulse–momentum methods or by work–energy methods can also be solved directly using Newton's second law. However, *when the principle of impulse–momentum or that of work–energy is suitable, these methods will often get the solution more quickly and with less labor.*

Not only are the work–energy and impulse–momentum methods not suitable for solving all problems, few real problems are specifically set up for either one. Generally, both methods will be used in different parts of the same problem. The maximum benefit is realized by choosing the particular method most suitable for a particular problem or part of a problem. In fact, it is often useful to combine all three methods— impulse–momentum, work–energy, and Newton's second law—to solve particular problems.

Many problems, such as Example Problem 19-5, involve several phases for which different principles are suitable. During the first phase, only conservative forces act and the work–energy equation is the easiest method to use to find the velocity of box A just prior to its impact with box B. During the impact phase, impulsive forces dominate, and the impulse–momentum principle is the most convenient method to use to find the velocities of the boxes just after the impact. During the final phase, relationships between forces, velocities, and position are again required and work–energy is the most convenient method to use. However, Newton's second law is used to find the normal force so that the work done by friction can be computed.

The vertical component of the linear impulse–momentum equation could also have been used to find the normal force, and it would have given the same result. Since there was no motion in the normal direction, however, the impulse–momentum method offered no advantage over the direct application of Newton's second law.

CONCEPTUAL EXAMPLE 19-1: IMPULSIVE FORCES

What is the difference between impulsive and nonimpulsive forces? How can one be sure that a force is nonimpulsive and can be neglected?

SOLUTION

Forces that are so large that they have a significant impulse

$$\text{Impulse} = \int_0^{\delta t} \mathbf{F}\, dt$$

even over very short time intervals ($\delta t \to 0$) are called impulsive forces. The forces of collision between two elastic bodies is a common example of impulsive forces. Since $m(\mathbf{v}_2 - \mathbf{v}_1) = \int \mathbf{F}\, dt$, impulsive forces give the appearance of an instantaneous change of velocity.

Forces whose impulse is small compared to that of impulsive forces are called nonimpulsive forces. Weight and friction are two examples of forces whose effect is generally negligible over very short intervals of time.

As an example of the difference between and the effect of impulsive and nonimpulsive forces, consider an experiment in which a small hammer is tapped on a hard surface. The 1-lb hammer is fitted with a force transducer between the hammer body and a plastic tip, and the weight of the handle of the hammer is negligible. When the hammer is tapped on a hard surface, the force transducer reports a force that rises rapidly to a peak force of F_p and falls just as rapidly back to zero. The area under the force–time graph is the impulse of the force and is generally between the area of the approximating triangle

$$\text{Area} = \frac{F_p T_d}{2}$$

and the area of a half sine wave

$$\text{Area} = \frac{2 F_p T_d}{\pi}$$

For one impact of the plastic end against a steel surface, the maximum force and impact duration were recorded as $F_p = 335$ lb and $T_d = 0.00044$ s, respectively. Approximating the area under the graph as midway between that of the approximating triangle [$\frac{1}{2}(335)(0.00044) = 0.074$ lb $\cdot$ s] and the half sine wave [$2(335)(0.00044)/\pi = 0.094$ lb $\cdot$ s] gives an impulse of about 0.085 lb $\cdot$ s. By comparison, the weight of the hammer ($W = 1$ lb) is constant, and during the 0.00044 s of the impact its impulse is only $W\,\delta t = 0.00044$ lb $\cdot$ s. The error introduced in this case by neglecting the impulse of the weight of the hammer is only about one-half of one percent.

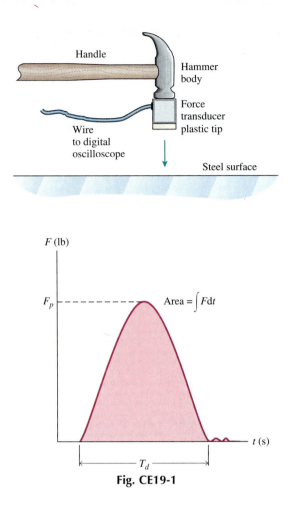

Area $= \int F dt$

Fig. CE19-1

If the velocity of the hammer immediately before the impact were 1.5 ft/s toward the surface, then the impulse–momentum principle gives

$$\frac{1}{32.2}(-1.5) + 0.085 = \frac{1}{32.2}v_f$$

and the velocity of the hammer immediately after the impact (just 0.00044 s later) would be 1.25 ft/s away from the surface. The coefficient of restitution of this impact would be (see Section 19-4)

$$e = -\frac{1.25 - 0}{(-1.5) - 0} \cong 0.8$$

When the same hammer was tapped against a soft surface such as a hand, the peak force and impact duration recorded were $F_p = 55$ lb and $T_d = 0.01$ s, respectively. Approximating the area under the graph as midway between that of the approximating triangle $[\frac{1}{2}(55)(0.01) = 0.275$ lb $\cdot$ s] and the half sine wave $[2(55)(0.01)/\pi = 0.350$ lb $\cdot$ s] gives an impulse of about 0.31 lb $\cdot$ s. This time, the impulse of the weight of the hammer during the 0.01 s of the impact is $W \delta t = 0.01$ lb $\cdot$ s. Although the error introduced in this case by neglecting the impulse of the weight of the hammer is about 3 percent, it is still no worse an approximation than that of the impulse of the impact force.

If there is doubt about whether a force is impulsive or not, either the force should be included in the calculation or the error introduced by not including the force should be estimated.

A 2500-lb car is initially at rest on the deck of a ferry, which is tied to a dock as shown in Fig. 19-8. The ferry weighs 25,000 lb.

a. If the car accelerates uniformly from rest to 20 mi/h in 4 s, determine the average tension in the cable during this time.

b. At $t = 4$ s, the cable between the ferry and the dock breaks, the driver of the car applies the brakes, and the car comes to a stop relative to the ferry. Neglect the friction between the ferry and the water, and determine the speed at which the boat will hit the dock.

Fig. 19-8

SOLUTION

a. The free-body diagram of the car and boat is shown in Fig. 19-9. The forces between the wheels of the car and the boat are internal forces; they are not needed in the linear impulse–momentum equation and are not shown on the diagram. The only force that has an impulse in the horizontal direction is the tension T, and the x-component of Eq. 19-3

$$(\Sigma \mathbf{L})_i + \Sigma \int \mathbf{R}\, dt = (\Sigma \mathbf{L})_f$$

is

$$0 + T_{avg}\,(4) = \frac{2500}{32.2}\,(29.333)$$

where $v_c = 20$ mi/h $= 29.333$ ft/s. Therefore,

$$T_{avg} = 569 \text{ lb} \qquad \text{Ans.}$$

b. After the cable to the dock breaks, there are no forces in the x-direction, so momentum is conserved in the x-direction. Therefore, Eq. 19-3 gives

$$\frac{2500}{32.2}\,29.333 = \frac{27,500}{32.2}\,v_f$$

and

$$v_f = 2.667 \text{ ft/s} = 1.818 \text{ mi/h} \qquad \text{Ans.}$$

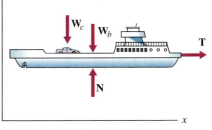

Fig. 19-9

The average force is the constant force which would give the same impulse. That is,

$$\text{Impulse} = \int T\, dt = \int T_{avg}\, dt$$

Then, since the average force is constant (by definition), it can be taken outside of the integral and

$$\text{Impulse} = T_{avg} \int dt = \\ T_{avg}(t_f - t_i) = \\ T_{avg}(4)$$

A 3-kg cannonball is fired with an initial velocity of $v_0 = 150$ m/s and $\theta_0 = 60°$ as shown in Fig. 19-10. At the peak of its trajectory, the ball explodes and splits into two pieces. The 1-kg piece hits the ground at $x = 500$ m, and $y = 2500$ m when $t = 35$ s.

a. Determine when and where the 2-kg piece hits the ground.
b. Determine the average magnitude of the explosive force F_{avg} if the duration of the explosion is $\Delta t = 0.005$ s.

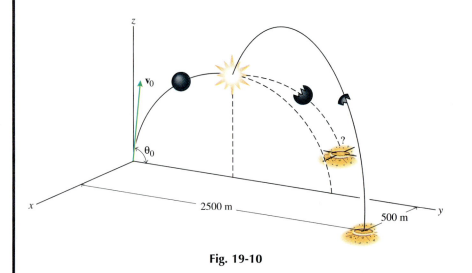

Fig. 19-10

SOLUTION

a. The only force acting on the system of two particles is gravity, so the motion of the mass center is given by

$$\mathbf{a}_G = -9.81\mathbf{k} \text{ m/s}^2$$
$$\mathbf{v}_G = 75\mathbf{j} + (129.9 - 9.81t)\mathbf{k} \text{ m/s}$$
$$\mathbf{r}_G = 75t\mathbf{j} + (129.9t - 4.905t^2)\mathbf{k} \text{ m}$$

The peak of the trajectory corresponds to the time when $dz_G/dt = 129.9 - 9.81t = 0$, which gives $t = 13.24$ s. Then

$$x = 0 \text{ m} \qquad y = 993.2 \text{ m} \qquad z = 860.1 \text{ m}$$

at the instant of the explosion.

   After the explosion, the 1-kg mass is acted on only by gravity, so its motion is given by

$$\mathbf{a}_1 = -9.81\mathbf{k} \text{ m/s}^2$$
$$\mathbf{v}_1 = v_{10x}\mathbf{i} + v_{10y}\mathbf{j} + [v_{10z} - 9.81(t - 13.24)]\mathbf{k} \text{ m/s}$$
$$x_1 = v_{10x}(t - 13.24) \text{ m}$$
$$y_1 = 993.2 + v_{10y}(t - 13.24) \text{ m}$$
$$z_1 = 860.1 + v_{10z}(t - 13.24) - 4.905(t - 13.24)^2 \text{ m}$$

where the constants of integration ($x_{10} = 0$ m, $y_{10} = 993.2$ m, and $z_{10} = 860.1$ m) were chosen to match the known position immediately after the explosion. The remaining constants of integration are determined by using the known impact time and position.

$$v_{10x} = \frac{500}{(35 - 13.24)} = 22.98 \text{ m/s}$$

$$v_{10y} = \frac{2500 - 993.2}{35 - 13.24} = 69.26 \text{ m/s}$$

$$v_{10z} = \frac{-860.1 + 4.905(35 - 13.24)^2}{35 - 13.24} = 67.21 \text{ m/s}$$

During the explosion, the only external force that has an impulse is gravity, so Eq. 19-3

$$(\Sigma \mathbf{L})_i + \Sigma \int \mathbf{R} \, dt = (\Sigma \mathbf{L})_f$$

gives

$$(3)(75\mathbf{j}) + (3)(-9.81\mathbf{k})(0.005) = (1)(22.98\mathbf{i} + 69.26\mathbf{j} + 67.21\mathbf{k}) + (2)\mathbf{v}_{20} \quad (a)$$

and therefore

$$\mathbf{v}_{20} = -11.49\mathbf{i} + 77.87\mathbf{j} - 33.68\mathbf{k} \text{ m/s}$$

Finally, the only force acting on the 2-kg particle after the explosion is gravity, so its motion is given by

$$\mathbf{a}_2 = -9.81\mathbf{k} \text{ m/s}^2$$
$$\mathbf{v}_2 = -11.49\mathbf{i} + 77.87\mathbf{j} - [33.68 + 9.81(t - 13.24)]\mathbf{k} \text{ m/s}$$

and

$$x_2 = -11.49(t - 13.24) \text{ m}$$
$$y_2 = 993.2 + 77.87(t - 13.24) \text{ m}$$
$$z_2 = 860.1 - 33.68(t - 13.24) - 4.905(t - 13.24)^2 \text{ m}$$

This particle hits the ground when $z_2 = 0$, which gives

$$t = 23.49 \text{ s} \qquad \text{Ans.}$$
$$x_2 = -117.7 \text{ m} \qquad \text{Ans.}$$
$$y_2 = 1791 \text{ m} \qquad \text{Ans.}$$

b. Equation 19-3 can also be applied to each of the pieces individually over the duration of the impact. For the 1-kg piece, Eq. 19-3 gives

$$(1)(75\mathbf{j}) + \mathbf{F}(0.005) = (1)[22.98\mathbf{i} + 69.26\mathbf{j} + 67.21\mathbf{k}] \quad (b)$$

Therefore, the average force exerted on the 1-kg piece by the explosion is

$$\mathbf{F}_{avg} = 4596\mathbf{i} - 1148\mathbf{j} + 13{,}442\mathbf{k} \text{ N}$$

and a force of equal magnitude but opposite direction is exerted on the 2-kg piece. The average magnitude of the explosive force exerted on each piece is then

$$F_{avg} = 14{,}250 \text{ N} \qquad \text{Ans.}$$

Note that the effect of the impulse of gravity in Eq. *a* is to change the *z*-component of $\mathbf{v}_{20}$ by only 0.074 m/s or about 0.2 percent. The weight force ($W = 29.43$ N) is very small compared to the explosive force ($F_{avg} = 14{,}250$ N). That is, gravity is not an impulsive force, and it would usually be left out of Eq. *a*.

Recognizing that the weight force is non-impulsive and has very little effect, it has been left out of Eq. *b*. Its impulse is only (1)(9.81)(0.005) = 0.0491 N · s and, if included, it would decrease the *z*-component of $\mathbf{F}_{avg}$ by only 9.81 N or about 0.07 percent.

A 20-lb box A slides down a frictionless ramp and strikes a 10-lb box B (Fig. 19-11). As a result of the impact, the two boxes become hooked together and slide as a single unit on the rough surface ($\mu_k = 0.6$). Determine

a.   The velocity of the boxes immediately after impact.
b.   The distance d that the boxes will slide before coming to rest.

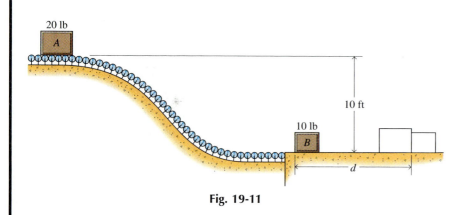

20 lb

10 ft

10 lb

d

Fig. 19-11

SOLUTION

a.   **Work–Energy from 1→2:** The free-body diagram of box A as it slides down the ramp is shown in Fig. 19-12a on the next page. The normal force **N** acts perpendicular to the motion and so does no work. The weight force has a potential, so the work–energy equation (Eq. 17-23)

$$T_1 + V_1 + U^{(o)}_{1\to2} = T_2 + V_2$$

gives

$$0 + (20)(10) + 0 = \frac{1}{2}\left(\frac{20}{32.2}\right)v^2_{A2} + 0$$

and

$$v_{A2} = 25.38 \text{ ft/s}$$

Linear Impulse–Momentum from 2→3: Box B is initially at rest ($v_{B2} = 0$), and after the impact both boxes have the same velocity ($v_{A3} = v_{B3} = v_3$). Over the brief duration of the impact there are no external impulsive forces in the x-direction, so the x-component of linear momentum is conserved:

$$\Sigma(mv_x)_i = \Sigma\,(mv_x)_f$$

$$\left(\frac{20}{g}\right)(25.38) + \left(\frac{10}{g}\right)(0) = \left(\frac{20}{g}\right)v_3 + \left(\frac{10}{g}\right)v_3$$

which gives

$$v_3 = 16.92 \text{ ft/s} \qquad\qquad \text{Ans.}$$

Many problems consist of separate and distinct phases. In each phase, the method which is best suited to that phase should be used to relate the information given and the results desired. For example, the first phase of this problem (as box A slides down the ramp) requires a relationship between the position of the box, the speed of the box, and the forces acting on the box. This type of problem is most easily solved using the principle of work and energy. The second phase, which consists of the collision of the two boxes, is most easily solved using the principle of impulse and momentum. The final phase again requires a relationship between the position of the boxes, the speed of the boxes, and the forces acting on the boxes, and it is solved using the principle of work and energy. During this last phase, Newton's second law is also needed to determine the normal and frictional forces acting on the boxes.

b. **Newton's Second Law:** The free-body diagram of the pair of boxes following the impact is shown in Fig. 19-12b. During this phase of the motion, the boxes slide along a straight horizontal line and there is no acceleration in the vertical direction. Therefore,

$$\uparrow \Sigma F = N - 30 = 0$$

and $N = 30$ lb = constant. Then the friction force is

$$F = 0.6(30) = 18 \text{ lb} = \text{constant}$$

Work–Energy from 3→4: The free-body diagram of Fig. 19-12b still applies. Both the normal force and the weight force act perpendicular to the motion and therefore do no work. The work done by the constant friction force is computed

$$U^{(o)}_{1 \to 2} = Fd = -18d$$

so the work–energy equation (Eq. 17-23)

$$T_1 + V_1 + U^{(o)}_{1 \to 2} = T_2 + V_2$$

gives

$$\frac{1}{2}\left(\frac{30}{32.2}\right)(16.92)^2 + 0 - 18d = 0 + 0$$

and

$$d = 7.41 \text{ ft} \qquad\qquad \text{Ans.}$$

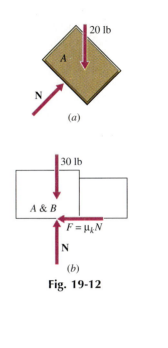

(a)

(b)

Fig. 19-12

PROBLEMS

Introductory Problems

19-19* At some instant of time, the mass center of three particles weighing $W_3 = 3$ lb, $W_5 = 5$ lb, and $W_1 = 1$ lb, is located at $\mathbf{r}_G = 8\mathbf{i} + 5\mathbf{j} + 3\mathbf{k}$ ft and has a velocity given by $\mathbf{v}_G = 5\mathbf{i} - 12\mathbf{k}$ ft/s. At this same instant, the position of the 3-lb particle is $\mathbf{r}_3 = 5\mathbf{i}$ ft and its velocity is $\mathbf{v}_3 = 3\mathbf{i} + 8\mathbf{k}$ ft/s, and the position of the 1-kg particle is $\mathbf{r}_1 = 8\mathbf{j} + 3\mathbf{k}$ ft and its velocity is $\mathbf{v}_1 = 3\mathbf{i} - 3\mathbf{j}$ ft/s. Determine the position and velocity of the 5-lb particle at this instant.

19-20* At some instant of time, the position and velocity of three particles are given by

	Particle		
	1	2	3
m, kg	1	2	3
x, m	3	8	5
y, m	4	3	7
v_x, m/s	10	0	-2
v_y, m/s	-5	5	3

Find the location and velocity of the mass center at this instant.

19-21 An 800-lb cannon fires a 5-lb cannonball in a horizontal direction as shown in Fig. P19-21. The cannon sits on a skid; the kinetic coefficient of friction between the skid and the horizontal surface is $\mu_k = 0.25$. If the cannonball has a speed of 650 ft/s when it leaves the cannon, determine

a. The recoil velocity of the cannon.
b. The distance that the cannon will slide before coming to rest.

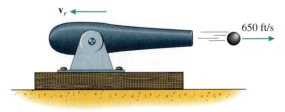

Fig. P19-21

19-22 A 0.05-kg bullet is moving to the right with a speed of 450 m/s when it strikes a 20 kg box of sand that is also moving to right with a speed of 5 m/s (Fig. P19-22). The kinetic coefficient of friction between the box and the horizontal surface is $\mu_k = 0.30$. Determine

a. The speed of the bullet–box unit when the bullet becomes embedded in the box of sand.
b. The distance that the unit moves before coming to rest.

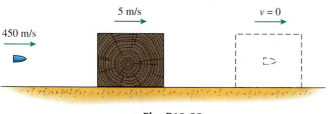

Fig. P19-22

19-23* Two disks of putty slide on a frictionless horizontal surface and collide as shown in Fig. P19-23. Disk A weighs 1 lb and has a velocity $\mathbf{v}_A = 15\mathbf{i}$ ft/s; disk B weighs 2 lb and has a velocity $\mathbf{v}_B = 10\mathbf{i} + 24\mathbf{j}$ ft/s. After the impact the disks move as a single unit with a velocity $\mathbf{v}$ oriented at an angle θ relative to the horizontal axis. Determine the magnitude of $\mathbf{v}$ and the angle θ.

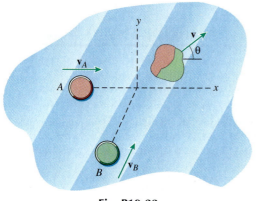

Fig. P19-23

19-24 A 25-kg sled is sliding along a flat, horizontal, frictionless surface with a speed of 5 m/s when a 60-kg man jumps onto it. If the initial speed of the man was 2 m/s at right angles to the motion of the sled, determine the final (common) velocity of the sled and man.

Intermediate Problems

19-25 A 4-lb box is released from rest at position A of the smooth, circular ramp, which has a radius of 20 ft (Fig. P19-25). When the box leaves the ramp at B, it lands on the 40-lb cart, which is initially at rest, and they roll to the right as a single unit. Determine the (common) speed of the cart and box when they move as a single unit.

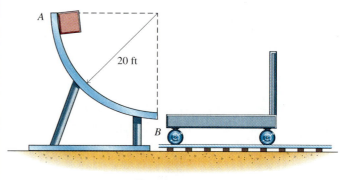

Fig. P19-25

19-26* Two cars collide at an intersection as shown in Fig. P19-26. Car A has a mass of 1000 kg and an initial speed of $v_A = 25$ km/h, whereas car B has a mass of 1500 kg. If the cars become entangled and move as a single unit at an angle of $\theta = 30°$ after the collision, determine the speed v_B of car B just prior to the collision.

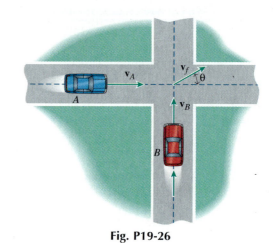

Fig. P19-26

19-27* A 2-lb particle is sliding along a flat, horizontal, frictionless surface at $v_i = 10$ ft/s as shown in Fig. P19-27. When the particle is 20 ft from the wall, it explodes and splits into two equal pieces. One piece hits the wall at $y_A = 5$ ft while the other piece hits the wall at $y_B = 10$ ft. Determine

a. The impulse exerted on particle A by the explosion.
b. The velocity $\mathbf{v}_{A/B}$ of particle A relative to particle B immediately after the explosion.
c. The time difference between when particle A hits the wall and when particle B hits the wall.

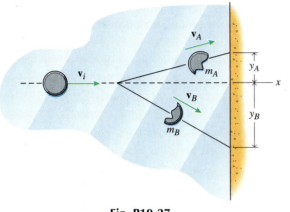

Fig. P19-27

19-28 A 5-kg particle is sliding along a flat, horizontal, frictionless surface at $v_i = 10$ m/s as shown in Fig. P19-27. When the particle is 20 m from the wall, it explodes and splits into two equal pieces. One piece m_A hits the wall at $y_A = 5$ m, whereas the other piece m_B hits the wall at $y_B = 10$ m. If both particles hit the wall at the same time, determine

a. The size of the two particles m_A and m_B.
b. The impulse exerted on particle A by the explosion.
c. The velocity $\mathbf{v}_{A/B}$ of particle A relative to particle B immediately after the explosion.

19-29 A 25-lb box A slides down a ramp ($\theta = 25°$) and strikes a 10-lb box B, which is attached to a spring of stiffness $k = 10$ lb/in. (Fig. P19-29). On impact, the two boxes become hooked and slide as a single unit on the rough ($\mu_k = 0.4$) surface. If box A starts from rest with $d = 20$ ft, determine

a. The velocity of the boxes immediately after the impact.
b. The maximum compression in the spring during the resulting motion.
c. The acceleration of the boxes at the instant of maximum compression.

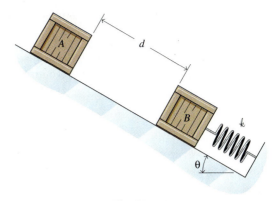

Fig. P19-29

19-30* A 0.30-kg block of wood is attached to a spring with $k = 7500$ N/m (Fig. P19-30). The block is at rest on a rough ($\mu_k = 0.4$) horizontal surface when it is struck by a 0.030-kg bullet with an initial speed $v_i = 150$ m/s. On impact, the bullet becomes embedded in the wood. Determine

a. The speed of the block and the bullet immediately after the impact.
b. The distance that the block will slide before coming to rest again.

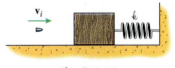

Fig. P19-30

Challenging Problems

19-31 The 180-lb man and the 1500-lb flatcar shown in Fig. P19-31 are initially at rest. At $t = 0$ the man starts running with a constant speed of 20 ft/s (relative to the flatcar) from B to A. Determine

a. The speed of the flatcar when the man runs off the end at A.
b. The absolute distance (the distance relative to the ground) traveled by the man as he runs from B to A.

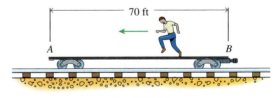

Fig. P19-31

19-32* A 5-kg cannonball is fired with an initial velocity of $v_0 = 125$ m/s and $\theta_0 = 75°$ in the yz-plane as shown in Fig. P19-32. At the peak of its trajectory, the ball explodes and splits into two pieces. A 2-kg piece hits the ground at $x = 50$ m and $y = 350$ m when $t = 25$ s. Determine

a. When and where the 3-kg piece hits the ground.
b. The impulse exerted on the 2-kg piece by the explosion.
c. The average magnitude of the explosive force F_{avg} exerted on the 2-kg piece by the explosion if the duration of the explosion is $\Delta t = 0.003$ s.

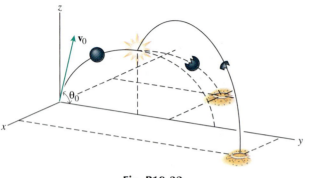

Fig. P19-32

19-33* A 10-lb cannonball is fired with an initial velocity of $v_0 = 450$ ft/s and $\theta_0 = 50°$ in the yz-plane as shown in Fig. P19-32. When $t = 5$ s, the ball explodes and splits into two pieces. A 6-lb piece hits the ground at $x = 1000$ ft and $y = 7000$ ft when $t = 25$ s. Determine

a. When and where the 4-lb piece hits the ground.
b. The impulse exerted on the 6-lb piece by the explosion.
c. The average magnitude of the explosive force F_{avg} exerted on the 6-lb piece by the explosion if the duration of the explosion is $\Delta t = 0.001$ s.

19-34 A 15-kg block of wood is attached to a spring with $k = 4500$ N/m (Fig. P19-34). The block is at rest on a rough ($\mu_k = 0.3$) horizontal surface with the spring undeformed when it is struck by a 0.03-kg bullet. On impact, the bullet becomes embedded in the wood. Determine the maximum initial speed v_i of the bullet for which the spring does not rebound.

Fig. P19-34

Computer Problems

C19-35 Two cars collide at an intersection as shown in Fig. P19-35. Car A weighs 2200 lb and has an initial speed of $v_A = 30$ mi/h, whereas car B weighs 3500 lb. On impact, the wheels of both cars lock and both cars slide ($\mu_k = 0.3$). After impact, the cars become entangled and move as a single unit. Calculate and plot

a. The (common) speed v_f of the two cars immediately after the collision as a function of the initial speed of car B for 10 mi/h $< v_B <$ 50 mi/h.
b. The location (y versus x) where the cars will come to rest for 10 mi/h $< v_B <$ 50 mi/h. (Let the x-axis coincide with the initial path of car A and the y-axis coincide with the initial path of car B.)

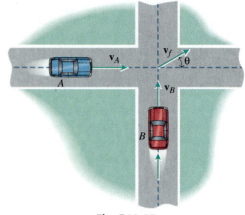

Fig. P19-35

C19-36 A ballistic pendulum consists of a 3-kg box of sand suspended from a light, 1.5-m-long cord (Fig. P19-36). Plot a calibration curve for a 30-g bullet striking the pendulum. That is, if a 30-g bullet strikes the box and becomes embedded in the sand, calculate and plot the initial speed of the bullet v_i as a function of θ, the maximum angle through which the pendulum swings after the impact, for $5° < \theta < 75°$.

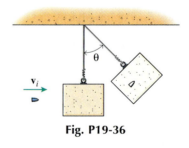

Fig. P19-36

19-4   COLLISION OF ELASTIC BODIES

An *impact* (collision between two bodies) is an event that usually occurs in a very brief interval of time. It is usually accompanied by relatively large reaction forces between the two bodies and correspondingly large changes of velocity of one or both bodies. The large reaction forces also result in considerable deformation of the impacting bodies and the consequent conversion of mechanical energy into sound and heat.

Impact events are categorized according to the relative location of the mass centers of the bodies, the relative velocity of the mass centers, and the *line of impact:* the straight line normal to the contacting surfaces at the point of impact (Fig. 19-13). If the mass centers of both bodies are on the line of impact, the impact is called a *central impact* (Fig. 19-14a–d). When the mass center of one or both bodies does not lie on the line of impact, the impact is called an *eccentric impact* (Fig. 19-14e,f). Obviously, only central impact can occur between particles because the size and shape of particles is not supposed to affect the calculation of their motion.

Further categorization is based on the orientation of the velocities of the bodies relative to the line of impact. A collision for which the initial velocities of the impacting bodies are along the line of impact is called a *direct impact* (Fig. 19-14a,b,e). A direct impact is a head-on collision. A collision for which the initial velocities of the impacting bodies are not along the line of impact is called an *oblique impact* (Fig. 19-14c,d,f).

The impact of two bodies consists of two phases—a deformation or compression phase followed by a restoration or restitution phase—and is accompanied by the generation of heat and sound. During the first phase, which lasts from the instant of contact to the instant of max-

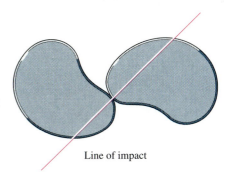

Line of impact

Figure 19-13   The *line of impact* is the straight line normal to the contacting surface at the point of impact.

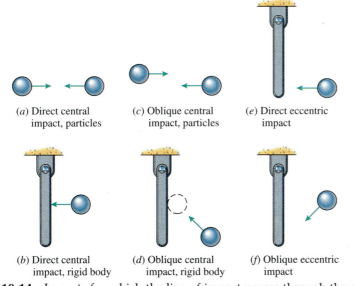

(*a*) Direct central impact, particles

(*c*) Oblique central impact, particles

(*e*) Direct eccentric impact

(*b*) Direct central impact, rigid body

(*d*) Oblique central impact, rigid body

(*f*) Oblique eccentric impact

Figure 19-14   Impacts for which the line of impact passes through the mass centers of both bodies are called *central impacts*. Impacts for which the initial velocities of both bodies are along the line of impact are called *direct impacts*.

imum deformation, the two bodies are compressed by the large inter-action force. At the end of this phase, the bodies are neither coming closer together nor moving apart: The relative velocity along the line of impact is zero. In the second phase, which lasts from the instant of maximum deformation until the instant of separation, the bodies move apart as the forces within the bodies act to restore the bodies to their original shapes. Generally, not all of the deformation is recoverable, however. Because of the permanent deformation of the bodies and the sound vibrations that are produced, some of the initial mechanical energy is dissipated in the collision.

Considerable study has been done on the relationship between the forces of impact and the resulting deformation when two bodies collide. The deformation of the bodies is found to depend on the rate of deformation as well as on the temperature and the material that the bodies are made of. Fortunately, the details of the impact can often be avoided; the linear impulse–momentum equation can be used to give a simple relation between the relative velocities of the bodies before and after the impact.

It must be remembered, however, that the impact of two bodies is a complex event. Although the simple analysis that follows allows the solution of many impact problems that could not otherwise be solved, always keep in mind that the results of these impact calculations are only approximate.

19-4.1   Direct Central Impact

Consider the motion of two particles A and B along a common line (the line of impact) as in Fig. 19-15a. It is assumed that the speed v_{Ai} is greater than the speed v_{Bi} so that particle A will eventually overtake and collide with particle B. Also, in accordance with general observations, it is commonly assumed that during the brief impact interval, $\Delta t = t_f - t_i$:

1.   The velocity of one or both particles may change greatly.
2.   The positions of the particles do not change significantly.
3.   Nonimpulsive forces may be neglected.
4.   Friction forces between the two bodies are negligible.

Furthermore, since all motion and forces act along the line of impact, only the component of the impulse–momentum equation along the line of impact need be considered. The positive direction along this line will be taken to the right: forces and velocities to the right will be positive; forces and velocities to the left, negative.

During the collision, there are no impulsive, external forces acting on the pair of particles and thus linear momentum is conserved for the pair of particles

$$m_A v_{Ai} + m_B v_{Bi} = m_A v_{Af} + m_B v_{Bf} \qquad (19\text{-}6)$$

On the other hand, if the particles are examined individually, the internal force is impulsive and must be included in the linear impulse–momentum equation. During the deformation phase of the impact (Fig. 19-15b), the linear impulse–momentum equation gives

$$m_A v_{Ai} - \int_{t_i}^{t_c} F_d \, dt = m_A v_c \qquad (a)$$

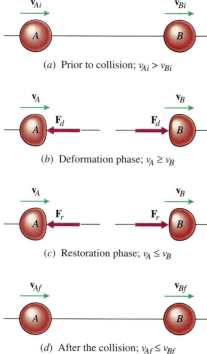

(a) Prior to collision; $v_{Ai} > v_{Bi}$

(b) Deformation phase; $v_A \geq v_B$

(c) Restoration phase; $v_A \leq v_B$

(d) After the collision; $v_{Af} \leq v_{Bf}$

Figure 19-15   The impact event can be divided into two separate phases. The *deformation phase*, which lasts from the moment of first contact until the two particles are moving together with a common speed, and the *restoration phase*, which lasts from the end of the deformation phase until the particles separate.

for particle A, where F_d is the interaction force on particle A during the deformation phase, v_c is the common velocity of the two particles at the end of the deformation phase, and t_c is the time at the end of the deformation phase. During the restoration phase (Fig. 19-15c), the linear impulse–momentum equation gives

$$m_A v_c - \int_{t_c}^{t_f} F_r \, dt = m_A v_{Af} \qquad (b)$$

for particle A, where F_r is the interaction force on particle A during the restoration phase and v_{Af} is the final velocity of particle A—the velocity of A after the collision is over.

The magnitude of the deformation impulse $\int_{t_i}^{t_c} F_d \, dt$ is generally larger than the magnitude of the restoration impulse $\int_{t_c}^{t_f} F_r \, dt$. The *coefficient of restitution e* is defined as the ratio of these two impulses:

$$e = \frac{\int_{t_c}^{t_f} F_r \, dt}{\int_{t_i}^{t_c} F_d \, dt} = \frac{m_A v_c - m_A v_{Af}}{m_A v_{Ai} - m_A v_c} = \frac{v_c - v_{Af}}{v_{Ai} - v_c} \qquad (c)$$

A similar analysis of particle B alone gives a similar equation for the coefficient of restitution:

$$e = \frac{\int_{t_c}^{t_f} F_r \, dt}{\int_{t_i}^{t_c} F_d \, dt} = \frac{m_B v_{Bf} - m_B v_c}{m_B v_c - m_B v_{Bi}} = \frac{v_{Bf} - v_c}{v_c - v_{Bi}} \qquad (d)$$

Eliminating the unknown velocity v_c between these two equations gives

$$e = -\frac{v_{Bf} - v_{Af}}{v_{Bi} - v_{Ai}} = -\frac{(v_{B/A})_f}{(v_{B/A})_i} \qquad (19\text{-}7)$$

That is, the coefficient of restitution e is the negative of the ratio of the relative velocity of the two particles after impact and the relative velocity of the two particles before impact. This ratio is a measure of the elastic properties of the particles and must be measured experimentally.

Equations 19-6 and 19-7 together give two equations for determining the velocities of the two bodies following the impact given the two velocities before the impact.

An impact for which $e = 1$ is called a *perfectly elastic impact*. In this case, Eq. 19-7 gives

$$v_{Ai} + v_{Af} = v_{Bi} + v_{Bf} \qquad (e)$$

But Eq. 19-6 can be rewritten

$$m_A(v_{Ai} - v_{Af}) = m_B(v_{Bf} - v_{Bi}) \qquad (f)$$

and multiplying these two equations together gives

$$m_A(v_{Ai}^2 - v_{Af}^2) = m_B(v_{Bf}^2 - v_{Bi}^2) \qquad (g)$$

Finally, rearranging and dividing by 2 gives

$$\frac{1}{2}m_A v_{Ai}^2 + \frac{1}{2} m_B v_{Bi}^2 = \frac{1}{2}m_A v_{Af}^2 + \frac{1}{2} m_B v_{Bf}^2 \qquad (h)$$

Therefore, for a perfectly elastic impact, $e = 1$, the kinetic energy of the pair of particles is conserved.

On the other hand, an impact for which $e = 0$ is called a *perfectly plastic impact*. In this case, the relative velocity of the two particles after impact is zero and the two particles move together at the same speed. This corresponds to the maximum loss of kinetic energy for the impact. The coefficient of restitution for all real objects always lies between these two limiting cases, $0 \le e < 1$.

The coefficient of restitution is not a property of a single body or a single type of material. Rather, it depends on the types of material of both colliding bodies. Although it is frequently considered a constant, the coefficient of restitution actually varies considerably with the impact velocity as well as with the sizes, shapes, and temperatures of the colliding bodies. Though handbook values for e may be used in the absence of better data, they are generally unreliable.

19-4.2   Oblique Central Impact

The analysis of the oblique central impact of two particles is a simple extension of the analysis of direct central impact. For the case of perfectly smooth and frictionless particles, oblique central impact will be seen to be merely the superposition of a uniform motion in the direction perpendicular to the line of impact and a direct central impact along the line of impact.

Coordinate axes are chosen along (n-axis) and perpendicular (t-axis) to the line of impact as shown in Fig. 19-16. The impulse–momentum equation applies equally to the combined system of particles (Fig. 19-16) and to each particle individually (Fig. 19-17). During the brief period of the impact, the only impulsive forces that act on either particle are the internal forces that act along the line of impact (n-axis).

Since there is no impulsive force on either particle in the t-direction, the t-component of linear momentum is conserved for each particle individually:

$$m_A(v_{Ai})_t = m_A(v_{Af})_t \qquad (i)$$
$$m_B(v_{Bi})_t = m_B(v_{Bf})_t \qquad (j)$$

Therefore, the t-components of the particles' velocities (the components perpendicular to the line of impact) are unchanged by the impact:

$$(v_{Ai})_t = (v_{Af})_t \qquad (19\text{-}8a)$$
$$(v_{Bi})_t = (v_{Bf})_t \qquad (19\text{-}8b)$$

For the system consisting of the pair of particles, there are no impulsive, external forces in any direction, and hence linear momentum is conserved in both the n- and t-directions:

$$m_A(v_{Ai})_t + m_B(v_{Bi})_t = m_A(v_{Af})_t + m_B(v_{Bf})_t \qquad (19\text{-}9a)$$
$$m_A(v_{Ai})_n + m_B(v_{Bi})_n = m_A(v_{Af})_n + m_B(v_{Bf})_n \qquad (19\text{-}9b)$$

Equation 19-9a is the sum of Eqs. i and j and contributes no additional information to the problem. Therefore, still another equation is needed

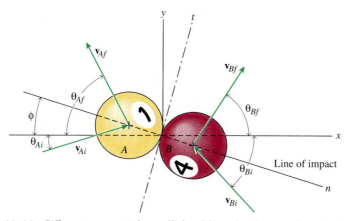

Figure 19-16   When two particles collide obliquely, the coefficient of restitution only relates the components of the motion along the line of impact.

to solve for the remaining two unknowns: $(v_{Af})_n$ and $(v_{Bf})_n$. Repeating the analysis of Section 19-4.1 for the n-direction gives

$$e = -\frac{(v_{Bf})_n - (v_{Af})_n}{(v_{Bi})_n - (v_{Ai})_n} = -\frac{(v_{B/A})_{fn}}{(v_{B/A})_{in}}$$ **(19-9c)**

where $(v_{B/A})_{fn}$ and $(v_{B/A})_{in}$ are the n-components of the final and initial velocity of B relative to A, respectively. Equation 19-9b and 19-9c give the final two equations necessary to find the normal components of velocity $(v_{Af})_n$ and $(v_{Bf})_n$.

Once the four final velocity components $(v_{Af})_t$, $(v_{Af})_n$, $(v_{Bf})_t$, and $(v_{Bf})_n$ are found, the magnitudes and directions of the final velocities $\mathbf{v}_{Af}$ and $\mathbf{v}_{Bf}$ may be easily determined. Though the n- and t-axes are required to solve for the velocity components, these directions seldom have significance in the overall problem being solved. Therefore, the final results should be reported relative to standard horizontal and vertical axes or relative to the initial directions of $\mathbf{v}_A$ and $\mathbf{v}_B$ and not relative to the n- and t-axes.

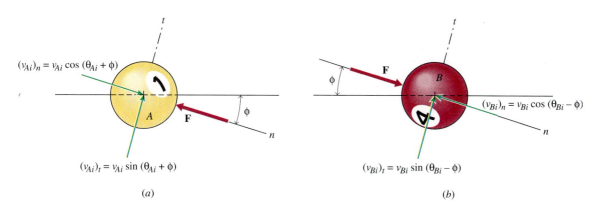

Figure 19-17   When the impulse–momentum equation is applied to the particles individually, the impact forces must be included.

19-4.3 Constrained Impact

In the foregoing analysis it was assumed that both particles moved freely except for the impact. That is, no external impulsive forces acted on either particle. When one or both of the colliding particles are constrained against motion in some direction as in Example Problem 19-8, the constraint force is likely to be just as impulsive as the internal force. Therefore, linear momentum for the system consisting of the pair of particles may not be conserved in either the n- or t-directions. Also, for the constrained particle, linear momentum in the t-direction will probably not be conserved, and the t-component of its velocity will likely change through the impact.

The equation for the coefficient of restitution, Eq. 19-9c, may still be used to relate the relative velocities along the line of impact. However, the other equations must be replaced with a combination of the following:

1. Conservation of linear momentum for the system of particles in the direction perpendicular to the constraint.
2. Conservation of linear momentum for the unconstrained particle in the t-direction.
3. Kinematic constraints on the direction of the velocity of the constrained particle.

These equations will obviously depend on the exact nature of the constraint and will have to be derived specifically for each particular problem being solved.

EXAMPLE PROBLEM 19-6

Two masses slide on a horizontal frictionless rod as shown in Fig. 19-18. Slider A has a mass of 2 kg and is sliding to the right at 3 m/s, whereas slider B has a mass of 0.75 kg and is sliding to the left at 1 m/s. If the coefficient of restitution for the sliders is 0.6, determine

a. The velocity of each mass after they collide.
b. The percentage decrease in energy due to the collision.

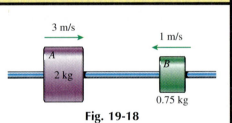

Fig. 19-18

SOLUTION

a. This is a direct central impact problem with the line of impact along the rod. For the system consisting of the pair of masses, there are no impulsive forces in the direction of the rod. Therefore, for the pair of masses, the component of linear momentum along the rod is conserved (Eq. 19-6)

$$m_A v_{Ai} + m_B v_{Bi} = m_A v_{Af} + m_B v_{Bf}$$
$$(2)(3) + (0.75)(-1) = 2v_{Af} + 0.75v_{Bf}$$

Also, the definition of the coefficient of restitution (Eq. 19-7)

$$e = -\frac{v_{Bf} - v_{Af}}{v_{Bi} - v_{Ai}}$$

gives

$$0.6 = -\frac{v_{Bf} - v_{Af}}{(-1) - (3)}$$

Solving these two equations simultaneously gives

$$v_{Af} = 1.255 \text{ m/s} \qquad \text{Ans.}$$
$$v_{Bf} = 3.65 \text{ m/s} \qquad \text{Ans.}$$

b. The sum of the kinetic energy of the two particles before the collision was

$$T_i = \Sigma \frac{1}{2}mv^2 = \frac{1}{2}(2)(3)^2 + \frac{1}{2}(0.75)(1)^2 = 9.375 \text{ J}$$

After the collision, the sum of the kinetic energy is

$$T_f = \frac{1}{2}(2)(1.255)^2 + \frac{1}{2}(0.75)(3.65)^2 = 6.571 \text{ J}$$

The percentage decrease in kinetic energy is then

$$\frac{9.375 - 6.571}{9.375}(100) = 29.9\% \qquad \text{Ans.}$$

Note that the collision has caused slider A to slow down but it is still moving to the right after the collision. Although both sliders move to the right after the collision, slider B is moving faster than slider A. Therefore, the distance between the two sliders will continually increase and the two sliders will not collide again.

EXAMPLE PROBLEM 19-7

Two pucks of equal radius sliding on a smooth horizontal surface collide obliquely as shown in Fig. 19-19. Puck A weighs 5 lb and is traveling to the right at 6 ft/s, whereas puck B weighs 2 lb and is traveling to the left at 3 ft/s. If the coefficient of restitution for the collision is 0.7 and the duration of the contact is 0.001 s, determine

a. The velocities of the pucks immediately after they collide.
b. The percentage energy loss due to the collision.
c. The average interaction force of puck B on A.

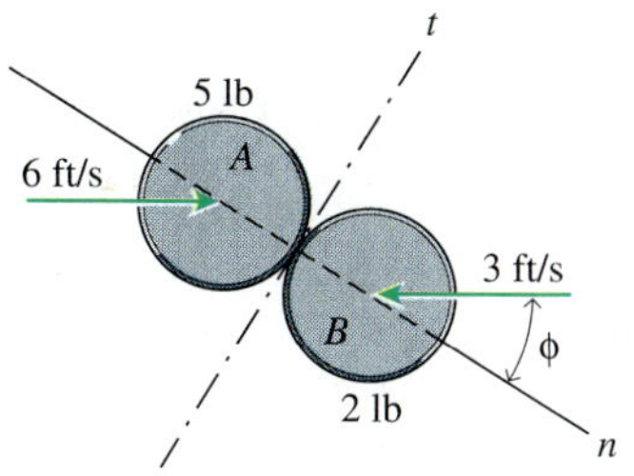

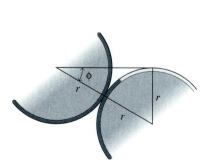

Fig. 19-19

SOLUTION

a. First, coordinates n and t are drawn along and perpendicular to the line of impact as shown in Fig. 19-19. As shown in Fig. 19-20a, the vertical distance between the centers is r, the radius of the pucks, and the slant distance between the centers is $2r$. Therefore the angle ϕ between the horizontal and the line of impact is given by

$$\phi = \sin^{-1}\frac{r}{2r} = 30°$$

Next, the initial velocities are resolved into components along and perpendicular to the line of impact.

$$(v_{Ai})_t = 6 \sin 30° = 3.00 \text{ ft/s}$$
$$(v_{Ai})_n = 6 \cos 30° = 5.196 \text{ ft/s}$$
$$(v_{Bi})_t = -3 \sin 30° = -1.500 \text{ ft/s}$$
$$(v_{Bi})_n = -3 \cos 30° = -2.598 \text{ ft/s}$$

(a)

Fig. 19-20

457

Since the only impulsive force acting on the system is the internal reaction force (which acts along the n-direction), linear momentum in the t-direction is conserved for each particle, and their t-components of velocity are unchanged by the impact:

$$(v_{Af})_t = 3.000 \text{ ft/s} \quad \text{and} \quad (v_{Bf})_t = -1.500 \text{ ft/s}$$

Next, considering the pair of particles as a system, there is no external impulsive force in any direction, and therefore linear momentum is conserved in every direction. In particular, for the n-direction, conservation of linear momentum (Eq. 19-6)

$$m_A v_{Ai} + m_B v_{Bi} = m_A v_{Af} + m_B v_{Bf}$$

gives

$$\left(\frac{5}{g}\right)(5.196) + \left(\frac{2}{g}\right)(-2.598) = \left(\frac{5}{g}\right)(v_{Af})_n + \left(\frac{2}{g}\right)(v_{Bf})_n$$

This equation is then combined with the definition of the coefficient of restitution (Eq. 19-7)

$$e = -\frac{(v_{Bf})_n - (v_{Af})_n}{(v_{Bi})_n - (v_{Ai})_n}$$

$$0.7 = -\frac{(v_{Bf})_n - (v_{Af})_n}{(-2.598) - (5.196)}$$

to get

$$(v_{Af})_n = 1.410 \text{ ft/s} \quad \text{and} \quad (v_{Bf})_n = 6.866 \text{ ft/s}$$

Finally, the final velocities are expressed relative to their initial horizontal directions. The magnitude of the final velocity of puck A is (Fig. 19-20b)

$$v_{Af} = \sqrt{(1.410)^2 + (3.000)^2} = 3.31 \text{ ft/s}$$

and its direction relative to the horizontal is given by

$$\tan(\theta_{Af} + 30°) = 3.00/1.410$$

or $\theta_{Af} = 34.8°$. Therefore, the final velocity of puck A is

$$\mathbf{v}_{Af} = 3.31 \text{ ft/s} \angle 34.8° \qquad \text{Ans.}$$

For puck B (Fig. 19-20c)

$$v_{Bf} = \sqrt{(1.500)^2 + (6.866)^2} = 7.03 \text{ ft/s}$$
$$\tan(\theta_{Bf} - 30°) = 1.500/6.866$$
$$\theta_{Bf} = 42.3°$$

and

$$\mathbf{v}_{Bf} = 7.03 \text{ ft/s} \searrow 42.3° \qquad \text{Ans.}$$

b. The sum of the kinetic energy of the two particles before the collision was

$$T_i = \Sigma \frac{1}{2}mv^2 = \frac{1}{2}\left(\frac{5}{32.2}\right)(6)^2 + \frac{1}{2}\left(\frac{2}{32.2}\right)(3)^2 = 3.075 \text{ lb} \cdot \text{ft}$$

After the collision, the sum of the kinetic energy is

$$T_f = \frac{1}{2}\left(\frac{5}{32.2}\right)(3.31)^2 + \frac{1}{2}\left(\frac{2}{32.2}\right)(7.03)^2 = 2.385 \text{ lb} \cdot \text{ft}$$

The percentage decrease in kinetic energy is then

$$\frac{3.075 - 2.385}{3.075}(100) = 22.4\% \qquad \text{Ans.}$$

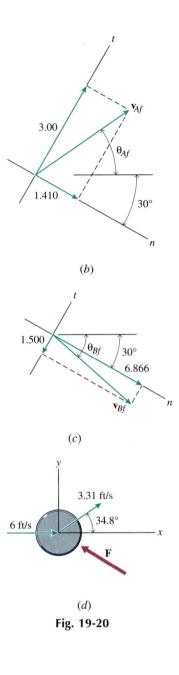

(b)

(c)

(d)

Fig. 19-20

It is generally more useful to express the final velocity in terms of the external world than to give its components along and perpendicular to the line of impact. The final velocity should either be written relative to standard horizontal and vertical coordinates or relative to the initial velocity of the particle.

c. Applying the linear impulse–momentum equation to puck A (Fig. 19-20d) gives

$$\frac{5}{32.2}(6\mathbf{i}) + .001\mathbf{F} = \frac{5}{32.2}(3.31)(\cos 34.8° \ \mathbf{i} + \sin 34.8° \ \mathbf{j})$$

which gives

$$\mathbf{F} = -509.6\mathbf{i} + 293.3\mathbf{j} \ \text{lb} \qquad \text{Ans.}$$

or

$$\mathbf{F} = 588 \ \text{lb} \ \text{\textbackslash} 30° \qquad \text{Ans.}$$

Note that the direction of the impact force is indeed along the line of impact.

EXAMPLE PROBLEM 19-8

A 3-kg sphere B is hanging at the end of a 1.5-m-long inextensible cord when it is struck by a 2-kg sphere A of the same material (Fig. 19-21). Sphere A is initially just touching the cord and drops 1 m before striking B. If the coefficient of restitution is 0.8 and the duration of contact is $\Delta t = 0.01$ s, determine

a. The velocity of each sphere immediately after the collision.
b. The average tensile force in the cord due to the impact.
c. The maximum angle θ that sphere B will swing as a result of the collision.

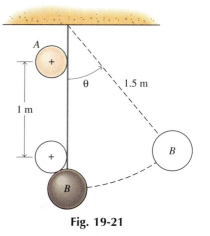

Fig. 19-21

SOLUTION

a. **Work–Energy:** First, use the work–energy principle to determine the velocity of sphere A just prior to the impact. Taking the initial position of A as the zero for gravitational potential energy,

$$T_1 + V_{g1} + U_{1\to2}^{(o)} = T_2 + V_{g2}$$
$$0 + 0 + 0 = \frac{1}{2}m_A v_{A2}^2 - m_A(9.81)(1)$$

which gives

$$\mathbf{v}_{A2} = 4.429 \ \text{m/s}\downarrow$$

where $\mathbf{v}_{A2} = \mathbf{v}_{Ai}$ is the velocity of sphere A just prior to the impact.

Linear Impulse–Momentum: Next, coordinates n and t are drawn along and perpendicular to the line of impact as shown in Fig. 19-22a. The horizontal distance between the centers of the spheres is r_A, and the slant distance between the centers is $r_A + r_B$. Therefore, the angle ϕ between the vertical and the line of impact is given by

$$\phi = \sin^{-1} \frac{r_A}{r_A + r_B}$$

But the spheres are made of the same material and thus have the same density $\rho = \text{mass/volume}$

$$\rho = \frac{2 \ \text{kg}}{\frac{4}{3}\pi r_A^3} = \frac{3 \ \text{kg}}{\frac{4}{3}\pi r_B^3}$$

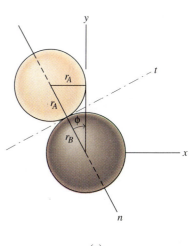

(a)

Fig. 19-22

459

Therefore,

$$r_B = \sqrt[3]{3/2}\,r_A = 1.145 r_A$$

and

$$\phi = 27.79°$$

Next, the velocities are resolved into components along and perpendicular to the line of impact

$$(v_{Ai})_t = -4.429 \sin 27.79° = -2.065 \text{ m/s}$$
$$(v_{Ai})_n = 4.429 \cos 27.79° = 3.918 \text{ m/s}$$
$$(v_{Bi})_t = (v_{Bi})_n = 0 \text{ m/s}$$

Furthermore, since the cord is inextensible, sphere B cannot move downward after impact, so

$$\mathbf{v}_{Bf} = v_{Bf}\,\mathbf{i}$$

and

$$(v_{Bf})_t = v_{Bf} \cos 27.79° = 0.8847 v_{Bf}$$
$$(v_{Bf})_n = v_{Bf} \sin 27.79° = 0.4662 v_{Bf}$$

Because the cord constrains the motion of sphere B, the tension in the cord is just as impulsive as the internal reaction force, and linear momentum for the pair of spheres is not conserved in either the n- or the t-direction. Instead, linear impulse–momentum equations will be written for each of the spheres separately. Referring to the free-body diagrams of Figs. 19-22b and 19-22c, the t- and n-components of these equations are

$$2(-2.065) = 2(v_{Af})_t \qquad (a)$$
$$2(3.918) - F\,\Delta t = 2(v_{Af})_n \qquad (b)$$
$$0 + T\,\Delta t \sin 27.79° = 3(0.8847 v_{Bf}) \qquad (c)$$
$$0 - T\,\Delta t \cos 27.79° + F\,\Delta t = 3(0.4662 v_{Bf}) \qquad (d)$$

where F is the average impact force between the spheres, T is the average tensile force in the cord, and $\Delta t = 0.01$ s is the duration of the impact. Equations a through d are combined with the definition of the coefficient of restitution

$$0.8 = -\frac{0.4662 v_{Bf} - (v_{Af})_n}{0 - (3.918)} \qquad (e)$$

and solved to get

$$(v_{Af})_t = -2.065 \text{ m/s} \qquad (v_{Af})_n = -2.242 \text{ m/s}$$
$$\mathbf{v}_{Bf} = 1.915 \text{ m/s} \rightarrow \qquad \text{Ans.}$$

The final velocity of A still needs to be expressed relative to the horizontal direction. The magnitude of the final velocity of sphere A is (Fig. 19-22d)

$$v_{Af} = \sqrt{(-2.065)^2 + (-2.242)^2} = 3.048 \text{ ft/s}$$

and its direction relative to the horizontal is given by

$$\tan(\theta_{Af} + 27.79°) = 2.242/2.065$$

or $\theta_{Af} = 19.56°$. Therefore, the final velocity of sphere A is

$$\mathbf{v}_{Af} = 3.05 \text{ ft/s} \, \angle \, 19.56° \qquad \text{Ans.}$$

b. Now that the velocity components have been found, Eq. c gives the average tension in the cord

$$T = 1090 \text{ N} \qquad \text{Ans.}$$

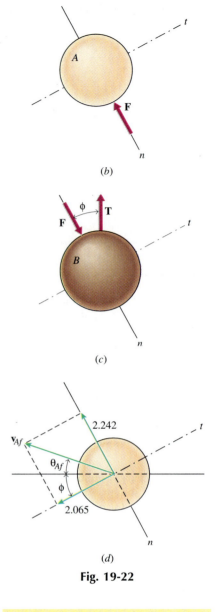

(b)

(c)

(d)

Fig. 19-22

Since the cord is assumed inextensible, sphere B cannot move downward after the impact, and the velocity of sphere B immediately after the impact must be in the horizontal direction. Since the impulsive force $\mathbf{F}$ is trying to stretch the cord holding sphere B, the tension in the cord must also be an impulsive force to prevent the cord from stretching.

c. Work–Energy: Finally, work–energy is used again to find the angle through which sphere B will swing after the collision. Using the upper end of the cord as the zero for gravitational potential energy,

$$T_1 + V_{g1} + U^{(o)}_{1 \to 2} = T_2 + V_{g2}$$

or

$$\frac{1}{2}m_B(1.915)^2 - m_B(9.81)(1.5) + 0 = 0 - m_B(9.81)(1.5 \cos \theta)$$

which gives

$$\theta = 28.9° \qquad\qquad \text{Ans.}$$

PROBLEMS

Introductory Problems

19-37–19-42 Two beads are sliding freely on a horizontal rod as shown in Fig. P19-37. For the conditions specified, determine

a. The final velocity of both beads.

b. The percentage of the initial kinetic energy lost as a result of the collision.

c. The average interaction force between the beads if the duration of the impact is 0.005 s.

Problem	m_A	v_A	m_B	v_B	e
19-37*	9 lb	3 ft/s	2 lb	0 ft/s	0.3
19-38*	0.5 kg	2 m/s	5 kg	0 m/s	0.7
19-39	7 lb	5 ft/s	3 lb	2 ft/s	0.7
19-40*	3 kg	1 m/s	1 kg	−3 m/s	0.9
19-41	6 lb	3 ft/s	1 lb	−2 ft/s	0.3
19-42	2 kg	3 m/s	3 kg	2 m/s	0.5

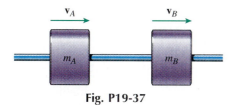

Fig. P19-37

19-43* Three beads are sliding freely on a horizontal rod as shown in Fig. P19-43. Initially, beads B and C are at rest and bead A is moving to the right at 5 ft/s. If the coefficient of restitution is $e = 0.8$ for all collisions, determine

a. The final velocity of each of the beads after all collisions have taken place.

b. The percentage of the initial kinetic energy that is lost as a result of the collisions.

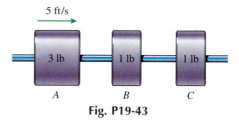

Fig. P19-43

19-44* Three beads are sliding freely on a horizontal rod as shown in Fig. P19-44. Initially, bead B is at rest, bead A is moving to the right at 3 m/s, and bead C is moving to the left at 2 m/s. If the coefficient of restitution is $e = 0.8$ for all collisions and the first collision occurs between beads B and C, determine

a. The final velocity of each of the beads after all collisions have taken place.

b. The percentage of the initial kinetic energy that is lost as a result of the collisions.

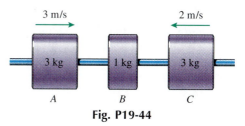

Fig. P19-44

19-45 A 2-lb sphere falls and bounces on a 10-lb plate that is resting on the ground (Fig. P19-45a). If the sphere starts from rest 6 ft above the plate and rebounds to a height of 5 ft after the impact, determine

a. The coefficient of restitution for the collision.
b. The height that the same sphere would rebound if the 10-lb plate were resting on two springs, each having $k = 120$ lb/ft (Fig. P19-45b).

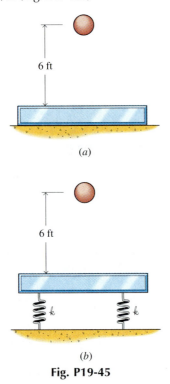

(a)

6 ft

(b)

Fig. P19-45

Intermediate Problems

19-46* Two spheres are hanging from cords as shown in Fig. P19-46. The distance from the ceiling to the center of each sphere is 2 m, and the coefficient of restitution is $e = 0.75$. If sphere A ($m_A = 2$ kg) is drawn back 60° and released from rest, determine

a. The maximum angle θ_B that sphere B ($m_B = 3$ kg) will swing as a result of the impact.
b. The angle θ_A that sphere A will rebound as a result of the impact.

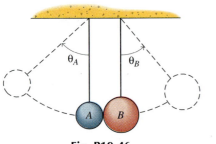

Fig. P19-46

19-47* A 1-lb sphere A is attached to a light inextensible cord that is 3 ft long as shown in Fig. P19-47. Sphere A is released from rest with the cord in the horizontal position and strikes the 2-lb stationary sphere B when the cord is vertical. If the coefficient of restitution between the spheres is $e = 0.9$, determine the distance b to where sphere B hits the horizontal surface.

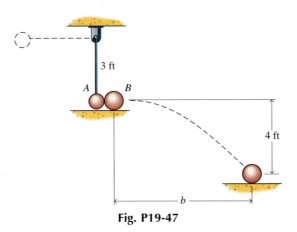

Fig. P19-47

19-48 The 4-kg collar A is moving to the right along a frictionless rod with a speed of 10 m/s when it strikes the 2-kg collar B that is at rest as shown in Fig. P19-48. If the coefficient of restitution between the collars is $e = 0.8$, determine the maximum deflection of the spring ($k = 1.75$ kN/m).

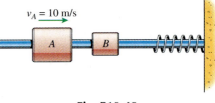

Fig. P19-48

19-49 A small sphere is released from rest at A and moves along the smooth tube shown in Fig. P19-49. At B the sphere collides with an identical sphere C, which is attached to a light inextensible cord. If the coefficient of restitution between the spheres is $e = 0.8$, determine the height h to which sphere C rises.

462

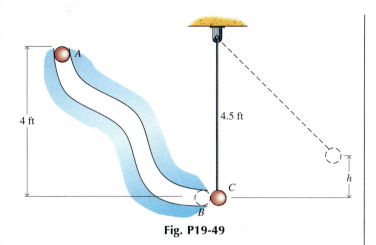

4 ft

4.5 ft

h

Fig. P19-49

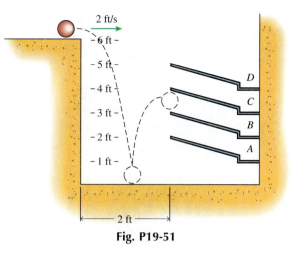

2 ft/s

– 6 ft –
– 5 ft –
– 4 ft –
– 3 ft –
– 2 ft –
– 1 ft –

D
C
B
A

2 ft

Fig. P19-51

19-50* The 2-kg sphere of Fig. P19-50 is released from rest when $\theta_A = 60°$ and swings down, striking the 5-kg box B. The distance from the ceiling to the center of the sphere is 1 m, the coefficient of restitution for the collision is $e = 0.7$, and the kinetic coefficient of friction between the box and the floor is $\mu_k = 0.1$. Determine

a. The velocity of the box immediately after the impact.
b. The distance that the box will slide before coming to rest again.

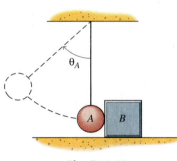

θ_A

A   B

Fig. P19-50

19-51 Balls are to be sorted according to their coefficient of restitution by rolling them off a 6-ft-high step and bouncing them into sorting bins as shown in Fig. P19-51. If the speed of the balls as they leave the step is 2 ft/s, determine the range of coefficients that will be collected in bin B, whose opening is from 3 ft to 4 ft above the floor.

19-52 Two identical pucks are sliding on an air hockey table as shown in Fig. P19-52. If puck A has an initial velocity of 5 m/s to the right, puck B is initially at rest, and the coefficient of restitution is $e = 0.9$, determine the final velocities (magnitudes and directions) of both pucks.

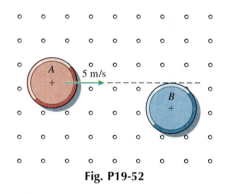

A   5 m/s   B

Fig. P19-52

19-53* Two identical pucks are sliding on an air hockey table as shown in Fig. P19-53. If the coefficient of restitution is $e = 0.9$, determine the final velocities (magnitudes and directions) of both pucks.

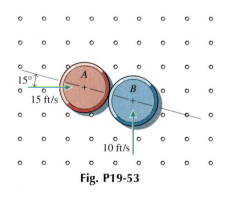

15°
15 ft/s
A
B
10 ft/s

Fig. P19-53

19-54* In a pool shot, the cue ball knocks the 1-ball into the corner pocket as shown in Fig. P19-54. If the coefficient of restitution is $e = 0.95$, determine the velocity of the cue ball after the collision.

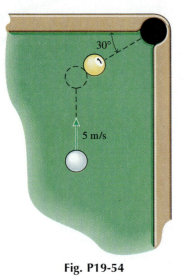

Fig. P19-54

19-55 Two identical spheres collide as shown in Fig. P19-55. If the coefficient of restitution is $e = 0.7$ and the velocity of sphere A after the collision is in the vertical direction as shown, determine the velocity of sphere B after the collision.

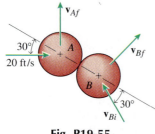

Fig. P19-55

19-56 Two pucks (of different size) collide on an air table as shown in Fig. P19-56. If the coefficient of restitution is $e = 0.7$ and the final velocity of each puck is 90° from its initial direction, determine the magnitudes of the final velocities and the mass of puck B.

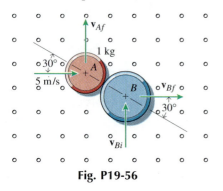

Fig. P19-56

Challenging Problems

19-57* A device designed to accept or reject small balls according to their coefficient of restitution is shown in Fig. P19-57. Determine the distance d to where the box used to catch the balls should be placed if the device is to accept or reject balls with a coefficient of restitution of $e = 0.7$.

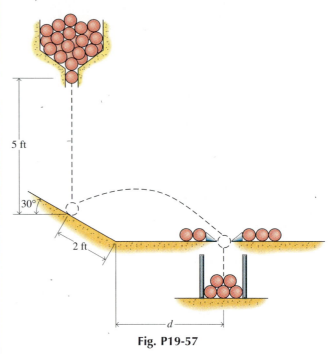

Fig. P19-57

19-58* A ball falls on a hard surface and bounces over a vertical wall as shown in Fig. P19-58. If the coefficient of restitution is $e = 0.8$, the ball starts from rest with $h = 1$ m, and the ball just clears the wall at the peak of its bounce, determine the distances b, c, and d in the figure.

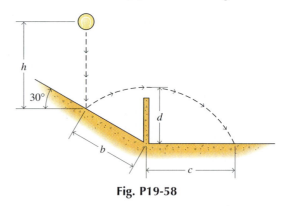

Fig. P19-58

19-59 A boy standing at a distance $d = 4$ m from the bottom of a building throws a ball against the wall of the building (Fig. P19-59). He releases the ball with a velocity of 15 m/s at an angle of 30° with respect to the horizontal. If the coefficient of restitution between the ball and the building is $e = 0.4$, determine the velocity of the ball immediately after it bounces off the wall.

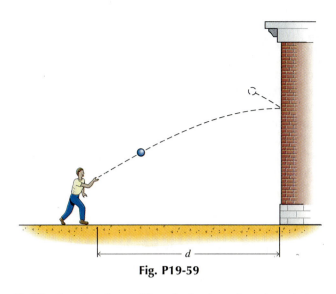

Fig. P19-59

19-60 An attention-getting device in a department store window consists of an air gun that repeatedly bounces a ball off a wall as shown in Fig. P19-60. If the coefficient of restitution is $e = 0.9$ and the velocity of the ball as it leaves the air gun is 3 m/s, determine

a. The distance x that the device must be placed away from the wall.
b. The distance y to where the ball will strike the wall.

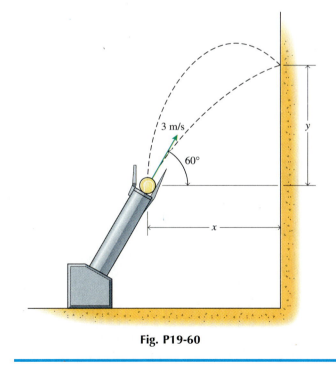

Fig. P19-60

19-61* A 2-lb sphere B is at rest on a ledge when it is struck by an identical sphere (Fig. P19-61). The distance from the ceiling to the center of sphere A is 3 ft, and the coefficient of restitution is $e = 0.7$. At the moment of impact, the cord is vertical and the center of sphere A is level with the bottom of sphere B. If sphere A is released from rest with $\theta_A = 60°$, determine the distance x traveled by sphere B before it bounces on the horizontal surface.

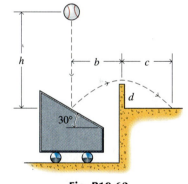

Fig. P19-61

19-62 A 0.5-kg ball falls and bounces off a 2-kg cart as shown in Fig. P19-62. The cart is free to roll in the horizontal direction, and the coefficient of restitution of the collision is $e = 0.8$. If the ball starts from rest with $h = 1$ m and just clears the wall at the peak of its bounce, determine distances b, c, and d in the figure.

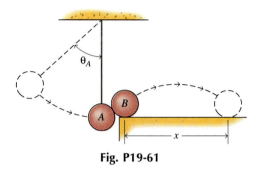

Fig. P19-62

19-5 ANGULAR IMPULSE AND ANGULAR MOMENTUM OF A PARTICLE

The angular impulse–momentum principle, which is developed in the next several sections, relates the moments of the forces acting on a particle and the velocity of the particle when the moments are known as functions of time. Again combining Newton's second law with the principles of kinematics, the angular impulse–momentum principle is particularly useful for solving problems in which several or all of the forces always act through a fixed point.

19-5.1 Angular Momentum

The *angular momentum* $\mathbf{H}_O$ of a particle P about a fixed point O is defined as the moment of the linear momentum $\mathbf{L}$ about the point O. If $\mathbf{r}_{P/O}$ is the position vector from point O to the particle P of mass m and velocity $\mathbf{v}$, then

$$\mathbf{H}_O = \mathbf{r}_{P/O} \times (m\mathbf{v}) \tag{19-10}$$

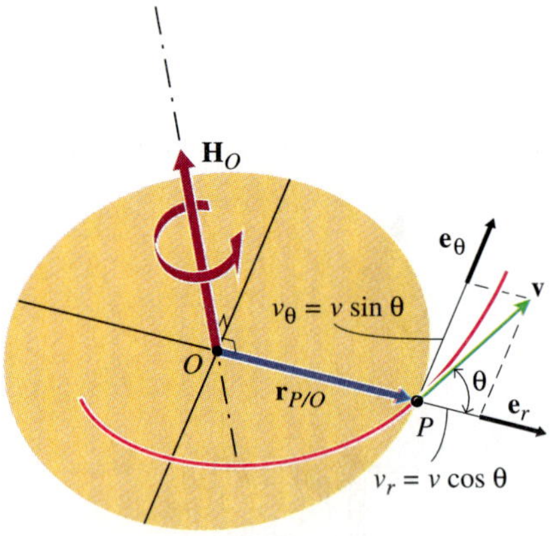

The magnitude of the angular momentum is equal to $r_{P/O}mv \sin\theta$, where v is the speed of the particle and θ is the angle between the position vector $\mathbf{r}_{P/O}$ and the velocity $\mathbf{v}$. The direction of the angular momentum vector will be perpendicular to the plane formed by the vectors $\mathbf{r}_{P/O}$ and $\mathbf{v}$ (Fig. 19-23).

Expressing the velocity in terms of polar coordinates in the plane of $\mathbf{r}_{P/O}$ and $\mathbf{v}$ (Fig. 19-23), the angular momentum becomes

$$\mathbf{H}_O = \mathbf{r}_{P/O} \times m(v_r\mathbf{e}_r + v_\theta\mathbf{e}_\theta) \tag{a}$$

But the first term on the right-hand side is zero because the cross product of two parallel vectors is zero. Therefore, only the component of the velocity perpendicular to $\mathbf{r}_{P/O}$ contributes to the angular momentum, and the angular momentum represents a rotation of the particle about an axis through O along the angular momentum vector.

In the SI system of measurement, the units of angular momentum are $kg \cdot m^2/s$ or, equivalently, $N \cdot m \cdot s$. In the U.S. Customary system of measurement, they are $slug \cdot ft^2/s = lb \cdot ft \cdot s$.

Figure 19-23 The angular momentum of a particle about a fixed point is the moment of the linear momentum about the point.

19-5.2 Angular Impulse

The *angular impulse* of the resultant force about a fixed point O is defined as the impulse of the moment

$$\int_{t_i}^{t_f} \mathbf{r}_{P/O} \times \mathbf{R}\, dt = \int_{t_i}^{t_f} \mathbf{M}_O\, dt \tag{b}$$

where $\mathbf{r}_{P/O}$ is again the position of the particle P relative to the fixed point O and $\mathbf{R}$ is the resultant of all forces acting on the particle. In the SI system of measurement the magnitude of the angular impulse is expressed in $N \cdot m \cdot s = kg \cdot m^2/s$, which is the same as the units of angular momentum. If U.S. Customary units are used, the angular impulse is expressed in $lb \cdot ft \cdot s = slug \cdot ft^2/s$.

19-5.3 Principle of Angular Impulse and Momentum

Differentiating the angular momentum with respect to time gives

$$\frac{d\mathbf{H}_O}{dt} = \frac{d\mathbf{r}_{P/O}}{dt} \times (m\mathbf{v}) + \mathbf{r}_{P/O} \times \left(m\frac{d\mathbf{v}}{dt} \right)$$

$$= \mathbf{v} \times (m\mathbf{v}) + \mathbf{r}_{P/O} \times (m\mathbf{a}) \qquad (c)$$

But the first term $\mathbf{v} \times (m\mathbf{v}) = \mathbf{0}$, since the cross product of any vector with itself is zero. In the second term, Newton's second law can be used to replace the $m\mathbf{a}$ with the resultant force $\mathbf{R}$. Then the second term is just $\mathbf{r}_{P/O} \times \mathbf{R} = \mathbf{M}_O$, the moment of the resultant force about point O. Therefore

$$\frac{d}{dt}\mathbf{H}_O = \mathbf{M}_O \qquad (19\text{-}11)$$

That is, *the time rate of change of the angular momentum of a particle about a fixed point O is equal to the resultant moment about O of all forces acting on the particle.*

If the moments of the forces are known as functions of time, Eq. 19-11 can be integrated from some initial time t_i to some final time t_f to get the Principle of Angular Impulse and Momentum:

$$\mathbf{H}_{Oi} + \int_{t_i}^{t_f} \mathbf{M}_O \, dt = \mathbf{H}_{Of} \qquad (19\text{-}12)$$

That is, *the final angular momentum $\mathbf{H}_{Of}$ of a particle about a fixed point O is the vector sum of its initial angular momentum $\mathbf{H}_{Oi}$ about O and the angular impulse $\int \mathbf{M}_O \, dt$ about O of the resultant of all forces acting on the particle during the time interval.*

Like the linear impulse–momentum principle, Eqs. 19-11 and 19-12 are vector equations representing three scalar components. Since $\mathbf{M}_O$ may vary in both magnitude and direction, rectangular Cartesian components are usually the most convenient to use. The three components can be applied independently of one another.

19-5.4 Conservation of Angular Momentum

It is unusual to know how the resultant moment about O, $\mathbf{M}_O = \mathbf{r}_{P/O} \times \mathbf{R}$, varies with time except in a few special cases. For example, if all of the forces acting on a particle pass through a single point O, then the sum of moments about that point will be zero and the initial and final angular momentum about O will be the same:

$$\mathbf{H}_{Oi} = \mathbf{H}_{Of} \qquad (d)$$

This behavior is called the Principle of Conservation of Angular Momentum.

19-5.5 Systems of Particles

For a system of interacting particles, the angular impulse–momentum equations can be written for each particle separately and the equations added together. For example, for the set of particles in Fig. 19-6, the angular impulse–momentum equations (Eq. 19-11) are

$$\frac{d}{dt}(\mathbf{r}_{1/O} \times m_1\mathbf{v}_1) = \mathbf{r}_{1/O} \times (\mathbf{R}_1 + \mathbf{f}_{12} + \mathbf{f}_{13} + \cdots + \mathbf{f}_{1i} + \cdots)$$

$$\frac{d}{dt}(\mathbf{r}_{2/O} \times m_2\mathbf{v}_2) = \mathbf{r}_{2/O} \times (\mathbf{R}_2 + \mathbf{f}_{21} + \mathbf{f}_{23} + \cdots + \mathbf{f}_{2i} + \cdots)$$

$$\vdots$$

$$\frac{d}{dt}(\mathbf{r}_{\ell/O} \times m_\ell\mathbf{v}_\ell) = \mathbf{r}_{\ell/O} \times (\mathbf{R}_\ell + \mathbf{f}_{\ell 1} + \mathbf{f}_{\ell 2} + \cdots + \mathbf{f}_{\ell i} + \cdots)$$

and so on. Adding these equations together gives

$$\sum_\ell \left(\frac{d}{dt}\mathbf{H}_{\ell/O}\right) = \sum_\ell \left(\mathbf{r}_{\ell/O} \times \mathbf{R}_\ell\right) + (\mathbf{r}_{1/O} \times \mathbf{f}_{12} + \mathbf{r}_{2/O} \times \mathbf{f}_{21})$$
$$+ (\mathbf{r}_{1/O} \times \mathbf{f}_{13} + \mathbf{r}_{3/O} \times \mathbf{f}_{31}) + \cdots$$

But the internal forces always occur in pairs having equal magnitude and the same line of action but opposite direction along the line of action (Fig. 19-24). Therefore, the sum of moments about O for each pair of forces is zero, and for the system of particles

$$\frac{d}{dt}\mathbf{H}_O = \sum_{\ell=1}^{N} \mathbf{M}_{\ell/O} \qquad (19\text{-}13)$$

where $\mathbf{H}_O = \Sigma\mathbf{H}_{\ell/O}$ is the total angular momentum about O of the system of particles, $\Sigma\mathbf{M}_{\ell/O}$ is the sum of moments about O of all the external forces acting on the system of particles, and the moments of the internal forces need not be considered.

Of course, Eq. 19-13 can be integrated with respect to time from t_i to t_f to get the *Principle of Angular Impulse and Momentum for a System of Interacting Particles:*

$$(\mathbf{H}_O)_i + \int_{t_i}^{t_f} \sum_{\ell=1}^{N} \mathbf{M}_{\ell/O}\, dt = (\mathbf{H}_O)_f \qquad (19\text{-}14)$$

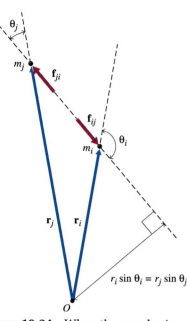

That is, *the final angular momentum* $(\mathbf{H}_O)_f$ *of a system of particles about a fixed point O is the vector sum of their initial angular momentum* $(\mathbf{H}_O)_i$ *about the point O and the angular impulse* $\int \Sigma\mathbf{M}_{\ell/O}\, dt$ *about O of all external forces acting on the system of particles.*

A common requirement for a system of particles is to compute the moments and angular momentum about the mass center of the system rather than about a fixed point O. The angular momentum of the system of particles about the mass center G is defined as the moment of the linear momentum

$$\mathbf{H}_G = \Sigma\,\mathbf{H}_{\ell/G} = \Sigma\,\mathbf{r}_{\ell/G} \times (m_\ell\mathbf{v}_\ell) \qquad (e)$$

Figure 19-24   When the angular impulse–momentum equations for a pair of interacting particles are added together, the moments of the internal forces cancel out. The internal forces have equal magnitudes and opposite directions, and they are collinear. Therefore, their moments about any point will add to zero.

where $\mathbf{r}_{\ell/G}$ is the position of the ℓth particle relative to the mass center G (Fig. 19-25) and $\mathbf{v}_\ell = \dot{\mathbf{r}}_{\ell/O}$ is the absolute velocity of the ℓth particle. The absolute velocity can be replaced using the relative velocity equation $\mathbf{v}_\ell = \mathbf{v}_G + \mathbf{v}_{\ell/G}$, where $\mathbf{v}_G$ is the velocity of the mass center of the system of particles, to get

$$\mathbf{H}_G = \Sigma\mathbf{r}_{\ell/G} \times m_\ell(\mathbf{v}_G + \mathbf{v}_{\ell/G})$$
$$= (\Sigma m_\ell\mathbf{r}_{\ell/G}) \times \mathbf{v}_G + \Sigma\mathbf{r}_{\ell/G} \times (m_\ell\mathbf{v}_{\ell/G}) \qquad (f)$$

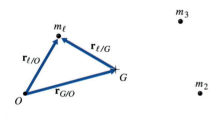

Figure 19-25   The positions with respect to the mass center G and with respect to a fixed point O are related by the triangle law of addition.

The velocity of the mass center $\mathbf{v}_G$ is the same for every particle and can be taken outside the summation in the first term. But then the quantity inside the parentheses is zero by the definition of the mass center, since the position vectors $\mathbf{r}_{\ell/G}$ are measured from the mass center. Therefore,

$$\mathbf{H}_G = \Sigma \mathbf{r}_{\ell/G} \times (m_\ell \mathbf{v}_\ell) = \Sigma \mathbf{r}_{\ell/G} \times (m_\ell \mathbf{v}_{\ell/G}) \qquad (19\text{-}15)$$

That is, the angular momentum of the system of particles about the mass center G can be computed using either the absolute velocity $\mathbf{v}_\ell = \dot{\mathbf{r}}_{\ell/O}$ or the velocity relative to the mass center $\mathbf{v}_{\ell/G}$.

Taking the time derivative of $\mathbf{H}_G$ gives

$$\begin{aligned}
\frac{d\mathbf{H}_G}{dt} &= \sum \left(\frac{d\mathbf{r}_{\ell/G}}{dt} \times m_\ell \mathbf{v}_\ell + \mathbf{r}_{\ell/G} \times m_\ell \frac{d\mathbf{v}_\ell}{dt} \right) \\
&= \Sigma \mathbf{v}_{\ell/G} \times m_\ell \mathbf{v}_\ell + \Sigma \mathbf{r}_{\ell/G} \times m_\ell \mathbf{a}_\ell \\
&= \Sigma \mathbf{v}_{\ell/G} \times m_\ell (\mathbf{v}_G + \mathbf{v}_{\ell/G}) + \Sigma \mathbf{r}_{\ell/G} \times m_\ell \mathbf{a}_\ell \\
&= (\Sigma m_\ell \mathbf{v}_{\ell/G}) \times \mathbf{v}_G + \Sigma \mathbf{v}_{\ell/G} \times m_\ell \mathbf{v}_{\ell/G} + \Sigma \mathbf{r}_{\ell/G} \times m_\ell \mathbf{a}_\ell
\end{aligned}$$

But the summation in the first term is again zero by the definition of the mass center since $\sum m_\ell \mathbf{v}_{\ell/G} = \dfrac{d}{dt} \sum m_\ell \mathbf{r}_{\ell/G}$ and the position vectors $\mathbf{r}_{\ell/G}$ are measured from the mass center G. Also, every term in the second summation of this equation is zero, since the cross product of any vector with itself is always zero. Finally, using Newton's second law to replace the factors $m_\ell \mathbf{a}_\ell$ as in the derivation of Eq. 19-13 gives

$$\frac{d\mathbf{H}_G}{dt} = \Sigma \mathbf{M}_{\ell/G} \qquad (19\text{-}16)$$

where $\Sigma \mathbf{M}_{\ell/G}$ is the sum of moments of the external forces about the mass center G and the moments of the internal forces need not be considered. Integrating Eq. 19-16 with respect to time gives

$$(\mathbf{H}_G)_i + \int_{t_i}^{t_f} \Sigma \mathbf{M}_{\ell/G}\, dt = (\mathbf{H}_G)_f \qquad (19\text{-}17)$$

Note that the form of the angular momentum principle for a system of particles is the same whether moments are summed around a fixed point O or about the moving mass center G. However, the form of the equations will not be the same if moments are summed about an arbitrary moving point P. In this case, the terms involving $\Sigma m_\ell \mathbf{r}_{\ell/P}$ will not drop out and there will be extra terms in Eqs. 19-16 and 19-17.

Finally, it should be noted that Eqs. 19-13 through 19-17 have been derived for a general system of particles. They apply equally to a system of independently moving particles and a system of particles that make up a rigid body.

A 500-lb satellite is in a circular orbit 100 mi above the earth and is to be moved to another circular orbit 1000 mi above the earth using a rocket engine having a thrust of 750 lb (Fig. 19-26). The orbital transfer is effected using an elliptic transfer orbit by firing the maneuvering engine first at A and then at B. If the required velocity at A in the transfer orbit is 26,874 ft/s, determine the length of the engine burns at A and B necessary to effect the orbit change.

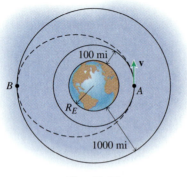

Fig. 19-26

SOLUTION

In the circular orbits, the acceleration of the satellite is

$$a = \frac{v^2}{r}$$

directed toward the center of the earth, and the only force acting on the satellite is

$$F = \frac{GMm}{r^2} = \frac{gmR_E^2}{r^2}$$

also directed toward the center of the earth. Then Newton's law gives

$$\frac{gmR_E^2}{r^2} = \frac{mv^2}{r}$$

which gives the velocity in the circular orbits as

$$v_c = R_E\sqrt{\frac{g}{r}}$$

Therefore, using $R_E = 3960$ mi,

$$v_{100c} = (3960)(5280)\sqrt{\frac{32.2}{(4060)(5280)}} = 25,626 \text{ ft/s}$$

for the lower circular orbit, and

$$v_{1000c} = (3960)(5280)\sqrt{\frac{32.2}{(4960)(5280)}} = 23,185 \text{ ft/s}$$

for the higher circular orbit.

At altitudes greater than about 90 mi above the surface of the earth, atmospheric drag is negligible. The only forces acting on the satellite are the gravitational attraction of the earth and the thrust of the satellite's maneuvering engines. (The gravitational forces exerted on the satellite by the sun, by the moon, and by other planets are negligible compared to the gravitational force exerted on the satellite by the earth.) Since the gravitational force always acts through the center of the earth, it will never give an angular impulse about an axis through the center of the earth. If the engine thrust is zero (or also acts through the center of the earth), then the angular momentum about an axis through the center of the earth will be conserved.

Since the velocity in a circular orbit is always perpendicular to the radius, the angular momentum about the axis perpendicular to the plane of the circular orbit is just $H = rmv$;

$$H_{100c} = (4060)(5280)\left(\frac{500}{32.2}\right)(25,626) = 8.530(10^{12}) \text{ lb} \cdot \text{ft} \cdot \text{s}$$

$$H_{1000c} = (4960)(5280)\left(\frac{500}{32.2}\right)(23,185) = 9.428(10^{12}) \text{ lb} \cdot \text{ft} \cdot \text{s}$$

At A in the elliptic orbit, the velocity is also perpendicular to the radius, so the angular momentum of the satellite in the elliptical orbit at A is also $H = rmv$;

$$H_{100e} = (4060)(5280)\left(\frac{500}{32.2}\right)(26,874) = 8.946(10^{12}) \text{ lb} \cdot \text{ft} \cdot \text{s}$$

Then applying the angular momentum principle (Eq. 19-12) across the duration of the burn at A gives (gravity acts through the axis and thus has no moment about the axis)

$$H_{100c} + r_A T \, \Delta t_A = H_{100e}$$
$$8.530(10^{12}) + (4060)(5280)(750) \, \Delta t_A = 8.946(10^{12})$$

or

$$\Delta t_A = 25.8 \text{ s} \qquad\qquad \text{Ans.}$$

Applying the angular momentum principle across the duration of the burn at B gives

$$H_{100e} + r_B T \, \Delta t_B = H_{1000c}$$
$$8.946(10^{12}) + (4960)(5280)(750) \, \Delta t_B = 9.428(10^{12})$$

or

$$\Delta t_B = 24.5 \text{ s} \qquad\qquad \text{Ans.}$$

Angular momentum is the moment of linear momentum, $\mathbf{H} = \mathbf{r} \times m\mathbf{v}$. If the velocity vector $\mathbf{v}$ is perpendicular to the radial vector $\mathbf{r}$, then the angular momentum will be $H = rmv$ about an axis perpendicular to the plane of $\mathbf{r}$ and $\mathbf{v}$.

A 0.6-kg mass slides on a smooth horizontal surface at the end of an inextensible string (Fig. 19-27). The other end of the string passes through a hole in the surface and is attached to a spring having $k = 100$ N/m. The spring is unstretched when $\ell = 0$. If $v = 10$ m/s and $\ell = 0.5$ m at the instant shown, determine the minimum and maximum values of ℓ in the resulting motion.

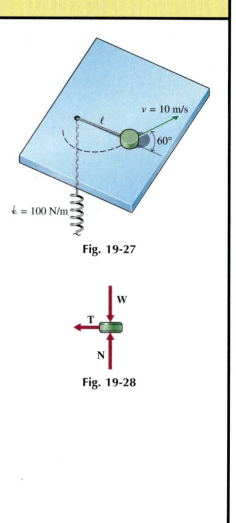

Fig. 19-27

SOLUTION

Angular momentum about a vertical axis through the hole is conserved since none of the three forces acting on the mass has a moment about the axis (Fig. 19-28). Both the weight **W** and the normal force **N** act parallel to the axis and thus have no moment about the axis, while the string tension acts through the axis and has no moment about the axis.

At the instant shown, the angular momentum is

$$H_{Oi} = (0.5)(0.6)(10 \sin 60) \text{ N} \cdot \text{m} \cdot \text{s}$$

When the string is at its minimum or maximum length, the velocity of the mass is perpendicular to the string and the angular momentum is

$$H_{Of} = \ell(0.6)v$$

Therefore, conservation of angular momentum about a vertical axis through the hole gives

$$(0.5)(0.6)(10 \sin 60) = \ell(0.6)v \qquad \qquad (a)$$

A second equation relating the length of the string and the velocity is obtained from the work–energy equation. Neither the weight **W** nor the normal force **N** do work, and the work done by the spring has a potential. Therefore,

$$\frac{1}{2}(0.6)(10)^2 + \frac{1}{2}(100)(0.5)^2 = \frac{1}{2}(0.6)v^2 + \frac{1}{2}(100)\ell^2 \qquad (b)$$

Solving equations a and b simultaneously gives

$$\ell_{max} = 0.828 \text{ m} \qquad \qquad \text{Ans.}$$
$$\ell_{min} = 0.405 \text{ m} \qquad \qquad \text{Ans.}$$

Fig. 19-28

PROBLEMS

Introductory Problems

19-63* Two identical 1-lb disks are moving on a frictionless horizontal surface with the velocities shown in Fig. P19-63. For the system consisting of the two disks, determine

a. The linear momentum of the system.
b. The angular momentum of the system with respect to point O.
c. The angular momentum of the system with respect to point C.
d. The kinetic energy of the system.

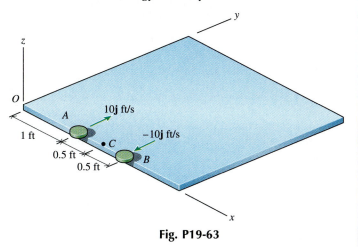

Fig. P19-63

19-64* Two 1-kg disks A and B are moving along a smooth horizontal surface as shown in Fig. P19-64. At the instant shown, the velocity of A is $\mathbf{v}_A = -5\mathbf{j}$ m/s and the velocity of B is $\mathbf{v}_B = 5\mathbf{i}$ m/s. For the system consisting of the two disks, determine

a. The angular momentum of the system with respect to O.
b. The angular momentum of the system with respect to the mass center of the system.

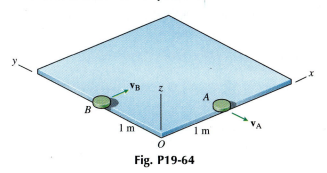

Fig. P19-64

19-65 The vertical shaft of Fig. P19-65 is rotating with an initial angular velocity of 20 rad/s when the 0.5-lb collar A starts to slide slowly outward along the lightweight horizontal arm. Determine the decrease in the angular velocity of the shaft as the collar A slides from 3 in. out to 24 in. from the axis of the shaft.

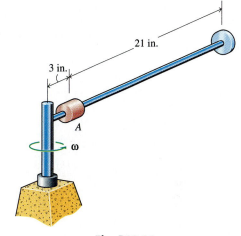

Fig. P19-65

19-66* A 250-g particle slides in a circular path on a smooth horizontal surface at the end of an inextensible string (Fig. P19-66). The other end of the cord is drawn very slowly through the central hole, reducing the radius of the circular path from 500 mm to 200 mm. If the initial velocity of the particle is 5 m/s, determine the velocity when the radius is 200 mm.

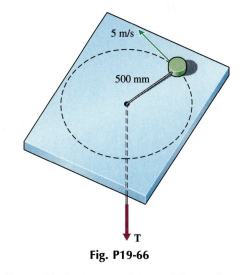

Fig. P19-66

19-67 Prove Kepler's second law: "The radius vector from the sun to a planet sweeps equal areas in equal time." That is, prove that $dA/dt = $ constant, where dA is the shaded area in Fig. P19-67.

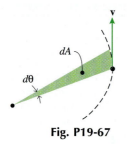

Fig. P19-67

19-68 A satellite is in an elliptical earth orbit with semimajor axis $a = 17{,}000$ km and semiminor axis $b = 13{,}725$ km (Fig. P19-68). If the velocity of the satellite is 9500 m/s at A, determine the velocity of the satellite at B and at C.

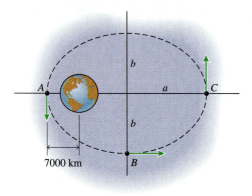

Fig. P19-68

Intermediate Problems

19-69* A 1-lb disk A is attached to a lightweight, inextensible, 2-ft-long cord as shown in Fig. P19-69. The cord is rotating in a horizontal plane with a constant angular velocity of $\pi\mathbf{k}$ rad/s. A 2-lb disk B with a velocity $\mathbf{v}_B = 15\mathbf{j}$ ft/s collides with disk A when the cord is parallel to the x-axis. If the collision is perfectly plastic ($e = 0$), determine the angular velocity of the cord immediately after the collision. Neglect the effects of friction.

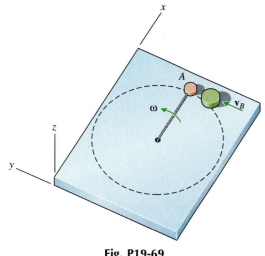

Fig. P19-69

19-70* Two 2-kg balls A and B are connected by a lightweight, rigid, 400-mm-long rod that is free to rotate in the xy-plane about a ball-and-socket joint at O as shown in Fig. P19-70. The rod is initially stationary. If a 2-kg ball with a velocity of $5\mathbf{j}$ m/s collides with ball A (the collision is perfectly plastic, $e = 0$) and causes the system of three balls to rotate about the z-axis, determine the angular velocity of the rod immediately after the collision.

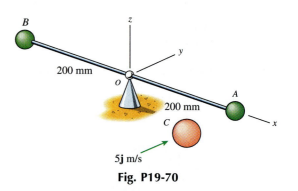

Fig. P19-70

19-71 A 5-lb ball is swinging at the end of a 2-ft-long, inextensible cord (Fig. P19-71). At the instant shown, the velocity is in a horizontal plane with $v = 6$ ft/s and $\theta = 60°$. Determine the minimum angle θ in the resulting motion of the ball.

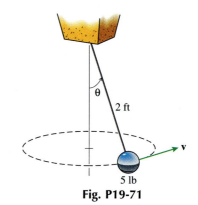

Fig. P19-71

19-72 For a system of n particles, show that the total angular momentum about a fixed point O can be written

$$\mathbf{H}_O = \mathbf{H}_G + \mathbf{r}_G \times m\mathbf{v}_G$$

where $\mathbf{H}_O = \Sigma \mathbf{r}_{i/O} \times m_i\mathbf{v}_i$, $\mathbf{H}_G = \Sigma \mathbf{r}_{i/G} \times m_i\mathbf{v}_i$, $m = \Sigma m_i$, and $\mathbf{r}_G$ and $\mathbf{v}_G$ are the position and velocity, respectively, of the mass center of the system of particles relative to the fixed point O.

19-73 For a system of n particles, show that the total angular momentum about a moving point P can be written

$$\mathbf{H}_P = \mathbf{H}_G + \mathbf{r}_{G/P} \times m\mathbf{v}_G$$

where $\mathbf{H}_P = \Sigma \mathbf{r}_{i/P} \times m_i\mathbf{v}_i$, $\mathbf{H}_G = \Sigma \mathbf{r}_{i/G} \times m_i\mathbf{v}_i$, $m = \Sigma m_i$, and $\mathbf{v}_G$ is the absolute velocity of the mass center of the system of particles.

Challenging Problems

19-74   A marble rolls freely on the inside of a 30° cone as shown in Fig. P19-74. At the instant shown the velocity of the marble is horizontal, with $z = 500$ mm. Determine the maximum height to which the marble will rise if the initial velocity is $v_i = 4$ m/s. Repeat for $v_i = 1$ m/s.

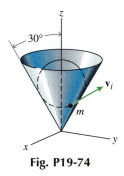

Fig. P19-74

19-75*   A marble rolls freely on the inside of a cone as shown in Fig. P19-75. At the instant shown the velocity of the marble is horizontal, with $z = 24$ in. If the minimum height in the resulting motion is 12 in., determine the initial velocity of the marble v_i and the velocity of the marble at its lowest point.

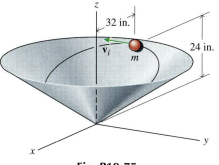

Fig. P19-75

19-76*   A 2-kg particle sliding on a smooth horizontal surface is attached to the end of an elastic cord (Fig. P19-76). The other end of the cord, which has an unstretched length of 400 mm and an elastic constant $k = 250$ N/m, is attached at A. At its closest approach to A ($d = 200$ mm), the particle has a speed of 5 m/s. Determine

a. The velocity of the particle (speed v and direction θ) when the length of the cord is 750 mm and the particle is moving away from A.

b. The length of the elastic cord and the velocity of the particle when the particle is at its farthest from A.

c. The velocity of the particle when the length of the cord is 600 mm and the particle is moving toward A.

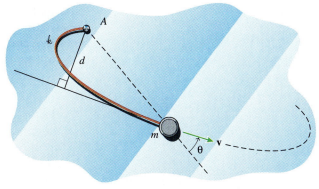

Fig. P19-76

19-77   A 2-lb particle slides on a smooth horizontal surface at the end of an elastic cord (Fig. P19-77). The other end of the cord, which has an unstretched length of 18 in. and an elastic constant of $k = 8$ lb/ft, is attached at A. If $v = 10$ ft/s, $\theta = 40°$, and $\ell = 27$ in. at the instant shown, determine

a. The velocity of the particle (speed v and direction θ) when the tension in the cord is zero.

b. The distance d of closest approach to point A.

c. The length of the elastic cord and the velocity of the particle when the particle is at its farthest from A.

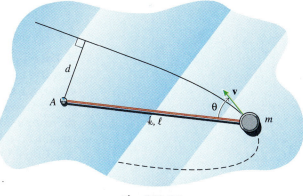

Fig. P19-77

19-78   For a system of n particles, show that the total external moment about an arbitrary point P can be written

$$\Sigma \mathbf{M}_P = \dot{\mathbf{H}}_G + \mathbf{r}_{G/P} \times m\mathbf{a}_G$$

where $\Sigma \mathbf{M}_P = \Sigma \mathbf{r}_{i/P} \times \mathbf{F}_i$, $\mathbf{H}_G = \Sigma \mathbf{r}_{i/G} \times m_i \mathbf{v}_i$, $m = \Sigma m_i$, and $\mathbf{a}_G$ is the absolute acceleration of the mass center of the system of particles.

19-6   SYSTEMS WITH VARIABLE MASS

The kinetics principles developed in the past several chapters apply only to constant systems of particles—systems that neither gain nor lose particles. Many important dynamics problems, however, consist of large systems of particles in which individual particles are not easily identified (such as fluid flow). For these types of problems, it is often more convenient to study the particles in a fixed region of space—a *control volume*—than to study a fixed system of particles. Two of the more common types of variable mass problems are

1. **Steady Flow of Mass.**   In many fluid flow problems, fluid particles enter and leave a control volume at the same rate. Although the mass (total number of particles) of fluid in the control volume at any time is constant, the particles that make up this mass are constantly changing. Since the particles entering have different momenta than the particles exiting the control volume, external forces must be exerted on the control volume even though the total momentum of the particles within the control volume does not change with time. This type of problem is considered in detail in Section 19-6.1.

2. **Systems Gaining or Losing Mass.**   In many problems encountered in dynamics, particles leave (or enter) a control volume at a constant rate over some interval of time. For example, in rocket propulsion, the control volume may consist of the rocket shell and the unburned fuel. As the engine burns fuel, the fuel is expelled from the control volume. Not only does the system decrease in mass as particles are taken away from the system, but the particles are expelled at some velocity relative to the rest of the system. Therefore, the total momentum of the system may vary even in the absence of any external applied force. This type of problem is considered in detail in Section 19-6.3.

19-6.1   Steady Flow of Mass

A knowledge of the forces exerted on fan and turbine blades by a steadily moving fluid stream is important in the analysis of many machines. A complete analysis of such problems will be presented in a course in fluid mechanics. The presentation here is just to illustrate how the momentum principles developed in this chapter apply to such steady flow problems.

Consider the problem of finding the force exerted on a fixed reducing bend in a pipe as a steady stream of fluid passes through it as shown in Fig. 19-29. Fluid enters the bend with some velocity $\mathbf{v}_1$, pressure p_1, and density (mass per unit volume) ρ_1, which are assumed to be constant over the inlet area A_1. The fluid then leaves the bend, with velocity $\mathbf{v}_2$, pressure p_2, and density ρ_2 also assumed constant over the exit area A_2. The flow is assumed steady; that is, there is no increase or decrease of fluid inside the bend. Therefore, the rate at which fluid leaves the bend is exactly the same as the rate at which fluid enters the bend.

A control volume $\mathcal{CV}$ is drawn that encloses a region of fluid bounded by the surface on which the force is desired and surfaces on which the forces are known or can be determined. Furthermore, the

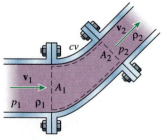

Figure 19-29   Although the total mass of fluid within the pipe bend is constant, the particles that make up this mass are constantly changing.

surfaces bounding the control volume are chosen so that the rate of fluid flow across the surface is either zero or is known or can be easily determined. The system of particles enclosed by the control volume is a variable mass system, since it continually gains particles flowing in and loses an equal number of particles flowing out. Therefore, the momentum principles developed earlier in this chapter for fixed systems of particles do not apply directly to the mass that makes up the control volume.

In order to get a fixed system of particles to which the momentum principles do apply, consider the larger group of particles shown in Fig. 19-30a. This system consists of the particles in the original control volume at time t (having total mass m_t) plus the particles that will enter the control volume in the time interval Δt (having total mass Δm_1). Since all particles within a distance $\Delta s_1 = v_1 \Delta t$ will enter the bend in the time Δt, the volume of the additional region is $V_1 = A_1 \Delta s_1$.[1] The total mass $\mathcal{M}$ of this larger group of particles is

$$\mathcal{M} = m_t + \Delta m_1 = m_t + \rho_1 V_1 = m_t + \rho_1 A_1 v_1 \, \Delta t \qquad (a)$$

At time $t + \Delta t$ this same system of particles will occupy the region shown in Fig. 19-30b. This region consists of the particles in the original control volume at time $t + \Delta t$ (having total mass $m_{t+\Delta t}$) plus those particles that have left the control volume during the time Δt (having total mass Δm_2). The mass of the fixed system of particles is now given by

$$\mathcal{M} = m_{t+\Delta t} + \Delta m_2 = m_{t+\Delta t} + \rho_2 V_2 = m_{t+\Delta t} + \rho_2 A_2 v_2 \, \Delta t \qquad (b)$$

But by the assumption of steady flow, the mass of the particles inside the original control volume is the same at all times; that is, $m_t = m_{t+\Delta t}$. Therefore, combining Eqs. a and b gives in the limit as $\Delta t \to 0$

$$\rho_1 A_1 v_1 = \rho_2 A_2 v_2 \qquad (19\text{-}18)$$

which verifies the earlier statement that the rate at which fluid leaves the bend is exactly the same as the rate at which fluid enters the bend.

Equation 19-18 expresses the principle of *conservation of mass*. The terms in the equation are called the *mass flow rate* $\dot{m} = \rho v A$ and represent the rate at which mass is entering or leaving the control volume. In SI units, the mass flow rate has dimensions of kg/s. In U.S. customary units, the mass flow rate has dimensions of slug/s = lb · s/ft. For the flow of incompressible fluids (fluids for which the density is constant) as well as other constant density flows, the *volume flow rate*

$$Q = \frac{\dot{m}}{\rho} = A_1 v_1 = A_2 v_2 \qquad (c)$$

is often used instead of the mass flow rate. The volume flow rate has dimensions of m³/s in the SI system of units and ft³/s in the U.S. customary system of units.

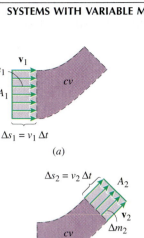

(a)

(b)

Figure 19-30   Since the flow is steady, fluid does not accumulate within the bend. Therefore, the amount of fluid entering the bend in any given amount of time must equal the amount of fluid that leaves the bend in the same amount of time.

---

[1]It is assumed here that the area A_1 was chosen perpendicular to the velocity so that the region is a right cylinder. If the velocity is not perpendicular to the area, then the formula for the volume will include the *sine* of the angle between the velocity and the plane of the area.

At time t the fixed system of particles identified on the previous page has linear momentum

$$\mathbf{L}_t = \Delta m_1 \mathbf{v}_1 + (\mathbf{L}_{\mathcal{CV}})_t = (\dot{m}\,\Delta t)\mathbf{v}_1 + (\mathbf{L}_{\mathcal{CV}})_t \qquad (d)$$

where $(\mathbf{L}_{\mathcal{CV}})_t$ is the linear momentum of all particles in the control volume at time t and $\Delta m_1 \mathbf{v}_1$ is the linear momentum of the particles that are about to enter the control volume at time t. At time $t + \Delta t$ the same system of particles will have linear momentum

$$\mathbf{L}_{t+\Delta t} = (\mathbf{L}_{\mathcal{CV}})_{t+\Delta t} + \Delta m_2 \mathbf{v}_2 = (\mathbf{L}_{\mathcal{CV}})_{t+\Delta t} + (\dot{m}\,\Delta t)\mathbf{v}_2 \qquad (e)$$

But because the flow is steady, $(\mathbf{L}_{\mathcal{CV}})_t = (\mathbf{L}_{\mathcal{CV}})_{t+\Delta t}$ and the linear momentum equation (Eq. 19-3)

$$\mathbf{L}_t + \Sigma(\textstyle\int \mathbf{F}\,dt) = \mathbf{L}_{t+\Delta t} \qquad (f)$$

gives in the limit as $\Delta t \to 0$

$$(\dot{m}\,\Delta t)\mathbf{v}_1 + \Sigma\mathbf{F}\,\Delta t = (\dot{m}\,\Delta t)\mathbf{v}_2 \qquad (g)$$

or

$$\Sigma\mathbf{F} = \dot{m}\,(\mathbf{v}_2 - \mathbf{v}_1) \qquad (19\text{-}19)$$

where $\Sigma\mathbf{F}$ is the sum of all external forces acting on the system of particles inside the control volume.

In Eq. 19-19 it is important to include *all* external forces acting on the system of particles inside the control volume. Therefore, a correct free-body diagram is just as important for these fluid flow problems as it is for any other particle or rigid body problem. Referring to the free-body diagram of the control volume (Fig. 19-31),

$$\Sigma\mathbf{F} = \mathbf{F}_p + \mathbf{W} - p_1 A_1 \mathbf{n}_1 - p_2 A_2 \mathbf{n}_2 \qquad (h)$$

where $\mathbf{F}_p$ is the force exerted by the pipe bend on the fluid in the control volume (the fluid exerts an equal but opposite force back on the pipe bend); $\mathbf{W}$ is the weight of the fluid in the control volume; $\mathbf{n}_1$ and $\mathbf{n}_2$ are the outward-pointing unit normals to the surfaces A_1 and A_2, respectively, and $-p_1 A_1 \mathbf{n}_1$ and $-p_2 A_2 \mathbf{n}_2$ are the forces exerted on the fluid in the control volume by the adjoining portions of the fluid.

An analogous result can be obtained using the angular momentum principle Eq. 19-14 (or Eq. 19-17). Taking the angular momenta and the moments of all external forces with respect to some arbitrary fixed point O (or about the mass center G), gives

$$\mathbf{r}_1 \times [(\dot{m}\,\Delta t)\mathbf{v}_1] + \Sigma\mathbf{M}_O\,\Delta t = \mathbf{r}_2 \times [(\dot{m}\,\Delta t)\mathbf{v}_2] \qquad (i)$$

or

$$\Sigma\mathbf{M}_O = \dot{m}(\mathbf{r}_2 \times \mathbf{v}_2 - \mathbf{r}_1 \times \mathbf{v}_1) \qquad (19\text{-}20)$$

where $\Sigma\mathbf{M}_O = \Sigma(\mathbf{r} \times \mathbf{F})$ is the sum of the moments of all external forces acting on the fluid inside the control volume, and $\mathbf{r}_1$ and $\mathbf{r}_2$ are the position vectors of the centers of the areas A_1 and A_2, respectively. The angular momenta and moments of all forces are to be computed relative to the same fixed point O (or relative to the mass center G).

19-6.2   Common Applications of Steady Flow

Equations 19-19 and 19-20 can be used to solve a wide variety of fluid flow problems, including the reducing elbow of Fig. 19-29 and the sit-

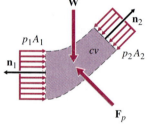

Figure 19-31   The force $\mathbf{F}_p$ shown on the free-body diagram represents the force exerted on the fluid within the pipe bend by the surfaces in contact with the fluid.

uations depicted in Fig. 19-32. Some considerations for the application of these principles to some of the more common classes of flows are as follows:

1. **Enclosed Flows.** In flows through pipes (Fig. 19-29), pipe bends (Fig. 19-32a), and nozzles (Fig. 19-32b), the cross-sectional area of the flow is presumed known at any section needed, and the fluid velocities can be obtained from the flow rate $\dot{m}$ or Q. The fluid pressure in the pipe will, in general, not be negligible or constant and must either be given or be determined from other fluid mechanics principles. The weight of the fluid is usually negligible compared to other forces in the problem unless the control volume is very large. If the pipe length is very long and/or if the pipe diameter is very small, fluid friction on the side walls of the pipe (not shown in the free-body diagram of Fig. 19-31) may also have to be included.

2. **Channel Flows.** In flows of water under a sluice gate (Fig. 19-32c) or over a weir (Fig. 19-32d), the fluid pressure is very important and cannot be neglected. It is shown in fluid mechanics that in a region of the flow where the streamlines are straight and parallel, the fluid pressure increases linearly with depth ($p = \rho g h$, where h is the depth of a point below the surface of the fluid). Although the weight would be needed to find the force on the bottom of the channel, it is not needed to find the force on vertical surfaces such as the sluice gate or the weir shown in Figs. 19-32c and 19-32d. Fluid friction on the bottom or sides or both of the channel is seldom significant compared to the other forces in the problem.

3. **Free Jet Flows.** It is shown in fluid mechanics that the pressure in free jets (fluid flow that is not contained by pipe or channel walls) is the same as that of the surrounding fluid. For the fluid being deflected by the fixed vane of Fig. 19-32e, this means that the pressure in the jet of water approaching the vane and the pressure in the jet of water leaving the vane are both zero. It is further shown for flows such as this that the speed of the jet leaving the vane is the same as that of the jet entering the vane. The fluid weight and frictional forces on the vane are seldom significant.

4. **Stationary Fans.** When air or water passes through a fan, the velocity increases from one side of the blades to the other. Except for a region near the fan, the fluid approaching the fan and the jet of fluid leaving the fan of Fig. 19-32f can be treated as free jets: The pressure can be neglected. While the jet of fluid expelled by the fan (called the slipstream) is usually concentrated and of uniform velocity, the air approaching the fan is usually more dispersed: The intake area is very large, and the intake velocity can be neglected.

5. **Moving Vanes and Propellers.** Flows around moving vanes and moving propellers are not steady flows, and the foregoing equations do not directly apply. However, if the vanes or propellers are moving in a straight line with a constant speed, the flow will appear steady to an observer moving with the vane or propeller. Therefore, these problems can be converted to a steady flow situation and the foregoing equations applied by choosing a coordinate system moving with the vane or propeller. The velocities (and therefore the flowrates!) must be expressed relative to the moving

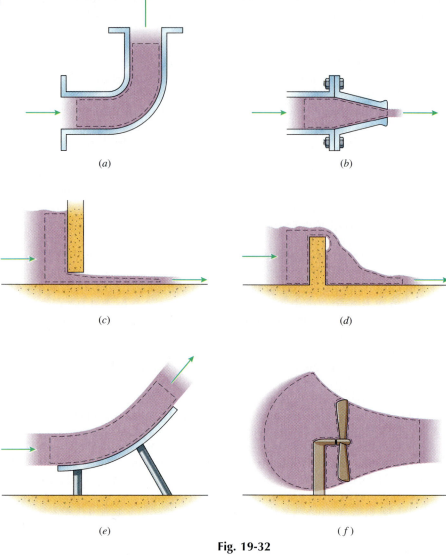

(a)   (b)

(c)   (d)

(e)   (f)

Fig. 19-32

coordinate system. If the vane or propeller is not moving in a
straight line with a constant speed, then alternative equations must
be developed that properly take into account the acceleration of
the flow/coordinate system.

19-6.3   Systems Gaining or Losing Mass

The other type of variable mass system that will be analyzed here is
the system that gains mass by collecting particles (such as a moving
container being filled with water or grain) or that loses mass by ex-
pelling particles (such as a rocket burning fuel). The general procedure
will be developed for a system that is acquiring mass but will apply
equally to both situations.

Consider a body that is absorbing a stream of particles as shown
in Fig. 19-33. The body is obviously a variable mass system and the
momentum principles developed earlier in this chapter do not directly

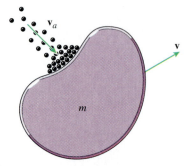

Figure 19-33   The mass of the moving body increases as particles strike
and stick to it.

apply to this body. Instead, define a system of particles consisting of the body (which at time t has mass m and velocity $\mathbf{v}$) and the particle(s) (having total mass Δm and velocity $\mathbf{v}_a$) that will be absorbed in the time interval Δt (Fig. 19-34). This larger system is a fixed system of particles over the interval Δt, and the momentum principles can be used to relate the forces acting on the system and the change in momentum of the system.

At time t the momentum of the system is just the sum of the momenta of the parts (Fig. 19-34a)

$$\mathbf{L}_t = m\mathbf{v} + \Delta m\, \mathbf{v}_a \qquad (j)$$

while the momentum after the particle(s) has (have) been absorbed and all the mass is moving as a single object of mass $m + \Delta m$ and velocity $\mathbf{v} + \Delta\mathbf{v}$ (Fig. 19-34b) is

$$\mathbf{L}_{t+\Delta t} = (m + \Delta m)(\mathbf{v} + \Delta\mathbf{v}) = m\mathbf{v} + \Delta m\, \mathbf{v} + m\, \Delta\mathbf{v} + \Delta m\, \Delta\mathbf{v} \qquad (k)$$

Then if $\mathbf{R}$ is the resultant of all *external* forces acting on the system, the linear momentum principle (Eq. 19-3) gives

$$m\mathbf{v} + \Delta m\, \mathbf{v}_a + \int \mathbf{R}\, dt = m\mathbf{v} + \Delta m\, \mathbf{v} + m\, \Delta\mathbf{v} + \Delta m\, \Delta\mathbf{v} \qquad (l)$$

Dividing through by Δt and rearranging this expression gives

$$\frac{1}{\Delta t}\int_t^{t+\Delta t} \mathbf{R}\, dt = m\frac{\Delta\mathbf{v}}{\Delta t} - \frac{\Delta m}{\Delta t}(\mathbf{v}_a - \mathbf{v}) + \frac{\Delta m\, \Delta\mathbf{v}}{\Delta t} \qquad (m)$$

which in the limit as $\Delta t \to 0$ becomes

$$\mathbf{R} = m\mathbf{a} - \dot{m}(\mathbf{v}_a - \mathbf{v}) = m\mathbf{a} - \dot{m}\mathbf{v}_{a/m} \qquad (19\text{-}21)$$

where $\mathbf{a} = \dot{\mathbf{v}} = \lim\limits_{\Delta t \to 0} \Delta\mathbf{v}/\Delta t$ is the acceleration of the body due to the action of the external forces and the absorbed particles; $\dot{m} = \lim\limits_{\Delta t \to 0} \Delta m/\Delta t$ is the rate at which the body is absorbing mass from the particle stream; $\mathbf{v}_{a/m} = \mathbf{v}_a - \mathbf{v}$ is the relative velocity of the absorbed particles with respect to the body; and $\lim\limits_{\Delta t \to 0} \Delta m\, \Delta\mathbf{v}/\Delta t = \lim\limits_{\Delta t \to 0} \dot{m}\Delta\mathbf{v} = \mathbf{0}$.

The force $\mathbf{R}$ represents the resultant of *all external forces* that act on the system—the body and the piece of mass being absorbed. However, this force resultant *does not include* the forces of action and reaction $\mathbf{P}$ (Fig. 19-35) between the body and the mass being absorbed, since these are internal to the system.

It is interesting to compare Eq. 19-21 with Newton's second law of motion. For example, Newton's second law of motion might be written in the form (the resultant force equals the change in linear momentum)

$$\mathbf{R} = \frac{d(m\mathbf{v})}{dt} = \dot{m}\mathbf{v} + m\dot{\mathbf{v}} \qquad (n)$$

Equation 19-21 agrees with Newton's law in this form only if $\mathbf{v}_a = \mathbf{0}$; that is, the acquired mass is at rest before it is picked up.

If Eq. 19-21 is compared with Newton's second law of motion written in its usual form, $\Sigma\mathbf{F} = m\mathbf{a}$, it is convenient to rearrange the terms of Eq. 19-21

$$\mathbf{R} + \dot{m}\mathbf{v}_{a/m} = m\mathbf{a} \qquad (19\text{-}22)$$

That is, the effect on the body of the particles being absorbed is the same as that of a force in the direction of the relative velocity of mag-

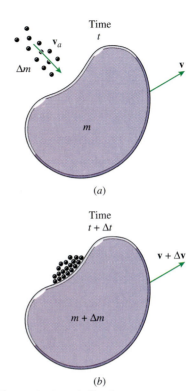

Figure 19-34   A fixed mass system can be obtained by considering the body and the particles that will be absorbed in the time interval Δt as a single large system.

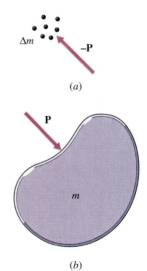

Figure 19-35   As the particles are absorbed by the body, they will exert a force $\mathbf{P}$ on the body. The body will exert an "equal but opposite" force on the particles.

nitude $\dot{m}v_{a/m}$. In fact, applying the linear momentum principle to the single particle Δm (Fig. 19-35a) gives

$$\Delta m \, \mathbf{v}_a + \int (-\mathbf{P}) \, dt = \Delta m(\mathbf{v} + \Delta \mathbf{v}) \qquad (p)$$

which after rearranging, dividing through by Δt, and letting $\Delta t \to 0$ gives

$$\mathbf{P} = \dot{m}(\mathbf{v}_a - \mathbf{v}) = \dot{m}\mathbf{v}_{a/m} \qquad (19\text{-}23)$$

This "effective force" will tend to accelerate the body if the particles are added "from behind" or will tend to decelerate the body if the particles are added "from in front" of the body.

Equation 19-21 may also be used for a body expelling mass such as a rocket burning fuel. In this case, the mass flow rate $\dot{m}$ is negative.[2] Then, by Eqs. 19-22 and 19-23, the effect on the body of the particles being expelled will be the same as that of a force in the direction opposite that of the relative velocity and of magnitude $P = |\dot{m}\mathbf{v}_{a/m}|$. That is, particles expelled "from the rear" will tend to accelerate the body whereas particles expelled "from the front" will tend to decelerate the body. This is the mechanism of propulsion by rockets.

19-6.4   Special Cases of Systems Gaining or Losing Mass

Equations 19-21 through 19-23 may be applied to a wide variety of systems gaining or losing mass from rockets to hoisting cables. The equations simplify for some special cases:

1. **A Rocket Sled.**   When a rocket sled accelerates horizontally along a straight track, the weight and track reaction forces will be normal to the velocity and relative velocity. The drag force due to aerodynamic forces is generally proportional to the square of the speed of the rocket kv^2. Therefore, the component of Eq. 19-22 in the direction of the sled's motion is

$$(m_0 - bt) \, \dot{v} = bu - kv^2 \qquad (19\text{-}24)$$

where b is the constant rate at which the rocket is burning fuel and u is the velocity of the burned gases relative to the sled. If the rocket thrust P is known rather than the mass flow rate and relative velocity, Eq. 19-23 can be combined with Eq. 19-22 to give

$$(m_0 - bt) \, \dot{v} = P - kv^2 \qquad (19\text{-}25)$$

Equation 19-24 or 19-25 is then solved to find the speed of the sled as a function of time.

2. **All External Forces Are Zero.**   When a space ship travels in outer space, it encounters no air resistance. If the spaceship is also far from any planets or stars, then any gravitational force acting on the space ship is also negligible, so $\mathbf{R} = \mathbf{0}$ and Eq. 19-21 becomes

$$m\mathbf{a} = \dot{m}\mathbf{v}_{a/m} \qquad (q)$$

---

[2]That is, for a rocket burning fuel at a constant rate b the rate of change of mass $\dot{m} = -b$. Then, if the initial mass of the rocket is m_0, the mass of the rocket at time t will be $m = m_0 + \dot{m}t = m_0 - bt$. Furthermore, if the velocity of the burned gases with respect to the rocket is $\mathbf{u} = \mathbf{v}_{a/m}$, then the thrust on the rocket, $\mathbf{P} = \dot{m}\mathbf{v}_{a/m} = -b\mathbf{u}$ will be in the direction opposite the relative velocity $\mathbf{u}$.

EXAMPLE PROBLEM 19-11

Water is being expelled from a nozzle at a constant rate of 500 gal/min as shown in Fig. 19-36. The nozzle is attached to a 4-in.-diameter pipe with six bolts and has an exit diameter of 2 in. If the pressure measured in the pipe is 16.45 lb/in.², determine the force in each bolt. (The specific weight of water is 62.4 lb/ft³ and 7.481 gal = 1 ft³.)

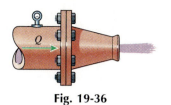

Fig. 19-36

SOLUTION

The flow rate is given as

$$Q = \frac{500 \text{ gal/min}}{(7.481 \text{ gal/ft}^3)(60 \text{ s/min})}$$
$$= 1.114 \text{ ft}^3/\text{s}$$
$$= v_1 A_1 = v_2 A_2$$

from which the velocities are determined to be

$$v_1 = 12.76 \text{ ft/s}$$

and

$$v_2 = 51.06 \text{ ft/s}$$

Applying the x-component of Eq. 19-19

$$\Sigma \mathbf{F} = \dot{m}(\mathbf{v}_2 - \mathbf{v}_1)$$

to the free-body diagram of the water contained in the nozzle (Fig. 19-37a) gives

$$p_1 A_1 - F_n = \rho Q(v_{2x} - v_{1x})$$

or

$$(16.45)\frac{\pi}{4}(4)^2 - F_n = \frac{62.4}{32.2}(1.114)(51.06 - 12.76)$$

which gives

$$F_n = 124.03 \text{ lb}$$

as the force exerted on the water by the nozzle. The water exerts an equal but opposite force back on the nozzle. Then from the free-body diagram of the nozzle (Fig. 19-37b), equilibrium gives the tension in the bolts as

$$T = \frac{F_n}{6} = 20.67 \text{ lb} \qquad\qquad \text{Ans.}$$

Air on the outside of the nozzle exerts a constant pressure of p_{atm} on all external surfaces of the nozzle. Water pressure inside the pipe is 16.45 lb/in.² above p_{atm}. At every point of the nozzle, the air pressure on the outside of the nozzle cancels with the p_{atm} portion of the water pressure inside the nozzle. These canceling forces are removed from the calculations by calling the air pressure zero and referencing all pressures to atmospheric pressure. Since the pressure in the free jet is the same as that of the surrounding atmosphere, $p_2 = 0$.

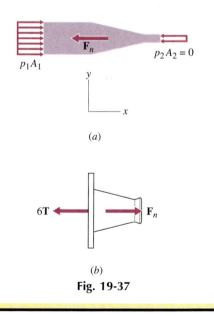

(a)

(b)

Fig. 19-37

Water having a density $\rho = 1000 \text{ kg/m}^3$ flows under a sluice gate as shown in Fig. 19-38. The width of the channel is 2 m and the flowrate is $Q = 3.10 \text{ m}^3/\text{s}$. Compare the force exerted on the gate by the flowing water with the force the water would exert on the gate if it were not moving.

3 m   0.5 m   0.3 m

Fig. 19-38

SOLUTION

The free-body diagram of the water pressing on the gate (Fig. 19-39a) includes the pressure forces of the adjacent water in the channel $\mathbf{F}_1$ and $\mathbf{F}_2$. The magnitudes of these forces are equal to the triangular areas under the pressure-loading diagrams (Fig. 19-39b, 19-39c).

$$F_1 = 0.5[(1000)(9.81)(3)](3)(2) = 88{,}290 \text{ N}$$
$$F_2 = 0.5[(1000)(9.81)(0.3)](0.3)(2) = 882.9 \text{ N}$$

The mass flowrate and velocities of the water are obtained from the volume flowrate $\dot{m} = \rho Q = \rho v_1 A_1 = \rho v_2 A_2$ giving

$$\dot{m} = (1000)(3.10) = 3100 \text{ kg/s}$$
$$v_1 = \frac{3.10}{(3)(2)} = 0.5167 \text{ m/s}$$

and

$$v_2 = \frac{3.10}{(0.3)(2)} = 5.167 \text{ m/s.}$$

Applying the x-component of Eq. 19-19

$$\Sigma \mathbf{F} = \dot{m}(\mathbf{v}_2 - \mathbf{v}_1)$$

to the free-body diagram of the water pressing on the gate (Fig. 19-39a) gives

$$88{,}290 - 882.9 - F_g = (3100)(5.167 - 0.5167)$$

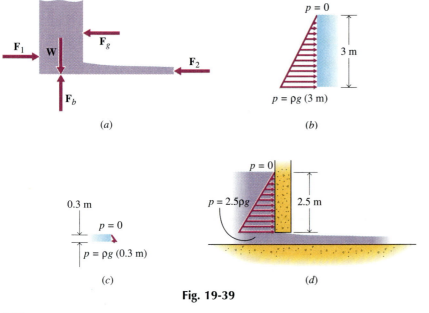

(a)

$p = 0$

3 m

$p = \rho g \text{ (3 m)}$

(b)

0.3 m

$p = 0$

$p = \rho g \text{ (0.3 m)}$

(c)

$p = 0$

$p = 2.5\rho g$

2.5 m

(d)

Fig. 19-39

which gives the force exerted on the water by the sluice gate

$$\mathbf{F}_g = 73{,}000 \text{ N}\leftarrow \qquad \text{on the water}$$

The water exerts an equal and opposite force back on the sluice gate:

$$\mathbf{F}_g = 73.0 \text{ kN} \rightarrow \qquad \text{on the gate} \qquad\qquad \text{Ans.}$$

If the fluid were not moving, it would exert a pressure force on the gate that increases linearly with depth as shown in Fig. 19-39d. The magnitude of this force is equal to the area under the pressure-loading diagram

$$\begin{aligned} F_{gs} &= 0.5[(1000)(9.81)(2.5)](2.5)(2) \\ &= 61{,}300 \text{ N}\rightarrow \qquad \text{on the gate} \qquad\qquad \text{Ans.}\end{aligned}$$

EXAMPLE PROBLEM 19-13

A jet of water ($\gamma = \rho g = 62.4$ lb/ft³) is deflected by a turning vane as shown in Fig. 19-40. The water jet has an absolute velocity of 30 ft/s and a diameter of 1 in. If the turning angle of the vane is 50°, determine the horizontal force $\mathbf{P}$ required to move the vane to the left at a steady rate of 10 ft/s.

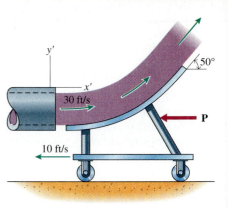

Fig. 19-40

SOLUTION

In a coordinate system moving to the left with the vane (Fig. 19-41a), the flow is steady and Eq. 19-19 applies. In this coordinate system, the water appears to be approaching the vane at a speed of 40 ft/s and the mass flow rate is

$$\dot{m} = \rho Q = \rho v_1 A_1 = \left(\frac{62.4}{32.2}\right)(40)\left(\frac{\pi}{4}\right)\left(\frac{1}{12}\right)^2 = 0.4228 \text{ slug/s}$$

In free-jet problems such as this, the speed of the water leaving the vane is the same as the speed of the water approaching the vane, 40 ft/s. Therefore, applying the x-component of Eq. 19-19

$$\Sigma \mathbf{F} = \dot{m}(\mathbf{v}_2 - \mathbf{v}_1)$$

to the free-body diagram of Fig. 19-41b gives the force exerted on the water by the vane

$$\begin{aligned}-F_{wx} &= \rho Q(v_{2x} - v_{1x}) = (0.4228)(40\cos 50 - 40) \\ \mathbf{F}_{wx} &= 6.04 \text{ lb}\leftarrow \qquad \text{on the water}\end{aligned}$$

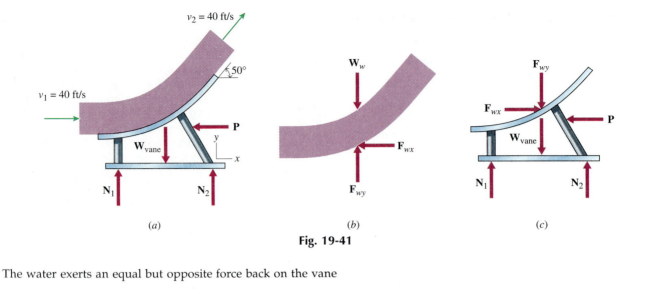

$v_2 = 40$ ft/s

$50°$

$v_1 = 40$ ft/s

$\mathbf{P}$

$\mathbf{W}_{vane}$

y

x

$\mathbf{N}_1$ $\mathbf{N}_2$

(a)

$\mathbf{W}_w$

$\mathbf{F}_{wx}$

$\mathbf{F}_{wy}$

(b)

$\mathbf{F}_{wy}$

$\mathbf{F}_{wx}$ $\mathbf{P}$

$\mathbf{W}_{vane}$

$\mathbf{N}_1$ $\mathbf{N}_2$

(c)

Fig. 19-41

The water exerts an equal but opposite force back on the vane

$$\mathbf{F}_{wx} = 6.04 \text{ lb} \rightarrow \quad \text{on the vane}$$

Finally, applying the equilibrium equation to the vane (Fig. 19-41c) gives the force required to keep the vane moving to the left at a constant speed

$$\mathbf{P} = 6.04 \text{ lb} \leftarrow \quad \text{on the vane} \qquad \text{Ans.}$$

EXAMPLE PROBLEM 19-14

A 1000-kg rocket sled accelerates from rest along a horizontal track (Fig. 19-42). The rocket motor burns fuel at the rate of 15 kg/s, and the velocity of the exhaust gas relative to the sled is 3500 m/s. If 200 kg of the rocket's initial weight is fuel, ignore aerodynamic drag and determine

a. The initial acceleration of the sled.
b. The thrust exerted on the sled by the rocket motor.
c. The velocity and acceleration of the sled an instant before the rocket motor burns out.

Fig. 19-42

SOLUTION

a. Since there are no external forces on the rocket sled in the horizontal direction, the horizontal component of Eq. 19-21 gives

$$R_x = 0 = ma_x - \dot{m}v_{a/m}$$

where $m = 1000 + \dot{m}t$, $\dot{m} = -15$ kg/s, and $v_{a/m} = -3500$ m/s. Solving for the initial acceleration ($t = 0$) gives

$$a_x = \frac{-15}{1000}(-3500) = 52.5 \text{ m/s}^2$$
$$= 5.35 \ g \qquad \text{Ans.}$$

If the rocket sled continued to accelerate at the initial rate of $a_x = 5.35g$ (that is, if the thrust of the motor and the mass of the sled were both constant), the speed of the sled at burnout (after 13.33 s) would be 700 m/s.

486

b. The thrust on the rocket sled is given by Eq. 19-23

$$P = \dot{m}v_{a/m} = (-15)(-3500)$$
$$= 52{,}500 \text{ N} \qquad \text{Ans.}$$

c. The mass of the rocket sled at the instant that the fuel is burnt up will be 800 kg. The acceleration will then be

$$a_x = \frac{-15}{800}(-3500) = 65.6 \text{ m/s}^2$$
$$= 6.69 \ g \qquad \text{Ans.}$$

The velocity of the rocket sled is obtained by integrating the acceleration

$$\frac{dv}{dt} = a_x = \frac{-15}{1000 - 15t}(-3500)$$

giving

$$v(t) = 3500 \ln\left(\frac{1000}{1000 - 15t}\right)$$

The rocket will burn up the 200 kg of fuel in 200/15 = 13.33 s. Therefore, the speed of the sled at burnout will be

$$v(13.33) = 781 \text{ m/s} \qquad \text{Ans.}$$

> The rocket motor expels mass at a constant rate of 3500 m/s relative to the rocket sled. Therefore, $v_{a/m} = -3500$ m/s and the thrust on the sled $P = 52.5$ kN are both independent of the speed of the rocket sled. As the motor burns and expels fuel, the mass of the sled decreases and the constant thrust causes the acceleration of the sled to increase.

PROBLEMS

Introductory Problems

19-79* A 2-in.-diameter jet of water ($\gamma = \rho g = 62.4 \text{ lb/ft}^3$) is deflected through an angle of 120° by a fixed blade as shown in Fig. P19-79. If the water is moving with a horizontal velocity of 20 ft/s as it enters the blade, determine the horizontal and vertical components of the force exerted on the blade by the stream of water.

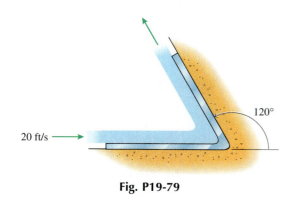

Fig. P19-79

19-80* A 50-mm-diameter jet of water ($\rho = 1000 \text{ kg/m}^3$) strikes the middle of a fixed plate as shown in Fig. P19-80.

If the incoming jet has a speed of 7 m/s and the water leaving the plate flows radially outward, determine the force exerted on the plate by the stream of water.

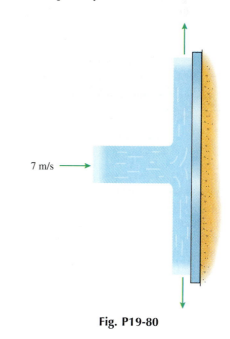

7 m/s

Fig. P19-80

19-81 A child's toy consists of a clown head attached to a garden hose that shoots a jet of water ($\gamma = \rho g = 62.4$ lb/ft^3) vertically upward, suspending a cone-shaped hat as shown in Fig. P19-81. The weight of the hat is 0.5 lb, and the water coming out of the hat makes an angle of 30° with the vertical. If the diameter of the jet entering the hat is 0.5 in., determine the flowrate required to support the hat.

Fig. P19-81

19-82 A 2300-kg helicopter has a slipstream diameter of 10 m. Determine the speed of the air ($\rho = 1.225$ kg/m^3) in the slipstream when the helicopter is hovering.

19-83* A 40,000-lb railcar is being filled with grain at the rate of 5000 lb/s (Fig. P19-83). If the grain is falling straight downward, determine the horizontal force **P** necessary to keep the car moving horizontally at a constant speed of 1 ft/s.

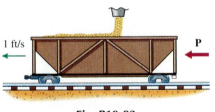

Fig. P19-83

19-84 Sand ($\rho = 1860$ kg/m^3) is flowing out of a dump truck at a rate of 0.7 m^3/s (Fig. P19-84). The initial mass of the truck and the sand is 20,000 kg. If the sand is discharged at an angle of 40° to the horizontal and a speed of 5 m/s relative to the truck, determine the force (magnitude and direction) necessary to keep the truck moving forward at a constant speed of 0.6 m/s.

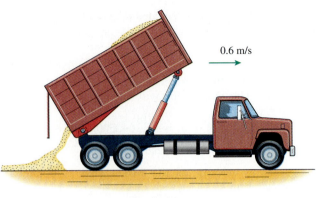

Fig. P19-84

Intermediate Problems

19-85* Water ($\gamma = \rho g = 62.4$ lb/ft^3) is being expelled from a nozzle at a constant rate of 500 gal/min (7.481 gal = 1 ft^3) as shown in Fig. P19-85. The nozzle is attached to a 4-in.-diameter pipe with six bolts and has an exit diameter of 2 in. If the pressure measured in the pipe is 16.45 lb/in.^{2}, determine the force in each of the six bolts holding the nozzle to the pipe. (This is the same nozzle as in Example Problem 19-11 extended and turned through 180°.)

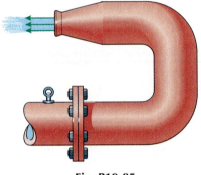

Fig. P19-85

19-86* Water ($\rho = 1000$ kg/m^3) flows over a sharp crested weir at a constant rate of $Q = 3.33$ m^3/s as shown in Fig. P19-86. If the channel is 5 m wide, determine the horizontal component of the force exerted on the weir by the water.

Fig. P19-86

19-87 Water ($\gamma = \rho g = 62.4$ lb/ft³) flows under an inclined sluice gate at a constant rate of $Q = 125$ ft³/s as shown in Fig. P19-87. The width of the channel is 10 ft. Given that the force of the water on the gate is perpendicular to the gate, determine the magnitude of the force exerted on the gate by the water.

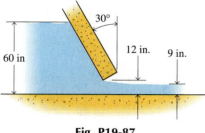

Fig. P19-87

19-88* A 25-mm-diameter jet of water ($\rho = 1000$ kg/m³) is deflected through an angle of 50° by a turning vane as shown in Fig. P19-88. The combined mass of the vane and its base is 10 kg. If the coefficient of static friction between the base and the floor is $\mu_s = 0.25$, determine the maximum velocity of the jet of water for which the vane will not move.

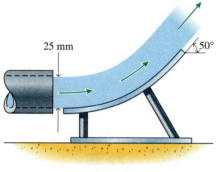

Fig. P19-88

19-89 The table fan of Fig. P19-89 exhausts a 10-in.-diameter jet of air ($\gamma = \rho g = 0.0749$ lb/ft³) having a speed of 20 ft/s. If the fan weighs 5 lb, determine the minimum coefficient of friction between the fan and the table for which the fan will not slide.

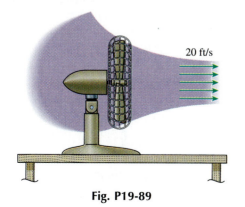

Fig. P19-89

19-90 A 2200-kg spacecraft fires its retrorocket to decrease its orbital speed. The engine burns fuel at the rate of 10 kg/s, and the velocity of the exhaust gas relative to the spacecraft is 2700 m/s. If the initial speed of the spacecraft is 8000 m/s, determine

a. The initial thrust exerted on the spacecraft by the engine.
b. The initial deceleration of the spacecraft.
c. The velocity of the spacecraft after 800 kg of fuel have been burned.

19-91* A small rocket weighs 175 lb when empty, carries 1300 lb of fuel, burns fuel at the rate of 20 lb/s, and has an exhaust velocity of 7000 ft/s. If this rocket is used to launch a 50-lb payload (vertically upward), determine the initial thrust exerted on the payload and the maximum velocity attained by the payload.

19-92 A 6-m-long chain having a total mass of 3 kg is being raised at a constant rate of 3 m/s by a force **F** (Fig. P19-92). Determine the magnitude F of the force when

a. $y = 1$ m.
b. $y = 2$ m.
c. $y = 4$ m.

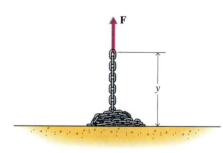

Fig. P19-92

Challenging Problems

19-93* The rocket of Problem 19-91 is to be replaced with a two-stage rocket having the same total shell weight and carrying the same amount of fuel. The first stage of the two-stage rocket weighs 100 lb when empty, carries 750 lb of fuel, burns fuel at the rate of 20 lb/s, and has an exhaust velocity of 7000 ft/s. When the first stage's fuel is used up, its shell is discarded and the second stage is ignited. The second stage weighs 75 lb when empty, carries 550 lb of fuel, burns fuel at the rate of 15 lb/s, and also has an exhaust velocity of 7000 ft/s. If this rocket is used to launch a 50-lb payload (vertically upward), determine

a. The initial thrust exerted on the payload by the engine.
b. The velocity of the rocket when the first stage burns out and is discarded.
c. The maximum velocity attained by the payload.

19-94* The rocket sled shown in Fig. P19-94 has a mass of 8000 kg (which includes 3000 kg of fuel). The rocket motor burns fuel at the rate of 800 kg/s, and the velocity of the exhaust gas relative to the sled is 3000 m/s. The horizontal resistance (due to the track and air) acting on the sled is given by $F_R = 400v$, where F_R is in newtons and v is the velocity of the sled in meters per second. Determine the maximum velocity attained by the sled.

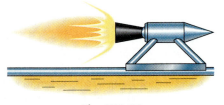

Fig. P19-94

19-95 A 14-ft-long chain having a total weight of 21 lb is being raised by a constant force $F = 9$ lb (Fig. P19-95). If the chain starts from rest when $y = 1$ ft, determine

a. The speed of the chain when $y = 5$ ft.
b. The maximum speed attained by the chain.
c. The maximum height attained by the chain.

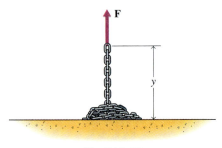

Fig. P19-95

19-96* A 6-m-long chain having a total mass of 3 kg is loosely coiled on the floor (Fig. P19-96). The upper end of the chain is attached to a lightweight cord, which passes over a small frictionless pulley. The other end of the cord is tied to a 1.5-kg block. If the system is released from rest with $y = 1$ m, determine

a. The maximum upward velocity $\dot{y}$ attained by the upper end of the chain.
b. The maximum height y_{max} reached by the upper end of the chain.

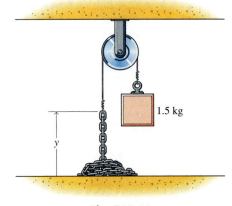

Fig. P19-96

19-97 A 24-ft-long chain having a total weight of 12 lb is loosely coiled on the floor (Fig. P19-97). The upper end of the chain is attached to a lightweight cord, which passes over a small frictionless pulley 20 ft above the floor. The other end of the cord is tied to an 8-lb block. If the system is released from rest with $h = 15$ ft and $y = 1$ ft, determine the speed of the block just before it hits the floor.

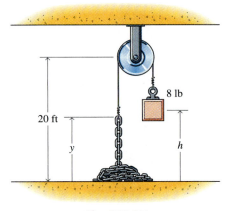

Fig. P19-97

19-98 A long, uniform chain having a density of 0.25 kg/m is piled on a horizontal surface as shown in Fig. P19-98. If one end of the chain falls through a hole, determine the speed $\dot{y}$ of the end of the chain when $y = 3$ m. (Assume that all links are at rest until the moment that they fall through the hole.)

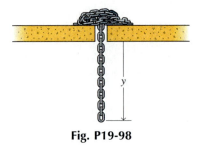

Fig. P19-98

SUMMARY

All of the study of kinetics is based on Newton's second law of motion. In Chapters 15 and 16, Newton's second law was used directly to relate the forces acting on particles and rigid bodies and the resulting acceleration of the particles and the rigid bodies. When information about the acceleration or the value of a force at an instant is desired, Newton's second law is usually the easiest method to use.

In Chapters 17 and 18, Newton's second law was integrated with respect to position to get the principle of work and energy. Since the principle of work and energy is just a combination of Newton's second law and the principles of kinematics, any problem that can be solved using the principle of work and energy can also be solved using Newton's second law. However, the principle of work and energy is particularly useful for solving problems in which the speeds of a body for two different positions of its motion are to be related and the forces involved can be expressed as functions of the position of the body.

In this chapter, Newton's second law was integrated with respect to time to get the principle of impulse and momentum. The resulting equations are useful for solving problems in which the velocity of a body for two different instants of time are to be related and the forces involved can be expressed as functions of time. The principles of impulse and momentum were found to be particularly useful for solving problems involving collisions between bodies and problems involving variable mass systems.

The principle of linear impulse and momentum is expressed by

$$(m\mathbf{v})_i + \int_{t_i}^{t_f} \mathbf{R} \, dt = (m\mathbf{v})_f \tag{19-1}$$

The final linear momentum $\mathbf{L}_f = (m\mathbf{v})_f$ *of a particle is the vector sum of its initial linear momentum* $\mathbf{L}_i = (m\mathbf{v})_i$ *and the impulse* $\int \mathbf{R} \, dt$ *of the resultant of all forces acting on the particle.* This equation is a vector equation representing three scalar equations. The three scalar equations are completely independent of each other.

The principle of angular impulse and momentum is expressed by

$$\mathbf{H}_{Oi} + \int_{t_i}^{t_f} \mathbf{M}_O \, dt = \mathbf{H}_{Of} \tag{19-12}$$

The final angular momentum $\mathbf{H}_{Of} = [\mathbf{r}_{P/O} \times (m\mathbf{v})]_f$ *of a particle about a fixed point O is the vector sum of its initial angular momentum* $\mathbf{H}_{Oi} = [\mathbf{r}_{P/O} \times (m\mathbf{v})]_i$ *about O and the angular impulse* $\int \mathbf{M}_O \, dt$ *about O of the resultant of all forces acting on the particle during the time interval.* Like the principle of linear impulse and momentum, Eq. 19-12 is a vector equation representing three scalar equations. In planar problems, only the component perpendicular to the plane gives useful information.

An impact or collision between two bodies is an event that occurs in a very brief interval of time. It is usually accompanied by relatively large reaction forces between the two bodies and correspondingly large changes of velocity of one or both bodies.

When two particles collide head-on, the relative velocities of the particles before and after the impact are related by the coefficient of restitution

$$e = -\frac{v_{Bf} - v_{Af}}{v_{Bi} - v_{Ai}} = -\frac{(v_{B/A})_f}{(v_{B/A})_i} \qquad (19\text{-}7)$$

When two particles collide obliquely, Eq. 19-7 relates the normal components of the relative velocities.

Neither the work–energy method nor the impulse–momentum method is suitable for solving all problems. The maximum benefit from these methods is realized by choosing the particular method most suitable for a particular problem or part of a problem. It is often useful to combine all three methods—impulse–momentum, work–energy, and Newton's second law—to solve particular problems.

REVIEW PROBLEMS

19-99* A 25-lb shell is fired from a 3000-lb gun as shown in Fig. P19-99. The shell leaves the barrel of the gun with a velocity of 1500 ft/s at an angle of 30° to the horizontal. Assume that the barrel is rigidly attached to the frame of the gun and that the gun is free to move horizontally. If the shell leaves the end of the barrel 0.005 s after firing, determine

a.   The resultant vertical force exerted on the gun by the ground.
b.   The recoil velocity of the gun.

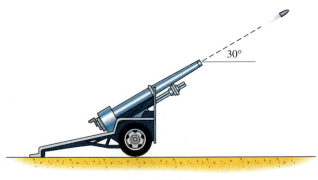

Fig. P19-99

19-100* A 5-kg particle is sliding along a flat, horizontal, frictionless surface at $v_i = 4$ m/s as shown in Fig. P19-100. When the particle is 10 m from the wall, it explodes and splits into two pieces $m_A = 3$ kg and $m_B = 2$ kg. If the 3-kg piece hits the wall 3 s after the explosion at $y_A = 7.5$ m, determine

a.   The impulse exerted on particle A by the explosion.
b.   The velocity $\mathbf{v}_{A/B}$ of particle A relative to particle B immediately after the explosion.
c.   The position y_B at which particle B hits the wall.
d.   The time difference between when particle A hits the wall and when particle B hits the wall.

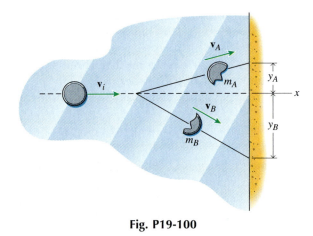

Fig. P19-100

19-101 The soccer ball shown in Fig. P19-101 weighs 1 lb and has a velocity of 20 ft/s at 40° to the horizontal prior to striking the soccer player's head. After the soccer player hits the ball, it has a velocity of 30 ft/s at 20° to the horizontal. If the duration of the impact is 0.15 s, determine the average force (magnitude and direction) exerted on the soccer ball by the soccer player's head.

20 ft/s

40°

20°

30 ft/s

Fig. P19-101

19-102 Two cars collide at an intersection (Fig. P19-102). Car A has a mass of 1200 kg and car B has a mass of 1500 kg. On impact, the wheels of both cars lock and the cars slide ($\mu_k = 0.2$). After impact, the cars become entangled and move as a single unit a distance of 10 m at an angle of $\theta = 60°$. Determine the speeds of the cars v_A and v_B just prior to the impact.

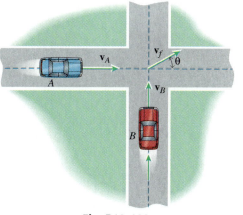

$\mathbf{v}_A$

$\mathbf{v}_f$

θ

A

$\mathbf{v}_B$

B

Fig. P19-102

19-103* A 3200-lb car moving at 45 mi/h collides head-on with a stationary 2200-lb car. On impact, the wheels of the cars lock and slide ($\mu_k = 0.5$). If the cars become entangled and move as a single unit after the impact,

a. Estimate the distance that the cars move after the collision.

b. Determine the amount of kinetic energy lost in the collision.

19-104* Water ($\rho = 1000 \text{ kg/m}^3$) flows through a 40-mm-diameter pipe at a constant rate of $Q = 0.15 \text{ m}^3/\text{min}$ (Fig. P19-104). A nozzle attached to the pipe has an exit diameter of 20 mm and bends the flow through a 30° angle. If the

nozzle is attached to the pipe using four bolts and the pressure in the pipe just before the nozzle is 29.68 kN/m², determine the average tension in the four bolts. (Assume that the bend is in a horizontal plane and neglect the weight of water in the nozzle.)

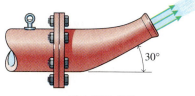

30°

Fig. P19-104

19-105 In a pool shot, the cue ball knocks the 1-ball into the corner pocket as shown in Fig. P19-105. If the coefficient of restitution is $e = 0.9$, determine the distance d to where the cue ball will strike the cushion.

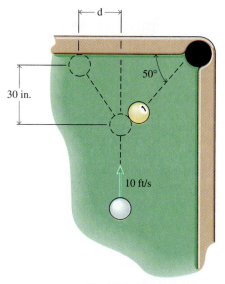

d

30 in.

50°

10 ft/s

Fig. P19-105

19-106* A ballistic pendulum consists of a 3-kg box of sand suspended from a lightweight cord 2 m long (Fig. P19-106). A 0.05-kg bullet strikes the box and becomes embedded in the sand. If the maximum angle of swing of the pendulum following the impact is 25°, determine

a. The common speed of the sand and the bullet immediately after the impact.

b. The initial speed v_i of the bullet.

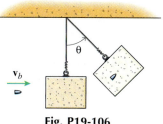

$\mathbf{v}_b$

θ

Fig. P19-106

493

19-107 A 2-lb sphere A is hanging motionless at the end of an inextensible cord as shown in Fig. P19-107 when it is struck by an identical sphere B rolling along the horizontal surface. The distance from the ceiling to the center of sphere A is 3 ft, and the center of the sphere is initially level with the horizontal surface. If the coefficient of restitution of the collision is $e = 0.8$, determine

a. The angle that the final velocity of sphere B makes with the horizontal.
b. The maximum angle θ_A through which sphere A will swing as a result of the impact.

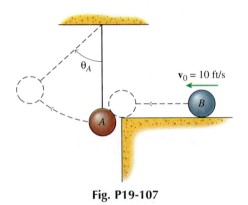

Fig. P19-107

19-108 The orbit maneuvering engine of an 1800-kg spacecraft burns fuel at the rate of 10 kg/s, and the velocity of the exhaust gas relative to the spacecraft is 2700 m/s. Initially, the spacecraft is in a circular orbit 160-km above the surface of the earth with a velocity of 7810 m/s. If the satellite uses its engine to increase its speed to 8190 m/s, determine

a. The initial thrust exerted on the spacecraft by the engine.
b. The length of time that the engine must be on if the mass of the spacecraft and the thrust are assumed constant.
c. The length of time that the engine must be on, taking into account the decrease in mass of the spacecraft as it burns fuel.

19-109* A small bag of sand weighing 10 lb swings in a vertical xz-plane at the end of a 5-ft-long rope as shown in Fig. P19-109. When the bag is at the end of its swing ($\theta = 20°$ and $\dot{\theta} = 0$), it is struck by a 2.1-oz bullet traveling at 825 ft/s in the y-direction. If the bullet embeds in the sand and the rope remains straight, determine for the ensuing motion

a. The maximum angle θ_{max} that the rope will make with the vertical.
b. The velocity of the sandbag when $\theta = \theta_{max}$.
c. The tension in the rope when $\theta = \theta_{max}$.

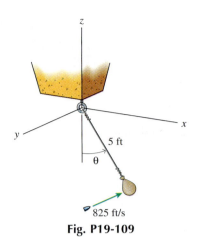

Fig. P19-109

19-110 Two collars are sliding freely on a horizontal rod as shown in Fig. P19-110. Show that the final speeds of the two collars are given by

$$v_{Af} = v_{Ai} - \frac{(1 + e)m_B}{m_A + m_B}(v_{Ai} - v_{Bi})$$

$$v_{Bf} = v_{Bi} + \frac{(1 + e)m_A}{m_A + m_B}(v_{Ai} - v_{Bi})$$

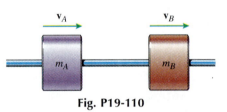

Fig. P19-110

19-111 Two collars are sliding freely on a horizontal rod as shown in Fig. P19-110. Show that the maximum decrease in kinetic energy of the system (which corresponds to $e = 0$) is

$$\Delta T = \frac{1}{2}\left(\frac{m_A m_B}{m_A + m_B}\right)(v_{Ai} - v_{Bi})^2$$

19-112 An 18,000-kg railcar is being filled with grain at the rate of 2000 kg/s (Fig. P19-112). If the grain enters the car at an angle of 60° and a speed of 6 m/s, determine the initial acceleration of the railcar.

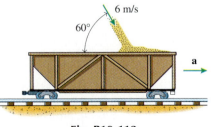

Fig. P19-112

19-113* A 1-lb block of wood is at rest on a rough ($\mu_k = 0.25$) horizontal surface when it is struck by a 0.25-oz bullet (Fig. P19-113). On impact, the bullet becomes embedded in the wood. If the block slides 25 ft before coming to rest again, determine

a. The speed of the block and the bullet immediately after impact.
b. The initial speed v_i of the bullet.

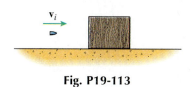

Fig. P19-113

19-114* A 50,000-kg railcar A rolls with an initial speed of $v_A = 3$ m/s along a straight and level track. Car A collides with and couples to a second railcar B of mass 80,000 kg, which has an initial speed of $v_B = 2$ m/s (Fig. P19-114). Determine $\mathbf{v}_f$, the common final velocity of the two cars, if car B was initially moving

a. In the same direction as car A (as shown).
b. In the opposite direction.

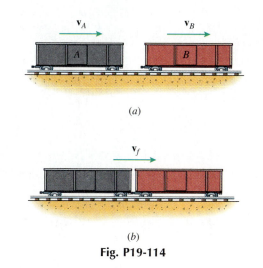

(a)

(b)

Fig. P19-114

In the absence of external moments, the angular momentum vector of the spinning top will be constant. The axis of the top will remain vertical and the rotation rate of the top will remain constant.

Chapter 20

KINETICS OF RIGID BODIES: IMPULSE AND MOMENTUM

20-1   INTRODUCTION

The linear and angular momentum principles for particle motion were derived in the last chapter. It was seen that these equations are first integrals of the equations of motion with respect to time. The resulting equations relate the forces acting on the particle, the velocity of the particle, and time. Therefore, these principles are particularly useful for solving problems in which the velocities of a body at two different instants of time are to be related and the forces involved can be expressed as functions of time.

The impulse–momentum principles were also applied to an arbitrary system of interacting particles. All of those results apply immediately to a rigid body because a rigid body is just a system of interacting particles. All that remains is to simplify the general results using the relative velocity equation that relates the velocities of particles in a rigid body.

20-2   LINEAR IMPULSE AND MOMENTUM OF A RIGID BODY

The linear momentum of a system of particles—rigid or non-rigid—was defined in Section 19-3 as the sum of the linear momenta of the individual particles. Using the definition of the mass center, this was written

$$\mathbf{L} = \sum_{\ell} \mathbf{L}_{\ell} = \sum_{\ell} (m\mathbf{v})_{\ell} = m\mathbf{v}_G \qquad (a)$$

For a continuous rigid body, the sum must be replaced with an integral

$$\mathbf{L} = \int d\mathbf{L} = \int \mathbf{v}\, dm = m\mathbf{v}_G \qquad (b)$$

Then the Linear Impulse–Momentum Principle was written (Eq. 19-5)

$$m(\mathbf{v}_G)_i + \sum_\ell \int_{t_i}^{t_f} \mathbf{R}_\ell\, dt = m(\mathbf{v}_G)_f \qquad \textbf{(20-1)}$$

where $\sum_\ell \int_{t_i}^{t_f} \mathbf{R}_\ell dt$ is the impulse of all external forces acting on the system of particles and the impulses due to internal forces have no effect and can be ignored. But the system of particles is arbitrary, and Eq. 20-1 applies equally to a system of independent, interacting particles and to rigid bodies.

20-3   ANGULAR IMPULSE AND MOMENTUM OF A RIGID BODY IN PLANE MOTION

The angular momentum of a system of particles—rigid or non-rigid—was also defined in Section 19-5 as the sum of the angular momenta of the individual particles. For the general system of interacting particles of Section 19-5, this led to the differential statements of the Angular Impulse–Momentum Principles (Eqs. 19-13 and 19-16):

$$\sum_\ell \mathbf{M}_{O\ell} = \frac{d}{dt}\mathbf{H}_O \qquad \text{and} \qquad \sum_\ell \mathbf{M}_{G\ell} = \frac{d}{dt}\mathbf{H}_G \qquad \textbf{(20-2)}$$

and the integral statements of the Angular Impulse–Momentum Principle (Eqs. 19-14 and 19-17):

$$(\mathbf{H}_O)_i + \int_{t_i}^{t_f} \sum_\ell \mathbf{M}_{O\ell}\, dt = (\mathbf{H}_O)_f \qquad \textbf{(20-3a)}$$

$$(\mathbf{H}_G)_i + \int_{t_i}^{t_f} \sum_\ell \mathbf{M}_{G\ell}\, dt = (\mathbf{H}_G)_f \qquad \textbf{(20-3b)}$$

where $\mathbf{H} = (\sum \mathbf{r}_\ell \times m\mathbf{v}_\ell)$, O is a fixed point, and G is the mass center of the system of particles. Again, the system of particles is arbitrary and Eqs. 20-2 and 20-3 apply equally to an arbitrary system of particles and in the system of particles that make up a rigid body.

The angular momentum of a particle can be calculated relative to any point, fixed or moving. For an arbitrary system of interacting particles, the particles move independently and the expression of the Angular Impulse–Momentum Principle relative to a fixed point O is usually the most useful. For rigid bodies, however, velocities of points in the body are related by the angular velocity, and the expression of the Angular Impulse–Momentum Principle relative to the mass center is usually the most useful.

Let $\mathbf{r}_{\ell/G}$ be the position of the element of mass dm relative to the mass center G of the rigid body shown in Fig. 20-1. If the absolute velocity of dm is denoted by $\mathbf{v}_\ell = \dot{\mathbf{r}}_{\ell/O}$, the angular momentum of dm about G is the moment of the linear momentum

$$d\mathbf{H}_G = (\mathbf{r}_{\ell/G} \times \mathbf{v}_\ell)\, dm \qquad (c)$$

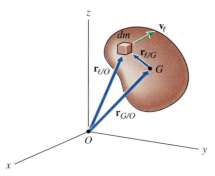

Figure 20-1   The angular momentum of a rigid body is the sum of the angular momenta of the particles dm that make up the body.

Then the angular momentum of the entire rigid body is

$$\mathbf{H}_G = \int d\mathbf{H}_G = \int (\mathbf{r}_{\ell/G} \times \mathbf{v}_\ell)\, dm$$

$$= \int \mathbf{r}_{\ell/G} \times (\mathbf{v}_G + \mathbf{v}_{\ell/G})\, dm \qquad (d)$$

where the absolute velocity has been replaced using the relative velocity equation $\mathbf{v}_\ell = \mathbf{v}_G + \mathbf{v}_{\ell/G} = \mathbf{v}_G + \boldsymbol{\omega} \times \mathbf{r}_{\ell/G}$ and $\mathbf{v}_G$ is the velocity of the mass center of the rigid body.

20-3.1 Planar Motion of a Rigid Body

For the planar motion of a rigid body, the angular velocity is perpendicular to the plane of the motion $\boldsymbol{\omega} = \omega\mathbf{k}$. Therefore, the angular momentum of a rigid body about an axis through its mass center G can be written

$$\mathbf{H}_G = \int \mathbf{r}_{\ell/G} \times (\mathbf{v}_G + \omega\mathbf{k} \times \mathbf{r}_{\ell/G})\, dm$$

$$= \left(\int \mathbf{r}_{\ell/G}\, dm \right) \times \mathbf{v}_G + \int \mathbf{r}_{\ell/G} \times (\omega\mathbf{k} \times \mathbf{r}_{\ell/G})\, dm \qquad (e)$$

In the first term, the velocity of the mass center $\mathbf{v}_G$ has been taken outside the integral since it is the same for every element of mass dm. But the integral $\int \mathbf{r}_{\ell/G}\, dm$ is zero by the definition of the mass center, since the position vector $\mathbf{r}_{\ell/G}$ is measured from the mass center.

In the remaining term let $\mathbf{r}_{\ell/G} = x\mathbf{i} + y\mathbf{j} + z\mathbf{k}$. Then, taking the constant ω outside the integral and expanding the triple cross product gives

$$\mathbf{H}_G = -\omega \int xz\, dm\, \mathbf{i} - \omega \int yz\, dm\, \mathbf{j} + \omega \int (x^2 + y^2)\, dm\, \mathbf{k}$$

$$= -\omega I_{Gxz}\mathbf{i} - \omega I_{Gyz}\mathbf{j} + \omega I_{Gz}\mathbf{k} \qquad (20\text{-}4)$$

where $I_{Gz} = \int (x^2 + y^2)\, dm$ is the moment of inertia with respect to the z-axis and $I_{Gxz} = \int xz\, dm$ and $I_{Gyz} = \int yz\, dm$ are the products of inertia of the rigid body with respect to planes through the mass center.[1] If the rigid body is symmetric about the plane of motion (e.g., a slab of uniform thickness in the z-direction or a cylinder with axis parallel to the z-axis) or if the z-axis through G is an axis of symmetry,[2] then the products of inertia I_{Gxz} and I_{Gyz} will be zero and

$$\mathbf{H}_G = \omega I_G \mathbf{k} \qquad (20\text{-}5)$$

where $I_G = I_{Gz}$ is the moment of inertia of the rigid body with respect to an axis through the mass center G and perpendicular to the plane

[1]Note that if the angular momentum is calculated relative to a fixed point O rather than the mass center G, then $\int \mathbf{r}_{\ell/O}\, dm = m\mathbf{r}_{G/O} \neq \mathbf{0}$ and the first term will reduce to $m\mathbf{r}_{G/O} \times \mathbf{v}_G$ rather than zero. Also, the position vectors in the second term will be measured from different points and

$$\int \mathbf{r}_{G/O} \times (\omega\mathbf{k} \times \mathbf{r}_{\ell/G})\, dm$$

will not give the moments of inertia in Eq. 20-4.

[2]These two situations are special cases of the more general situation of the z-axis through G being a principal axis of inertia.

of motion. Then, substitution of Eq. 20-5 in Eq. 20-3*b* gives the differential form of the Angular Impulse–Momentum equation for a rigid body in plane motion

$$\sum \mathbf{M}_G = \frac{dI_G\omega}{t}\,\mathbf{k} = I_G\alpha\mathbf{k} \qquad (20\text{-}6)$$

where $\sum \mathbf{M}_G$ is the sum of moments of the external forces about the mass center G. For planar motion in which the forces lie in the plane of motion, the moment $\mathbf{M}_G = M_G\mathbf{k}$, so the x- and y-components of Eq. 20-6 are satisfied identically.[3] Then the z-component of Eq. 20-6 can be written

$$\sum M_G = \frac{dI_G\omega}{dt} = I_G\alpha \qquad (20\text{-}7)$$

which is just one of the general equations of motion for rigid bodies (Eq. 16-19*b*). Finally, integration of Eq. 20-7 with respect to time gives the integral form of the Angular Impulse–Momentum equation for a rigid body

$$(I_G\omega)_i + \int_{t_i}^{t_f} \sum M_G\, dt = (I_G\omega)_f \qquad (20\text{-}8)$$

Equation 19-17 (which applies to any system of interacting particles—rigid or non-rigid) and Eq. 20-8 (which applies to a rigid body) both state that the angular impulse $\int_{t_i}^{t_f} \sum M_G\, dt$ acting on a system of particles is equal to the change in the angular momentum $(H_G)_f - (H_G)_i$ of the system of particles. The only difference between Eqs. 19-17 and 20-8 is in the manner in which the initial and final angular momentum is calculated. Therefore, the use of Eq. 20-8 requires only that the body behave rigidly at the initial time t_i and at the final time t_f so that H_G can be computed using $I_G\omega$ at these times. Between the initial and final instants of time, parts of the body may move relative to each other and $(I_G)_i$ may not be the same as $(I_G)_f$.

20-3.2   Rotation About a Fixed Axis

A common type of problem encountered in dynamics is the rotation of a rigid body about a fixed axis (Fig. 20-2). In this case, all particles in the body travel in circular paths in planes perpendicular to the axis. Therefore, the motion is a plane motion and Eqs. 20-1 through 20-8 apply. Although the foregoing equations apply directly to this type of problem, the Angular Impulse–Momentum Principle can be simplified by combining it with the Linear Impulse–Momentum Principle.

Select a coordinate system with the z-axis along the axis of rotation and origin O where the axis of rotation penetrates the plane of motion (the plane containing the mass center). Then the mass center G will travel in a circular path about the axis of rotation with a velocity $\mathbf{v}_G = \omega\mathbf{k} \times \mathbf{r}_{G/O}$, where $\mathbf{r}_{G/O} = x_G\mathbf{i} + y_G\mathbf{j}$. Therefore, the sum of the an-

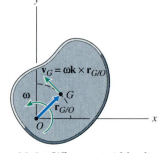

Figure 20-2 When a rigid body rotates about a fixed axis, the mass center travels along a circular path about the fixed axis.

---

[3]Note that if the products of inertia I_{Gxz} and I_{Gyz} are not zero, then the angular momentum will also have x- and y-components even for planar motion. This means that moment components in the x- and/or y-direction will be required to keep the motion in the xy-plane if the angular velocity is changing.

gular momentum about G and the moment of the linear momentum about O gives

$$\mathbf{H}_G + \mathbf{r}_{G/O} \times m\mathbf{v}_G = I_G\omega\mathbf{k} + \mathbf{r}_{G/O} \times m(\omega\mathbf{k} \times \mathbf{r}_{G/O})$$
$$= I_G\omega\mathbf{k} + m(x_G^2 + y_G^2)\omega\mathbf{k} = I_O\omega\mathbf{k} = \mathbf{H}_O \qquad (f)$$

Similarly, the sum of the angular impulse about G and the moment of the linear impulse about the fixed point O gives

$$\int_{t_i}^{t_f} \sum_{\ell} M_{G\ell}\mathbf{k}\, dt + \mathbf{r}_{G/O} \times \int_{t_i}^{t_f} \sum \mathbf{R}_\ell\, dt$$

$$= \int_{t_i}^{t_f} \left(\sum_{\ell} (\mathbf{r}_{\ell/G} \times \mathbf{R}_\ell) + \left(\mathbf{r}_{G/O} \times \sum \mathbf{R}_\ell \right) \right) dt$$

$$= \int_{t_i}^{t_f} \sum_{\ell} \left((\mathbf{r}_{\ell/G} + \mathbf{r}_{G/O}) \times \mathbf{R}_\ell \right) dt$$

$$= \int_{t_i}^{t_f} \sum_{\ell} M_{O\ell}\mathbf{k}\, dt \qquad (g)$$

Therefore, adding the moment of Eq. 20-1 about the fixed point O to Eq. 20-8 gives

$$(I_O\omega)_i + \int_{t_i}^{t_f} \sum_{\ell} M_{O\ell}\, dt = (I_O\omega)_f \qquad \textbf{(20-9)}$$

That is, for a rigid body rotating about a fixed axis, the change in angular momentum about the axis of rotation is equal to the angular impulse about the axis.

20-3.3 Graphical Representation of the Linear and Angular Momentum Principles

Together, Eqs. 20-1 and 20-8 say that the momentum of the particles making up the rigid body can be replaced with an equivalent "force-couple" system at the mass center G. The equivalent "force" is equal to the linear momentum vector $\mathbf{L} = m\mathbf{v}_G$, and the equivalent "couple" is equal to the angular momentum vector $\mathbf{H}_G = I_G\omega\mathbf{k}$. Then the results of Eqs. 20-1 and 20-8 can be summarized graphically as shown in the kinetic diagrams of Fig. 20-3. That is, the sum of the equivalent "force-couple" of the momentum at time t_i (Fig. 20-3a) and the equivalent force-couple of the impulses (Fig. 20-3b) is equal to the equivalent "force-couple" of the momentum at time t_f (Fig. 20-3c).

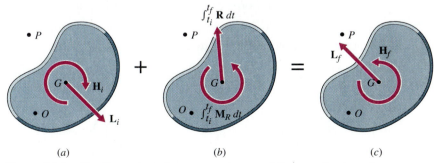

(a)   (b)   (c)

Figure 20-3 The linear impulse–momentum and the angular impulse–momentum principles are both contained in this graphical representation. The linear momenta and linear impulse act through the mass center; the angular momenta and angular impulse are with respect to the mass center.

The graphical representation of Fig. 20-3 also includes the special case of rotation about a fixed axis. If O is a point on the axis of rotation as described in Section 20-3.2, then computing the moment of the equivalent force-couple systems in each part of the figure gives Eq. 20-9. For example, in Fig. 20-3a the velocity of the mass center $\mathbf{v}_G = \omega \times \mathbf{r}_{G/O}$ is perpendicular to $\mathbf{r}_{G/O}$, and the moment of the force-couple system about O is

$$H_G + r_{G/O}mv_G = I_G\omega + r_{G/O}mr_{G/O}\omega$$
$$= (I_G + mr_{G/O}^2)\omega = I_O\omega \qquad (h)$$

as was obtained in Section 20-3.2.

The graphical representation of Fig. 20-3 can also be used to write the Angular Impulse–Momentum Principle about an arbitrary fixed point P. However, if the body is not rotating about an axis through P, then $\mathbf{v}_G \neq \omega \times \mathbf{r}_{G/P}$ and the sum of H_G and the moment of the linear momentum will *not* reduce to $I_P\omega$.

20-3.4   Center of Percussion

Just as a force and couple can be reduced to its simplest form (its resultant), it is possible to reduce the linear and angular momentum vectors of Fig. 20-3 to a single, equivalent linear momentum vector. The "resultant" will be equal to the linear momentum of the mass center $m\mathbf{v}_G$ and will act along a line in the direction of the linear momentum $m\mathbf{v}_G$ and located a distance

$$d = \frac{I_G\omega}{mv_G} \qquad (i)$$

away from the mass center (Fig. 20-4).

In particular, for a body rotating about a fixed axis through O, the system linear and angular momenta at the mass center G (Fig. 20-5a) are equivalent to the system linear momentum $m\mathbf{v}_G$ at point P (Fig. 20-5b). The linear momentum is clearly the same for both kinetic diagrams. The angular momentum will also be the same if the position of P is chosen such that

$$r_P(mv_G) = I_G\omega + r_G(mv_G) \qquad (j)$$

The point P located in this manner is called the *center of percussion*.

Note that the location of the center of percussion depends on the motion of the body as well as on the size, shape, and mass distribution of the body. Since the body of Fig. 20-5 is rotating about a fixed axis, $v_G = r_G\omega$, and therefore

$$r_P(mr_G\omega) = mk_G^2\omega + r_G(mr_G\omega) \qquad (k)$$

where k_G is the radius of gyration of the body relative to an axis through

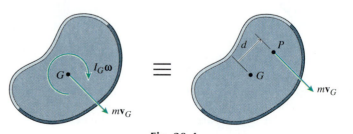

Fig. 20-4

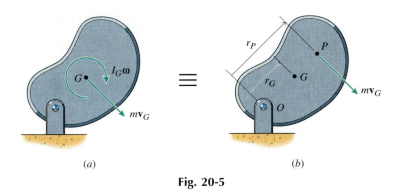

(a)　　　　　　　(b)

Fig. 20-5

its mass center parallel to the axis of rotation. Dividing through by the common factor $m\omega$ gives

$$r_P r_G = k_G^2 + r_G^2 \qquad (l)$$

or

$$(r_P - r_G)r_G = k_G^2 \qquad \textbf{(20-10)}$$

That is, the distance between the center of percussion and the mass center $d = r_P - r_G$ is equal to the ratio of k_G^2, which is a constant for a given body, and r_G, which depends on the location of the axis of rotation.

20-4 SYSTEMS OF RIGID BODIES

As has already been pointed out, the use of Eq. 20-8 requires only that the body behave rigidly at the initial time t_i and at the final time t_f so that the angular momentum H_G can be computed using $I_G\omega$ at these times. Between the initial and final instants of time, parts of the body may move relative to each other and $(I_G)_i$ may not be the same as $(I_G)_f$. If parts of the body are moving relative to each other at the initial and/or final instants of time, then a separate angular impulse momentum equation should be written for each part that does behave in a rigid manner and the equations added together. If the moments of momentum and moments of forces are all written relative to the same point in each equation, then the moments of the joint forces holding the various parts together will cancel in pairs and need not be computed. This is most easily accomplished using the graphical representation of Fig. 20-6. In the first and last parts of the figure the momentum of each rigid body has been replaced with an equivalent "force-couple" system at its own mass center. In the middle part of the figure, the joint forces are internal forces and need not be shown.

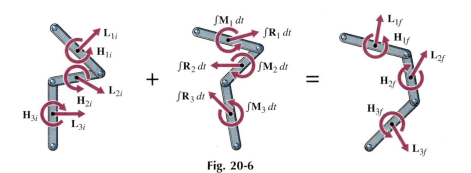

Fig. 20-6

How does a figure skater control her spinning by extending her arms or pulling her arms in close to her body?

SOLUTION

According to the angular impulse–momentum principle

$$(I\omega)_i + \int_i^f \sum M \, dt = (I\omega)_f$$

the angular rotation rate of a body can be changed by a moment about the axis of rotation. However, the moments produced by air resistance and by ice friction are so small that they have little effect over short time periods such as 30 s to 60 s. Therefore, the skater's angular momentum is conserved

$$(I\omega)_i = (I\omega)_f$$

If the skater's body configuration does not change ($I_i = I_f$), then the skater's rotation rate will not change either ($\omega_i = \omega_f$). If the skater's body configuration does change, then her rotation rate must also change.

The figure skater starts her spin about a vertical axis with both arms and one leg outstretched. Although her arms and leg contribute only about 20 percent of her total body mass, the moment of inertia of a piece of mass dm is proportional to the square of its distance from the axis of rotation, $dI = r^2 \, dm$. By pulling her arms and leg in next to her body, the skater reduces her moment of inertia to about one-half of its initial value and nearly doubles her rate of rotation.

(a)   *(b)*

Fig. CE20-1

EXAMPLE PROBLEM 20-1

A 20-lb uniform disk rotates about an axis through its center (Fig. 20-7). The radius of the disk is 9 in., and the initial angular velocity of the disk is 600 rev/min clockwise. A counterclockwise torque $M = 10 \sin nt$ acts on the disk, where M is in lb · ft, t is in seconds, and $n = 1$ rad/s. Neglect any friction between the bearings and the axle of the disk, and determine the angular velocity of the disk after 1 s; 3 s; 5 s.

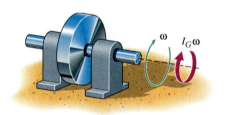

Fig. 20-7

SOLUTION

The Impulse–Momentum Principle is shown in graphical form in Fig. 20-8, in which the initial and final linear momentum of the mass center are both zero. The initial angular velocity of the disk is

$$\omega_0 = 600 \text{ rev/min} \left(\frac{2\pi \text{ rad/rev}}{60 \text{ s/min}} \right) = 62.83 \text{ rad/s}$$

and the moment of inertia about the axle (the mass center) is

$$I_G = \frac{1}{2} mR^2 = \frac{1}{2} \frac{20}{32.2} \left(\frac{9}{12} \right)^2$$
$$= 0.17469 \text{ lb} \cdot \text{ft} \cdot \text{s}^2$$

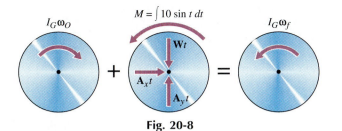

Fig. 20-8

Then the sum of moments about the axle (which is also the mass center) gives

$$\downarrow +H_G: \quad -(0.17469)(62.83) + \int_0^\tau 10 \sin t \, dt = 0.17469\omega_f$$

or

$$\omega_f = 57.24 \, (1 - \cos \tau) - 62.83 \text{ rad/s } \downarrow$$

Therefore

$$\omega_f(1 \text{ s}) = -36.52 \text{ rad/s}$$
$$\cong 36.5 \text{ rad/s } \downarrow \qquad \text{Ans.}$$
$$\omega_f(3 \text{ s}) \cong 51.1 \text{ rad/s } \downarrow \qquad \text{Ans.}$$
$$\omega_f(5 \text{ s}) = -21.83 \text{ rad/s}$$
$$\cong 21.8 \text{ rad/s } \downarrow \qquad \text{Ans.}$$

Since $\cos \tau$ varies between $+1$ and -1, the angular velocity varies between 51.65 rad/s counterclockwise and 62.83 rad/s clockwise. Although the torque and the motion of the disk both repeat every $t = 2\pi$ s, the torque is a maximum when $t = \pi/2$ s, $3\pi/2$ s, $5\pi/2$ s, ..., while the angular speed of the disk is a maximum when $t = 0$ s, π s, 2π s, ...

EXAMPLE PROBLEM 20-2

A bowling ball is modeled as a 7-kg uniform sphere 300 mm in diameter (Fig. 20-9). The ball is released on a horizontal wood floor with an initial velocity of $v_0 = 6$ m/s and an initial angular velocity of $\omega_0 = 0$. If the coefficient of kinetic friction between the ball and the floor is $\mu_k = 0.1$, determine

a.   The time t_f at which the ball begins to roll without slipping.
b.   The velocity v_f and angular velocity ω_f of the ball at time t_f.

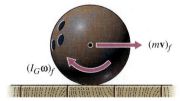

Fig. 20-9

SOLUTION

The Impulse–Momentum Principle is shown in graphical form in Fig. 20-10a, in which the moment of inertia of the sphere about an axis through its mass center is

$$I_G = \frac{2}{5}mR^2 = \frac{2}{5}(7)(0.15)^2 = 0.0630 \text{ kg} \cdot \text{m}^2$$

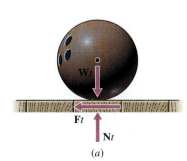

(a)

Fig. 20-10

505

The initial momentum of the sphere is replaced with an equivalent "force-couple" at the mass center consisting of a "force" in the x-direction of magnitude $L_i = (7 \text{ kg})(6 \text{ m/s}) = 42 \text{ kg} \cdot \text{m/s}$ and a "moment" about the mass center of $(I_G\omega)_i = 0$. Similarly, the final momentum of the sphere is replaced with a "force" in the x-direction of magnitude $L_f = mv_f$ acting through the mass center and a "moment" about the mass center of $I_G\omega_f$. Then, with reference to the figure, the linear and angular momentum principles give

$$+ \rightarrow L_x: \qquad\qquad 42 - Ft = 7\,v_f \qquad\qquad (a)$$
$$+ \uparrow L_y: \qquad 0 + Nt - (7)(9.81)t = 0 \qquad\qquad (b)$$
$$\curvearrowright + H_G: \qquad\qquad 0 + (0.150)Ft = (0.0630)\omega_f \qquad (c)$$

Equation b gives $N = 68.67$ N for any $0 < t < t_f$ (or even for any $t > t_f$). Since the ball slides between $t = 0$ and $t = t_f$, the friction force on the ball is

$$F = \mu_k N = 0.1(68.67) = 6.867 \text{ N}$$

Finally, kinematics is used to relate the linear and angular velocity at time t_f (Fig. 20-10b)

$$v_f = 0.150\omega_f \qquad\qquad (d)$$

Then Eqs. a, c, and d give

$$t_f = 1.747 \text{ s} \qquad v_f = 4.286 \text{ m/s} \rightarrow \qquad \omega_f = 28.57 \text{ rad/s} \downarrow \qquad \text{Ans.}$$

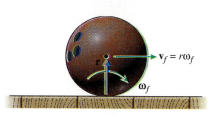

This same result is trivially obtained using Newton's second law. Since there is no motion in the y-direction, $a_y = 0$ and $\Sigma F_y = 0$ or $N - (7)(9.81) = 0$.

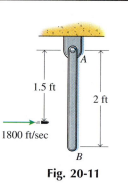

(b)

Fig. 20-10

EXAMPLE PROBLEM 20-3

A 2-ft-long uniform rod weighing 3 lb hangs from a frictionless pin at A (Fig. 20-11). A bullet weighing 0.05 lb and traveling with an initial speed of 1800 ft/s strikes the rod and becomes embedded in it. Determine the angular velocity of the rod immediately after the bullet becomes embedded in it.

1.5 ft

2 ft

1800 ft/sec

B

Fig. 20-11

SOLUTION

Since the rod is in fixed-axis rotation about an axis through A, Eq. 20-9 will be used. The appropriate momentum and free-body diagrams are shown in Fig. 20-12. The initial linear momentum of the bullet is

$$L_{ib} = \frac{0.05}{32.2}1800 = 2.795 \text{ lb} \cdot \text{s}$$

and its moment about point A is $1.5L_{ib} = 4.193 \text{ ft} \cdot \text{lb} \cdot \text{s}$. The initial linear and angular momentum of the rod are both zero. Therefore, the total angular momentum about A of the system just before the bullet strikes the rod is $H_{Ai} = 4.193 \text{ ft} \cdot \text{lb} \cdot \text{s}$.

When the momenta and impulses are added together, the interaction force of the bullet and the rod is an internal force and need not be shown on the free-body diagram. None of the other three forces have a moment about A, and so the total angular impulse about A is zero.

506

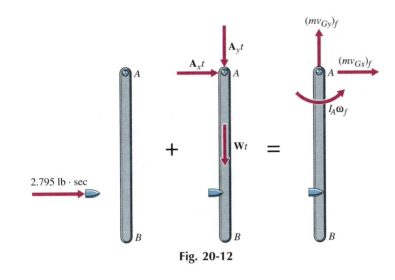

Fig. 20-12

Immediately after the bullet lodges in the rod, the rod and bullet rotate as a single unit about the fixed pin A. The final angular momentum of the system is $H_{Af} = (I_A\omega)_f$ where

$$I_A = \frac{1}{3}\frac{3}{32.2}(2)^2 + \frac{0.05}{32.2}(1.5)^2$$
$$= 0.12772 \text{ lb} \cdot \text{ft} \cdot \text{s}^2$$

is the moment of inertia of both the rod and the embedded bullet about the axis through A.

Finally then, the angular impulse–momentum principle (Eq. 20-9) gives

$$\downarrow + H_A: \qquad 4.193 + 0 = 0.12772\omega_f$$

or

$$\omega_f = 32.8 \text{ rad/s} \downarrow \qquad\qquad \text{Ans.}$$

> Before the bullet strikes the rod, the kinetic energy of the bullet is $T_i = 2516$ lb · ft and the kinetic energy of the rod is zero. After the bullet becomes embedded in the rod, the kinetic energy of the system is only $T_f = 68.8$ lb · ft. Most of the "lost energy" is expended in deforming the rod and the bullet as the bullet becomes embedded in the rod.

PROBLEMS

Introductory Problems

20-1* Figure P20-1 shows a radial-arm saw that has an operating speed of 1500 rev/min. The blade and motor armature have a combined weight of 3 lb and a centroidal radius of gyration of $k_G = 1$ in. When the saw is turned off, bearing friction and a magnetic brake exert a constant braking torque **T** on the blade and motor.

a. Determine the length of time that the blade rotates before coming to rest if $T = 0.015$ lb · in. (bearing friction only).

b. Determine the torque **T** necessary to stop the blade in just 0.25 s.

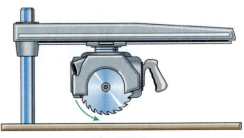

Fig. P20-1

20-2* The grinding wheel shown in Fig. P20-2 coasts to rest from an angular velocity of 2400 rev/min in 150 s. The grinding wheel and motor armature have a combined mass of 3 kg and a radius of gyration with respect to the centroidal axis of rotation of $k_G = 100$ mm. Determine the average frictional moment exerted on the flywheel by the bearings of the motor.

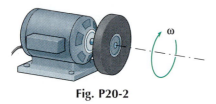

Fig. P20-2

20-3* The starting moment of the electric motor shown in Fig. P20-2 is given by $M_0 e^{-t}$ where M_0 is a constant and the bearings exert a constant resisting moment of 0.006 ft · lb. The grinding wheel and motor armature have a combined weight of 5 lb and a radius of gyration with respect to the axis of rotation of $k_G = 2.3$ in. If the motor reaches its operating speed of 3000 rev/min in 3 s, determine the value of M_0.

20-4 The platters and hub of the computer disk drive shown in Fig. P20-4 have a mass of 650 g and a centroidal radius of gyration of 70 mm. Determine the average motor torque necessary to attain an operating speed of 3600 rev/min in just 2 s after the drive is turned on.

Fig. P20-4

20-5 The turntable platter shown in Fig. P20-5 weighs 1.5 lb and has a centroidal radius of gyration of $k_G = 5$ in. If the belt drive exerts a torque of 0.10 lb · in. on the turntable, determine the time required to attain its operating speed of $33\frac{1}{3}$ rev/min.

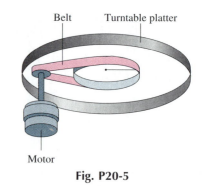

Fig. P20-5

20-6 The speed of a rotating system is controlled with a brake as shown in Fig. P20-6. The rotating parts of the system have a mass of 300 kg and a radius of gyration with respect to the axis of rotation of $k_G = 200$ mm. The kinetic coefficient of friction between the brake pad and brake drum is $\mu_k = 0.50$. If a 500-N force **P** is being applied to the brake lever, determine the time t required to reduce the speed of the system from 1000 rev/min to rest.

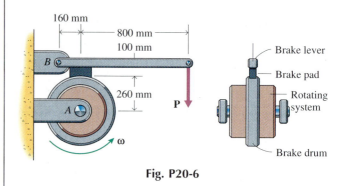

Fig. P20-6

20-7* The 20-lb uniform wheel of Fig. P20-7 is rotating at 3000 rev/min when a force of $P = 40(1 - e^{-0.5t})$ lb is applied to the handle of the brake arm. If the kinetic coefficient of friction between the brake arm and the wheel is $\mu_k = 0.1$, determine the length of time until the wheel stops rotating

a. If it is rotating clockwise.
b. If it is rotating counterclockwise.

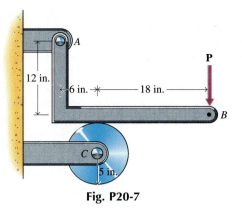

Fig. P20-7

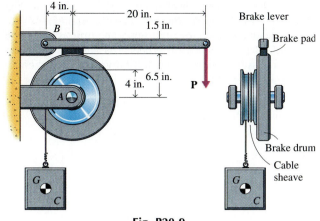

Fig. P20-9

20-8 A force of $P = 50$ N is applied to the end of a rope wrapped around the outside of a hollow drum as shown in Fig. P20-8a. The radius of gyration of the 20-kg drum is $k_G = 175$ mm and axle friction may be neglected.

a. If the drum is released from rest, determine the downward velocity of point A on the rope after 10 s.

b. Repeat part a if the 50-N force is replaced with a 50-N weight as shown in Fig. P20-8b.

20-10* The 20-kg stepped wheel of Fig. P20-10 has a radius of gyration of $k_G = 150$ mm and an initial angular velocity of 3000 rev/min counterclockwise. If the kinetic coefficient of friction between the brake arm and the wheel is $\mu_k = 0.2$, determine the length of time

a. Until the wheel stops rotating.

b. Until the rotation rate of the wheel is 3000 rev/min clockwise.

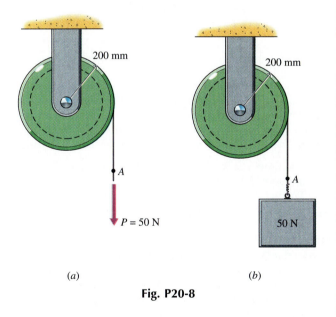

(a)   (b)

Fig. P20-8

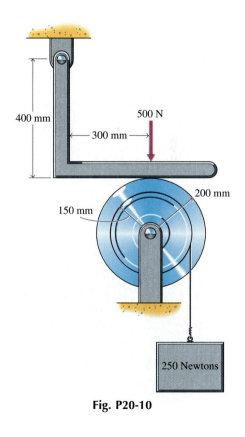

Fig. P20-10

Intermediate Problems

20-9* A brake for regulating the descent of a body is shown in Fig. P20-9. The rotating parts of the brake (cable sheave and brake drum) weigh 250 lb and have a radius of gyration with respect to their axis of rotation of $k_G = 4.25$ in. The kinetic coefficient of friction between the brake pad and the brake drum is $\mu_k = 0.50$. Determine the magnitude of the force **P** that must be applied to the brake lever to reduce the speed of the 500-lb body C from 10 ft/s to 5 ft/s in a time of 3 s.

20-11 The uniform wheel A ($W_A = 4$ lb) shown in Fig. P20-11 is initially raised and given a counterclockwise angular velocity of 4500 rev/min while the uniform wheel B ($W_B = 9$ lb) remains at rest. Wheel A is then released and allowed to spin against wheel B. If the kinetic coefficient of friction between the two wheels is $\mu_k = 0.1$, determine

a. The length of time before the wheels begin to rotate together without slipping.
b. The final angular velocities of both wheels.

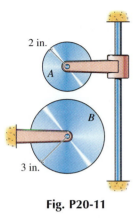

Fig. P20-11

20-12 The two uniform wheels A and B shown in Fig. P20-12 are initially at rest when a constant counterclockwise moment $M = 2.5$ N · m is suddenly applied to wheel B. Wheel A has a 200-mm diameter and a 10-kg mass, whereas wheel B has a 300-mm diameter and a 25-kg mass. If the coefficients of static and kinetic friction between the two wheels are $\mu_s = 0.2$ and $\mu_k = 0.1$, determine the angular velocities of both wheels at $t = 15$ s.

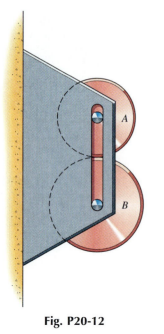

Fig. P20-12

20-13* The two uniform wheels A and B shown in Fig. P20-12 are initially rotating together without slipping when wheel B is suddenly stopped. Wheel A has an 8-in. diameter, weighs 20 lb, and has an initial angular velocity of 4500 rev/min counterclockwise; wheel B has a 12-in. diameter and weighs 45 lb. If the kinetic coefficient of friction between the two wheels is $\mu_k = 0.2$, determine the length of time before wheel A stops rotating.

20-14 A 300-mm-diameter, 5-kg homogeneous sphere is lowered onto a horizontal surface with an initial angular velocity of $\omega_0 = 3000$ rev/min and zero linear velocity, $v_0 = 0$ (Fig. P20-14). If the kinetic coefficient of friction between the sphere and the surface is $\mu_k = 0.15$, determine

a. The time t_f at which the sphere will begin to roll without slipping.
b. The velocity v_f of the mass center at t_f.
c. The angular velocity ω_f of the sphere at t_f.

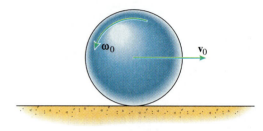

Fig. P20-14

20-15* A 14-in.-diameter, 16-lb homogeneous sphere is lowered onto a horizontal surface with an initial angular velocity of $\omega_0 = 3000$ rev/min and a linear velocity of $v_0 = 20$ ft/s (Fig. P20-14). If the kinetic coefficient of friction between the sphere and the surface is $\mu_k = 0.15$, determine

a. The time t_f at which the sphere will begin to roll without slipping.
b. The velocity v_f of the mass center at t_f.
c. The angular velocity ω_f of the sphere at t_f.

20-16* A 200-mm-diameter, 5-kg homogeneous sphere is lowered onto an inclined surface with an initial angular velocity of $\omega_0 = 3000$ rev/min (Fig. P20-16). If the kinetic coefficient of friction is $\mu_k = 0.20$ and $\theta = 15°$, determine the smallest initial velocity v_0 for which the sphere will stop sliding and stop rotating at the same instant.

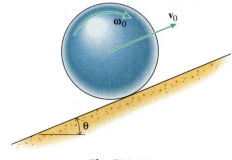

Fig. P20-16

20-17 A 14-in.-diameter, 16-lb homogeneous sphere is lowered onto an inclined surface with an initial angular velocity of $\omega_0 = 3000$ rev/min and zero linear velocity $v_0 = 0$ (Fig. P20-16). If the kinetic coefficient of friction is $\mu_k = 0.25$ and $\theta = 20°$, determine

a. The time t_1 at which the sphere will begin to roll without slipping.
b. The velocity v_1 of the mass center and the angular velocity ω_1 at t_1.

20-18 The uniform slender rod AB ($m = 3$ kg, $\ell = 800$ mm) is resting on a frictionless horizontal surface when it is struck with an impulse $P\Delta t = 20$ N · s as shown in Fig. P20-18. If $b = 650$ mm, determine

a. The angular velocity ω of the rod immediately after the impact.
b. The velocity v_A of end A immediately after the impact.

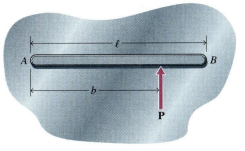

Fig. P20-18

20-19* The uniform slender rod AB ($W = 3$ lb, $\ell = 2$ ft) is resting on a frictionless horizontal surface when it is struck with an impulse $P\Delta t = 2$ lb · s as shown in Fig. P20-18. Determine the distance b for which end A is an instantaneous center of zero velocity (the velocity of A is zero immediately after the impact).

20-20* The uniform slender rod AB ($m = 3$ kg, $\ell = 800$ mm) is resting on a frictionless horizontal surface when it is struck with an impulse $P\Delta t = 5$ N · s as shown in Fig. P20-20. If $b = 300$ mm and the duration of the impact is $\Delta t = 0.002$ s, determine

a. The angular velocity of the rod immediately after the impact.
b. The average magnitude of the force exerted on the rod by the frictionless pin at A.

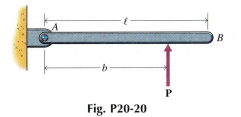

Fig. P20-20

20-21 The uniform slender rod AB ($W = 3$ lb, $\ell = 2$ ft) is resting on a frictionless horizontal surface when it is struck with an impulse $P\Delta t = 2$ lb · s as shown in Fig. P20-20. If the duration of the impact is $\Delta t = 0.002$ s, determine the distance b for which the average force exerted on the rod by the frictionless pin at A is zero.

20-22 Determine the location of the center of percussion for a slender rod rotating about a frictionless pin if the pin is

a. Located at on end of the rod.
b. Located at a point $\ell/4$ from one end of the rod.
c. Located at the middle of the rod.

Challenging Problems

20-23* A 3-lb uniform slender rod AB 4 ft long is rotating in a vertical plane about a frictionless pin through its center as shown in Fig. P20-23. A small piece of putty ($W_P = 0.4$ lb) falls and strikes the rod when it is horizontal. If the initial rotation of the rod is counterclockwise at 120 rev/min and the putty starts from rest at $h = 5$ ft, determine

a. The rotation rate of the rod and putty immediately after the impact.
b. The average force of contact between the rod and the putty for an impact duration of $\Delta t = 0.005$ s.
c. The average magnitude of the force exerted on the rod by the frictionless pin for an impact duration of $\Delta t = 0.005$ s.
d. The total system energy lost in the collision.

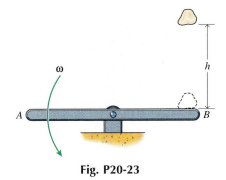

Fig. P20-23

20-24* A 3-kg uniform slender rod *AB* 800 mm long hangs in a vertical plane from a frictionless pivot when a 0.03-kg bullet strikes the rod and becomes embedded in it (Fig. P20-24). If the initial velocity of the bullet is $v_0 = 350$ m/s, determine

a. The rotation rate of the rod and bullet immediately after the impact.
b. The average force of contact between the rod and the bullet for an impact duration of $\Delta t = 0.001$ s.
c. The average magnitude of the force exerted on the rod by the frictionless pin at *A* for an impact duration of $\Delta t = 0.001$ s.
d. The total system energy lost in the impact.
e. The maximum angle through which the rod will swing after the impact.

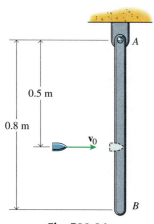

Fig. P20-24

20-25 When an LP record drops onto a spinning turntable (Fig. P20-25), it slips for a short time before attaining its final speed of $33\frac{1}{3}$ rev/min. The record has a diameter of 12 in. and weighs 2.0 oz. If the coefficient of friction between the record and the turntable is $\mu_k = 0.2$, determine the length of time that the record will slip. (Assume that the contact force between the turntable and the record is uniformly distributed over the surface of the record.)

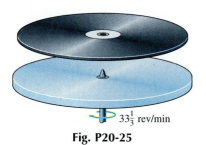

Fig. P20-25

20-26 A 1.5-kg uniform slender rod *AB* hangs in a vertical plane from a frictionless pivot as shown in Fig. P20-26. A 125-g arrow traveling at 100 m/s strikes the rod at a point 500 mm below the pivot. The arrow may be modeled as a uniform slender rod 800 mm long. If the arrowhead becomes embedded in the rod, determine

a. The angular velocity ω of the system immediately after the impact.
b. The average moment exerted on the arrow by the bar *AB* for an impact duration of $\Delta t = 0.01$ s.

Fig. P20-26

Computer Problems

C20-27 The uniform slender rod *AB* ($W = 3$ lb, $\ell = 2$ ft) is resting on a frictionless horizontal surface when it is struck with an impulse $P\Delta t = 2$ lb · s as shown in Fig. P20-27. Calculate and plot

a. The angular velocity ω of the rod immediately after the impact as a function of *b* the location of the impulse for $0 < b < 2$ ft.
b. The velocity v_A of end *A* immediately after the impact as a function of *b* the location of the impulse for $0 < b < 2$ ft.

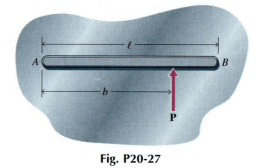

Fig. P20-27

C20-28 The uniform slender rod AB ($m = 3$ kg, $\ell = 800$ mm) is resting on a frictionless horizontal surface when it is struck with an impulse $P\Delta t = 5$ N · s as shown in Fig. P20-28. If the duration of the impact is $\Delta t = 0.002$ s, calculate and plot

a. The angular velocity ω of the rod immediately after the impact as a function of b the location of the impulse for $0 < b < 2$ ft.
b. The average magnitude of the force exerted on the rod by the frictionless pin at A as a function of b the location of the impulse for $0 < b < 2$ ft.

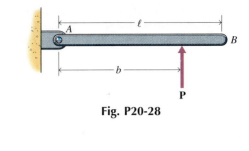

Fig. P20-28

20-5 IMPACT OF RIGID BODIES

The impact of colliding bodies was discussed in Section 19-4. For particle collisions, only the cases of direct central and oblique central impact were applicable, and these cases were developed in detail there. In the present section, the additional cases of eccentric impact will be developed.

Even for the relatively simple case of particle collisions, the impact phenomena were seen to be complex events. Fortunately, however, the details of the impact can often be avoided; the linear impulse–momentum equation can be used to give a simple relation between the relative velocities of the bodies before and after the impact. Although the impulse–momentum method of approach is only an approximation of a very complex event and should be applied with care, the method allows the solution of some impact problems that would not otherwise be simply solvable.

20-5.1 Impulsive Forces and Impulsive Motion

Although impact events occur in a relatively short time interval, it is observed that the velocities and angular velocities of bodies can change significantly. The changes in momentum and angular momentum then require an impulse that does not go to zero for the brief impact times. Forces characterized by very large magnitudes such that they produce a significant change in momentum (a large impulse) even for very short time periods are called *impulsive forces*. Motions that result from impulsive forces are called *impulsive motions*. The forces generated when one body strikes another is an example of an impulsive force.

Forces that produce a negligible change in momentum (a small impulse) for small time periods are called *nonimpulsive forces*. Examples of nonimpulsive forces are the weight of a body, friction forces, and spring forces. The magnitudes of nonimpulsive forces are always small compared to the magnitudes of impulsive forces. When the principle of impulse and momentum is applied over a short time interval, the impulse of nonimpulsive forces may often be neglected compared to the impulse of the impulsive forces.

It is usually not known ahead of time whether unknown reaction forces are impulsive or not. Generally, the reaction force of any support that acts to prevent motion in some direction is just as impulsive as the forces trying to cause motion in that direction.

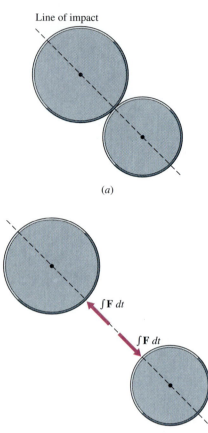

Line of impact

(a)

(b)

$\int \mathbf{F}\,dt$

$\int \mathbf{F}\,dt$

Figure 20-13 In a central impact of
two bodies, the contact forces of the
impact pass through the mass centers
of the bodies.

The final decision of whether or not the impulse of a force can be
ignored must be based on the required accuracy of the result and on
the estimated effect that the term has on the equation.

20-5.2 Assumptions for Impact Problems

By their very nature, impact events occur in very brief intervals of time.
Based on observations of many impact events, it is assumed that during the brief impact interval $\Delta t = t_f - t_i$:

1. The positions of the impacting bodies do not change appreciably.
2. The velocities and/or angular velocities of one or both of the im-
    pacting bodies may change greatly.
3. Nonimpulsive forces and moments may be neglected.
4. Friction forces (forces tangent to the plane of impact) may be ne-
    glected.[4]

20-5.3 Eccentric Impact of Rigid Bodies

The analysis of particle collision problems in Section 19-4 illustrated
the case of central impact. In central impact the line of impact coincides with the line connecting the mass centers. Therefore the contact
forces of impact pass through the mass centers of the bodies (Fig.
20-13). These problems were solved using the concept of conservation
of linear momentum in conjunction with the coefficient of restitution
e, which compares the relative velocity of separation of the points of
contact (after the collision) to their relative velocity of approach (prior
to the collision).

The impact problem for rigid bodies is quite similar to that for particles, but it is complicated slightly by the fact that the line of impact
usually does not pass through the mass centers of the two bodies (Fig.
20-14). As noted in Section 19-4, such a collision is called an *eccentric
impact*.

An additional complication arises if the coefficient of restitution is
defined as the ratio of the restitution impulse and the deformation impulse, as in Section 19-4. An analysis similar to that in Section 19-4
again gives the coefficient of restitution in terms of the relative velocity of separation of the points of contact (after the collision) to their relative velocity of approach (prior to the collision). However, the velocity of the body at the point of impact is usually different from the
velocity of its mass center. Therefore, when eccentric impact is involved, the relative velocity equations must be used to relate the velocities of the contact points in the equation for the coefficient of restitution and the velocities of the mass centers in the linear and angular
momentum principles.

Consider the impact of the rigid bodies shown in Fig. 20-14. Points
A and B are the mass centers of the bodies and points C and D are the
actual points of contact. Coordinates t and n are chosen in and perpendicular to the plane of contact as shown. The coefficient of restitution is again defined as the ratio of the restitution impulse and the de-

---

[4]This last is often *not* a good approximation. For such problems, the reader is referred
to Raymond M. Brach, *Mechanical Impact Dynamics: Rigid Body Collisions* (New York: Wiley, 1991).

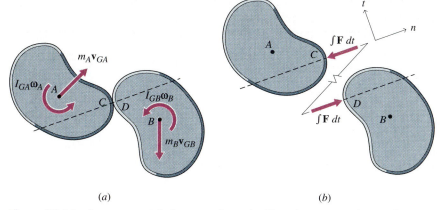

(a)   (b)

Figure 20-14   In an eccentric impact of two bodies, the contact forces do not pass through the mass center of at least one of the two bodies.

formation impulse. Then, an analysis similar to that of Section 19-4 gives the coefficient of restitution

$$e = -\frac{(v_{Df})_n - (v_{Cf})_n}{(v_{Di})_n - (v_{Ci})_n} \tag{20-11}$$

where $(v_{Ci})_n$ and $(v_{Di})_n$ are the initial components of the velocities of points C and D (before impact) and $(v_{Cf})_n$ and $(v_{Df})_n$ are the final velocities of the points (after impact). The components of the velocities of points C and D are related to the velocities of the mass centers A and B by the relative velocity equations

$$\mathbf{v}_C = \mathbf{v}_A + \boldsymbol{\omega}_A \times \mathbf{r}_{C/A} \tag{a}$$

and

$$\mathbf{v}_D = \mathbf{v}_B + \boldsymbol{\omega}_B \times \mathbf{r}_{D/B} \tag{b}$$

The result of combining Eqs. 20-11, a, and b gives one scalar equation relating the velocities $\mathbf{v}_A$ and $\mathbf{v}_B$ and the angular velocities $\boldsymbol{\omega}_A$ and $\boldsymbol{\omega}_B$ after the impact. Four additional scalar equations (two vector equations) can be obtained from applying the Linear Impulse–Momentum Principle to each body separately. Finally, the Principles of Angular Impulse–Momentum can be applied about the center of mass of each body, giving two more scalar equations for a total of seven equations. These equations are used to solve for the seven unknowns $(v_{Cf})_n$, $(v_{Cf})_t$, $(v_{Df})_n$, $(v_{Df})_t$, ω_{Af}, ω_{Bf}, and the magnitude of the impulsive contact force F exerted between the two bodies. As shown in Fig. 20-14, the force $\mathbf{F}$ is directed along the normal to the plane of contact.

   If one or both of the colliding bodies is constrained to rotate about a fixed point or points, an impulsive reaction will be exerted at the fixed point(s). The impulse of these reactions must also be included in the equations of linear and angular impulse–momentum.

A 1.5-kg uniform rod 800 mm long is at rest on a frictionless horizontal surface when it is struck by a 0.5-kg disk as shown in Fig. 20-15. If the disk strikes the rod 200 mm from the end and the coefficient of restitution of the collision is 0.4, determine

a. The velocity of the disk after the collision.
b. The velocity of the mass center of the rod after the collision.
c. The angular velocity of the rod after the collision.
d. The location of a point on the rod that is instantaneously at rest during the collision.

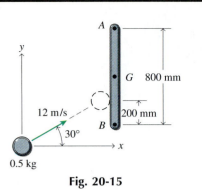

Fig. 20-15

SOLUTION

a. The kinetic diagrams of the rod and the disk are shown in Fig. 20-16, in which the moment of inertia of the rod about an axis through its mass center is

$$I_G = \frac{1}{12}(1.5)(0.8)^2 = 0.0800 \text{ kg} \cdot \text{m}^2$$

Since friction between the disk and rod is neglected, there is no component of linear impulse in the y-direction on either the rod or the disk. Therefore, the y-component of the linear momentum principle applied to the rod and disk separately gives (Fig. 20-16)

rod: $\qquad [0] + 0 = [1.5v_{Gy}]$
disk: $\qquad [(0.5)(12 \sin 30°)] + 0 = [0.5v_{dy}]$

or

$$v_{Gy} = 0 \qquad \text{and} \qquad v_{dy} = 6 \text{ m/s}.$$

For the combined system of the rod and the disk, there is no linear impulse in the x-direction or moment of impulse about the mass center of the rod. Therefore, the x-component of the linear momentum principle applied to the combined system gives (Fig. 20-16)

$$[(0.5)(12 \cos 30°)] + 0 = [0.5v_{dx} + 1.5v_{Gx}] \qquad (a)$$

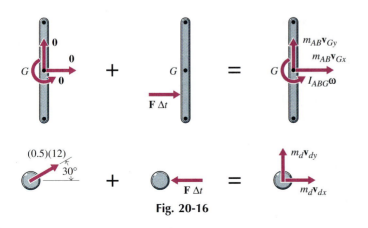

Fig. 20-16

and the angular momentum principle applied to the combined system gives

$$[(0.2)(0.5)(12 \cos 30°) + 0] + 0 = [(0.2)(0.5)v_{dx} + 0.08\omega] \qquad \textbf{(b)}$$

Finally, the coefficient of restitution relates the x-components of velocities of the contact points before and after the collision:

$$e = -\frac{(v_{Gx} + 0.2\omega) - v_{dx}}{0 - 12 \cos 30°} = 0.4 \qquad \textbf{(c)}$$

Solving Eqs. *a*, *b*, and *c* simultaneously gives $v_{dx} = 1.203$ m/s, $v_{Gx} = 3.06$ m/s, and $\omega = 11.49$ rad/s. Then combining the x- and y-components of the velocity of the disk after the collision gives

$$\mathbf{v}_d = 6.12 \text{ m/s} \angle 78.7° \qquad \text{Ans.}$$

b. Combining the x- and y-components of the velocity of the mass center of the rod after the collision gives

$$\mathbf{v}_G = 3.06 \text{ m/s} \rightarrow \qquad \text{Ans.}$$

c. The angular velocity of the rod after the collision is

$$\omega = 11.49 \text{ rad/s } \downharpoonleft \qquad \text{Ans.}$$

d. If C is an instantaneous center of zero velocity, then (Fig. 20-17)

$$3.06 = 11.49d$$

and the instantaneous center is

$$d = 0.266 \text{ m} \qquad \text{Ans.}$$

away from the mass center G on the side opposite the point of contact.

Although the x-component of the disk's velocity is reduced by the collision, the disk is still moving to the right after the collision. Since the velocity of the disk after the collision is up and to the right (at 78.7°), the disk will eventually collide with the rod again.

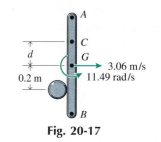

Fig. 20-17

Note that if the rod were fixed at C with a frictionless pin, then the point of contact would be the center of percussion for the rod. That is, the radius of gyration of the rod is $k_G = \sqrt{0.08/1.5}$ and $0.2d = k_G^2$ which is in agreement with Eq. 20-10.

EXAMPLE PROBLEM 20-5

A 25-lb uniform rod 3 ft long is attached to a frictionless hinge at A (Fig. 20-18). The rod starts from rest in the vertical position shown, falls against the bumper C, and rebounds upward. If the coefficient of restitution of the collision is 0.6, determine

a. The maximum angle θ_{max} that the bar will make with the horizontal after the collision.
b. The average magnitude of the support reaction at A for an impact duration of 0.01 s.
c. The amount of total system energy lost during the collision.

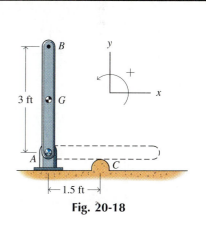

Fig. 20-18

SOLUTION

a. **Work–energy:** From the time t_0 (when the bar is vertical) until the time t_1 (just before the bar collides with the bumper C), the only force that does work on the bar is gravity. Therefore, the Work–Energy Principle can be used to determine the motion of the bar at time t_1. The initial kinetic energy of the bar at t_1 can be written $T = \frac{1}{2}I\omega_{A1}^2$ where $I_A = \frac{1}{3}m\ell^2 = 2.329$ lb · ft · s². Then the Work–Energy Principle

$$0 + (25)(1.5) = \frac{1}{2}(2.329) \omega_1^2 + 0$$

gives $\omega_1 = 5.675$ rad/s $\downarrow$, and the velocity of the mass center at t_1 is $v_{G1} = 1.5\omega_1 = 8.512$ ft/s$\downarrow$.

Impact: The coefficient of restitution is used to determine the change in motion across the impact

$$e = 0.6 = -\frac{v_{G2} - 0}{(-8.512) - 0}$$

where the velocity of the bumper is zero both before and after the impact. Therefore, $v_{G2} = 5.107$ ft/s$\uparrow$ and the angular velocity of the rod after the impact is $\omega_2 = v_{G2}/1.5 = 3.405$ rad/s$\downarrow$.

Work-Energy: From time t_2 (just after the collision) until time t_3 (when the bar is at its maximum angle θ_{max}), the only force that does work on the bar is again gravity. At the maximum angle, the kinetic energy of the bar is zero and the Work–Energy Principle

$$\frac{1}{2}(2.329)(3.405)^2 + 0 = 0 + (25)(1.5 \sin \theta_{max})$$

gives

$$\theta_{max} = 21.1° \qquad \text{Ans.}$$

b. **Impulse–Momentum:** Referring to the kinetic diagram of Fig. 20-19, the x-component of the Linear Impulse–Momentum Principle

$$0 + A_x(0.01) = 0$$

gives $A_x = 0$. Using the same kinetic diagram (in which $I_G = \frac{1}{12}m\ell^2 = 0.5823$ lb · ft · s^2, the Angular Impulse–Momentum Principle

$$(0.5823)(-5.675) + (1.5)A_y(0.01) = (0.5823)(3.405)$$

gives $A_y = 352$ lb. Therefore, the average magnitude of the support reaction at A is

$$A = 352 \text{ lb} \qquad \text{Ans.}$$

c. Since the kinetic energy at times t_0 and t_3 are both zero, the total system energy at these times is just the potential energies

$$E_0 = (25)(1.5) = 37.5 \text{ lb · ft}$$
$$E_3 = (25)(1.5 \sin 21.1°) = 13.50 \text{ lb · ft}$$

The energy lost is then

$$\frac{37.50 - 13.50}{37.50}(100) = 64.0\% \qquad \text{Ans.}$$

Note that the weight force has been left off of Fig. 20-19 since it is not an impulsive force. Even if it had been included, it would not affect the calculation of A_y since the weight acts through the mass center and has no moment about the mass center. Referring again to the kinetic diagram of Fig. 20-19, the y-component of the Linear Impulse-Momentum Principle gives that $C = 1410$ lb. If the weight force were included here, the bumper force would be $C = 1435$ lb or a little less than 2 percent larger.

The bar bounces off of the bumper because the bar and the bumper both deform slightly and then recover (spring back to their original shape). If the collision were perfectly elastic ($e = 1$), no energy would be lost in the deformation and recovery process. For real collisions, however, some energy is always lost (and $e < 1$). The "lost energy" is expended generating sound waves, generating heat as the molecules slide past one another, and creating a permanent shape change in either the rod or the bumper.

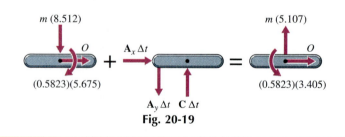

$m\,(8.512)$ $m\,(5.107)$

$\mathbf{A}_x \Delta t$

$(0.5823)(5.675)$ $(0.5823)(3.405)$

$\mathbf{A}_y \Delta t$ $\mathbf{C} \Delta t$
Fig. 20-19

PROBLEMS

Introductory Problems

20-29* The uniform rod of Fig. P20-29 falls horizontally and strikes the rigid corner with its right end. If the velocity of the rod just before the impact is v_0 and the impact is perfectly elastic ($e = 1$), determine the angular velocity ω and the velocity of the mass center $\mathbf{v}_G$ immediately after the impact.

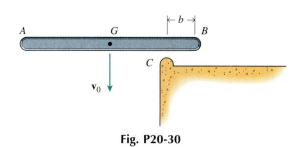

Fig. P20-29

20-30 The uniform rod of Fig. P20-30 falls horizontally and strikes the rigid corner at a distance $b = \ell/3$ from its right end. If the velocity of the rod just before the impact is v_0 and the impact is perfectly elastic ($e = 1$), determine the angular velocity ω and the velocity of the mass center $\mathbf{v}_G$ immediately after the impact.

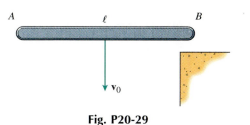

Fig. P20-30

20-31 A slender uniform rod ($\ell = 21$ in., $W_{AB} = 10$ lb) is at rest on a frictionless horizontal surface when it is struck by a small disk ($W_d = m_d g = 2$ lb) as shown in Fig. P20-31. If $b = 3$ in., $e = 0.6$, and the initial velocity of the disk is $v_0 = 15$ ft/s at an angle of $\theta = 60°$, determine

a. The velocity of the disk after the collision.
b. The velocity of the mass center of the rod after the collision.
c. The angular velocity of the rod after the collision.
d. The location of a point on the rod that is instantaneously at rest during the collision.

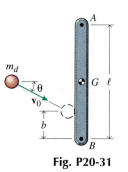

Fig. P20-31

20-32* A slender uniform rod ($\ell = 900$ mm, $m_{AB} = 5$ kg) is at rest on a frictionless horizontal surface when it is struck by a small disk as shown in Fig. P20-31. The disk strikes the rod at a point $b = 250$ mm from B with an initial velocity of $v_0 = 10$ m/s at an angle of $\theta = 40°$. If $e = 0.5$, determine the mass of the disk m_d for which the disk will have no component of velocity along the line of impact after the collision.

Intermediate Problems

20-33* A slender uniform rod ($\ell = 36$ in., $W_{AB} = 12$ lb) hangs motionless from a frictionless hinge at A as shown in Fig. P20-33. A small ball ($W_b = m_b g = 2$ lb) strikes the rod at a point $d = 4$ in. from the bottom end. If $e = 0.5$ and the initial velocity of the ball is $v_0 = 30$ ft/s at an angle of $\theta = 40°$ to the horizontal, determine

a. The velocity of the ball after the collision.
b. The angular velocity of the rod after the collision.
c. The average magnitude of the support reaction at A for an impact duration of 0.005 s.
d. The maximum angle of swing of rod AB after the collision.

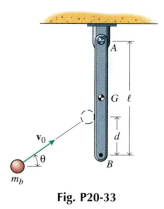

Fig. P20-33

20-34 A slender uniform rod ($\ell = 600$ mm, $m_{AB} = 5$ kg) is attached to a frictionless hinge at A as shown in Fig. P20-34. The rod is released from rest with $\phi_0 = 60°$ and a small ball ($m_b = 0.8$ kg) strikes the rod at a point $d = 100$ mm from the bottom end when it is vertical $\phi = 0°$. The coefficient of restitution is $e = 0.7$ and the initial velocity of the ball v_0 makes an angle of $\theta = 50°$ with the horizontal. If the maximum swing of the rod after the impact is 30°, determine

a. The initial velocity of the ball v_0.
b. The velocity of the ball after the collision.
c. The average magnitude of the support reaction at A for an impact duration of 0.008 s.

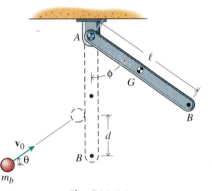

Fig. P20-34

20-35* A slender uniform bar ($\ell = 24$ in., $W_{AB} = 5$ lb) is attached to a frictionless hinge at A as shown in Fig. P20-35. The bar is released from rest with $\phi_0 = 90°$ and strikes the bumper at C ($d = 8$ in.). If the coefficient of restitution is $e = 0.7$, determine

a. The angular velocity of the bar immediately after the collision with the bumper.
b. The maximum angle of rebound of the bar.
c. The average magnitude of the support reaction at A for an impact duration of 0.005 s.
d. The amount of total system energy lost in the collision.

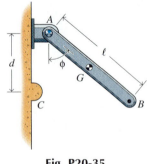

Fig. P20-35

20-36 The uniform slender bar CD of Fig. P20-36 is attached to a frictionless pin at E and rests on a frictionless support at F. The bar is 800 mm long with a mass of 4 kg and is initially at rest. A 2-kg ball is dropped from a height of $h = 2.5$ m and strikes the left end of the bar as shown. If the coefficient of restitution is $e = 0.8$, determine

a. The velocity of the ball immediately after the impact.
b. The angular velocity of the bar CD immediately after the impact.
c. The average magnitude of the support reaction at E for an impact duration of 0.003 s.

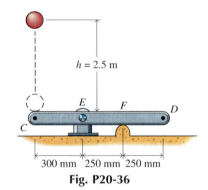

Fig. P20-36

20-37 A slender uniform rod ($\ell = 30$ in., $W_{AB} = 4$ lb) is released from rest at an angle of $\theta = 70°$ to the horizontal and strikes a hard horizontal surface as shown in Fig. P20-37. If the initial height of the rod is $h = 60$ in. and the coefficient of restitution is $e = 0.7$, determine

a. The angular velocity of the rod immediately after the impact.
b. The velocity of the mass center of the rod immediately after the impact.
c. If end B of the rod will strike the surface as the rod rotates immediately after the impact.

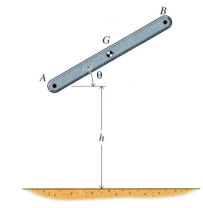

Fig. P20-37

20-38* A slender uniform rod ($\ell = 800$ mm, $m_{AB} = 2$ kg) is released from rest from an initial height of $h = 2$ m and strikes a hard horizontal surface as shown in Fig. P20-37. If the coefficient of restitution is $e = 0.7$ and end B just clears the surface as the rod rotates immediately after the impact, determine

a. The angle θ at which the bar was released.
b. The angular velocity of the rod immediately after the impact.
c. The velocity of the mass center of the rod immediately after the impact.

Challenging Problems

20-39* Bar AB of Fig. P20-39a is attached to a frictionless pin at A; bar CD is attached to a frictionless pin at E and rests on a frictionless support at F. Both AB and CD are uniform slender bars 36 in. long and weighing 5 lb. Both bars are initially at rest when a slight disturbance causes bar AB to fall to the right and strike bar CD as shown in Fig. P20-39b. If the coefficient of restitution is $e = 0.6$, determine

a. The angular velocities of both bars immediately after the impact.
b. The maximum angle of rebound of bar AB after the impact.
c. The average magnitude of the support reaction at E for an impact duration of 0.005 s.

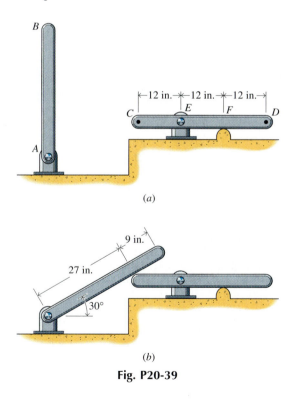

(a)

(b)

Fig. P20-39

20-40* The uniform slender bar CD of Fig. P20-40 is attached to a frictionless pin at E and rests on a frictionless

support at F. The bar is 800 mm long with a mass of 4 kg and is initially at rest. The uniform slender bar AB is 500 mm long with a mass of 3 kg. Bar AB is released from rest with $h = 2.5$ m and strikes bar CD as shown. If the coefficient of restitution is $e = 0.6$, determine

a. The angular velocities of both bars immediately after the impact.
b. The average magnitude of the support reaction at E for an impact duration of 0.003 s.

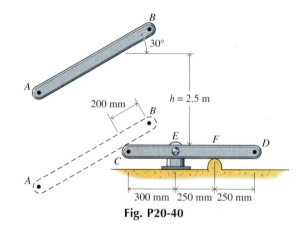

Fig. P20-40

20-41 A slender uniform bar ($\ell = 24$ in., $W_{AB} = 5$ lb) swings from a frictionless pin at A as shown in Fig. P20-41. Pin A is not attached to the support but simply rests on a horizontal ledge. The bar is released from rest with $\phi_0 = 90°$ and strikes the bumper at C ($d = 10$ in.). If the coefficient of restitution is $e = 0.7$, determine the angular velocity ω and the velocity of the mass center $\mathbf{v}_G$ immediately after the collision with the bumper.

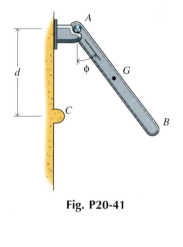

Fig. P20-41

20-42 A slender uniform rod ($\ell = 750$ mm, $m_{AB} = 10$ kg) hangs motionless from a frictionless pin at A as shown in Fig. P20-42. Pin A is not attached to the support but simply rests on a horizontal ledge. A small ball ($m_b = 2$ kg) strikes the rod at a point $d = 350$ mm from the bottom end. If the coefficient of restitution is $e = 0.8$ and the initial velocity of the ball is $v_0 = 20$ m/s at an angle of $\theta = 40°$ to the horizontal, determine

a. The velocity of the ball immediately after the impact.
b. The angular velocity of the rod immediately after the impact.
c. The velocity of the mass center of the rod immediately after the impact.

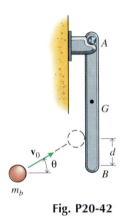

Fig. P20-42

20-43 Bar AB of Fig. P20-43a is attached to a frictionless pin at A; bar CD rests on frictionless supports at E and F. Both AB and CD are uniform slender bars 36 in. long and weighing 5 lb. Both bars are initially at rest when a slight disturbance causes bar AB to fall to the right and strike bar CD as shown in Fig. P20-43b. If the coefficient of restitution is $e = 0.6$, determine

a. The angular velocities of both bars immediately after the impact.
b. The velocity of the mass center of bar CD immediately after the impact.

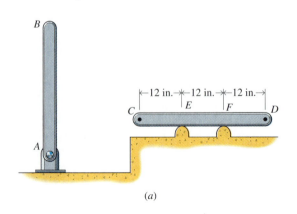

(a)

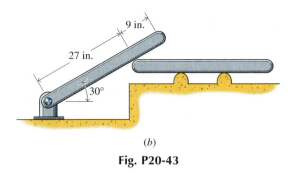

(b)

Fig. P20-43

20-44* The uniform slender bar CD of Fig. P20-44 rests on frictionless supports at E and F. The bar is 800 mm long with a mass of 4 kg and is initially at rest. The uniform slender bar AB is 500 mm long with a mass of 3 kg. Bar AB is released from rest with $h = 2.5$ m and strikes bar CD as shown. If the coefficient of restitution is $e = 0.6$, determine

a. The angular velocities of both bars immediately after the impact.
b. The velocity of the mass center of bar CD immediately after the impact.

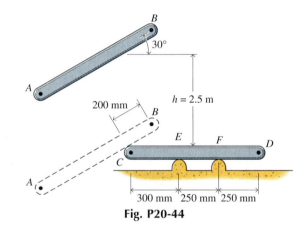

Fig. P20-44

Computer Problems

C20-45 A slender uniform rod ($\ell = 21$ in., $W_{AB} = 10$ lb) is at rest on a frictionless horizontal surface when it is struck by a small disk ($W_d = m_d g = 2$ lb) as shown in Fig. P20-45. If the coefficient of restitution is $e = 0.6$ and the initial velocity of the disk is $v_0 = 15$ ft/s at an angle of $\theta = 60°$, calculate and plot

a. The angular velocity of the rod after the collision (ω_{AB}) as a function of the impact location (b) for $0 < b < 21$ in.
b. The velocity of the mass center of the rod after the collision (v_G) as a function of the impact location (b) for $0 < b < 21$ in.
c. The velocity of end A of the rod immediately after the collision (v_A) as a function of the impact location (b) for $0 < b < 21$ in.

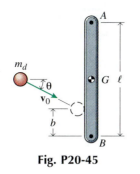

Fig. P20-45

C20-46 A slender uniform bar ($\ell = 750$ mm, $m_{AB} = 8$ kg) is attached to a frictionless hinge at A as shown in Fig. P20-46. The bar is released from rest with $\phi_0 = 60°$ and strikes the bumper at C. If the coefficient of restitution is $e = 0.7$, calculate and plot

a. The angular velocity of the bar immediately after the collision with the bumper (ω_{AB}) as a function of the location of the bumper (d) for $0 < d < 750$ mm.

b. The average magnitude of the support reaction at A for an impact duration of 0.005 s as a function of the location of the bumper (d) for $0 < d < 750$ mm.

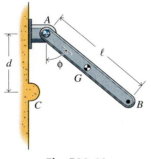

Fig. P20-46

C20-47 A slender uniform rod ($\ell = 36$ in., $W_{AB} = 12$ lb) hangs motionless from a frictionless pin at A as shown in Fig. P20-47. Pin A is not attached to the support but simply rests on a horizontal ledge. A small ball ($W_b = m_b g = 2$ lb) strikes the rod at a point d from the bottom end. If the coefficient of restitution is $e = 0.5$ and the initial velocity of the ball is $v_0 = 30$ ft/s at an angle of $\theta = 40°$ to the horizontal, determine

a. The angular velocity of the rod after the collision (ω_{AB}) as a function of the impact location (d) for $0 < d < 36$ in.

b. The average magnitude of the support reaction at A for an impact duration of 0.005 s as a function of the impact location (d) for $0 < d < 36$ in.

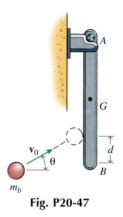

Fig. P20-47

C20-48 A slender uniform rod ($\ell = 800$ mm, $m_{AB} = 3$ kg) is released from rest from an initial height of $h = 2$ m and strikes a hard horizontal surface as shown in Fig. P20-48. If the coefficient of restitution is $e = 0.7$ and $\theta = 70°$, calculate and plot the height h_A of end A, the height h_G of the mass center, and the height h_B of end B, all as functions of time t from $t = 0$ (the time when end A first hits the horizontal surface) until one of the ends again hits the horizontal surface.

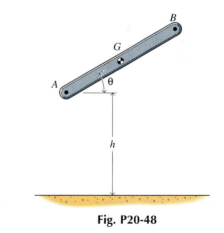

Fig. P20-48

20-6   ANGULAR IMPULSE AND MOMENTUM OF A RIGID BODY IN THREE-DIMENSIONAL MOTION

The general forms of the impulse–momentum equations stated earlier in this chapter (Eqs. 20-1, 20-2, and 20-3) apply equally to an arbitrary system of particles and to the system of particles that make up a rigid body. The general forms of the equations also apply equally to two-dimensional or three-dimensional motion. In fact, not only is the form of the linear impulse–momentum equation (Eq. 20-1) exactly the same for an arbitrary system of interacting particles, a rigid body in planar motion, and a rigid body in general three-dimensional motion, but the terms in the equation are also calculated in exactly the same manner for all three cases.

Calculation of the angular momentum terms in Eqs. 20-2 and 20-3, however, depends on whether the particles move independently or form a rigid body. For the planar motion of a symmetrical rigid body, the angular momentum was simply (Eq. 20-5)

$$H_G\mathbf{k} = I_G\omega\mathbf{k}$$

For the three-dimensional motion of a rigid body, however, the angular momentum has a number of additional components that are absent in plane motion.

20-6.1   Angular Momentum

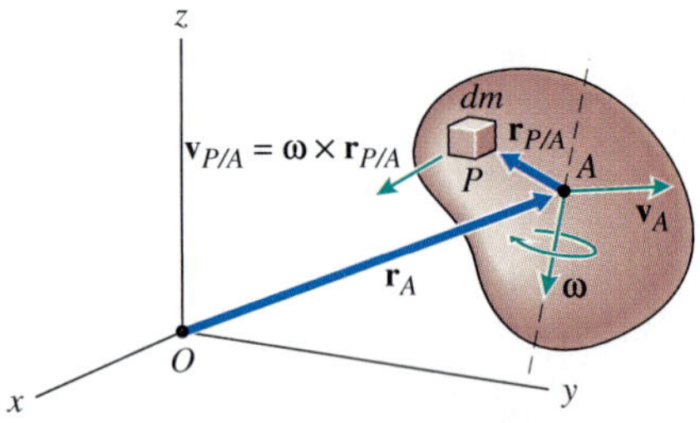

The angular momentum of a particle about a point is just the moment of the linear momentum of the particle about that point. Let A be an arbitrary point in a rigid body (Fig. 20-20). Then the angular momentum of the particle P of mass dm about point A is given by

$$d\mathbf{H}_A = \mathbf{r}_{P/A} \times \mathbf{v}_P \, dm = \mathbf{r}_{P/A} \times (\mathbf{v}_A + \mathbf{v}_{P/A}) \, dm$$
$$= \mathbf{r}_{P/A} \times [\mathbf{v}_A + (\boldsymbol{\omega} \times \mathbf{r}_{P/A})] \, dm \qquad (a)$$

where $\mathbf{v}_A$ and $\mathbf{v}_P = \mathbf{v}_A + \mathbf{v}_{P/A}$ are the absolute velocities of point A and particle P; $\mathbf{r}_{P/A}$ and $\mathbf{v}_{P/A} = \boldsymbol{\omega} \times \mathbf{r}_{P/A}$ are the position and velocity of particle P relative to point A; and $\boldsymbol{\omega}$ is the angular velocity of the rigid body. Integration of Eq. a over all the particles in the rigid body gives the angular momentum of the entire rigid body about A as

$$\mathbf{H}_A = \int \mathbf{r}_{P/A} \times [\mathbf{v}_A + (\boldsymbol{\omega} \times \mathbf{r}_{P/A})] \, dm$$
$$= \left(\int \mathbf{r}_{P/A} \, dm\right) \times \mathbf{v}_A + \int \mathbf{r}_{P/A} \times (\boldsymbol{\omega} \times \mathbf{r}_{P/A}) \, dm \qquad (b)$$

where $\mathbf{v}_A$ is independent of dm and has been taken outside of the integral in the first term. Then by the definition of the mass center, the first integral can be written $\int \mathbf{r}_{P/A} \, dm = m\mathbf{r}_{G/A}$. Therefore,

$$\mathbf{H}_A = \mathbf{r}_{G/A} \times (m\mathbf{v}_A) + \int \mathbf{r}_{P/A} \times (\boldsymbol{\omega} \times \mathbf{r}_{P/A}) \, dm \qquad (c)$$

Equation c can be simplified even further for special choices of the point A. For example, if point A is a fixed point about which the body is rotating, then $\mathbf{v}_A = \mathbf{0}$ and Eq. c becomes

$$\mathbf{H}_A = \int \mathbf{r}_{P/A} \times (\boldsymbol{\omega} \times \mathbf{r}_{P/A}) \, dm \qquad (d)$$

Figure 20-20   The angular momentum of a rigid body is the sum of the angular momenta of the particles dm that make up the body.

Similarly, if point A is the center of mass G, then $\mathbf{r}_{G/A} = \mathbf{r}_{G/G} = \mathbf{0}$, and Eq. c becomes

$$\mathbf{H}_G = \int \mathbf{r}_{P/G} \times (\boldsymbol{\omega} \times \mathbf{r}_{P/G})\, dm \qquad (e)$$

Even when point A is an arbitrary point, Eq. c can be written in a slightly more convenient form. Using the substitution $\mathbf{r}_{P/A} = \mathbf{r}_{P/G} + \mathbf{r}_{G/A}$ in Eq. c gives

$$\mathbf{H}_A = \mathbf{r}_{G/A} \times (m\mathbf{v}_A) + \int (\mathbf{r}_{P/G} + \mathbf{r}_{G/A}) \times [\boldsymbol{\omega} \times (\mathbf{r}_{P/G} + \mathbf{r}_{G/A})]\, dm \qquad (f)$$

But $\mathbf{r}_{G/A}$ and $\boldsymbol{\omega}$ are independent of dm and can be taken outside of the integral. Therefore,

$$\mathbf{H}_A = \mathbf{r}_{G/A} \times (m\mathbf{v}_A) + \int (\mathbf{r}_{P/G} \times (\boldsymbol{\omega} \times \mathbf{r}_{P/G})\, dm$$

$$+ \left(\int (\mathbf{r}_{P/G}\, dm) \times (\boldsymbol{\omega} \times \mathbf{r}_{G/A}) + \mathbf{r}_{G/A} \times \left(\boldsymbol{\omega} \times \int \mathbf{r}_{P/G}\, dm \right) \right)$$

$$+ \mathbf{r}_{G/A} \times (\boldsymbol{\omega} \times \mathbf{r}_{G/A}) \int dm \qquad (g)$$

But the first integral in Eq. g is just $\mathbf{H}_G$ (Eq. f), the second and third integrals are zero because $\int \mathbf{r}_{P/G}\, dm = m\mathbf{r}_{G/G} = \mathbf{0}$ by the definition of the mass center, and the last term is $\mathbf{r}_{G/A} \times m\mathbf{r}_{G/A}$. Finally, combining the first and last terms and using the relative velocity equation $\mathbf{v}_G = \mathbf{v}_A + \mathbf{v}_{G/A}$ gives

$$\mathbf{H}_A = \mathbf{r}_{G/A} \times (m\mathbf{v}_G) + \mathbf{H}_G \qquad (20\text{-}12)$$

That is, momentum properties of a rigid body may be represented by the equivalent "force-couple" system shown in the kinetic diagram of Fig. 20-21. Although the resultant angular momentum vector $\mathbf{H}_G$ is a free vector, it is represented as acting about the mass center G for convenience. The resultant linear momentum vector $\mathbf{L} = m\mathbf{v}_G$ acts through the mass center G.

Equations d and e have similar structure and can be developed simultaneously by writing $\mathbf{r} = x\mathbf{i} + y\mathbf{j} + z\mathbf{k}$ and $\boldsymbol{\omega} = \omega_x\mathbf{i} + \omega_y\mathbf{j} + \omega_z\mathbf{k}$. For Eq. d, $\mathbf{H} = \mathbf{H}_A$, $\mathbf{r} = \mathbf{r}_{P/A}$, and the positions x, y, and z are measured relative to coordinate axes centered at the fixed point A. For Eq. e, $\mathbf{H} = \mathbf{H}_G$, $\mathbf{r} = \mathbf{r}_{P/G}$, and the positions x, y, and z are measured relative to coordinate axes centered at the mass center G.

Now, expanding the triple vector product in Eqs. d and e gives

$$\mathbf{H} = \left(\omega_x \int (y^2 + z^2)\, dm - \omega_y \int xy\, dm - \omega_z \int xz\, dm \right)\mathbf{i}$$

$$+ \left(-\omega_x \int xy\, dm + \omega_y \int (x^2 + z^2)\, dm - \omega_z \int yz\, dm \right)\mathbf{j}$$

$$+ \left(-\omega_x \int xz\, dm - \omega_y \int yz\, dm + \omega_z \int (x^2 + y^2)\, dm \right)\mathbf{k} \qquad (h)$$

where the components of the angular velocity are also independent of dm and have again been taken outside of the integrals. The integrals in Eq. h represent the mass moments of inertia and products of inertia

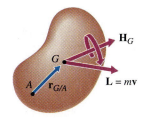

Figure 20-21 The momentum properties of a rigid body may be represented by an "equivalent force-couple" at the mass center of the body.

of the body with respect to the *xyz*-axes:

$$I_x = \int (y^2 + z^2)\, dm \qquad I_{xy} = \int xy\, dm = I_{yx}$$

$$I_y = \int (x^2 + z^2)\, dm \qquad I_{yz} = \int yz\, dm = I_{zy} \tag{i}$$

$$I_z = \int (x^2 + y^2)\, dm \qquad I_{xz} = \int xz\, dm = I_{zx}$$

The calculation of the moments and products of inertia was covered in the volume on *Statics*. Much of that material has been repeated in Appendix A of this volume.

Substituting the moments of inertia from Eqs. *i* into Eq. *h* gives the angular momentum of the body about a fixed point *A* or about its mass center *G*:

$$\begin{aligned}
\mathbf{H} = &(I_x\omega_x - I_{xy}\omega_y - I_{xz}\omega_z)\mathbf{i} \\
&+ (-I_{yx}\omega_x + I_y\omega_y - I_{yz}\omega_z)\mathbf{j} \\
&+ (-I_{zx}\omega_x - I_{zy}\omega_y + I_z\omega_z)\mathbf{k}
\end{aligned} \tag{20-13}$$

where the moments and products of inertia are relative to axes through the fixed point *A* for $\mathbf{H}_A$ or relative to axes through the mass center *G* for $\mathbf{H}_G$. Equation 20-13 is valid for a particular position of the body. Since the orientation of the coordinate axes is fixed, the moments of inertia and products of inertia will, in general, change as the body rotates relative to *xyz*.

Equation 20-13 appears rather complicated, but it can be simplified considerably for special orientations of the coordinate axes. The ideal set of axes is the principal axes of inertia. If the coordinate axes coincide with the principal axes of inertia, then all the products of inertia are zero $I_{xy} = I_{yx} = I_{yz} = I_{zy} = I_{xz} = I_{zx} = 0$. Then Eq. 20-13 becomes (for that instant only, in most cases)

$$\mathbf{H} = (I_x\omega_x)\mathbf{i} + (I_y\omega_y)\mathbf{j} + (I_z\omega_z)\mathbf{k} \tag{20-14}$$

where I_x, I_y, and I_z are the principal moments of inertia. Even though use of the principal axes simplifies the expression for the angular momentum, it is not always convenient for geometric reasons to use these axes to compute $\mathbf{H}$.

Finally, it must be noted that the angular momentum vector $\mathbf{H}$ and the angular velocity $\boldsymbol{\omega}$ will have different directions unless $\boldsymbol{\omega}$ is directed along a principal axes of inertia. For example, for the plane motion of a rigid body that is symmetrical with respect to the *xy*-plane, the *z*-axis is a principal direction, I_z is a principal moment of inertia, $\boldsymbol{\omega} = \omega_z\mathbf{k}$, and Eq. 20-14 gives

$$\mathbf{H} = (I_z\omega_z)\mathbf{k} = I_z(\omega_z\mathbf{k}) = I_z\boldsymbol{\omega} \tag{20-15}$$

Therefore, the vectors $\mathbf{H}$ and $\boldsymbol{\omega}$ are collinear. Actually, if the three principal moments of inertia are equal, $I_x = I_y = I_z = I$, then Eq. 20-14 gives

$$\mathbf{H} = I(\omega_x\mathbf{i} + \omega_y\mathbf{j} + \omega_z\mathbf{k}) = I\boldsymbol{\omega} \tag{20-16}$$

and the vectors $\mathbf{H}$ and $\boldsymbol{\omega}$ are also collinear. However, if the three principal moments of inertia are equal, then every axis is a principal axis. Therefore, Eq. 20-16 is just a special case of Eq. 20-15, since no matter what direction $\boldsymbol{\omega}$ is in, it will coincide with a principal direction.

20-6.2 Angular Impulse–Momentum Principle

The Angular Impulse–Momentum Principle for a system of particles (Eq. 20-3)

$$(\mathbf{H}_O)_i + \int_{t_i}^{t_f} \sum \mathbf{M}_{\ell/O}\, dt = (\mathbf{H}_O)_f \qquad \text{(20-17a)}$$

$$(\mathbf{H}_G)_i + \int_{t_i}^{t_f} \sum \mathbf{M}_{\ell/G}\, dt = (\mathbf{H}_G)_f \qquad \text{(20-17b)}$$

applies to any system of particles whether it consists of independently moving, interacting particles (where $\mathbf{H}$ is computed by summing $\mathbf{r}_\ell \times m\mathbf{v}_\ell$ for all the particles) or the particles that make up a rigid body. Restricting the point about which moments and angular momentum are calculated to either the mass center G or a fixed point O about which the rigid body is rotating allows the angular momentum $\mathbf{H}$ to be computed using Eq. 20-13. If the rigid body is rotating about a fixed point O, then Eq. 20-17a can be used, with the moments of inertia in Eq. 20-13 calculated relative to coordinate axes centered at the fixed point O. Otherwise, Eq. 20-17b must be used, with the moments of inertia in Eq. 20-13 calculated relative to coordinate axes centered at the mass center G.

The angular impulse–momentum equations (Eq. 20-17) are particularly useful when the moment of external forces about a specified axis is known. For example, if no external force acting on a body has a moment about a particular axis, then the angular momentum about that axis is constant. Usually, when the body is acted on by impulsive forces, only the moments of the impulsive forces need be considered.

When a rigid body is rotating about a fixed point O that is not the center of mass, the impulse of the reaction must be included in the analysis if the mass center G is used as a reference (Eq. 20-17b). In such cases, it is generally more convenient to use the fixed point O for reference because the reaction at O has no moment about O and will not enter into Eq. 20-17a.

As was the case with the application of the angular impulse momentum principle to a rigid body in planar motion, the body need move rigidly only at the initial and final instants of time for Eq. 20-13 to be used to calculate the angular momentum. Between the initial and final instants, parts of the body may move freely relative to each other.

20-6.3 Graphical Representation of the Linear and Angular Momentum Principles

The linear and angular momentum (linear and angular impulse) are analogous to a force and a couple. The impulse momentum principles expressed by Eqs. 20-1 and 20-17b

$$m(\mathbf{v}_G)_i + \int_{t_i}^{t_f} \mathbf{R}\, dt = m(\mathbf{v}_G)_f$$

$$(\mathbf{H}_G)_i + \int_{t_i}^{t_f} \sum \mathbf{M}_{\ell/G}\, dt = (\mathbf{H}_G)_f$$

can be represented by the kinetic diagrams of Fig. 20-22, in which the linear and angular momenta have been replaced with an equivalent

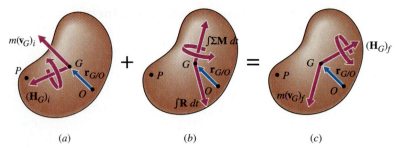

(a)   (b)   (c)

Figure 20-22   The linear impulse–momentum and the angular impulse–momentum principles are both contained in this graphical representation. The linear momenta and linear impulse act through the mass center; the angular momenta and angular impulses are with respect to the mass center.

"force-couple" $\mathbf{L} = m\mathbf{v}_G$ and $\mathbf{H}_G$ applied at the mass center. That is, the sum of the equivalent "force-couple" of the momenta at time t_1 (Fig. 20-22a) and the equivalent force-couple of the impulses (Fig. 20-22b) is equal to the equivalent "force-couple" of the momenta at time t_f (Fig. 20-22c).

The graphical representation of Fig. 20-22 can also be used to write the Angular Impulse–Momentum Principle relative to an arbitrary fixed point P as well as relative to a fixed point O about which the body is rotating. The former statement is immediately verified by comparing the moments about P of the equivalent "force-couple" in each part of the figure with the sum of $\mathbf{r}_{G/P} \times$ (Eq. 20-1) and Eq. 20-17b. The latter statement is similarly verified by adding $\mathbf{r}_{G/O} \times$ (Eq. 20-1) to Eq. 20-17b and making use of the parallel-axis theorems for moments and products of inertia.

20-6.4   Systems of Rigid Bodies

As has already been pointed out, the use of Eq. 20-17 requires only that the body behave rigidly at the initial time t_i and at the final time t_f so that the angular momentum $\mathbf{H}$ can be calculated using Eq. 20-13 at these times. Between the initial and final instants of time, parts of the body may move relative to each other. If parts of the body are moving relative to each other at the initial and/or the final instants of time, then a separate angular impulse momentum equation should be written for each part that does behave in a rigid manner and the equations added together. If the moments of momentum and moments of forces

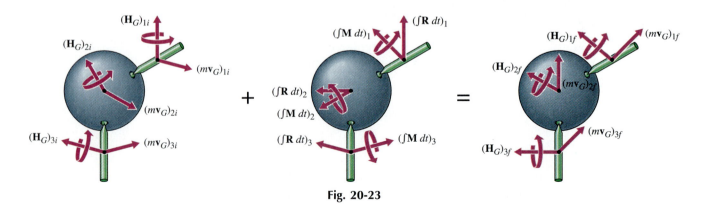

Fig. 20-23

are all written relative to the same point in each equation, then the moments of the joint forces holding the various parts together will cancel in pairs and need not be computed. This is most easily accomplished using the graphical representation of Fig. 20-23. In the first and last parts of the figure, the momentum of each rigid body has been replaced with an equivalent "force-couple" system at its own mass center. In the middle part of the figure, the joint forces are internal forces and need not be shown.

EXAMPLE PROBLEM 20-6

A homogeneous wheel of diameter 800 mm, thickness 50 mm, and mass 40 kg is rigidly attached to an axle OG of length 1200 mm, diameter 50 mm, and mass 10 kg as shown in Fig. 20-24. The axle is pivoted at the fixed point O, and the wheel rolls without slipping on a horizontal floor. If the center of the wheel G has a speed of 2 m/s as the wheel rolls, determine for the instant shown

a.   The angular velocity $\boldsymbol{\omega}$ of the wheel and axle system.
b.   The angular momentum $\mathbf{H}_O$ about O of the wheel and axle system.
c.   The angle between $\boldsymbol{\omega}$ and $\mathbf{H}_O$.

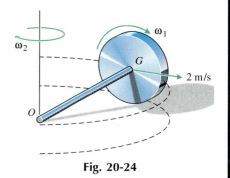

Fig. 20-24

SOLUTION

a.   Set up a coordinate system with the x-axis along OG, the y-axis in the vertical plane containing OG, and the z-axis in the horizontal plane (Fig. 20-25a). These axes are principal axes for the wheel and axle system. As the wheel rotates about the axle OG at a rate of ω_1, it also rotates with the axle about a vertical axis at a rate ω_2. The total angular velocity of the system is therefore

$$\boldsymbol{\omega} = \omega_1 \mathbf{i} - \omega_2(\sin\theta\,\mathbf{i} + \cos\theta\,\mathbf{j})$$

where $\theta = \tan^{-1}(400/1200) = 18.43°$. As the wheel rotates through an angle ϕ, the axle rotates through an angle γ and the arc lengths $(1.2/\cos\theta)\gamma$ and $(0.4)\phi$ are equal (Fig. 20-25b). Therefore, $(1.2/\cos\theta)\omega_2 = (0.4)\omega_1$ or $\omega_1 = 3.162\omega_2$. But since O is fixed,

$$\begin{aligned}\mathbf{v}_G = 2.0\mathbf{k} &= \mathbf{v}_O + \mathbf{v}_{G/O} = \mathbf{0} + \boldsymbol{\omega} \times \mathbf{r}_{G/O} \\ &= [(3.162 - \sin\theta)\omega_2\mathbf{i} - (\cos\theta)\omega_2\mathbf{j}] \times (1.2\mathbf{i})\end{aligned}$$

Therefore, $\omega_2 = 1.7568$ rad/s, $\omega_1 = 5.5556$ rad/s, and

$$\boldsymbol{\omega} = 5.00\mathbf{i} - 1.667\mathbf{j} \text{ rad/s} \qquad\qquad \text{Ans.}$$

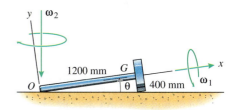

(a)

b.   Then, since the xyz-axes are principal axes, the angular momentum of the system about O is

$$\mathbf{H}_O = (I_x\omega_x)\mathbf{i} + (I_y\omega_y)\mathbf{j} + (I_z\omega_z)\mathbf{k}$$

where

$$\begin{aligned}I_x &= \frac{1}{2}(10)(0.025)^2 + \frac{1}{2}(40)(0.400)^2 \\ &= 3.203 \text{ kg} \cdot \text{m}^2\end{aligned}$$

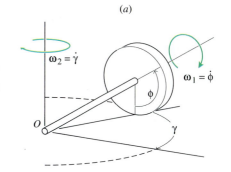

(b)

Fig. 20-25

and

$$I_y = I_z = \frac{1}{4}(10)(0.025)^2 + \frac{1}{3}(10)(1.200)^2$$

$$+ \frac{1}{4}(40)(0.400)^2 + \frac{1}{12}(40)(0.050)^2 + (40)(1.225)^2$$

$$= 66.43 \text{ kg} \cdot \text{m}^2$$

Therefore,

$$\mathbf{H}_O = 16.03\mathbf{i} - 110.7\mathbf{j} \text{ kg} \cdot \text{m}^2/\text{s} \qquad \text{Ans.}$$

c. The angle between $\boldsymbol{\omega}$ and $\mathbf{H}_O$ is given by

$$\cos^{-1}\left(\frac{\boldsymbol{\omega} \cdot \mathbf{H}_O}{\omega H_O}\right) = \cos^{-1}\left(\frac{264.6}{(5.270)(111.88)}\right)$$

$$= 63.33° \qquad \text{Ans.}$$

As the wheel rotates through an angle ϕ, the contact point moves a distance $s = 0.4\,\phi$ along the surface and the axle rotates about the vertical axis by an amount $\gamma = s/(1.2/\cos\theta)$ (Fig. 20-25b). Therefore, the rotations of the axle and the disk are related by $0.4\,\phi = (1.2/\cos\theta)\gamma$. The time derivative of this relationship gives the relationship between the angular speeds, $0.4\,\omega_1 = (1.2/\cos\theta)\,\omega_2$.

EXAMPLE PROBLEM 20-7

A 2-lb uniform sign is suspended by two wires as shown in Fig. 20-26. The thickness of the sign is negligible compared to its surface dimensions. If a bullet weighing 0.05 lb and traveling with a speed of 500 ft/s in the negative x-direction strikes the sign at the corner C and becomes embedded in it, determine the velocity $\mathbf{v}_G$ of the mass center of the sign and the angular velocity $\boldsymbol{\omega}$ of the sign immediately after the impact.

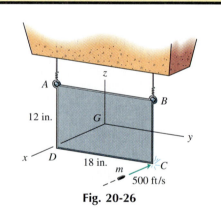

Fig. 20-26

SOLUTION

Separate linear and angular impulse–momentum equations will be written for the bullet and the sign and then added together. When the equations are added together, the linear and angular impulses of the force of interaction between the bullet and the sign will cancel. The combined kinetic diagrams are then shown in Fig. 20-27, in which the impulse of the force of interaction between the bullet and the sign has been omitted.

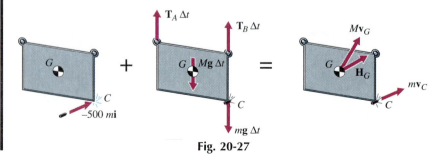

Fig. 20-27

530

Linear Momentum: Initially, the sign is at rest and has no linear momentum $(\mathbf{L}_s)_i = \mathbf{0}$. After the impact, the linear momentum of the sign is just $(\mathbf{L}_s)_f = M\mathbf{v}_G$ where $M = 2/32.2$ slugs is the mass of the sign and $\mathbf{v}_G$ is the velocity of the mass center of the sign immediately after the impact. The initial and final linear momenta of the bullet are

$$(\mathbf{L}_b)_i = -500m\mathbf{i} = -0.7764\mathbf{i} \text{ lb} \cdot \text{s}$$
$$(\mathbf{L}_b)_f = m[\mathbf{v}_G + (\omega_x\mathbf{i} + \omega_y\mathbf{j} + \omega_z\mathbf{k}) \times (0.75\mathbf{j} - 0.50\mathbf{k})]$$
$$= m[\mathbf{v}_G - (0.5\omega_y + 0.75\omega_x)\mathbf{i} + 0.5\omega_x\mathbf{j} + 0.75\omega_x\mathbf{k}]$$

where $m = 0.05/32.2$ slugs is the mass of the bullet, $\boldsymbol{\omega} = \omega_x\mathbf{i} + \omega_y\mathbf{j} + \omega_z\mathbf{k}$ is the angular velocity of the sign immediately after the impact, and the velocity of the bullet has been replaced using the relative velocity equation $\mathbf{v} = \mathbf{v}_G + \boldsymbol{\omega} \times \mathbf{r}_{C/G}$. Then the x-, y- and z-components of the linear impulse–momentum equation are

$$-500m = (m + M)v_{Gx} - m(0.5\omega_y + 0.75\omega_z) \qquad (a)$$
$$0 = (m + M)v_{Gy} + 0.5\,m\omega_x \qquad (b)$$
$$(T_A + T_B - 2.05)\Delta t = (m + M)v_{Gz} + 0.75\,m\omega_x \qquad (c)$$

Angular Momentum About G**:** Initially, the sign is at rest and has no angular momentum $(\mathbf{H}_{Gs})_i = \mathbf{0}$. Since the x-, y-, and z-axes are principal axes, the angular momentum of the sign after the impact is just $(\mathbf{H}_{Gs})_f = I_x\omega_x\mathbf{i} + I_y\omega_y\mathbf{j} + I_z\omega_z\mathbf{k}$ where

$$I_x = \frac{1}{12}\left(\frac{2}{32.2}\right)(1.5^2 + 1.0^2) = 0.016822 \text{ lb} \cdot \text{ft} \cdot \text{s}^2$$
$$I_y = \frac{1}{12}\left(\frac{2}{32.2}\right)(1.0^2) = 0.005176 \text{ lb} \cdot \text{ft} \cdot \text{s}^2$$
$$I_z = \frac{1}{12}\left(\frac{2}{32.2}\right)(1.5^2) = 0.011646 \text{ ft} \cdot \text{lb} \cdot \text{s}^2$$

The angular momentum of the bullet about G before the impact is the moment of its linear momentum

$$(\mathbf{H}_{Gb})_i = (0.75\mathbf{j} - 0.50\mathbf{k}) \times m(-500\mathbf{i})$$
$$= 0.3882\mathbf{j} + 0.5823\mathbf{k} \text{ lb} \cdot \text{ft} \cdot \text{s}$$

Similarly,

$$(\mathbf{H}_{Gb})_f = (0.75\mathbf{j} - 0.50\mathbf{k}) \times m[\mathbf{v}_G + \boldsymbol{\omega} \times (0.75\mathbf{j} - 0.50\mathbf{k})]$$

Then, the x-, y-, and z-components of the angular impulse–momentum equation are

$$(T_B - T_A - 0.05)(0.75)\Delta t = I_x\omega_x + 0.75mv_{Gz} + 0.5mv_{Gy} + (0.75^2 + 0.5^2)m\omega_x \quad (d)$$
$$0.3882 = I_y\omega_y - 0.5mv_{Gx} + 0.5m(0.5\omega_y + 0.75\omega_z) \qquad (e)$$
$$0.5823 = I_z\omega_z - 0.75mv_{Gx} + 0.75m(0.5\omega_y + 0.75\omega_z) \qquad (f)$$

Equations a through f are to be solved subject to these additional constraints. The wires are inextensible (neither A nor B can have a component of velocity in the negative z-direction immediately after the impact); the tensions T_A and T_B must not be negative (the wires cannot withstand compressive forces); and if either A or B has a component of velocity in the positive z-direction immediately after impact, then the corresponding tension T_A or T_B must be zero. Guessing that the wires both remain taut (T_A and T_B both remain positive) and that $v_{Gz} = \omega_x = 0$ gives the solution

$$v_{Gx} = -10.64 \text{ ft/s} \qquad \qquad \text{Ans.}$$
$$v_{Gy} = v_{Gz} = \omega_x = 0 \qquad \qquad \text{Ans.}$$
$$\omega_y = 63.8 \text{ rad/s} \qquad \qquad \text{Ans.}$$
$$\omega_z = 42.6 \text{ rad/s} \qquad \qquad \text{Ans.}$$

Substitution of $\mathbf{v}_G$ and $\boldsymbol{\omega}$ into Eqs. c and d gives $T_A = 1.00$ lb and $T_B = 1.05$ lb, thus verifying that both wires remain taut. Furthermore, it is easily verified that $\mathbf{v}_A = \mathbf{v}_G + \mathbf{v}_{A/G} = 53.2\mathbf{i}$ ft/s and $\mathbf{v}_B = \mathbf{v}_G + \mathbf{v}_{B/G} = -10.64\mathbf{i}$ ft/s so that neither A nor B has a z-component of velocity immediately after the impact.

PROBLEMS

Introductory Problems

20-49–20-57 For each of the objects shown, determine the angular momentum $\mathbf{H}_O$ (where O is the origin of the coordinate system) and the angle between the angular velocity and the angular momentum vectors at the instant shown. Neglect the mass of the shafts to which the spheres, plates, and so on are mounted.

20-49* The slender bent rod of Fig. P20-49 weighs 0.2 lb/ft.

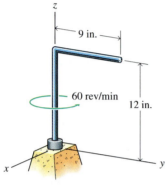

Fig. P20-49

20-50* The slender branched rod of Fig. P20-50 has a mass of 0.25 kg/m.

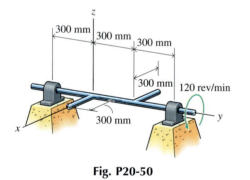

Fig. P20-50

20-51 The three identical spheres of Fig. P20-51 each weigh 2 lb and have a diameter of 4 in. The centers of the spheres are 10 in. from the center of the shaft, and they are symmetrically located around the shaft.

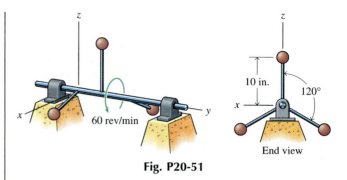

Fig. P20-51

20-52* The 1.5-kg cylinder AB of Fig. P20-52 has a diameter of 50 mm and a length of 200 mm. It is mounted on a thin circular disk having a radius of 400 mm and a mass of 0.5 kg. The distance between the axis of the cylinder and the axis of the shaft is 300 mm.

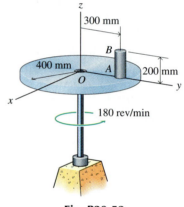

Fig. P20-52

20-53 The bent rod of Fig. P20-53 has a diameter of 0.5 in., weighs 0.2 lb/ft, and is 18 in. long.

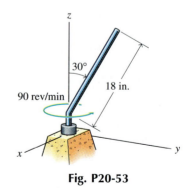

Fig. P20-53

20-54 The 100-mm-long rod of Fig. P20-54 has a diameter of 20 mm, a mass of 3 kg, and is mounted at an angle of 30° to the axis of the shaft.

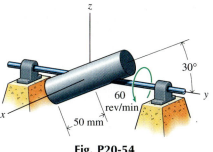

Fig. P20-54

20-55* The thin circular plate of Fig. P20-55 has a radius of 9 in. and weighs 1.5 lb. The plane of the plate makes an angle of 60° with the axis of the shaft.

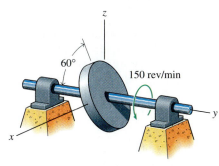

Fig. P20-55

20-56 The thin rectangular plate (300 mm by 800 mm) of Fig. P20-56 has a mass of 5 kg and rotates about a shaft along its diagonal.

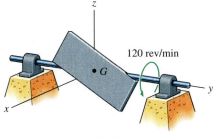

Fig. P20-56

20-57* The 12-in.-long circular cylinder of Fig. P20-57 has a diameter of 6 in., weighs 4 lb, and rotates about a shaft along its "diagonal" as shown.

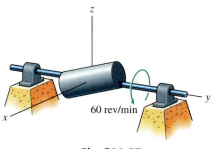

Fig. P20-57

Intermediate Problems

20-58* The thin rectangular plate of Fig. P20-58 is 300 mm tall, 800 mm wide, and has a mass of 5 kg. The plate is balanced on edge when it is struck by an impulse of $\mathbf{F}\,\Delta t = -10\mathbf{i}$ N · s at corner C. Assume that Δt is sufficiently small that nonimpulsive forces can be neglected, and determine

a. The velocity of the mass center of the plate immediately after the impact.
b. The angular velocity of the plate immediately after the impact.
c. The angle between the angular velocity and the angular momentum vectors immediately after the impact.

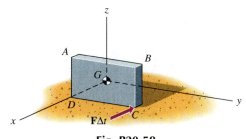

Fig. P20-58

20-59 The thin plate of Fig. P20-59 is an equilateral triangle 18 in. on a side and weighs 5 lb. The plate is balanced on edge when it is struck by an impulse of $\mathbf{F}\,\Delta t = -0.15\mathbf{i}$ lb · s at corner A. Assume that Δt is sufficiently small that nonimpulsive forces can be neglected and determine

a. The velocity of the mass center of the plate immediately after the impact.
b. The angular velocity of the plate immediately after the impact.
c. The angle between the angular velocity and the angular momentum vectors immediately after the impact.

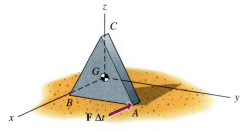

Fig. P20-59

20-60* The thin circular plate of Fig. P20-60 has a radius of 300 mm and a mass of 2 kg. The plate is suspended on a wire when it is struck by an impulse of $\mathbf{F}\,\Delta t = -1.4\mathbf{i}$ N · s at point A. Assume that Δt is sufficiently small that nonimpulsive forces can be neglected and determine

a. The angle between the angular velocity and the angular momentum vectors immediately after the impact.
b. The velocity of point A immediately after the impact.

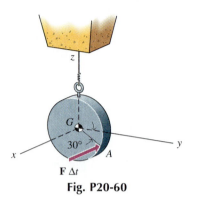

Fig. P20-60

20-61* The thin plate of Fig. P20-61 is a square 6 in. on a side and weighs 2 lb. The plate is suspended on a wire attached to the middle of side AB when it is struck by an impulse of $\mathbf{F}\,\Delta t = -0.05\mathbf{i}$ lb · s at corner C. Assume that Δt is sufficiently small that nonimpulsive forces can be neglected and determine

a. The angle between the angular velocity and the angular momentum vectors immediately after the impact.
b. The velocity of corner C immediately after the impact.

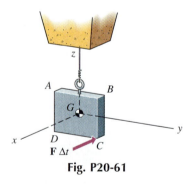

Fig. P20-61

20-62 The thin plate of Fig. P20-62 is a square 200 mm on a side and has a mass of 1.5 kg. The plate is suspended on a wire attached to corner A when it is struck by an impulse of $\mathbf{F}\,\Delta t = -2.5\mathbf{i}$ N · s at E (the midpoint of side AD). Assume that Δt is sufficiently small that nonimpulsive forces can be neglected and determine

a. The angle between the angular velocity and angular momentum vectors immediately after the impact.
b. The velocity of corner C immediately after the impact.

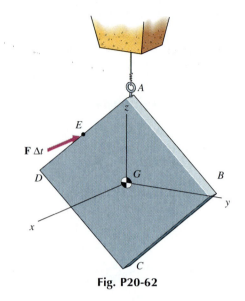

Fig. P20-62

20-63 The thin plate of Fig. P20-63 is a rectangle 9 in. high and 18 in. wide and weighs 2.5 lb. The plate is suspended on a wire attached to the corner A when it is struck by an impulse of $\mathbf{F}\,\Delta t = -0.1\mathbf{i}$ lb · s at corner D. Assume that Δt is sufficiently small that nonimpulsive forces can be neglected and determine

a. The angle between the angular velocity and angular momentum vectors immediately after the impact.
b. The velocity of corner D immediately after the impact.

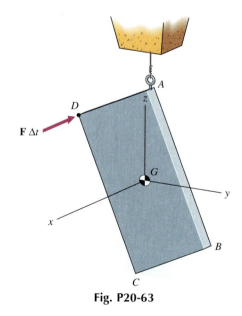

Fig. P20-63

20-64* The thin plate of Fig. P20-64 has a mass of 1.2 kg. The plate is suspended on a wire attached to its edge when it is struck by an impulse of $\mathbf{F}\,\Delta t = 0.5\mathbf{i} + \mathbf{j} - 0.8\mathbf{k}$ N · s at corner C. Assume that Δt is sufficiently small that nonimpulsive forces can be neglected and determine

a. The angle between the angular velocity and the angular momentum vectors immediately after the impact.
b. The velocity of corner C immediately after the impact.

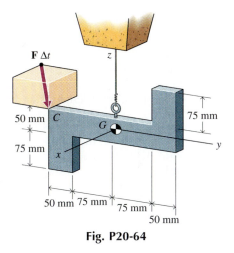

Fig. P20-64

20-65 The assembly of Fig. P20-65 is made by joining five identical slender rods (12 in. long and 0.5 lb each). The assembly is suspended on a wire attached to the midpoint of segment CD when it is struck by an impulse of $\mathbf{F}\,\Delta t = 0.1\mathbf{j} - 0.05\mathbf{k}$ lb · s at A. Assume that Δt is sufficiently small that nonimpulsive forces can be neglected and determine

a. The angle between the angular velocity and the angular momentum vectors immediately after the impact.
b. The velocity of end A immediately after the impact.

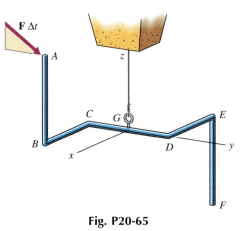

Fig. P20-65

20-66 The assembly of Fig. P20-66 is made by joining five identical slender rods (300 mm long and 0.25 kg each). The assembly is suspended on a wire attached to the midpoint of segment CD when it is struck by an impulse of $\mathbf{F}\,\Delta t = 0.5\mathbf{j} - 0.25\mathbf{k}$ N · s at A. Assume that Δt is sufficiently small that nonimpulsive forces can be neglected and determine

a. The angle between the angular velocity and the angular momentum vectors immediately after the impact.
b. The velocity of end A immediately after the impact.

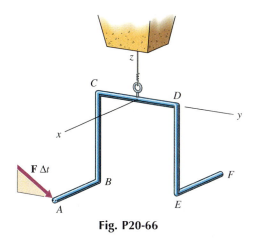

Fig. P20-66

20-67* The assembly of Fig. P20-67 is made by joining six identical slender rods (15 in. long and 0.6 lb each) to another slender rod (DK: 30 in. long and 1.2 lb). The assembly is suspended on a wire attached to joint K when it is struck by an impulse of $\mathbf{F}\,\Delta t = 0.15\mathbf{j} - 0.1\mathbf{k}$ lb · s at A. Assume that Δt is sufficiently small that nonimpulsive forces can be neglected and determine

a. The angle between the angular velocity and the angular momentum vectors immediately after the impact.
b. The velocity of end A immediately after the impact.

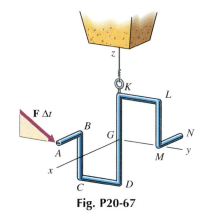

Fig. P20-67

20-68* A 50-g bullet traveling with a speed of 150 m/s in the negative x-direction strikes the rectangular plate of Problem 20-58 at corner C. If the bullet becomes embedded in the plate, determine

a. The velocity of the mass center of the plate immediately after the impact.
b. The angular velocity of the plate immediately after the impact.
c. The velocity of corner C immediately after the impact.

20-69 A 0.08-lb bullet traveling with a speed of 600 ft/s in the negative x-direction strikes the triangular plate of Problem 20-59 at corner A. If the bullet becomes embedded in the plate, determine

a. The velocity of the mass center of the plate immediately after the impact.
b. The angular velocity of the plate immediately after the impact.
c. The velocity of corner A immediately after the impact.

20-70 A 30-g bullet traveling with a speed of 180 m/s in the negative x-direction strikes the circular plate of Problem 20-60 at point A. If the bullet becomes embedded in the plate, determine

a. The magnitude of the impulse exerted on the plate by the bullet.
b. The angular momentum $\mathbf{H}_G$ of the plate immediately after the impact.
c. The velocity of point A immediately after the impact.

20-71* A 0.05-lb bullet traveling with a speed of 800 ft/s in the negative x-direction strikes the square plate of Problem 20-61 at corner C. If the bullet becomes embedded in the plate, determine

a. The magnitude of the impulse exerted on the plate by the bullet.
b. The angular momentum $\mathbf{H}_G$ of the plate immediately after the impact.
c. The velocity of corner C immediately after the impact.

Challenging Problems

20-72* A 125-g arrow traveling with a speed of 100 m/s in the negative x-direction strikes the rectangular plate of Problem 20-58 at corner C. (The arrow may be modeled as a uniform slender rod 800 mm long.) If the arrowhead becomes embedded in the plate, determine

a. The velocity of the mass center of the plate immediately after the impact.
b. The angular velocity of the plate immediately after the impact.
c. The velocity of corner C immediately after the impact.

20-73 A 0.2-lb arrow traveling with a speed of 300 ft/s in the negative x-direction strikes the triangular plate of Problem 20-59 at corner A. (The arrow may be modeled as a uniform slender rod 32 in. long.) If the arrowhead becomes embedded in the plate, determine

a. The velocity of the mass center of the plate immediately after the impact.
b. The angular velocity of the plate immediately after the impact.
c. The velocity of corner A immediately after the impact.

20-74 A 100-g arrow traveling with a speed of 150 m/s in the negative x-direction strikes the circular plate of Problem 20-60 at the point A. (The arrow may be modeled as a uniform slender rod 800 mm long.) If the arrowhead becomes embedded in the plate, determine

a. The magnitude of the impulse exerted on the plate by the arrow.
b. The angular momentum $\mathbf{H}_G$ of the plate immediately after the impact.
c. The velocity of point A immediately after the impact.

20-75* A 0.25-lb arrow traveling with a speed of 400 ft/s in the negative x-direction strikes the square plate of Problem 20-61 at corner C. (The arrow may be modeled as a uniform slender rod 32 in. long.) If the arrowhead becomes embedded in the plate, determine

a. The magnitude of the impulse exerted on the plate by the arrow.
b. The angular momentum $\mathbf{H}_G$ of the plate immediately after the impact.
c. The velocity of corner C immediately after the impact.

SUMMARY

The linear and angular momentum principles are integrals of the equations of motion with respect to time. They are particularly useful for solving problems in which the velocities of a body at two different instants of time are to be related and the forces involved can be expressed as functions of time.

The linear momentum of a system of particles whether rigid or non-rigid is the product of its mass and the velocity of its mass center $\mathbf{L} = m\mathbf{v}_G$. Therefore, the Linear Impulse–Momentum Principle as expressed by Eq. 20-1

$$m(\mathbf{v}_G)_i + \sum_\ell \int_{t_i}^{t_f} \mathbf{R}_\ell \, dt = m(\mathbf{v}_G)_f \tag{20-1}$$

applies equally to a system of independent, interacting particles and to rigid bodies.

The angular momentum of a particle can be calculated relative to any point, fixed or moving. For an arbitrary system of interacting particles, the particles move independently, and the expression of the Angular Impulse–Momentum Principle relative to a fixed point O is usually the most useful. For rigid bodies, however, velocities of points in the body are related by the angular velocity and the expression of the Angular Impulse–Momentum Principle relative to the mass center is usually the most useful.

For planar motion, $\boldsymbol{\omega} = \omega\mathbf{k}$, the angular momentum of a rigid body is

$$\mathbf{H}_G = -\omega I_{Gxz}\mathbf{i} - \omega I_{Gyz}\mathbf{j} + \omega I_{Gz}\mathbf{k} \tag{20-4}$$

If the products of inertia I_{Gxz} and I_{Gyz} are not zero, then the angular momentum will also have x- and y-components even for planar motion. This means that moment components in the x- and/or y-direction will be required to keep the motion in the xy-plane if the magnitude of the angular velocity is changing.

If a rigid body is symmetric about the plane of motion or if the z-axis through its mass center G is an axis of symmetry, then the products of inertia will be zero. During a planar motion of this body, $\boldsymbol{\omega} = \omega\mathbf{k}$, the angular momentum of the body will have only a z-component

$$\mathbf{H}_G = \omega I_G\mathbf{k} \tag{20-5}$$

where $I_G = I_{Gz}$ is the moment of inertia of the rigid body with respect to an axis through the mass center G and perpendicular to the plane of motion. For the planar motion of this body, the Angular Impulse–Momentum Principle is expressed by Eq. 20-8

$$(I_G\omega)_i + \int_{t_i}^{t_f} \sum M_G \, dt = (I_G\omega)_f \tag{20-8}$$

Between the initial and final instants of time, parts of the body may move relative to each other and $(I_G)_i$ may not be the same as $(I_G)_f$.

When rigid bodies collide, the coefficient of restitution relates the relative velocity of the points of contact before and after the collision. Since the principles of linear and angular impulse–momentum involve

the velocities of the mass centers of the rigid bodies, the velocities of
the contact points must be related to the velocities of the mass centers
of the rigid bodies using the relative velocity equations.

REVIEW PROBLEMS

20-76 A student sits on a chair that is free to rotate about
a vertical axis and holds a dumbbell in each hand as shown
in Fig. P20-76. The combined mass of the student and chair
is 80 kg, and the radius of gyration about the vertical axis
of rotation is $k_G = 140$ mm. Each dumbbell has a mass of
7 kg, and initially the student holds the dumbbells outstretched (at a distance of 900 mm from the axis of rotation) as in Fig. P20-76a. If the initial angular velocity of the
student is 10 rev/min, determine the angular velocity after the dumbbells have been pulled in close to the body (to
a distance of 200 mm from the axis of rotation) as shown
in Fig. P20-76b.

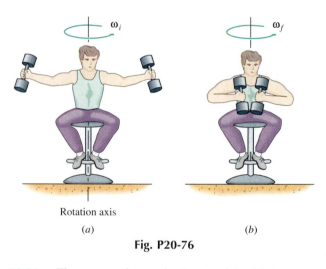

(a)   (b)

Fig. P20-76

20-77* The two uniform wheels A and B of Fig. P20-77
are initially at rest when a constant counterclockwise moment $M = 3$ lb · ft is suddenly applied to wheel A. Wheel
A has a 10-in. diameter and weighs 20 lb. Wheel B has a
16-in. diameter and weighs 30 lb. If the static and kinetic
coefficients of friction between the two wheels are $\mu_s = 0.2$
and $\mu_k = 0.1$, respectively, determine the angular velocities
of both wheels at $t = 5$ s, 15 s, and 25 s.

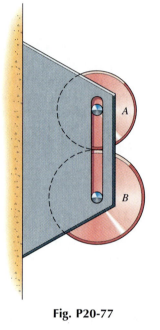

Fig. P20-77

20-78* Each of the four wheels of the cart shown in Fig.
P20-78 is a uniform disk 500 mm in diameter having a mass
of 5 kg. If the cart and its load add an additional 30 kg to
the total mass, determine the speed that the cart will attain
in 10 s after starting from rest.

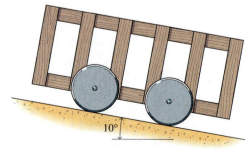

Fig. P20-78

20-79 The uniform slender rod AB ($W_{AB} = 3$ lb, $\ell_{AB} = 4$ ft) is hanging from a 16-in.-long wire when it is struck by a bullet as shown in Fig. P20-79. If the initial speed of the 0.08-lb bullet is $v_0 = 500$ ft/s, determine

a. The angular velocity ω of the rod immediately after the impact.
b. The velocity $\mathbf{v}_A$ of end A of the rod immediately after the impact.
c. The velocity $\mathbf{v}_B$ of end B of the rod immediately after the impact.

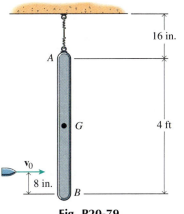

Fig. P20-79

20-80 A 200-mm-diameter, 5-kg homogeneous sphere is lowered onto a horizontal surface with an initial angular velocity of $\omega_0 = 3000$ rev/min (Fig. P20-80). If the kinetic coefficient of friction is $\mu_k = 0.15$, determine the initial velocity v_0 for which the sphere will end up at rest (the angular velocity and the linear velocity will both be zero when the sphere stops sliding).

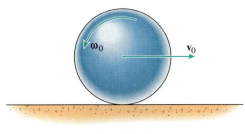

Fig. P20-80

20-81* A square crate sliding across a frictionless floor strikes a small obstacle A as shown in Fig. P20-81. If the crate rotates about A after impact, determine

a. The minimum initial speed v_0 for which the crate will tip all the way over.
b. The velocity $\mathbf{v}_G$ and the angular velocity ω of the crate immediately after the impact.

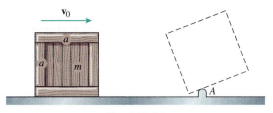

Fig. P20-81

20-82* Determine the height h at which the cue stick shown in Fig. P20-82 must strike the billiard ball if the ball is to roll without slipping and without the aid of friction. (Assume that the cue stick exerts only a horizontal force on the ball.)

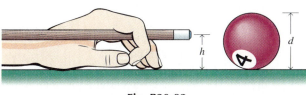

Fig. P20-82

20-83 The rear drive wheels of the tractor shown in Fig. P20-83 are 5 ft in diameter. The two wheels rotate as a unit having a combined weight of 2000 lb and a centroidal radius of gyration about the axle of $k_G = 1.5$ ft. The rest of the tractor weighs an additional 5000 lb, and its center of mass is 1.5 ft above the rear axle. If the tractor accelerates from zero to 20 mi/h in 3 s, determine

a. The average frictional force exerted on the drive tires by the ground.
b. The average torque applied to the axle of the rear wheels.
c. The minimum distance a between the rear axle and the center of gravity of the tractor body for which the tractor will not overturn.

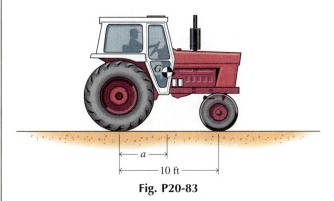

Fig. P20-83

20-84 The uniform slender rod AB ($m_{AB} = 3$ kg, $\ell_{AB} = 800$ mm) is resting on a frictionless horizontal surface when it is struck by a bullet as shown in Fig. P20-84. If the initial speed of the 0.03-kg bullet is $v_0 = 350$ m/s, determine

a. The angular velocity ω of the rod immediately after the impact.
b. The velocity $\mathbf{v}_A$ of end A of the rod immediately after the impact.
c. The velocity $\mathbf{v}_B$ of end B of the rod immediately after the impact.

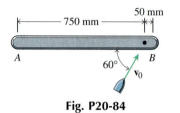

Fig. P20-84

20-85* Sphere A of Fig. P20-85 is rolling without slipping on a horizontal surface when it collides head-on with an identical stationary sphere B. If the kinetic coefficient of friction between the spheres and the horizontal surface is $\mu_k = 0.4$ and the collision is perfectly elastic ($e = 1$), determine

a. The linear and angular velocities of both spheres immediately after the impact.
b. The velocities of both spheres when they both roll without slipping after the impact.

Fig. P20-85

20-86 A 2-m-long slender rod of mass 5 kg is initially balanced on end as shown in Fig. P20-86. When the rod is disturbed, it first falls against the sharp corner C and then rotates about C. Determine

a. The angular velocity of the rod ω and the velocity of the mass center of the rod $\mathbf{v}_G$ immediately after the impact with the corner C.
b. The angular velocity of the rod when B strikes the horizontal surface.

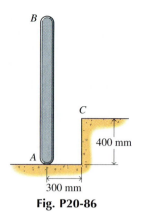

Fig. P20-86

20-87 The uniform rod of Fig. P20-87 falls horizontally and strikes the two rigid corners A and B. Corner A is slightly lower than corner B, so the rod strikes corner B first. If the velocity of the rod just before the impact is v_0 and the impact is perfectly elastic ($e = 1$), determine the angular velocity ω and the velocity of the mass center $\mathbf{v}_G$ of the rod

a. Immediately after striking corner B.
b. Immediately after striking corner A.

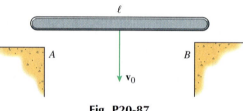

Fig. P20-87

20-88 The uniform slender rod AB ($m_{AB} = 2.5$ kg, $\ell_{AB} = 1.2$ m) is balanced on end on a horizontal surface when it is struck by an arrow as shown in Fig. P20-88. The arrow may be modeled as a uniform slender rod 800 mm long. If the initial speed of the 125-g arrow is $v_0 = 100$ m/s, determine

a. The angular velocity ω of the rod immediately after the impact.
b. The velocity $\mathbf{v}_A$ of end A of the rod immediately after the impact.
c. The velocity $\mathbf{v}_B$ of end B of the rod immediately after the impact.

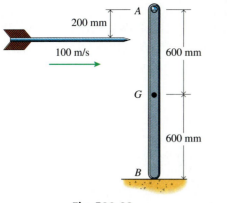

200 mm

100 m/s

A

600 mm

G

600 mm

B

Fig. P20-88

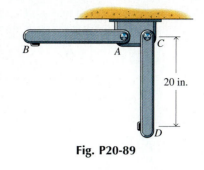

B

A

C

20 in.

D

Fig. P20-89

20-89* The two uniform, slender bars shown in Fig. P20-89 rotate in a vertical plane. Bar *AB* (5 lb, 2 ft long) is released from rest when it is horizontal and impacts bar *CD* (8 lb, 2 ft long) when it is vertical. If the coefficient of restitution is $e = 0.8$, determine

a.   The angular velocity of both bars immediately after the impact.
b.   The maximum angle through which *CD* will swing after the impact.
c.   The rebound angle of bar *AB*.

20-90* An 8-kg uniform sphere 400 mm in diameter rolls without slipping on a horizontal surface and strikes a 100-mm-tall step as shown in Fig. P20-90. The collision with the step is perfectly plastic ($e = 0$), and the sphere rotates about the corner of the step after the impact. Determine

a.   The angular velocity ω and the velocity of the mass center v_G of the sphere immediately after the collision if the initial speed of the sphere is $v_0 = 2.5$ m/s.
b.   The kinetic energy lost in the collision.
c.   The minimum initial speed v_0 for which the sphere will rotate all the way over to the upper level.

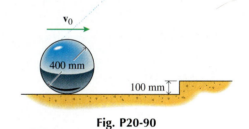

v_0

400 mm

100 mm

Fig. P20-90

If rotating machinery is not carefully balanced, it will vibrate causing excessive wear of its bearings, damaging its supports, and generating irritating noise.

MECHANICAL VIBRATIONS

21-1 INTRODUCTION

A mechanical vibration is the repeated oscillation of a particle or rigid body about an equilibrium position. In many devices vibratory motions are desirable and are deliberately generated; for example, a pendulum used to control a clock, a plucked string on a guitar or piano, the vibrator used to compact concrete in a form, and so on. The task of the engineer in such problems is to create and to control the vibrations. However, most vibrations in rotating machinery and in structures are undesirable. If rotating machine parts are not carefully balanced, they will vibrate. The vibrations can cause discomfort to the machine operator as well as damage to the machine or its support. Vibrations in structures due to earthquakes or to traffic of vehicles nearby may cause damage to or even collapse of the structure. In these cases the task of the engineer is to eliminate the vibrations (or at least to reduce the effect of the vibrations as much as possible) by appropriate design.

When a particle or rigid body in stable equilibrium is displaced by the application of an additional force, a mechanical vibration will result. Some common examples are the following:

1. The horizontal oscillation of a body attached to a spring (Fig. 21-1a) when it is displaced from its equilibrium position and then released.
2. The vertical oscillation of a flexible board or rod (Fig. 21-1b) when it is displaced from its equilibrium position and then released.
3. The rotational oscillation of a pendulum bob supported by an inextensible cord of negligible weight (Fig. 21-1c) when it is displaced from its equilibrium position and then released.

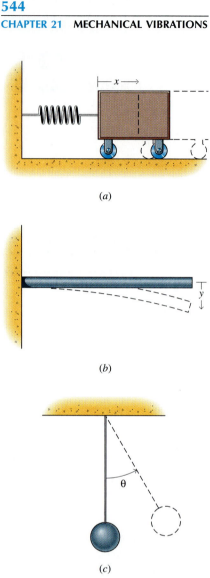

Figure 21-1 Masses attached to springs, flexible rods, and pendulums are all examples of vibrating systems.

The common characteristic in each of these examples is that restoring forces act on the body, causing it to return to its equilibrium position (Fig. 21-2a). However, when the body reaches its equilibrium position, it has a nonzero velocity and passes through it (Fig. 21-2b). The process is then repeated as the restoring forces again act to cause the body to return to its equilibrium position (Fig. 21-2c). The motion is repeated over and over again as the body moves back and forth through its equilibrium position.

In many cases, the position or motion of a body can be completely specified by one coordinate (e.g., x in Fig. 21-1a; θ in Fig. 21-1c; etc.). Such bodies have one *degree of freedom*. In other cases, a body can vibrate independently in two directions (Fig. 21-3a), or two bodies can be connected together but each can vibrate independently in a single direction (Fig. 21-3b). Since two coordinates are required to completely specify the position or motion of these systems, they have two degrees of freedom. Only single-degree-of-freedom systems are covered in this first course in dynamics.

Figure 21-4 shows typical graphs of the displacement (x or y or θ) from the equilibrium position versus time. Oscillations that repeat uniformly as in Figs. 21-4a and 21-4b are called *periodic*; oscillations that do not repeat uniformly (Fig. 21-4c) are called *aperiodic* or *random vibrations*. Random vibrations are not covered in this first course in Dynamics.

One of the more important characteristics of a periodic oscillation is the *period* τ, which is the minimum amount of time before the motion repeats itself. The motion completed in one period is called a *cycle*. The period is expressed in *seconds per cycle* or just seconds. The *frequency* f of an oscillation is the reciprocal of the period

$$f = \frac{1}{\tau} \tag{21-1}$$

or the number of cycles that occur per unit of time. The customary unit of frequency, *cycles per second (cps)*, is also called *hertz (Hz)*. The *amplitude* A of an oscillation is the maximum displacement of the body from its equilibrium position.

Finally, it must be noted that the study of vibrations is merely an application of the principles developed earlier. In the earlier chapters the acceleration was usually obtained only for a particular position of the body and at a particular instant of time. Here, the acceleration will be obtained for an arbitrary position of the body and then integrated to get the velocity and position for all future times.

21-2 UNDAMPED FREE VIBRATIONS

Mechanical vibrations are generally categorized as either *free vibrations* (also called *natural vibrations*) or *forced vibrations*. A free vibration is produced and maintained by forces such as elastic and gravitational forces, which depend only on the position and motion of the body. A forced vibration is produced and maintained by an externally applied periodic force, which does not depend on the position or motion of the body.

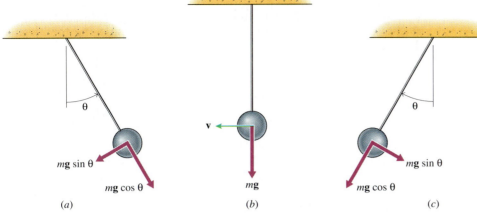

Figure 21-2 When a pendulum is displaced from its equilibrium position, a restoring force is created that tries to return the pendulum to its equilibrium position.

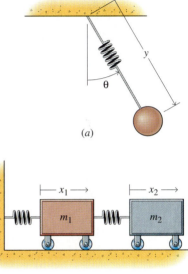

Free vibrations and forced vibrations may be further categorized as either damped or undamped. When forces that oppose the restoring force (friction, air resistance, etc.) are negligible, the vibration is called *undamped*. When resisting forces are not negligible, the vibration is called *damped*. An undamped free vibration will repeat itself indefinitely; a damped free vibration will eventually die out.

Of course, all real systems contain frictional forces, which will eventually stop a free vibration. In many systems, however, the energy loss due to air resistance, the internal friction of springs, or other friction forces are small enough that an analysis based on negligible damping often gives quite satisfactory engineering results. In particular, the frequency and period of vibration obtained for a freely vibrating system are very close to the values obtained for a system that has a small amount of damping.

Figure 21-3 Systems that require two coordinates to completely locate all parts of the system are said to have *two degrees of freedom*.

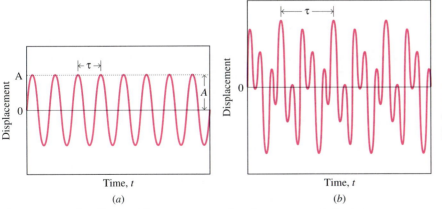

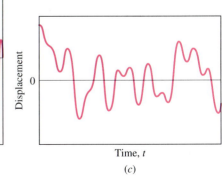

Figure 21-4 Oscillations that repeat exactly after some period of time τ are called *periodic*. Oscillations that do not repeat exactly over some long period of time are called *aperiodic* or *random vibrations*.

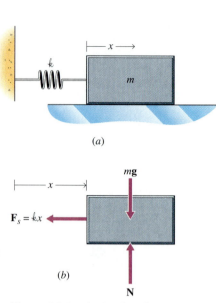

(a)

(b)

Figure 21-5 A simple vibrating system consists of a single mass and a spring. The force in the spring is proportional to the amount of displacement from the equilibrium position and is always directed toward the equilibrium position.

21-2.1   Undamped Free Vibration of Particles

Consider a block of mass m sliding on a frictionless horizontal surface as shown in Fig. 21-5a. Vibration is induced by displacing the block a distance x_0 and then releasing it with an initial velocity of $\dot{x}_0 = v_0$.

The free-body diagram of the block is shown in Fig. 21-5b, in which the block has been displaced an arbitrary amount in the positive coordinate direction. The elastic restoring force of the spring $F_s = kx$ is always directed toward the equilibrium position, whereas the acceleration $a_x = d^2x/dt^2 = \ddot{x}$ acts in the direction of *positive displacement*. It is important to remember that since acceleration is the second time derivative of displacement, both the displacement x and the acceleration $\ddot{x}$ must be measured positive in the same direction. Applying Newton's second law ($\Sigma F_x = ma_x = m\ddot{x}$) to the block gives the differential equation of motion for the block

$$-kx = m\ddot{x} \qquad \text{or} \qquad \ddot{x} = -\frac{k}{m}x \qquad (a)$$

Therefore, when the block is to the right of the equilibrium position (x is positive), its acceleration is to the left ($\ddot{x}$ is negative) or toward the equilibrium position. Similarly, when the block is to the left of the equilibrium position (x is negative), its acceleration is to the right ($\ddot{x}$ is positive), also toward the equilibrium position. That is, the acceleration of the block is proportional to its displacement from the equilibrium position and is directed toward that position.

21-2.2   Simple Harmonic Motion

Equation a describes *simple harmonic motion:* a motion for which the acceleration is proportional to the displacement from a fixed point and is directed toward the fixed point. Most of the vibrations encountered in engineering applications may be represented by a simple harmonic motion. Many other vibrations may be closely approximated by simple harmonic motion. A thorough knowledge of this concept is most helpful when analyzing such systems.

Equation a is a standard type of differential equation (a homogeneous, second-order, linear differential equation with constant coefficients) and is usually written in the form

$$\ddot{x} + \omega_n^2 x = 0 \qquad (21\text{-}2a)$$

The coefficient ω_n, which has units of rad/s, is related to the frequency of the oscillation and is called the *natural circular frequency*[1]

$$\omega_n = \sqrt{k/m} \qquad (21\text{-}2b)$$

The general solution of Eq. 21-2 is[2]

$$x(t) = B \cos \omega_n t + C \sin \omega_n t \qquad (21\text{-}3)$$

---

[1]While the natural circular frequency ω_n is often equal to $\sqrt{k/m}$ as in the present example, it is not always so. More generally, ω_n^2 is the ratio of the *effective spring constant* (the coefficient of the x term) and the *effective mass* (the coefficient of the $\ddot{x}$ term) in the differential equation of motion.

[2]It is easily verified by direct substitution that the solution (Eq. 21-3) satisfies the differential equation (Eq. 21-2) for any values of the constants B and C.

in which B and C are constants of integration to be determined from the initial conditions of the problem ($x = x_0$ and $\dot{x} = v_0$ when $t = 0$).

The solution (Eq. 21-3) can also be written as either

$$x(t) = A \cos (\omega_n t - \phi_c) \qquad (21\text{-}4a)$$

or

$$x(t) = A \sin (\omega_n t - \phi_s) \qquad (21\text{-}4b)$$

To verify that Eq. 21-4a is equal to Eq. 21-3, first expand Eq. 21-4a to get

$$x(t) = A (\cos \omega_n t \cos \phi_c + \sin \omega_n t \sin \phi_c) \qquad (b)$$

Then, setting Eq. b equal to Eq. 21-3 gives

$$(B - A \cos \phi_c) \cos \omega_n t + (C - A \sin \phi_c) \sin \omega_n t = 0 \qquad (c)$$

But if Eq. 21-3 is truly equal to Eq. 21-4a, then Eq. c must hold for any and all values of t. In particular, when $t = 0$, $\cos \omega_n t = 1$ and $\sin \omega_n t = 0$, so

$$B = A \cos \phi_c \qquad (d)$$

Similarly, when $t = \pi/2\omega_n$, $\cos \omega_n t = 0$ and $\sin \omega_n t = 1$, so

$$C = A \sin \phi_c \qquad (e)$$

Therefore, Eqs. 21-3 and 21-4a will be equal if

$$A = \sqrt{B^2 + C^2} \qquad \text{and} \qquad \tan \phi_c = \frac{C}{B} \qquad (f)$$

(The equality of Eqs. 21-3 and 21-4b is verified in a similar fashion.) Since $\cos (\omega_n t - \phi_c)$ oscillates between -1 and $+1$, the amplitude of the oscillation is $A = \sqrt{B^2 + C^2}$. The *phase angle* ϕ_c (or ϕ_s) is the amount by which the solution must be shifted to make it a simple cosine (or sine) curve.

The velocity and acceleration of the block are obtained by differentiating Eq. 21-3 or 21-4 with respect to time. For example, the velocity of the block is

$$
\begin{aligned}
v(t) = \dot{x}(t) &= -\omega_n B \sin \omega_n t + \omega_n C \cos \omega_n t & (21\text{-}5a)\\
&= -\omega_n A \sin (\omega_n t - \phi_c) & (21\text{-}5b)\\
&= \omega_n A \cos (\omega_n t - \phi_s) & (21\text{-}5c)
\end{aligned}
$$

and the acceleration of the block is

$$
\begin{aligned}
a(t) = \ddot{x}(t) &= -\omega_n^2 B \cos \omega_n t - \omega_n^2 C \sin \omega_n t & (21\text{-}6a)\\
&= -\omega_n^2 A \cos (\omega_n t - \phi_c) & (21\text{-}6b)\\
&= -\omega_n^2 A \sin (\omega_n t - \phi_s) & (21\text{-}6c)
\end{aligned}
$$

Since the cosine curve (Eq. 21-4a) and the sine curve (Eq. 21-4b) repeat whenever their argument increases by an angle of 2π rad, the period of the oscillation is given by $\omega_n \tau_n = 2\pi$ or

$$\tau_n = \frac{2\pi}{\omega_n} \qquad (21\text{-}7)$$

where the natural circular frequency ω_n is obtained from the differential equation of motion. The natural frequency of the oscillation in *hertz* (cycles per second) is then

$$f_n = \frac{1}{\tau_n} = \frac{\omega_n}{2\pi} \qquad (21\text{-}8)$$

and the natural frequency f_n and the natural circular frequency ω_n are related by $\omega_n = 2\pi f_n$. That is, a natural frequency of $f_n = 1$ Hz is equivalent to a natural circular frequency of $\omega_n = 2\pi$ rad/s.

It must be pointed out that the results obtained in this section are not limited to the vibration of a particle on a horizontal surface. They may be used to analyze the vibrational motion of a particle whenever the equations of motion reduce to the form (Eq. 21-2)

$$\ddot{x} + \omega_n^2 x = 0$$

which characterizes simple harmonic motion.

On the other hand, if the equations of motion do not reduce to the form of Eq. 21-2, the motion may still be an oscillatory motion but it will not be simple harmonic motion. In this case new expressions for the period, frequency, and so on must be obtained from solving the differential equation of motion.

21-2.3   Displaced Equilibrium Position

Simple harmonic motion also occurs if the block is suspended from the spring (Fig. 21-6a) rather than sliding on a frictionless surface so long as the y-coordinate is measured from the equilibrium position of the system. To see that this is so, draw the free-body diagrams of the block in its equilibrium position (Fig. 21-6b) and in an arbitrary displaced position (Fig. 21-6c). In the equilibrium position (before the block has been displaced and released), the sum of forces acting on the block must add to zero

$$mg - k\delta_{eq} = 0 \qquad (g)$$

where δ_{eq} is the static deformation of the spring (the elongation of the spring in the static equilibrium position $y = 0$). Therefore, the static deformation of the spring is $\delta_{eq} = mg/k$.

When the block has been displaced downward (in the positive y-direction) some amount y, the spring will be stretched a total amount of $y + \delta_{eq}$ and the force exerted on the block will be $k(y + \delta_{eq}) = k(y + mg/k)$ upward. Writing Newton's second law gives

$$mg - k(y + mg/k) = m\ddot{y} \qquad (h)$$

or

$$m\ddot{y} + ky = 0 \qquad (i)$$

which is again the equation for simple harmonic motion and has the solution

$$y = B \cos \omega_n t + C \sin \omega_n t \qquad (j)$$

The circular frequency ω_n, the natural frequency f_n, the period of vibration τ_n, and other vibrational characteristics of the block are then obtained as in Section 21-2.2.

If the position of the block were measured from the position where the spring is unstretched ($\hat{y} = 0$ when the spring is unstretched) rather

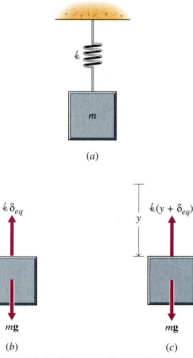

(a)

$k\delta_{eq}$

$k(y + \delta_{eq})$

y

mg

(b)

mg

(c)

Figure 21-6   The coordinate describing the position of the mass should always be measured from the equilibrium position rather than from the position where the spring is unstretched.

than from the equilibrium position, then the force in the spring would be ky and Eq. (i) would become

$$m\ddot{y} + ky = mg \qquad (k)$$

But the solution to Eq. k is just a constant plus Eq. 21-3

$$\hat{y}(t) = \frac{mg}{k} + B \cos \omega_n t + C \sin \omega_n t$$
$$= \delta_{eq} + B \cos \omega_n t + C \sin \omega_n t \qquad (l)$$

where $\hat{y}(t) = y(t) + \delta_{eq}$. That is, the oscillation consists of simple harmonic motion about the equilibrium position $\hat{y} = \delta_{eq}$.

21-2.4 Approximately Simple Harmonic Motion

If the equations of motion do not reduce to the form of Eq. 21-2,

$$\ddot{x} + \omega_n^2 x = 0$$

then the motion is not simple harmonic motion. Many motions, however, are well approximated by Eq. 21-2 as long as the amplitude of the motion is small. Such motions can be approximated as simple harmonic motions, and all the results of Section 21-2.2 directly apply.

For example, consider the oscillation of the simple pendulum shown in Fig. 21-7a. The pendulum consists of a particle of mass m swinging at the end of an inextensible, lightweight cord of length ℓ. The pendulum is released with an initial angle θ_0 and an initial speed $\dot{\theta}_0 = \omega_0$. Since the cord is inextensible, the particle will travel along a circular path with an acceleration

$$\mathbf{a} = \ell \ddot{\theta} \mathbf{e}_t + \ell \dot{\theta}^2 \mathbf{e}_n \qquad (m)$$

where the normal direction is toward the suspension point and the tangential direction is in the direction of increasing θ (Fig. 21-7b). The tangential component of Newton's second law, $\Sigma F_t = ma_t$ then gives the differential equation of motion

$$-mg \sin \theta = m\ell \ddot{\theta} \qquad (n)$$

or

$$\ddot{\theta} = -\frac{g}{\ell} \sin \theta \qquad (p)$$

As long as the angle θ is small, $\sin \theta \cong \theta$ (where θ is in radians) and Eq. p becomes

$$\ddot{\theta} = -\frac{g}{\ell} \theta \qquad (q)$$

Therefore, the pendulum has simple harmonic motion with a natural circular frequency $\omega_n = \sqrt{g/\ell}$ and period $\tau_n = 2\pi/\omega_n = 2\pi\sqrt{\ell/g}$. If the angle θ does not remain small, the resulting motion will still be an oscillatory motion, but it will not be a simple harmonic motion. The solution in this case must be obtained from solving the differential equation of motion Eq. p.[3]

---

[3]If the angle θ does not remain small, the solution to Eq. p can still be found, although it cannot be written in terms of simple functions such as polynomials or trigonometric functions. For $\theta_{max} = 5°$ the difference between the exact and approximate solutions is only about 0.05% in the value of the period; $\theta_{max} = 10°$, 0.19%; $\theta_{max} = 20°$, 0.76%; and $\theta_{max} = 40°$, 3.15%.

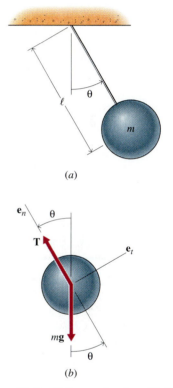

(a)

(b)

Figure 21-7 If the angle of oscillation is small, then the motion can be approximated by a simple harmonic motion.

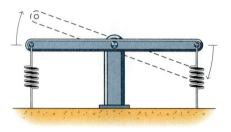

(a)

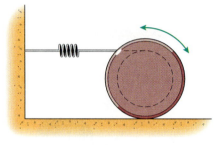

(b)

Fig. 21-8

21-2.5   Undamped Free Vibration of Rigid Bodies

A rigid body oscillating about a fixed axis (Fig. 21-8a) and a wheel oscillating on a flat surface (Fig. 21-8b) are also one-degree-of-freedom vibrating systems. The analysis of such rigid-body systems is essentially the same as for a particle. First, the free-body diagram is drawn for an arbitrary position of the rigid body. Next, the equations of motion are written. Finally, the principles of kinematics are used to reduce the equations of motion to a single differential equation involving a single variable that describes the position and motion of the rigid body. If the resulting differential equation can be written in the form of Eq. 21-2

$$\ddot{x} + \omega_n^2 x = 0$$

then the motion of the rigid body is a simple harmonic motion and all the results of Section 21-2.2 apply. If the equation of motion cannot be written in the form of Eq. 21-2, the resulting motion may still be an oscillatory motion, but it will not be a simple harmonic motion. The solution in this case must be obtained from solving the differential equation of motion.

EXAMPLE PROBLEM 21-1

A 10-lb cart is attached to three springs and rolls on an inclined surface as shown in Fig. 21-9. The elastic modulus of the springs are $k_1 = k_2 = 5$ lb/ft and $k_3 = 15$ lb/ft. If the cart is moved 3 in. up the incline from its equilibrium position and released with an initial velocity of 15 in./s up the incline when $t = 0$, determine

a. The period τ_n, the frequency f_n, and the circular frequency ω_n of the resulting vibration.
b. The position of the cart as a function of time.
c. The amplitude A of the resulting vibration.

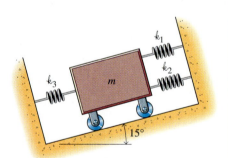

Fig. 21-9

SOLUTION

a. The free-body diagram of the cart is drawn in Fig. 21-10a in which the x-coordinate measures the position of the cart along the incline with $x = 0$ at the equilibrium position. In the equilibrium position (before the cart has been disturbed), the spring forces are proportional to their deformation ($F_1 = k_1\delta_{eq1}$, $F_2 = k_2\delta_{eq2}$, and $F_3 = k_3\delta_{eq3}$), so equilibrium gives

$$k_1\delta_{eq1} + k_2\delta_{eq2} - k_3\delta_{eq3} - mg \sin 15° = 0 \qquad (a)$$

When the cart is at an arbitrary (positive) location x, the stretch in springs 1 and 2 will be reduced ($F_1 = k_1[\delta_{eq1} - x]$ and $F_2 = k_2[\delta_{eq2} - x]$)

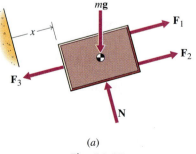

(a)

Fig. 21-10

and the stretch in spring 3 will be increased ($F_3 = k_3[\delta_{eq3} + x]$). Therefore, Newton's second law of motion ($\Sigma F_x = m\ddot{x}$) gives

$$k_1(\delta_{eq1} - x) + k_2(\delta_{eq2} - x) - k_3(\delta_{eq3} + x) - mg\sin 15° = m\ddot{x}$$

or

$$(k_1\delta_{eq1} + k_2\delta_{eq2} - k_3\delta_{eq3} - mg\sin 15°) - (k_1 + k_2 + k_3)x = m\ddot{x}$$

The constant terms in the first set of parentheses add to zero by Eq. *a*, however, so the differential equation of motion is

$$m\ddot{x} + (k_1 + k_2 + k_3)x = 0$$

or

$$\ddot{x} + 80.50x = 0 \qquad (b)$$

The natural circular frequency, the natural frequency, and the period are

$$\omega_n = \sqrt{80.50} = 8.972 \text{ rad/s} \qquad \text{Ans.}$$

$$f_n = \frac{\omega_n}{2\pi} = 1.428 \text{ Hz} \qquad \text{Ans.}$$

$$\tau_n = \frac{1}{f_n} = 0.700 \text{ s} \qquad \text{Ans.}$$

b. The displacement and velocity of the cart can be written in the form

$$x(t) = B\cos 8.972t + C\sin 8.972t$$
$$\dot{x}(t) = -8.972B\sin 8.972t + 8.972C\cos 8.972t$$

But at $t = 0$, $x = B = 3$ in. and $\dot{x} = 8.972C = 15$ in./s. Therefore, $B = 3$ in. and $C = 1.672$ in., and

$$x(t) = 3\cos 8.972t + 1.672\sin 8.972t \text{ in.} \qquad \text{Ans.}$$

This solution is shown in Fig. 21-10*b*.

Alternatively, the position and velocity of the cart can be written in the form

$$x(t) = A\cos(8.972t - \phi_c)$$
$$\dot{x}(t) = -8.972A\sin(8.972t - \phi_c)$$

Then, applying the initial conditions $x(0) = A\cos\phi_c = 3$ in. and $\dot{x}(0) = -8.972A(-\sin\phi_c) = 15$ in./s gives $A = 3.43$ in. and $\phi_c = 29.13° = 0.508$ rad. Therefore, the equation describing the position of the cart is

$$x(t) = 3.43\cos(8.972t - 0.508) \text{ in.} \qquad (c)$$

The displacement and velocity of the cart can also be written in the form

$$x(t) = A\sin(8.972t - \phi_s)$$
$$\dot{x}(t) = 8.972A\cos(8.972t - \phi_s)$$

Then, applying the initial conditions $x(0) = A(-\sin\phi_s) = 3$ in. and $\dot{x}(0) = 8.972A\cos\phi_s = 15$ in./s gives $A = 3.43$ in. and $\phi_s = -60.87° = -1.062$ rad. Therefore, the equation describing the position of the cart is

$$x(t) = 3.43\sin(8.972t + 1.062) \text{ in.} \qquad (d)$$

(The solution described by Eqs. *c* and *d* is exactly the same as that shown in Fig. 21-10*b*. The phase angles $\phi_c = 0.508$ rad and $\phi_s = -1.062$ are also indicated on Fig. 21-10*b*.)

c. Since the maximum value of the cosine function is 1, the amplitude of the vibration is

$$A = 3.43 \text{ in.} \qquad \text{Ans.}$$

Since it is not known how much the springs were stretched or compressed before being attached to the cart, it is not possible to determine values for the static deformations δ_{eq1}, δ_{eq2}, and δ_{eq3} individually. However, Eq. *b* does give a relationship between the static deformations and the weight of the cart.

If the displacement x is not measured from the equilibrium position of the system, the constant terms will not cancel out and there will be a constant on the right-hand side of Eq. *b*. Then the solution will have a constant term in addition to the sine and cosine terms, and the solution will oscillate about some non-zero value rather than oscillating about zero.

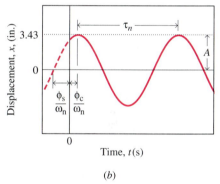

(b)

Fig. 21-10

The pendulum of Fig. 21-11 consists of a 2-kg uniform bar 0.8 m long suspended from a frictionless pin at one end. Determine the natural frequency and the period of the resulting oscillation. (Assume a small angle of oscillation.)

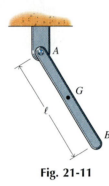

Fig. 21-11

SOLUTION

The free-body diagram of the pendulum is drawn in Fig. 21-12. Since the motion is a rotation about a fixed axis, Newton's second law can be written in the form $\Sigma M_A = I_A \alpha = I_A \ddot{\theta}$, which gives

$$-\frac{\ell}{2}(mg \sin \theta) = \left(\frac{1}{3}m\ell^2\right)\ddot{\theta}$$

But if the angle of oscillation is small, $\sin \theta \cong \theta$ (in radians), so the differential equation of motion of the pendulum is

$$\ddot{\theta} + \frac{3g}{2\ell}\theta = 0$$

Therefore, the natural circular frequency, the natural frequency, and the period of the oscillation are

$$\omega_n = \sqrt{(3)(9.81)/(2)(0.8)} = 4.289 \text{ rad/s} \qquad \text{Ans.}$$

$$f_n = \frac{\omega_n}{2\pi} = 0.683 \text{ Hz} \qquad \text{Ans.}$$

$$\tau_n = \frac{1}{f_n} = 1.465 \text{ s} \qquad \text{Ans.}$$

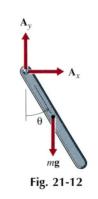

Fig. 21-12

Note that the natural circular frequency $\omega_n = \sqrt{3g/2\ell}$ contains neither a spring constant nor mass. The natural circular frequency is the square root of the coefficient that appears in the differential equation of motion when it is written in the standard form of Eq. 21-2a.

These results can be compared with the results for a simple pendulum in which all of the mass is concentrated at the end of a massless rod or string. In Section 21-2.4 the natural circular frequency of the simple pendulum was found to be $\omega_n = \sqrt{g/\ell} = \sqrt{9.81/0.8} = 3.502 \text{ rad/s}$. Therefore, the natural frequency and period of the simple pendulum would be $f_n = \omega_n/2\pi = 0.557 \text{ Hz}$ and $\tau_n = 1/f_n = 1.794 \text{ s}$, respectively.

A 5-lb uniform cylinder 12 in. in diameter rolls without slipping on an inclined plane as shown in Fig. 21-13. A linear spring ($k = 24$ lb/ft) is attached to the cylinder at point A (which is $e = 3$ in. from the center of the cylinder), and the spring is unstretched in the position shown. If the cylinder is released from rest in the position shown, determine

a. The period τ_n, the frequency f_n, and the circular frequency ω_n of the resulting vibration.

b. The position of the center of mass of the cylinder as a function of time.

Fig. 21-13

SOLUTION

a. Figure 21-14*a* shows the free-body diagram of the cylinder in its equilibrium position. To get from its initial position to the equilibrium position, the cylinder has to roll counterclockwise through an angle θ_{eq}, the center of mass of the cylinder has to move down the incline a distance x_{Geq}, and the spring is stretched by the amount δ_{eq}. If θ_{eq} is small, then $\sin \theta_{eq} \cong \theta_{eq}$ (in radians), $\cos \theta_{eq} \cong 1$, $\theta_{eq} \cong \tan \theta_{eq} \cong \dfrac{\delta_{eq}}{(r + e)} \cong \dfrac{x_{Geq}}{r}$ (Fig. 21-14*b*), and the spring force $k\delta_{eq}$ remains parallel to the surface. Then the equilibrium equations

$$\Sigma F_x = 0: \qquad mg \sin 10° - F - k\delta_{eq} = 0$$
$$\Sigma M_G = 0: \qquad Fr - k\delta_{eq}\, e \cos \theta_{eq} = 0$$

can be combined to get

$$mgr \sin 10° - k(r + e)\delta_{eq} = 0 \qquad\qquad (a)$$

Equation *a* gives $\delta_{eq} = 0.02412$ ft $= 0.2894$ in., from which $x_{Geq} = 0.01608$ ft $= 0.1929$ in. and $\theta_{eq} = 0.03216$ rad $= 1.842°$ can be found.

Next, the free body diagram of the cylinder is drawn (Fig. 21-14*c*) for an arbitrary position in which the center of mass has moved an additional distance x_G down the incline, the cylinder has rotated through an additional angle θ, and (still assuming small angles) the spring has been stretched an additional amount $[(r + e)/r]x_G$. Then the equations of motion for the cylinder are

$$\Sigma F_x = ma_{Gx}: \qquad mg \sin 10° - F - k\left(\delta_{eq} + \frac{9}{6}x_G\right) = m\ddot{x}_G$$

$$\Sigma M_G = I_G\alpha: \qquad Fr - k\left(\delta_{eq} + \frac{9}{6}x_G\right)e \cos (\theta_{eq} + \theta) = \frac{1}{2}mr^2\ddot{\theta}$$

> The equilibrium position of the cylinder is only $x_{Geq} = 0.1929$ in. lower than the initial position. In the equilibrium position, the spring is stretched $\delta_{eq} = 0.2894$ in. and the cylinder has rotated only $\theta_{eq} = 0.03216$ rad $\cong 1.842°$. As a check of the small angle approximations, note that $\sin \theta_{eq} = 0.03215 \cong \theta_{eq}$ and $\cos \theta_{eq} = 0.9995 \cong 1$.

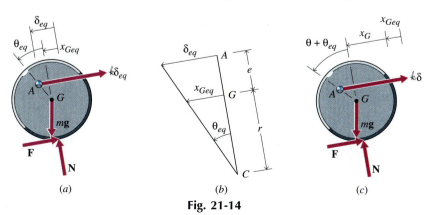

(*a*)   (*b*)   (*c*)

Fig. 21-14

To get the differential equation describing the vibration, replace $\cos(\theta_{eq} + \theta)$ with 1; multiply the first equation by r and add it to the second to get

$$[mgr \sin 10° - k(r + e)\delta_{eq}] - k(r + e)\left(\frac{9}{6}x_G\right) = mr\ddot{x}_G + \frac{1}{2}mr^2\ddot{\theta}$$

The term in the brackets is zero by Eq. a and the accelerations are related by $\ddot{x}_G = r\ddot{\theta}$. Therefore, the differential equation of motion is

$$mr\ddot{x}_G + \frac{1}{2}mr\ddot{x}_G + k(r + e)\left(\frac{9}{6}x_G\right) = 0$$

or

$$0.11646\ddot{x}_G + 27.00x_G = 0$$

and the natural circular frequency, natural frequency, and period of the vibration are

$$\omega_n = \sqrt{(27.00)/(0.11646)} = 15.23 \text{ rad} \qquad \text{Ans.}$$

$$f_n = \frac{\omega_n}{2\pi} = 2.423 \text{ Hz} \qquad \text{Ans.}$$

$$\tau_n = \frac{1}{f_n} = 0.413 \text{ s} \qquad \text{Ans.}$$

b. The position and velocity of the center of mass of the cylinder can be written in the form

$$x_G(t) = B \cos 15.23t + C \sin 15.23t$$
$$\dot{x}_G(t) = -15.23B \sin 15.23t + 15.23C \cos 15.23t$$

When $t = 0$, $x_G = B = -x_{Geq} = -0.1929$ in. and $\dot{x}_G = 15.23C = 0$. Therefore, the position of the center of mass of the cylinder is

$$x_G(t) = -0.1929 \cos 15.23t \text{ in.} \qquad \text{Ans.}$$

As a final check that the small angle of oscillation approximation remains valid, note that the amplitude of the oscillation is 0.1929 in. Therefore, the maximum angle of rotation from the initial position is $\theta_{max} = \theta_{eq} + (0.1929/6) = 0.06431$ rad $= 3.685°$. But $\sin \theta_{max} = 0.06428 \cong \theta_{max}$ and $\cos \theta_{max} = 0.9979 \cong 1$.

PROBLEMS

Introductory Problems

21-1–21-6 The following equations represent the position of a particle in simple harmonic motion. For each equation, plot the position, velocity, and acceleration of the particle versus time for two complete cycles of the oscillation.

21-1* $x(t) = 8 \cos \pi t$ in.

21-2* $x(t) = 5 \sin \pi t/4$ mm

21-3 $x(t) = 3 \cos(\pi t/2 - \pi/4)$ in.

21-4* $x(t) = 10 \sin(3\pi t/4 + \pi/8)$ mm

21-5 $x(t) = 4 \cos 5t - 3 \sin 5t$ in.

21-6 $x(t) = 5 \sin 3t + 12 \cos 3t$ mm

21-7–21-12 The following equations represent the position of a particle in simple harmonic motion. For each equation

a. Write the equation for the particle motion in the form $x(t) = A \cos(\omega_n t - \phi_c)$.
b. Find the maximum velocity and the particle position when it occurs.
c. Find the maximum acceleration and the particle position when it occurs.

21-7* $x(t) = 3 \cos \pi t - 4 \sin \pi t$ in.

21-8 $x(t) = 12 \cos \pi t/2 + 5 \sin \pi t/2$ mm

21-9* $x(t) = 8 \cos 10t + 6 \sin 10t$ in.

21-10* $x(t) = 10 \cos 3\pi t/4 - 24 \sin 3\pi t/4$ mm

21-11 $x(t) = 5 \sin \pi t$ in.

21-12 $x(t) = 4 \sin (3t + \pi/3)$ mm

21-13–21-18   The following equations represent the position of a particle in simple harmonic motion. For each equation

a.   Write the equation for the particle motion in the form $x(t) = A \sin (\omega_n t - \phi_s)$.
b.   Find the earliest value of t for which the particle's position is zero.
c.   Find the earliest value of t for which the particle's velocity is zero.

21-13* $x(t) = 5 \cos \pi t - 12 \sin \pi t$ in.

21-14* $x(t) = 4 \cos \pi t/2 + 3 \sin \pi t/2$ mm

21-15 $x(t) = 8 \cos 3\pi t/4 + 6 \sin 3\pi t/4$ in.

21-16 $x(t) = 5 \cos 10t - 5 \sin 10t$ mm

21-17* $x(t) = 5 \cos \pi t$ in.

21-18 $x(t) = 8 \cos (3\pi t/2 + 2\pi/3)$ mm

21-19   An instrument used to measure the vibration of a particle indicates a simple harmonic motion with a natural frequency of 5 Hz and a maximum acceleration of 160 ft/s². Determine the amplitude and maximum velocity of the vibration.

21-20*   A particle vibrates with a simple harmonic motion. When the particle passes through the equilibrium position, its velocity is 2 m/s. When the particle is 20 mm away from its equilibrium position, its acceleration is 50 m/s². Determine the magnitude of the velocity at this position.

Intermediate Problems

21-21*   A block that has a mass m slides on a frictionless horizontal surface as shown in Fig. P21-21. Determine the modulus k of the single spring that could replace the two springs shown without changing the frequency of vibration of the block.

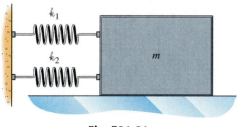

Fig. P21-21

21-22*   A block that has a mass m slides on a frictionless horizontal surface as shown in Fig. P21-22. Determine the modulus k of the single spring that could replace the two springs shown without changing the frequency of vibration of the block.

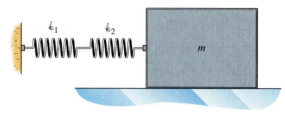

Fig. P21-22

21-23   A 2-lb mass is suspended in a vertical plane by three springs as shown in Fig. P21-23. If the mass is displaced 1.5 in. below its equilibrium position and released with an upward velocity of 10 in./s when $t = 0$, determine

a.   The differential equation governing the motion.
b.   The period and amplitude of the resulting vibration.
c.   The position of the mass as a function of time.
d.   The earliest time $t_1 > 0$ when the mass passes through its equilibrium position.

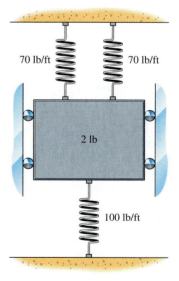

Fig. P21-23

21-24 A 4-kg mass is suspended in a vertical plane as shown in Fig. P21-24. The two springs remain in tension at all times, and the two pulleys are both small and frictionless. If the mass is displaced 15 mm above its equilibrium position and released with a downward velocity of 750 mm/s when $t = 0$, determine

a. The differential equation governing the motion.
b. The period and amplitude of the resulting vibration.
c. The position of the mass as a function of time.
d. The earliest time $t_1 > 0$ when the velocity of the mass is zero.

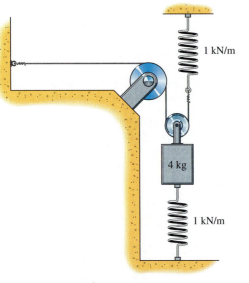

Fig. P21-24

21-25 A 20-lb block slides on a frictionless horizontal surface as shown in Fig. P21-25. The two springs remain in tension at all times and the two pulleys are both small and frictionless. If the block is displaced 3 in. to the left of its equilibrium position and released with a velocity of 50 in./s to the right when $t = 0$, determine

a. The differential equation governing the motion.
b. The period and amplitude of the resulting vibration.
c. The position of the block as a function of time.
d. The earliest time $t_1 > 0$ when the velocity of the block is zero.

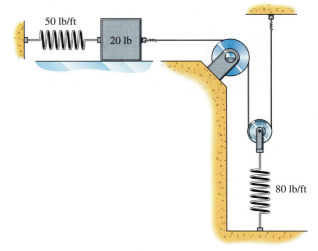

Fig. P21-25

21-26* An 8-kg mass slides on a frictionless horizontal surface as shown in Fig. P21-26. The two springs remain in tension at all times, and the pulleys are small and frictionless. If the mass is displaced 25 mm to the right of its equilibrium position and released with a velocity of 800 mm/s to the right when $t = 0$, determine

a. The differential equation governing the motion.
b. The period and amplitude of the resulting vibration.
c. The position of the mass as a function of time.
d. The earliest time $t_1 > 0$ when the acceleration of the mass is zero.

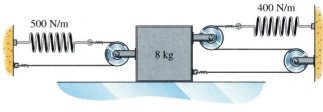

Fig. P21-26

21-27 The 10-lb block of Fig. P21-27 slides on a frictionless horizontal surface while the 5-lb block moves in a vertical plane. The springs remain in tension at all times, and the two pulleys are both small and frictionless. Write the differential equation of motion for $x(t)$, the position of the 10-lb block, and determine the frequency and period of the resulting vibratory motion.

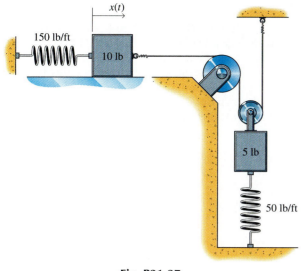

Fig. P21-27

21-28* The two masses of Fig. P21-28 each slide on a frictionless horizontal surface. The springs remain in tension at all times, and the pulleys are small and frictionless. Write the differential equation of motion for $x(t)$, the position of the 10-kg block, and determine the frequency and period of the resulting vibratory motion.

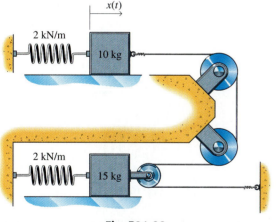

Fig. P21-28

21-29* The two blocks of Fig. P21-29 hang in a vertical plane from a massless bar, which is horizontal in the equilibrium position. If the springs remain in tension at all times, write the differential equation of motion for $y(t)$, the position of the 10-lb block, and determine the frequency and period of the resulting vibratory motion. (Assume small oscillations.)

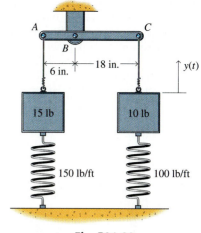

Fig. P21-29

21-30 The two masses of Fig. P21-30 each slide on a frictionless horizontal surface. The bar ABC is vertical in the equilibrium position and has negligible mass. If the springs remain in tension at all times, write the differential equation of motion for $x(t)$, the position of the 10-kg mass, and determine the frequency and period of the resulting vibratory motion. (Assume small oscillations.)

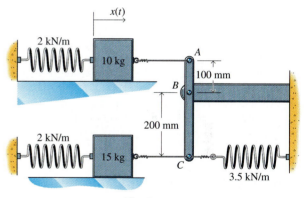

Fig. P21-30

21-31* The 5-lb block of Fig. P21-31 slides on a frictionless horizontal surface while the 3-lb block hangs in a vertical plane. The bar ABC has negligible mass and arm AB is horizontal in the equilibrium position. If the springs remain in tension at all times, write the differential equation of motion for $y(t)$, the position of the 3-lb block, and determine the frequency and period of the resulting vibratory motion. (Assume small oscillations.)

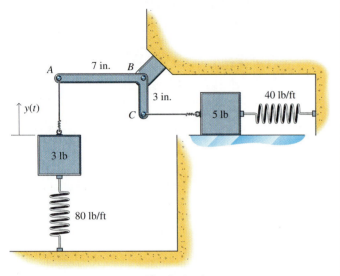

Fig. P21-31

21-32 A 0.5-kg plunger is at rest in a vertical frictionless guide when a 0.3-kg ball rolls off a 4-m-high step and bounces off it as shown in Fig. P21-32. If the collision is perfectly elastic ($e = 1$) and the spring modulus is $k = 200$ N/m, determine $y(t)$, the position of the plunger, as a function of the time after the ball strikes it.

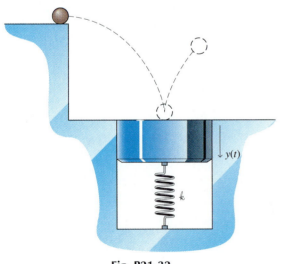

Fig. P21-32

Challenging Problems

21-33 A 7-lb uniform cylinder rolls without slipping on a horizontal surface as shown in Fig. P21-33. The two springs are connected to a small frictionless pin at G the center of the 8-in.-diameter cylinder. Write the differential equation of motion for $x_G(t)$, the position of the center of mass of the cylinder, and determine the frequency and period of the resulting vibratory motion.

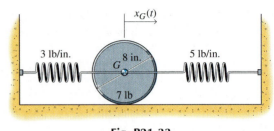

Fig. P21-33

21-34* A 4-kg uniform cylinder is suspended in the vertical plane in the loop of a lightweight cord as shown in Fig. P21-34. If the 500-mm-diameter cylinder does not slip on the cord, write the differential equation of motion for $y_G(t)$, the position of the mass center of the cylinder, and determine the frequency and period of the resulting vibratory motion.

Fig. P21-34

21-35* An 18-lb stepped cylinder rolls without slipping on a horizontal surface as shown in Fig. P21-35. The two springs are attached to cords, which are wrapped securely around the 12-in.-diameter central hub. If the radius of gyration of the stepped cylinder is 9 in., write the differential equation of motion for $x_G(t)$, the position of the mass center of the cylinder, and determine the frequency and period of the resulting vibratory motion.

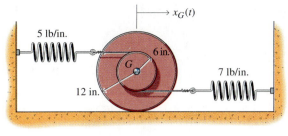

Fig. P21-35

21-36 A 2-kg thin circular disk ($r = 200$ mm) hangs from a small frictionless pin on its rim as shown in Fig. P21-36. Write the differential equation of motion for $\theta(t)$, the angular position of the disk, and determine the frequency and period of the resulting vibratory motion.

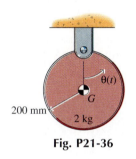

Fig. P21-36

21-37* A 15-lb thin rectangular plate (18 in. by 12 in.) hangs from a small frictionless pin at the middle of its long edge as shown in Fig. P21-37. Write the differential equation of motion for $\theta(t)$, the angular position of the plate, and determine the frequency and period of the resulting vibratory motion.

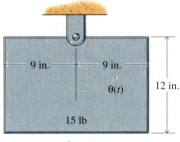

Fig. P21-37

21-38 The circular disk of Problem 21-36 is replaced with a thin circular ring of the same mass ($m = 2$ kg) and radius ($r = 200$ mm) as shown in Fig. P21-38. Write the differential equation of motion for $\theta(t)$, the angular position of the ring, and determine the frequency and period of the resulting vibratory motion.

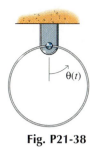

Fig. P21-38

21-39 A 3-lb uniform slender bar 5 ft long is attached to a frictionless pivot at A as shown in Fig. P21-39. The bar is horizontal in the equilibrium position. If end C is pulled down 5 in. and released from rest, determine

a. The differential equation of motion for $\theta(t)$, the angular position of the bar.
b. The maximum velocity of end C in the resulting vibratory motion.

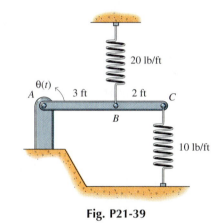

Fig. P21-39

21-40* A 2-kg uniform slender bar 500 mm long is attached to a frictionless pivot at B as shown in Fig. P21-40. The bar is horizontal in the equilibrium position. If end C is pulled down 15 mm and released from rest, determine

a. The differential equation of motion for $\theta(t)$, the angular position of the bar.
b. The maximum velocity of end C in the resulting vibratory motion.

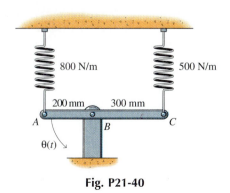

Fig. P21-40

21-41 Two slender uniform bars are welded together as shown in Fig. P21-41. Bar ABC weighs 2 lb and is horizontal in the equilibrium position, bar BD weighs 3 lb and is vertical in the equilibrium position, and the pivot is frictionless. If end D is pulled to the left 3 in. and released from rest, determine

a. The differential equation of motion for $\theta(t)$, the angular position of the bar.
b. The maximum velocity of end D in the resulting vibratory motion.

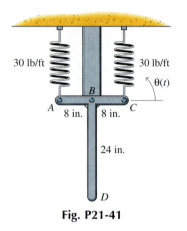

Fig. P21-41

21-42* A brass (8750 kg/m³) paperweight in the form of a half cylinder (75 mm long, 100 mm in diameter) rests on a flat horizontal surface as shown in Fig. P21-42. If the cylinder is tipped slightly to one side and rolls without slipping, write the differential equation of motion for $\theta(t)$, the angular position of the paperweight, and determine the frequency and period of the resulting vibratory motion.

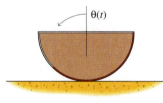

Fig. P21-42

21-43* The lightweight cord attached to the 10-lb block of Fig. P21-43 is wrapped around a 7-lb uniform cylinder. If the cord does not slip on the cylinder, write the differential equation of motion for $y(t)$, the position of the 10-lb block, and determine the frequency and period of the resulting vibratory motion.

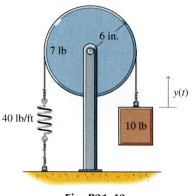

Fig. P21-43

21-44 A 5-kg uniform slender bar 400 mm long is rigidly attached to an 8-kg uniform cylinder 300 mm in diameter as shown in Fig. P21-44. If the cylinder rolls without slipping on the horizontal surface, write the differential equation of motion for $\theta(t)$, the angular position of the cylinder, and determine the frequency and period of the resulting vibratory motion.

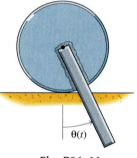

Fig. P21-44

Computer Problems

C21-45 A 2-lb piston is initially at rest on two identical springs ($k = 6$ lb/ft each) when a 0.5-lb ball of putty is dropped onto it (Fig. P21-45). If the collision is perfectly plastic ($e = 0$) and $h = 2$ ft, calculate and plot

a. The position $y(t)$ of the piston as a function of time t for one complete cycle of the resulting oscillatory motion.
b. The force exerted on the putty by the piston as a function of time t for one complete cycle of the resulting oscillatory motion.

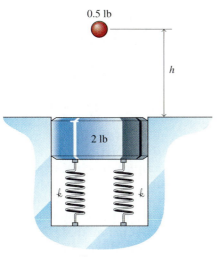

0.5 lb

h

2 lb

k k

Fig. P21-45

C21-46 A simple pendulum consists of a concentrated mass m on the end of a lightweight rod AB as shown in Fig. P21-46. If the hinge at A is frictionless, the differential equation of motion for the pendulum is given by

$$\ell\ddot{\theta} + g\sin\theta = 0 \qquad (a)$$

The solution of this equation is only approximated by simple harmonic motion when the angle θ is small so that $\sin\theta \cong \theta$. If $\ell = 1.3$ m, $m = 1$ kg, and the pendulum is released from rest when $\theta = \theta_0$,

a. Use the Euler method of solving differential equations to solve Eq. a for the angle θ as a function of time for various initial angles θ_0 ($10° \le \theta_0 \le 120°$).

b. Then for $\theta_0 = 80°$, plot θ as a function of time t through one complete cycle of the oscillation. On the same graph, plot the solution using the simple harmonic approximation.

c. For each initial angle $\theta_0 = 10°, 20°, 30°, \ldots, 120°$, determine τ, the period of the oscillation. For example, determine the elapsed time for 10 swings through $\theta = 0$ and divide by 5.

d. Plot Err, the percent relative error in the period obtained by using the small angle approximation as a function of θ_0 ($10° \le \theta_0 \le 120°$), where $Err = [(\tau - \tau_n)/\tau] \times 100$, and $\tau_n = \sqrt{\ell/g}$ is the natural period of the simple harmonic motion solution.

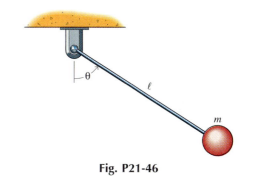

θ

ℓ

m

Fig. P21-46

21-3 DAMPED FREE VIBRATIONS

The analysis of undamped free vibrations in the previous sections is only an idealization of real systems because it does not account for the energy lost to friction. Once set in motion, such idealized systems vibrate forever with a constant amplitude. All real systems, however, lose energy to friction and will eventually stop unless there is a source of energy to keep them going. When the amount of energy loss in the system is small, the results of the previous sections are often in good agreement with real systems, at least for short intervals of time. For longer intervals of time or for cases in which the energy loss is not small, the effects of friction forces must be included.

There are several types of friction forces that can remove mechanical energy from a vibrating system. Some of the more common friction forces are: *fluid friction* (also called *viscous damping force*), which arises when bodies move through viscous fluids; *dry friction* (also called *coulomb friction*), which arises when a body slides across a dry surface; and *internal friction*, which arises when a solid body is deformed. Damping caused by fluid friction is quite common in engineering work, and only linear viscous damping is considered in this first course in Dynamics.

Dashpot

Figure 21-15   A *dashpot* (or *viscous damper*) is an idealization of physical devices such as automobile shock absorbers and door closers. A dashpot always resists the motion with a force that is proportional to the speed of the motion.

21-3.1   The Linear Viscous Damper

Viscous damping occurs naturally when mechanical systems such as pendulums vibrate in air or water. Viscous damping is also exhibited by devices called *dashpots* (represented symbolically by Fig. 21-15), which are intentionally added to mechanical systems to limit or control vibration. A typical dashpot consists of a piston moving in a cylinder filled with a viscous fluid. Movement of the piston is opposed by the fluid, which must either flow through small holes in the piston or flow through a narrow gap around the piston. Devices such as door closers and shock absorbers are examples of actual dashpots. In addition, symbolic dashpots are sometimes used to represent the frictional loss in systems that do not have distinct damping devices. The mass of a dashpot, like the mass of a spring, is generally neglected.

The viscous dampers considered here are linear. That is, the magnitude of the viscous damping force is directly proportional to the speed with which the damper is being extended or compressed

$$F = c\dot{x} \tag{21-9}$$

The constant of proportionality c is called the *coefficient of viscous damping*. Typical units for c are $\text{N} \cdot \text{s/m}$ in the SI system of units or $\text{lb} \cdot \text{s/ft}$ in the U.S. Customary system of units. The direction of the viscous damping force is always opposite to the direction of the velocity.

21-3.2   Viscous Damped Free Vibration

To illustrate viscous damped free vibration, damping is added to the block-spring system of Fig. 21-5a as shown in Fig. 21-16a. The free-body diagram of the block is now shown in Fig. 21-16b, in which the block again has been displaced an arbitrary amount in the positive coordinate direction. The elastic restoring force of the spring $F_s = kx$ is still directed toward the equilibrium position (the negative coordinate direction). Since the positive directions for the velocity $\dot{x}$ and the acceleration $\ddot{x}$ are the same as the positive coordinate direction, the damping force $F_d = c\dot{x}$ also acts in the negative coordinate direction. Applying Newton's second law ($\Sigma F = ma_x = m\ddot{x}$) to the block gives the differential equation of motion for the block

$$-kx - c\dot{x} = m\ddot{x} \tag{21-10a}$$

or

$$m\ddot{x} + c\dot{x} + kx = 0 \tag{21-10b}$$

which is again a second-order linear differential equation with constant coefficients.

It is well known from the theory of ordinary differential equations that the solution of any linear, ordinary differential equation with constant coefficients is always of the form

$$x(t) = De^{\lambda t} \tag{a}$$

where the constants D and λ must be chosen to satisfy the differential equation and the initial conditions. Substituting Eq. *a* into Eq. 21-10b gives

$$De^{\lambda t}(m\lambda^2 + c\lambda + k) = 0 \tag{b}$$

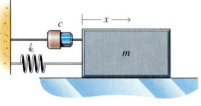

(a)

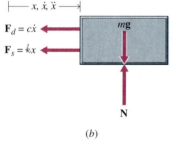

(b)

Fig. 21-16

If the constant D is zero, then Eq. a gives the trivial solution $x = 0$, which is of no interest. Clearly, the exponential $e^{\lambda t}$ is never zero. Therefore, the factor $De^{\lambda t}$ is never zero and it is permissible to divide Eq. b by it, which gives the characteristic equation

$$m\lambda^2 + c\lambda + k = 0 \qquad (c)$$

which has roots

$$\lambda_{1,2} = \frac{-c \pm \sqrt{c^2 - 4mk}}{2m} \qquad (21\text{-}11)$$

The displacement of the block is then given by[4]

$$x(t) = D_1 e^{\lambda_1 t} + D_2 e^{\lambda_2 t} \qquad (21\text{-}12)$$

where the constants D_1 and D_2 are determined from the initial conditions (at $t = 0$; $x = D_1 + D_2 = x_0$ and $\dot{x} = D_1\lambda_1 + D_2\lambda_2 = v_0$) and λ_1 and λ_2 are given by Eq. 21-11.

Before discussing the solution, however, the roots (Eq. 21-11) will be rewritten in terms of more convenient variables. The dimensionless combination of constants

$$\zeta = \frac{c}{2\sqrt{mk}} = \frac{c}{2m\omega_n} \qquad (21\text{-}13)$$

is called the *damping ratio*.[5] In terms of the damping ratio ζ and the natural circular frequency ω_n, Eq. 21-11 becomes

$$\lambda_{1,2} = -\zeta\omega_n \pm \omega_n\sqrt{\zeta^2 - 1} \qquad (21\text{-}14)$$

The behavior of the system depends on whether the quantity under the radical in Eq. 21-14 is positive, zero, or negative. The value of c that makes the radical zero is called the *critical damping coefficient* c_{cr}. Therefore,

$$c_{cr} = 2m\omega_n = 2\sqrt{mk} \qquad (21\text{-}15)$$

The solution (Eq. 21-12) will have three totally distinct types of behavior depending on whether the actual system damping c is greater than, equal to, or less than c_{cr}.[6] Each possibility is analyzed separately in the next three sections.

[4]When the quantity under the radical in Eq. 21-11 is zero, the two roots are identical: $\lambda_1 = \lambda_2 = \lambda = -c/2m = -\sqrt{k/m} = -\omega_n$ (since $c = 2\sqrt{mk}$). In this case the general solution of Eq. 21-10 is

$$x(t) = (B + Ct)e^{\lambda t}$$

as can be easily verified by direct substitution.

[5]In the present example, the constants that appear in the definitions of the damping ratio ζ and the natural circular frequency ω_n are the actual system mass m, damping coefficient c, and spring constant k. In general, however, they should be interpreted as the *effective mass* of the system (the coefficient of the $\ddot{x}$ term in Eq. 21-10), the *effective damping coefficient* (the coefficient of the $\dot{x}$ term in Eq. 21-10), and the *effective spring constant* (the coefficient of the x term in Eq. 21-10), respectively.

[6]As always, the constants m, c, and k may or may not be the actual system values. Rather, they should be interpreted as the coefficients of the differential equation of motion when written in the standard form—Eq. 21-10b.

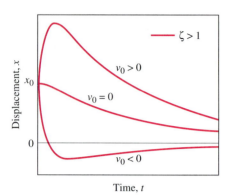

Figure 21-17 An *overdamped system* does not oscillate. The system returns to rest at its equilibrium position.

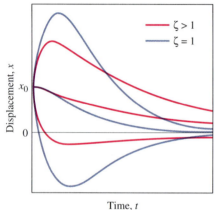

Figure 21-18 A *critically damped system* returns to rest more quickly than any other system starting from the same initial position and initial velocity.

21-3.3 Overdamped Systems

When the damping coefficient c is greater than c_{cr}, then the damping ratio ζ is greater than one, the radical Eq. 21-14 is real, and the two roots λ_1 and λ_2 are both real and are unequal. Furthermore, since $\sqrt{\zeta^2 - 1} < \zeta$, both roots will be negative. Therefore, the displacement (Eq. 21-12) simply decreases to zero as t increases, and the motion is nonvibratory.

The displacement given by Eq. 21-12 is shown in Fig. 21-17 for representative initial conditions. The damping is so severe in this case that an overdamped system returns slowly to its equilibrium position. Since the system does not actually oscillate, there is no period or frequency associated with *overdamped* or *supercritically damped* motions.

21-3.4 Critically Damped Systems

When the damping coefficient c is equal to c_{cr}, then the damping ratio ζ is equal to one, the radical in Eq. 21-14 is zero, and the two roots $\lambda_1 = \lambda_2 = -\omega_n$ are equal and negative. The solution in this case has the special form

$$x(t) = (B + Ct)e^{-\omega_n t} \tag{21-16}$$

Again, the displacement (Eq. 21-12) simply decreases to zero as t increases, and the motion is nonvibratory.

Qualitatively, the motion described by Eq. 21-16 for critical damping is the same as the motion for supercritical damping. Critical damping is of special importance only because it is the dividing point between nonvibratory motions and damped oscillatory motions. That is, critical damping is the smallest amount of damping for which a system will not oscillate. In addition, a critically damped system will come to rest in less time than any other system starting from the same initial conditions.[7] Figure 21-18 shows representative displacement curves for both an overdamped and a critically damped system starting from the same initial displacement and the same initial velocity.

21-3.5 Underdamped Systems

When the damping coefficient c is less than c_{cr}, then the damping ratio ζ is less than one, the radical in Eq. 21-14 is imaginary, and the two roots λ_1 and λ_2 are complex conjugates,

$$\lambda_1 = -\zeta\omega_n + i\omega_d \tag{d}$$
$$\lambda_2 = -\zeta\omega_n - i\omega_d \tag{e}$$

where $i = \sqrt{-1}$ and

$$\omega_d = \omega_n\sqrt{1 - \zeta^2} \tag{21-17}$$

When these values are substituted back into Eq. 21-12, the equation for the displacement becomes

$$x(t) = e^{-\zeta\omega_n t}(D_1 e^{i\omega_d t} + D_2 e^{-i\omega_d t}) \tag{f}$$

---

[7]Strictly speaking, the system described by Eq. 21-16 does not really come to rest for any finite value of time. Practically speaking, however, the motion will become imperceptible after some finite time and the system may be said to be at rest.

By making use of the Euler formula, $e^{ix} = \cos x + i \sin x$, Eq. f can be rewritten

$$
\begin{aligned}
x(t) &= e^{-\zeta\omega_n t}[(D_1 + D_2) \cos \omega_d t + i(D_1 - D_2) \sin \omega_d t] \\
&= e^{-\zeta\omega_n t}(B \cos \omega_d t + C \sin \omega_d t) \quad\quad\quad\quad\text{(21-18a)} \\
&= A e^{-\zeta\omega_n t} \cos (\omega_d t - \phi_c) \quad\quad\quad\quad\quad\quad\text{(21-18b)}
\end{aligned}
$$

where the constants $B = D_1 + D_2$ and $C = i(D_1 - D_2)$ or $A = \sqrt{B^2 + C^2}$ and $\phi_c = \tan^{-1} B/C$ are to be determined from the initial conditions. A typical displacement curve given by Eq. 21-18 is shown in Fig. 21-19. Just as in the preceding cases, the displacement goes to zero as t goes to infinity. However, here the response oscillates within the bounds of the exponential decay curves $A e^{-\zeta\omega_n t}$ and $-A e^{-\zeta\omega_n t}$ as it goes to zero.

The motion described by Eq. 21-18 is called *time-periodic*. The motion oscillates about the equilibrium position, but the amplitude $A e^{-\zeta\omega_n t}$ decreases because the exponent $-\zeta\omega_n = -c/2m$ is negative. Since the amplitude of a damped oscillation decreases monotonically with time, the oscillation will never repeat itself exactly. Therefore, a damped oscillation does not have a period in the same sense as defined for free undamped vibrations. Because of the similarity of Eqs. 21-18b and 21-4a, however, it is customary to call the constant $\omega_d = \omega_n\sqrt{1 - \zeta^2}$ the *damped natural circular frequency*. Since $0 < \zeta < 1$ for underdamped vibrations, the damped natural circular frequency ω_d will always be less than the undamped natural circular frequency ω_n. Also, by analogy with free, undamped vibrations, a *damped natural frequency* f_d and a *damped period* τ_d may be defined as

$$
f_d = \frac{\omega_d}{2\pi} = \frac{\omega_n\sqrt{1 - \zeta^2}}{2\pi} \quad\quad\quad\text{(21-19a)}
$$

$$
\tau_d = \frac{2\pi}{\omega_d} = \frac{2\pi}{\omega_n\sqrt{1 - \zeta^2}} \quad\quad\quad\text{(21-19b)}
$$

The period as defined by Eq. 21-19b is seen to be the time interval between two successive points where the curve of Eq. 21-18 touches one of the limiting curves shown in Fig. 21-19 or twice the time interval between two successive passings through equilibrium. It is interesting to note that the damped period τ_d and the damped natural frequencies f_d and ω_d are constant (independent of time) even though the amplitude is not.

The linear viscous damper in many physical systems is not a real physical element but is merely a mathematical concept used to explain the energy dissipation. For this and other reasons, it is usually necessary to determine the value of the viscous damping ratio ζ experimentally. The determination is easily accomplished by measuring the displacement at two successive "peaks" of the motion; for example, at x_1 and x_2 in Fig. 21-19. Since $\cos (\omega_d t - \phi) = 1$ at both t_1 and t_2, the ratio of these two amplitudes is

$$
\frac{x_1}{x_2} = \frac{A e^{-\zeta\omega_n t_1}}{A e^{-\zeta\omega_n (t_1 + \tau_d)}} = e^{\zeta\omega_n \tau_d} \quad\quad\quad\text{(g)}
$$

Then taking the natural logarithm of both sides and defining the *logarithmic decrement* $\delta = \ln (x_1/x_2)$ gives

$$
\delta = \zeta\omega_n \tau_d = \zeta\omega_n \frac{2\pi}{\omega_n\sqrt{1 - \zeta^2}} = \frac{2\pi\zeta}{\sqrt{1 - \zeta^2}} \quad\quad\quad\text{(21-20)}
$$

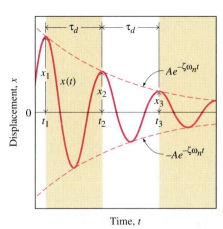

Figure 21-19 An *underdamped system* oscillates about its equilibrium position many times before coming to rest. The amplitude of the oscillations decrease monotonically with time.

Note that δ depends only on the damping ratio ζ and not on t_1 or t_2. That is, the logarithmic decrement does not depend on which two successive peaks are used to measure it. Finally, solving for ζ gives

$$\zeta = \frac{\delta}{\sqrt{(2\pi)^2 + \delta^2}} \qquad (21\text{-}21)$$

When the amount of damping in the system is small, the displacements x_1 and x_2 will be nearly equal $x_1 \cong x_2$, so $\delta = \ln(x_1/x_2)$ will be very small. Then $\sqrt{(2\pi)^2 + \delta^2} \cong 2\pi$, so $\zeta \cong \delta/2\pi$ or $\delta \cong 2\pi\zeta$.

EXAMPLE PROBLEM 21-4

A 5-kg block slides on an inclined frictionless surface as shown in Fig. 21-20. The elastic modulus of the springs are $k_1 = k_2 = 2$ kN/m, and the viscous damping coefficients are $c_1 = c_2 = 25$ N·s/m. If the cart is moved 50 mm up the incline from its equilibrium position and released with an initial velocity of 1.25 m/s down the incline when $t = 0$, determine

a. The damped period τ_d, the damped frequency f_d, and the damped circular frequency ω_d of the resulting vibration.
b. The position of the block as a function of time.
c. The time t_1 at which the amplitude is reduced to 1 percent of its initial value.

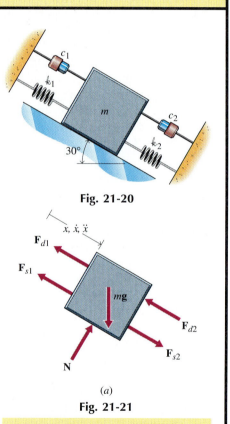

Fig. 21-20

SOLUTION

a. The free-body diagram of the block is drawn in Fig. 21-21a, in which the x-coordinate measures the position of the block along the incline with $x = 0$ at the equilibrium position. In the equilibrium position (before the block has been disturbed), the spring forces are proportional to their static deformation ($F_{s1} = k_1\delta_{eq1}$, $F_{s2} = k_2\delta_{eq2}$). Since $\dot{x} = 0$ in the static equilibrium position, $F_{d1} = F_{d2} = 0$. Therefore, equilibrium gives

$$mg \sin 30° - k_1\delta_{eq1} + k_2\delta_{eq2} = 0 \qquad (a)$$

When the block is at an arbitrary (positive) location x, the stretch in spring 1 will be increased, $F_1 = k_1[\delta_{eq1} + x]$, and the stretch in spring 2 will be reduced, $F_2 = k_2[\delta_{eq2} - x]$, from their equilibrium values. Both dashpots resist the motion and exert forces $F_{d1} = c_1\dot{x}$ and $F_{d2} = c_2\dot{x}$ in the negative coordinate direction (when $\dot{x}$ is positive). Therefore, Newton's second law of motion ($\Sigma F_x = m\ddot{x}$) gives

$$mg \sin 30° - k_1(\delta_{eq1} + x) + k_2(\delta_{eq2} - x) - (c_1 + c_2)\dot{x} = m\ddot{x}$$

or

$$mg \sin 30° - k_1\delta_{eq1} + k_2\delta_{eq2} = m\ddot{x} + (c_1 + c_2)\dot{x} + (k_1 + k_2)x \qquad (b)$$

The left side of Eq. b is zero by Eq. a, however, so the differential equation of motion is

$$m\ddot{x} + (c_1 + c_2)\dot{x} + (k_1 + k_2)x = 0$$
$$5\ddot{x} + 50\dot{x} + 4000x = 0$$

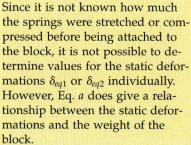

(a)

Fig. 21-21

Since it is not known how much the springs were stretched or compressed before being attached to the block, it is not possible to determine values for the static deformations δ_{eq1} or δ_{eq2} individually. However, Eq. a does give a relationship between the static deformations and the weight of the block.

Then the natural circular frequency, the damping ratio, the damped circular frequency, the damped natural frequency, and the damped period are

$$\omega_n = \sqrt{4000/5} = 28.284 \text{ rad/s}$$

$$\zeta = \frac{50}{2(5)(28.284)} = 0.17678$$

$$\omega_d = \omega_n\sqrt{1 - (0.17678)^2} = 27.84 \text{ rad/s} \qquad \text{Ans.}$$

$$f_d = \frac{\omega_d}{2\pi} = 4.431 \text{ Hz} \qquad \text{Ans.}$$

$$\tau_d = \frac{1}{f_d} = 0.2257 \text{ s} \qquad \text{Ans.}$$

b. The displacement and velocity of the block can be written in the form

$$x(t) = e^{-5.000t}(B \cos 27.84t + C \sin 27.84t)$$
$$\dot{x}(t) = -5.000e^{-5.000t}(B \cos 27.84t + C \sin 27.84t)$$
$$+ e^{-5.000t}(-27.84B \sin 27.84t + 27.84C \cos 27.84t)$$

But at $t = 0$, $x = B = -50 \text{ mm}$ and $\dot{x} = -5B + 27.84C = 1250 \text{ mm/s}$. Therefore, $B = -50 \text{ mm}$, $C = 35.92 \text{ mm}$, and

$$x(t) = e^{-5.000t}(-50 \cos 27.84t + 35.92 \sin 27.84t) \text{ mm} \qquad \text{Ans.}$$

This solution is shown in Fig. 21-21b. Also shown, for comparison, are the two parts of the damped solution $x_u(t) = -50 \cos 27.84t + 35.92 \sin 27.84t$ and the amplitude $Ae^{-5.00t}$, where $A = \sqrt{B^2 + C^2} = 61.56 \text{ mm}$.

c. The initial amplitude of the oscillation is just A. Therefore, at time t_1

$$Ae^{-5.000t_1} = 0.01A$$

which gives

$$t_1 = 0.921 \text{ s} \qquad \text{Ans.}$$

(or a little over 4 cycles of the oscillation).

Both dashpots resist the motion with forces proportional to the speed of the motion $F_d = c\dot{x}$. If the motion is to the right ($\dot{x} > 0$), then $c\dot{x}$ will be positive and the dashpot force F_d acts to the left as shown on the free-body diagram. If the motion is to the left ($\dot{x} < 0$), then $c\dot{x}$ will be negative indicating that the dashpot force F_d acts to the right.

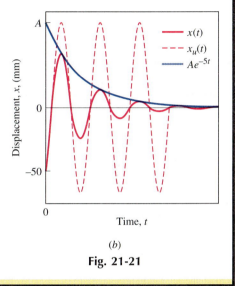

(b)

Fig. 21-21

EXAMPLE PROBLEM 21-5

A 20-lb cart rolls on a flat, horizontal surface as shown in Fig. 21-22. The cart is pulled 15 in. to the right and released with a velocity of 15 ft/s to the left at $t = 0$. If the spring constant $k = 40$ lb/ft and the damping coefficient c corresponds to critical damping, determine

a. The value of the damping coefficient c.
b. If the cart will overshoot the equilibrium position before coming to rest.

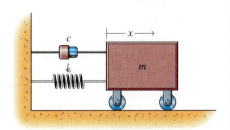

Fig. 21-22

SOLUTION

a. The free-body diagram of the cart is drawn in Fig. 21-23a for an arbitrary (positive) location x. Applying Newton's second law of motion ($\Sigma F = m\ddot{x}$) gives

$$-c\dot{x} - kx = m\ddot{x}$$

or

$$\frac{20}{32.2}\ddot{x} + c\dot{x} + 40x = 0$$

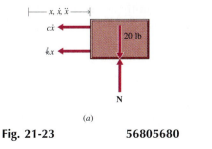

(a)

Fig. 21-23

Then, the natural circular frequency is

$$\omega_n = \sqrt{\frac{40}{20/32.2}} = 8.025 \text{ rad/s}$$

and the damping ratio is

$$\zeta = \frac{c_{cr}}{2\dfrac{20}{32.2}8.025} = 1$$

Therefore, the damping coefficient is

$$c = c_{cr} = 9.97 \text{ lb} \cdot \text{s/ft} \qquad\qquad \text{Ans.}$$

b. For the case of critical damping, the displacement and velocity of the cart are given by

$$x(t) = (B + Ct)e^{-\omega_n t} = (B + Ct)e^{-8.025t} \text{ in.}$$
$$\dot{x}(t) = [C - 8.025(B + Ct)]e^{-8.025t} \text{ in./s}$$

But at $t = 0$; $x = B = 15$ in. and $\dot{x} = C - 8.025B = -180$ in./s. Therefore, $B = 15$ in., $C = -59.63$ in./s, and

$$x(t) = (15 - 59.63t)e^{-8.025t} \text{ in.}$$

Since B and C have opposite signs, there will exist a time $t_1 = 15/59.63 = 0.252$ s at which the position of the cart will be zero. Before t_1 the cart will be on one side of the equilibrium position, and after t_1 it will be on the other side. Therefore, the cart will pass through the equilibrium position before coming to rest. Ans.
The position and velocity of the cart are shown in Fig. 21-23b for $0 \le t \le 1.5$ s.

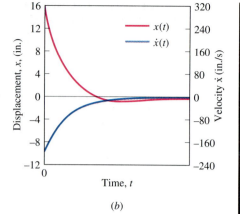

(b)

Fig. 21-23b

A 3-kg uniform slender bar 150 mm long is in equilibrium in the horizontal position shown in Fig. 21-24. When E is displaced down a small amount and released, the amplitude of each peak of the oscillation is observed to be 0.9 of the amplitude of the previous peak. If the spring constant is $k = 400$ N/m, determine

a. The value of the damping coefficient c.
b. The damped period τ_d, the damped frequency f_d, and the damped circular frequency ω_d of the resulting vibration.

Fig. 21-24

SOLUTION

a. The logarithmic decrement is determined from the ratio of successive amplitudes $\delta = \ln(x_1/x_2) = \ln(1/0.9) = 0.10536$. Then the damping ratio is

$$\zeta = \frac{\delta}{\sqrt{(2\pi)^2 + \delta^2}} = 0.01677 = \frac{c_{eff}}{2\sqrt{m_{eff}k_{eff}}} \qquad (a)$$

where m_{eff}, c_{eff}, and k_{eff} are the coefficients in the differential equation of motion.
The free-body diagram of the bar is drawn in Fig. 21-25 in which θ measures the angular position of the bar, with θ positive counterclockwise and $\theta = 0$ at the equilibrium position. In the equilibrium position (before

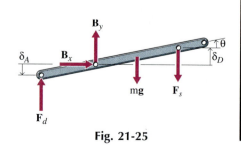

Fig. 21-25

the bar has been disturbed), the dashpot force is $F_d = 0$ and the spring force is $F_s = k\delta_{eq}$, where δ_{eq} is the extension of the spring in the equilibrium position. Therefore, the moment equilibrium equation

$$\downarrow + \Sigma M_B = 0: \quad -0.075 k \delta_{eq} - 0.025mg = 0 \quad (b)$$

gives $\delta_{eq} = -24.53$ mm.

When the bar is rotated counterclockwise (in the positive θ direction), the extension in the spring is $\delta_{eq} + \delta_D$, where for small angles of rotation $\delta_D \cong 0.075\,\theta$. Similarly, the dashpot is compressed at the rate of $\dot{\delta}_A \cong 0.050\,\dot{\theta}$. Therefore, the differential equation of motion is

$$\downarrow + \Sigma M_B = 0: \quad -0.025mg - 0.075k(\delta_{eq} + \delta_D) - 0.050c\dot{\delta}_A = I_B\ddot{\theta}$$

or

$$I_B\ddot{\theta} + (0.050)^2 c\dot{\theta} + (0.075)^2 k\theta = -0.075k\delta_{eq} - 0.025mg \quad (c)$$

where $I_B = \dfrac{1}{12}(3)(0.150)^2 + (3)(0.025)^2 = 7.5(10^{-3})$ kg $\cdot$ m². But the right-hand side of Eq. c is zero by Eq. b, so

$$\ddot{\theta} + 0.3333c\dot{\theta} + 300\theta = 0$$

Substituting the coefficients $m_{eff} = 1$, $c_{eff} = 0.3333c$, and $k_{eff} = 300$ into Eq. a gives the viscous damping coefficient

$$c = \frac{0.01677}{0.3333}(2)\sqrt{300} = 1.743 \text{ N} \cdot \text{s/m} \qquad \text{Ans.}$$

b. Then the natural circular frequency is $\omega_n = \sqrt{300} = 17.321$ rad/s and

$$\omega_d = \omega_n\sqrt{1 - \zeta^2} = 17.318 \text{ rad/s} \qquad \text{Ans.}$$

$$f_d = \frac{\omega_d}{2\pi} = 2.756 \text{ Hz} \qquad \tau_d = \frac{1}{f_d} = 0.363 \text{ s} \qquad \text{Ans.}$$

It is tempting to skip the analysis of the system in the equilibrium position and simply assume that any constants in the differential equation of motion can be discarded. However, students are strongly encouraged to properly account for all forces acting on the system including those forces which act to hold the system in its equilibrium position. The failure of these forces to drop out of the differential equation of motion will signal an error in the analysis.

PROBLEMS

Introductory Problems

21-47–21-60 The following equations represent the position of a particle in damped vibratory motion. For each equation,

a. Classify the motion as underdamped, overdamped, or critically damped.
b. Plot the position, velocity, and acceleration of the particle versus time from $t = 0$ until the amplitude has decreased to 5 percent of its initial value or for three cycles of the oscillation, whichever comes first.

21-47* $x(t) = 10e^{-0.1t} \cos(5t - 1.2)$ in.

21-48* $x(t) = (5 + 3t)e^{-2t}$ mm

21-49 $x(t) = 10e^{-0.5t} - 8e^{-1.5t}$ rad

21-50* $x(t) = e^{-0.05t}(8\cos 3t - 6\sin 3t)$ mm

21-51 $x(t) = 8e^{-0.5t} - 8e^{-2t}$ in.

21-52* $x(t) = (-2 + 5t)e^{-1.5t}$ rad

21-53 $x(t) = e^{-0.02t}(12\sin 12t - 5\cos 12t)$ in.

21-54 $x(t) = -8e^{-0.02t}\sin(15t + 2.5)$ mm

21-55* $x(t) = -(5 + 10t)e^{-0.2t}$ in.

21-56 $x(t) = 7e^{-2t} + 5e^{-3t}$ rad

21-57* $x(t) = (4 - t)e^{-1.2t}$ rad

21-58* $x(t) = 6e^{-0.15t}\sin(10t - 2.5)$ mm

21-59 $x(t) = 3e^{-0.06t}\cos(8t + 1.8)$ in.

21-60 $x(t) = 5e^{-0.5t} - 8e^{-1.5t}$ mm

21-61–21-68 The following differential equations and initial conditions represent the motion of a particle in damped vibratory motion. For each equation,

a. Classify the motion as underdamped, overdamped, or critically damped.
b. Plot the position, velocity, and acceleration of the particle versus time from $t = 0$ until the amplitude has decreased to 5 percent of its initial value or for three cycles of the oscillation, whichever comes first.

21-61* $0.5\ddot{x} + 5\dot{x} + 40x = 0;$ x, in.
$x(0) = 3$ in.; $\dot{x}(0) = 15$ in./s

21-62* $3\ddot{x} + 60\dot{x} + 240x = 0;$ x, mm
$x(0) = -30$ mm; $\dot{x}(0) = 150$ mm/s

21-63 $0.25\ddot{x} + 5\dot{x} + 25x = 0;$ x, in.
$x(0) = -5$ in.; $\dot{x}(0) = 50$ in./s

21-64* $2\ddot{x} + 4\dot{x} + 40x = 0;$ x, mm
$x(0) = 100$ mm; $\dot{x}(0) = 150$ mm/s

21-65 $0.1\ddot{x} + 5\dot{x} + 5x = 0;$ x, in.
$x(0) = 8$ in.; $\dot{x}(0) = 25$ in./s

21-66 $4\ddot{x} + 100\dot{x} + 200x = 0;$ x, mm
$x(0) = -100$ mm; $\dot{x}(0) = -250$ mm/s

21-67* $0.2\ddot{x} + 2\dot{x} + 5x = 0;$ x, in.
$x(0) = -15$ in; $\dot{x}(0) = 0$ in./s

21-68 $5\ddot{x} + 10\dot{x} + 50x = 0;$ x, mm
$x(0) = 0$ mm; $\dot{x}(0) = 500$ mm/s

Intermediate Problems

21-69* A block that has a mass m slides on a frictionless horizontal surface as shown in Fig. P21-69. Determine the damping coefficient c of the single dashpot that could replace the two dashpots shown without changing the frequency of vibration of the block.

Fig. P21-69

21-70 A block that has a mass m slides on a frictionless horizontal surface as shown in Fig. P21-70. Determine the damping coefficient c of the single dashpot that could replace the two dashpots shown without changing the frequency of vibration of the block.

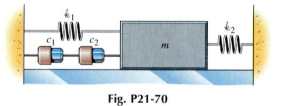

Fig. P21-70

21-71 A 10-lb block is suspended in a vertical plane by two springs and a dashpot as shown in Fig. P21-71. If the block is displaced 7 in. above its equilibrium position and released with an upward velocity of 150 in./s when $t = 0$, determine

a. The differential equation governing the motion.
b. The period of the resulting vibration.
c. The position of the block as a function of time.
d. The earliest time $t_1 > 0$ when the block passes through its equilibrium position.

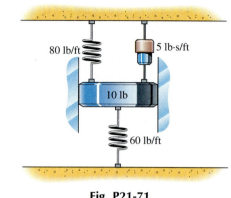

Fig. P21-71

21-72* A 2-kg mass is suspended in a vertical plane by two springs and a dashpot as shown in Fig. P21-72. If the mass is displaced 5 mm below its equilibrium position and released with an upward velocity of 250 mm/s when $t = 0$, determine

a. The differential equation governing the motion.
b. The period of the resulting vibration.
c. The position of the block as a function of time.
d. The earliest time $t_1 > 0$ when the block passes through its equilibrium position.

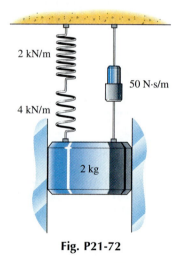

Fig. P21-72

21-73 A 20-lb block slides on a frictionless horizontal surface as shown in Fig. P21-73. The two springs remain in tension at all times and the two pulleys are both small and frictionless. If the block is displaced 3 in. to the left of its equilibrium position and released with a velocity of 50 in./s to the right when $t = 0$, determine

a. The differential equation governing the motion.
b. The period of the resulting vibration.
c. The position of the block as a function of time.
d. The earliest time $t_1 > 0$ when the velocity of the block is zero.

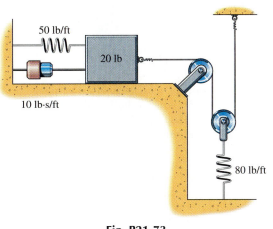

Fig. P21-73

21-74* A 4-kg mass is suspended in a vertical plane as shown in Fig. P21-74. The spring remains in tension at all times and the two pulleys are both small and frictionless. If the mass is displaced 15 mm above its equilibrium position and released with a downward velocity of 750 mm/s when $t = 0$, determine

a. The differential equation governing the motion.
b. The period of the resulting vibration.
c. The position of the mass as a function of time.
d. The earliest time $t_1 > 0$ when the velocity of the mass is zero.

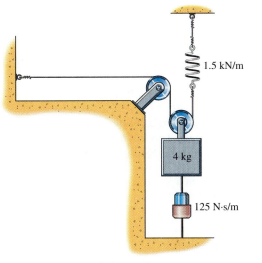

Fig. P21-74

21-75* The two blocks of Fig. P21-75 hang in a vertical plane from a massless bar, which is horizontal in the equilibrium position. If $a = 6$ in., assume small oscillations and determine

a. The damping ratio ζ.
b. The type of motion (underdamped, critically damped, or overdamped).
c. The frequency and period of the motion (if any exists).
d. The value of a that gives critical damping.

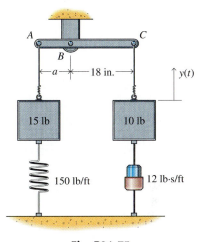

Fig. P21-75

21-76 The two masses of Fig. P21-76 each slide on a frictionless horizontal surface. The bar ABC is vertical in the equilibrium position and has negligible mass. If $a = 100$ mm, assume small oscillations and determine

a. The damping ratio ζ.
b. The type of motion (underdamped, critically damped, or overdamped).
c. The frequency and period of the motion (if any exists).
d. The value of a that gives critical damping.

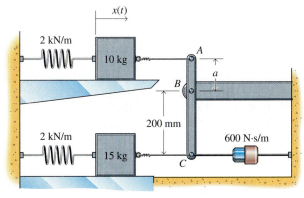

Fig. P21-76

21-77* The 5-lb block of Fig. P21-77 slides on a friction-less horizontal surface while the 3-lb block hangs in a vertical plane. The bar ABC has negligible mass and arm AB is horizontal in the equilibrium position. If $c = 15$ lb · s/ft, assume small oscillations and determine

a. The damping ratio ζ.
b. The type of motion (underdamped, critically damped, or overdamped).
c. The frequency and period of the motion (if any exists).
d. The value of c that gives critical damping.

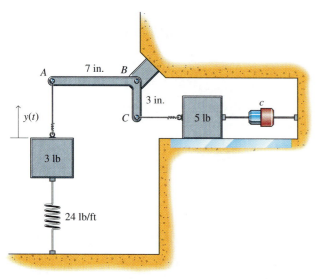

Fig. P21-77

21-78 The two masses of Fig. P21-76 each slide on a frictionless horizontal surface. The bar ABC is vertical in the equilibrium position and has negligible mass. At $t = 0$ the 10-kg mass is x_0 mm to the left of its equilibrium position. If the system is released with zero initial velocity, determine the amount of time and/or the number of cycles required to reduce the amplitude of the motion to 1 percent of its initial value for $a = 500$ mm.

21-79 The damping coefficient c of a dashpot is to be determined by observing the oscillation of a 10-lb block suspended from it as shown in Fig. P21-79. When the block is pulled downward and released, the amplitude of the resulting vibration is observed to decrease from 5 in. to 3 in. in 20 cycles of oscillation. Determine the value of c if the 20 cycles are completed in 5 s.

Fig. P21-79

21-80* The damping coefficient c of a dashpot is to be determined by observing the oscillation of a block suspended from it as shown in Fig. P21-79. When the block is pulled downward and released, the amplitude of the resulting vibration is observed to decrease from 75 mm to 20 mm in 10 cycles of oscillation. Determine the value of c if the spring constant is $k = 1.5$ kN/m and the 10 cycles are completed in 8 s.

Challenging Problems

21-81* A 7-lb uniform cylinder rolls without slipping on a horizontal surface as shown in Fig. P21-81. The spring and the dashpot are connected to a small frictionless pin at G the center of the 8-in.-diameter cylinder. For this system, determine

a. The damping ratio ζ.
b. The type of motion (underdamped, critically damped, or overdamped).
c. The frequency and period of the motion (if any exists).

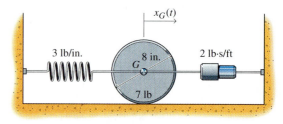

Fig. P21-81

21-82 A 5-kg uniform cylinder rolls without slipping on an inclined surface as shown in Fig. P21-82. The spring is attached to a lightweight inextensible cord, which is wrapped around the cylinder, and the dashpot is connected to a small frictionless pin at G the center of the 400-mm-diameter cylinder. For this system, determine

a. The damping ratio ζ.
b. The type of motion (underdamped, critically damped, or overdamped).
c. The frequency and period of the motion (if any exists).

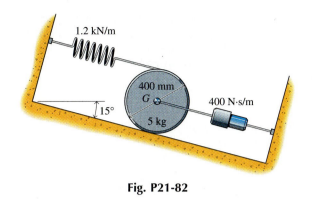

Fig. P21-82

21-83* A 3-lb uniform slender bar 5 ft long is attached to a frictionless pivot at A as shown in Fig. P21-83. The bar is horizontal in the equilibrium position. For this system, determine

a. The damping ratio ζ.
b. The type of motion (underdamped, critically damped, or overdamped).
c. The frequency and period of the motion (if any exists).

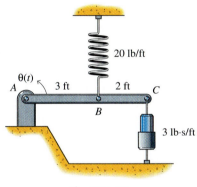

Fig. P21-83

21-84* A 2-kg uniform slender bar 500 mm long is attached to a frictionless pivot at B as shown in Fig. P21-84. The bar is horizontal in the equilibrium position. For this system, determine

a. The damping ratio ζ.
b. The type of motion (underdamped, critically damped, or overdamped).
c. The frequency and period of the motion (if any exists).

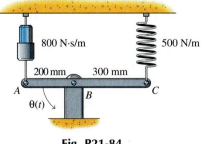

Fig. P21-84

21-85 Two slender uniform bars are welded together as shown in Fig. P21-85. Bar ABC weighs 2 lb and is horizontal in the equilibrium position, bar BD weighs 3 lb and is vertical in the equilibrium position, and the pivot is frictionless. For this system, determine

a. The damping ratio ζ.
b. The type of motion (underdamped, critically damped, or overdamped).
c. The frequency and period of the motion (if any exists).

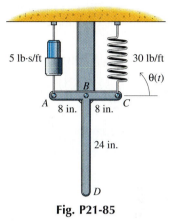

Fig. P21-85

21-86 The 5-kg uniform flywheel shown in Fig. P21-86 rotates about a frictionless pivot. A rope wrapped around the outside of the flywheel is attached to a pair of springs ($k = 1.2$ kN/m, each). If the rope does not slip on the flywheel and the springs remain in tension at all times, determine

a. The damping ratio ζ.
b. The type of motion (underdamped, critically damped, or overdamped).
c. The frequency and period of the motion (if any exists).

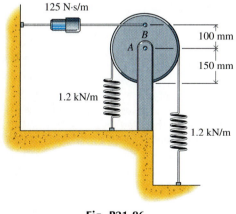

Fig. P21-86

Computer Problems

C21-87 A 10-lb block is suspended in a vertical plane by two springs and a dashpot as shown in Fig. P21-87. If the block is displaced 7 in. above its equilibrium position and released with an upward velocity of 150 in./s when $t = 0$, calculate and plot

a. The position of the block as a function of time for the first two cycles of the oscillation.

b. The force exerted on the upper surface by the spring F_s and by the dashpot F_d as a function of time for the first two cycles of the oscillation.

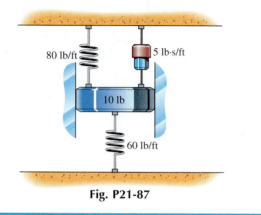

80 lb/ft

5 lb·s/ft

10 lb

60 lb/ft

Fig. P21-87

C21-88 A 10-kg block slides on a smooth horizontal surface as shown in Fig. P21-88. At time $t = 0$ the position and velocity of the block are $x_0 = 0.175$ m and $v_0 = 3$ m/s, respectively. If $k = 1000$ N/m and $c = 15$ N·s/m, calculate and plot

a. The position x of the block as a function of time t ($0 \leq t \leq 5$ s).

b. The velocity v of the block as a function of its position x ($0 \leq t \leq 5$ s).

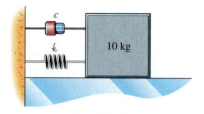

c

k

10 kg

Fig. P21-88

21-4 FORCED VIBRATIONS

A forced vibration is produced and maintained by an externally applied periodic force that does not depend on the position or motion of the body. The force may be applied directly to the body, as in the force that keeps the pendulum of a clock moving. The force may be generated when the support to which the body is attached oscillates, as in the force applied by car springs to a car when the wheels roll over a bumpy road. The force may also be generated internally by the motion of unbalanced rotating parts, as in the force transmitted to an axle when a wheel rotates about an axis that does not pass through its center of mass.

Forced vibrations will occur whenever a periodically varying force is applied to a body. Since any nonharmonic, periodic function of time may be expressed by a Fourier series (a series of simple harmonic functions), a harmonic function of time

$$P = P_0 \sin \Omega t \qquad \text{or} \qquad P = P_0 \cos \Omega t \qquad (a)$$

will be considered here. The constants P_0 and Ω are the amplitude and frequency (rad/s), respectively, of the driving force.

21-4.1 Harmonic Excitation Force

To illustrate viscous damped forced vibration, a harmonic excitation force $P_0 \sin \Omega t$ is added to the block–spring–dashpot system of Fig. 21-16a as shown in Fig. 21-26a. The free-body diagram of the block is shown in Fig. 21-26b, in which the block has been displaced an arbitrary amount in the positive coordinate direction. The elastic restoring force of the spring $F_s = kx$ is directed toward the equilibrium position (the negative coordinate direction), and the damping force $F_d = c\dot{x}$ acts opposite the velocity (also in the negative coordinate direction). Applying Newton's second law of motion ($\Sigma F = ma_x = m\ddot{x}$) to the block gives the differential equation of motion for the block

$$-c\dot{x} - kx + P_0 \sin \Omega t = m\ddot{x}$$

or

$$m\ddot{x} + c\dot{x} + kx = P_0 \sin \Omega t \tag{21-22}$$

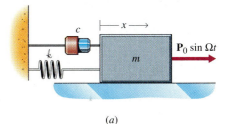

(a)

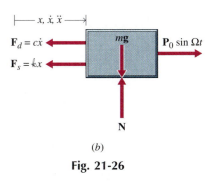

(b)

Fig. 21-26

Equation 21-22 is a nonhomogeneous, linear, second-order differential equation with constant coefficients. Its general solution consists of two parts, a *particular solution* and a *complementary solution*. The particular solution is any function $x_p(t)$ that satisfies the differential equation as written. The complementary solution is the function $x_c(t)$ that satisfies the homogeneous part of the differential equation, which is identical to Eq. 21-10. The complementary solution, therefore, is given by Eqs. 21-12, 21-16, or 21-18, depending on the value of the damping ratio ζ. The general solution to Eq. 21-22 is then

$$x(t) = x_c(t) + x_p(t) \tag{21-23}$$

The complementary part of the solution has already been discussed in considerable detail in Section 21-3. Therefore, the complementary solution will not be considered further except to note that

1. Whether the system is overdamped, underdamped, or critically damped, $x_c(t)$ contains two constants that must be chosen to satisfy the initial conditions. In evaluating the constants, however, care must be taken to include the particular solution. That is, if the initial position and velocity are $x(0) = x_0$ and $\dot{x}(0) = v_0$, respectively, then $x_c(0) = x_0 - x_p(0)$ and $\dot{x}_c(0) = v_0 - \dot{x}_p(0)$.
2. No real system is completely frictionless. Therefore, the complementary solution $x_c(t)$ will always decay away with time. Since the complementary solution is visible only for some (usually short) period of time after the motion is started, it is called a *transient solution*.

The particular part of the solution is any function $x_p(t)$ that satisfies Eq. 21-22. Since the periodic exciting force is harmonic, it seems reasonable to guess that $x_p(t)$ is also harmonic

$$\begin{aligned} x_p(t) &= D \sin (\Omega t - \psi_s) \\ &= D \sin \Omega t \cos \psi_s - D \sin \psi_s \cos \Omega t \end{aligned} \tag{21-24}$$

where the constants D and ψ_s are to be chosen to make the solution $x_p(t)$ satisfy the differential equation (Eq. 21-22). Taking the appropriate derivatives and substituting them into the differential equation gives

$$D[(k - m\Omega^2) \cos \psi_s + c\Omega \sin \psi_s] \sin \Omega t$$
$$+ D[c\Omega \cos \psi_s - (k - m\Omega^2) \sin \psi_s] \cos \Omega t = P_0 \sin \Omega t \qquad (b)$$

But the solution (Eq. 21-24) is supposed to satisfy the differential equation at every instant of time. Therefore, Eq. *b* must hold for every instant of time. In particular, when $t = 0$, $\sin \Omega t = 0$ and $\cos \Omega t = 1$ so that[8]

$$\tan \psi_s = \frac{c\Omega}{k - m\Omega^2} = \frac{2\zeta\Omega/\omega_n}{1 - (\Omega/\omega_n)^2} \qquad (21\text{-}25)$$

The *phase angle* ψ_s represents the amount by which the response $D \sin (\Omega t - \psi_s)$ lags the applied force $P_0 \sin \Omega t$. That is, the response hits its peak ψ_s/Ω seconds later than the applied force hits its peak.

When $\Omega t = \pi/2$, $\sin \Omega t = 1$ and $\cos \Omega t = 0$, so Eq. *b* gives

$$D = \frac{P_0}{(k - m\Omega^2) \cos \psi_s + c\Omega \sin \psi_s} \qquad (c)$$

where (with reference to Fig. 21-27)

$$\sin \psi_s = \frac{c\Omega}{\sqrt{(k - m\Omega^2)^2 + (c\Omega)^2}} \qquad (d)$$

$$\cos \psi_s = \frac{k - m\Omega^2}{\sqrt{(k - m\Omega^2)^2 + (c\Omega)^2}} \qquad (e)$$

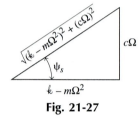

Fig. 21-27

Therefore, the amplitude of the particular solution is

$$D = \frac{P_0}{\sqrt{(k - m\Omega^2)^2 + (c\Omega)^2}}$$
$$= \frac{P_0/k}{\sqrt{[1 - (\Omega/\omega_n)^2]^2 + (2\zeta\Omega/\omega_n)^2}} \qquad (21\text{-}26)$$

Since the amplitude of the particular solution is constant, the particular solution is called the *steady-state* vibration. That is, after the transient part of the solution x_c has decayed away, the system will oscillate according to $x_p(t) = D \sin (\Omega t - \psi_s)$ as long as the driving force $P_0 \sin \Omega t$ is applied.

Note, however, that $\delta_p = P_0/k$ is the deflection of the spring that would result if the force P_0 were applied statically to the spring.[9] Then the ratio D/δ_p represents the factor by which the magnitude of the dynamic oscillation is greater than the static deflection. This ratio is called the *dynamic magnification factor* and is given by

$$\frac{D}{\delta_p} = \frac{D}{P_0/k} = \frac{1}{\sqrt{[1 - (\Omega/\omega_n)^2]^2 + (2\zeta\Omega/\omega_n)^2}} \qquad (21\text{-}27)$$

[8]As always, the coefficients m, k, c, and P_0 in the solutions Eqs. 21-25 through 21-27 must be interpreted as the coefficients of the differential equation Eq. 21-22. They may or may not refer to actual system values of mass, spring constant, etc.

[9]The static deflection δ_p should not be confused with the equilibrium deflection δ_{eq} of Sections 21-2 and 21-3. The static deflection δ_p describes the deflection that would occur if the force P_0 were applied statically to the spring and has nothing to do with the equilibrium position of the system.

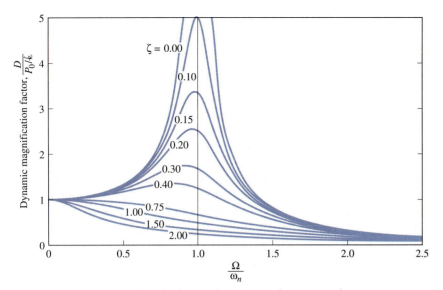

Figure 21-28 When a lightly damped system is driven at a frequency near its natural frequency ($\Omega/\omega_n \cong 1$), the amplitude of the vibration can become very large.

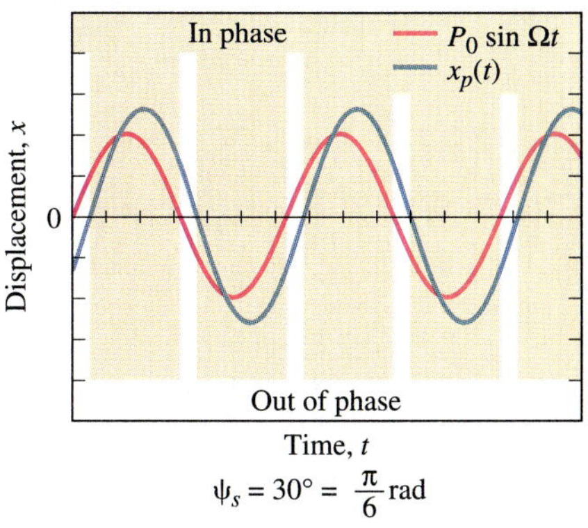

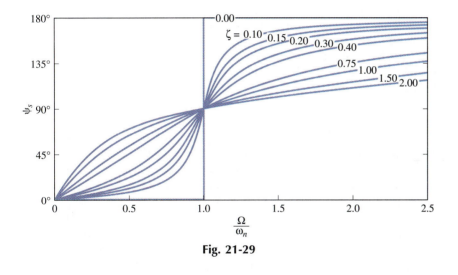

Fig. 21-29

$\psi_s = 30° = \dfrac{\pi}{6}$ rad

(a)

Fig. 21-30

Figures 21-28 and 21-29 show the variation of the magnification factor D/δ_p and the phase angle ψ_s with the frequency ratio Ω/ω_n for various values of the damping ratio ζ. When the disturbing force $P_0 \sin \Omega t$ is applied at low frequencies ($\Omega/\omega_n < 1$), the steady-state response is mostly *in phase* with the disturbing force ($0 < \psi_s < 90°$). That is, the disturbing force generally acts to the right ($P_0 \sin \Omega t > 0$) when the block is to the right of its equilibrium position ($x_p > 0$) and vice versa (Fig. 21-30a). In fact, for very low frequencies ($\Omega/\omega_n \cong 0$), the system is essentially in static equilibrium; the phase angle is nearly zero ($\psi_s \cong 0$), the magnification factor is approximately one ($D/\delta_p \cong 1$), and the steady-state response is $x_p(t) \cong (P_0 \sin \Omega t)/k$.

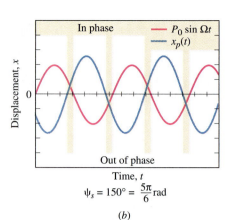

$$\psi_s = 150° = \frac{5\pi}{6} \, \text{rad}$$

(b)

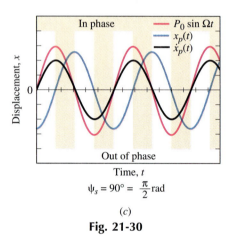

$$\psi_s = 90° = \frac{\pi}{2} \, \text{rad}$$

(c)

Fig. 21-30

When the disturbing force is applied at high frequencies ($\Omega/\omega_n > 1$), the steady-state response is mostly *out of phase* with the disturbing force ($90° < \psi_s < 180°$). That is, the disturbing force generally acts to the right ($P_0 \sin \Omega t > 0$) when the block is to the left of its equilibrium position ($x_p < 0$) and vice verse (Fig. 21-30b). For very high frequencies ($\Omega/\omega_n \gg 1$), the response is nearly totally out of phase with the disturbing force ($\psi_s \cong 180°$), and the magnification factor is approximately zero independent of the damping ratio. The block is held essentially stationary by the inertial resistance of the block.

When the disturbing force is applied at a frequency close to the natural frequency of the system ($\Omega/\omega_n \cong 1$) and the damping is light ($\zeta \cong 0$), the amplitude of the vibration is magnified substantially. In fact, if the system possesses no damping and is excited by a harmonic force with a frequency very close to the natural frequency $\Omega \cong \omega_n$, the amplitude of the vibration becomes very large according to Eq. 21-26.[10] This condition is called *resonance*. Figure 21-28 suggests that the amplitude of an oscillation may be controlled either by avoiding the condition of resonance or (if resonance cannot be avoided) by increasing the damping ζ.

When the frequency of the disturbing force matches the natural frequency of the system ($\Omega/\omega_n = 1$), as Fig. 21-29 shows, the response lags 90° behind the disturbing force independent of the damping ratio. Therefore the displacement $x_p(t) = D \sin (\Omega t - \pi/2) = -D \cos \Omega t$ is maximum when the disturbing force $P_0 \sin \Omega t$ is zero and vice versa (Fig. 21-30c). However, the velocity $\dot{x}_p(t) = D\Omega \cos (\Omega t - \pi/2) = D\Omega \sin \Omega t$ is in phase with the disturbing force $P_0 \sin \Omega t$. Also, when $\Omega/\omega_n = 1$, Eq. 21-27 shows that the magnification factor $D/\delta_p = 1/2\zeta$. These features are often used to determine the natural frequency and the damping ratio of a system experimentally.

It must be noted, however, that except for $\zeta = 0$, the magnification factor curves (and hence the amplitude of the vibration) do not peak at exactly $\Omega/\omega_n = 1$. Increasing the amount of damping decreases the resonant frequency—the frequency at which the magnification curve peaks. When $\zeta = \sqrt{1/2}$, the maximum amplitude occurs at $\Omega = 0$. When $\zeta \geq \sqrt{1/2}$, the amplitude of vibration D is less than the static displacement δ_p for all frequencies $\Omega > 0$. The exact location of the resonant frequency for any given value of ζ can be calculated by setting the derivative of the magnification factor with respect to Ω/ω_n equal to zero.

In summary then, the complete solution consists of two superimposed vibrations $x(t) = x_c(t) + x_p(t)$. For underdamped systems $\zeta < 1$, the displacement is

$$x(t) = Ae^{-\zeta\omega_n t} \cos (\omega_d t - \phi_c) + D \sin (\Omega t - \psi_s) \quad \text{(21-28)}$$

The first term in Eq. 21-28 represents a free vibration of the system. The frequency of this vibration depends only on properties of the system (the spring constant k, the damping coefficient c, and the mass m)

---

[10]Of course, all real systems possess some damping, so the amplitude of the vibration cannot become infinite. In addition, physical constraints such as the length of the spring also limit the amplitude of the vibration. Still, resonance is a dangerous condition and should always be avoided.

and is independent of the applied disturbing force. The amplitude of the free vibration (or transient vibration) term decays with time due to friction (damping) forces. The constants A and ϕ_c are chosen to fit the complete solution to the initial conditions.

The last term in Eq. 21-28, which represents the steady-state vibration of the system, is the part of the solution that is usually of primary interest. The frequency of the steady-state vibration is the same as that of the applied disturbing force, and its amplitude depends on the frequency ratio Ω/ω_n.

21-4.2 Harmonic Motion of Foundation

The cause of forced vibrations need not be a periodic force applied directly to the mass in the system. In many systems such as car suspensions, the forced vibrations are caused by the periodic movement of the system's support rather than by a directly applied force. It will be shown that the periodic movement of the support is directly equivalent to a periodic disturbing force. So long as the coefficients m, k, c, and P_0 in the solutions, Eqs. 21-25 through 21-28, are interpreted as the coefficients of the system's differential equation of motion, those solutions apply to this case as well.

For example, suppose that the support to which the spring of Fig. 21-16a is attached is given a periodically varying displacement $x_b(t) = b \sin \Omega t$ as shown in Fig. 21-31a. The free-body diagram of the block is shown in Fig. 21-31b, in which the block has been displaced an arbitrary amount in the positive coordinate direction. The elongation of the spring is the difference between the displacements of the block and the movable support $x_b(t) - x(t) = b \sin \Omega t - x(t)$. Therefore, the elastic restoring force of the spring is $F_s = k(b \sin \Omega t - x)$ to the right (the spring is stretched and pulls on the block whenever $b \sin \Omega t > x$). Since the dashpot is attached to a fixed support, the rate of extension of the dashpot is just $\dot{x}(t)$ and the damping force $F_d = c\dot{x}$ acts opposite the velocity (in the negative coordinate direction). Applying Newton's second law of motion ($\Sigma F = ma_x = m\ddot{x}$) to the block gives the differential equation

$$-c\dot{x} + k(b \sin \Omega t - x) = m\ddot{x}$$

or

$$m\ddot{x} + c\dot{x} + kx = kb \sin \Omega t \qquad (f)$$

But by comparison, Eq. f can be transformed into Eq. 21-22 simply by replacing P_0 with kb. Therefore, Eq. f has the same solution as Eq. 21-22. That is, the solutions defined by Eqs. 21-25 through 21-28 also describe the motion of the block when subjected to the support displacement $x_b(t) = b \sin \Omega t$ when the constants m, c, k, and P_0 are interpreted as the coefficients of the differential equation of motion when written in the form of Eq. 21-22.

21-4.3 Rotating Imbalance

Another common source of forced vibrations is an imbalance in a rotating machine piece. For example, the small mass m_s of Fig. 21-32a rotates with angular frequency Ω about an axis fixed in the larger block of mass M. When the block is displaced an arbitrary amount $x(t)$ in the

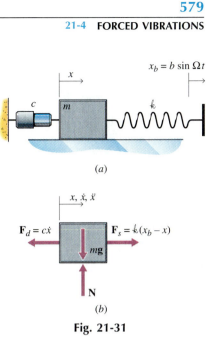

(a)

(b)

Fig. 21-31

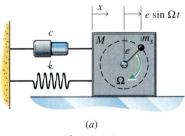

(a)

Fig. 21-32

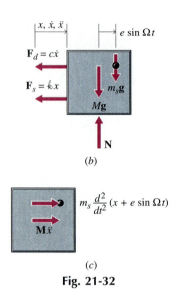

(b)

(c)

Fig. 21-32

positive coordinate direction, the position of the small mass is $x(t) + e \sin \Omega t$. In the free-body diagram of Fig. 21-32b, the internal forces between the mass and block need not be shown. The elastic restoring force of the spring $F_s = kx$ is directed toward the equilibrium position (the negative coordinate direction). The damping force $F_d = c\dot{x}$ acts opposite the velocity—also in the negative coordinate direction. Then, referring to the freebody diagram of Fig. 21-32b and the kinetic diagram of Fig. 21-32c, Newton's second law of motion ($\Sigma F = ma_x = m\ddot{x}$) gives the differential equation

$$-c\dot{x} - kx = M\ddot{x} + m_s \frac{d^2(x + e \sin \Omega t)}{dt^2}$$

or

$$(M + m_s)\ddot{x} + c\dot{x} + kx = em_s\Omega^2 \sin \Omega t \qquad (g)$$

Again by comparison, Eq. g can be transformed into Eq. 21-22 simply be replacing P_0 with $em_s\Omega^2$ and m with $M + m_s$. That is, the solutions defined by Eqs. 21-25 through 21-28 also describe the motion of the block when subjected to the rotational imbalance of the small mass m_s when the constants m, c, k, and P_0 are interpreted as the coefficients of the differential equation of motion when written in the form of Eq. 21-22.

CONCEPTUAL EXAMPLE 21-1: STRUCTURAL VIBRATIONS

What do masses bouncing on springs have to do with real structural problems? Why do structural designers need to worry about dynamics and vibrations?

SOLUTION

In *Strength of Materials* it will be shown that the deflection of beams caused by concentrated loads is proportional to the load applied. In particular, for a simply supported beam subjected to a single concentrated load P at its middle, the deflection at the center of the beam is given by

$$\delta = \frac{PL^3}{48EI} \qquad (a)$$

where E is a constant (called Young's Modulus) characterizing the elasticity or flexibility of the material that the beam is made of and I is the second moment of area of the beam's cross-sectional shape. Equation a can be rearranged to get

$$P = \frac{48EI}{L^3}\delta = k\delta \qquad (b)$$

That is, the beam acts like a spring of modulus

$$k = \frac{48EI}{L^3} \qquad (c)$$

For a 20-ft long and 1-ft-deep I-beam of structural steel, the equivalent spring modulus is about $k = 275$ kip/ft.

Suppose that a 500-lb motor, generator, or some other piece of equipment is supported by this structural steel I-beam. The effect is the same as if the ma-

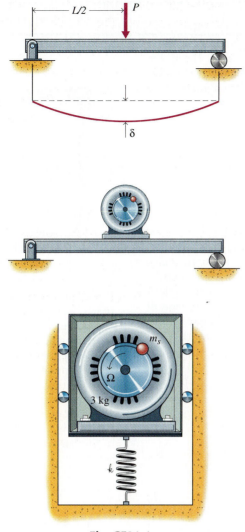

Fig. CE21-1

chine is simply sitting on a very stiff spring (k = 275 kip/ft). When the machine is at rest, the spring (that is, the beam) will deflect only $\delta = W/k = 500/275,000 = 0.00182$ ft $= 0.0218$ in. However, all machines have a slight imbalance and will vibrate a little. Therefore, it is important to check the natural frequency of the system and to verify that the operating speed of the machine is not too close to the natural frequency.

It is easily verified that the natural frequency of this simple mass-spring system is

$$\omega_n = \sqrt{k/m} = \sqrt{kg/W}$$
$$= \sqrt{(275,000)(32.2)/(500)} = 133.08 \text{ rad/s} \qquad (d)$$

or $f_n = \omega_n/2\pi = 21.18$ Hz $= 1271$ rev/min. Although this natural frequency is probably well away from the normal operating speed of the machine, the machine will have to go through this frequency every time it is started or stopped. If the machine does not go through this frequency fast enough, the resonance of the system could cause large deflections and significant structural damage to the support.

A 24-lb block slides on a frictionless surface as shown in Fig. 21-33. The spring is unstretched when bar AB is vertical and bar BC is horizontal. The weights of bars AB and BC may be neglected. Assume that oscillations remain small and determine

a. The range of frequencies Ω for which the angular steady-state motion of bar AB is less than $\pm 5°$.
b. The position of the block as a function of time if the block is pulled 2 in. to the right and released from rest when $t = 0$ and $\Omega = 25$ rad/s.

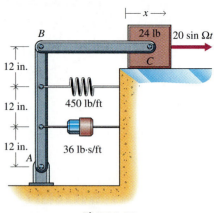

Fig. 21-33

SOLUTION

a. Free-body diagrams of the block and of the bar AB are drawn in Figs. 21-34a and 21-34b in which the block has been displaced an arbitrary distance in the positive coordinate direction (to the right). When the block moves a distance x to the right, the bar AB rotates clockwise through an angle θ. If oscillations remain small, then $\sin \theta \cong \theta$, $\cos\theta \cong 1$, the compression of the spring will be $2x/3$, and the rate of compression of the dashpot will be $\dot{x}/3$. Since the mass of the bar is negligible, the moment of inertia of the bar is negligible and

$$\zeta \Sigma M_A = \left(36\frac{\dot{x}}{3}\right) + 2\left(450\frac{2x}{3}\right) - 3T = 0$$

or

$$T = 4\dot{x} + 200x \qquad (a)$$

Then, applying Newton's second law of motion ($\Sigma F = ma_x = m\ddot{x}$) to the block gives

$$20 \sin \Omega t - T = \frac{24}{32.2}\ddot{x} \qquad (b)$$

Adding Eqs. a and b gives the differential equation of motion of the block

$$0.7453\ddot{x} + 4\dot{x} + 200x = 20 \sin \Omega t$$

Therefore, the natural circular frequency and damping ratio of the system are

$$\omega_n = \sqrt{200/0.7453} = 16.381 \text{ rad/s}$$
$$\zeta = \frac{4}{2(0.7453)(16.381)} = 0.1638$$

Since it is desired to keep the angular motion of bar AB less than $5° = 0.08727$ rad, the maximum amplitude of the steady-state vibration of the block is (Eq. 21-26)

$$D \cong (3 \text{ ft})(0.08727 \text{ rad}) = \frac{20}{\sqrt{(200 - 0.7453\Omega^2)^2 + (4\Omega)^2}}$$

which corresponds to the limiting frequencies

$$\Omega = 14.12 \text{ rad/s} \quad \text{or} \quad 17.57 \text{ rad/s}$$

Frequencies between these two values give amplitudes that are too great, so the allowable range of frequencies is

$$0 < \Omega < 14.12 \text{ rad/s} \qquad 17.57 \text{ rad/s} < \Omega \qquad \text{Ans.}$$

If oscillations remain small, then the vertical motion of pin B will be negligible ($3 - 3 \cos \theta \cong 0$), and the spring and the dashpot will remain essentially horizontal. By similar triangles, the compression in the spring will be two-thirds of the horizontal movement of pin B, and the rate of compression of the dashpot will be one-third the rate of the horizontal movement of pin B.

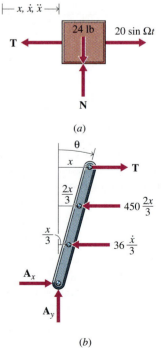

Fig. 21-34

b. When $\Omega = 25$ rad/s, the equation of motion of the block is

$$x(t) = Ae^{-\zeta\omega_n t} \cos(\omega_d t - \phi_c) + D \sin(\Omega t - \psi_s)$$

where

$$\omega_d = 16.381\sqrt{1 - (0.1638)^2} = 16.16 \text{ rad/s}$$

$$D = \frac{20}{\sqrt{[200 - 0.7453(25)^2]^2 + [4(25)]^2}}$$

$$= 0.07042 \text{ ft} = 0.8451 \text{ in.}$$

and

$$\psi_s = \tan^{-1} \frac{(4)(25)}{200 - (0.7453)(25)^2} = 159.4° = 2.782 \text{ rad}$$

But at $t = 0$

$$x(0) = 2 = A \cos \phi_c - 0.8451 \sin 159.4°$$
$$\dot{x}(0) = A[16.16 \sin \phi_c - (0.1638)(16.381) \cos \phi_c]$$
$$+ (0.8451)(25) \cos 159.4° = 0$$

Therefore $A = 2.803$ in., $\phi_c = 34.94° = 0.610$ rad, and

$$x(t) = 2.803e^{-2.68t} \cos(16.16t - 0.610) + 0.845 \sin(25t - 2.782) \text{ in.} \quad \textbf{Ans.}$$

This solution is sketched in Fig. 21-34c. For comparison, a unit loading curve ($\sin 25t$) and the steady-state portion of the response $x_p(t) = 0.845 \sin(25t - 2.782)$ are also shown in the figure.

After just one cycle of the oscillation, the displacement $x(t)$ is in phase with (reaches its maximum at the same time as) the steady state portion of the displacement. After three cycles of the oscillation (about 0.75 s), the difference between the displacement and the steady state portion of the displacement is nearly gone.

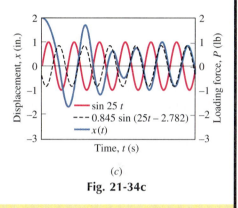

(c)

Fig. 21-34c

EXAMPLE PROBLEM 21-8

A 3-kg motor sits on a spring ($k = 150$ kN/m) and a dashpot ($c = 120$ N · s/m) as shown in Fig. 21-35. A small mass ($m_s = 0.5$ kg) is attached to the edge of the motor's pulley ($e = 25$ mm). Determine the maximum amplitude of the resulting forced vibration of the motor.

SOLUTION

The free-body diagram of the motor is shown in Fig. 21-36a for an arbitrary (positive) location y. The downward force in the spring is $F_s = k(y + \delta_{eq})$, where δ_{eq} is the stretch of the spring in the equilibrium position (where $y = 0$). In the equilibrium position (before the motor is started running), $y = \dot{y} = 0$ and the vertical component of equilibrium ($\uparrow \Sigma F_y = 0$)

$$-(3 + 0.5)(9.81) - 150,000\delta_{eq} = 0 \quad (a)$$

gives the static deflection of the spring $\delta_{eq} = -2.289(10^{-4})$ m $= -0.2289$ mm. After the motor is started running, the free-body diagram of Fig. 21-36a still applies. The kinetic diagrams of the motor and the mass are shown in Fig. 21-36b. Adding the y-components of Newton's second law of motion ($\Sigma F_y = m\ddot{y}$) for the motor and the mass gives

$$-(3 + 0.5)(9.81) - 150,000(y + \delta_{eq}) - 120\dot{y} \quad (b)$$
$$= 3\frac{d^2 y}{dt^2} + 0.5\frac{d^2}{dt^2}(y + 0.025 \sin \Omega t)$$

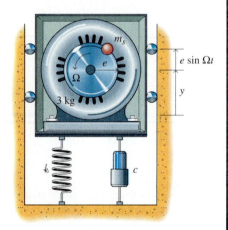

Fig. 21-35

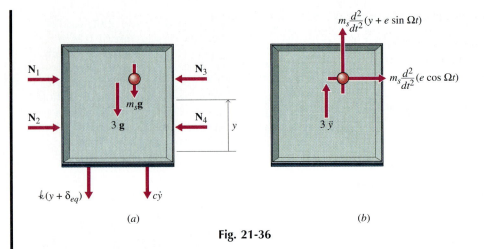

(a)   (b)

Fig. 21-36

Then, substituting $\delta_{eq} = -2.289(10^{-4})$ m into Eq. *b* or equivalently subtracting Eq. *a* from Eq. *b* gives the differential equation of motion for the motor

$$3.5\ddot{y} + 120\dot{y} + 150{,}000y = 0.0125\Omega^2 \sin \Omega t$$

Therefore, the natural circular frequency and damping ratio of the motion are

$$\omega_n = \sqrt{150{,}000/3.5} = 207.0 \text{ rad/s}$$

$$\zeta = \frac{120}{2(3.5)(207.0)} = 0.08282$$

and the amplitude of the steady-state vibration is (Eq. 21-26)

$$D = \frac{0.0125\Omega^2/150{,}000}{\sqrt{[1 - (\Omega/207.0)^2]^2 + [2(0.08282)\Omega/207.2]^2}}$$

To find the value of Ω that gives the maximum amplitude, set the derivative $dD/d\Omega = 0$, which gives $\Omega = 208.4$ rad/s. Then

$$D_{max} = 0.02163 \text{ m} = 21.63 \text{ mm} \qquad\qquad \text{Ans.}$$

PROBLEMS

Introductory Problems

21-89* A 10-lb block is suspended in a vertical plane by two springs and a dashpot as shown in Fig. P21-89. An upward force $P(t) = 70 \sin 30t$ lb is applied to the block. If the block has an upward velocity of 150 in./s and is displaced 7 in. above its equilibrium position when $t = 0$, determine

a. The differential equation governing the motion.
b. The position of the block as a function of time.

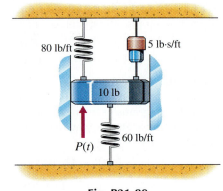

Fig. P21-89

21-90 A 2-kg mass is suspended in a vertical plane by two springs and a dashpot as shown in Fig. P21-90. A downward force $P(t) = 600 \sin 20t$ N is applied to the mass. If the mass has an upward velocity of 250 mm/s and is displaced 5 mm below its equilibrium position when $t = 0$, determine

a. The differential equation governing the motion.
b. The position of the mass as a function of time.

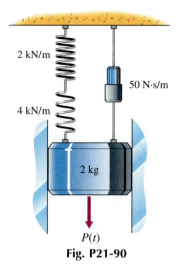

Fig. P21-90

21-91* A 20-lb block slides on a frictionless horizontal surface as shown in Fig. P21-91. The two springs remain in tension at all times, and the two pulleys are both small and frictionless. A force $P(t) = 40 \sin 12t$ lb to the right is applied to the block. If the block has a velocity of 50 in./s to the right and is displaced 3 in. to the left of its equilibrium position when $t = 0$, determine

a. The differential equation governing the motion.
b. The position of the block as a function of time.

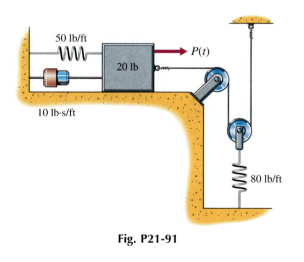

Fig. P21-91

21-92* A 4-kg mass is suspended in a vertical plane as shown in Fig. P21-92. The spring remains in tension at all times, and the two pulleys are both small and frictionless. A downward force $P(t) = 150 \sin 18t$ N is applied to the mass. If the mass has a downward velocity of 750 mm/s and is displaced 15 mm above its equilibrium position when $t = 0$, determine

a. The differential equation governing the motion.
b. The position of the mass as a function of time.

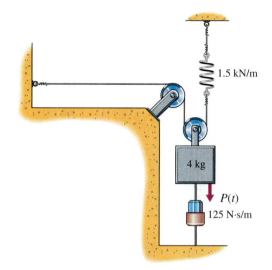

Fig. P21-92

21-93 The two blocks of Fig. P21-93 hang in a vertical plane from a massless bar which is horizontal in the equilibrium position. If an upward force $P(t) = 4 \sin \Omega t$ lb is applied at point D of the bar, determine

a. The maximum amplitude of the steady-state oscillation of the 10-lb block.
b. The range of frequencies Ω that must be avoided if the amplitude of the steady-state oscillation of the 10-lb block is not to exceed 1.5 in.

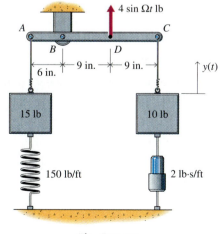

Fig. P21-93

21-94* The two masses of Fig. P21-94 each slide on a frictionless horizontal surface. The bar ABC is vertical in the equilibrium position and has negligible mass. If a force $P(t) = 50 \sin \Omega t$ N is applied at point D of the bar, determine

a. The maximum amplitude of the steady-state oscillation of the 10-kg block.
b. The range of frequencies Ω that must be avoided if the amplitude of the steady-state oscillation of the 10-kg block is not to exceed 25 mm.

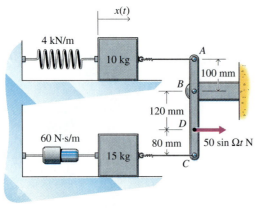

Fig. P21-94

21-95 Determine the maximum magnification factor given by Eq. 21-27 and the frequency ratio (Ω/ω_n) at which it occurs as a function of the damping ratio ζ.

21-96 The particular solution to the differential equation of motion

$$m\ddot{x} + c\dot{x} + kx = P_0 \cos \Omega t$$

can be written in the form

$$x_p(t) = D \cos(\Omega t - \psi_c)$$

Determine expressions for D and ψ_c similar to Eqs. 21-25 and 21-26 for this case.

Intermediate Problems

21-97* A 10-lb block is suspended in a vertical plane by two springs and a dashpot as shown in Fig. P21-97. Determine the amplitude of the resulting steady-state oscillation when the lower support oscillates vertically according to $y_\ell = 7 \sin 30t$ in.

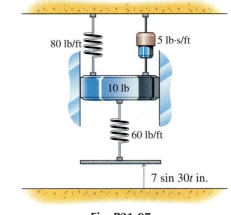

Fig. P21-97

21-98* A 20-kg block slides on a frictionless surface as shown in Fig. P21-98. The spring ($k = 500$ N/m) and the dashpot ($c = 40$ N · s/m) are attached to an oscillating wall. Determine

a. The differential equation governing the motion of the block.
b. The particular solution in the form $x_p(t) = D \sin (\Omega t - \psi_s)$.

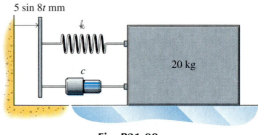

Fig. P21-98

21-99 A 20-lb block slides on a frictionless horizontal surface as shown in Fig. P21-99. The two springs remain in tension at all times, and the two pulleys are both small and frictionless. Determine the amplitude of the resulting steady-state oscillation when the lower support oscillates vertically according to $y_\ell = 15 \sin 12t$ in.

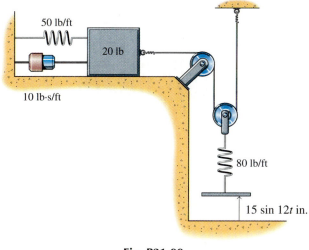

Fig. P21-99

21-100 A 4-kg mass is suspended in a vertical plane as shown in Fig. P21-100. The spring remains in tension at all times, and the two pulleys are both small and frictionless. Determine the amplitude of the resulting steady-state oscillation when the lower support oscillates vertically according to $y_\ell = 200 \cos 18t$ mm.

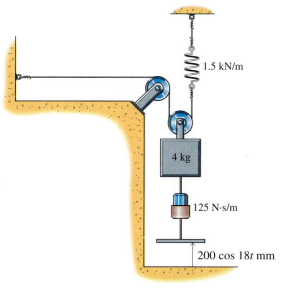

Fig. P21-100

21-101 The two blocks of Fig. P21-101 hang in a vertical plane from a massless bar which is horizontal in the equilibrium position. Determine the amplitude of the resulting steady-state oscillation of the 10-lb block when $a = 9$ in. and the lower right support oscillates vertically according to $y_r = 4 \cos 9t$ in.

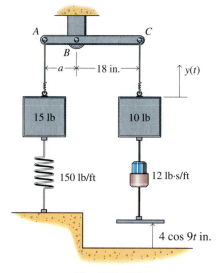

Fig. P21-101

21-102* The two masses of Fig. P21-102 each slide on a frictionless horizontal surface. The bar ABC is vertical in the equilibrium position and has negligible mass. Determine the amplitude of the resulting steady-state oscillation of the 10-kg block when $a = 150$ mm and the lower left support oscillates horizontally according to $x_\ell = 5 \sin 8t$ mm.

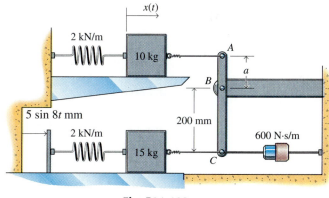

Fig. P21-102

Challenging Problems

21-103 A 10-lb block is suspended in a vertical plane by two springs and a dashpot as shown in Fig. P21-103. If a 2-lb weight rotating around a 6-in.-radius circle at a rate of $\Omega = 30$ rad/s is added to the 10-lb block, determine the amplitude of the resulting steady-state oscillation.

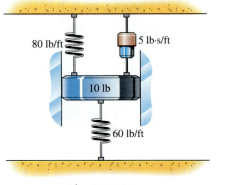

Fig. P21-103

21-104* A 2-kg mass is suspended in a vertical plane by two springs and a dashpot as shown in Fig. P21-104. If a 0.6-kg mass rotating around a 150-mm radius circle at a rate of $\Omega = 20$ rad/s is added to the 2-kg mass, determine the amplitude of the resulting steady-state oscillation.

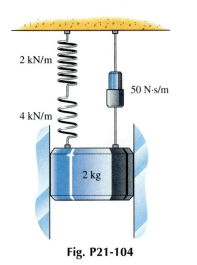

Fig. P21-104

21-105* A 20-lb block slides on a frictionless horizontal surface as shown in Fig. P21-105. The two springs remain in tension at all times and the two pulleys are both small and frictionless. If a 1.5-lb weight rotating around an 18-in.-radius circle at a rate of $\Omega = 12$ rad/s is added to the 20-lb block, determine the amplitude of the resulting steady-state oscillation.

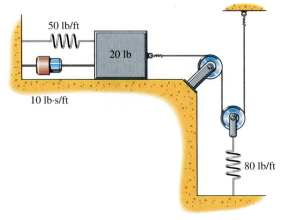

Fig. P21-105

21-106 A 4-kg mass is suspended in a vertical plane as shown in Fig. P21-106. The spring remains in tension at all times and the two pulleys are both small and frictionless. If a 0.8-kg mass rotating around a 400-mm-radius circle at a rate of $\Omega = 25$ rad/s is added to the 4-kg mass, determine the amplitude of the resulting steady-state oscillation.

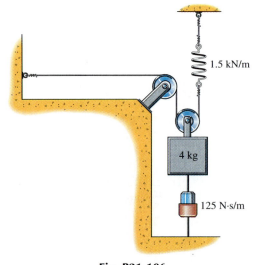

Fig. P21-106

Computer Problems

C21-107 When the forcing frequency Ω of a forced oscillation is close to the natural frequency ω_n of the system, the amplitude of the oscillation varies sinusoidally at a frequency $|\Omega - \omega_n|$. This phenomenon is known as *beating*.

Consider the 25-lb block that slides on a smooth horizontal surface as shown in Fig. P21-107. At time $t = 0$ the position and velocity of the block are $x_0 = 0$ ft and $v_0 = 8$ ft/s, respectively. If $k = 40$ lb/ft, $P_0 = 10$ lb, and $\Omega = 8$ rad/s, calculate and plot the position x of the block as a function of time t ($0 < t < 25$ s). (Try some other values of k; say, $k = 48$ lb/ft or $k = 50$ lb/ft.)

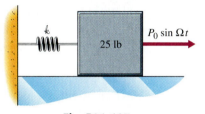

Fig. P21-107

C12-108 A 50-g coin sits on top of a 2-kg piston as shown in Fig. P21-108. The bottom end of the spring is oscillated according to $\delta = \delta_0 \sin \Omega t$. At time $t = 0$ the position and velocity of the piston are both zero ($x_0 = 0$ m and $v_0 = 0$ m/s). If $k = 205$ N/m, $\delta_0 = 20$ mm, and $\Omega = 8$ rad/s,

a. Calculate and plot the force F_s that must be exerted on the lower end of the spring to produce the motion as a function of time t ($0 < t < 10$ s).
b. Calculate and plot the force F_c exerted on the coin by the piston as a function of t ($0 < t < 10$ s).
c. Determine the maximum value of δ_0 for which the coin stays in contact with the piston; that is, for which $F_c > 0$ always.

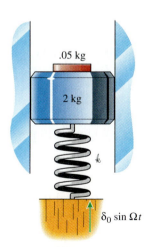

Fig. P21-108

C21-109 A 25-lb block slides on a smooth horizontal surface as shown in Fig. P21-109. At time $t = 0$, the position and velocity of the block are $x_0 = 6$ in. and $v_0 = 0$ in./s, respectively. If $k = 40$ lb/ft, $c = 1$ lb $\cdot$ s/ft, $\Omega = 8$ rad/s, and $P_0 = 10$ lb, calculate and plot

a. The position x of the block as a function of time t ($0 < t < 10$ s). On the same graph, plot the steady-state part of the solution.
b. The velocity v of the block as a function of x ($0 < t < 5$ s). On the same graph, plot the steady-state part of the solution.

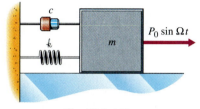

Fig. P21-109

C21-110 A 5-kg block slides on a smooth horizontal surface as shown in Fig. P21-109. At time $t = 0$, the position and velocity of the block are $x_0 = 25$ mm and $v_0 = 0$ mm/s, respectively. If $k = 125$ N/m, $c = 5$ N $\cdot$ s/m, $\Omega = 8$ rad/s, and $P_0 = 1000$ N, calculate and plot

a. The ratio F/P_0 (where F is the force exerted by the system on the wall) as a function of time t ($0 < t < 10$ s).
b. The ratio $(F/P_0)_{max}$ for the steady-state part of the solution as a function of Ω/ω_n ($0.1 < \Omega/\omega_n < 3$) for $c = 5$ N $\cdot$ s/m, 15 N $\cdot$ s/m, . . . , 50 N $\cdot$ s/m.

21-5 ENERGY METHODS

The approach taken in the first few sections of this chapter has been to get the differential equations of motion by applying Newton's second law of motion to the appropriate free-body diagram(s). These differential equations were then solved to get the frequency, period, and amplitude of vibration, as well as the equations of position, velocity, and acceleration of the system. In this direct approach all forces (including internal connection forces and friction or viscous damping forces) must be included in the analysis.

If no friction or viscous damping forces act on the system, however, the work–energy principle as described in Chapters 17 and 18 may be a simpler approach. When the only forces acting on the system are conservative (as in the undamped free vibration of particles and rigid bodies), the work–energy principle reduces to the *conservation of energy: the total mechanical energy of the system is constant*

$$T + V = \text{constant} \qquad (a)$$

The conservation of energy principle can be manipulated to give both the differential equation of motion and the natural frequency of the vibration.

Although all real systems lose energy to friction, the damping in many real systems is light. Only very small errors in determining the systems' natural frequency and period result from approximating these systems as undamped. The work–energy approach is particularly suited to problems involving systems of particles connected by rigid links and systems of connected rigid bodies. Using the work–energy method, the system need not be taken apart and the motion of the individual pieces need not be considered separately.

21-5.1 Differential Equation of Motion by Energy Methods

Consider the block of Fig. 21-37, which slides on a frictionless, horizontal surface. When the block is displaced a distance x in the positive coordinate direction, the kinetic energy of motion is $T = \frac{1}{2}mv^2 = \frac{1}{2}m\dot{x}^2$ and the potential energy of the elastic spring force will be $V = \frac{1}{2}kx^2$. Differentiating the conservation of energy principle with respect to time gives

$$\frac{d}{dt}(T + V) = \frac{d}{dt}\left(\frac{1}{2}kx^2 + \frac{1}{2}m\dot{x}^2\right) = (kx + m\ddot{x})\dot{x} = 0 \qquad (b)$$

Since the velocity $\dot{x}$ is not zero at every instant of time, Eq. b gives the differential equation of motion

$$m\ddot{x} + kx = 0 \qquad (c)$$

which is the same as Eq. 21-2a. Then the natural frequency ω_n, period τ_n, and so on all follow from the differential equation as in Section 21-2.

21-5.2 Frequency of Vibration by Energy Methods

The natural frequency and period of vibration can also be determined using the conservation of energy principle without first deriving the

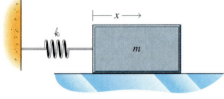

(a)

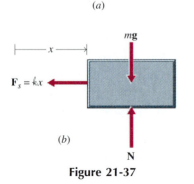

(b)

Figure 21-37

differential equation of motion. In Section 21-2 it was seen that for a system that vibrates with simple harmonic motion about its equilibrium position (where $x = 0$), the position and velocity of the system can be written (Eqs. 21-4 and 5)

$$x(t) = A \sin (\omega_n t - \phi_s)$$
$$v(t) = \dot{x}(t) = A\omega_n \cos (\omega_n t - \phi_s)$$

But it is noted from these two expressions that the position is a maximum ($x_{max} = A$) when the velocity is zero. That is, the potential energy is a maximum when the kinetic energy is zero. Likewise, the maximum velocity ($v_{max} = A\omega_n = \omega_n x_{max}$) occurs where the position is zero, and hence the kinetic energy is a maximum when the potential energy is zero. Therefore, the total mechanical energy of the system is

$$T + V = T_{max} + 0 = 0 + V_{max} = \frac{1}{2}m\dot{x}_{max}^2 = \frac{1}{2}kx_{max}^2 \qquad (d)$$

$$\frac{1}{2}m(\omega_n A)^2 = \frac{1}{2}kA^2 \qquad (e)$$

Solving for the natural angular frequency ω_n gives

$$\omega_n = \sqrt{k/m} \qquad (f)$$

which is again the same as Eq. 21-2b.

EXAMPLE PROBLEM 21-9

A 10-lb cart is attached to three springs and rolls on an inclined surface as shown in Fig. 21-38. The elastic moduli of the springs are $k_1 = k_2 = 5$ lb/ft and $k_3 = 15$ lb/ft. Determine the differential equation of motion for the cart using the energy method.

SOLUTION

The free-body diagram of the cart (Fig. 21-39) shows that four of the five forces acting on the cart are conservative and the fifth does no work. Therefore, the differential equation can be obtained using conservation of energy.

Before the motion is started, the cart is in its static equilibrium position, and equilibrium ($\Sigma F_x = 0$) gives that

$$k_1 \delta_{eq1} + k_2 \delta_{eq2} - k_3 \delta_{eq3} - mg \sin 15° = 0 \qquad (a)$$

where δ_{eq1}, δ_{eq2}, and δ_{eq3} are the deformation of the springs in the static equilibrium position ($x = 0$). Although the individual values of δ_{eq1}, δ_{eq2}, δ_{eq3} cannot be determined, Eq. a relates them to the weight of the cart.

For the arbitrary position shown in Fig. 21-39, the kinetic energy of the cart is

$$T = \frac{1}{2}mv^2 = \frac{1}{2}m\dot{x}^2$$

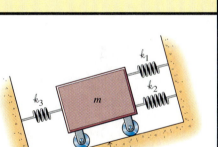

Fig. 21-38

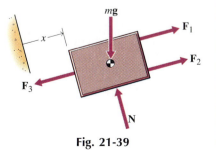

Fig. 21-39

When the cart moves to the right, its center of gravity rises, so the gravitational potential energy of the cart is

$$V_g = mgx \sin 15°$$

Also when the cart moves to the right, the stretches in springs 1 and 2 are decreased and the stretch in spring 3 is increased. Therefore, the elastic potential energies of the three spring forces are

$$V_{e1} = \frac{1}{2}k_1(\delta_{eq1} - x)^2 \qquad V_{e2} = \frac{1}{2}k_2(\delta_{eq2} - x)^2$$

$$V_{e3} = \frac{1}{2}k_3(\delta_{eq3} + x)^2$$

and the conservation of energy equation becomes

$$T + V = \frac{1}{2}m\dot{x}^2 + mgx \sin 15° + \left[\frac{1}{2}k_1(\delta_{eq1} - x)^2\right]$$

$$+ \left[\frac{1}{2}k_2(\delta_{eq2} - x)^2\right] + \left[\frac{1}{2}k_3(\delta_{eq3} + x)^2\right]$$

$$= \text{constant} \qquad (b)$$

Taking the time derivative of Eq. b gives

$$\frac{d}{dt}(T + V) = [m\ddot{x} + mg \sin 15° - k_1(\delta_{eq1} - x)$$

$$- k_2(\delta_{eq2} - x) + k_3(\delta_{eq3} + x)]\dot{x} = 0 \qquad (c)$$

Since the velocity of the cart $\dot{x}$ is not always zero, the term inside the square brackets must be zero. Finally, adding Eqs. a and c gives the differential equation of motion of the cart

$$m\ddot{x} + (k_1 + k_2 + k_3)x = 0 \qquad \text{Ans.}$$

which is exactly the same as derived in Example Problem 21-1.

> Although the differential equation has been obtained using energy methods, it has involved considerably more effort than was needed in Example Problem 21-1. For a single particle moving in rectilinear motion, the force-mass-acceleration (Newton's second law) approach is generally more direct. Energy methods generally have an advantage for systems of connected particles, for single particles moving in curvilinear motion, and for rigid bodies. For these types of problems, using energy methods eliminates the need to combine several equations to eliminate unknown forces from the differential equation of motion.

EXAMPLE PROBLEM 21-10

A 5-kg block slides on a horizontal frictionless surface as shown in Fig. 21-40. The elastic moduli of the springs are $k_1 = 600$ N/m and $k_2 = 300$ N/m. Assume that the cords remain in tension and determine the natural frequency of the free, undamped vibration of the block using the energy method.

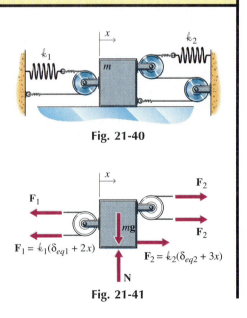

Fig. 21-40

Fig. 21-41

SOLUTION

The free-body diagram of the block (Fig. 21-41) shows that two of the forces (mg and $\mathbf{N}$) do no work and the other forces are due to the springs and are conservative. Therefore, the block oscillates with simple harmonic motion, and the natural frequency of the vibration can be obtained using conservation of energy.

Before the motion is started, the block is in its static equilibrium position, and equilibrium ($\Sigma F_x = 0$) gives that

$$3k_2\delta_{eq2} - 2k_1\delta_{eq1} = 0 \qquad (a)$$

where δ_{eq1} and δ_{eq2} are the deformation of the springs in the static equilibrium position ($x = 0$). When the block has moved to the right an amount x, the stretch in spring 1 is increased by $2x$ and the stretch in spring 2 is decreased by $3x$. Therefore, the elastic potential energy of the spring forces is

$$V_e = \frac{1}{2}k_1(\delta_{eq1} + 2x)^2 - \frac{1}{2}k_1\delta_{eq1}^2 + \frac{1}{2}k_2(\delta_{eq2} - 3x)^2 - \frac{1}{2}k_2\delta_{eq2}^2 \qquad (b)$$

where the constants have been subtracted so that the zero of potential energy is at the equilibrium position. Expanding Eq. b and simplifying using Eq. a gives

$$V_e = \frac{1}{2}(4k_1x^2 + 4k_1x\delta_{eq1} + 9k_2x^2 - 6k_2x\delta_{eq2}) = \frac{1}{2}(4k_1 + 9k_2)x^2$$

The kinetic energy of the block is just

$$T = \frac{1}{2}mv^2 = \frac{1}{2}m\dot{x}^2$$

For a body oscillating in simple harmonic motion, however, the position and velocity can be written

$$x = A \sin(\omega_n t - \phi_s)$$
$$\dot{x} = A\omega_n \cos(\omega_n t - \phi_s)$$

Therefore, when the position is zero ($x = 0$), the potential energy is zero ($V = 0$), the velocity is a maximum ($\dot{x} = \dot{x}_{max} = A\omega_n$), and the kinetic energy is also a maximum ($T = T_{max} = \frac{1}{2}mA^2\omega_n^2$). On the other hand, when the position is a maximum ($x = x_{max} = A$), the potential energy is also a maximum [$V = V_{max} = \frac{1}{2}(4k_1 + 9k_2)A^2$], the velocity is zero ($\dot{x} = 0$), and the kinetic energy is also zero ($T = 0$). Writing the conservation of energy equation between these two positions ($T_{max} + 0 = 0 + V_{max}$) gives

$$\frac{1}{2}mA^2\omega_n^2 = \frac{1}{2}(4k_1 + 9k_2)A^2 \qquad (c)$$

Finally, solving Eq. c for the natural circular frequency gives

$$\omega_n = \sqrt{\frac{4k_1 + 9k_2}{m}} = 39.5 \text{ rad/s} \qquad \text{Ans.}$$

EXAMPLE PROBLEM 21-11

The two blocks shown in Fig. 21-42 slide on horizontal, frictionless surfaces. The connecting links have negligible weight, and ABC is vertical in the equilibrium position. Assume small oscillations and use the energy method to determine

a. The differential equation of motion of the 15-lb block.
b. The natural frequency of the oscillation.

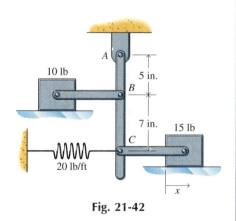

Fig. 21-42

SOLUTION

a. The free-body diagram of the assembly is shown in Fig. 21-43. Since the connecting links are all rigid, the work done by the forces at the connections need not be considered. Then the only force that does work is the spring force, and it is conservative. Therefore, the energy method may be used to determine the differential equation of motion and the natural frequency of vibration.

When the 15-lb block moves a distance x to the right, the 10-lb block moves a distance $5x/12$ to the right. Therefore, the kinetic energy of the system is

$$T = \frac{1}{2}\frac{15}{32.2}\dot{x}^2 + \frac{1}{2}\frac{10}{32.2}\left(\frac{5}{12}\dot{x}\right)^2 = 0.2599\dot{x}^2$$

The elastic potential energy of the spring force is

$$V = \frac{1}{2}20x^2$$

Taking the time derivative of the total mechanical energy of the system ($T + V$ = constant) gives

$$\frac{d}{dt}(T + V) = (0.5198\ddot{x} + 20x)\dot{x} = 0$$

Since the velocity $\dot{x}$ is not zero at every instant of time, the quantity inside the parentheses must be zero, which gives the differential equation of motion of the 15-lb block

$$0.5198\ddot{x} + 20x = 0 \qquad\qquad \text{Ans.}$$

b. Once the differential equation of motion is determined, the natural frequency of vibration is just

$$\omega_n = \sqrt{20/0.5198} = 38.5 \text{ rad/s} \qquad\qquad \text{Ans.}$$

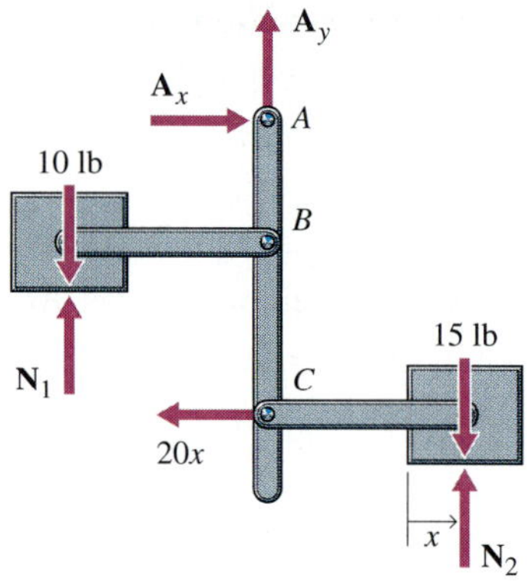

Fig. 21-43

Solving this problem using the force-mass-acceleration (Newton's second law) approach would require separate free-body diagrams of the two blocks and the vertical link ABC. The internal link forces would be eliminated by combining the equations of motion of the various pieces. Since the connecting links are rigid, however, the internal forces drop out of the work-energy equation of the system, and the differential equation of motion is obtained more directly.

PROBLEMS

Introductory Problems

21-111–21-116 For each of the following problems, determine the differential equation of motion using the energy method.

21-111 Problem 21-25

21-112* Problem 21-26

21-113 Problem 21-27

21-114* Problem 21-28

21-115* Problem 21-29

21-116 Problem 21-30

21-117–21-122 For each of the following problems, determine the natural frequency of vibration ω_n using the energy method.

21-117 Problem 21-25

21-118* Problem 21-26

21-119 Problem 21-27

21-120* Problem 21-28

21-121* Problem 21-29

21-122 Problem 21-30

SUMMARY

A mechanical vibration is the repeated oscillation of a particle or rigid body about an equilibrium position. In many devices vibratory motions are desirable and are deliberately generated. The task of the engineer in such problems is to create and to control the vibrations. However, most vibrations in rotating machinery and in structures are undesirable. In these cases the task of the engineer is to eliminate the vibrations (or at least reduce the effect of the vibrations as much as possible) by appropriate design.

The study of vibrations is a direct application of the principles developed earlier. In the earlier chapters, the acceleration was usually obtained only for a particular position of the body and at a particular instant of time. In this chapter, the acceleration was obtained for an arbitrary position of the body and then integrated using principles of ordinary differential equations to get the velocity and position for all future times.

An undamped free vibration will repeat itself indefinitely. Once set in motion, such idealized systems vibrate forever with a constant amplitude. Of course, all real systems contain frictional forces that will eventually stop a free vibration. In many systems, however, the energy loss due to air resistance, the internal friction of springs, or other friction forces are small enough that an analysis based on negligible damping often gives quite satisfactory engineering results. In particular, the frequency and period of vibration obtained for a freely vibrating system are very close to the values obtained for a system that has a small amount of damping.

A forced vibration is produced and maintained by an externally applied periodic force that does not depend on the position or motion of the body. A damped, forced vibration is maintained as long as the periodic force that produces the vibration is applied.

When a periodic force is applied to a body, the body will begin to oscillate with a combination of free and forced vibrations. Because some friction is present in all real systems, however, the free vibration part of the motion will always decay away. Therefore, the free vibration part of the motion is called the *transient* motion. The frequency of steady-state forced vibrations is the same as that of the applied disturbing force and is independent of the natural frequency and other characteristics of the vibrating body. The amplitude of steady-state forced vibrations, however, does depend on the natural frequency of the system as well as on the frequency of the applied load.

When the disturbing force is applied at a frequency close to the natural frequency of the system and the damping is light, the amplitude of the vibration is magnified substantially. This condition is called *resonance*. The amplitude of an oscillation may be controlled either by avoiding the condition of resonance or (if resonance cannot be avoided) by increasing the damping ζ.

In all cases, the constants m, c, k, and P_0 appearing in the solutions are to be interpreted as the coefficients in the differential equation of motion rather than the actual system mass, the actual damping coefficient, and so on.

REVIEW PROBLEMS

21-135* A 60-lb child bounces up and down on a pair of elastic bands as shown in Fig. P21-135. If the amplitude of the oscillation is observed to decrease by 3 percent every 5 cycles and the 5 cycles take 6.5 s, determine the elastic modulus k and damping coefficient c of the elastic bands.

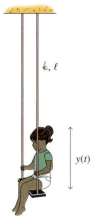

Fig. P21-135

21-136* The pendulum shown in Fig. P21-136 consists of a 5-kg mass on the end of a 0.9-m-long lightweight stick. The other end of the stick oscillates along a horizontal guide. Assume small oscillations and determine

a. The differential equation of motion for θ, the angular position of the pendulum.
b. The amplitude of the steady-state oscillation.

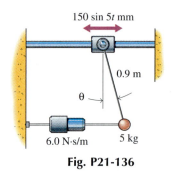

Fig. P21-136

21-137 A small particle slides along the bottom of a 10-in.-radius circular bowl. Neglect friction, and assume small oscillations. If the particle has a speed of 15 in./s when it is at the bottom of the bowl, determine

a. The differential equation governing the motion.
b. The period and amplitude of the resulting vibration.
c. The position of the particle as a function of time.

21-138* A 10-kg mass slides on a frictionless horizontal surface as shown in Fig. P21-138. At $t = 0$ the mass passes through its equilibrium position with a speed of 2.5 m/s to the right. If $k = 1.2$ kN/m and $c = 180$ N · s/m, determine

a. The force F_k exerted on the mass by the spring when it reaches its maximum extension.
b. The force F_c exerted on the mass by the dashpot when the mass returns to its equilibrium position.

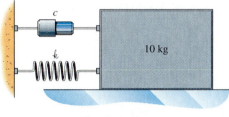

Fig. P21-138

21-139 The particular solution Eq. 21-24 does not satisfy the differential equation of motion Eq. 21-22 when the frequency of the applied force is exactly the same as the system natural frequency.

a. Show that the particular solution has the form

$$x_p(t) = Dt \sin\left(\Omega t - \frac{\pi}{2}\right)$$

when $\Omega = \omega_n$ and $c = 0$.
b. Determine the value of D in terms of the system parameters m, k, ω_n, and P_0.

21-140 A 10-kg mass slides on a frictionless horizontal surface as shown in Fig. P21-140. If $k = 800$ N/m, $c = 30$ N · s/m, $\Omega = 1.5$ Hz, and $P_0 = 80$ N, determine

a. The amplitude of the steady-state oscillation.
b. The dynamic magnification factor.
c. The magnitude of the total force transmitted to the wall F_w.
d. The transmissibility—the ratio of F_w and P_0 (the amplitude of the applied force).

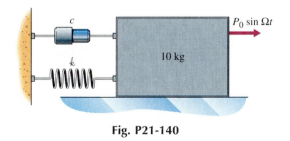

Fig. P21-140

21-141* A 2-lb piston is initially at rest on two identical springs ($k = 6$ lb/ft each) when a 0.5-lb ball of putty is dropped onto it (Fig. P21-141). If the collision is perfectly plastic ($e = 0$) and $h = 16$ ft, determine

a. The differential equation governing the motion of the piston.
b. The period and amplitude of the resulting vibration.
c. The force exerted on the putty by the piston when the springs are at maximum compression.
d. The force exerted on the putty by the piston when the system passes through equilibrium on the way up.
e. The maximum height h_{max} from which the putty could be dropped and not lose contact with the piston during the subsequent oscillation.

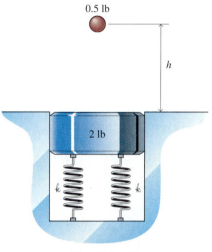

Fig. P21-141

21-142 A 4-kg mass hangs from an elastic band as shown in Fig. P21-142. The unstretched length of the band is 1.5 m, and the equilibrium length is 2.0 m. If the elastic band is to remain taut when the upper support is oscillated according to $\delta = a \sin \Omega t$, determine

a. The maximum amplitude a_{max} when $\Omega = 4$ rad/s.
b. The allowed range of frequencies Ω when $a = 0.7$ m.

Fig. P21-142

21-143* A 1-in-diameter marble ($W = 1$ oz) rolls without slipping in the bottom of a 12-in.-radius circular bowl. Assume small oscillations. If the marble has a speed of 15 in./s when it is at the bottom of the bowl, determine

a. The differential equation governing the motion.
b. The frequency and amplitude of the resulting vibration.
c. The position of the marble as a function of time.

21-144* When the system shown in Fig. P21-144 is in equilibrium, spring 1 ($k_1 = 1.2$ kN/m) is stretched 50 mm and spring 2 ($k_2 = 1.8$ kN/m) is stretched 90 mm. If the mass m is pulled down a distance δ and released from rest, determine

a. The differential equation governing the motion.
b. The maximum distance δ_{max} so that all cord remain in tension.
c. The frequency and amplitude of the resulting vibration.
d. The position of the mass as a function of time.

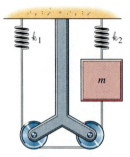

Fig. P21-144

21-145 A 0.1-lb coin sits on top of a 2-lb piston as shown in Fig. P21-145. If the bottom end of the spring is oscillated according to $\delta = a \sin \Omega t$, where $a = 1.5$ in. and $\Omega = 2\pi$ rad/s, determine

a. The differential equation governing the motion of the piston.
b. The amplitude of the resulting vibration.
c. The force exerted on the coin by the piston when the spring is at maximum compression.
d. The force exerted on the coin by the piston when the spring is at maximum extension.

e. The maximum amplitude a_{max} when $\Omega = 10$ rad/s if the coin is to remain in contact with the piston.
f. The allowed range of frequencies Ω when $a = 2$ in. if the coin is to remain in contact with the piston.

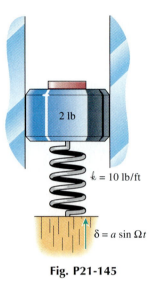

Fig. P21-145

21-146 A 6-kg mass is suspended from a cord, which is wrapped around the outside of a 10-kg cylinder 600 mm in diameter (Fig. P21-146). The system is in equilibrium with point A 200 mm directly above the frictionless axle. If the mass is pulled down 50 mm and the system released from rest, determine

a. The differential equation governing the vertical motion of the mass.
b. The frequency and amplitude of the resulting vibration.
c. The position of the mass as a function of time.

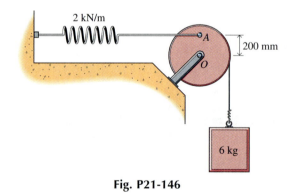

Fig. P21-146

MOMENTS AND PRODUCTS OF INERTIA

A-1 MOMENTS OF INERTIA

In analyses of the motion of rigid bodies, expressions are often encountered that involve the product of the mass of a small element of the body and the square of its distance from a line of interest. This product is called the second moment of the mass of the element or more frequently the *moment of inertia* of the element. Thus, the moment of inertia dI of an element of mass dm about the axis OO shown in Fig. A-1 is defined as

$$dI = r^2 \, dm$$

The moment of inertia of the entire body about axis OO is defined as

$$I = \int_m r^2 \, dm \qquad \text{(A-1)}$$

Since both the mass of the element and the distance squared from the axis to the element are always positive, the moment of inertia of a mass is always a positive quantity.

Moments of inertia have the dimensions of mass multiplied by length squared, ML^2. Common units for the measurement of moment of inertia in the SI system are $\text{kg} \cdot \text{m}^2$. In the U.S. Customary system, force, length, and time are selected as the fundamental quantities, and mass has the dimensions FT^2L^{-1}. Therefore, moment of inertia has the units $\text{lb} \cdot \text{s}^2 \cdot \text{ft}$. If the mass of the body W/g is expressed in slugs $(\text{lb} \cdot \text{s}^2/\text{ft})$, the units for measurement of moment of inertia in the U.S. Customary system are $\text{slug} \cdot \text{ft}^2$.

The moments of inertia of a body with respect to an xyz-coordinate system can be determined by considering an element of mass as shown in Fig. A-2. From the definition of moment of inertia,

$$dI_x = r_x^2 \, dm = (y^2 + z^2) \, dm$$

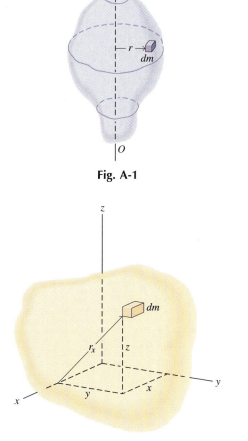

Fig. A-1

Fig. A-2

599

Similar expressions can be written for the y- and z-axes. Thus,

$$I_x = \int_m r_x^2 \, dm = \int_m (y^2 + z^2) \, dm$$

$$I_y = \int_m r_y^2 \, dm = \int_m (z^2 + x^2) \, dm \qquad \text{(A-2)}$$

$$I_z = \int_m r_z^2 \, dm = \int_m (x^2 + y^2) \, dm$$

A-1.1   Radius of Gyration

The definition of moment of inertia (Eq. A-1) indicates that the dimensions of moment of inertia are mass multiplied by a length squared. As a result, the moment of inertia of a body can be expressed as the product of the mass m of the body and a length k squared. This length k is defined as the *radius of gyration* of the body. Thus, the moment of inertia I of a body with respect to a given line can be expressed as

$$I = mk^2 \qquad \text{or} \qquad k = \sqrt{\frac{I}{m}} \qquad \text{(A-3)}$$

The radius of gyration of the mass of a body with respect to any axis can be viewed as the distance from the axis to the point where the total mass must be concentrated to produce the same moment of inertia with respect to the axis as does the actual (or distributed) mass.

The radius of gyration for masses is very similar to the radius of gyration for areas discussed in Section 10-2-3. The radius of gyration for masses is not the distance from the given axis to any fixed point in the body such as the mass-center. The radius of gyration of the mass of a body with respect to any axis is always greater than the distance from the axis to the mass center of the body. There is no useful physical interpretation for a radius of gyration; it is merely a convenient means of expressing the moment of inertia of the mass of a body in terms of its mass and a length.

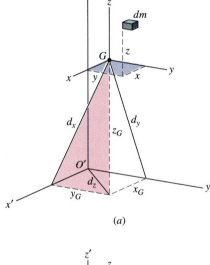

(a)

A-1.2   Parallel-Axis Theorem for Moments of Inertia

The parallel-axis theorem for moments of inertia is very similar to the parallel-axis theorem for second moments of area discussed in Section 10-2-1. Consider the body shown in Fig. A-3a, which has an xyz-coordinate system with its origin at the mass-center G of the body and a parallel $x'y'z'$-coordinate system with its origin at point O'. Observe in Fig. A-3b that

$$x' = x_G + x$$
$$y' = y_G + y$$
$$z' = z_G + z$$

The distance d_x (see Fig. A-3a) between the x'- and x-axes is

$$d_x = \sqrt{y_G^2 + z_G^2}$$

The moment of inertia of the body about an x'-axis that is parallel to

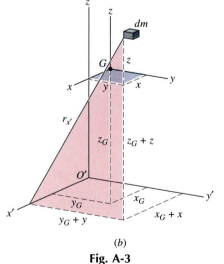

(b)

Fig. A-3

the x-axis through the mass center (see Fig. A-3b) is by definition

$$I_{x'} = \int_m r_{x'}^2 \, dm$$

$$= \int_m [(y')^2 + (z')^2] \, dm$$

$$= \int_m [(y_G + y)^2 + (z_G + z)^2] \, dm$$

$$= \int_m (y^2 + z^2) \, dm + y_G^2 \int_m dm + 2y_G \int_m y \, dm + z_G^2 \int_m dm + 2z_G \int_m z \, dm$$

However,

$$\int_m (y^2 + z^2) \, dm = I_{xG}$$

and, since the x- and y-axes pass through the mass center G of the body,

$$\int_m y \, dm = 0 \qquad \int_m z \, dm = 0$$

Therefore,

$$I_{x'} = I_{xG} + (y_G^2 + z_G^2)m = I_{xG} + d_x^2 m$$
$$I_{y'} = I_{yG} + (z_G^2 + x_G^2)m = I_{yG} + d_y^2 m \qquad \text{(A-4)}$$
$$I_{z'} = I_{zG} + (x_G^2 + y_G^2)m = I_{zG} + d_z^2 m$$

Equation A-4 is the parallel-axis theorem for moments of inertia. The subscript G indicates that the x-axis passes through the mass-center G of the body. Thus, if the moment of inertia of a body with respect to an axis passing through its mass-center is known, the moment of inertia of the body with respect to any parallel axis can be found, without integrating, by use of Eqs. A-4.

A similar relationship exists between the radii of gyration for the two axes. Thus, if the radii of gyration for the two parallel axes are denoted by k_x and $k_{x'}$, the foregoing equation may be written

$$k_{x'}^2 \, m = k_{xG}^2 \, m + d_x^2 m$$

Hence

$$k_{x'}^2 = k_{xG}^2 + d_x^2$$
$$k_{y'}^2 = k_{yG}^2 + d_y^2 \qquad \text{(A-5)}$$
$$k_{z'}^2 = k_{zG}^2 + d_z^2$$

Note: Equations A-4 and A-5 are valid only for transfers to or from xyz-axes passing through the mass center of the body. They are not valid for two arbitrary axes.

A-1.3   Moments of Inertia by Integration

When integration methods are used to determine the moment of inertia of a body with respect to an axis, the mass of the body can be divided into elements in various ways. Depending on how the element is chosen, single, double, or triple integration may be required. The geometry of the body usually determines whether Cartesian or polar coordinates are used. In any case, the elements of mass should always be selected, so that

1. All parts of the element are the same distance from the axis with respect to which the moment of inertia is to be determined.

2. If condition 1 is not satisfied, the element should be selected so that the moment of inertia of the element with respect to the axis about which the moment of inertia of the body is to be found is known. The moment of inertia of the body can then be found by summing the moments of inertia of the elements.

3. If the location of the mass-center of the element is known and the moment of inertia of the element with respect to an axis through its mass-center and parallel to the given axis is known, the moment of inertia of the element can be determined by using the parallel-axis theorem. The moment of inertia of the body can then be found by summing the moments of inertia of the elements.

When triple integration is used, the element always satisfies the first requirement, but this condition is not necessarily satisfied by elements used for single or double integration.

In some instances, a body can be regarded as a system of particles. The moment of inertia of a system of particles with respect to a line of interest is the sum of the moments of inertia of the particles with respect to the given line. Thus, if the masses of the particles of a system are denoted by $m_1, m_2, m_3, \ldots, m_n$, and the distances of the particles from a given line are denoted by $r_1, r_2, r_3, \ldots, r_n$, the moment of inertia of the system can be expressed as

$$I = \sum mr^2 = m_1 r_1^2 + m_2 r_2^2 + m_3 r_3^2 + \cdots + m_n r_n^2$$

Moments of inertia for thin plates are relatively easy to determine. For example, consider the thin plate shown in Fig. A-4. The plate has a uniform density ρ, a uniform thickness t, and a cross-sectional area A. The moments of inertia about the x-, y-, and z-axes are by definition

$$I_{xm} = \int_m y^2 \, dm = \int_V y^2 \rho \, dV = \int_A y^2 \rho t \, dA = \rho t \int_A y^2 \, dA = \rho t I_{xA}$$

$$I_{ym} = \int_m x^2 \, dm = \int_V x^2 \rho \, dV = \int_A x^2 \rho t \, dA = \rho t \int_A x^2 \, dA = \rho t I_{yA} \qquad \text{(A-6)}$$

$$I_{zm} = \int_m (x^2 + y^2) \, dm = \rho t I_{yA} + \rho t I_{xA} = \rho t (I_{yA} + I_{xA})$$

where the subscripts m and A denote moments of inertia and second moments of area, respectively. Since the equations for the moments of inertia of thin plates contain the expressions for the second moments of area, the results listed in Appendix B (Table B-3) for second moments of areas can be used for moments of inertia by simply multiplying the results listed in the table by ρt.

For the general three-dimensional body, moments of inertia with respect to x-, y-, and z-axes are

$$I_x = \int_m r_x^2 \, dm = \int_m (y^2 + z^2) \, dm$$

$$I_y = \int_m r_y^2 \, dm = \int_m (z^2 + x^2) \, dm \qquad \text{(A-2)}$$

$$I_z = \int_m r_z^2 \, dm = \int_m (x^2 + y^2) \, dm$$

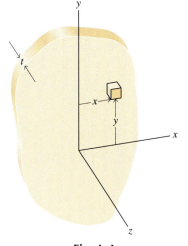

Fig. A-4

If the density of the body is uniform, the element of mass dm can be expressed in terms of the element of volume dV of the body as $dm = \rho\, dV$. Equations A-2 then become

$$I_x = \rho \int_V (y^2 + z^2)\, dV$$

$$I_y = \rho \int_V (z^2 + x^2)\, dV \qquad \text{(A-7)}$$

$$I_z = \rho \int_V (x^2 + y^2)\, dV$$

If the density of the body is not uniform, it must be expressed as a function of position and retained within the integral sign.

The specific element of volume to be used depends on the geometry of the body. For the general three-dimensional body, the differential element $dV = dx\, dy\, dz$, which requires a triple integration, is usually used. For bodies of revolution, circular plate elements, which require only a single integration, can be used. For some problems, cylinder elements and polar coordinates are useful. Procedures for determining moments of inertia are illustrated in the following example problems.

EXAMPLE PROBLEM A-1

Determine the moment of inertia of a homogeneous right circular cylinder with respect to the axis of the cylinder.

SOLUTION

The moment of inertia of the cylinder can be determined from the definition of moment of inertia (Eq. A-1) by selecting a cylindrical tube type of element as shown in Fig. A-5a. Thus,

$$dI_{zm} = r^2\, dm = r^2(\rho\, dV) = r^2\rho(2\pi rh\, dr) = 2\pi\rho hr^3\, dr$$

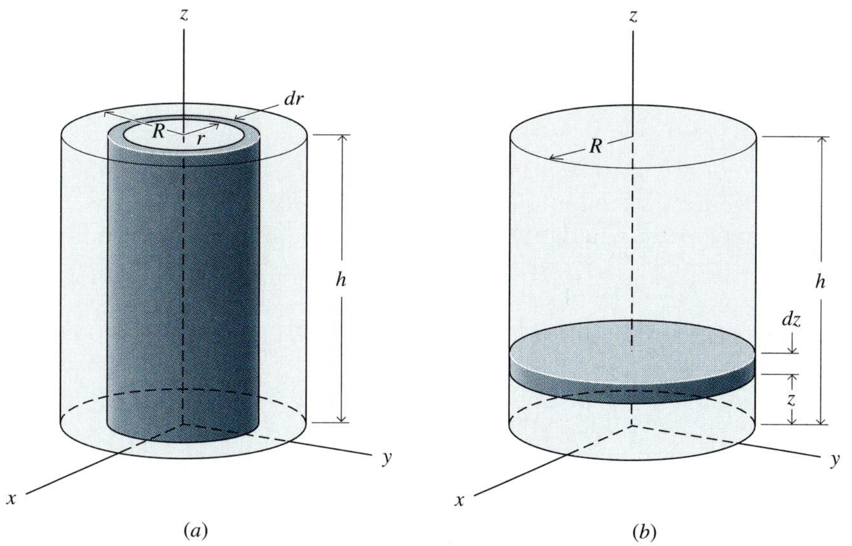

(a)   (b)

Fig. A-5

Therefore,

$$I_z = \int_m dI_{zm} = \int_0^R 2\pi\rho h r^3 \, dr = \left[\frac{\pi\rho h r^4}{2} \right]_0^R = \frac{1}{2}\pi\rho h R^4$$

Alternatively, a thin circular plate type of element, such as the one shown in Fig. A-5b, can be used. The moment of inertia for this type of element is given by Eq. A-6 as

$$dI_{zm} = \rho t(I_{yA} + I_{xA})$$

Substituting the second moments for a circular area from Table B-3 yields

$$dI_{zm} = \rho\left(\frac{\pi R^4}{4} + \frac{\pi R^4}{4} \right) dz = \frac{1}{2}\pi\rho R^4 \, dz$$

Therefore,

$$I_z = \int_m dI_{zm} = \int_0^h \frac{1}{2}\pi\rho R^4 \, dz = \left[\frac{1}{2}\pi\rho R^4 z \right]_0^h = \frac{1}{2}\pi\rho h R^4$$

The mass of the cylinder is

$$m = \rho V = \rho(\pi R^2 h) = \rho\pi R^2 h$$

Therefore,

$$I_z = \frac{1}{2}(\rho\pi R^2 h)R^2 = \frac{1}{2}mR^2 \qquad\qquad \text{Ans.}$$

EXAMPLE PROBLEM A-2

Determine the moment of inertia for the homogeneous rectangular prism shown in Fig. A-6a with respect to

a. Axis y through the mass-center of the prism.
b. Axis y' along an edge of the prism.
c. Axis x through the centroid of an end of the prism.

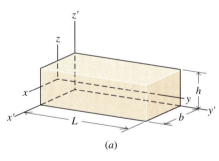

(a)

SOLUTION

a. A thin rectangular plate type of element, such as the one shown in Fig. A-6b, will be used. The moment of inertia for this type of element is given by Eq. A-6 as

$$dI_{ym} = \rho t(I_{zA} + I_{xA})$$

Substituting the second moments for a rectangular area from Table B-3 yields

$$dI_{ym} = \rho\left(\frac{hb^3}{12} + \frac{bh^3}{12} \right) dy = \rho\frac{bh}{12}(b^2 + h^2) \, dy$$

Therefore,

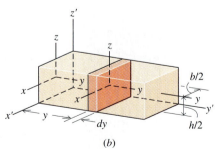

(b)

Fig. A-6

$$I_y = \int_m dI_{ym} = \int_0^L \rho\frac{bh}{12}(b^2 + h^2) \, dy$$

$$= \rho\frac{bh}{12}\left[(b^2 + h^2)\, y \right]_0^L = \frac{\rho bh L}{12}(b^2 + h^2)$$

The mass of the prism is

$$m = \rho V = \rho(bhL) = \rho bhL$$

Therefore,

$$I_y = \frac{\rho bhL}{12}\,(b^2 + h^2) = \frac{1}{12}\,m\,(b^2 + h^2) \qquad\qquad \text{Ans.}$$

b. The parallel-axis theorem (Eq. A-4) can be used to determine the moment of inertia about the y'-axis along an edge of the prism. Thus,

$$\begin{aligned}
I_{y'} &= I_{yG} + (z_G^2 + x_G^2)m \\
&= \frac{1}{12}\,m(b^2 + h^2) + \left(\frac{h^2}{4} + \frac{b^2}{4}\right) m \\
&= \frac{1}{3}\,m(b^2 + h^2) \qquad\qquad \text{Ans.}
\end{aligned}$$

c. The moment of inertia about an x-axis through the mass-center of the thin rectangular plate type of element shown in Fig. A-6b is given by Eq. A-6 as

$$dI_{zm} = \rho t\, I_{xA}$$

Substituting the second moment for a rectangular area from Table B-3 yields

$$dI_{xG} = \rho\, \frac{bh^3}{12}\, dy = \frac{\rho bh^3}{12}\, dy$$

The parallel-axis theorem (Eq. A-4) with $d_x = y$ then gives the moment of inertia for the thin rectangular plate element about the x-axis shown in Fig. A-6b as

$$dI_x = dI_{xG} + d_x^2 m = \frac{\rho bh^3}{12}\, dy + y^2(\rho bh\, dy) = \frac{\rho bh}{12}\,(h^2 + 12y^2)\, dy$$

$$\begin{aligned}
I_x &= \int_m dI_x = \int_0^L \frac{\rho bh}{12}\,(h^2 + 12y^2)\, dy \\
&= \frac{\rho bh}{12}\left[h^2 y + 4y^3\right]_0^L = \frac{\rho bhL}{12}\,(h^2 + 4L^2)
\end{aligned}$$

But

$$m = \rho bhL$$

Therefore,

$$I_x = \frac{\rho bhL}{12}\,(h^2 + 4L^2) = \frac{1}{12}\,m\,(h^2 + 4L^2) \qquad\qquad \text{Ans.}$$

A-1.4   Moment of Inertia of Composite Bodies

Frequently in engineering practice, a body of interest can be broken up into a number of simple shapes, such as cylinders, spheres, plates, and rods, for which the moments of inertia have been evaluated and tabulated. The moment of inertia of the *composite body* with respect to any axis is equal to the sum of the moments of inertia of the separate parts of the body with respect to the specified axis. For example,

$$I_x = \int_m (y^2 + z^2) \, dm$$

$$= \int_{m_1} (y^2 + z^2) \, dm_1 + \int_{m_2} (y^2 + z^2) \, dm_2 + \cdots + \int_{m_n} (y^2 + z^2) \, dm_n$$

$$= I_{x1} + I_{x2} + I_{x3} + \cdots + I_{xn}$$

When one of the component parts is a hole, its moment of inertia must be subtracted from the moment of inertia of the larger part to obtain the moment of inertia for the composite body. A listing of the moments of inertia for some frequently encountered shapes such as rods, plates, cylinders, spheres, and cones is presented in Appendix B (Table B-5).

EXAMPLE PROBLEM A-3

Determine the moment of inertia of the cast-iron flywheel shown in Fig. A-7 with respect to the axis of rotation of the flywheel. The specific weight of the cast iron is 460 lb/ft³.

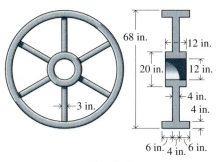

Fig. A-7

SOLUTION

The rim and hub of the flywheel are hollow cylinders, and the spokes are rectangular prisms. The density of the cast iron is

$$\rho = \frac{w}{g} = \frac{460}{32.2} = 14.29 \text{ slugs/ft}^3$$

With all dimensions converted to feet, the moment of inertia of the rim is

$$I_R = \frac{1}{2} m_o R_o^2 - \frac{1}{2} m_i R_i^2$$

$$= \frac{1}{2} \left[\pi \left(\frac{34}{12} \right)^2 \left(\frac{16}{12} \right) (14.29) \right] \left(\frac{34}{12} \right)^2 - \frac{1}{2} \left[\pi \left(\frac{30}{12} \right)^2 \left(\frac{16}{12} \right) (14.29) \right] \left(\frac{30}{12} \right)^2$$

$$= 1929 - 1169 = 760 \text{ slug} \cdot \text{ft}^2$$

The moment of inertia of the hub is

$$I_H = \frac{1}{2} m_o R_o^2 - \frac{1}{2} m_i R_i^2$$

$$= \frac{1}{2} \left[\pi \left(\frac{10}{12} \right)^2 \left(\frac{12}{12} \right) (14.29) \right] \left(\frac{10}{12} \right)^2 - \frac{1}{2} \left[\pi \left(\frac{6}{12} \right)^2 \left(\frac{12}{12} \right) (14.29) \right] \left(\frac{6}{12} \right)^2$$

$$= 10.82 - 1.40 = 9.42 \text{ slug} \cdot \text{ft}^2$$

The moment of inertia of each spoke is

$$I_S = I_G + d^2m$$
$$= \frac{1}{12}\left[\frac{3}{12}\left(\frac{4}{12}\right)\left(\frac{20}{12}\right)(14.29)\right]\left[\left(\frac{3}{12}\right)^2 + \left(\frac{20}{12}\right)^2\right]$$
$$+ \left(\frac{20}{12}\right)^2\left[\frac{3}{12}\left(\frac{4}{12}\right)\left(\frac{20}{12}\right)(14.29)\right]$$
$$= 0.4698 + 5.5131 = 5.9829 = 5.98 \text{ slug} \cdot \text{ft}^2$$

The total moment of inertia for the flywheel is

$$I = I_R + I_H + 6I_S$$
$$= 760 + 9.42 + 6(5.98) = 8.05 \text{ slug} \cdot \text{ft}^2 \qquad \text{Ans.}$$

A-2 PRODUCT OF INERTIA

In analyses of the motion of rigid bodies, expressions are sometimes encountered that involve the product of the mass of a small element and the coordinate distances from a pair of orthogonal coordinate planes. This product, which is similar to the mixed second moment of an area, is called the *product of inertia* of the element. For example, the product of inertia of the element shown in Fig. A-8 with respect to the *xz*- and *yz*-planes is by definition

$$dI_{xy} = xy\,dm \qquad (A-8)$$

The sum of the products of inertia of all elements of mass of the body with respect to the same orthogonal planes is defined as the product of inertia of the body. The three products of inertia for the body shown in Fig. A-8 are

$$I_{xy} = \int_m xy\,dm$$
$$I_{yz} = \int_m yz\,dm \qquad (A-9)$$
$$I_{zx} = \int_m zx\,dm$$

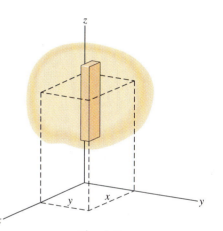

Fig. A-8

Products of inertia, like moments of inertia, have the dimensions of mass multiplied by a length squared, ML^2. Common units for the measurement of product of inertia in the SI system are $kg \cdot m^2$. In the U.S. Customary system, common units are $slug \cdot ft^2$.

The product of inertia of a body can be positive, negative, or zero, since the two coordinate distances have independent signs. The product of inertia will be positive for coordinates with the same sign and negative for coordinates with opposite signs. The product of inertia will be zero if either of the planes is a plane of symmetry, since pairs of elements on opposite sides of the plane of symmetry will have positive and negative products of inertia that will add to zero in the summation process.

The integration methods used to determine moments of inertia apply equally well to products of inertia. Depending on the way the element is chosen, single, double, or triple integration may be required. Moments of inertia for thin plates were related to second moments of area for the same plate. Likewise, products of inertia can be related to the mixed second moments for the plates. If the plate has a uniform density ρ, a uniform thickness t, and a cross-sectional area A, the products of inertia are by definition

$$I_{xym} = \int_m xy \, dm = \int_V xy\rho \, dV = \int_A xy\rho t \, dA = \rho t \int_A xy \, dA = \rho t I_{xyA}$$

$$I_{yzm} = \int_m yz \, dm = 0 \tag{A-10}$$

$$I_{zxm} = \int_m zx \, dm = 0$$

where the subscripts m and A denote products of inertia of mass and mixed second moments of area, respectively. The products of inertia I_{yzm} and I_{zxm} for a thin plate are zero because the x- and y-axes are assumed to lie in the midplane of the plate (plane of symmetry).

A parallel-axis theorem for products of inertia can be developed that is very similar to the parallel-axis theorem for mixed second moments of area discussed in Section 10-2-5. Consider the body shown in Fig. A-9, which has an xyz-coordinate system with its origin at the mass-center G of the body and a parallel $x'y'z'$-coordinate system with its origin at point O'. Observe in the figure that

$$x' = x_G + x$$
$$y' = y_G + y$$
$$z' = z_G + z$$

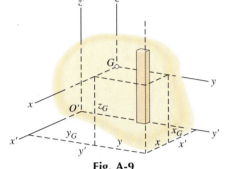

Fig. A-9

The product of inertia $I_{x'y'}$ of the body with respect to the $x'z'$- and $y'z'$-planes is by definition

$$I_{x'y'} = \int_m x'y' \, dm = \int_m (x_G + x)(y_G + y) \, dm$$

$$= \int_m x_G y_G \, dm + \int_m x_G y \, dm + \int_m y_G x \, dm + \int_m xy \, dm$$

Since x_G and y_G are the same for every element of mass dm,

$$I_{x'y'} = x_G y_G \int_m dm + x_G \int_m y \, dm + y_G \int_m x \, dm + \int_m xy \, dm$$

However,

$$\int_m xy \, dm = I_{xy}$$

and, since the x- and y-axes pass through the mass center G of the body,

$$\int_m y \, dm = 0 \qquad \int_m z \, dm = 0$$

Therefore,

$$I_{x'y'} = I_{xyG} + x_G y_G m$$
$$I_{y'z'} = I_{yzG} + y_G z_G m \tag{A-11}$$
$$I_{z'x'} = I_{zxG} + z_G x_G m$$

Equations A-11 are the parallel-axis theorem for products of inertia. The subscript G indicates that the x- and y-axes pass through the mass-center G of the body. Thus, if the product of inertia of a body with respect to a pair of orthogonal planes that pass through its mass-center is known, the product of inertia of the body with respect to any other pair of parallel planes can be found, without integrating, by use of Eqs. A-11.

Procedures for determining products of inertia are illustrated in the following example problems.

EXAMPLE PROBLEM A-4

Determine the product of inertia I_{xy} for the homogeneous quarter cylinder shown in Fig. A-10a.

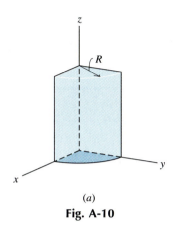

(a)

Fig. A-10

SOLUTION

All parts of the element of mass dm, shown in Fig. A-10b, are located at the same distances x and y from the xz- and yz-planes; therefore, the product of inertia dI_{xy} for the element is by definition

$$dI_{xy} = xy \, dm$$

Summing the elements for the entire body yields

$$\begin{aligned}
I_{xy} &= \int_m dI_{xy} = \int_m xy \, dm = \int_V xy \, \rho dV \\
&= \int_0^R \int_0^{\sqrt{R^2 - x^2}} \rho xy(h \, dy \, dx) \\
&= \int_0^R \rho hx \left[\frac{y^2}{2} \right]_0^{\sqrt{R^2 - x^2}} dx \\
&= \int_0^R \frac{1}{2} \rho h(R^2 x - x^3) \, dx \\
&= \frac{1}{2} \rho h \left[\frac{R^2 x^2}{2} - \frac{x^4}{4} \right]_0^R = \frac{1}{2} \rho h \left(\frac{R^4}{4} \right) = \frac{1}{8} \rho R^4 h
\end{aligned}$$

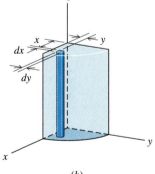

(b)

Alternatively, the thin-plate element shown in Fig. A-10c could be used to determine I_{xy}. From Eq. A-10 and the data from Table B-5,

$$dI_{xym} = \rho t \, dI_{xyA} = \frac{1}{8} \rho R^4 \, dz$$

Therefore,

$$I_{xym} = \rho t \int_A dI_{xyA} = \int_0^h \frac{1}{8} \rho R^4 \, dz = \frac{1}{8} \rho R^4 h$$

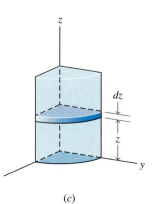

(c)

Since the mass of the body is

$$m = \rho V = \rho\left(\frac{1}{4}\pi R^2 h\right) = \frac{1}{4}\rho\pi R^2 h$$

the product of inertia I_{xy} can be written as

$$I_{xy} = \frac{1}{2\pi}\left(\frac{1}{4}\rho\pi R^2 h\right)R^2 = \frac{1}{2\pi}mR^2 \qquad \text{Ans.}$$

EXAMPLE PROBLEM A-5

Determine the products of inertia I_{xy}, I_{yz}, and I_{zx} for the homogeneous flat-plate steel ($\rho = 7870$ kg/m³) washer shown in Fig. A-11. The hole is located at the center of the plate.

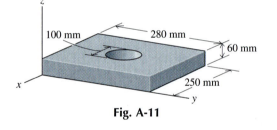

Fig. A-11

SOLUTION

The products of inertia are zero for the planes of symmetry through the mass-centers of the plate and hole. Since the xy-, yz-, and zx-planes shown in Fig. A-11 are parallel to these planes of symmetry, the parallel-axis theorem for products of inertia (Eqs. A-11) can be used to determine the required products of inertia. The masses of the plate, hole, and washer are

$$m_P = \rho V = \rho bht = 7870(0.280)(0.250)(0.060) = 33.05 \text{ kg}$$
$$m_H = \rho V = \rho\pi R^2 t = 7870\pi(0.050)^2(0.060) = 3.71 \text{ kg}$$
$$m_W = m_P - m_H = 33.05 - 3.71 = 29.34 \text{ kg}$$

From Eqs. A-11,

$$\begin{aligned} I_{zy} &= I_{xyG} + x_G y_G\, m \\ &= 0 + (-0.125)(0.140)(29.34) = -0.513 \text{ kg} \cdot \text{m}^2 \end{aligned} \qquad \text{Ans.}$$

$$\begin{aligned} I_{yz} &= I_{yzG} + y_G z_G \\ &= 0 + (0.140)(0.030)(29.34) = 0.1232 \text{ kg} \cdot \text{m}^2 \end{aligned} \qquad \text{Ans.}$$

$$\begin{aligned} I_{zx} &= I_{zxG} + z_G x_G\, m \\ &= 0 + (0.030)(-0.125)(29.34) = -0.1100 \text{ kg} \cdot \text{m}^2 \end{aligned} \qquad \text{Ans.}$$

A-3 PRINCIPAL MOMENTS OF INERTIA

In some instances, in the dynamic analysis of bodies, maximum and minimum moments of inertia, and principal axes which are similar to maximum and minimum second moments of an area, must be determined. Again, the problem is one of transforming known or easily calculated moments and products of inertia with respect to one coordinate system (such as an *xyz*-coordinate system along the edges of a rectangular prism) to a second *x'y'z'*-coordinate system through the same origin *O* but inclined with respect to the *xyz*-system.

For example, consider the body shown in Fig. A-12, where the *x'*-axis is oriented at angles $\theta_{x'x}$, $\theta_{x'y}$, and $\theta_{x'z}$ with respect to the *x*-, *y*-, and *z*-axes, respectively. The moment of inertia $I_{x'}$ is by definition

$$I_{x'} = \int_m r^2 \, dm$$

The distance *d* from the origin of coordinates to the element *dm* is given by the expression

$$d^2 = x^2 + y^2 + z^2 = x'^2 + y'^2 + z'^2 = x'^2 + r^2$$

Therefore,

$$r^2 = x^2 + y^2 + z^2 - x'^2$$

and since

$$x' = x \cos \theta_{x'x} + y \cos \theta_{x'y} + z \cos \theta_{x'z}$$
$$r^2 = x^2 + y^2 + z^2 - (x \cos \theta_{x'x} + y \cos \theta_{x'y} + z \cos \theta_{x'z})^2$$

Recall that

$$\cos^2\theta_{x'x} + \cos^2\theta_{x'y} + \cos^2\theta_{x'z} = 1$$

Therefore,

$$r^2 = (x^2 + y^2 + z^2)(\cos^2\theta_{x'x} + \cos^2\theta_{x'y} + \cos^2\theta_{x'z})$$
$$- (x \cos \theta_{x'x} + y \cos \theta_{x'y} + z \cos \theta_{x'z})^2$$

which reduces to

$$r^2 = (y^2 + z^2) \cos^2\theta_{x'x} + (z^2 + x^2) \cos^2\theta_{x'y} + (x^2 + y^2) \cos^2\theta_{x'z}$$
$$- 2xy \cos \theta_{x'x} \cos \theta_{x'y} - 2yz \cos \theta_{x'y} \cos \theta_{x'z} - 2zx \cos \theta_{x'z} \cos \theta_{x'x}$$

Therefore

$$I_{x'} = \int_m r^2 \, dm = \cos^2\theta_{x'x} \int_m (y^2 + z^2) \, dm + \cos^2\theta_{x'y} \int_m (z^2 + x^2) \, dm$$
$$+ \cos^2\theta_{x'z} \int_m (x^2 + y^2) \, dm - \cos \theta_{x'x} \cos \theta_{x'y} \int_m 2xy \, dm$$
$$- \cos \theta_{x'y} \cos \theta_{x'z} \int_m 2yz \, dm - \cos \theta_{x'z} \cos \theta_{x'x} \int_m 2zx \, dm$$

From Eqs. A-2 and A-9,

$$I_x = \int_m (y^2 + z^2) \, dm \qquad I_{xy} = \int_m xy \, dm$$

$$I_y = \int_m (z^2 + x^2) \, dm \qquad I_{yz} = \int_m yz \, dm$$

$$I_z = \int_m (x^2 + y^2) \, dm \qquad I_{zx} = \int_m zx \, dm$$

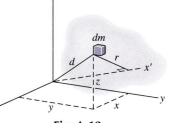

Fig. A-12

Therefore,

$$I_{x'} = I_x \cos^2\theta_{x'x} + I_y \cos^2\theta_{x'y} + I_z \cos^2\theta_{x'z} - 2I_{xy} \cos\theta_{x'x} \cos\theta_{x'y}$$
$$- 2I_{yz} \cos\theta_{x'y} \cos\theta_{x'z} - 2I_{zx} \cos\theta_{x'z} \cos\theta_{x'x} \quad \text{(A-12a)}$$

In a similar fashion the product of inertia

$$I_{x'y'} = \int_m x'y' \, dm$$

can be expressed in terms of I_x, I_y, I_z, I_{xy}, I_{yz}, and I_{zx} as

$$I_{x'y'} = -I_x \cos\theta_{x'x} \cos\theta_{y'x} - I_y \cos\theta_{x'y} \cos\theta_{y'y} - I_z \cos\theta_{x'z} \cos\theta_{y'z}$$
$$+ I_{xy}(\cos\theta_{x'x} \cos\theta_{y'y} + \cos\theta_{x'y} \cos\theta_{y'x})$$
$$+ I_{yz}(\cos\theta_{x'y} \cos\theta_{y'z} + \cos\theta_{x'z} \cos\theta_{y'y})$$
$$+ I_{zx}(\cos\theta_{x'z} \cos\theta_{y'x} + \cos\theta_{x'x} \cos\theta_{y'z}) \quad \text{(A-12b)}$$

If the original xyz-axes are principal axes (such as those shown for the figures in Table B-5),

$$I_{xy} = I_{yz} = I_{zx} = 0$$

and Eqs. A-12 reduce to

$$I_{x'} = I_x \cos^2\theta_{x'x} + I_y \cos^2\theta_{x'y} + I_z \cos^2\theta_{x'z} \quad \text{(A-13a)}$$

and

$$I_{x'y'} = -I_x \cos\theta_{x'x} \cos\theta_{y'x} - I_y \cos\theta_{x'y} \cos\theta_{y'y}$$
$$- I_z \cos\theta_{x'z} \cos\theta_{y'z} \quad \text{(A-13b)}$$

Equation A-12a for moments of inertia is the three-dimensional equivalent of Eq. 10-14 for second moments of area. By using a similar but much more complicated procedure than the one used with Eq. 10-14 to locate principal axes and determine maximum and minimum second moments of area, principal axes can be located and maximum and minimum moments of inertia can be determined. The procedure yields the following equations

$$(I_x - I_P) \cos\theta_{Px} - I_{xy} \cos\theta_{Py} - I_{zx} \cos\theta_{Pz} = 0$$
$$(I_y - I_P) \cos\theta_{Py} - I_{yz} \cos\theta_{Pz} - I_{xy} \cos\theta_{Px} = 0 \quad \text{(A-14)}$$
$$(I_z - I_P) \cos\theta_{Pz} - I_{zx} \cos\theta_{Px} - I_{yz} \cos\theta_{Py} = 0$$

This set of equations has a nontrivial solution only if the determinant of the coefficients of the direction cosines is equal to zero. Expansion of the determinant yields the following cubic equation for determining the principal moments of inertia of the body for the particular origin of coordinates being used:

$$I_P^3 - (I_x + I_y + I_z)I_P^2 + (I_xI_y + I_yI_z + I_zI_x - I_{xy}^2 - I_{yz}^2 - I_{zx}^2)I_P$$
$$- (I_xI_yI_z - I_xI_{yz}^2 - I_yI_{zx}^2 - I_zI_{xy}^2 - 2I_{xy}I_{yz}I_{zx}) = 0 \quad \text{(A-15)}$$

Equation A-15 yields three values I_1, I_2, and I_3 for the principal moments of inertia. One value is the maximum moment of inertia of the body for the origin of coordinates being used, a second value is the minimum moment of inertia of the body for the origin of coordinates being used, and the third value is an intermediate value of the moment of inertia of the body that has no particular significance.

The direction cosines for the principal inertia axes can be obtained by substituting the three values I_1, I_2, and I_3 obtained from Eq. A-15, in turn, into Eqs. A-14 and using the additional relation

$$\cos^2\theta_{Px} + \cos^2\theta_{Py} + \cos^2\theta_{Pz} = 1$$

Equations A-14 and A-15 are valid for bodies of any shape. The procedure for locating principal axes and determining maximum and minimum moments of inertia is illustrated in the following example problem.

EXAMPLE PROBLEM A-6

Locate the principal axes and determine the maximum and minimum moments of inertia for the rectangular steel ($w = 490$ lb/ft^3) block shown in Fig. A-13.

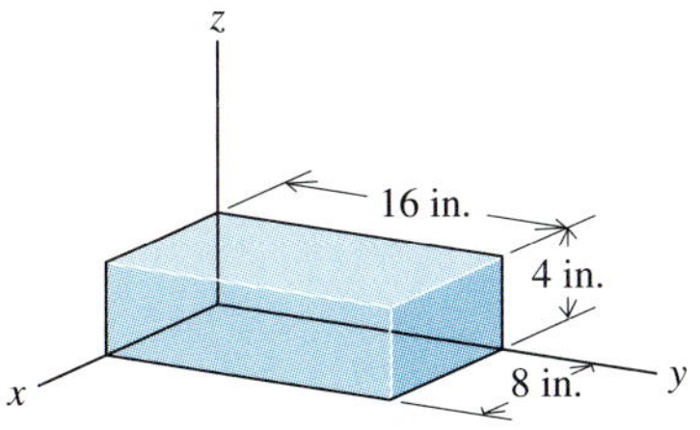

Fig. A-13

SOLUTION

The moments and products of inertia for the block are given by Eqs. A-4 and A-11 as

$$\begin{aligned}
I_x &= I_{xG} + (y_G^2 + z_G^2)m & I_{xy} &= I_{xyG} + x_G y_G m \\
I_y &= I_{yG} + (z_G^2 + x_G^2)m & I_{yz} &= I_{yzG} + y_G z_G m \\
I_z &= I_{zG} + (x_G^2 + y_G^2)m & I_{zx} &= I_{zxG} + z_G x_G m
\end{aligned}$$

The mass of the block is

$$m = \rho V = \frac{w}{g}bhL = \frac{490}{32.2}\left(\frac{8}{12}\right)\left(\frac{4}{12}\right)\left(\frac{16}{12}\right) = 4.509 \text{ slugs}$$

Thus, from the results listed in Table B-5,

$$\begin{aligned}
I_x &= \frac{1}{12}m(b^2 + h^2) + \left[\left(\frac{h}{2}\right)^2 + \left(\frac{b}{2}\right)^2\right]m \\
&= \frac{1}{3}m(b^2 + h^2) = \frac{1}{3}(4.509)\left[\left(\frac{8}{12}\right)^2 + \left(\frac{4}{12}\right)^2\right] = 0.835 \text{ slug} \cdot \text{ft}^2 \\
I_y &= \frac{1}{12}m(b^2 + L^2) + \left[\left(\frac{b}{2}\right)^2 + \left(\frac{L}{2}\right)^2\right]m \\
&= \frac{1}{3}m(b^2 + L^2) = \frac{1}{3}(4.509)\left[\left(\frac{8}{12}\right)^2 + \left(\frac{16}{12}\right)^2\right] = 3.340 \text{ slug} \cdot \text{ft}^2 \\
I_z &= \frac{1}{12}m(h^2 + L^2) + \left[\left(\frac{h}{2}\right)^2 + \left(\frac{L}{2}\right)^2\right]m \\
&= \frac{1}{3}m(h^2 + L^2) = \frac{1}{3}(4.509)\left[\left(\frac{4}{12}\right)^2 + \left(\frac{16}{12}\right)^2\right] = 2.839 \text{ slug} \cdot \text{ft}^2 \\
I_{xy} &= 0 + \left(\frac{L}{2}\right)\left(\frac{h}{2}\right)m = \frac{1}{4}mLh = \frac{1}{4}(4.509)\left(\frac{16}{12}\right)\left(\frac{4}{12}\right) = 0.501 \text{ slug} \cdot \text{ft}^2 \\
I_{yz} &= 0 + \left(\frac{h}{2}\right)\left(\frac{b}{2}\right)m = \frac{1}{4}mhb = \frac{1}{4}(4.509)\left(\frac{4}{12}\right)\left(\frac{8}{12}\right) = 0.251 \text{ slug} \cdot \text{ft}^2 \\
I_{zx} &= 0 + \left(\frac{b}{2}\right)\left(\frac{L}{2}\right)m = \frac{1}{4}mbL = \frac{1}{4}(4.509)\left(\frac{8}{12}\right)\left(\frac{16}{12}\right) = 1.002 \text{ slug} \cdot \text{ft}^2
\end{aligned}$$

Once the moments and products of inertia have been determined, the principal moments of inertia can be determined by using Eq. A-15. Thus,

$$\begin{aligned}
I_P^3 - (I_x + I_y &+ I_z)I_P^2 + (I_x I_y + I_y I_z + I_z I_x - I_{xy}^2 - I_{yz}^2 - I_{zx}^2)I_P \\
&- (I_x I_y I_z - I_x I_{yz}^2 - I_y I_{zx}^2 - I_z I_{xy}^2 - 2I_{xy}I_{yz}I_{zx}) = 0
\end{aligned}$$

Substituting values for the moments and products of inertia gives

$$I_P^3 - 7.014I_P^2 + 13.324I_P - 3.548 = 0$$

which has the solution

$$I_1 = I_{max} = 3.451 \text{ slug} \cdot \text{ft}^2 \qquad \text{Ans.}$$
$$I_2 = I_{int} = 3.246 \text{ slug} \cdot \text{ft}^2$$
$$I_3 = I_{min} = 0.317 \text{ slug} \cdot \text{ft}^2 \qquad \text{Ans.}$$

The principal directions are obtained by substituting the principal moments of inertia, in turn, into Eqs. A-14. With $I_P = I_1 = I_{max} = 3.451$ slug $\cdot$ ft^2:

$$(I_x - I_P) \cos \theta_{Px} - I_{xy} \cos \theta_{Py} - I_{zx} \cos \theta_{Pz} = 0$$
$$(I_y - I_P) \cos \theta_{Py} - I_{yz} \cos \theta_{Pz} - I_{xy} \cos \theta_{Px} = 0$$
$$(I_z - I_P) \cos \theta_{Pz} - I_{zx} \cos \theta_{Px} - I_{yz} \cos \theta_{Py} = 0$$

$$-2.616 \cos \theta_{1x} - 0.501 \cos \theta_{1y} - 1.002 \cos \theta_{1z} = 0$$
$$-0.501 \cos \theta_{1x} - 0.111 \cos \theta_{1y} - 0.251 \cos \theta_{1z} = 0 \qquad (a)$$
$$-1.002 \cos \theta_{1x} - 0.251 \cos \theta_{1y} - 0.612 \cos \theta_{1z} = 0$$

Equations a together with the required relationship for direction cosines

$$\cos^2\theta_{1x} + \cos^2\theta_{1y} + \cos^2\theta_{1z} = 1$$

have the solution

$$\cos \theta_{1x} = 0.0891 \qquad \text{or} \qquad \theta_{1x} = 84.9°$$
$$\cos \theta_{1y} = -0.9643 \qquad\qquad\quad \theta_{1y} = 164.6° \qquad \text{Ans.}$$
$$\cos \theta_{1z} = 0.2495 \qquad\qquad\quad \theta_{1z} = 75.6°$$

With $I_P = I_2 = I_{int} = 3.246$ slug $\cdot$ ft^2:

$$-2.411 \cos \theta_{2x} - 0.501 \cos \theta_{2y} - 1.002 \cos \theta_{2z} = 0$$
$$-0.501 \cos \theta_{2x} + 0.094 \cos \theta_{2y} - 0.251 \cos \theta_{2z} = 0$$
$$-1.002 \cos \theta_{2x} - 0.251 \cos \theta_{2y} - 0.407 \cos \theta_{2z} = 0$$
$$\cos^2\theta_{2x} + \cos^2\theta_{2y} + \cos^2\theta_{2z} = 1$$

the solution becomes

$$\cos \theta_{2x} = -0.4105 \qquad \text{or} \qquad \theta_{2x} = 114.2°$$
$$\cos \theta_{2y} = 0.1925 \qquad\qquad\quad \theta_{2y} = 78.9° \qquad \text{Ans.}$$
$$\cos \theta_{2z} = 0.8913 \qquad\qquad\quad \theta_{2x} = 27.0°$$

With $I_P = I_3 = I_{min} = 0.317$ slug $\cdot$ ft^2:

$$0.518 \cos \theta_{3x} - 0.501 \cos \theta_{3y} - 1.002 \cos \theta_{3z} = 0$$
$$-0.501 \cos \theta_{3x} + 3.023 \cos \theta_{3y} - 0.251 \cos \theta_{3z} = 0$$
$$-1.002 \cos \theta_{3x} - 0.251 \cos \theta_{3y} + 2.522 \cos \theta_{3z} = 0$$
$$\cos^2\theta_{3x} + \cos^2\theta_{3y} + \cos^2\theta_{3z} = 1$$

the solution becomes

$$\cos \theta_{3x} = 0.9075 \qquad \text{or} \qquad \theta_{3x} = 24.8°$$
$$\cos \theta_{3y} = 0.1818 \qquad\qquad\quad \theta_{3y} = 79.5° \qquad \text{Ans.}$$
$$\cos \theta_{3z} = 0.3786 \qquad\qquad\quad \theta_{3z} = 67.8°$$

Thus, the unit vectors associated with the three principal directions are

$$\mathbf{n}_1 = 0.0891\mathbf{i} - 0.9643\mathbf{j} + 0.2495\mathbf{k} \qquad \mathbf{n}_1 \cdot \mathbf{n}_2 = 0$$
$$\mathbf{n}_2 = -0.4105\mathbf{i} + 0.1925\mathbf{j} + 0.8913\mathbf{k} \qquad \mathbf{n}_2 \cdot \mathbf{n}_3 = 0$$
$$\mathbf{n}_3 = 0.9075\mathbf{i} + 0.1818\mathbf{j} + 0.3786\mathbf{k} \qquad \mathbf{n}_3 \cdot \mathbf{n}_1 = 0$$

which verifies that the three principal axes are orthogonal.

Appendix B

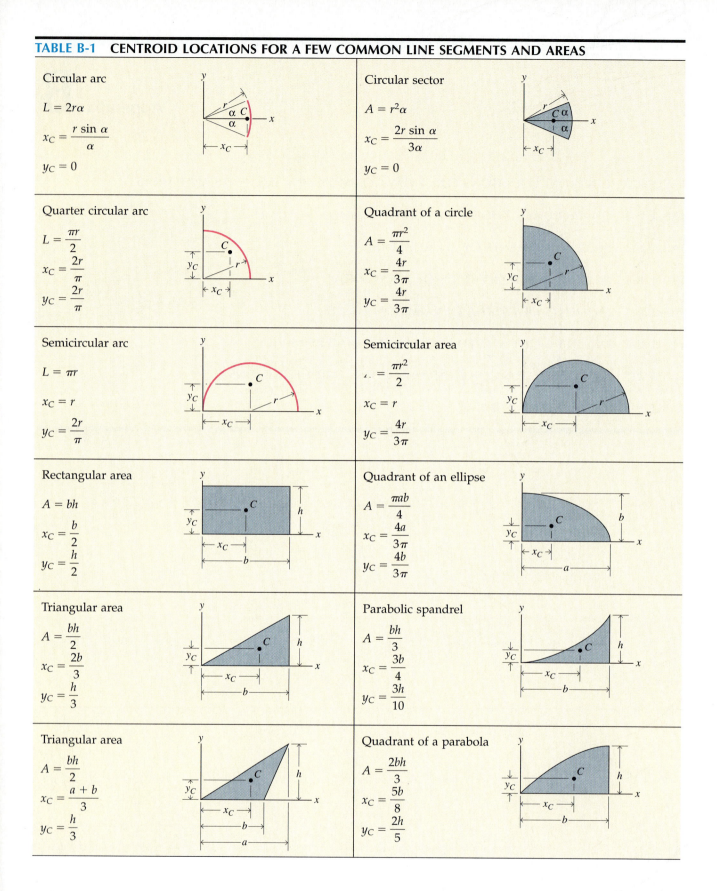

Circular arc

$L = 2r\alpha$

$x_C = \dfrac{r \sin \alpha}{\alpha}$

$y_C = 0$

Circular sector

$A = r^2\alpha$

$x_C = \dfrac{2r \sin \alpha}{3\alpha}$

$y_C = 0$

Quarter circular arc

$L = \dfrac{\pi r}{2}$

$x_C = \dfrac{2r}{\pi}$

$y_C = \dfrac{2r}{\pi}$

Quadrant of a circle

$A = \dfrac{\pi r^2}{4}$

$x_C = \dfrac{4r}{3\pi}$

$y_C = \dfrac{4r}{3\pi}$

Semicircular arc

$L = \pi r$

$x_C = r$

$y_C = \dfrac{2r}{\pi}$

Semicircular area

$A = \dfrac{\pi r^2}{2}$

$x_C = r$

$y_C = \dfrac{4r}{3\pi}$

Rectangular area

$A = bh$

$x_C = \dfrac{b}{2}$

$y_C = \dfrac{h}{2}$

Quadrant of an ellipse

$A = \dfrac{\pi ab}{4}$

$x_C = \dfrac{4a}{3\pi}$

$y_C = \dfrac{4b}{3\pi}$

Triangular area

$A = \dfrac{bh}{2}$

$x_C = \dfrac{2b}{3}$

$y_C = \dfrac{h}{3}$

Parabolic spandrel

$A = \dfrac{bh}{3}$

$x_C = \dfrac{3b}{4}$

$y_C = \dfrac{3h}{10}$

Triangular area

$A = \dfrac{bh}{2}$

$x_C = \dfrac{a + b}{3}$

$y_C = \dfrac{h}{3}$

Quadrant of a parabola

$A = \dfrac{2bh}{3}$

$x_C = \dfrac{5b}{8}$

$y_C = \dfrac{2h}{5}$

TABLE B-2   CENTROID LOCATIONS FOR A FEW COMMON VOLUMES

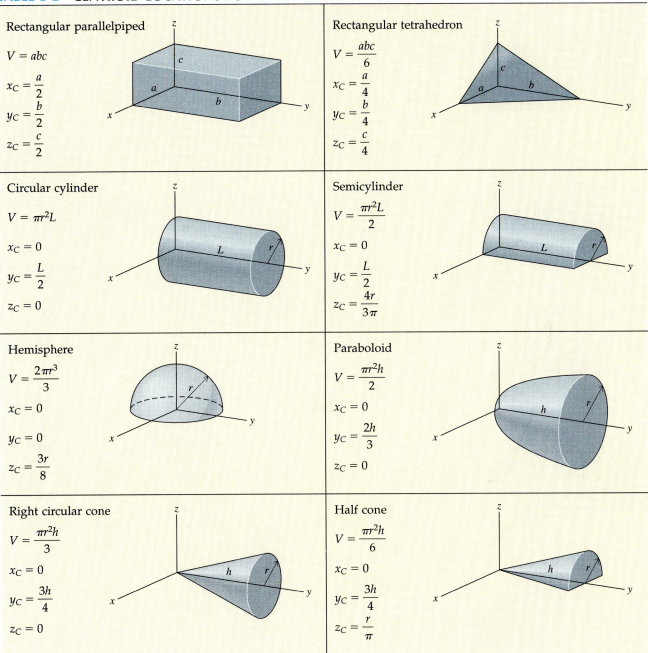

Rectangular parallelpiped

$V = abc$

$x_C = \dfrac{a}{2}$

$y_C = \dfrac{b}{2}$

$z_C = \dfrac{c}{2}$

Rectangular tetrahedron

$V = \dfrac{abc}{6}$

$x_C = \dfrac{a}{4}$

$y_C = \dfrac{b}{4}$

$z_C = \dfrac{c}{4}$

Circular cylinder

$V = \pi r^2 L$

$x_C = 0$

$y_C = \dfrac{L}{2}$

$z_C = 0$

Semicylinder

$V = \dfrac{\pi r^2 L}{2}$

$x_C = 0$

$y_C = \dfrac{L}{2}$

$z_C = \dfrac{4r}{3\pi}$

Hemisphere

$V = \dfrac{2\pi r^3}{3}$

$x_C = 0$

$y_C = 0$

$z_C = \dfrac{3r}{8}$

Paraboloid

$V = \dfrac{\pi r^2 h}{2}$

$x_C = 0$

$y_C = \dfrac{2h}{3}$

$z_C = 0$

Right circular cone

$V = \dfrac{\pi r^2 h}{3}$

$x_C = 0$

$y_C = \dfrac{3h}{4}$

$z_C = 0$

Half cone

$V = \dfrac{\pi r^2 h}{6}$

$x_C = 0$

$y_C = \dfrac{3h}{4}$

$z_C = \dfrac{r}{\pi}$

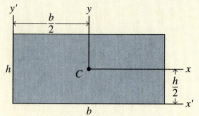

$$I_x = \frac{bh^3}{12}$$

$$I_{x'} = \frac{bh^3}{3}$$

$$A = bh$$

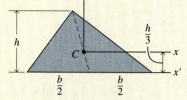

$$I_x = \frac{bh^3}{36}$$

$$I_{x'} = \frac{bh^3}{12}$$

$$A = \frac{1}{2}bh$$

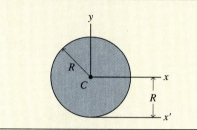

$$I_x = \frac{\pi R^4}{4}$$

$$I_{x'} = \frac{5\pi R^4}{4}$$

$$A = \pi R^2$$

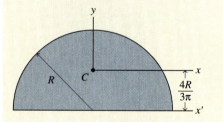

$$I_x = \frac{\pi R^4}{8} - \frac{8R^4}{9\pi}$$

$$I_y = \frac{\pi R^4}{8}$$

$$I_{x'} = \frac{\pi R^4}{8}$$

$$A = \frac{1}{2}\pi R^2$$

$$I_x = \frac{\pi R^4}{16} - \frac{4R^4}{9\pi}$$

$$I_{x'} = \frac{\pi R^4}{16}$$

$$A = \frac{1}{4}\pi R^2$$

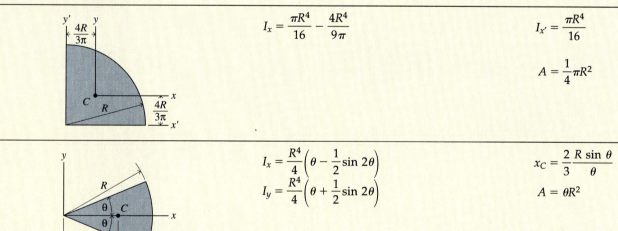

$$I_x = \frac{R^4}{4}\left(\theta - \frac{1}{2}\sin 2\theta\right)$$

$$I_y = \frac{R^4}{4}\left(\theta + \frac{1}{2}\sin 2\theta\right)$$

$$x_C = \frac{2}{3}\frac{R \sin \theta}{\theta}$$

$$A = \theta R^2$$

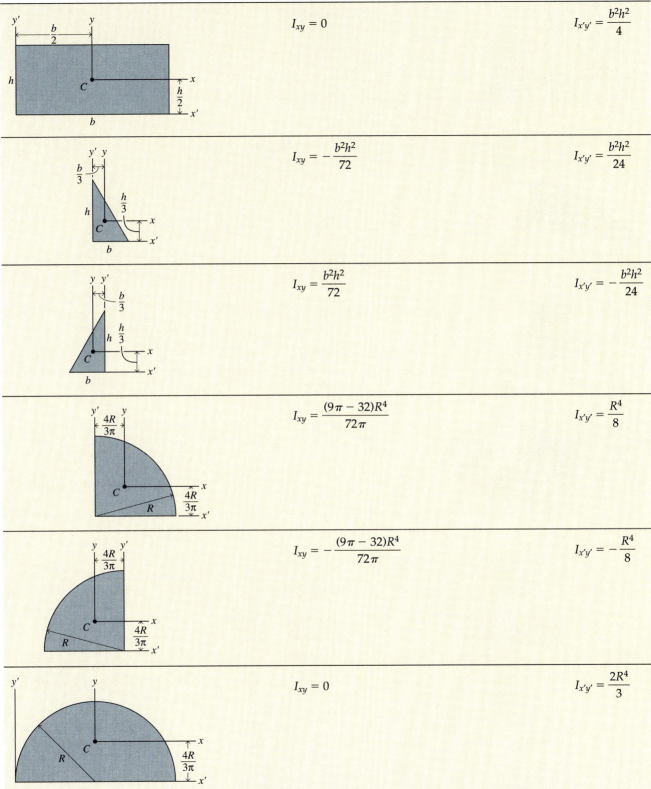

$$I_{xy} = 0 \qquad\qquad I_{x'y'} = \frac{b^2h^2}{4}$$

$$I_{xy} = -\frac{b^2h^2}{72} \qquad\qquad I_{x'y'} = \frac{b^2h^2}{24}$$

$$I_{xy} = \frac{b^2h^2}{72} \qquad\qquad I_{x'y'} = -\frac{b^2h^2}{24}$$

$$I_{xy} = \frac{(9\pi - 32)R^4}{72\pi} \qquad\qquad I_{x'y'} = \frac{R^4}{8}$$

$$I_{xy} = -\frac{(9\pi - 32)R^4}{72\pi} \qquad\qquad I_{x'y'} = -\frac{R^4}{8}$$

$$I_{xy} = 0 \qquad\qquad I_{x'y'} = \frac{2R^4}{3}$$

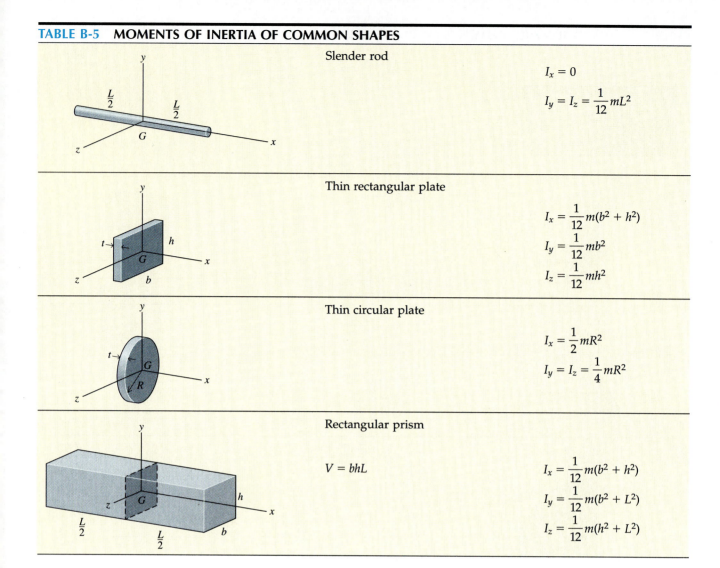

Slender rod

$$I_x = 0$$

$$I_y = I_z = \frac{1}{12}mL^2$$

Thin rectangular plate

$$I_x = \frac{1}{12}m(b^2 + h^2)$$

$$I_y = \frac{1}{12}mb^2$$

$$I_z = \frac{1}{12}mh^2$$

Thin circular plate

$$I_x = \frac{1}{2}mR^2$$

$$I_y = I_z = \frac{1}{4}mR^2$$

Rectangular prism

$$V = bhL$$

$$I_x = \frac{1}{12}m(b^2 + h^2)$$

$$I_y = \frac{1}{12}m(b^2 + L^2)$$

$$I_z = \frac{1}{12}m(h^2 + L^2)$$

Right circular cone

$$V = \frac{1}{3}\pi R^2 h$$

$$x_G = \frac{3}{4}h$$

$$I_x = \frac{3}{10}mR^2$$

$$I_y = I_z = \frac{3}{20}m(R^2 + 4h^2)$$

$$I_{yG} = I_{zG} = \frac{3}{80}m(4R^2 + h^2)$$

Circular cylinder

$$V = \pi R^2 L$$

$$I_x = \frac{1}{2}mR^2$$

$$I_y = I_z = \frac{1}{12}m(3R^2 + L^2)$$

Hemisphere

$$V = \frac{2}{3}\pi R^3$$

$$x_G = \frac{3}{8}R$$

$$I_x = I_y = I_z = \frac{2}{5}mR^2$$

$$I_{yG} = I_{zG} = \frac{83}{320}mR^2$$

Sphere

$$V = \frac{4}{3}\pi R^3$$

$$I_x = I_y = I_z = \frac{2}{5}mR^2$$

TABLE B-6 DENSITY ρ OF SELECTED ENGINEERING MATERIALS

	kg/m^3	$slug/ft^3$
Solids		
Aluminum	2,770	5.35
Brass	8,750	16.91
Concrete	2,410	4.66
Cast iron	7,370	14.24
Copper	8,910	17.21
Earth (wet)	1,760	3.40
(dry)	1,280	2.47
Glass	2,590	5.00
Gold	19,300	37.29
Lead	11,370	21.97
Steel	7,870	15.21
Wood (soft pine)	480	0.93
(hard oak)	800	1.55
Liquids		
Ice	900	1.74
Mercury	13,570	26.22
Oil	900	1.74
Water (fresh)	1,000	1.94
(sea)	1,030	1.99
Gases		
Air	1.225	$2.377(10^{-3})$

TABLE B-7 MISCELLANEOUS CONVERSION FACTORS AND DEFINITIONS

Length
I ft = 12 in.
1 mi = 5280 ft
1 nautical mile = 6080 ft
1 in. = 25.40 mm
1 mi = 1.609 km

Volume
1 cup = 8 fl oz (fluid ounce)
1 pint = 2 cup = 16 fl oz
1 quart = 2 pint = 32 fl oz
1 gal = 4 quart = 128 fl oz = 231 in.^{3}
1 ft^3 = 7.48 gal
1 barrel (petroleum) = 42 gal
1 liter = 10^{-3} m^3 = (100 mm)^3

Mass
1 metric ton = 1000 kg
1 slug = 14.59 kg

Force
1 lb = 16 oz
1 kip (kilo-pound) = 1000 lb
1 ton = 2000 lb
1 lb = 4.448 Newton

Energy
1 BTU (British Thermal Unit) = 778 ft · lb

Power
1 hp (horsepower) = 550 ft · lb/s
1 ft · lb/s = 1.356 watt

TABLE B-8   ASTRONOMICAL DATA

Universal Gravitational Constant

$G = 6.673(10^{-11})$ m^3/(kg · s^2) $= 3.439(10^{-8})$ ft^4/(lb · s^4)

The Sun

Mass	1.990(10^{30}) kg	1.364(10^{29}) lb · s^2/ft
Mean radius	696,000 km	432,000 mi

The Earth

Mass	5.976(10^{24}) kg	4.095(10^{23}) lb · s^2/ft
Mean radius	6370 km	3960 mi
Rotation rate	23.93 hr	

The Moon

Mass	7.350(10^{22}) kg	5.037(10^{21}) lb · s^2/ft
Mean radius	1740 km	1080 mi
Mean distance to the Earth (center to center)		
	384,000 km	239,000 mi
Eccentricity (e)	0.055	

The Solar system

Planet	Mean Distance to Sun A.U.*	e	Mean Diameter (relative to Earth)	Mass (relative to Earth)
Mercury	0.387	0.206	0.380	0.05
Venus	0.723	0.007	0.975	0.81
Earth	1.000	0.017	1.000	1.00
Mars	1.524	0.093	0.532	0.11
Jupiter	5.203	0.048	11.27	317.8
Saturn	9.539	0.056	9.49	95.2

*Astronomical Unit (A.U.) is equal to the mean distance from the earth to the sun = 149.6(10^6) km = 92.96(10^6) mi.

ANSWERS TO SELECTED PROBLEMS

CHAPTER 13

13-1 $v(t) = 10t - 30$ ft/s; $a(t) = 10$ ft/s²
$x = -5$ ft; $v = 20$ ft/s; $a = 10$ ft/s²
$s = 65.0$ ft

13-2 $v(t) = -4$ m/s; $a(t) = 0$ m/s²
$x = -5$ m; $v = -4$ m/s;
$a = 0$ m/s²
$s = 20$ m

13-3 $v(t) = -e^{-t/3}$ ft/s;
$a(t) = (1/3)e^{-t/3}$ ft/s²
$x = 0.567$ ft; $v = -0.1888$ ft/s;
$a = 0.0630$ ft/s²
$s = 2.43$ ft

13-7 $x(t) = 60 + 48t - 8t^2$ ft;
$a(t) = -16$ ft/s²
$x = -68$ ft; $v = -80$ ft/s;
$a = -16$ ft/s²
$s = 108$ ft

13-8 $x(t) = -4t^3/3 + 20t^2 - 70t + 20$ m
$a(t) = -8t + 40$ m/s²
$x = 57.3$ m; $v = -6$ m/s;
$a = -24$ m/s²
$s = 55.5$ m

13-9 $x(t) = 53.1 - 90e^{-t/3}$ ft;
$a(t) = -10e^{-t/3}$ ft/s²
$x = 46.9$ ft; $v = 2.08$ ft/s;
$a = -0.695$ ft/s²
$s = 10.75$ ft

13-13 $x(t) = -40 - 30t + 9t^2 - 0.5t^3$ ft
$v(t) = -30 + 18t - 1.5t^2$ ft/s
$x = -62.5$ ft; $v = 10.50$ ft/s;
$a = 9$ ft/s²
$s = 102.5$ ft

13-14 $x(t) = -19.62 + 44.62t$
$\quad - 4.905t^2$ m
$v(t) = 44.62 - 9.81t$ m/s
$x = 70.1$ m; $v = 15.19$ m/s;
$a = -9.81$ m/s²
$s = 70.2$ m

13-15 $x(t) = 432e^{-t/6} + 16.98t - 245$ ft
$v(t) = 16.98 - 72e^{-t/6}$ ft/s
$x = 68.3$ ft; $v = -26.7$ ft/s;
$a = 7.28$ ft/s²
$s = 63.3$ ft

13-19 $a = 7.33$ ft/s²; $x = 238$ ft
13-22 $a = -0.333$ m/s²; $x = 18$ m
13-23 $x = 2108$ ft
13-26 $a = -5.21$ m/s²
13-28 $a = -1.257$ m/s²
13-29 $v = 3.00$ ft/s
13-31 $v = 5.74$ ft/s
13-34 $x = 1$ m

13-36 $v = \dfrac{1}{\sqrt{333.33\, e^{(421.1-y)/166.67} - 3270}}$ m/s
$h = 40.5$ m

13-37 $v = 50 - 0.50x$ ft/s; $v = 12.5$ ft/s
13-39 $v = \sqrt{2000(1 - e^{-x/2500})}$ ft/s
$v = 44.7$ ft/s
13-41 $v = 20$ ft/s; $t = 1.833$ s
13-42 $v = 8.73$ ft/s; $t = 85.0$ s
13-44 $a = -2.10$ m/s²; $a = -2.68$ m/s²
13-46 $x = 23.0$ m; $v = 10.5$ m/s
$x = 70$ m; $v = 12.0$ m/s
13-47 $t = 72.2$ s
13-50 $7:03:16.2$; $7:08:32.4$; $7:13:48.6$
13-51 $\mathbf{v}_{A/B} = 140$ mi/h $\rightarrow$;
$\mathbf{v}_{B/A} = 140$ mi/h $\leftarrow$
13-52 $\mathbf{v}_{A/B} = 35$ m/s $\downarrow$; $\mathbf{v}_{B/A} = 35$ m/s $\uparrow$
13-54 $t = 1$ h 6.7 min; $t = 2$ h 13.3 min
13-56 8:44 PM; 533 km (from where
A starts)
13-57 $t = 100$ s; $x = 2500$ ft
13-59 39.1 mi (from first town)
13-60 1:30 PM; 12.54 km (from 1st
town)
13-62 $b = 22.2$ m
13-65 $t = 90$ s; $t = 18$ s
13-67 $x = 843$ ft; $v = 75.1$ mi/h
13-70 $v = 1$ m/s $\uparrow$; $a = 0.1$ m/s² $\uparrow$
13-72 $v = 2$ m/s $\nearrow$; $a = 0.4$ m/s² $\nearrow$
13-73 $v = 18$ ft/s $\downarrow$; $a = 1.5$ ft/s² $\uparrow$
13-75 $v_C = 20$ ft/s $\leftarrow$; $a_C = 2$ ft/s² $\rightarrow$
$v_A = 6.67$ ft/s $\rightarrow$;
$a_A = 0.667$ ft/s² $\leftarrow$
13-77 $x = 813$ ft
13-80 $v = 3$ m/s $\downarrow$
13-82 $v = 3.75$ m/s $\uparrow$; $a = 2.53$ m/s² $\uparrow$
13-83 $v = 6.25$ ft/s $\leftarrow$; $a = 3.52$ ft/s² $\leftarrow$
13-87 $\theta = 33.87°$ or $82.69°$
13-88 $x = 3.37$ km
13-90 $v_0 = 24.9$ m/s
13-93 $v_0 = 20.1$ ft/s
$v_x = 20.1$ ft/s $\rightarrow$; $v_y = 32.2$ ft/s $\downarrow$
13-94 $v_0 = 134.8$ km/h; $\theta = 34.99°$ $\nwarrow$
13-97 $a = 2.92$ ft; $\theta_0 = 2.796°$
$h = 0.732$ ft (above initial)
13-99 $h = 3.63$ ft; $t = 2.91$ s; $d = 268$ ft
13-103 $v = 1761$ ft/s $\searrow$ $81.66°$;
$a = 325$ ft/s² $\searrow$ $80.58°$
13-104 $v = 46.5$ mm/s $\searrow$ $38.92°$;
$a = 330$ mm/s² $\nearrow$ $39.76°$
13-106 $v = 262$ mm/s $\searrow$ $52.24°$;
$a = 2023$ mm/s² $\searrow$ $69.42°$
13-107 $v_r = 347$ ft/s; $v_\theta = -85.8$ ft/s
$a_r = 30.9$ ft/s²;

13-110 $a_\theta = -4.71$ ft/s²
$r = 210$ mm
$v_r = -108.5$ mm/s;
$v_\theta = 420$ mm/s
$a_r = -1186$ mm/s²;
$a_\theta = -434$ mm/s²
13-111 $v_r = -11.78$ in./s; $v_\theta = 46.1$ in./s
$a_r = -1685$ in./s²;
$a_\theta = -370$ in./s²
13-113 $\dot{\theta}_{max} = 24$ rad/s
13-116 $\dot{\theta}_{max} = 5.48$ rad/s
13-119 $v = 40.6$ mi/h
13-120 $a = 10.86$ m/s² $\measuredangle$ $62.60°$
13-123 $v = 29.7$ mi/h
13-124 $a = 11.78$ m/s² $\measuredangle$ $34.89°$
13-128 $v_{max} = 200$ m/s; $a_{max} = 89.4$ m/s²
13-129 $a = 5.86$ m/s² $\measuredangle$ $57.38°$
13-132 $a = 4.12$ m/s² $\measuredangle$ $42.91°$
13-133 $\mathbf{v} = 8\,\mathbf{i} + 4\,\mathbf{j}$ ft/s; $\mathbf{a} = 4\,\mathbf{i}$ ft/s²
$\mathbf{v} = 8.94\,\mathbf{e}_t$ ft/s;
$\mathbf{a} = 3.58\,\mathbf{e}_t + 1.789\,\mathbf{e}_n$ ft/s²
$\rho = 44.72$ ft
13-134 $v = 361$ mm/s $\measuredangle$ $85.24°$;
$a = 180$ mm/s² $\uparrow$
$a_t = 179.4$ mm/s²;
$a_n = 14.95$ mm/s²
13-137 $\mathbf{v} = 650\,\mathbf{e}_r + 1637\,\mathbf{e}_\theta$ ft/s
$\quad = 1761$ ft/s $\searrow$ $81.66°$
$\mathbf{a} = 114.3\,\mathbf{e}_r + 304\,\mathbf{e}_\theta$ ft/s²
$\quad = 325$ ft/s² $\searrow$ $80.58°$
$\mathbf{v} = 1761\,\mathbf{e}_t$ ft/s;
$\mathbf{a} = 325\,\mathbf{e}_t + 6.08\,\mathbf{e}_n$ ft/s²
13-140 $v_x = -0.728$ m/s;
$v_y = -2.91$ m/s
$a_n = 0.257$ m/s²; $a_t = 5$ m/s²;
$\theta = 2.95°$
13-141 $v_r = -11.78$ in./s; $v_\theta = 46.1$ in./s
$a_t = 58.9$ in./s²;
$a_n = 1724$ in./s²; $\theta = 88.05°$
13-143 $T = 96.4$ s; $\phi = 19.47°$
13-144 $v = 6.51$ m/s $\searrow$ $50.19°$
13-147 $x = 32.9$ ft;
$\mathbf{v}_{B/A} = 23.3$ ft/s $\searrow$ $48.82°$
13-149 $d = 623$ ft; $\dot{d} = 20.8$ ft/s
13-150 $\phi = 73.69°$; $v = 138.8$ km/h
13-152 $\mathbf{r}_{B/A} = 10.00\,\mathbf{i} + 8.66\,\mathbf{j}$ km
$\mathbf{v}_{B/A} = -500\,\mathbf{i} - 346\,\mathbf{j}$ km/h
$d = 13.23$ km; $T = 4.30$ min
13-155 $v_B = 29.4$ ft/s; $s = 35.8$ ft
$v_{B/A} = 31.1$ ft/s $\searrow$ $58.18°$
13-159 $\mathbf{v} = 10t\,\mathbf{i} + 3\,\mathbf{j} + 45t^2\,\mathbf{k}$ ft/s
$\mathbf{a} = 10\,\mathbf{i} + 90t\,\mathbf{k}$ ft/s²

14-165 $\alpha_A = -2.62$ rad/s^2;
$\alpha_B = 1.047$ rad/s^2
$\omega_A = 375$ rev/min $\downarrow$

14-166 $\alpha_{BC} = 87.2$ rad/s^2 $\downarrow$;
$a_C = 44.4$ m/s^2 $\rightarrow$

14-169 $\mathbf{v}_A = -21.3\ \mathbf{i} + 41.5\ \mathbf{j}$
$+ 4.33\ \mathbf{k}$ ft/s
$\mathbf{a}_A = 10\ \mathbf{i} - 45.0\ \mathbf{j} - 0.99\ \mathbf{k}$ ft/s^2

14-170 $v_C = 1.047$ m/s $\leftarrow$;
$a_C = 0.0533$ m/s^2 $\rightarrow$

CHAPTER 15

15-1 $P = 422$ lb; $P = 242$ lb
15-2 $F = 2000$ N
15-4 $F = 84.4$ N; $\mu_k = 0.573$
15-7 $v = 73.7$ ft/s; $x = 184.3$ ft
$v = 35.9$ ft/s; $x = 89.8$ ft
15-8 $a = 2.41$ m/s^2; $t = 3.53$ s;
$v = 6.94$ m/s
15-12 $F_p = 5580$ N
15-13 $a = 6.25$ ft/s^2 $\uparrow$; $T = 1194$ lb
15-14 $F_D = 67.8$ N
15-17 $a_B = 38.1$ ft/s^2 $\leftarrow$;
$v_A = 397$ ft/s $\rightarrow$
15-18 $A = 10.69$ kN $\uparrow$; $C = 1.781$ kN $\uparrow$;
$T = 12.02$ kN
15-21 $a_B = 12.53$ ft/s^2 $\searrow$; $T = 3.07$ lb
15-24 $T_A = 210$ N; $T_B = 420$ N
15-25 $a_B = 2.22$ ft/s^2 $\downarrow$; $W_B = 44.4$ lb;
$T = 41.3$ lb
15-28 $a_A = 0.1369$ m/s^2 $\rightarrow$; $T = $
17.49 N; $x_A = 1.711$ m $\rightarrow$
15-31 $t = 5.49$ s; $x = 63.1$ ft $\leftarrow$;
$x_{10} = 0$ ft
15-32 $a = -4.20g$ $\leftarrow$; $t = 2.15$ s;
$x = 60.3$ m
15-36 $h_{max} = 4590$ m; $h_{max} = 1036$ m;
$v = 86.6$ m/s $\downarrow$
15-40 $v = 0.957$ m/s $\downarrow$
15-41 $v_B = 5.68$ ft/s $\downarrow$; $a_B = 10.73$ ft/s^2
15-45 $a_A = 3.16$ ft/s^2 $\rightarrow$;
$a_B = 3.42$ ft/s^2 $\rightarrow$
$T = 19.62$ lb
$a_A = 3.52$ ft/s^2 $\rightarrow$;
$a_B = 3.20$ ft/s^2 $\rightarrow$
$T = 21.9$ lb
15-49 $x = 5.40$ mi; $t = 43.2$ s
15-50 $R = 4990$ m; $t = 25.5$ s
15-54 $x = 0.1529$ m
15-55 $F = 8.17$ lb $\searrow$ 50.66°
15-58 $F_b = 30.65$ N
15-59 $T = 88.7$ lb; $P = 8.87$ lb
15-64 $\theta = 65.53°$; $P = 11.84$ N
15-65 $T_A = 20.2$ lb; $T_B = 5.75$ lb;
$\omega = 3.72$ rad/s
15-67 $T = 5.22$ lb; $v = 18.39$ ft/s
$T = 30.8$ lb; $v = 31.6$ ft/s
15-68 $T = 41.6$ N; $T = 20.8$ N;
$T = 46.7$ N
15-72 $N = 31.2$ N; $d = 20.0$ m
15-73 $N = 3.02$ lb $\measuredangle$ 26.57°;

15-79 $v_C = 23,200$ ft/s; $\tau = 118.3$ min
15-80 $v_C = 7120$ m/s; $v_P = 10,070$ m/s
15-82 $e = 0.0669$; $\tau = 107.1$ min
$v_P = 7810$ m/s; $v_a = 6830$ m/s
15-86 $d = 2770$ km $= 24.9°$
15-87 $h = 929$ mi
15-90 $h_P = 2630$ km; $\tau = 218$ min
$v_P = 7440$ m/s; $v_a = 4470$ m/s
15-93 $\Delta v = 307$ ft/s; $t = 65.8$ min
15-94 $e = 0.1803$; $\tau = 347$ min
$\Delta v_A = 512$ m/s; $\Delta v_B = 467$ m/s
15-97 $h_P = 250$ mi; $h_a = 831$ mi
$h_P = 120$ mi; $h_a = 388$ mi
$h_P = 120$ mi; $h_a = 388$ mi
15-98 $v_r = 2080$ m/s; $v_\theta = 5540$ m/s
15-102 $v = 1681$ m/s $\measuredangle$ 87.48° to
radial dir
$\Delta v = 36.5$ m/s
15-103 $a = 1.175$ ft/s^2; $T = 268$ lb;
$s = 58.8$ ft
15-104 $T = 62.5$ N; $\omega = 27.8$ rev/min
15-106 $a = 507$ m/s^2 $\nearrow$ 58.744°
$N = 241$ N $\nearrow$ 57.02°
15-109 $v = 13.37$ ft/s; $y = 40$ ft
15-111 $v = 88.3$ ft/s; $N_B = 3640$ lb;
$N_C = 941$ lb
$v_{min} = 35.9$ ft/s
15-114 $\Delta v_A = 168.3$ m/s;
$\Delta v_C = 164.7$ m/s; $\phi = 11.18°$
15-117 $N = 54.9$ lb $\nearrow$ 85.66°;
$B = 17.25$ lb $\searrow$ 30°
$N = 11.47$ lb $\searrow$ 34.34°;
$B = 18.02$ lb $\measuredangle$ 30°
15-118 $e = 0.372$; $h_P = 409$ km;
$\tau = 172.2$ min
$v_P = 8980$ m/s; $v_a = 4110$ m/s
15-121 $e = 1.155$; $h = 3710$ mi;
$v = 27,400$ ft/s
15-124 $v = 211$ m/s; $y = 6340$ m
$v = 197.7$ m/s; $y = 6130$ m
15-127 $v_B = 13.57$ ft/s $\uparrow$; $T = 25.7$ lb

CHAPTER 16

16-1 $A = 275$ lb $\searrow$ 78.69°;
$B = 643$ lb $\searrow$ 78.69°
$a = 2.50$ ft/s^2 $\rightarrow$
16-2 $P_{max} = 736$ N
16-4 $A = 1659$ N $\searrow$ 81.47°;
$B = 2110$ N $\searrow$ 81.47°
$a = 0.372$ m/s^2 $\rightarrow$
16-7 $A_n = 403$ lb $\searrow$; $A_f = 120.8$ lb $\measuredangle$
$B_n = 116.9$ lb $\searrow$; $B_f = 35.1$ lb $\measuredangle$
16-9 $t = 6.47$ s
16-11 $a = 21.4$ ft/s^2 $\rightarrow$
16-12 $a = 2.45$ m/s^2; $A_n = 971$ N;
$B_n = 1482$ N
16-15 $\ddot{\theta} = 2.42$ rad/s^2 $\downarrow$; $T_A = 102.4$ lb;
$T_B = 205$ lb
16-16 $L_A = 3.79$ N (C); $L_B = 52.8$ N (T)
$a_G = 8.50$ m/s^2 $\nearrow$ 60°

16-19 $W_B = 499$ lb
16-20 $t = 5.92$ s
16-22 $m_B = 150$ kg
16-27 $a_G = 24.2$ ft/s^2 $\downarrow$; $A = 5$ lb $\uparrow$
16-28 $\alpha = 9.10$ rad/s^2 $\downarrow$;
$A = 531$ N $\nearrow$ 22.12°
16-31 $a_G = 21.6$ ft/s^2 $\nearrow$ 63.44°;
$A = 50$ lb $\searrow$ 53.13°
16-32 $a_G = 7.11$ m/s^2 $\nearrow$ 67.00°;
$A = 343$ N $\searrow$ 49.67°
16-36 $\alpha = 11.41$ rad/s^2 $\downarrow$; $T = 126.6$ N
16-37 $\alpha = 2.84$ rad/s^2 $\downarrow$; $T_A = 54.4$ lb;
$T_B = 84.7$ lb
16-41 $a_B = 0.595$ ft/s^2 $\uparrow$; $T = 50.9$ lb
16-42 $B_x = 2180$ N $\rightarrow$; $B_y = 3860$ N $\downarrow$;
$t = 2.22$ s
16-43 $\omega = 8.26$ rad/s $\downarrow$; $A = 57.9$ lb
$\omega = 11.68$ rad/s $\downarrow$; $A = 155.9$ lb
16-46 $\theta = 166.50°$; $\omega = 6.83$ rad/s γ;
$A = 262$ N
16-47 $A_x = 77.6$ lb $\leftarrow$; $A_y = 10.25$ lb $\uparrow$
16-51 $a_G = 10.80$ ft/s^2 $\nearrow$ 28°; $\mu = 0.152$
16-52 $\alpha = 18.43$ rad/s^2 $\downarrow$; $\mu = 0.155$
16-54 $a_G = 4$ m/s^2 $\rightarrow$; $\alpha = 30.7$ rad/s^2 $\downarrow$
$a_G = 4.60$ m/s^2 $\rightarrow$;
$\alpha = 23.0$ rad/s^2 $\downarrow$
16-55 $\alpha = 46.4$ rad/s^2 γ; $T = 6.40$ lb
16-57 $C_n = 19.39$ lb $\searrow$; $C_f = 1.932$ lb $\measuredangle$
$\alpha = 23.2$ rad/s^2 $\downarrow$
16-61 $a_B = 4.34$ ft/s^2 $\searrow$; $T = 9.63$ lb;
$\mu = 0.0021$
16-62 $a_B = 0.755$ m/s^2 $\downarrow$; $T = 453$ N
16-65 $a_B = 5.54$ ft/s^2 $\downarrow$;
$\alpha = 12.13$ rad/s^2 $\downarrow$
$T = 16.56$ lb
16-66 $a_B = 2.09$ m/s^2 $\downarrow$; $T_A = 545$ N;
$T_B = 386$ N
16-69 $F = 6.32$ lb $\measuredangle$; $B = 10.96$ lb $\uparrow$
16-72 $\alpha = 6.37$ rad/s^2 γ; $A = 63.7$ N $\leftarrow$;
$B = 12.26$ N $\uparrow$
16-74 $t = 1.487$ s; $s = 8.78$ m
16-75 $s = 18.00$ ft
16-80 $\alpha_{AB} = 9.44$ rad/s^2 $\downarrow$; $T_B = 60.3$ N
16-85 $A = -219$ $\mathbf{k}$ lb; $B = 249$ $\mathbf{k}$ lb
16-86 $A = -2940$ $\mathbf{k}$ N; $B = -1462$ $\mathbf{k}$ N
16-91 $A = -850$ $\mathbf{i} - 1907$ $\mathbf{j} + 43$ $\mathbf{k}$ lb
$B = -1918$ $\mathbf{i} - 850$ $\mathbf{j}$ lb;
$T = 9.12$ lb · ft
16-92 $A = -4310$ $\mathbf{i} - 9730$ $\mathbf{j}$
$+ 161.9$ $\mathbf{k}$ N
$B = -9740$ $\mathbf{i} - 4350$ $\mathbf{j}$ N;
$T = -5.25$ N · m
16-94 $A = -1957$ $\mathbf{k}$ N; $B = 2450$ $\mathbf{k}$ N
16-95 $A = 339$ $\mathbf{k}$ lb; $B = -279$ $\mathbf{k}$ lb
16-101 $A = 339$ $\mathbf{k}$ lb; $B = -279$ $\mathbf{k}$ lb
16-102 $A = -1957$ $\mathbf{k}$ N; $B = 2450$ $\mathbf{k}$ N
16-104 $A = -0.400$ $\mathbf{i} - 27.4$ $\mathbf{j}$ N;
$B = 0.400$ $\mathbf{i} + 105.8$ $\mathbf{j}$ N
16-107 $A = -665$ $\mathbf{i} + 9.59$ $\mathbf{j}$ lb
$C = -180.7$ $\mathbf{j} + 19.55$ $\mathbf{k}$ lb · ft
16-108 $A = -44.7$ $\mathbf{i}$ kN; $B = -2.13$ $\mathbf{i}$ kN
$N = 0.491$ kN; $F = 0$ kN
16-111 $a_C = 2.53$ ft/s^2 $\downarrow$; $T = 922$ lb

$B_x = 553$ lb $\rightarrow$; $B_y = 958$ lb $\downarrow$

16-112 $F = 467$ N; $A_f = 83.3$ N

16-115 $N_r = 570$ lb; $N_f = 2530$ lb; $\mu = 0.770$

16-116 $\alpha = 19.03$ rad/s^2 $\downarrow$; $C = 383$ N $\angle$ $79.56°$

16-121 $\omega_{AB} = 1.925$ rad/s $\downarrow$; $\alpha_{AB} = 13.63$ rad/s^2 $\downarrow$; $B = 88.0$ lb $\angle$ $58.18°$

16-122 $\alpha_A = 12.19$ rad/s^2 $\downarrow$; $T = 13.33$ N $C_n = 256$ N; $D_n = 169.2$ N

16-124 $L_C = 459$ N (T); $L_D = 56.4$ N (T); $F = 147.2$ N $\searrow$ $30°$

16-126 $\alpha = 16.39$ rad/s^2 $\downarrow$; $a_G = 2.41$ m/s^2 $\nearrow$ $24.16°$ $A_n = 221$ N $\uparrow$; $A_f = 57.3$ N $\leftarrow$

CHAPTER 17

17-1 $U_F = -94.0$ lb $\cdot$ ft; $U_W = 85.5$ lb $\cdot$ ft

17-2 $U_F = -6.00$ kJ; $U_W = -25.0$ kJ

17-5 $U_T = 200(10^6)$ lb $\cdot$ ft; $U_W = 0$ lb $\cdot$ ft

17-8 $U\mu = -0.405$ kJ; $U_W = -1.472$ kJ; $U_F = 2.60$ kJ

17-9 $U_5 = 3.75$ lb $\cdot$ ft; $U_F = -0.750$ lb $\cdot$ ft; $U_s = -5.63$ lb $\cdot$ ft

17-10 $U_{F5} = -2.94$ J; $U_s = 135.0$ J $U_{W10} = -22.5$ J; $U_{F10} = -3.78$ J

17-13 $U_s = 20.8$ lb $\cdot$ ft; $U_W = -0.220$ lb $\cdot$ ft

17-14 $U = -11.10$ J

17-17 $U_W = -3.00$ lb $\cdot$ ft; $U_{12} = 8.76$ lb $\cdot$ ft

17-18 $U = 73.6$ J

17-21 $\ell = 2160$ ft

17-22 $s = 50.3$ m

17-25 $P = 71{,}300$ lb

17-26 $s = 17.69$ m; $s = 31.5$ m

17-28 $v = 6.65$ m/s; $d = 11.27$ m

17-29 $v_0 = 14.02$ ft/s

17-31 $v_{max} = 21.4$ ft/s; $x_{max} = 21.1$ in.; $F = 110.6$ lb

17-34 3.86 m/s $< v_0 < 3.96$ m/s

17-37 $v = 11.81$ ft/s

17-38 $v_{max} = 0.900$ m/s; $\delta = 158.2$ mm; $\delta_{max} = 277$ mm

17-41 $\theta = 29.10°$

17-42 $h = 375$ mm; $N = 1.472$ N

17-49 $v_A = 16.05$ ft/s $\rightarrow$; $v_B = 8.02$ ft/s $\downarrow$

17-50 $v_A = 2.56$ m/s; $b = 1.333$ m

17-52 $v_A = 2.17$ m/s; $v_A = 2.40$ m/s

17-55 $b = 0.988$ ft

17-58 $k = 157.0$ N/m

17-59 $v_{max} = 1.311$ ft/s; $\delta = 0.2$ ft; $\delta_{max} = 0.4$ ft Will rebound

17-62 $b = 0.236$ m

17-63 $a = 1.234$ ft

17-66 $v_2 = 8.17$ m/s; $x_{max} = 0.896$ m

17-69 $\mu_k = 0.318$

17-70 $\delta = 0.262$ m; 6.53 m left of init. pos.

17-73 $v_{10} = 15.37$ ft/s; $\delta = 1.400$ in.

17-74 $v_A = 6.49$ m/s; $v_B = 5.85$ m/s

17-77 $v = 11.81$ ft/s

17-78 $v_{max} = 0.900$ m/s; $\delta = 158.2$ mm; $\delta_{max} = 277$ mm

17-81 $v = 18.58$ ft/s

17-82 $v_0 = 2.30$ m/s

17-85 $v = 1.139$ ft/s

17-86 $v = 9.64$ km/h; $\mathcal{P} = 856$ w

17-89 $v = 2.64$ ft/s; $x_{max} = 7.42$ in.

17-90 $v_2 = 8.17$ m/s; $x_{max} = 0.896$ m

17-96 2.02 m/s $< v_0 < 2.12$ m/s

17-97 $T_A = 45.3$ lb; $T_B = 21.3$ lb

17-101 $\mathcal{P} = 28.5$ hp; $\mathcal{P} = 43.7$ hp

17-103 $b_{min} = 9.89$ in.; $v_C = 6.95$ ft/s

17-104 $\mathcal{P} = 31.0$ kw; $v = 1.333$ m/s

17-106 $\delta_{max} = 0.292$ m; $\delta_{eq} = 0.0491$ m

17-109 $\mathcal{P} = 21.7$ hp; $v = 51.6$ mi/h

CHAPTER 18

18-1 $\theta = 16.68$ rev

18-2 $M = 616$ N $\cdot$ m

18-3 $W_A = 96.0$ lb

18-7 $\omega = 9.10$ rad/s; $B = 9.13$ lb $\angle$ $63.92°$

18-8 $\omega = 5.42$ rad/s; $A = 2470$ N $\searrow$ $51.65°$

18-11 $\theta_s = 51.98°$

18-12 $\theta_m = 53.13°$

18-14 $d = 18.09$ m; $d_P = 12.06$ m

18-17 $\theta_C = 55.15°$; $\theta_h = 60.00°$; $\theta_s = 53.97°$

18-21 $\omega = 22.8$ rad/s; $v_G = 15.22$ ft/s

18-22 $F = 0$ N; $N = 285$ N

18-25 $v_C = 7.65$ ft/s; $\omega = 15.30$ rad/s; $v_A = 15.30$ ft/s

18-26 $v_C = 4.15$ m/s; $\omega = 27.7$ rad/s; $v_A = 5.53$ m/s

18-28 $y_{max} = 0.239$ m $\downarrow$; $y_{max} = 0.218$ m $\uparrow$ $v_C = 0.515$ m/s; $\omega = 3.43$ rad/s; $v_A = 1.029$ m/s

18-31 $\omega = 11.79$ rad/s $\circlearrowright$; $v_A = 59.0$ ft/s $\rightarrow$

18-32 $\omega = 9.38$ rad/s $\circlearrowright$; $v_A = 19.60$ m/s $\angle$ $19.43°$

18-35 $v_A = 15.66$ ft/s $\downarrow$

18-36 $v_A = 5.48$ m/s $\downarrow$

18-39 $\theta_C = 48.19°$; $\omega = 1.638$ rad/s; $v_B = 13.10$ ft/s

18-42 $\omega_C = 11.50$ rad/s $\downarrow$; $v_A = 0.415$ m/s $\rightarrow$

18-43 $\omega_C = 6.71$ rad/s $\downarrow$; $v_B = 7.12$ ft/s $\searrow$ $45.00°$

18-46 $h_{max} = 1.559$ m; $N = 94.2$ N $\uparrow$ $\omega = 1.657$ rad/s $\downarrow$;

$v_C = 1.134$ m/s $\rightarrow$

18-49 $\omega_{AB} = 5.62$ rad/s $\downarrow$; $v_C = 4.33$ ft/s $\rightarrow$ $\omega_{AB} = 6.49$ rad/s $\downarrow$; $v_C = 8.66$ ft/s $\leftarrow$

18-50 $\omega_{AB} = 123.5$ rad/s $\downarrow$; $v_C = 18.52$ m/s $\rightarrow$ $\omega_{AB} = 206$ rad/s $\downarrow$; $v_C = 21.2$ m/s $\rightarrow$

18-57 $T = 85.3$ lb $\cdot$ ft

18-58 $T = 199.4$ J

18-60 $T = 137.4$ J

18-63 $T = 171.56$ lb $\cdot$ ft

18-65 $T = 34.4$ lb $\cdot$ ft

18-66 $T = 9.36$ J

18-68 $\omega = 9.22$ rad/s

18-71 $T = 15.94$ lb $\cdot$ ft

18-73 $T = 7.28$ lb $\cdot$ ft

18-74 $T = 213$ J

18-77 $d = 149.2$ ft; $d = 134.3$ ft

18-78 $T = 8.57$ J

18-80 $v = 4.45$ m/s

18-83 $v_C = 15.36$ ft/s; $v_s = 15.90$ ft/s; $v_h = 13.77$ ft/s

18-85 $\omega = 75.7$ rad/s

18-86 $\omega = 1.147$ rad/s; $A = 405$ N $\angle$ $29.01°$

18-89 $v = 6.95$ ft/s; $v = 6.38$ ft/s

18-90 $R = 102.6$ N $\searrow$ $89.98°$; $R = 68.6$ N $\searrow$ $87.83°$

18-92 $v = 4.27$ m/s; $P = 491$ N; $s = 1.038$ m

18-93 $\omega = 82.1$ rad/s; $\omega = 43.9$ rad/s

CHAPTER 19

19-1 $F = 0.0828$ lb; $\mu_k = 0.207$

19-2 $t = 14.16$ s

19-4 $F = 185.7$ N; $\mu_k = 0.183$

19-7 $v_a = 51.5$ ft/s; $v_b = 25.8$ ft/s

19-9 $v = 369$ mi/h

19-10 $v_5 = 15.90$ m/s $\angle$ $21.19°$ $v_{10} = 19.27$ m/s $\angle$ $26.57°$ $v_{15} = 15.90$ m/s $\angle$ $21.19°$

19-13 $P = 187.6$ lb $\angle$ $16.54°$

19-16 $t_1 = 7.85$ s; $t_f = 19.59$ s $t_m = 14.11$ s; $v_{max} = 7.50$ m/s

19-17 $t_1 = 6.01$ s; $t_f = 25.6$ s $t_m = 15.81$ s; $v_{max} = 74.0$ ft/s

19-19 $\mathbf{r}_5 = 11.40$ i $+ 7.40$ j $+ 4.80$ k ft $\mathbf{v}_5 = 9.60$ i $+ 0.60$ j $- 26.4$ k ft/s

19-20 $\mathbf{r}_G = 5.67$ i $+ 5.17$ j m $\mathbf{v}_G = 0.667$ i $+ 2.33$ j m/s

19-23 $v = 19.80$ ft/s $\angle$ $53.90°$

19-26 $v_B = 9.62$ km/h

19-27 $\mathbf{F}\delta t = 0.1035$ i $+ 0.1035$ j lb $\cdot$ s $\mathbf{v}_{A/B} = 6.67$ i $+ 6.67$ j ft/s $\Delta t = 1.500$ s

19-30 $v = 13.64$ m/s; $s = 0.0903$ m

19-32 $\mathbf{r}_3 = -31.7$ i $+ 1079$ j m; $t = 24.4$ s

F$\delta t = 7.88\,\mathbf{i} - 72.3\,\mathbf{j} + 7.43\,\mathbf{k}$ N · s; $F = 24.4$ kN

19-33 $\mathbf{r}_4 = -877\,\mathbf{i} + 5030\,\mathbf{j}$ ft; $t = 16.70$ s
F$\delta t = 9.32\,\mathbf{i} - 2.15\,\mathbf{j} + 13.46\,\mathbf{k}$ lb · s; $F = 16,510$ lb

19-37 $v_A = 2.29$ ft/s, $v_B = 3.19$ ft/s $\Delta T = 16.54$ %; $F = 39.6$ lb

19-38 $v_A = -1.091$ m/s, $v_B = 0.309$ m/s $\Delta T = 46.4\%; F = 309$ N

19-40 $v_A = -0.900$ m/s, $v_B = 2.70$ m/s $\Delta T = 19.00\%; F = 1140$ N

19-43 $v_A = 1.816$ ft/s, $v_B = 3.48$ ft/s $v_C = 6.08$ ft/s, $\Delta T = 21.5\%$

19-44 $v_A = -1.216$ m/s, $v_B = 1.616$ m/s $v_C = 1.677$ m/s, $\Delta T = 60.3\%$

19-46 $\theta_A = 2.87°; \theta_B = 40.97°$
19-47 $b = 4.39$ ft
19-50 $v_B = 1.521$ m/s; $d = 1.180$ m
19-53 $v_A = 4.25$ ft/s $\nwarrow 80.93°$; $v_B = 16.71$ ft/s $\measuredangle 20.31°$
19-54 $v_C = 4.33$ m/s $\nwarrow 60.83°$
19-57 $d = 4.16$ ft
19-58 $b = 0.630$ m; $c = 1.031$ m; $d = 0.438$ m
19-61 $x = 1.839$ ft
19-63 $\mathbf{H}_O = -0.311\,\mathbf{k}$ lb · ft · s; $\mathbf{H}_C = -0.311\,\mathbf{k}$ lb · ft · s $T = 3.11$ lb · ft
19-64 $\mathbf{H}_O = -10\,\mathbf{k}$ N · m · s; $\mathbf{H}_G = -5\,\mathbf{k}$ N · m · s
19-66 $v = 12.5$ m/s
19-69 $\dot\theta = 6.05$ rad/s
19-70 $\dot\theta = 8.33$ rad/s
19-75 $v_i = 4.63$ ft/s; $v = 9.27$ ft/s
19-76 $v = 3.11$ m/s, $\theta = 25.37°$ $r_{max} = 0.834$ m; $v = 1.199$ m/s $v = 4.47$ m/s, $\theta = 158.12°$
19-79 $F_x = 25.4$ lb →; $F_y = 14.65$ lb ↓
19-80 $F = 96.2$ N →
19-83 $P = 155.3$ lb ←
19-85 $F_b = 57.4$ lb
19-86 $F = 384$ kN →
19-88 $v = 17.36$ m/s
19-91 $P = 4350$ lb; $v_{max} = 11,300$ ft/s
19-93 $P = 4350$ lb; $v = 3530$ ft/s; $v_{max} = 14,160$ ft/s
19-94 $v = 1257$ m/s
19-96 $\dot{y} = 2.37$ m/s; $y_{max} = 4.62$ m
19-99 $F = 116,500$ lb ↑; $v_r = 10.83$ ft/s ←
19-100 **F**$\delta t = -2.00\,\mathbf{i} + 7.50\,\mathbf{j}$ N · s $\mathbf{v}_{A/B} = -1.666\,\mathbf{i} + 6.25\,\mathbf{j}$ m/s $y_B = 7.50$ m; $\Delta t = 1.000$ s
19-103 $d = 47.5$ ft; $\Delta T = 40.7$ %
19-104 $F_b = 6.26$ N
19-106 $v = 1.917$ m/s; $v_i = 117.0$ m/s
19-109 $\theta_{max} = 48.84°; v = 4.86$ ft/s; $T = 2.62$ lb
19-113 $v = 20.1$ ft/s; $v_i = 1304$ ft/s
19-114 $v = 2.38$ m/s →; $v = 0.077$ m/s ←

CHAPTER 20

20-1 $t = 81.3$ s; $T = 4.88$ lb · in.
20-2 $M = 0.0503$ N · m
20-3 $M_0 = 1.905$ lb · ft
20-7 $t = 12.16$ s; $t = 9.76$ s
20-9 $P = 90.1$ lb
20-10 $t = 4.96$ s; $t = 19.77$ s
20-13 $t = 12.20$ s
20-15 $t_f = 12.02$ s; $\omega_f = 65.3$ rad/s $\downarrow$; $v_f = 38.1$ ft/s ←
20-16 $v_0 = 4.27$ m/s $\measuredangle$
20-19 $b = 1.333$ ft
20-20 $\omega = 2.34$ rad/s $\downarrow$; $A = 1094$ N
20-23 $\omega = 6.41$ rad/s $\downarrow$; $B = 76.5$ lb $P = 76.5$ lb; $\Delta T = 69.7\%$
20-24 $\omega = 8.11$ rad/s $\downarrow$; $F = 10.38$ kN; $A = 649$ N $\Delta T = 98.8\%; \theta = 141.78°$
20-29 $\omega_f = 3v_0/\ell$ $\downarrow$; $v_{Gf} = v_0/2$ ↓
20-32 $m_d = 1.570$ kg
20-33 $v_b = 19.36$ ft/s $\nwarrow 84.88°$; $\omega = 3.66$ rad/s $\downarrow$ $A = 103.0$ lb; $\theta = 54.28°$
20-35 $\omega = 4.86$ rad/s $\downarrow$; $\phi = 59.34°$ $A = 367$ lb; $\Delta T = 51.0\%$
20-38 $\theta = 66.2°; \omega = 21.7$ rad/s $\downarrow$; $v_G = 0.890$ m/s ↑
20-39 $\omega_{AB} = 0.432$ rad/s $\downarrow$; $\omega_{CD} = 5.13$ rad/s $\downarrow$ $\theta = 30.38°; E = 256$ lb $\nwarrow 68.95°$
20-40 $\omega_{AB} = 12.13$ rad/s $\downarrow$; $\omega_{CD} = 15.56$ rad/s $\downarrow$ $E = 6930$ N $\nwarrow 68.61°$
20-44 $\omega_{AB} = 11.05$ rad/s $\downarrow$; $\omega_{CD} = 14.17$ rad/s $\downarrow$ $v_G = 2.23$ m/s $\measuredangle 39.37°$
20-49 $\mathbf{H}_O = -0.01098\,\mathbf{j} + 0.00549\,\mathbf{k}$ ft · lb · s; $\theta = 63.43°$
20-50 $\mathbf{H}_O = 0.0424\,\mathbf{i} + 0.0566\,\mathbf{j}$ N ·m · s; $\theta = 36.87°$
20-52 $\mathbf{H}_O = -0.848\,\mathbf{j} + 3.31\,\mathbf{k}$ N · m · s; $\theta = 14.38°$
20-55 $\mathbf{H}_O = 0.1801\,\mathbf{j} + 0.0446\,\mathbf{k}$ ft · lb · s; $\theta = 13.90°$
20-57 $\mathbf{H}_O = 0.0350\,\mathbf{j} + 0.0211\,\mathbf{k}$ ft · lb · s; $\theta = 31.16°$
20-58 $\mathbf{v}_G = -2\,\mathbf{i}$ m/s; $\boldsymbol{\omega} = 40\,\mathbf{j} + 15\,\mathbf{k}$ rad/s; $\theta = 48.89°$
20-60 $\theta = 0°; \mathbf{v}_A = -3.50\,\mathbf{i}$ m/s
20-61 $\theta = 0°; \mathbf{v}_C = -5.64\,\mathbf{i}$ ft/s
20-64 $\theta = 54.09°$; $\mathbf{v}_C = 3.18\,\mathbf{i} + 1.010\,\mathbf{j} - 0.882\,\mathbf{k}$ m/s
20-67 $\theta = 25.57°$; $\mathbf{v}_A = 2.44\,\mathbf{i} + 3.44\,\mathbf{j} - 2.83\,\mathbf{k}$ ft/s
20-68 $\mathbf{v}_G = -1.402\,\mathbf{i}$ m/s; $\boldsymbol{\omega} = 28.0\,\mathbf{j} + 10.51\,\mathbf{k}$ rad/s $\mathbf{v}_C = -9.81\,\mathbf{i}$ m/s
20-71 $F\Delta t = 1.057$ lb · s; $\mathbf{H}_G = 0.264\,\mathbf{j} + 0.264\,\mathbf{k}$ lb · ft · s $\mathbf{v}_C = -119.2\,\mathbf{i}$ ft/s
20-72 $\mathbf{v}_G = -2.20\,\mathbf{i} - 0.1491\,\mathbf{j} + 0.316\,\mathbf{k}$ m/s

$\boldsymbol{\omega} = 0.790\,\mathbf{i} + 26.2\,\mathbf{j} + 14.99\,\mathbf{k}$ rad/s $\mathbf{v}_C = -12.12\,\mathbf{i} - 0.0306\,\mathbf{j} + 0.632\,\mathbf{k}$ m/s
20-75 $F\Delta t = 2.68$ lb · s; $M\Delta t = 0.865$ lb · ft · s $\mathbf{H}_G = 0.0469\,\mathbf{j} + 0.0469\,\mathbf{k}$ lb · ft · s $\mathbf{v}_C = -60.6\,\mathbf{i} - 5.37\,\mathbf{j} + 5.37\,\mathbf{k}$ ft/s
20-77 $\omega_A = 201$ rad/s $\downarrow$; $\omega_B = 32.2$ rad/s $\downarrow$ $\omega_A = 603$ rad/s $\downarrow$; $\omega_B = 96.6$ rad/s $\downarrow$ $\omega_A = 1005$ rad/s $\downarrow$; $\omega_B = 161.0$ rad/s $\downarrow$
20-78 $v = 14.20$ m/s
20-81 $v_0 = 1.051\sqrt{ag}$; $v_G = 0.5574\sqrt{ag}$ $\measuredangle 45°$ $\omega = 0.7882\sqrt{g/a}$
20-82 $h = 7r/5$
20-85 $v_A = 0; \omega_A = v_0/r; v_B = v_0; \omega_B = 0$ $v_A = 2v_0/7; v_B = 5v_0/7$
20-89 $\omega_{AB} = 0.748$ rad/s $\downarrow$; $\omega_{CD} = 4.81$ rad/s $\downarrow$ $\theta_{AB} = 8.73°; \theta_{CD} = 58.62°$
20-90 $\omega = 8.04$ rad/s $\downarrow$; $v_G = 1.607$ m/s $\measuredangle 60°$ $\Delta T = 58.7\%; v_0 = 1.841$ m/s →

CHAPTER 21

21-1 $\dot{x}(t) = -8\pi \sin \pi t$ in./s $\ddot{x}(t) = -8\pi^2 \cos \pi t$ in./s²
21-2 $\dot{x}(t) = (5\pi/4) \cos \pi t/4$ mm/s $\ddot{x}(t) = -(5\pi^2/16) \sin \pi t/4$ mm/s²
21-4 $\dot{x}(t) = (30\pi/4) \cos (3\pi t/4 + \pi/8)$ mm/s $\ddot{x}(t) = -(90\pi^2/16) \sin(3\pi t/4 + \pi/8)$ mm/s²
21-7 $x(t) = 5 \cos (\pi t + 0.9273)$ in. $v_{max} = 5\pi$ in./s at $x = 0$ in. $a_{max} = 5\pi^2$ in./s² at $x = -5$ in.
21-9 $x(t) = 10 \cos (10t - 0.6435)$ in. $v_{max} = 100$ in./s at $x = 0$ in. $a_{max} = 1000$ in./s² at $x = -10$ in.
21-10 $x(t) = 26 \cos (3\pi t/4 + 1.1760)$ mm $v_{max} = 61.3$ mm/s at $x = 0$ mm. $a_{max} = 144.3$ mm/s² at $x = -26$ mm
21-13 $x(t) = 13 \sin (\pi t + 2.7468)$ in. $t = 0.1257$ s; $t = 0.626$ s
21-14 $x(t) = 5 \sin (\pi t/2 + 0.9273)$ mm $t = 1.410$ s; $t = 0.410$ s
21-17 $x(t) = 5 \sin (\pi t + \pi/2)$ in. $t = 0.5$ s; $t = 0$ s
21-20 $v = 1.732$ m/s
21-21 $k = k_1 + k_2$

21-22 $k = \dfrac{k_1 k_2}{k_1 + k_2}$

21-26 $\ddot{x} + 700x = 0$; $\tau_n = 0.237$ s;
$A = 39.2$ mm
$x(t) = 30.2 \sin 26.458t$
$\qquad + 25.0 \cos 26.458t$ mm
$t_1 = 0.0926$ s

21-28 $\ddot{x} + 181.8x = 0$; $\omega_n = 13.48$ rad/s;
$\tau_n = 0.466$ s

21-29 $\ddot{y} + 322y = 0$; $\omega_n = 17.94$ rad/s;
$\tau_n = 0.350$ s

21-31 $\ddot{y} + 718y = 0$; $\omega_n = 26.8$ rad/s;
$\tau_n = 0.235$ s

21-34 $\ddot{y}_G + 533y_G = 0$; $\omega_n = 23.1$ rad/s;
$\tau_n = 0.272$ s

21-35 $\ddot{x}_G + 178.6x_G = 0$;
$\omega_n = 13.36$ rad/s; $\tau_n = 0.470$ s

21-37 $\ddot{\theta} + 30.9\theta = 0$;
$\omega_n = 5.56$ rad/s; $\tau_n = 1.130$ s

21-40 $\ddot{\theta} + 1650\,\theta = 0$;
$v_{C\text{max}} = 0.609$ m/s

21-42 $\ddot{\theta} + 127.9\,\theta = 0$;
$\omega_n = 11.31$ rad/s; $\tau_n = 0.556$ s

21-43 $\ddot{y} + 95.4y = 0$;
$\omega_n = 9.77$ rad/s; $\tau_n = 0.643$ s

21-47 underdamped
$\dot{x}(t) = e^{-0.1t}[-\cos(5t - 1.2)$
$\qquad - 50\sin(5t - 1.2)]$ in./s
$\ddot{x}(t) = e^{-0.1t}[-250\cos(5t - 1.2)$
$\qquad + 10\sin(5t - 1.2)]$ in./s^2

21-48 critically damped
$\dot{x}(t) = (-7 - 6t)e^{-2t}$ mm/s
$\ddot{x}(t) = (8 + 12t)e^{-2t}$ mm/s^2

21-50 underdamped
$\dot{x}(t) = e^{-0.05t}[-18.4\cos 3t$
$\qquad - 23.7\sin 3t]$ mm/s
$\ddot{x}(t) = e^{-0.05t}[-70.2\cos 3t$
$\qquad + 56.4\sin 3t]$ mm/s^2

21-52 critically damped
$\dot{x}(t) = (8 - 7.5t)e^{-1.5t}$ rad/s
$\ddot{x}(t) = (-19.5 + 11.25t)e^{-1.5t}$
$\qquad$ rad/s^2

21-55 critically damped
$\dot{x}(t) = (-9 + 2t)e^{-0.2t}$ in./s
$\ddot{x}(t) = (3.8 - 0.4t)e^{-0.2t}$ in./s^2

21-57 critically damped
$\dot{x}(t) = (-5.8 + 1.2t)e^{-1.2t}$ rad/s
$\ddot{x}(t) = (8.16 - 1.44t)e^{-1.2t}$ rad/s^2

21-58 underdamped
$\dot{x}(t) = e^{-0.15t}[-0.9\sin(10t -$
$\qquad 2.5) + 60\cos(10t - 2.5)]$
$\qquad$ mm/s
$\ddot{x}(t) = e^{-0.15t}[-600\sin(10t -$
$\qquad 2.5) - 18\cos(10t - 2.5)]$ mm/s^2

21-61 underdamped
$x(t) = e^{-5t}[3\cos 7.416t$
$\qquad + 4.05\sin 7.416t]$ in.
$\dot{x}(t) = e^{-5t}[15\cos 7.416t$
$\qquad - 42.5\sin 7.416t]$ in./s
$\ddot{x}(t) = e^{-5t}[-390\cos 7.416t$
$\qquad + 101.1\sin 7.416t]$ in./s^2

21-62 overdamped
$x(t) = -31.8e^{-5.53t}$
$\qquad + 1.771e^{-14.47\,t}$ mm
$\dot{x}(t) = 175.6e^{-5.53t}$
$\qquad - 25.6e^{-14.47t}$ mm/s
$\ddot{x}(t) = -971e^{-5.53t}$
$\qquad + 371e^{-14.47t}$ mm/s^2

21-64 underdamped
$x(t) = e^{-t}[100\cos 4.359t$
$\qquad + 57.4\sin 4.359t]$ mm
$\dot{x}(t) = e^{-t}[150\cos 4.359t$
$\qquad - 493\sin 4.359t]$ mm/s
$\ddot{x}(t) = e^{-t}[-2300\cos 4.359t$
$\qquad - 160.6\sin 4.359t]$ mm/s^2

21-67 critically damped
$x(t) = (-15 - 75t)e^{-5t}$ in.
$\dot{x}(t) = 375te^{-5t}$ in./s
$\ddot{x}(t) = (375 - 1875t)e^{-5t}$ in./s^2

21-69 $c = c_1 + c_2$

21-72 $\ddot{y} + 25\dot{y} + 667y = 0$; $\tau_d = 0.278$ s;
$t_1 = 0.0240$ s
$y(t) = e^{-12.5t}[-5\cos 22.592t$
$\qquad + 8.30\sin 22.592t]$ mm

21-74 $4\ddot{y} + 125\dot{y} + 6000y = 0$;
$\tau_d = 0.1773$ s; $t_1 = 0.0552$ s
$y(t) = e^{-15.63t}[15\cos 35.438t$
$\qquad - 14.55\sin 35.438t]$ mm

21-75 $\zeta = 2.44$; overdamped;
$a = 12.18$ in.
No frequency or period exists

21-77 $\zeta = 0.806$; underdamped;
$c = 18.61$ lb · s/ft
$\omega_d = 8.31$ rad/s; $\tau_d = 0.756$ s

21-80 $c = 8.03$ N · s/m

21-81 $\zeta = 0.292$; underdamped
$\omega_d = 10.05$ rad/s; $\tau_d = 0.625$ s

21-83 $\zeta = 3.17$; overdamped
No frequency or period exists

21-84 $\zeta = 11.04$; overdamped
No frequency or period exists

21-89 $0.311\ddot{y} + 5\dot{y} + 140y = 70\sin 30t$
$y(t) = e^{-8.05t}[10\cos 19.647t$
$\qquad + 16\sin 19.647t]$
$\qquad + 4.10\sin(30t - 2.320)$ in.

21-91 $0.621\ddot{x} + 10\dot{x} + 70x = 40\sin 12t$
$x(t) = e^{-8.05t}[0.897\cos 6.921t$
$\qquad + 9.36\sin 6.921t]$
$\qquad + 3.95\sin(12t - 1.7314)$ in.

21-92 $4\ddot{y} + 125\dot{y} + 6000y = 150\sin 18t$
$y(t) = e^{-15.63t}[-2.59\cos 35.438t$
$\qquad + 6.84\sin 35.438t]$
$\qquad + 28.8\sin(18t - 0.4462)$ mm

21-94 $D_{\max} = 34.0$ mm;
$5.30 \le \Omega \le 8.63$ rad/s

21-97 $D = 2.05$ in.

21-98 $20\ddot{x} + 40\dot{x} + 500x = 2.50\sin 8t$
$\qquad - 1.600\cos 8t$
$x(t) = 3.52\sin(8t - 2.183)$ mm

21-102 $D = 1.46$ mm

21-104 $D = 34.5$ mm

21-105 $D = 0.983$ in.

21-112 $\ddot{x} + 700x = 0$

21-114 $\ddot{x} + 181.8x = 0$

21-115 $\ddot{y} + 322y = 0$

21-118 $\omega_n = 26.5$ rad/s

21-120 $\omega_n = 13.48$ rad/s

21-121 $\omega_n = 17.94$ rad/s

21-124 $\ddot{y}_G + 533y_G = 0$

21-126 $\ddot{\theta} + 1650\,\theta = 0$

21-127 $\ddot{y} + 95.4y = 0$

21-129 $\omega_n = 13.36$ rad/s

21-130 $\omega_n = 5.72$ rad/s

21-133 $\omega_n = 9.77$ rad/s

21-135 $k = 21.8$ lb/ft;
$c = 0.00873$ lb · s/ft

21-136 $4.5\,\ddot{\theta} + 5.4\,\dot{\theta} + 49.1\,\theta =$
$19.28\sin(5t - 0.2355)$
$D = 0.280$ rad

21-138 $F_k = 114.3$ N; $F_C = 4.86$ N

21-141 $\ddot{y} + 154.5y = 0$; $\tau_n = 0.505$ s;
$A = 6.22$ in.
$F_P = 1.743$ lb; $F_P = 0.5$ lb;
$h = 2.5$ ft

21-143 $\ddot{\theta} + 24\,\theta = 0$; $\omega_n = 4.90$ rad/s;
$A = 0.266$ rad
$\theta(t) = 0.266\sin 4.899t$ rad

21-144 $\ddot{y} + 289y = 0$; $\omega_n = 16.99$ rad/s;
$A = \delta$
$\delta_{\max} = 50$ mm;
$y(t) = -50\cos 16.986t$ mm

INDEX

PHOTO CREDITS

Chapter 12 Opener: G.V. Foint/The Image Bank.

Chapter 13 Opener: Reza Estakhrian/Tony Stone Images/New York, Inc.

Chapter 14 Opener: D. Lynn Waldron/The Image Bank. Page 131: Courtesy Alice Halliday.

Chapter 15 Opener: Robin Smith/Tony Stone Images/New York, Inc. Page 194: Ken Levine/Allsport.

Chapter 16 Opener: Peter Newton/Tony Stone Images/New York, Inc.

Chapter 17 Opener: Jean-François Causse/Tony Stone Images/New York, Inc. Page 334: Hartwell/Sygma.

Chapter 18 Opener: Arthur Tilley/FPG International.

Chapter 19 Opener: Washnik/The Stock Market.

Chapter 20 Opener: Chris Rogers/The Stock Market.

Chapter 21 Opener: Louis Bencze/Tony Stone Images/New York, Inc.

TABLE 1-6 CONVERSION FACTORS BETWEEN THE SI AND U.S. CUSTOMARY SYSTEMS

Quantity	U. S. Customary to SI	SI to U. S. Customary
Length	1 in. = 25.40 mm	1 m = 39.37 in.
	1 ft = 0.3048 m	1 m = 3.281 ft
	1 mi = 1.609 km	1 km = 0.6214 mi
Area	$1 \text{ in.}^2 = 645.2 \text{ mm}^2$	$1 \text{ m}^2 = 1550 \text{ in.}^2$
	$1 \text{ ft}^2 = 0.0929 \text{ m}^2$	$1 \text{ m}^2 = 10.76 \text{ ft}^2$
Volume	$1 \text{ in.}^3 = 16.39(10^3) \text{ mm}^3$	$1 \text{ mm}^3 = 61.02(10^{-6}) \text{ in.}^3$
	$1 \text{ ft}^3 = 0.02832 \text{ m}^3$	$1 \text{ m}^3 = 35.31 \text{ ft}^3$
	$1 \text{ gal} = 3.785 \text{ L}^a$	$1 \text{ L} = 0.2642 \text{ gal}$
Velocity	1 in./s = 0.0254 m/s	1 m/s = 39.37 in./s
	1 ft/s = 0.3048 m/s	1 m/s = 3.281 ft/s
	1 mi/h = 1.609 km/h	1 km/h = 0.6214 mi/h
Acceleration	$1 \text{ in./s}^2 = 0.0254 \text{ m/s}^2$	$1 \text{ m/s}^2 = 39.37 \text{ in./s}^2$
	$1 \text{ ft/s}^2 = 0.3048 \text{ m/s}^2$	$1 \text{ m/s}^2 = 3.281 \text{ ft/s}^2$
Mass	1 slug = 14.59 kg	1 kg = 0.06854 slug
Second moment of area	$1 \text{ in.}^4 = 0.4162(10^6) \text{ mm}^4$	$1 \text{ mm}^4 = 2.402(10^{-6}) \text{ in.}^4$
Force	1 lb = 4.448 N	1 N = 0.2248 lb
Distributed load	1 lb/ft = 14.59 N/m	1 kN/m = 68.54 lb/ft
Pressure or stress	1 psi = 6.895 kPa	1 kPa = 0.1450 psi
	1 ksi = 6.895 MPa	1 MPa = 145.0 psi
Bending moment or torque	$1 \text{ ft} \cdot \text{lb} = 1.356 \text{ N} \cdot \text{m}$	$1 \text{ N} \cdot \text{m} = 0.7376 \text{ ft} \cdot \text{lb}$
Work or energy	$1 \text{ ft} \cdot \text{lb} = 1.356 \text{ J}$	$1 \text{ J} = 0.7376 \text{ ft} \cdot \text{lb}$
Power	$1 \text{ ft} \cdot \text{lb/s} = 1.356 \text{ W}$	$1 \text{ W} = 0.7376 \text{ ft} \cdot \text{lb/s}$
	1 hp = 745.7 W	1 kW = 1.341 hp

[a]Both L and l are accepted symbols for liter. Because "l" can be easily confused with the numeral "1," the symbol "L" is recommended for United States use by the National Institute of Standards and Technology (see NIST special publication 811, September 1991).

TABLE B-8 ASTRONOMICAL DATA

Universal Gravitational Constant

$G = 6.673(10^{-11}) \text{ m}^3/(\text{kg} \cdot \text{s}^2) = 3.439(10^{-8}) \text{ ft}^4/(\text{lb} \cdot \text{s}^4)$

The Sun

Mass	$1.990(10^{30})$ kg	$1.364(10^{29}) \text{ lb} \cdot \text{s}^2/\text{ft}$
Mean radius	696,000 km	432,000 mi

The Earth

Mass	$5.976(10^{24})$ kg	$4.095(10^{23}) \text{ lb} \cdot \text{s}^2/\text{ft}$
Mean radius	6370 km	3960 mi
Rotation rate	23.93 hr	

The Moon

Mass	$7.350(10^{22})$ kg	$5.037(10^{21}) \text{ lb} \cdot \text{s}^2/\text{ft}$
Mean radius	1740 km	1080 mi
Mean distance to the Earth (center to center)	384,000 km	239,000 mi
Eccentricity (e)	0.055	

The Solar System

Planet	Mean Distance to Sun A.U.[a]	e	Mean Diameter (relative to Earth)	Mass (relative to Earth)
Mercury	0.387	0.206	0.380	0.05
Venus	0.723	0.007	0.975	0.81
Earth	1.000	0.017	1.000	1.00
Mars	1.524	0.093	0.532	0.11
Jupiter	5.203	0.048	11.27	317.8
Saturn	9.539	0.056	9.49	95.2

[a]Astronomical Unit (A.U.) is equal to the mean distance from the Earth to the sun = $149.6(10^6)$ km = $92.96(10^6)$ mi.